Heavy Metals - Recent Advances

Edited by Basim A. Almayyahi

Published in London, United Kingdom

Heavy Metals - Recent Advances
http://dx.doi.org/10.5772/intechopen.104133
Edited by Basim A. Almayyahi

Contributors
Anamika Kalita Deka, Sunshri Basumatary, Kushwaha Jashvant Kumar, Ahmad Akhavan, Mautin Lawrence Ogun, Olajide Solomon Anagun, Olasunkanmi Kayode Awote, Surukite Opeolu Oluwole, Sesi Christiana Kappo, Faith Oseremi Alonge, Sanchari Biswas, Selina Acheampong, Iram Liaqat, Noor Virk, Nazish Mazhar Ali, Alberto Sánchez Calvo, María del Carmen Blanco Lopez, Muhammad Mudassir Usman, Shamsu Shuaibu Bala, K. Hamza, Samir Ghannem, Sonia Dhaouadi, Samir Touaylia, Sara Mohamed Yonues, Abdou Saad El-Tabl, Bongoua-Devisme Affi Jeanne, Kouakou Sainte Adélaïde Ahya Edith, Hien Marie Paule, Ndoye Fatou, Diouf Diégane, Guety Thierry, Iyioluwa Busuyi Raji, Lobina Gertrude Palamuleni, Carlos Montaño, Javier Montaño, Samuel A. Seriki, Charles C. Mfem, Mohammad Velayatzadeh, Jacqueline A. Malvestiti, Rodrigo P. Cavalcante, Renato Falcão Dantas, Valdemar Luiz Tornisielo, Opaluwa Obaje Daniel, Augustina Pruteanu, Ahmad Mohammadi, Fahimeh Mahmoudnia, Afrasiab Ur Rehman, Abdul Hakim Shahb, Atta Ur Rahmanb, Fida Ur Rahmanb, Sher Ali, Atta Ur Rehmana, Raza Ullaha, Ikram Ullaha, Muhammad Fayaza, Keying Shi, Nguyen Trung Minh, Seong-Taek Yun, Doan Thu Tra, Jang-Soon Kwon, Teodor Vintila, Eniko Gaspar, Maria Mihaela Antofie, Luca Magagnin, Adina Berbecea, Isidora Radulov, Reward Kokah Douglas, Majid Abdulhameed Ibrahim, Mahmood Hashim, Anfas Okash, Faizan Ahmad, Asima Shafi, Sadaf Zaidi, Adama Diarrassouba Tuo, Issiaka Ben Chérif Traoré, Albert Trokourey, Arturo Aguirre Gómez, Margarita Eugenia Gutiérrez Ruiz, Huanwen Chen, Jiaquan Xu, Lavinia Lupa, Laura Cocheci, Awonke Mbangi, Thandile Mdlambuzi, Pardon Muchaonyerwa, Andress Edowaye Odiko, Sarah Gnanasekaran, S. Amal Raj, Isaac Kow Tetteh, Albert Kwame Teye, David Ebuka Arthur, Karimatu Abdullahi, Michael Abatyough, Chidinma Chinelo Arthur, Dedik Budianta, Adipati Napoleon, Nanthi Bolan, Youssef F. F Lawgali, Abdelkarem A. S. Elgazali, Hatem Fawzi Gharour

First published in London, United Kingdom, 2023 by IntechOpen
IntechOpen is the global imprint of INTECHOPEN LIMITED, registered in England and Wales, registration number: 11086078, 5 Princes Gate Court, London, SW7 2QJ, United Kingdom

British Library Cataloguing-in-Publication Data
A catalogue record for this book is available from the British Library

Additional hard and PDF copies can be obtained from orders@intechopen.com

Heavy Metals - Recent Advances
Edited by Basim A. Almayyahi
p. cm.
Print ISBN 978-1-83768-514-1
Online ISBN 978-1-83768-515-8
eBook (PDF) ISBN 978-1-83768-516-5

For EU product safety concerns:
IN TECH d.o.o., Prolaz Marije Krucifikse Kozulić 3, 51000 Rijeka, Croatia,
info@intechopen.com or visit our website at intechopen.com.

Meet the editor

Dr. Basim Almayyahi is currently a professor in the Faculty of Science, University of Kufa, Iraq. His interests include radiation biophysics, radiation medical physics, the natural radioactivity of the environment, biomarkers of heavy metal, treatment of prostate cancer using gold nanoparticles, synthesis and applications of silver nanoparticles on bacterial pathogens, biosensors for detecting radon and lead, radiation detection and measurements, radiation protection, statistical physics, environmental statistics, and the natural radioactivity of gamma rays, radon, and alpha particles in tissue, blood, teeth, bones, soil, water, air, food, fertilizer, and building materials. He was a lecturer in the Department of Physics at Babylon University, Iraq, from 2001 to 2004, before taking up a post as a teacher in the Atomic and Nuclear Lab at the University of Kufa from 2005 to 2023. He is head of the Department of Environment and Director of Scientific Affairs and Cultural Relations at the Faculty of Science, University of Kufa. He has mentored five master's theses and one Ph.D. dissertation. He is the editor and author of four books and a reviewer for more than 250 journals. He is an editorial board member of twenty-three boards and a member of various institutions. He has published 130 scientific papers in domestic and international journals and participated in forty-three scientific congresses and conferences.

Contents

Preface **XIII**

Section 1
Heavy Metals Pollution **1**

Chapter 1 **3**
Evaluate the Impact of Soil Contamination on Vegetables and Fruits
by Augustina Pruteanu

Chapter 2 **31**
Abattoirs: The Hidden Sources of Plants' Heavy Metals and Other Pollutants in Lagos, Nigeria
by Mautin Lawrence Ogun, Olajide Solomon Anagun, Olasunkanmi Kayode Awote, Surukite Opeolu Oluwole, Sesi Christiana Kappo and Faith Oseremi Alonge

Chapter 3 **49**
Heavy Metals in Soils Following 50 Years of Sewage Sludge Application
by Thandile Mdlambuzi, Pardon Muchaonyerwa and Awonke Mbangi

Chapter 4 **67**
Some Metals Found in Stored Canned Fish Products Sold in Nigeria
by Andress Edowaye Odiko

Chapter 5 **79**
Birds as Intrinsic Bio-Indicators for Probing Heavy Metal Contamination Signatures in Polluted Environmental Matrices
by Sanchari Biswas

Chapter 6 **95**
Heavy Metals in Surface Soils and Crops
by Mohammad Velayatzadeh

Chapter 7 **119**
The Efficiency of Phytoremediation of the Big-Sage Plant in Accumulating Some Heavy Metals in Their Tissues *In Vitro*
by Majid Ibrahim, Mahmood Hashim and Anfas Okash

Chapter 8 139
Perspective Chapter: Environmental Assessment on the Effect of Chemical Waste from Dyeing Industries in Zaria
by David Ebuka Arthur, Karimatu Abdullahi, Michael Abatyough and Chidinma Chinelo Arthur

Chapter 9 153
Combined Effects of Earthworms and Plant Growth-Promoting Rhizobacteria (PGPR) on the Phytoremediation Efficiency of *Acacia mangium* in Polluted Dumpsite Soil in Bonoua, Côte d'Ivoire
by Bongoua-Devisme Affi Jeanne, Kouakou Sainte Adélaïde Ahya Edith, Hien Marie Paule, Ndoye Fatou, Guety Thierry and Diouf Diégane

Chapter 10 177
Perspective Chapter: Uptake Capacity of Metals (Al, Cu, Pb, Sn, Zn) in Contaminated Water Metal Production Trade Village Dong Xam, Thai Binh, Vietnam by *Vetiveria zizanioides*
by Nguyen Trung Minh, Seong-Taek Yun, Jang-Soon Kwon and Doan Thu Tra

Chapter 11 189
Heavy Metals in Indonesian Paddy Soils
by Dedik Budianta, Adipati Napoleon and Nanthi Bolan

Chapter 12 207
Heavy Metal Bioaccumulation in Sediment and Benthic Biota
by Sarah Gnanasekaran and S. Amal Raj

Chapter 13 221
Use of Several Pollution Indices for Metal Contamination Assessment in Aquatic Ecosystems, A Case Study, Ebrié Lagoon-Côte d'Ivoire
by Adama Diarrassouba Tuo, Issiaka Ben Chérif Traoré and Albert Trokourey

Chapter 14 239
Heavy Metal Pollution Resulting from Informal E-Waste Recycling in the Greater Accra Region of Ghana
by Albert Kwame Teye and Isaac Kow Tetteh

Section 2
Removal of Heavy Metals 263

Chapter 15 265
Biological Treatment of Heavy Metals with Algae
by Ahmad Mohammadi and Fahimeh Mahmoudnia

Chapter 16 281
Metals as Catalysts for Ozonation
by Jacqueline A. Malvestiti, Rodrigo P. Cavalcante, Valdemar Luiz Tornisielo and Renato Falcão Dantas

Chapter 17 295
Advances in the Adsorption Capacity, Rupture Time and Saturation Curve of Natural Zeolites
by Carlos Montaño and Javier Montaño

Chapter 18 313
Biorefinery for Rehabilitation of Heavy Metals Polluted Areas
by Teodor Vintila, Eniko Gaspar, Maria Mihaela Antofie, Luca Magagnin, Adina Berbecea and Isidora Radulov

Chapter 19 339
Perspective Chapter: Removal of Heavy Metals and Salmonella Pathogens from Sewage Sludge Using a Novel Chelating Agent and Its Reuse as a Fertilizer
by Sara Mohamed Yonues and Abdou Saad El-Tabl

Chapter 20 357
New Advancements in the Field of Pollution Treatment, Including Contamination of the Soil and Water
by Ahmad Akhavan

Chapter 21 377
Heavy Metals Removal from Water and Wastewater
by Lavinia Lupa and Laura Cocheci

Chapter 22 407
Removal of Heavy Metals from Wastewater with Special Reference to Groundnut Shells: Recent Advances
by Asima Shafi, Faizan Ahmad and Sadaf Zaidi

Section 3
Toxic Heavy Metals 423

Chapter 23 425
Perspective Chapter: Rapid Measurement of Potentially Toxic Elements (PTEs) in Petroleum Hydrocarbons Polluted Soils by X-Ray Fluorescence (XRF) Spectroscopy
by Reward Kokah Douglas

Chapter 24 435
Heavy Metals' Poisoning in Farm Animals
by Selina Acheampong

Chapter 25 457
Toxic Heavy Metals in Soil and Plants from a Gold Mining Area, South Africa
by Iyioluwa Busuyi Raji and Lobina Gertrude Palamuleni

Chapter 26 485
Ameliorative Effects of Zinc and Vitamin E on Physiological Changes after Exposure to Heavy Metal
by Samuel A. Seriki and Charles C. Mfem

Section 4
Impact of Heavy Metals **503**

Chapter 27 505
Environmental Impact of Heavy Metals
by Muhammad Mudassir Usman, Shamsu Shuaibu Bala and K. Hamza

Chapter 28 517
Occurrence and Impact of Heavy Metals on Groundwater Sources: A Case Study of Two Communities in Nasarawa State, Nigeria
by Opaluwa Obaje Daniel

Chapter 29 531
Effect of Heavy Metal Pollution on Invertebrates
by Samir Ghannem, Sonia Dhaouadi and Samir Touaylia

Section 5
Detection of Heavy Metals **541**

Chapter 30 543
Electrochemical Techniques for the Detection of Heavy Metals
by Afrasiab Ur Rehman, Abdul Hakim Shah, Atta Ur Rahman, Fida Ur Rahman, Sher Ali, Atta Ur Rehman, Raza Ullah, Ikram Ullah, Muhammad Fayaz and Keying Shi

Chapter 31 557
A Study on the Methyl and Ethylmercury Artifacts in Biological Samples Using Sodium Tetra(n-Propyl)Borate as a Derivatizing Agent
by Abdelkarem A.S. Elgazali, Youssef F. Lawgali and Hatem Fawzi Gharour

Chapter 32 575
Recent Advances in Evaluating Insects as Bioindicators of Heavy Metal Pollution
by Iram Liaqat, Noor Virk and Nazish Mazhar Ali

Chapter 33 593
Electrochemical Detection of Heavy Metals Based on Nanostructured, or Film-Modified Paper Electrodes
by Alberto Sánchez Calvo and María del Carmen Blanco Lopez

Chapter 34 607
Heavy Metal Speciation, and the Evaluation and Remediation of Polluted Mine Wastes and Soils
by Arturo Aguirre Gómez and Margarita Eugenia Gutiérrez Ruiz

Chapter 35 **647**
Monitoring Strategies for Heavy Metals in Foods and Beverages: Limitations for Human Health Risks
by Anamika Kalita Deka, Kushwaha Jashvant Kumar and Sunshri Basumatary

Chapter 36 **671**
Sequential Speciation Analysis of Metals in Geological Samples by Mass Spectrometry
by Jiaquan Xu and Huanwen Chen

Preface

Heavy metals can be found in many places, and understanding their effects on human health is important for ensuring that we can live healthy lives. This book covers several topics related to heavy metals, including sources, health effects, removal, and prevention of exposure.

Heavy metals come from many different sources, including air pollution, water pollution, contaminated soil, and certain types of food. Understanding where heavy metals come from can help us avoid exposure and reduce the risk of negative health effects.

Exposure to heavy metals can have a range of negative health effects, depending on the specific metal and the level of exposure. Some of the potential health effects can include neurological damage, reproductive problems, and cancer.

If heavy metals are present in our bodies, it may be possible to remove them through various methods such as chelation therapy or dietary changes. Understanding how to remove heavy metals from our bodies can help us reduce the risk of negative health effects.

In addition to removing heavy metals from our bodies, it is also important to prevent exposure in the first place. This can involve taking steps like avoiding certain types of foods, filtering our water, or using protective equipment when working in environments where heavy metals are present.

This book on heavy metals is a useful resource for anyone looking to better understand the potential risks associated with exposure and how to protect themselves from these risks.

Basim A. Almayyahi
University of Kufa,
Najaf, Iraq

Section 1

Heavy Metals Pollution

Chapter 1

Evaluate the Impact of Soil Contamination on Vegetables and Fruits

Augustina Pruteanu

Abstract

In the chapter will be presented: scientific substantiation on the models used worldwide to evaluate the contamination of soil, respectively vegetables and fruits; development of theoretical models to evaluate the impact of soil contamination by heavy metals on vegetables and fruits; testing of theoretical models in real conditions based on data obtained from laboratory; development of mathematical models to evaluate the impact on soil contamination on vegetables and fruits and thus on consumers health. The research presented in the chapter aim to develop some original models on the correlation between the level of soil contamination, respectively the remanence of heavy metal in vegetables and fruits harvested for consumption in fresh state. The statistical mathematical models elaborated by the interpolation of the experimental data are models with practical applications in both scientific research and agricultural management.

Keywords: contamination of soil, vegetables and fruits, heavy metal, mathematical models, accumulation of heavy metals

1. Introduction

The rapid development of industry and urbanization in developing countries has led to the chaotic increase in levels of toxic heavy metals in the environment. In addition, heavy metal contamination of agricultural soils and crop plants in these countries, because of the use of industrial waste water it can have negative effects on human health. Other sources of heavy metals from agriculture include manure, fertilizers and pesticides, and contamination from the air due to excessive use of cars [1].

The most important pollution in agriculture is the accumulation of heavy metals which are very toxic to soil, water, plants and humans. Although heavy metals have an key role in nature for soil conservation, their concentration above certain limits can have toxic effects [2]. Therefore, in order to understand the phenomenon of accumulation in soil and plants, it is necessary to know the following definitions [3]:

- Bioamplification is the accumulation of toxic substances in the tissues of tolerant organisms as the trophic level increases. This increase can occur if: the substance cannot be decomposed by environmental processes; the concentration of the substance gradually increases as it moves into a food chain; the impossibility of internal degradation or elimination of the substance.

- Bioaccumulation means increasing the concentration of toxic substances in certain tissues of organisms, over a period of time, due to absorption from food and environment (air, water, soil). The longer the body's exposure to the toxic substance, the more serious the risk of intoxication, even if the level of intoxication is not very high.

- Bioconcentration is the process of accumulating a toxic substance in an aquatic organism, when its source is only water.

Bioconcentration and bioaccumulation occur within an organism, bioamplification occurs at trophic levels (food chain).

Excess accumulation of heavy metals in crops from contaminated agricultural soils it results in soil pollution and low quality of food. Soil is the key factor, it is the basis of the food chain that determines food safety. Vegetables and fruits are plants that are commonly used in food due to their content in nutrients (iron, calcium, proteins), vitamins, minerals, fiber, beneficial to health. Consumption of contaminated fruit and vegetables entail risks for health therefore many researchers have studied food safety in this regard [1, 4–8].

The key functions of metals in plants are involvement in redox reactions and an integral part of enzymes. The essential metals for plants Fe, Cu, Mo, Zn play a major role in the formation of enzymes, the transport of electrons, the sustaining metabolism. In the soil, metals are known as essential trace elements, others non-essential (Hg, Ag, Pb, Ni, etc) and as ultra-trace elements [2].

Heavy metals are present in the environment and they are pollutants in both aquatic and terrestrial ecosystems. The most hazardous heavy metals (Cr, Ni, Cu, Zn, Cd, Pb, Hg) in the environment have three characteristics: their persistence in the environment, toxicity to the soil, water, plants, and organisms, their bioaccumulation in the structure of the soil, the composition of water, the tissues and organs of plants, the body of organisms [9].

The negative effect of metals on the activity of microbes in the soil indirectly affects plant growth. Plants that grow under the stress of heavy metals consume more energy for their survival, what affects other physiological processes, such as: absorption of nutrients, photosynthesis, respiration, metabolism and reproduction, water balance. Due to metallic stress, a lot of reactive oxygen accumulates in the plant [10–12]. Plant reactions to heavy metal toxicity may include: necrosis, chlorosis, senescence and wilting, slowing growth, metabolic disorders, loss of yield, nutrient deficiency, reduced ability to fix atmospheric nitrogen, the small number of seeds and, finally, death [2].

The accumulation of soil metals in plants depends on a number of factors, such as: the structure of the plant, the life cycle of the plant, the vigor of the plant, the pH of the soil, the depth of the root system, temperature, partial pressure of oxygen, carbohydrate level, respiration rate, nutrient exchange and microbial activity [2, 13].

Generally, plants can be integrated into three categories, given their reaction against metals: exclusions, accumulators and indicators based on the mechanism of action for to survive under stress, as suggested [2, 14]. Exclusions react to entering of metals into the vegetative aerial parts making this impossible by stopping the metal in the roots. Accumulators are plants that accumulate metals in the vegetative aerial

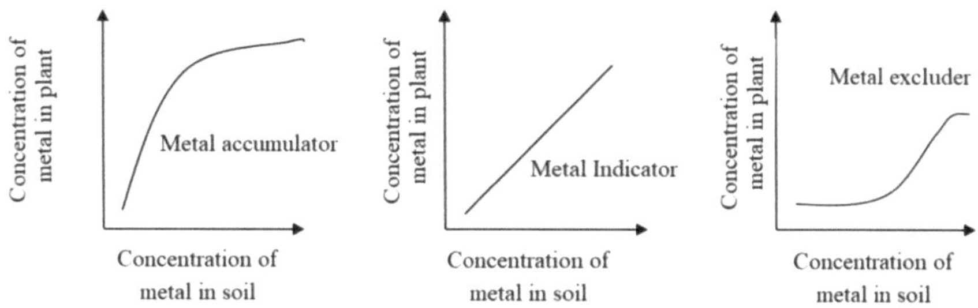

Figure 1.
The shape of curves for mathematical modeling of heavy metal concentrations in the soil-plant system [15].

parts greater than the metals in the soil. Indicators are species that continuously accumulate metals in vegetative aerial tissues tolerating metal concentrations and indicating the amount of metals in the soil. The behavior of plants in relation to the increase in the concentration of heavy metals in the soil is shown in **Figure 1** [15].

Hyperaccumulation of heavy metals is the process by which plants accumulate excess metals 0.1–1% of their dry weight, and concentrate them into roots, stems, or leaves [16]. A hyperaccumulating plant can accumulate and/or tolerate high levels amounts of metals. As an ecological adaptation, some plant species have the ability to grow in heavy-contaminated metal soils and accumulate them [2, 17]. As hyperaccumulating species, about 400 species of plants from 22 families are known. The Brassicaceae family contains a large number of these plants, which includes 87 species from 11 genera [15].

According to the paper [18] the soil factors that influence the absorption of metal by plants are: the concentration of metal in the soil, the processes and properties of the soil and the vegetable factors. These are: the refeeding of ions in the rhizosphere, the kinetic parameters that adjusts the absorption of metals from plants, the tolerance of metal by the plant.

For the assessment of phytoremediation of soils by means of plants, there are many specialized papers [12, 19–21], which uses the bioconcentration factor (BCF) and the translocation factor (TF) as parameters. Through BCF, the concentration of metals in plant tissues is determined in relation to their growth medium, while TF determines way metals are translocated in the aerial vegetative parts of plants [12, 19–21]. As plant species used to remove metals from the environment, they can be listed: peas [1, 22], lettuce [23], wood species (poplar, willow, ash) [24], oleander [25], flax and hemp [26], jute, rooster crest, field thyme [27], rapeseed and Indian mustard [28], sunflower and corn [29], cucumber [22, 30], cherry tomato [15], sweet pepper [31], cabbage and broccoli [32], spinach [33, 34].

The way plants tolerate, absorb, transport, capture, sequester and bioaccumulate metals differs depending on several factors, such as: plant species, phenology, type of metal, soil type and quality, climate, type of source that contaminates, chemical and physical behavior of the plant, environmental factors [12].

The following will be presented different parameters for the assessment of soils and plants with heavy metals, calculated with different mathematical formulas. This parameters have been used in many papers by researchers who have studied this major environmental issue currently existing globally at the moment.

- **The contamination factor (CF)** is the ratio of the metal tracked from the soil to the background value of the heavy metal, expressed in mg/kg^{-1} dry matter, as it is presented in the papers [35, 36] and is determined with the formula (1):

CF = Concentration of heavy metal in the sample/Background value of the heavy meta (1)

The contamination factor values are classified as such: CF < 1 (low contamination), 1 < CF < 3 (moderate contamination), 3 < CF < 6 (high contamination) and CF > 6 (very high contamination) [35, 36].

- **Geoaccumulation index (I_{geo})** can be used when desired effective environmental planning of pollution. It is used successfully to assess soil contaminated with heavy metals from natural or anthropogenic sources. It can be determined with the formula (2):

$$I_{geo} = \log_2 \left[\frac{C_n}{1.5 \bullet B_n} \right] \tag{2}$$

-where, Cn shows the measured value of the metal concentration, in mg kg^{-1} dry matter; n și Bn are the geochemical background value of the corresponding metal, measured in mg kg^{-1} dry matter and 1,5 is the compensation factor in the background concentration of heavy metals. In the study [35] shale values were used as background values.

In the papers [35, 36] geoaccumulation index has been classified into the following six categories: unpolluted environment ($I_{geo} \le 0$), unpolluted environment to moderately polluted environment ($0 < I_{geo} \le 1$), moderately polluted environment ($1 < I_{geo} \le 2$), moderately to heavily polluted environment ($2 < I_{geo} \le 3$), highly polluted environment ($3 < I_{geo} \le 4$), strongly to extremely polluted environment ($4 < I_{geo} \le 5$).

- **Enrichment factor (EF)** is also important to be able to assess the level of heavy metal pollution from anthropogenic sources, and it is calculated as the ratio of the concentration of the studied metal (the chosen reference metal must be in combination with very fine surface solids and occur naturally and evenly in the environment) and geochemical background, expressed also in mg $kg^{-}1$ dry matter, calculated with the formula (3):

$$EF = \text{sample metal/Background metal} \tag{3}$$

The values of enrichment factor were classified into: EF < 1 (soil without enrichment in Ref. metal), 1 < EF < 3 (soil less enriched in Ref. metal), 3 < EF < 5 (soil moderately enriched in Ref. metal), 5 < EF < 25 (soil severely enriched in Ref. metal), 25 < EF < 50 (soil very severely enriched in Ref. metal) si EF > 50 (soil extremely severely enriched in Ref. metal) [35, 37].

- **Degree of contamination (DC)** is presented in literature [35, 36] as a simple evaluation method for controlling anthropogenic pollution, as an indication of dangerous or not. A classification has been proposed for degree of contamination in three categories: DC < 6 (low degree of contamination), 6 < DC < 12 (moderate degree of contamination) and 12 < DC < 24 (considerable degree of contamination). It can be calculated as the sum of the contamination factor of each metal concerned, with the following formula (4):

$$DC = \sum nCF \tag{4}$$

- **Potential Ecological Risk** is a mathematical model that evaluates as the degree of soil pollution based on toxicity from heavy metals as well as assessment of environmental threats because of metals in response to environmental factors. The model is important because it shows which metal is more dangerous to the environment [38]. In the eqs. (5), E_{ri} is calculated for the determination of environmental risk and PERI (Eq. (6)) is calculated for determination as the sum of all the risk values presented by heavy metals in the soil.

$$E_{ri} = T_{ri} \bullet CF \tag{5}$$

where T_{ri} represents the value of the toxic or lethal response; and CF is the contamination factor.

$$PERI = \sum_{f=1}^{n} E_{ri} \tag{6}$$

The two formulas helps establish the degree of threat to soils because to heavy metals by the indication of limits and to measuring the environmental sensitivity for the metals concerned. Therefore, [36] classified the two parameters as follows: $E_{ri} < 40$ (low ecological risk), $40 < E_{ri} < 80$ (moderate ecological risk), $80 < E_{ri} < 160$ (considerable ecological risk), $160 < E_{ri} < 320$ (high ecological risk) and $E_{ri} > 320$ (very serious ecological risk). In the same way, PERI was classified as follows: PERI < 95 (low ecological risk), 95 < PERI <190 (moderate ecological risk), 190 < PERI < 380 (considerable ecological risk) and PERI >380 (very high ecological risk).

Soil pollution with each metal was determined in the paper [1, 39] using the **pollution load index, (PLI)**, calculated as the ratio of the concentration of heavy metal in polluted soils to the concentration of heavy metal in unpolluted soils.

In the paper [40] the level of chemical pollution of the soil was determined using the **anthropogenic coefficient (Kc)**, (Eq. (7)) of the concentration of a metal in a sample, calculated as the ratio of the content to the metal in a studied land (C) to the base level of the metal (CF).

$$K_c = C/CF \tag{7}$$

It is known that soil pollution due to anthropogenic activities may have different sources, therefore in the paper [40] the **total pollution index (Zc)**, (Eq. 8)) was calculated as a result of a group of heavy metals in a studied area.

$$Z_c = \sum_{i=1}^{n} K_c - (n-1) \tag{8}$$

where, K_c it is the anthropic coefficient of the concentration of a metal in a sample; n is the number of samples analyzed.

In mathematical modeling of the phenomenon of accumulation of metals in the soil-plant system, depending as a lot of factors such as was previously mentioned, it is necessary that the behavior of the plant be regularly monitored in order to study the prediction of dynamics. For this purpose, in the papers [40, 41] was determined the **biological absorption coefficient**, noted **(K_i^{bp})**, (Eq. (9)).

$$K_i^{bp} = C_i^r / C_i^p \tag{9}$$

C_i^r is the content of the i-th heavy metal in a plant, mg/kg; C_i^p is the content of the i-th heavy metal in soil, mg/kg.

In the paper [42] was determined the **normalized difference vegetation index (NDVI)**, (Eq. 10)), from mining areas where the soil is polluted with heavy metals. This index (NDVI) depend on the following factors: the topographic position index (TPI), wind speed (WP), precipitation (P), atmospheric dustfall (D), and surface temperature (W).

NDVI it is useful to be able to determine the cover of vegetation from an area, and it is calculated as the ratio of the value of the difference and the total value both near infrared bands, as well as visible infrared bands, and is calculated with the formula (10) [42]:

$$NDVI = NIR - red/NIR + red \tag{10}$$

under NDVI este normalized difference vegetation index, NIR arată the value of the near-infrared light, iar roșu arată the value of the visible infrared light.

The study [42] concluded that atmospheric dust was the main factor that increased the heavy metal content in the soil, being strongly influenced by wind speed and topography of the soil.

- **Mitotic index (MI),** helps to determine the number of cells in the leaves, as a growth parameter of the plant, and is determined as a ratio between the number of cells that divide and the total number of cells in the leaves, expressed as a percentage (%), and calculated with the formula (11) [31]:

$$MI = (\text{number of dividing cells}/\text{total number of cells})\ x100 \tag{11}$$

In a statistical analysis [30] which describes the modeling of heavy metals in the soil-plant system, rice was used in the study and was used as algorithms: multiple linear regression (MLR), support vector machines (SVM), random forest (RF), and cubist. They have helped predict the bioaccumulation coefficient of metals in rice and to identify the potential for transfer of metals into the tissues of rice plants. The flow diagram of the study [30] carried out in China is shown in **Figure 2**.

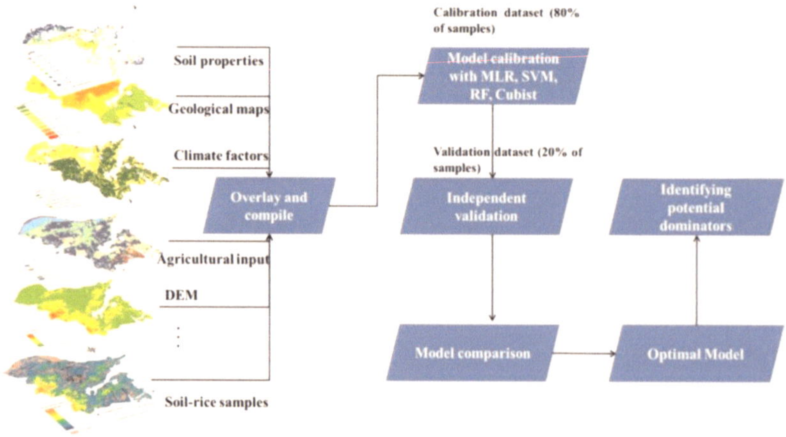

Figure 2.
Flowchart of the modeling of heavy metals in soil–rice system [30].

In order to establish the relationship between soil and the environment, in science and engineering, many forms of models have been used, such as they are: the Philip model, the Richard equation, the Hortonn model [43].

Modeling the distribution of heavy metals and their bioaccumulation in plants has been a major concern for many researchers because the concentration and distribution of heavy metals because the concentration and distribution of heavy metals concern both soil and crop quality. Therefore, the use of a precise model is essential for estimating the real values of heavy metals in the soil and their distribution for effective management in agriculture.

Three models (linear, logarithmic, polynomial, and quadratic models) have often been developed to test their adequacy to experimental data correlated to heavy metals [43, 44]. Validation of these models by direct analysis and comparison of the experimental data with the precise ones, indicates the percentage deviation (DV), determined with the relation (Eq. (12)):

$$D_v = \left[\frac{D_p + D_e}{D_e}\right] x100 \tag{12}$$

where, D_v is deviation, D_p is predicted values and D_e is experimented values [43].

Regression models has been widely used in the prediction of soil properties, especially the degree of pollution, because of their ease and wide use. These models show a global model of the problem studied and through a single regression equation can represent the process [43, 45].

Statistical models for predicting metal concentrations in plants have as main independent variables metal concentrations in the soil. Based on the available analytical data, the concentrations are total or only the bioavailable portion.

2. Experimentation of impact of soil contamination on vegetables and fruits

2.1 General principles of experimentation

It has been shown that when environmental conditions allow heavy metals to infiltrate the soil and, implicitly, groundwater, whole food chain is degraded and there is a direct risk of pollution. The dissemination of metals in food chains place as follows: contaminated soil with toxic metals - plants that absorb metals from the soil in the root, stem, leaves and fruit- humans and animals who eating plants contaminated with dangerous toxic metals and drinking water that can circulate through groundwater and surface water that has metals drains in them.

The danger of contamination of soil and plants with heavy metals depends primarily on the species of the plant and the properties of the soil, and second, the amount and concentration of the metal in the soil. But on the other hand the existence of chemical elements that can change the effect of metals and other substances, can reduce or amplify the absorption or adsorption processes in the soil. The adverse effect depends on the mobility and solubility of heavy metals in the soil. As control measures, there must be optimal conditions for passing metals from the soil solution into stable forms.

In plants, the concentration of metals varies depending on the species of the plant, its period of development, the vegetative organs (roots, stems, leaves, shoots, etc. branches, buds, flowers, fruits). There are plant species that can preferentially accumulate heavy metals according to their vegetative organs, as well as varying amounts. Therefore, in polluted areas, especially mining areas, it is not recommended, to ingest especially green vegetables, because toxic metals quickly get into the leaves due to foliar absorption.

Therefore, for areas at risk of soil pollution and implicitly of plants, the population must selectively consume the plants that grow on those contaminated soils. The selection of plants for consumption is based on the following criteria:

- the frequency with which a given species of plant is eaten;
- taxonomy, the family to which the genus or species of the plant belongs;
- the surface on which it grows and in what quantity a certain species is found to that place;
- the period of growth and vegetation of the plant;
- the vegetative part of the plant that can be eaten (only the fruit, only the leaves);
- the resistance of the plant to diseases and pests;
- the ability to spread the plant.

In this study, the following vegetables and fruits were used in a controlled environmental experimentation:

- vegetables - carrot (*Daucus carota*) grown as root, parsley (Petroselinum spp.) grown as leafy vegetables, cucumbers (*Cucumis sativus*), then increase the fresh lettuce over time;
- fruits - strawberries (Fragaria spp.); raspberry (*Rubus idaeus*).

These vegetables and fruits have been chosen because they are frequently eaten, and grown by the inhabitants of suburban areas of Romania, being essential and rich in micro-macronutrients, proteins, antioxidants and vitamins beneficial to the human body.

The selected vegetables also have a high capacity of accumulating metals without the phytotoxicity of the plants being observable by consumers. The vegetables studied had a relatively short life cycle (about 60 days) and developed well in pots, in the greenhouse, where a controlled environment was created.

2.2 Actual development of experiments

The heavy metals studied were copper, lead and zinc introduced into potted soils in four different concentrations for the solution of each metal, namely: 1.5%, 3.0%, 4.5% si 6.0%. The solutions were prepared individually, using distilled water, as a solvent and copper sulfate, lead acetate and zinc sulphate, as reagents. 250 ml solution of each metal (Cu, Pb, Zn) with the concentrations for each metal was placed, at 1 kg of soil.

Soil contamination with metal solutions and their concentrations was carried out by homogenizing them evenly in the soil prior to planting in pots, no additional until harvest.

In pots, carrot seeds, parsley, cucumber and salad seedlings, raspberry bushes and strawberry stems were planted, which were 1 year old at the time of planting.

At the same time, the reference samples were also made, which consisted of planting seeds, seedlings, bushes and stems in fertile, uncontaminated soil.

The basic properties of the soil used in the experiments were: PH 6.2, moisture 15.4%, particle elements <18 mm 85–95%, total phosphorus 0.8%, total potassium 1%, total nitrogen 2%.

The analysis of the metal content for samples from soils and plants was made on ash resulting from the samples. The methods and techniques used were consistent with the recommendations developed by [46, 47] using an atomic absorption spectrophotometer [48]. The method used was the spectrophotometric method (atomic absorption in flame) [48, 49].

2.3 Experimental results

Table 1 presents the content of metals in soil uncontaminated and in soli contamination with Cu, Pb and Zn separated concentrations.

The experimental data obtained for vegetables are presented in the **Table 2** and for fruits in the **Table 3**. Based on them, the mathematical modeling was done.

Also, aspects with the plants during the experimental research are shown in the **Figure 3**.

Metal	Concentrations of metals in soil, [%]				
	0	1.5	3.0	4.5	6.0
	Content initial of metals in soil (C_{is}), [mg/kg]				
Cu	17.6	58.9	267.2	525.1	680.8
Pb	6.75	48.7	84.7	117.7	285.2
Zn	39.8	202.7	534.8	921.7	1052.3

Table 1.
The content of metals in soil function of four concentrations.

Vegetable	Concentration of heavy metals, [%]	The contents of metals in vegetables (C_p), [mg/kg]		
		Cu	Pb	Zn
Carrot roots	0	7.6	5.4	20.2
	1.5	9.3	8.3	70.3
	3.0	8.5	9.0	73.5
	4.5	10.8	35.3	119.4
	6.0	10.5	44.0	110.5
Parsley leaves	0	4.5	6.1	32.4
	1.5	10.1	7.5	176.9
	3.0	8.1	14.9	185.0
	4.5	10.4	24.8	245.8
	6.0	10.7	45.2	349.9

Vegetable	Concentration of heavy metals, [%]	The contents of metals in vegetables (C_p), [mg/kg]		
		Cu	Pb	Zn
Cucumbers	0	5.2	2.7	34.1
	1.5	6.0	12.8	45.0
	3.0	8.5	12.0	65.3
	4.5	8.3	9.5	84.5
	6.0	10.0	11.8	106.3

Table 2.
The experimental data from vegetables.

Fruit	Concentration of heavy metals, [%]	The contents of metals in fruits (C_f), [mg/kg]		
		Cu	Pb	Zn
Strawberry	0	4.1	2.7	9.9
	1.5	17.5	—	—
	3.0	15.5	4.0	25.8
	4.5	—	7.1	—
	6.0	—	3.2	24.8
Raspberry	0	15.7	6.2	47.9
	1.5	13.8	—	—
	3.0	10.9	—	—
	4.5	19.8	5.3	35.4
	6.0	—	3.0	—

Table 3.
The experimental data from fruits.

3. Development of mathematical models

3.1 Study of the transfer coefficient of plants studied

The transfer coefficient reflects the heavy metal uptake ability from the soil by the plant as a function of the heavy metal concentration in the soil.

The definition formula is (Eq. (13)):

$$C_t = \frac{C_{fp}}{C_{is}} \tag{13}$$

For vegetables and fruits, the variation of the transfer coefficient (C_t) depending on the initial concentration of heavy metal in the plants (C_{fp}) and on the initial concentration of heavy metal in the soil (C_{is},) is represented in the following tables (**Table 4** for vegetables and **Table 5** for fruits).

Figure 3.
Aspects with the plants during the experimental research.

3.2 Statistical models regarding the phenomenon of accumulation of heavy metals in plants studied

Using the numerical data from the experiments, the statistical mathematical models were obtained by interpolation.

- Interpolation formulas for the amount of heavy metal accumulated at the end of vegetation period

The general form of the interpolation polynomials (degrees one - four) is (Eq. (14)):

$$C_p(C_{is}) = c_0 + c_1C_{is} + c_2C_{is}^2 + c_3C_{is}^3 + c_4C_{is}^4 \tag{14}$$

Vegetable	Concentration of heavy metals, [%]	The transfer coefficient (C_t)		
		Cu	Pb	Zn
Carrot roots	0	0.433	0.800	0.508
	1.5	0.157	0.169	0.347
	3.0	0.032	0.106	0.137
	4.5	0.020	0.299	0.130
	6.0	0.015	0.154	0.105
Parsley leaves	0	0.257	0.904	0.814
	1.5	0.172	0.155	0.873
	3.0	0.030	0.176	0.346
	4.5	0.020	0.211	0.267
	6.0	0.016	0.158	0.333
Cucumbers	0	0.297	0.404	0.857
	1.5	0.102	0.263	0.222
	3.0	0.032	0.142	0.122
	4.5	0.016	0.081	0.092
	6.0	0.015	0.041	0.101

Table 4.
The variation of the transfer coefficient from vegetables.

Fruit	Concentration of heavy metals, [%]	The contents of metals, [mg/kg]		
		Cu	Pb	Zn
Strawberry	0	0.233	0.400	0.249
	1.5	0.297	—	—
	3.0	0.058	0.047	0.048
	4.5	—	0.060	—
	6.0	—	0.011	0.024
Raspberry	0	0.892	0.919	1.204
	1.5	0.234	—	—
	3.0	0.041	—	—
	4.5	0.038	0.045	0.038
	6.0	—	0.011	—

Table 5.
The variation of the transfer coefficient from fruits.

3.2.1 Transfer coefficient interpolation

The transfer coefficient was determined according to (Eq. (13)). Similar to (Eq. (14)) and the transfer coefficient (Eq. 15))was determined:

$$C_t(C_{is}) = c_0 + c_1 C_{is} + c_2 C_{is}^2 + c_3 C_{is}^3 + c_4 C_{is}^4 \tag{15}$$

Table 6 shows the coefficients for the interpolation of the transfer coefficient Ct.

The experimental data for the transfer coefficient indicated a potential monotonic, asymptotic decrease, the mathematical modeling was done according to formula (16):

$$C_t(C_{is}) = \frac{c_0}{C_{is}} + c_1 \tag{16}$$

The coefficients of the polynomials (Eq. (15)), shown in **Table** 7, and those of the hyperbola (Eq. (16)), shown in **Table 6**.

Vegetable	Metal	Hyperbola coefficients	
		c0	c1
Carrot	Cu	7.55	$9.20\cdot10^{-3}$
	Pb	4.44	0.13
	Zn	15.41	0.14
Parsley leaves	Cu	4.34	0.02
	Pb	5.25	0.11
	Zn	18.87	0.39
Cucumbers	Cu	5.07	0.01
	Pb	2.10	0.10
	Zn	31.55	0.06

Table 6.
The hyperbolic interpolation equations corresponding to the transfer coefficient.

Vegetable	Metal	Coefficients of the interpolation polynomials					Error*,%
		c0	c1	c2	c3	c4	
Carrot	Cu	0.277	$-4.713\cdot10^{-4}$	0	0	0	175.587
		0.362	$-1.67\cdot10^{-3}$	$1.762\cdot10^{-6}$	0	0	122.469
		0.462	$-3.508\cdot10^{-3}$	$8.829\cdot10^{-6}$	$-6.731\cdot10^{-9}$	0	93.579
		0.598	−0.01	$5.342\cdot10^{-5}$	$-1.031\cdot10^{-7}$	$6.608\cdot10^{-11}$	$5.909\cdot10^{-13}$
	Pb	0.458	$-1.402\cdot10^{-3}$	0	0	0	158.923
		0.712	$-7.62\cdot10^{-3}$	$2.01\cdot10^{-5}$	0	0	113.81
		0.976	−0.028	$2.556\cdot10$–^{4}	$-5.917\cdot10^{-7}$	0	1.025
		0.972	−0.027	$2.419\cdot10^{-4}$	$-4.861\cdot10^{-7}$	$-2.259\cdot10^{-10}$	$3.623\cdot10^{-13}$
	Zn	0.446	$-3.642\cdot10^{-4}$	0	0	0	60.035
		0.545	$-1.094\cdot10^{-3}$	$6.61\cdot10^{-7}$	0	0	16.433

Vegetable	Metal	Coefficients of the interpolation polynomials					Error*,%
		c0	c1	c2	c3	c4	
		0.571	$-1.472\cdot10^{-3}$	$1.596\cdot10^{-6}$	$-5.799\cdot10^{-10}$	0	10.878
		0.546	$-9.403\cdot10^{-4}$	$-7.101\cdot10^{-7}$	$2.759\cdot10^{-9}$	$-1.534\cdot10^{-12}$	$1.934\cdot10^{-13}$
Parsley leaves	Cu	0.2	$-3.272\cdot10^{-4}$	0	0	0	112.486
		0.252	$-1.06\cdot10^{-3}$	$1.077\cdot10^{-6}$	0	0	47.272
		0.279	$-1.833\cdot10^{-3}$	$4.051\cdot10^{-6}$	$-2.832\cdot10^{-9}$	0	16.952
		0.303	$-2.759\cdot10^{-3}$	$1.014\cdot10^{-5}$	$-1.598\cdot10^{-8}$	$9.02\cdot10^{-12}$	$4.973\cdot10^{-13}$
	Pb	0.5	$-1.654\cdot10^{-3}$	0	0	0	171.259
		0.824	$-9.559\cdot10^{-3}$	$2.555\cdot10^{-5}$	0	0	106.208
		1.07	−0.028	$2.461\cdot10^{-4}$	$-5.541\cdot10^{-7}$	0	31.212
		1.184	−0.046	$6.82\cdot10^{-4}$	$-3.932\cdot10^{-6}$	$7.223\cdot10^{-9}$	$2.63\cdot10^{-12}$
	Zn	0.853	$-5.935\cdot10^{-4}$	0	0	0	49.634
		0.964	$-1.404\cdot10^{-3}$	$7.339\cdot10^{-7}$	0	0	39.635
		0.824	$6.817\cdot10^{-4}$	$-4.426\cdot10^{-6}$	$3.201\cdot10^{-9}$	0	23.812
		0.71	$3.178\cdot10^{-3}$	$-1.526\cdot10^{-5}$	$1.889\cdot10^{-8}$	$-7.205\cdot10^{-12}$	$7.119\cdot10^{-13}$
Cucumbers	Cu	0.19	$-3.136\cdot10^{-4}$	0	0	0	169.64
		0.245	$-1.094\cdot10^{-3}$	$1.148\cdot10^{-6}$	0	0	123.307
		0.286	$-2.288\cdot10^{-3}$	$5.738\cdot10^{-6}$	$-4.372\cdot10^{-19}$	0	99.395
		0.415	$-7.355\cdot10^{-3}$	$3.904\cdot10^{-5}$	$-7.634\cdot10^{-8}$	$4.936\cdot10^{-11}$	$4.653\cdot10^{-12}$
	Pb	0.31	$-1.142\cdot10^{-3}$	0	0	0	89.319
		0.435	$-4.183\cdot10^{-3}$	$9.829\cdot10^{-6}$	0	0	6.995
		0.433	$-4.033\cdot10^{-3}$	$8.062\cdot10^{-6}$	$4.44\cdot10^{-9}$	0	6.853
		0.418	$-1.798\cdot10^{-3}$	$-4.75\cdot10^{-5}$	$4.349\cdot10^{-7}$	$-9.206\cdot10^{-10}$	$7.649\cdot10^{-13}$
	Zn	0.587	$-5.594\cdot10^{-4}$	0	0	0	155.184
		0.831	$-2.355\cdot10^{-3}$	$1.626\cdot10^{-6}$	0	0	91.914
		1.013	$-5.077\cdot10^{-3}$	$8.362\cdot1^{0-6}$	$-4.179\cdot10^{-9}$	0	48.423
		1.136	$-7.765\cdot10^{-3}$	$2.002\cdot10^{-5}$	$-2.107\cdot10^{-8}$	$7.757\cdot10^{-12}$	$1.062\cdot10^{-12}$

Table 7.
The coefficients of the interpolation polynomials.

3.2.2 Study of the variation of the final concentration in fruit depending on the initial concentration of heavy metal injected into the soil

Based on **Table 3** which shows the variations in the content of heavy metals: Cu, Pb, Zn in strawberry and raspberries fruits grown in soil injected with heavy metal solutions separated by different concentrations, results the formula (Eq. (17)) and values from the interpolation equations presented in **Table 8**.

The general form of the interpolation polynomial (grades one - two) is:

$$C_f(C_{is}) = c_0 + c_1 C_{is} + c_2 C_{is}^2 \tag{17}$$

Fruit	Metal	Coefficients of the interpolation polynomials			
		c0	c1	c2	c3
Strawberries	Cu	4.14	14.02	−3.41	—
	Pb	2.71	−4.97	2.76	−0.32
	Zn	9.94	18.37	−6.05	0.56
Raspberry	Cu	15.7	1.91	−3.06	0.63
	Pb	4.14	14.02	−3.41	—
	Zn	47.9	−21.28	4.11	—

Table 8.
The interpolation equations corresponding to the fruits.

4. Experimental research on the accumulation of heavy metals over time in fresh lettuce

Experimental research focused on growing and monitoring a lettuce crop over a 68-day period, the optimal vegetation period is 45–50 days [50]. The lettuce culture was carried out this time again in a controlled environment, in pots and in the greenhouse. The soil was contaminated with zinc sulfate solution in three concentrations: 1.5%, 3.0% and 4.5%, respectively, prepared solutions similar to those in the previous experiment.

The soil was homogenized with each solution before planting the lettuce seedlings in pots, as shown in the **Figure 4**. One kg of soil was homogenized with 250 ml solution, under the same experimental conditions, without further addition. 11 samples were placed for each concentration of the soil in zinc, harvesting being made at distances of about seven days each.

The crop has also developed in the greenhouse, the environmental conditions being favorable for the development of plants and having slight variations.

Sampling of the vegetal samples was done in time up to 68 days after planting (**Figure 5**) and each time the lettuce was harvested, the soil sample was taken from the pot, after it was homogenized.

The determination of zinc from the contaminated soil and from the whole fresh lettuce plant (root and leaves) was performed by the spectrophotometric method (atomic absorption in the flame) [27, 28].

The variation in the zinc content of lettuce is shown graphically in **Figure 6**. It is generally observed that the concentration of zinc in lettuce leaves increases for each of the concentrations of contaminated soil.

Figure 4.
Planting the lettuce seedling.

Figure 5.
Pots with plants from the three types of crops harvested at several.

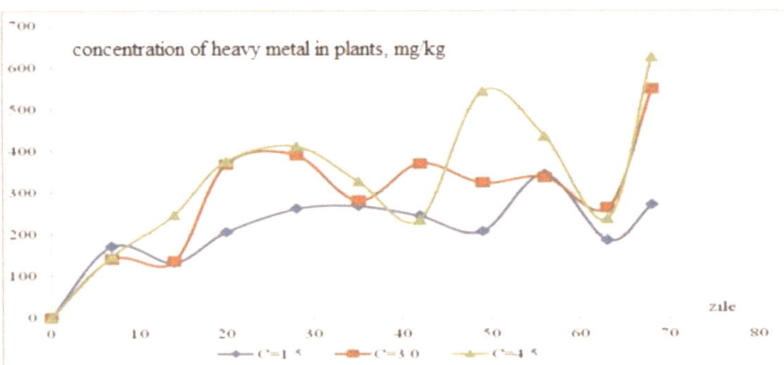

Figure 6.
Variation in time of the concentration of heavy metal (Zn) in lettuce.

At certain time periods, experimental data show a decrease in zinc in plants. If the phenomenon is real, and there was no error in the measurement, it can be concluded that the removal of metal from the plant was natural, in this regard, new experiments are indicated to be able to determine and which factors influenced this process.

If we count the zinc concentrations obtained in plants, we notice that: for soil contaminated with 4.5%, 7 of the 11 crops have the highest concentration of zinc; for soil contaminated with 3.0%, 3 of the 11 crops have the highest concentration of zinc; for soil contaminated with 1.5%, 1 of the 11 crops have the highest concentration of zinc. This observation is also supported by the average lettuce harvest, which recorded 228.9 mg/kg, for the crop developed in soil contaminated with 1.5% Zn, 315.3 mg/kg, for the crop developed in soil contaminated with 3.0% Zn si 357.9 mg/kg, for the crop developed in soil contaminated with 4.5% Zn.

The conclusion is that after the concentration of the soil in zinc the order of accumulation of metal in the lettuce is: 4.5% > 3.0% > 1.5%.

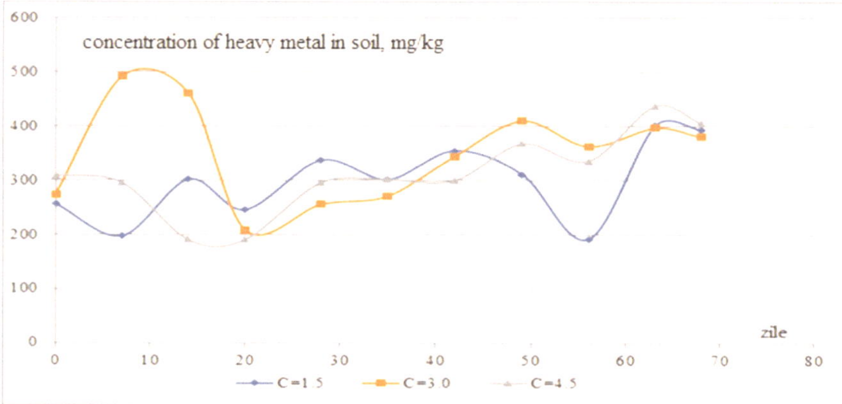

Figure 7.
Variation in time of the concentration of heavy metal (Zn) in soil.

The variation in the zinc content of lettuce is shown graphically in **Figure 7**. It is observed that, over time, the concentration of zinc oscillates, even slightly increasing. Again, it is questioned whether the phenomenon is real or whether measurement errors have occurred. If in the lettuce it can be believed that the metal could not be absorbed by the plant and that the mass has increased, in the case of soil, this conclusion is hard to say, although some soil chemicals can pass into plants as well.

5. Statistical modeling of experimental data on the accumulation of heavy metals over time in fresh lettuce

By interpolating the experimental data, functions were obtained that indicate the variation over time of the zinc concentrations from the soil into the plants. **Figure 8** showed linear interpolation, the results being similar to the increase in the concentration of metals in the soil-plant system. It has been observed that in soils with high concentrations of zinc there has been a large and rapid accumulation.

The interpolated functions obtained can help calculate other important parameters that influence the growth and development of plants on soils contaminated with zinc.

In **Figure 9** graphically represents the variations of Zn concentrations in lettuce grown on the soils of the three infestation categories, in the form interpolated by second-degree polynomials and in the form of experimental data. The same order is

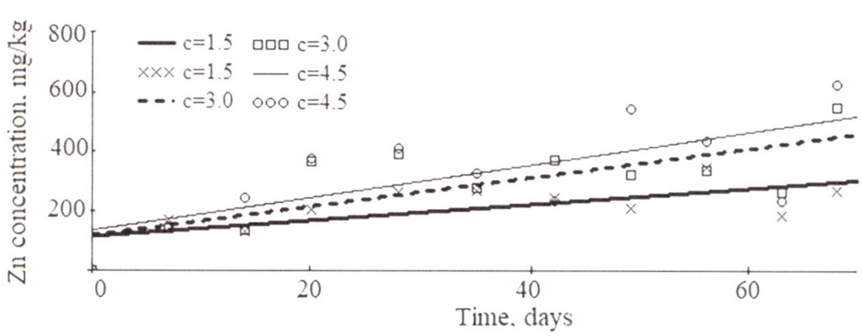

Figure 8.
Linear interpolation of the increase of Zn concentration in fresh lettuce.

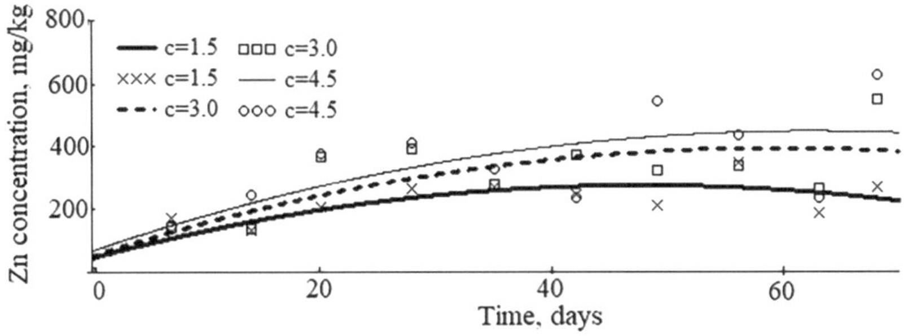

Figure 9.
Quadratic interpolation of the increase in heavy metal concentration in fresh lettuce.

easily visible between the heavy metal concentrations in the plants of the three categories of soil infestation, as in the case of linear interpolation.

From **Figure 9** it is observed that in plants the arrangement of curves of variation of heavy metal concentrations and their increasing monotony is permanent both for the third degree polynomial and for the separation of curves, without any common points between the three curves being observed.

The interpolation curves by third degree polynomials, for increasing the concentration of heavy metals in plants, are represented in **Figure 10** and for fourth degree polynomials, are represented in **Figure 11**.

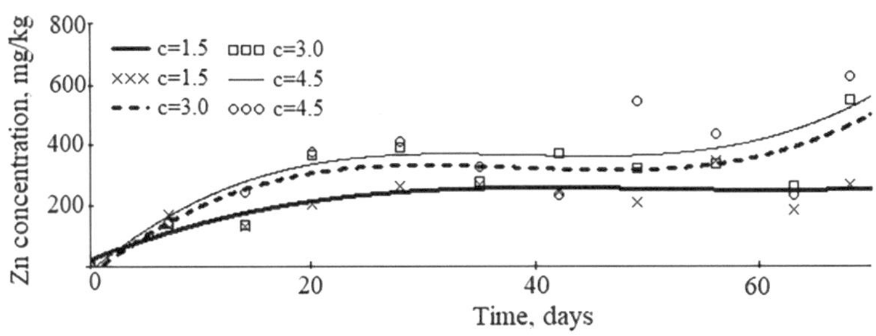

Figure 10.
Cubic interpolation of the increase in the concentration of heavy metal (Zn) in fresh lettuce.

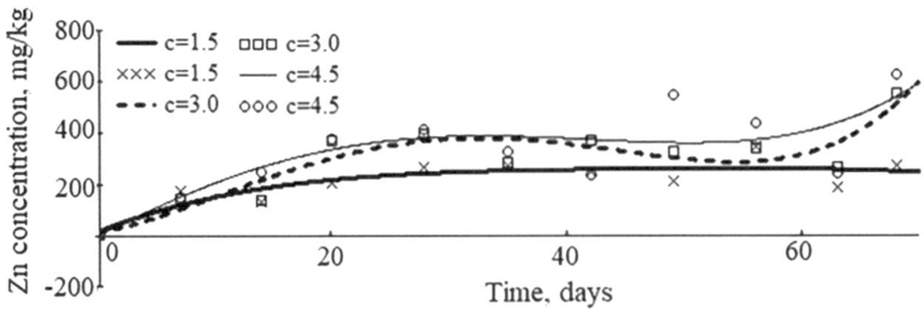

Figure 11.
Polynomial interpolation of the fourth degree for the increase of heavy metal (Zn) concentrations in fresh lettuce crops.

It is observed from **Figures 10** and **11** that, generally, the order and monotony of interpolation curves of the variation in the concentration of heavy metals in plants is observed by polynomial curves, with a few exceptions, the curve corresponding to the soil contaminated by 3% Zn slightly reaches the curve corresponding to the soil contaminated by 4.5% and it is very almost, toward the end of the study period, to the curve corresponding to plants grown in soil contaminated with 1.5% Zn.

6. Conclusions

6.1 Conclusions on the development of vegetables in contaminated soil

- In most cases of vegetables grown in soil contaminated with different concentrations of heavy metal have been observed a tendency to increase the amount of heavy metal accumulated in the plant as the amount of heavy metal in the soil increases.

- On vegetables, in general, the conclusion regarding the increasing monotony representing the variation of the heavy metal concentration in the plant remains valid. It is observed that the variation of the concentration of heavy metal in plants appear more often in the case of contaminations with copper, and less with lead and zinc.

- In the case of the transfer coefficient to vegetables, the trend is decreasing as the amount of heavy metal in the soil increases. The general conclusion is that for the vegetables studied (carrot, parsley and cucumber) bioaccumulation with, Pb and Zn is less as the amount of metals in the soil is higher. Therefore, the more contaminated the soil, the more difficult it is to phytoremediation with these plant species.

6.2 Conclusions on the development of fruits in contaminated soil

- Fruits have a longer vegetation period than vegetables, and experiments to accumulate metals in them are more complicated because, the general distribution of metals requires complex study, possibly over several seasons and on all vegetative plant organs (root, stem, branches, leaves, flowers, fruits and seeds). This way of working requires a long time and a large number of analyzes for a single plant. The analysis in this study focused on the concentration of Cu, Pb and Zn in fruits, ripe, therefore the conclusions for them are mediating.

- For experiments on strawberries with a single heavy metal (Cu, Pb, Zn), it was observed that the variation in the final content in the fruit has a maximum. Therefore, while the concentration in the soil increases, the absorption capacity of metal from the soil into strawberries decreases. It can be concluded that strawberries are developing mechanisms of protection.

- For experiments on raspberries with a single heavy metal (Cu, Pb, Zn), it was observed that the values of the transfer coefficient decrease as the concentration of metals in the soil increases. The order of absorption of metals in raspberry fruits being Zn > Cu > Pb.

- Also, in the case of strawberries and raspberries, the best absorption was observed in the low concentration of 1.5%.

6.3 Conclusions on mathematical modeling of vegetables and fruits

Mathematical modeling was performed based on the interpolation of experimental data obtained in this study.

- Polynomial equations are statistical models with practical applications in both research and agriculture. The formulas obtained can help with various calculations such as:

    - the number of plants corresponding to the implementation of the phytoremediation process of soils contaminated with heavy metals;

    - the compliance of plant production within the limits provided for human consumption;

    - the optimization of phytoremediation and crops with regard to the marketing and consumption.

- In this case, the first and second degree interpolation polynomials were the most used because they showed small variations between the experimental data. The curves obtained for third and fourth degree polynomials showed greater variations between the experimental data and their use may be at risk of showing errors, although it has been found that the fourth degree polynomial passes very close to experimental points when it is restricted to the set of experimental points only.

- Overall, the interpolation curves obtained in this study indicate the increase in the final concentration of Cu, Pb and Zn in the roots of carrot, parsley leaves, cucumbers and strawberry and raspberry fruits and the decrease in the transfer coefficient of the three metals studied at the end of the vegetation periods.

- For the transfer coefficient to vegetables, the experimental data presented in **Table 2** showed a hyperbolic decrease in this coefficient. The hyperbolic interpolation equation (Eq. (16)), proved effective, with few exceptions.

Statistical mathematical models determined with equations (Eqs.(14), (15) and Eq. (17)) whose coefficients are given in the **Tables 7–9**, for each case, they can be used for interpolation in the calculation of concentrations of heavy metals in the plant or calculation of transfer coefficients, only for the experimental range used in this study, for the initial concentration of each heavy metal in the soil.

- Mathematical modeling for fruits is poorly documented because at certain concentrations, plants have not developed. It is recommended for the two species of studied berries (strawberries and raspberries), interpolation curves of maximum second degree, because the experimental distributions are not linear. Interpolation was also attempted using third degree polynomial curves, which passed through experimental points but had a total zero error.

Vegetable	Metal	The coefficients of the interpolation polynomials					Error*,%
		c0	c1	c2	c3	c4	
Carrot	Cu	8.128	$3.867 \cdot 10^{-3}$	0	0	0	15.069
		8.119	$3.985 \cdot 10^{-3}$	$-1.745 \cdot 10^{-7}$	0	0	15.068
		8.406	$-4.221 \cdot 10^{-3}$	$3.138 \cdot 10^{-5}$	$-3.005 \cdot 10^{-8}$		14.225
		6.55	0.069	$4.497 \cdot 10^{-4}$	$1.01 \cdot 10^{-6}$	$-7.13 \cdot 10^{-10}$	$3.418 \cdot 10^{-13}$
	Pb	4.182	0.149	0	0	0	79.321
		−0.143	0.255	$-3.42 \cdot 10^{-4}$	0	0	74.097
		9.338	−0.465	$8.131 \cdot 10^{-3}$	$-2.129 \cdot 10^{-5}$	0	41.472
		−0.224	1.015	−0.029	$2.638 \cdot 10^{-4}$	$-6.097 \cdot 10^{-7}$	$2.269 \cdot 10^{-12}$
	Zn	33.157	0.083	0	0	0	37.801
		23.598	0.153	$-6.353 \cdot 10^{-5}$	0	0	33.614
		16.381	0.261	$-3.304 \cdot 10^{-4}$	$1.655 \cdot 10^{-7}$	0	31.78
		−6.277	0.759	$2.493 \cdot 10^{-3}$	$3.297 \cdot 10^{-6}$	$-1.438 \cdot 10^{-9}$	$1.107 \cdot 10^{-12}$
Parsley leaves	Cu	6.938	$5.925 \cdot 10^{-3}$	0	0	0	44.074
		6.603	0.011	$-6.986 \cdot 10^{-6}$	0	0	43.437
		6.084	0.026	$-6.411 \cdot 10^{-5}$	$5.441 \cdot 10^{-8}$	0	42.371
		0.882	0.231	$1.412 \cdot 10^{-3}$	$2.968 \cdot 10^{-6}$	$1.998 \cdot 10^{-9}$	$1.142 \cdot 10^{-12}$
	Pb	3.636	0.148	0	0	0	27.501
		2.875	0.166	$-6.016 \cdot 10^{-5}$	0	0	27.014
		6.903	−0.14	$3.54 \cdot 10^{-3}$	$-9.047 \cdot 10^{-6}$	0	0.7
		6.747	−0.116	$2.94 \cdot 10^{-3}$	$-4.396 \cdot 10^{-6}$	$-9.944 \cdot 10^{-9}$	$2.548 \cdot 10^{-13}$
	Zn	65.286	0.241	0	0	0	46.616
		60.405	0.277	$-3.245 \cdot 10^{-5}$	0	0	46.484
		−15.377	1.409	$-2.834 \cdot 10^{-3}$	$1.738 \cdot 10^{-6}$	0	8.212
		−30.093	1.733	$4.238 \cdot 10^{-3}$	$3.772 \cdot 10^{-6}$	$-9.342 \cdot 10^{-10}$	$2.966 \cdot 10^{-13}$
Cucumbers	Cu	5.647	$6.269 \cdot 10^{-3}$	0	0	0	19.454
		5.284	0.011	$-7.583 \cdot 10^{-6}$	0	0	17.047
		4.498	0.034	$-9.401 \cdot 10^{-5}$	$8.232 \cdot 10^{-8}$	0	3.285
		4.847	0.02	$3.612 \cdot 10^{-6}$	$-1.131 \cdot 10^{-7}$	$1.34 \cdot 10^{-10}$	$1.053 \cdot 10^{-13}$
	Pb	7.789	0.018	0	0	0	74.402
		4.457	0.1	$-2.635 \cdot 10^{-4}$	0	0	58.818
		0.164	0.426	$-4.1 \cdot 10^{-3}$	$9.641 \cdot 10^{-6}$	0	9.272
		−0.86	0.584	$8.039 \cdot 10^{-3}$	$4.016 \cdot 10^{-5}$	$-6.526 \cdot 10^{-8}$	$5.97 \cdot 10^{-13}$
	Zn	30.884	0.066	0	0	0	14.042
		33.759	0.045	$1.911 \cdot 10^{-5}$	0	0	12.642
		28.007	0.13	$-1.935 \cdot 10^{-4}$	$1.319 \cdot 10^{-7}$	0	7.389
		32.489	0.032	$2.342 \cdot 10^{-4}$	$-4.875 \cdot 10^{-7}$	$2.845 \cdot 10^{-10}$	$4.609 \cdot 10^{-13}$

Table 9.
The interpolation equations corresponding to the vegetables.

- It was observed that the variation of the heavy metal content in berries (fruit) does not recommend linear interpolation (linear regression), as these distributions have potential extreme points (minimum or maximum).

6.4 Conclusions on the development over time of fresh lettuce on soils with different concentrations of zinc

- The overall conclusion is accumulate zinc in the lettuce regardless of Zn concentration in the soil.

- At certain intervals, the concentration of zinc decreases in the lettuce.

- This phenomenon show that the removal of zinc from the lettuce can be done naturally, while the factors that influence this process remain to be studied.

- The accumulation of Zn in lettuce increases as the concentration of Zn in the soil increases (4.5% > 3.0% > 1.5%).

- At the higher concentration of zinc (4.5%) for contaminated soil, the accumulation of metal in lettuce is higher, in the chosen range of these experiments.

6.5 Conclusions on mathematical modeling for the development of fresh lettuce over time on soils with different concentrations of zinc

The mathematical modeling presented in this chapter was done through a mediation of experimental data.

Increased polynomic interpolation resulted in a good approximation of experimental data, but there were not enough reasons from the point of view of the phenomenon studied to get over the interpolation of the third degree polynomial. The functions determinated are particular and useful only in the experimental field chosen in this study.

Therefore, the statistical model can be used to validate the theoretical model, which is general both in terms of working conditions (soil type, metal type, metal concentrations, atmospheric conditions) and the plant species used (root species, leaf species, fruit species).

The results obtained from experiments and interpolation suggest indications for future research: New experiments on several plant vegetation cycles, possibly with modification of working parameters that can influence the bioaccumulation process of heavy metals in plants.

6.6 General conclusions

At the optimum vegetation period (final harvest, in the stage of ripening and human consumption), the plants (carrots, parsley, cucumbers, strawberries, Raspberries) have continuously accumulated (monotone growth) heavy metals (Cu, Pb, Zn) from contaminated soil.

At the end of the growing season, the transfer coefficient decreases as the concentration of heavy metals (Cu, Pb, Zn) in the soil increases.

The data presented in this chapter can be both the foundation of dynamic mathematical models that can simulate the life of a plant and the starting point for the development of future research, that can emphasize self-defense mechanisms developed by plants, adaptations to new environmental conditions, possibilities for restoring the qualities of some plants over time.

Acknowledgements

This paper was financed by support of Executive Agency for Higher Education, Research, Development and Innovation Funding, Exploratory Research Programme, PN-III-P4-ID-PCE-2016-0860, contr. 174/08.08.2017, Research on the development of some mathematical models to evaluate the impact of soil contamination on fruits and vegetables – CONTAMOD.

Conflict of interest

The authors declare no conflict of interest.

Author details

Augustina Pruteanu
The National Institute of Research – Development for Machines and Installations Designed for Agriculture and Food Industry – INMA, Bucharest, Romania

*Address all correspondence to: pruteanu_augustina@yahoo.com

References

[1] Galal TM, Hassan LM, Ahmed DA, Alamri SAM, Alrumman SA, Eid EM. Heavy metals uptake by the global economic crop (Pisum sativum L.) grown in contaminated soils and its associated health risks. PLoS One. 2021; **16**(6):e0252229. DOI: 10.1371/journal.pone.0252229

[2] Kumar V, Jogendra SJ, Kumar P. Heavy metals accumulation in crop plants: Sources, response mechanisms, stress tolerance and their effects. Contaminants in Agriculture and Environment: Health Risks and Remediation. 2019;**4**:38-57. DOI: 10.26832/AESA-2019-CAE-0161-04

[3] Sfiru R, Cardei P, Matache M. The role of mathematical modeling in research in the field of bioaccumulation of heavy metals. International Symposium ISB-INMATEH-Agricultural and Mechanical Engineering-Bucharest. 26-28 Oct 2017. ISSN:2344-4118. Available from: http://isb.pub.ro/wp-content/isbinmateh/2017/Volume_Symposium_2017.pdf

[4] Galal TM. Health hazards and heavy metals accumulation by summer squash (Cucurbita pepo L.) cultivated in contaminated soils. Environmental Monitoring and Assessment. 2016;**188**: 434. DOI: 10.1007/s10661-016-5448-3

[5] Shehata HS, Galal TM. Trace metal concentration in planted cucumber (Cucumis sativus L.) from contaminated soils and its associated health risks. Journal of Consumer Protection Food Safe. 2020;**15**:205-217

[6] Zeng X, Wang Z, Wang J, Guo J, Chen X, Zhuangs J. Health risk assessment of heavy metals via dietary intake of wheat grown in Tianjin sewage irrigation area. Ecotoxicology. 2015;**24**:2115-2124. DOI: 10.1007/s10646-015-1547-0

[7] Horiguchi H, Oguma E, Sasaki S, Miyamoto K, Ikeda Y, Machida M. Dietary exposure to cadmium at close to the current provisional tolerable weekly intake does not affect renal function among female Japanese farmers. Environmental Research. 2004;**95**:20-31. DOI: 10.1016/S0013-9351(03)00142-7

[8] Washington DC, Wang Y, Qiao M, Liu Y, Zhu Y. Health risk assessment of heavy metals in soils and vegetables from wastewater irrigated area, Beijing-Tianjin city cluster, China. Journal of Environmental Sciences. 2012;**24**:690-698

[9] Ali H, Khan E, Ilahi I. Environmental chemistry and ecotoxicology of hazardous heavy metals: Environmental persistence, toxicity, and bioaccumulation, Hindawi. Journal of Chemistry. 2019;**6730305**:14. DOI: 10.1155/2019/6730305

[10] Panda G, Das S, Bandopadhyay T, Guha A. Adsorption of nickel on husk of Lathyrus sativus: Behavior and binding mechanism. Colloids and Surfaces. B, Biointerfaces. 2007;**57**:135-142. DOI: 10.1016/j.colsurfb.2007.01.022

[11] Kumar R, Sharma RK, Singh AP. Cellulose based grafted biosorbents-journey from lignocellulose biomass to toxic metal ions sorption applications-a review. Journal of Molecular Liquids. 2017;**232**:62-93. DOI: 10.1016/j.molliq.2017.02.050

[12] Usman K, Al JH, Abu-Dieyeh MH, Alsafran MHSA. Comparative assessment of toxic metals bioaccumulation and the mechanisms of chromium (Cr) tolerance and uptake in Calotropis procera. Frontiers in Plant

Science. 2020;**11**:883. DOI: 10.3389/fpls.2020.00883

[13] Chen Y, Wang Y, Wu W, Lin Q, Xue S. Impacts of chelate-assisted phytoremediation on microbial community composition in the rhizosphere of a copper accumulator and non-accumulator. Science of the Total Environment. 2006;**356**(1–3):247-255

[14] Baker AJ. Accumulators and excluders-strategies in the response of plants to heavy metals. Journal of Plant Nutrition. 1981;**3**(1–4):643-654

[15] Budovich LS. Effects of heavy metals in soil and plants on ecosystems and the economy. Caspian Journal of Environmental Sciences. 2021;**19**(5): 991-997. DOI: 10.22124/CJES.2021.5331

[16] Baker AJ, Brooks R. Terrestrial higher plants which hyperaccumulate metallic elements: A review of their distribution, ecology and phytochemistry. Biorecovery. 1989;**1**(2):81-126

[17] Lombi E, Zhao FJ, Wieshammer G, Zhang G, McGrath SP. In situ fixation of metals in soils using bauxite residue: Biological effects. Environmental Pollution. 2002;**118**(3):445-452

[18] Cataldo DA, Wildung RE. Soil and plant factors influencing the accumulation of heavy metals by plants. Environmental Health Perspectives. 1978;**27**:149-159

[19] Usman K, Al-Ghouti MA, Abu-Dieyeh MH. The assessment of cadmium, chromium, copper, and nickel tolerance and bioaccumulation by shrub plant Tetraena qataranse. Scientific Reports. 2019;**9**:5658. DOI: 10.1038/s41598-019-42029-9

[20] Xia S, Song Z, Jeyakumar P, Shaheen SM, Rinklebe J, Ok YS. A critical review on bioremediation technologies for Cr (VI)-contaminated soils and wastewater. Critical Reviews in Environmental Science and Technology. 2019;**49**:1027-1078. DOI: 10.1080/10643389.2018.1564526

[21] Anjum NA, Hasanuzzaman M, Hossain MA, Thangavel P, Roychoudhury A, Gill SS. Jacks of metal/metalloid chelation trade in plants—An overview. Frontiers in Plant Science. 2015;**6**:192. DOI: 10.3389/fpls.2015.00192

[22] Eid EM, El-Bebany AF, Taher MA, Alrumman SA, Galal TM, Shaltout KH, et al. Heavy metal bioaccumulation, growth characteristics, and yield of Pisum sativum L. grown in agricultural soil-sewage sludge mixtures. Plants. 2020;**9**:1300. DOI: 10.3390/plants9101300

[23] Smical AI, Hotea V, Oros V, Juhasz J, Pop E. Studies on transfer and bioaccumulation of heavy metals from soil into lettuce. Environmental Engineering and Management Journal. 2008;7(5):609-615. Available from: http://omicron.ch.tuiasi.ro/EEMJ/

[24] Brkovic DL, Boskovic Rakocevic LS, Mladenovic JD, Simic ZB, Glisic RM, Grabovic FJ, et al. Metal bioaccumulation, translocation and phytoremediation potential of some woody species at mine tailings. Notulae Botanicae Horti Agrobotanici Cluj-Napoca. 2021;**49**(4):12487. DOI: 10.15835/nbha49412487. Available at: https://notulaebotanicae.ro/index.php/nbha/article/view/12487/9281

[25] Ibrahim N, El Afandi G. Phytoremediation uptake model of heavy metals (Pb, Cd and Zn) in soil using Nerium oleander. Heliyon. 2020;**6**: e04445. DOI: 10.1016/j.heliyon.2020.e04445

[26] Shehata SM, Badawy RK, Aboulsoud YIE. Phytoremediation of some heavy metals in contaminated soil. Bulletin of the National Research Centre. 2019;**43**:189. DOI: 10.1186/s42269-019-0214-7

[27] Osundiya MO, Ayejuyo OO, Olowu RA, Bamgboye OA, Ogunlola AO. Bioaccumulation of heavy metals in frequently consumed leafy vegetable grown along Nigeria-Benin Seme border. West Africa, Advances in Applied Science Research. 2014;**5**(1):1-7

[28] Turan M, Esringü A. Phytoremediation based on canola (Brassica napus L.) and Indian mustard (Brassica juncea L.) planted on spiked soil by aliquot amount of Cd, Cu, Pb, and Zn. Plant, Soil and Environment. 2007; **53**(1):7-15. DOI: 10.17221/3188-PSE

[29] Iram S, Basri R, Ahmad KS, Jaffri SB. Mycological assisted phytoremediation enhancement of bioenergy crops Zea mays and Helianthus annuus in heavy metal contaminated lithospheric zone. Soil and Sediment Contamination: An International Journal. 2019;**28**(4):411-430. DOI: 10.1080/15320383.2019.1597011

[30] Xie M, Li H, Zhu Y, Xue J, You Q, Jin B, et al. Predicting bioaccumulation of potentially toxic element in soil–rice systems using multi-source data and machine learning methods: A case study of an industrial city in Southeast China. Land. 2021;**10**:558. DOI: 10.3390/land10060558

[31] Khan S, Naz A, Asim M, Ahmad SS, Yousaf S, Muhammad S. Toxicity and bioaccumulation of heavy metals in spinach seedlings grown on freshly contaminated soil. Pakistan Journal of Botany. 2013;**45**(S1):501-508

[32] Antonious GF, Kochhar TS, Coolong T. Yield, quality, and concentration of seven heavy metals in cabbage and broccoli grown in sewage sludge and chicken manure amended soil. Journal of Environmental Science and Health, Part A. 2020;**47**:1955-1965. DOI: 10.1080/03601234.2012.676509

[33] Kumar V, Chopra AK, Srivastava S. Assessment of heavy metals in spinach (Spinacia oleracea L.) grown in sewage sludge-amended soil. Communications in Soil Science and Plant Analysis. 2016;**47**: 221-236. DOI: 10.1080/00103624.2015.1122799

[34] Eid EM, El-Bebany AF, Alrumman SA, Hesham A, Taher MA, Fawy KF. Effects of different sewage sludge applications on heavy metal accumulation, growth and yield of spinach (Spinacia oleracea L.). International Journal of Phytoremedicine. 2017;**19**:340-347. DOI: 10.1080/15226514.2016.1225286

[35] Sabir M, Baltrenaite-Gediene E, Ditta A, Ullah H, Kanwal A, Ullah S, et al. Bioaccumulation of heavy metals in a soil–plant system from an open dumpsite and the associated health risks through multiple routes. Sustainability. 2022;**14**: 13223. DOI: 10.3390/su142013223

[36] Hakanson L. An ecological risk index for aquatic pollution control. A sedimentological approach. Water Research. 1980;**14**:975-1001. DOI: 10.1016/0043-1354(80)90143-8

[37] Müller G. Schwermetalle in den Sedimenten des Rheins-Veranderungen seit 1971. Umschau. 1979;**79**:778-783

[38] Cao X, Li W, Song S, Wang C, Khan K. 2022, source apportionment and risk assessment of soil heavy metals around a key drinking water source area in northern China: Multivariate statistical analysis approach. Environmental Geochemistry and

Health. 2022;**34**:2365-2380. DOI: 10.1007/s10653-022-01251-7

[39] Liu WH, Zhao JZ, Ouyang ZY, Soderlund L, Liu GH. Impacts of sewage irrigation on heavy metals distribution and contamination in Beijing, China. Environment International. 2005;**31**: 805-812. DOI: 10.1016/j. envint.2005.05.042

[40] Popova E. Accumulation of heavy metals in the “soil-plant” system. AIP Conference Proceedings. 2016;**1772**: 050006. DOI: 10.1063/1.4964576

[41] Mirkin BM, Naumov LG, Solomesh AI. The Modern Science of Vegetation. Moscow, in Russian: Logos; 2000. pp. 517-669.

[42] Zeng Q, Shen L, Feng T, Hao R. Investigation of the distribution of heavy metals in the soil of the Dahuangshan mining area of the southern Junggar coalfield, Xinjiang, China. Minerals. 2022;**12**:1332. DOI: 10.3390/ min12101332

[43] Uzoije AP, Uzoigwe L, Otuonye E, Kamalu CO, Onunkwo-Akunne A. Modeling lateral distribution of heavy metal and bio-accumulation in earthworm in the varying acidic surface horizon of waste- polluted soil. International Journal of Energy Engineering. 2013;**3**(2):45-54. DOI: 10.5923/j.ijee.20130302.02

[44] Cârdei P, Tudora C, Vlăduț V, Pruteanu A, Găgeanu I, Cujbescu D, et al. Mathematical model to simulate the transfer of heavy metals from soil to plant. Sustainability. 2021;**13**(11):6157. DOI: 10.3390/su13116157

[45] Baltensweiller A, Stephan Zimmermann. Modeling soil acidicity in switerland using spatial statistics tools. In: Proceedings of the ESRI International Conference. 12-16 Jul 2010. p. 1493. Available at: file:///C:/Users/DI_OFF~1/ AppData/Local/Temp/MicrosoftEdge Downloads/669a1380-801a-4eca-b587- 0cef83e4a778/MODELLING_SOIL_ ACIDITY_IN_SWITZERLAND_USING_ SPATIA.pdf

[46] Order of the Minister of Agriculture, Food and Forests no. 223, M.O. 598/13. 2002. Available from: https://legislatie.just.ro/Public/ DetaliiDocument/37981

[47] Gergen I. Analytical Chemistry and Physical-Chemical Analysis. Timisoara: Mirton Publishing House; 1998

[48] Soil Quality — Determination of Cadmium, Chromium, Cobalt, Copper, Lead, Manganese, Nickel and Zinc — Flame and Electrothermal Atomic Absorption Spectrometric Methods. Available from: https://standards.iteh.ai/ catalog/standards/iso/0c26271c- 2975-495d-931c-bfb9014495ca/iso- 11047-1998

[49] Gergen I. Chemical and Physico-Chemical Methods in the Quality Control of Plant Agro-Food Products. Timişoara: Orizonturi Universitare Publishing House; 2003

[50] Sebastian B. Environmental Technologies of Vegetable Growing. Iasi, Romania: No. 2, “Terra Nostra” Publishing House; 2004

Chapter 2

Abattoirs: The Hidden Sources of Plants' Heavy Metals and Other Pollutants in Lagos, Nigeria

Mautin Lawrence Ogun, Olajide Solomon Anagun, Olasunkanmi Kayode Awote, Surukite Opeolu Oluwole, Sesi Christiana Kappo and Faith Oseremi Alonge

Abstract

Abattoirs are places where animals are slaughtered and processed for human consumption leading to the production of huge wastes. Abattoir wastes contain several pollutants, most of which have growth limiting effects on soil microbes, plants, animals, and the entire ecosystem. A larger fraction of these wastes contains heavy metals. Heavy metals present in abattoir wastes are often acquired by plants through bioaccumulation, biomagnification and bioconcentration and remain persistent via food chain in the ecosystem. Most abattoirs in the developing nations such as Nigeria (Lagos) lack good personnel, equipment, and healthy practices. These ineffective management practices often provide bedrock for the occurrence of several negative effects evident in disease, disruption of wellness and so on. To prevent this effects, good abattoir waste management such as burying, composting, rendering, anaerobic digestion, blood processing, incineration with proper policies, laws and regulations must be put in place and enforced by necessary government agencies especially in Lagos State, Nigeria to minimize the pollutants released into the ecosystem. .

Keywords: abattoirs, heavy metals, pollutants, soil, water, air, plants, bioaccumulation, biomagnification, diseases

1. Introduction

Lagos State is one of the thirty-six (36) states in Nigeria with population estimated at over 200 million people. It is one the major industrial and business hubs of Nigeria. It has twenty (20) Local Government areas which are subdivided into five main divisions—Ikeja, Badagry, Ikorodu, Lagos and Epe divisions [1]. However, being the cash-driven center of the nation, it is without its own peculiar challenges. One of the prevailing challenges of Lagos, without exceptions from other parts of the country and the world is pollution.

Pollution involves the introduction of any material (solid, liquid or gas) or types of energy (heat, sound, or radiation) that is hazardous to the ecosystem-plants, animals, and human health [2]. Also, it could further be seen as any reaction, by individuals or bodies, which changes the biochemistry, metabolically and/or environment of other living things in a wider or localized area, where the causal link is clearly established [3, 4]. Pollution may be natural like flood, drought, cyclones, and many others and artificially (human activities) mediated and could be grouped into air pollution (affecting the health of the atmosphere), soil pollution (reducing the biodiversity of the soil and its health status) and water pollution (affects the overall quality and safety of water bodies). Several substances have been attributed to the rise in pollution levels in the state and these substances are generally referred to as pollutant [5].

Pollutants are materials that when released into these ecosystems (air, soil, and water) makes them intolerable to the inhabitants. Examples of some pollutants include heavy metals—lead, methane, carbon monoxide, particulate matter, bad smell, plastic, volatile organic compound, and many others have been attributed to pollutions [6]. The majority of man-made fine particles come from domestic sources, specifically biomass-based cooking, and the use of fossil fuels to power homes and vehicles. Some areas near deserts in Africa and West Asia are also important contributors due to windblown dust [6].

However, several sources have been documented which could either be point sources like industries, refineries, mining and many others or non-point sources like car, busses, trains and many more. But abattoirs in Lagos State of Nigeria have also been documented to house several heavy metals and other pollutants [7]. One of these contaminants that is frequently present in trace amounts is heavy metals. Many of these metals, even when present in trace amounts, are harmful to the environment and can accumulate in the bodies of living things through a process called bioaccumulation or bio-concentration [8, 9]. Abattoirs produce a significant quantity of secondary waste materials in the form of massive animal feces [10]. The inappropriate disposal of animal parts, such as flesh, blood, and innards, has been linked to an increase in soil acidity [11, 12]. Soil microorganisms, not just plants, are negatively affected by heavy metals present due to improper treatment of abattoir wastes. When soil is highly concentrated in metals, it can bring heavy metals into the food chain, which can be problematic because of the toxicity of these elements to humans, especially when they bioaccumulate in the body through the ingestion of plants.

2. Abattoir as sources of environmental pollutants

The impacts of pollution on ecosystems are much more pronounced today than they were in the past. Many people assert this, citing various factors such as a decline in soil fertility, a weakening of water quality, several health issues (including those that can lead to metabolic disorder), an ecological impact, and so on [13–15]. A major contributor to these pollutions is the careless disposal of trash in these ecosystems, which disrupts the balance of nature. Since there are no well-established management rules on waste disposal, this is a typical practice in Nigeria. However, the sudden increase in the amount of communicable, non-communicable and zoonotic diseases, for example, cancer and tuberculosis in our localities today makes abattoir waste a disease surveillance focus [16]. Abattoir operations result in the release of various wastes and pathogenic organisms that pollutes the environment and pose serious threat to human health and quality of life [17]. Tragically enough, most abattoirs in Lagos, Nigeria are known for poor and obsolete structural design which is accompanied with deteriorating environment.

An abattoir is a facility specifically designed to carry out the inspection of animals, the sanitary slaughtering, processing, and proper preservation, and storage of meat products for human consumption and is registered and certified by the regulatory body. A meat processing plant is a specialized building authorized to receive animals for slaughter, hold them, and conduct quality control inspections on meat products prior to their release for sale. Abattoirs aim to improve the efficiency with which consumable parts of the meat processing cycle are recovered for human consumption [10]. However, substantial quantities of secondary waste, such as organic and inorganic substances that aren't fit for further exploitation, are produced anyway [18]. Land degradation, air pollution, water scarcity and contamination, loss of biodiversity (particularly plant life), and climate change are all blamed on animal production and the work of veterinary establishments like slaughterhouses [19, 20]. Abattoir waste is an emerging solid waste whose rate of generation is becoming alarming [21].

The soil and natural water bodies are often seriously threatened by the pollution caused by abattoir activities, which not only generate a large quantity of waste but are also a source of heavy metals, bacteria, and others that threaten plant health and human quality of life [17].

2.1 Types of abattoirs

2.1.1 Modern abattoir

They are the pinnacle of conventional abattoir design, equipment, and services, and are typically constructed and managed by centralized governments with the aid of foreign technical experts. These abattoirs are managed on factory lines and offer a variety of services, including cold storage, processing, by-product usage, and waste recycling. Being commercial or profit-driven establishments, most contemporary abattoirs have little interest in providing low-revenue services, such as direct slaughter for public consumption.

2.1.2 Slaughterhouses

Most public slaughters are performed in these establishments. These establishments essentially provide a place to slaughter animals in accordance with public health, inspection, and marketing regulations for a fee, and are used by licensed butchers and dealers. Typically, only operating in densely populated areas and larger cities, these businesses provide essential services under the watchful eye of state and regional authorities. Two types of slaughterhouses exist, they are:

2.1.3 Old slaughterhouses

These establishments just provide the necessary infrastructure for licensed butchers and traders to slaughter cattle in line with public health, inspection, and marketing standards, for the predetermined costs. They are service businesses supervised by city or state governments, typically catering to residents in densely populated urban regions. Most public slaughters are conducted by them.

2.1.4 Makeshift slaughterhouses

They are more often found in more rural or suburban settings (**Figure 1**). They could happen at the outskirts of major cities, though. In the second case, it is not recommended that they exist or continue to operate since their disregard for evident requirements in slaughterhouse architecture, equipment care, and hygiene is occasionally linked to illegal livestock dealing and the slaughter of sick and diseased animals. Animals and their products should only be permitted to leave these facilities if they are being inspected [22].

2.2 Processing of animal products

Animal slaughter involves many stages, and there are several procedures and legislative requirements involved however, several methods are used in processing animal products in abattoirs [23]. The type of method used is dependent on the desired products. The following methods of animal products processing is common in Lagos state, Nigeria.

Boiling: Animal skin (known as ponmo in the Nigeria) is dipped into hot water to soften it so the hair can be more easily removed. To further soften the cow skin, it is steeped in water for several hours after being shaved, which induces a brief period of fermentation. The result of this procedure is ponmo of the white variety.

Burning: This is one of the most prevalently used methods for cow skin. In this technique, the animal skin is first softened by being put into a fire, at which point the hair is singed off and the softening process may begin. Used tires are used to create the flame, and other petrochemical agents (such kerosene, diesel, or gasoline) are often added to increase the intensity of the fire (**Figure 2**). This procedure yields the brownish ponmo [24].

2.2.1 Evisceration

Evisceration involves the removal of the viscera. This includes the heart, lungs and kidneys from both cattle, sheep, and cow. If not properly inspected by the meat inspector, this method is mostly prone to conditions such as tapeworm (e.g. *Taenia saginata*), liver flukes, abscesses, cysts and tumors [23].

Figure 1.
A makeshift slaughterhouse.

Figure 2.
Processing of Ponmo.

2.2.2 Deboning

Deboning of meat is one of the several methods in the processing of animal meat, during which muscle, connective and or adipose tissue (meat), is removed from the bone content. This process is often done manually or using advance equipment.

2.3 Abattoir waste and its contamination potentials

Wastewater and solid waste from slaughterhouses can also be considered abattoir waste since they contain potentially harmful substances such feces, blood, fat, trimmings, paunch content, and urine [25]. Inadequate management or control of abattoir wastes can have negative consequences for human health, the environment, animal welfare, and the national economy. Solid, liquid, and gaseous waste categories exist for this category. Manure, feces, hair, horn, hoof, gallbladder, trimmings, internal organs, bones, condemned corpses or body parts, paper, carton, and plastics all fall under the category of solid wastes. Slaughterhouse liquid waste comprises of feces, blood, and wastewater. Slaughterhouse gaseous waste consists mostly of odors and emissions [26]. The contamination could be in.

2.3.1 Contamination of surface waters

There are oxygen-demanding substances in slaughterhouse scraps. Therefore, the quality of surrounding waterways can be impacted by runoff from abattoir waste piles. Fish mortality could arise from a lack of dissolved oxygen and the toxicity of ammonia in these streams. In addition, the nutrients (nitrogen and phosphorus) in abattoir effluent can promote eutrophication (excessive vegetative growth) in stream channels, which might diminish the size of receiving stream channels and lead to over-flooding and its concomitant damages. Abattoir effluent is known to degrade the physical and chemical quality of streams and may even provide a health risk to people who engage in water sports and other similar activities [27].

2.3.2 Contamination of underground water

Pollutants from abattoir wastes make their way into the earth and can degrade water quality [27]. Organic ground water pollutants manifest themselves through altered flavor, odor, foaming, or damage to irrigated crops. In their investigation, Elemile et al. [28] found that the quality of groundwater increased as distance from the slaughterhouse was decreased. Agbara abattoir in Lagos State was the subject of a separate study by Jimoh et al. [29], which examined the environmental effects of the wastes produced there. They found a high coliform level of 82.50 cfu/ml, and they also found residues of chromium.

2.3.3 Contamination of the abattoir environment

Abattoir wastes typically give off pungent odors that could be a source of localized air pollution and a nuisance to nearby residents. Certain odorous substances, such as sulphides, mercaptans, amines, organic acids, etc., are notoriously difficult to get rid of. They can adhere to fabric, last for extended periods of time, and be carried large distances [30].

2.3.4 Contamination of plants

Most abattoir waste runoff is a possible source of plant pollutants, especially heavy metals, because of rainfall. To make matters worse, the effluent flows and spreads to the surrounding habitats from most abattoirs in Lagos, Nigeria, which contain solid wastes, feces, corpse, horns, bits of tissue, etc. [5]. Animal wastes are known to include pathogenic organisms, producing salmonellosis, leptospirosis, tularemia, foot and mouth disease, hog cholera, etc. [31], and if abattoir effluent-polluted waterways are used to cultivate fruits and vegetables, transmission of illnesses is possible. This could cause heavy metals to build up in the food web.

3. Heavy metal contaminated abattoir wastes: mechanism of transfer in ecosystem

Abattoirs have repercussions for the environment because of the way they operate and dispose of their trash. Bleeding, treating wounds, removing the hide, eviscerating the animal (removing its internal organs), preparing the carcass (by cutting and boning it), and so on are all steps in the surgical. In many circumstances, disagreeable odors are produced alongside the massive amounts of solid waste and wastewaters with biochemical oxygen demands (BOD) that might be generated by any of these procedures. The effluents that are relevant and specific to abattoirs in Lagos include blood, excrement, hair, bones, and undigested stomach contents [32].

The processes involved in getting end-products in most abattoirs in Lagos State of Nigeria are unhealthy considering the type of abattoirs found, the personnel and the skills used in the processes [7]. However, these processes used in abattoirs in Lagos, often lead to the introduction, accumulation, and transfer of contaminants like heavy metals such as lead, cadmium, zinc etc. within the ecosystem [33].

Abattoir wastes such as organic and inorganic substances, as well as salts and chemicals added during processing, affects air, water, soil, plants, animals, and humans [33] (**Figure 3**). Heavy metals and other contaminants from the abattoirs

DOI: http://dx.doi.org/10.5772/intechopen.110339

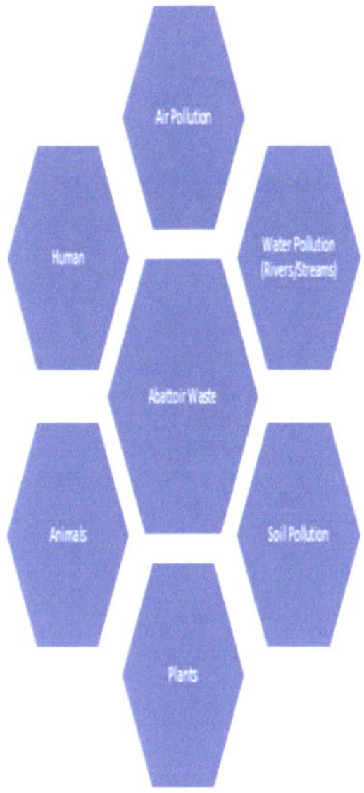

Figure 3.
Heavy metals' mechanism of transfer within the ecosystem from abattoir wastes.

firstly caused air pollution to the inhabitants and surrounding commuters very close to the abattoirs, leading to the inhalation of some virulent microorganisms which can cause air borne diseases [34] and irritation to the eyes and nose because of smokes and dust from animal skin (Ponmo) processing often done with burning of vehicular tyres.

Secondly, heavy metals in soil ecosystem (**Figure 3**), through discharge of abattoir effluents on soils thereby increasing heavy metal contaminants and other pollutants [12]. This effluent could kill the soil microbiota or increase the presence of virulent microbes which often causes diseases to plants, animals, and humans. The soil is the complex ecosystem of many plants. Thus, accumulation of metals has been reported in the soil and plants close to the abattoirs [7].

Thirdly, metals in surface and underground waters (**Figure 3**), this occurs through erosion and leaching of heavy metals in abattoir effluents. Improper discharge of effluents could lead to transfer of metals by rainwaters to nearby water bodies and increase the concentrations of metals in the aquatic bodies (**Figure 3**). This often discomfort the aquatic fauna and flora and increase the BOD of the water [28]. Also, leaching of this effluent containing heavy metals such as Pb and other contaminants could predispose the consumers of this water to toxins [35].

Finally, the mobilization of the metals into the air, soil, and water due to poor abattoir wastes management and personnel skills could not free plants, animals, and humans from the heavy metals' poisons [36, 37]. Plants that are contaminated with heavy metals because of bad abattoir practices accumulate these metals in their tissues, herbivores and humans consume these plants, and continuous consumptions of these herbivores and plants by humans leads to bioaccumulations of these metals in tissues and remain in food chain, thereby causing lots of health problems (**Figure 4**).

However, the continuous discharge of abattoir wastes and its poor management overtime, because of bioaccumulation, biomagnification and bioconcentration of heavy metals and other contaminants in the ecosystem in Lagos State, especially as it's involved the heavy metal transfer would pose great threat to air, water, soil, plants,

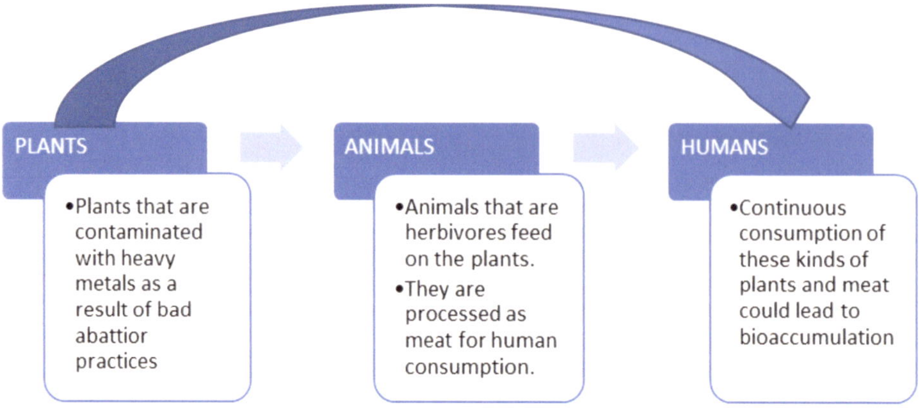

Figure 4.
Trend of pollutants' transfer within the food chain from abattoir wastes.

animals and human if not properly addressed (**Figure 4**). This in turn, may have negative effects on the economy of the country.

4. Effects of heavy metals on ecosystem

Most often found in greater concentrations in abattoir waste, heavy metals are hazardous to soils, plants, aquatic life, and human health. To exert their toxicity towards soil biota, heavy metals interfere with vital microbial processes, leading to a decrease in both the diversity and activity of soil microorganisms. The uptake of heavy metals by plants can be inhibited at low quantities, which can lead to their accumulation along the food chain and a possible threat to the health of animals and humans. However, at the level of aquatic systems, contaminants such as heavy metals encourage the creation of reactive oxygen species (ROS), which may harm fishes and other aquatic creatures.

4.1 Effects on soil

In industrialized countries, the issue of heavy metals such as copper, nickel, cadmium, zinc, lead, and thallium (etc.) in the soil is of critical relevance [38]. Soil microbial functions, such as respiration rate and enzyme activity, are important markers of soil contamination, and are typically impacted by an increase in metal concentration. Heavy metals have been shown to negatively impact soil's biological, physicochemical, and biochemical properties [39–41]. Heavy metal contamination can affect the size, composition, and activity of the microbial population, which can then have knock-on impacts on a variety of plant quality and yield indices [42]. Because of this, heavy metals are thought to be a significant contributor to soil pollution.

Toxic effects of heavy metals on soil biota include interference with vital microbial functions and a subsequent decrease in the diversity and abundance of soil microbes. As stated by Shun-Hong et al. [43], heavy metals can alter the composition of soil microbes, which in turn affects soil enzyme activity. But prolonged exposure to heavy metals can boost bacterial tolerance, which is useful for cleaning up polluted environments [44]. Heavy metals led to a decline in the bacterial population. The abundance and variety of bacteria living in polluted soils may have suffered because of this shift. The chemical affinities of enzymes in the soil system for various metals have been shown to have an impact on their respective activity, as described by Karaca et al. [45]. For instance, cadmium's (Cd) rapid mobility and low affinity for soil colloids makes it more hazardous to enzymes than lead (Pb). Lead (Pb) greatly reduces the activities of urease, catalase, invertase, and acid phosphatase; copper (Cu) lowers-glucosidase activity more than cellulose activity. Protease, urease, alkaline phosphatase, and arylsulfatase activities are negatively impacted by cadmium contamination, whereas invertase is unaffected [44]. As with enzymes, the sensitivity of different soil organisms to different metals varies.

4.2 Effects on plants

Some heavy metals, including arsenic (As), cadmium (Cd), mercury (Hg), lead (Pb), or selenium (Se), are not necessary for plant growth because they do not perform any recognized physiological function in plants, although others, like iron (Fe), copper (Cu), cobalt (Co), and zinc (Zn), are [46]. Chlorosis, stunted development, and decreased production are just some of the detrimental consequences that heavy metals can have on plants. Heavy metals can inhibit nitrogen fixation in plants, alter plant metabolism, and impair nutrient uptake. Temperature, pH, organic matter, moisture, and nutrient availability are just few of the variables that affect the uptake and accumulation of heavy metals in plant tissue. Root absorption is a significant pathway for heavy metals to enter the food chain and potentially affect animals and humans [46, 47]. Plant species and the efficiency of plant uptake of metals determine the accumulation of heavy metals in plants [48]. Lead (Pb) in soils can have a negative effect on plant productivity, and even trace amounts of lead can impede photosynthesis, cell division, and water uptake. Dark green leaves, elder leaves wilting, reduced growth, and brown, short roots are all possible manifestations of toxic effects [49].

4.3 Effects on aquatic environment

Extreme oxidative stress could be caused by even trace levels of heavy metals in aquatic species. Consequently, these contaminants are crucial to study in the field of ecotoxicology. Moreover, metals are frequently not degraded by microbes and hence persist in the marine environment indefinitely [50]. However, heavy metal contamination of a river could have catastrophic consequences for the aquatic ecosystem, reducing diversity of aquatic creatures and upsetting the delicate balance of the aquatic environment [51].

Particulate matter emitted into aquatic systems typically contains heavy metals, which settle and become a part of sediments. When it comes to metals and other pollutants in water, surface sediment is the most significant reservoir or sink. Aquatic macrophytes and other species with deep root systems can absorb sediment-bound contaminants [52]. Heavy metals could enter the food chain when an aquatic organism accumulates them. Many of the heavy metals used by carnivores like humans are found in marine life. The presence of fish makes this much more significant, as it could cause biomagnifications [51]. Increased formation of reactive oxygen species (ROS) due to the presence of heavy metals in aquatic systems is harmful to fish and other aquatic creatures [50]. Heavy metals are just one example of the environmental pollutants that could poison fish. Consequences for public health from these contaminants could be devastating. It is vital to be mindful of the sorts of fish you eat because of the potential health risks associated with heavy metal intake [53]. Mercury (Hg) is a significant contaminant due to the damage it could do to marine life and the health problems it can create for humans. Bacteria in watery sediments methylate organic mercury, forming a highly poisonous chemical compound known as methylmercury. Methylmercury makes up almost all the mercury found in fish muscles [53].

4.4 Effects on human health

Heavy metals in soil could be taken up by plants, and then by animals farther down the food chain, which could have serious consequences for human health. Growing plants in soil contaminated with heavy metals, such as that found near slaughterhouses, poses a threat since plant tissues can acquire these toxins [46]. When heavy metals are not broken down in the body, they accumulate in fat and muscle and become poisonous [54]. Negative impacts on human health are seen over long periods of time due to this buildup [48].

Toxic cadmium (Cd) has a specific gravity 8.65 times that of water, making it a heavy metal. Liver, placenta, kidneys, lungs, brain, and bones are particularly vulnerable to Cd poisoning [54]. It has been challenging to link morbidity and mortality to Cd's environmental exposures, even though exposure to Cd has been linked to a wide range of clinical conditions, including anosmia, cardiac failure cancers, cerebrovascular infarction, emphysema, osteoporosis, proteinuria, and cataract formation in the eyes [55].

When administered orally, zinc (Zn) is quite secure. Overexposure to Zn could cause systemic dysfunctions that limit growth and reproduction. Zinc poisoning has been associated with gastrointestinal symptoms, hemorrhagic cystitis, icterus (yellow mucus membrane), hepatic failure, renal failure, and anemia [56].

Copper (Cu) is a component of metalloenzymes where it could donate or take electrons, making it a critical element for mammalian nutrition. Diet and drinking water are the two most common routes of Cu exposure for people. Ingestion by mistake is the most common cause of acute Cu poisoning, while certain persons may be more vulnerable due to genetics or illness [57]. Mucosal irritation and corrosion, extensive capillary damage, hepatic and renal damage, central nervous system irritation, and depression may result from excessive Cu intake in humans. Necrotic abnormalities in the liver and kidney are also possible, in addition to severe gastrointestinal discomfort. When exposed to Ni, people could experience a wide range of symptoms, from skin irritation to problems with their lungs, nervous system, and mucous membranes [58].

Humans are extremely vulnerable to lead's (Pb) harmful effects on their physiology and nervous systems. Kidney, reproductive system, liver, and brain malfunction are possible outcomes of acute lead poisoning [59]. Even in trace amounts, Pb is the most dangerous element [60]. The synthesis of hemoglobin is inhibited by lead poisoning, and the cardiovascular system and the central nervous system (CNS) and the peripheral nervous system (PNS) are both damaged acutely and chronically (PNS). Anemia, exhaustion, gastrointestinal issues, and a lack of oxygen are some more long-term consequences. Low birth weights, hypertension, and muscle and joint pain are just some of the problems that lead exposure can bring [56, 59].

A strong oxidizing agent, caustic, soluble in alkaline and mildly acidic water, poisonous, and a possible carcinogen, chromium-Cr (VI) is harmful to plant and animal life [43, 61, 62]. Cr (VI) toxicity results from the fact that it can oxidize biological molecules despite being able to diffuse across cell membranes [61].

Mercury is poisonous and has no recognized biological or physiological function in humans. Inorganic mercury is linked to spontaneous abortion, congenital deformity, and gastrointestinal diseases (such as corrosive esophagitis and hematochezia) [56].

5. Management of abattoir wastes

Abattoir waste, like any other sort of waste, could be dangerous to humans and ecosystems if proper measures are not taken. Some slaughterhouse byproducts and wastes need to be repurposed as agricultural and industrial byproducts, but this requires recycling. Proper waste management techniques are necessary since this constitutes a significant threat to public health and is an annoyance in most slaughterhouses located in various market areas. This is because it pollutes the air, land, plant, and water and causes an infestation of flies and other disease vectors.

The risk of introducing enteric infections and extra nutrients into surface water is greatly exacerbated when slaughterhouse waste is disposed of unchecked into waterways. Furthermore, the extensive wastes produced by abattoir operations have been related to lower environmental air quality, potential patterns of transferable antibiotic resistance, plants contamination and various pathogenic pathogens with the potential to infect humans. But the following techniques are recommended for efficient abattoir waste management.

1. Burying: Most abattoirs use this approach, and it's the best option available. Landfills used for the disposal of abattoir waste must be covered soon after they are used, have a system in place to prevent wildlife from gaining access, and retain records of the locations and volumes of waste disposed of there.

2. Composting: In this, abattoir waste or carcasses are layered between absorbent carbon sources like wood chips, shavings, bark, barn animal bedding, hay, straw, etc. in a compost pile. Composting works best when the right amount of carbon and nitrogen sources are used. If the compost pile were built correctly, proper composting may be achieved without the need for manual turning or mechanical aeration.

3. Rendering: This involves extensive treatment of animal wastes into a more usable form.

4. Incineration: This process involves the combustion of substances present in the waste material. This method has the advantage of reduced waste per volume, lowered cost of waste management and in some cases a way to generate energy.

5. Anaerobic digestion: This is a series of event that occurred mostly in a bioreactor in which microorganisms utilize the materials present in the abattoir waste in the absence of oxygen leading to the formation of biogas. The abattoir waste is a protein-rich substrate and may result in sulphide formation during anaerobic digestion. The increased concentration of sulphides in the bioreactor can lead to higher concentrations of hydrogen sulphide in the biogas produced which may inhibit methanogens.

6. Blood processing: In this process, blood is removed from the body and sent to a treatment facility. Businesses use blood to create a range of goods that contain or could include beneficial nutrients.

Finally, proper policies, laws and regulations must be put in place and enforced by necessary government agencies especially in Lagos State, Nigeria to minimize the pollutants released into the ecosystem.

6. Conclusion

From the foregoing discourse, it safe to say that most abattoirs especially those in Lagos generate wastes that are loaded with heavy metals and other pollutants. However, these heavy metals are known to affect the ecosystem vis-à-vis disruption of soil health, excessive accumulation in plants and animals, depletion of water quality and reduction in atmospheric quality. These often have a great impact on the earth biota; thus, efforts should be made on mitigating the effects caused by heavy metal contaminated abattoir wastes through judicious abattoir waste management practices and regulations. Thus, abattoirs in Lagos are hidden sources of heavy metals in plants and other pollutants within the ecosystem.

Conflict of interest

The authors declare no conflict of interest.

Author details

Mautin Lawrence Ogun[1]*, Olajide Solomon Anagun[2], Olasunkanmi Kayode Awote[3], Surukite Opeolu Oluwole[1], Sesi Christiana Kappo[1] and Faith Oseremi Alonge[1]

1 Department of Botany, Lagos State University, Ojo, Nigeria

2 Department of Microbiology, Lagos State University, Ojo, Nigeria

3 Department of Biochemistry, Lagos State University, Ojo, Nigeria

*Address all correspondence to: mautin.ogun@lasu.edu.ng; mautinogun@gmail.com

References

[1] Odewunmi SG. Lagos State: Location, Extent and Constituents. Lagos State University- General Studies Book of Readings: Lagos and its Environs. Centre for General Nigerians Studies. Vol. 2. Ojo, Lagos State, Nigeria: Lagos State University Press; 2006. pp. 1-18

[2] Nathanson JA. Environmental Pollution. 2021. Available from: www.britannica.com/science/pollution-environment [Accessed: September 30, 2022]

[3] Khopkar SM. Environmental pollution monitoring and control. New Delhi, India: New Age International. 2007. pp. 32

[4] Vesilind PA, Peirce JJ, Weiner RF. Environmental pollution, and control. Amsterdam, Netherland: Elsevier; 2013. pp. 33-40

[5] Ezeoha SL, Ugwuishiwu BO. Status of abattoir wastes research in Nigeria. Nigerian Journal of Technology. 2011;**30**(2):143-148

[6] United Nations Environment Programme (UNEP). Emission Gas report 2022. Available from: https://www.unep.org/resources/emissions-gap-report-2022 [Accessed: October 20, 2022]

[7] Oluwole SO, Ogun ML, Oyetunji OS, Asokere SY, O'Daniel SC, Shittu AO, et al. Environmental assessment of heavy metal concentrations of *Colocasia esculenta* L., *Vernonia amygdalina* Del. and *Amaranthus spinosus* L. and soils collected around some abattoirs in Lagos State, Nigeria. *British*. Journal of Environmental Sciences. 2021;**9**(6):2054-2636

[8] Wong SC, Li XD, Zhang G, Qi SH, Min YS. 2002 heavy metals in agricultural soils of the Pearl River Delta. South China. Environmental Pollution. 2002;**19**:33-44

[9] Seaward MRD. The use of lichens for environmental impact assessment. Symbiosis. 2004;**37**:293-305

[10] RMAA (Red Meat Abattoir Association). Waste Management in red Meat Abattoir. Pretoria: South Africa; 2010. p. 40

[11] Jukna C, Jukna V, Suigzdaite J. Determination of heavy metals in viscera and muscles of cattle. Bulgarian Journal of Veterinary Medicine. 2006;**9**(1):35-41

[12] Chukwu UJ, Anucha SO. Impact of abattoir wastes on the physicochemical properties of soils within Port Harcourt Metropolis. The International Journal of Engineering and Science (IJES). 2016;**5**(6):17-21

[13] Daniel-Kalio LA, Braide SA. The effect of oil spill on a cultivated wetland area of Niger Delta. Nigerian Environmental Society Journal. 2004;**2**(2):153-158

[14] Yahaya MI, Mohammad S, Abdullahi BK. Seasonal variations of heavy metals concentration in abattoir dumping site soil in Nigeria. Journal of Applied Sciences and Environmental Management. 2009;**13**(4):9-13

[15] Okoye CO, Okwute GA. Heavy metal concentrations in food crops grown in crude oil impacted soils in Olomoro, delta state-Nigeria and their health implications. International Journal of Science and Engineering Invention. 2014;**3**(3):15-21

[16] Nwanta JA, Onunkwo JI, Ezenduka VE, Phil-Eze PO, Egege SC. Abattoir operations

and waste management in Nigeria: A review of challenges and prospects. Sokoto Journal of Veterinary Sciences. 2008;7(2):61-67

[17] Oruonye ED. Challenges of abattoir waste management in Jalingo Metropolis, Nigeria. International Journal of Research in Geography. 2015;**1**(2):22-31

[18] Steffen R, Kirsten I. Water and Wastewater Management in the Red Meat Industry Pretoria, South Africa. Pretoria: Water Research Commission; 1999. p. 36

[19] Steinfeld H, Gerber P, Wassenaar TD, Castel V, Rosales M, Rosales M, et al. Livestock's Long Shadow: Environmental Issues and Options. Rome: Food and Agriculture Organization of the United Nations; 2006. p. 270

[20] PSA (Philippine Sanitation Alliance). Slaughterhouses, Sanitation Technology Information Sheet. 2008:1-3

[21] Dán Á, Rónai Z, Széll Z, Sréter T. Prevalence, and genetic characterization of Echinococcus spp. in cattle, sheep, and swine in Hungary. Parasitology Research. 2018;**117**(9):3019-3022

[22] Clottey AJ. Manual for the Slaughter of Small Ruminants in Developing Countries. 3rd ed. Rome: FAO Consultant; 1995

[23] Hearle HJ. The Layout and Processes of an Abattoir. Food Safety: Meat Inspections; 2016. pp. 1-5

[24] Koleosho F. Exploring Exciting Culinary Experiences Using Indigenous Food Ingredient. Taste And Flavours of West Africa; 2015: Available from: www.funkekoleosho.blogspot.com [Accessed: September 30, 2022]

[25] Bandaw T, Herago T. Review on abattoir waste management. Global Veterinaria. 2017;**19**(2):517-524. DOI: 10.5829/idosi.gv.2017.517.524

[26] Estonilo LA. Standard Abattoir Guidelines of Manila. Laguna. Manila: Lake Development Authority (LLDA) Heightens Antipollution Campaign; 2006. p. 50

[27] Chukwu O, Mustapha HI, Abdul-Gafar HB. The effect of Minna abattoir wastes on surface water quality I. Environmental Research Journal. 2008;**2**(6):334-338

[28] Elemile OO, Raphael OD, Omole DO, Oloruntoba EO, Ajayi EO, Ohwavborua N. Assessment of the impact of abattoir effluent on the quality of groundwater in a residential area of Omu-Aran. Nigeria Environmental Sciences Europe. 2019;**31**:11-20

[29] Jimoh TY, Olayeri AA, Falebita TE, Reis GA, Olufowobi BI, Saba AO, et al. Environmental impacts of Agbara abattoir waste on the ecosystem of Ologe Lagoon, Nigeria. Journal of Advances in Biology and Biotechnology. 2022;**25**(5):11-17. DOI: 10.9734/JABB/2022/v25i530281

[30] Roberts H. Waste handling practices at red meat abattoirs in South Africa. 2011. Available from: http://www.sagepub.com/content/27/1/25 [Abstract retrieved 25th June 2022]

[31] Onwuka C, Eboatu AN, Ajiwe VIE, Morah EJ. Pollution studies on soils from crude oil producing areas of rivers state, Niger delta region, Nigeria. Open Access Library Journal. 2021;**8**(9):1-17

[32] Olawuni PO, Daramola OP, Soumah M. Environmental implications of abattoir waste generation and Management in Developing Countries: The case of Lagos state Abattior in Agege Nigeria. Greener Journal of Social Sciences. 2017;7(2):007-014

[33] Ediene VF, Iren OB, ldiong MM. Effects of abattoir effluent on

the physicochemical properties of surrounding soils in Calabar metropolis. International Journal of Advance Research. 2016;**4**:37-41

[34] Rabah AB, Oyeleke SB, Manga SB, Hassan LG, Ijah UJJ. Microbiological and physico-chemical assessment of soil contaminated with abattoir effluents in Sokoto metropolis, Nigeria. Science World Journal. 2010;**5**(3):1-4

[35] Ogbonnaya C. Analysis of groundwater pollution from abattoir waste in Minna, Nigeria. Resource Journal of Dairy Sciences. 2008;**2**(4):74-77

[36] Alonge DO. Textbook of Meat Hygiene in the Tropics. Ibadan, Nigeria: Farmcoe Press; 1991. pp. 58-57

[37] Alonge DO. Meat and Milk Hygiene in the Tropics. Ibadan, Nigeria: Farmose Press; 2005. pp. 77-86

[38] Hinojosa MB, Carreira JA, Ruız RG, Dick RP. Soil moisture pre-treatment effects on enzyme activities as indicators of heavy metalcontaminated and reclaimed soils. Soil Biology & Biochemistry. 2004;**36**:1559-1568

[39] Tóth G, Hermann T, Da Silva MR, Montanarella LJEI. Heavy metals in agricultural soils of the European Union with implications for food safety. Environment International. 2016;**88**:299-309

[40] Liu H, Xu F, Xie Y, Wang C, Zhang A, Li L, et al. Effect of modified coconut shell biochar on availability of heavy metals and biochemical characteristics of soil in multiple heavy metals contaminated soil. Science of the Total Environment. 2018;**645**:702-709

[41] Tang J, Zhang J, Ren L, Zhou Y, Gao J, Luo L, et al. Diagnosis of soil contamination using microbiological indices: A review on heavy metal pollution. Journal of Environmental Management. 2019;**242**:121-130

[42] Yao H, Xu J, Huang C. Substrate utilization pattern, biomass, and activity of microbial communities in a sequence of heavy metalpolluted paddy soils. Geoderma. 2003;**115**:139-148

[43] Shun-hong H, Bing P, Zhi-hui Y, Li-yuan C, Li-cheng Z. Chromium accumulation, microorganism population and enzyme activities in soils around chromium-containing slag heap of steel alloy factory. Transactions of Nonferrous Metals Society of China. 2009;**19**:241-248

[44] Singh J, Kalamdhad AS. Effects of heavy metals on soil, plants, human health, and aquatic life. International Journal of Research in Chemistry and Environment. 2011;**1**(2):15-21

[45] Karaca A, Cetin SC, Turgay OC, Kizilkaya R. Effects of heavy metals on soil enzyme activities. In: Sherameti I, Varma A, editors. Soil Heavy Metals, Soil Biology. Vol. 19. Berlin, Heidelberg: Springer; 2010. pp. 237-265

[46] Jordao CP, Nascentes CC, Cecon PR, Fontes RLF, Pereira JL. Heavy metal availability in soil amended with composted urban solid wastes. Environmental Monitoring and Assessment. 2006;**112**:309-326

[47] Sprynskyy M, Kosobucki P, Kowalkowski T, Buszewsk B. Influence of clinoptilolite rock on chemical speciation of selected heavy metals in sewage sludge. Journal of Hazardous Materials. 2007;**149**:310-316

[48] Khan S, Cao Q, Zheng YM, Huang YZ, Zhu YG. Health risks of heavy metals in contaminated soils and

food crops irrigated with wastewater in Beijing, China. Environmental Pollution. 2008;**152**:686-692

[49] Bhattacharyya P, Chakrabarti K, Chakraborty A, Tripathy S, Powell MA. Fractionation and bioavailability of Pb in municipal solid waste compost and Pb uptake by rice straw and grain under submerged condition in amended soil. Geosciences Journal. 2008;**12**(1):41-45

[50] Woo S, Yum S, Park HS, Lee TK, Ryu JC. Effects of heavy metals on antioxidants and stress-responsive gene expression in Javanese medaka (*Oryzias javanicus*). Comparative Biochemistry and Physiology, Part C. 2009;**149**:289-299

[51] Ayandiran TA, Fawole OO, Adewoye SO, Ogundiran MA. Bioconcentration of metals in the body muscle and gut of *Clarias gariepinus* exposed to sublethal concentrations of soap and detergent effluent. Journal of Cell and Animal Biology. 2009;**3**(8):113-118

[52] Peng K, Luo C, Luo L, Li X, Shena Z. Bioaccumulation of heavy metals by the aquatic plants *Potamogeton pectinatus* L. and *Potamogeton malaianus* Miq. and their potential use for contamination indicators and in wastewater treatment. Science of the Total Environment. 2008;**392**:22-29

[53] Soliman ZI. A study of heavy metals pollution in some aquatic organisms in Suez Canal in port- said harbour. Journal of Applied Sciences Research. 2006;**2**(10):657-663

[54] Sobha K, Poornima A, Harini P, Veeraiah K. A study on biochemical changes in the freshwater fish, catla catla (Hamilton) exposed to the heavy metal toxicant cadmium chloride. Kathmandu University Journal of Science, Engineering and Technology. 2007;**1**(4):1-11

[55] Lalor GC. Review of cadmium transfers from soil to humans and its health effects in the Jamaican environment. Science of the Total Environment. 2008;**400**:162-172

[56] Duruibe JO, Ogwuegbu MOC, Egwurugwu JN. Heavy metal pollution and human biotoxic effects. International Journal of Physical Sciences. 2007;**2**(5):112-118

[57] Stern BR, Solioz M, Krewski D, Aggett P, Aw TC, Baker S, et al. Copper, and human health: Biochemistry, genetics, and strategies for modeling dose response relationships. Journal of Toxicology and Environmental Health, Part B. 2007;**10**:157-222

[58] Argun ME, Dursun S, Ozdemir C, Karatas M. Heavy metal adsorption by modified oak sawdust: Thermodynamics and kinetics. Journal of Hazardous Materials. 2007;**141**:77-85

[59] Odum HT. Background of published studies on Lead and wetland. In: Odum HT, editor. Heavy Metals in the Environment Using Wetlands for their Removal. New York USA: Lewis Publishers; 2000. p. 32

[60] Kazemipour M, Ansari M, Tajrobehkar S, Majdzadeh M, Kermani HR. Removal of lead, cadmium, zinc, and copper from industrial wastewater by carbon developed from walnut, hazelnut, almond, pistachio shell, and apricot stone. Journal of Hazardous Materials. 2008;**150**:322-327

[61] Shaffer RE, Cross JO, Pehrsson SLR, Elam WT. Speciation of chromium in simulated soil samples using X-ray absorption spectroscopy and

multivariate calibration. Analytica Chimica Acta. 2001;**442**:295-304

[62] Jeyasingh J, Philip L. Bioremediation of chromium contaminated soil: Optimization of operating parameters under laboratory conditions. Journal of Hazardous Materials B. 2005;**118**:113-120

Chapter 3

Heavy Metals in Soils Following 50 Years of Sewage Sludge Application

Thandile Mdlambuzi, Pardon Muchaonyerwa and Awonke Mbangi

Abstract

Heavy metal contamination has increasingly become an environmental problem. While it is found in soils naturally through processes of weathering of parent materials, it is the anthropogenic activities that create the greatest threat. A study was conducted to investigate the vertical distribution of heavy metals in soils after over 50 years of sewage sludge application. Soil samples were collected at 10 cm intervals to a depth of 50 cm from five treated transects and a control. The soils were analyzed for zinc, copper, lead, nickel, cadmium, arsenic and chromium. The concentration of all the metals was higher in the treated soils compared to the control. The results were compared with two parameters: the total maximum thresholds (TMT) and maximum permissible limits (MPL). The TMT is the concentration of the metal beyond which the risk to the environment is unacceptable, while MPL is the concentration beyond which further waste disposal is prohibited. Zinc, chromium, lead and cadmium were above maximum permissible limits, in treated soils. High concentrations of all the metals, including Pb, and organic carbon were measured down to 40–50 cm depth. Only Cd (and Pb only in transect 2) was above the maximum permissible limits beyond the 20–30 cm depth.

Keywords: heavy metals, soils, sewage sludge, soil physicochemical properties, contamination

1. Introduction

Large quantities of sewage sludge generated globally present disposal challenges [1, 2]. High energy required for incineration and the scarcity of landfill space have made land application a major disposal option [3]. Land application of sewage sludge could benefit from the contents of organic matter and plant essential nutrients [4]. Sewage sludge from the Vlakplaas Wastewater Treatment Plant, South Africa, was found to contain 20–23% total carbon (C), 1.9–3.1% total nitrogen (N), 40–166 mg available phosphorus (P) kg^{-1} and 689–3804 mg potassium (K) kg^{-1} over a 4-year period [5]. Feasibility of using sewage sludge as a nutrient source could be limited by its composition of heavy metals, including cadmium (Cd), lead (Pb), chromium (Cr), mercury (Hg), arsenic (As), nickel (Ni), zinc (Zn) and copper (Cu) [1, 6]. The metals can be sorbed on soil colloids, lost through leaching to ground water or taken up by plants growing on contaminated sites. Soil conditions, sludge metal concentration and loading rates could determine the accumulation, mobility and fate of these metals, through interaction with soil colloids, pH and P.

A number of laboratory leaching tube and glasshouse studies have been conducted to determine effects of different soil properties, including pH, P and organic matter (OM) content, or other characteristics on mobility of selected heavy metals (HM) in soil [7–11]. Water solubility and phyto-availability of Zn, Cd and Pb were found to be reduced by P additions, with greater effects on Pb [10, 12]. Dissolved OM was found to enhance the mobility of Ni and Cu, whereas Zn mobility was not modified, in soil [13]. Leaching of Cu was found to increase with decline in pH, with the lowest mobility occurring at pH 5–7 [14, 15]. Kumpiene et al., [16] concluded that it was not feasible to make long-term predictions based on short-term standardized laboratory tests.

Long-term field experiments are impractical, and sampling and analysis of soils contaminated decades before, relative to adjacent uncontaminated soil, could be an alternative [16]. The longest studies of this nature were conducted on a site that had received a once-off treatment with sewage sludge 15–20 years before sampling [8, 17] reported no substantial vertical movement of Cd, Cu, Ni and Zn in soil, whereas calculated metal deficits suggested that there could have been leaching losses. Accumulation in the soil could result in high metal concentrations in tissues of volunteer and indigenous vegetables, with serious health risks. Indigenous and volunteer exotic vegetables grow on polluted sites and could be harvested and consumed. Addition of sewage sludge could modify soil pH, available P and OM, and their interactions with HM could determine the fate of the metals under field conditions. Limited studies have been conducted on effects of these soil properties (pH, OM and P) on the mobility of a mixture of HM under field conditions.

Darvill Waste Water Works (DWWW), South Africa, has continuously applied sewage sludge containing different HMs on a dedicated site for over 50 years. This long-term sewage sludge application site provides an opportunity to understand what happens to HMs in the soil under field conditions with time. The continuous application of sewage sludge could have resulted in increases in pH, OM, available P and HM concentrations in the soil. The forms of the HM, which depend on soil pH and their interaction with OM and P, could determine their fate in soil and their accumulation in plants, with risks on human health and plant tissue. It is therefore essential to investigate the effects of the long-term application of sewage sludge on the mobility and concentrations of HM with increasing soil depth. The objective of this study was to determine the effects of 50 years of sewage sludge application on the distribution of heavy metals, in the soil profile and selected physicochemical property composition on different transects of a loam soil.

1.1 Study area

The study was carried out at a dedicated sewage sludge application site at DWWW in Pietermaritzburg (PMB) (29.602500°S to 29.61139°S and 30.433900°E to 30.43861°E), KwaZulu-Natal, South Africa (**Figure 1**). The site has a mean annual rainfall of 680 mm and mean annual temperature of 18.4°C. The soil is formed from Ecca shale, a laminated carbonaceous sedimentary rock formed from the deposition of clastic sediments [18]. Over 250 m^3 of thickened sludge (about 3% solids) is produced per day and applied by sprinkler irrigation on 57 ha of land, per day, which translates to an average of about 48,000 kg solid sludge ha^{-1} $year^{-1}$.

The area treated with sludge was divided into five transects. A control transect, adjacent and upslope to the study field, was also included (**Figure 1**). Parts of

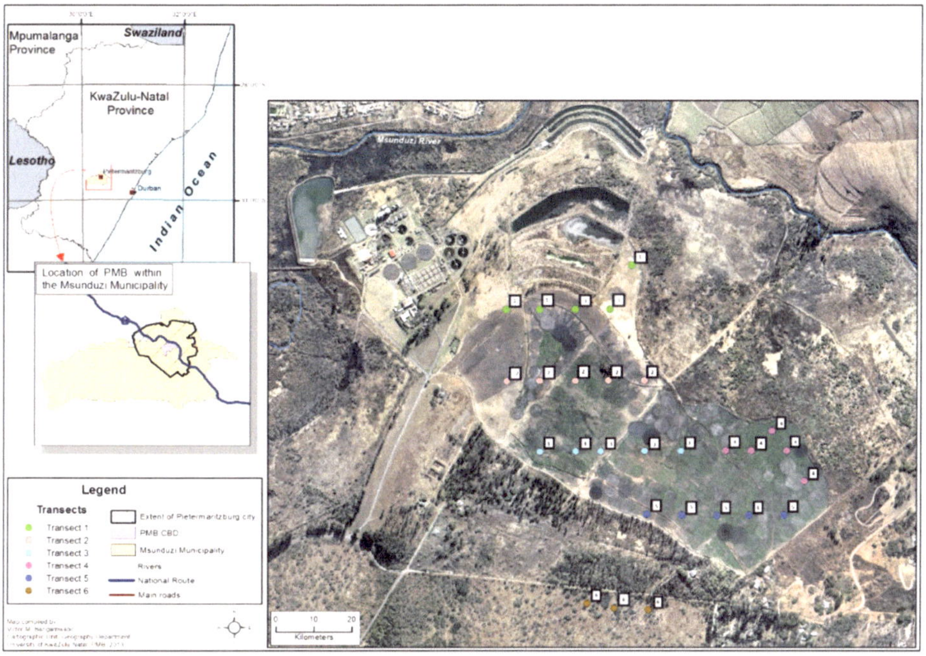

Figure 1.
Sampling points at the sewage sludge disposal site.

transects 1 and 4 were on lower slope positions, and transect 4 received drainage water from transects 3 and 5 and was always wet. Commercial turf grass production is currently practiced on the site, and the harvesting involves removal of the sludge-rich soil attached to the root system.

2. Materials and methods

Soil samples were collected from all transect points with three replicates around each point, at 0–10, 10–20, 20–30, 30–40 and 40–50 cm depths with a bucket auger. Samples from each depth per transect were mixed to get a composite sample and were oven-dried at 38°C for 72 h, ground with a mortar and pestle, and sieved (< 2 mm). Particle size analysis was only done for the 10–30 cm depth, after removal of the top 0–10 cm depth, which was mainly made up of OM. Particle size analysis was conducted using the double pipette procedure after removal of OM with hydrogen peroxide (H_2O_2) and dispersion with sodium hexametaphosphate solution [19]. Soil pH was determined in H_2O (1:5 soil: water). Total C was determined using a Leco TruMac CNS/NS Determinator (Leco Corporation). The available P was extracted with 0.25 M ammonium bicarbonate, EDTA disodium salt and 0.01 M ammonium fluoride (AMBIC) solution [20] and analyzed following the molybdenum blue calorimetric method [21], using the UV/VIS spectrophotometer Thermo Fisher Scientific model Genesys 20.

Total heavy metal concentrations were analyzed after extraction with aqua regia [22]. Soil samples (0.5 g) were treated with 12 ml concentrated HCl (32%) and 4 ml concentrated HNO_3 (55%) and digested in a microwave digester (EPA 3051H-HP500) at 175°C for 10 minutes. An aliquot of the digest (5 ml) was diluted to 50 ml with de-ionized water, in acid-washed glass test tubes, and analyzed for Zn, Cu, Ni, Cr, Pb, Cd and As with an inductively coupled plasma optical emission spectrometer (ICP-OES 720 Varian). The results were compared with total maximum thresholds (TMT) and maximum permissible limits (MPL) from the WRC South Africa [23]. The TMT is the concentration of the metal beyond which the risk to the environment is unacceptable, while MPL is the concentration beyond which further waste disposal is prohibited.

2.1 Statistical analysis

One-way analysis of variance (ANOVA) was used to compare pH, total C, available P and total heavy metal data across transects at each soil depth using Genstat Release 12.1 (Lawes Agricultural Trust, Rothamsted Experimental Station, Harpenden, UK).

3. Results and discussion

3.1 Soil physicochemical properties

Soils in all transects were loamy with 15–20% clay at all depths, except transect 1, which had 10–11% clay (**Table 1**). The silt contents ranged from 38 to 44%, while the sand content was 35–45%. The control soil was also loamy with 15% clay, 15% silt and 69% sand. The lower clay content and lower total C (**Table 2**) beyond the 0–10 cm depth of transect 1 could be explained by the removal of OM-rich soil during harvesting of turf grass for sale as an instant lawn.

Soil pH (H_2O) ranged from 4.6 to 5.4 at all depths in transect 1 and the control (4.9–5.1) and from 5.3 to 6.7 for all other transects, with pH 6.0–6.7 in the 0–10 cm depth (**Table 2**). Transect 1 and the control had lower pH than all other transects within the 0–30 cm depth, but there were no differences in the 30–40 cm depth. The total C was ≥9% in the 0–30 cm depth in transect 1 and in the top 30 cm of transects 2, 3, 4 and 5 ranged between 13 and 17%, while the control had 2% in the top 30 cm depth (**Table 2**). Transect 1 and the control had lower total C than all other transects, except the 0–10 cm depth, where all sludge-treated soils had similar levels. Total C rapidly decreased with depth beyond the 20–30 cm depth for all transects, with >2.5% at the 40–50 cm depth of transects 2, 3, 4 and 5. The available P was

Transect	Clay	Silt	Sand
		%	
1	11 ± 0.6	42 ± 2.2	45 ± 3.3
2	16 ± 2.1	44 ± 1.1	36 ± 4.1
3	17 ± 2.0	43 ± 2.3	41 ± 4.2
4	20 ± 1.2	41 ± 0.4	37 ± 1.3
5	15 ± 1.5	38 ± 10.5	35 ± 1.5
Control	15 ± 0.2	15 ± 1.1	69 ± 1.4

Table 1.
Soil particle size distribution of 10–30 cm depths of the different transects (mean ± standard deviation).

Parameter	Transect	0–10	10–20	20–30	30–40	40–50
pH	1	5.1	4.6	5.0	5.3	5.4
	2	6.0	5.8	5.6	5.7	5.9
	3	6.4	6.2	5.5	5.5	5.4
	4	6.4	6.0	5.7	5.8	6.4
	5	6.7	6.5	5.7	5.3	5.3
	Control	5.0	4.9	4.9	5.1	5.1
	LSD	0.63	0.47	0.64	ns	0.60
Total C (%)						
	1	15.7	7.8	4.5	2.2	1.7
	2	17.9	15.8	17.6	5.4	4.1
	3	13.6	15.6	17.3	5.6	5.7
	4	19.5	18.1	17.9	4.2	2.6
	5	14.5	13.6	13.3	4.9	3.5
	Control	2.4	2.2	2.0	1.9	1.8
	LSD	10.19	5.94	4.29	1.35	0.31
Available P (mg kg^{-1})						
	1	421	246	—	30	6
	2	444	402	284	23	16
	3	413	328	319	26	22
	4	308	396	394	29	22
	5	370	325	268	43	37
	Control	21	20	21	10	8
	LSD	67.4	45.0	43.6	2.0	6.1

Table 2.
Mean pH, total C contents and available P of soils used in the study.

>37 mg kg^{-1} in the 0–10 cm depth and declined with depth to 25–32 mg kg^{-1} in the top 30 cm of sludge-treated soil, except transect 4, where P increased from 30.8 to 39.6. The control had less than 22 mg P kg^{-1}, which did not change with depth, while for the other transects, the levels rapidly declined from the 20–30 cm depth to the 30–40 cm depth.

The higher pH values in transects with higher total C indicated that the sludge had a liming effect, which could affect the mobility of the HMs and the availability of P as seen in **Table 2**. Sewage sludge in South Africa has been found to have pH ranging from 6.4 to 6.7 [24], and 50 years of continuous application of large amounts of such sludge could have limed the soil from pH 5.0 to 6.0–6.7. The available P in the sludge-treated soil could have originated from food additives, dish washing and laundry detergents, personal care products [25], and human urine, which contains about 0.03% P [26]. The highly available P in soils treated with sewage sludge was in agreement with [5], who found that double the recommended rate of sewage sludge application increased the available P over a 4-year period (**Table 3**).

3.2 Heavy metal concentrations in soil

The HM concentrations in the sludge-treated soils than in the control, and their correlation with total C, suggested that the metals originated from the sludge. Variation in metal concentrations across transects treated with sewage sludge could be the result of clay content and slope position and non-uniform loading rates over the years. The increased HM concentration in soils on sewage sludge disposal land, compared with the control, was in agreement with [27]. The low Ni and the similar As between treated and control soil suggested that the long-term sludge application did not increase Ni and As levels, possibly due to low concentrations in the sludge. The high As even in control samples is an indication that local geology contributed significant quantities of As upon weathering of the rocks. The higher concentrations than the MPL of Cr (450 mg/kg), Zn (700 mg/kg), Pb (150 mg/kg) and Cd (5 mg/kg) posed a risk of leaching to ground water, toxicity to soil organisms and accumulation in plant tissue due to uptake. These risks depend on the mobility of these metals, which is affected by soil properties, including soil texture, organic matter content, pH and available P levels.

3.2.1 Chromium, zinc and copper

The strong positive correlations of pH, soil C and available P with Cr, Zn and Cu indicated that the mobility of these heavy metals was significantly affected by all three factors. The higher Cr, Zn and Cu (**Figures 2–4**) concentrations in the top 30 cm of transects amended with sludge, which was in agreement with [27], could be a result of sorption and precipitation due to the increase in soil pH. Metal cations are known to favor sorption and precipitation at high pH levels, while at low pH, they become more available [27]. Although soil Cu concentrations were lower than the MPL, transects treated with sludge had significantly higher levels than the control at all depths. The lower pH and Cr, Zn and Cu concentrations in the top 30 cm of transect 1 than the others, and the concentrations of these HM and pH with depth, beyond the 20–30 cm depth, emphasize the importance of sorption of the metals on more negatively charged colloids at high pH. Repeated harvesting of turf grass and the soil attached to the roots explains the lower total C, P and HM especially Cr, Zn and Cu. Gao et al., [7] reported that mobility of Zn and Cu declined with increase in pH from pH 4.5 to 6.5, while Cr in CrO_4^{3-} was not affected in soils enriched with OM. Formation of the hydrous chromium hydroxide (hydroxide of Cr^{3+}), the species to

Transect	Electrical conductivity ($mS\ m^{-1}$)				
	0–10	10–20	20–30	30–40	40–50
1	90.1	93.4	94.8	96.3	95.2
2	80.5	87.3	84.5	89.9	90.6
3	87.4	83.1	90.1	91.9	84.4
4	81.3	84.7	85.0	82.6	81.6
5	81.4	86.0	85.3	95.8	101.2
Control	96.8	98.1	99.3	97.9	96.8

Table 3.
Mean EC of soils used in the study.

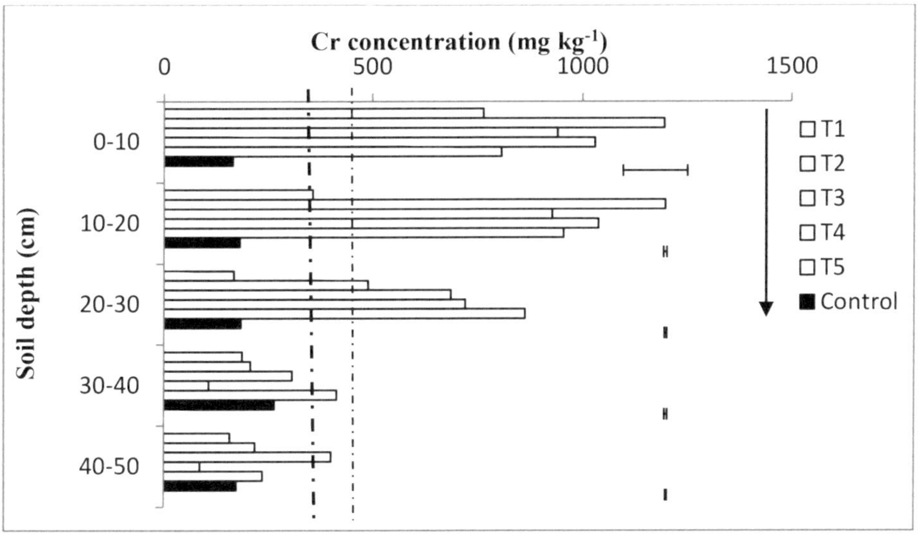

Figure 2.
Distribution of total chromium concentrations for all transects. Error bars represent least significant differences (LSD) at $p < 0.05$. TMT and MPL for Cr are 350 and 450 mg kg^{-1}, respectively [23].

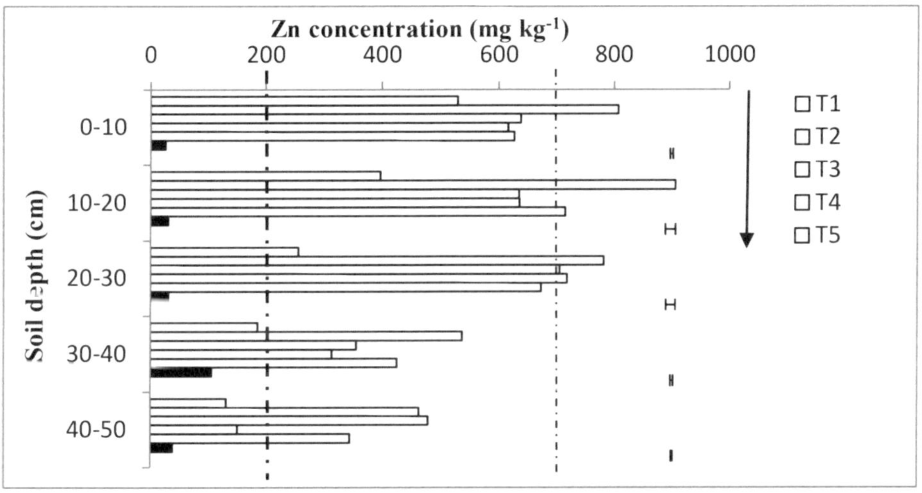

Figure 3.
Distribution of total zinc concentrations for all transects. Error bars represent least significant differences (LSD) at $p < 0.05$. TMT and MPL for Zn are 200 and 700 mg kg^{-1}, respectively [23].

which it rapidly converts under oxidizing soil conditions, could therefore have limited the mobility [16]. Whereas the effects of pH were evident, Cr and Zn concentrations were higher in transect 2 (>MPL) than in transect 5, which had higher pH, suggesting that other factors also contributed. Sorption to OM and clay minerals [27, 28] and formation of metal–organic matter complexes [27] could have resulted in the accumulation of Cr, Zn and Cu in the surface layers.

The accumulation of Cr in the surface layers (**Figure 2**) of the soil could be explained by complexation of the Cr^{3+} ion by OM, which was high in these layers, and

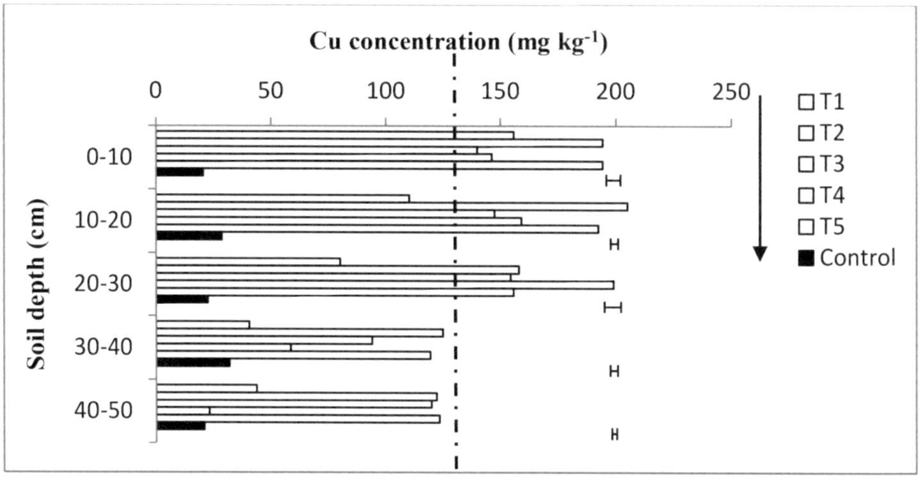

Figure 4.
Distribution of total copper concentrations for all transects. Error bars represent least significant differences (LSD) at $p < 0.05$. TMT and MPL for Cu are120 and 375 mg kg^{-1}, respectively [23].

chemisorption by silicate clays [27]. Copper and Zn tend to build up at the surface of contaminated soils with high OM, as a result of complex formation [27]. The lowest soil C and Cr, Zn concentrations in transect 1, while transect 2 had the highest than the other transects, together with the decline of these parameters with depth beyond the 30 cm depth, emphasized the close relationship of soil C with mobility of the metals. The high levels of these metals could have been toxic to soil microorganisms [29, 30], resulting in C sequestration in the soil and minimizing gaseous emissions. Any conditions that cause rapid oxidation of the >10% C in the top 30 cm depth of the soil could contribute significantly to CO_2 emissions.

The high available P in soils treated with sewage sludge suggested that the mobility of Cr, Zn and Cu could have been limited through precipitation as metal phosphates [10], leading to the accumulation of these metals in the top 30 cm [10, 12]. The high Cr, Zn and Cu levels in all transects treated with sewage sludge, and their drastic decline at deeper layers beyond the 0–30 cm depth for all transects, were clearly associated with trends in available P levels. For example, transect 2 had the highest available P and soil Cr, Zn (both >MPL) and Cu than the other transects in the top 30 cm. The exception was in the 0–10 cm depth of transect 1, where Cr, Zn and Cu concentrations were equal to, or higher than, those in transects 4 and 5, yetthe available P levels were lower, further suggesting that the available P alone does not fully explain the variations in these metals, emphasizing the role of OM and soil pH.

Although the metals declined beyond the 0–30 cm depth, the concentrations of Cr (except transects 1 and 4), Zn and Cu (except transect 4) remained higher than those of the control to a depth of 40–50 cm, suggesting that significant mobility occurred. Higher levels of organic C and total Cr, Zn and Cu (except transect 4 for Cr and Cu) at 40–50 cm depth could possibly be a result of the formation of mobile metal-soluble organic matter complexes [8, 31]. The displacement of dissolved organic matter from sorption sites by a high level of available P could have enhanced the mobility of the organic matter together with complexed metals to deeper layers. In the 40–50 cm depths of transects treated with sludge, soil C was more than twice that in the control, indicating mobility of C from the surface layers. Zhang and Zhang [11] reported that increased P application rates, to soil, resulted in elevated leaching of Zn and Cu, resulting in the formation of soluble Zn-dissolved organic matter complexes. Ashworth and Alloway [13] found that dissolved organic matter enhanced the mobility of Cu but not Zn. The findings were also contrary to [7] and [17].

Lead concentrations in the control ranged from 22 to 40 mg kg^{-1}. The concentrations were higher in soils treated with sewage sludge than those of the control at all depths. Transect 1 had the highest Pb concentration in the 0–10 cm depth, while transect 2 had the highest at all other depths (**Figure 5**). The highest Pb concentration (203 mg kg^{-1}) was in the 30–40 cm depth of transect 2. Whereas Pb remained high, with no clear trend with depth, it declined with depth for transect 1. Lead concentration was above the MPL (150 mg kg^{-1}) in transects 1 (0–10 cm), 2 and 4 (10–20 cm). Nickel concentrations in the control were in the range 7–15 mg kg^{-1}. The concentrations were higher in soils treated with sewage sludge than those of the control at all depths. Nickel accumulated in the top 30 cm and declined beyond that depth in all transects, except for transect 2 (**Figure 6**). Transect 2 had the highest concentration at all depths, with the greatest in the 40–50 cm depth. All transects had lower Ni than the MPL (200 mg kg^{-1}) (**Figure 6**).

The strong positive correlations of Ni, and the weak correlation of Pb, with soil C and available P, and not with soil pH, indicated that the mobility of these heavy metals was only affected by the former two factors. Although metal cations are known to favor sorption and precipitation at high pH levels, while at low pH they become more available [27], there was no clear trend between soil pH and Pb and Ni concentrations in our study. The accumulation of Pb and Ni levels in the surface layers of most transects could be explained by sorption to organic matter and clay minerals [27, 28] and was in agreement with McBride [27]. The highest soil C (similar to transect 4), Ni and Pb concentrations were in transect 2, throughout the depth of the profile, while transect 3 had generally increasing soil C, Ni and Pb within the top 30 cm, and they all declined beyond the 30 cm depth, emphasizing the close relationship of these parameters. In transect 4, soil C declined beyond the 20–30 cm depth, while soil Ni and Pb remained high down to the 30–40 cm depth. These trends emphasized the close relationship between soil C and concentrations of Ni and Ni.

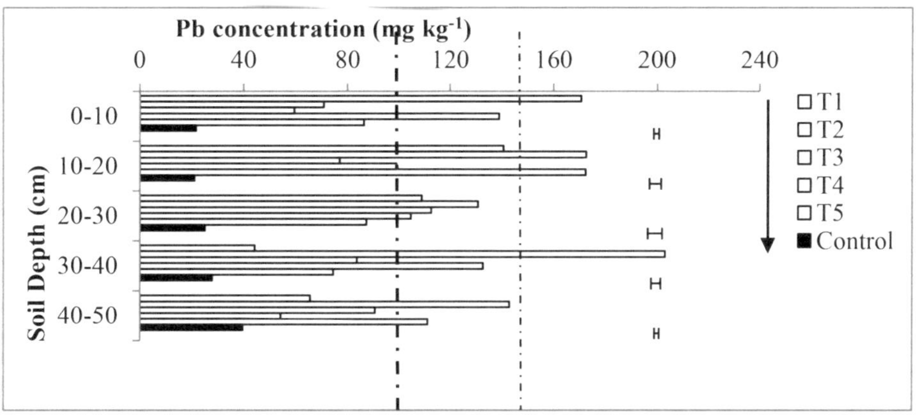

Figure 5.
Distribution of Pb concentrations for all transects. Error bars represent least significant differences (LSD) at $p < 0.05$. *TMT and MPL for Pb are 100 and 150 mg kg*$^{-1}$*, respectively [23].*

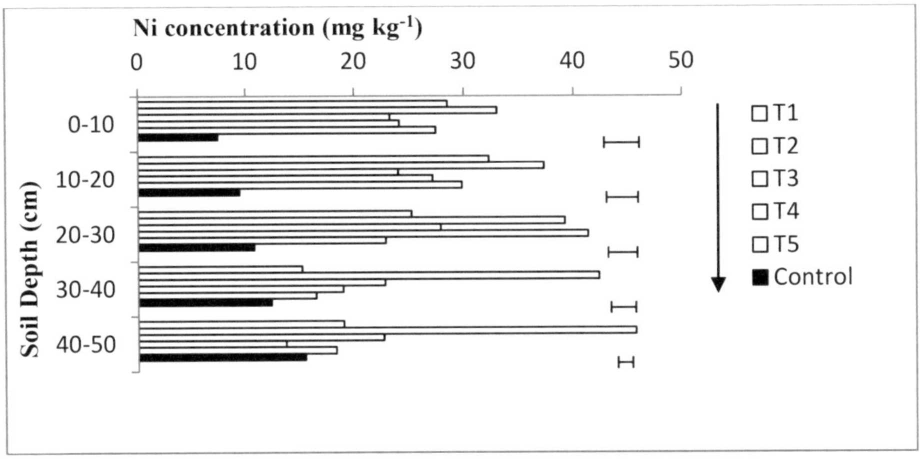

Figure 6.
Distribution of Ni concentrations for all transects. Error bars represent least significant differences (LSD) at $p < 0.05$. TMT and MPL for total Ni are 150 and 200 mg kg^{-1}, respectively [23].

The precipitation as metal phosphates [10] could also have contributed to moderating the mobility of Pb and N, which explains the accumulation of these metals in the top 30 cm of most treated transects [10, 12]. The available P and the total Pb and Ni were the highest in transects 1 (0–10 cm), 2 (10–30) and 4 (10–20 cm), which suggested the formation of Pb and Ni phosphate limited mobility. Solubility of Pb has been found to be reduced by the formation of insoluble Pb phosphate minerals [10, 32]. The accumulation of Pb in the surface soil could also have been toxic to soil microorganisms, resulting in sequestration of OM due to limited microbial degradation.

Contrary to the view that Pb tends to build up at the surface of contaminated soils with high organic matter, as a result of strong complex formation, with no downward movement [10, 27], the levels of Pb remained high beyond the 30 cm depth, which suggested that significant mobility had occurred. This was contrary to observations by [11]. The formation of mobile metal-soluble organic matter complexes could have resulted in the higher Pb, Ni and organic C in the 30–40 and 40–50 cm depths [8, 31]. This effect could have been enhanced by the displacement of dissolved organic matter from sorption sites by the high available P, resulting in the formation of soluble metal-dissolved organic matter complexes. Zhang and Zhang [11] reported that increased P application rates to soil resulted in elevated leaching of Ni. Beyond 30 cm, soil C, Pb and Ni concentrations were higher than in the control, and in transect 2, Pb concentrations were even higher than in MPL, while Ni continued to increase with depth, suggesting that factors other than soil C and available P also contributed.

3.2.2 Cadmium and arsenic

Cadmium concentration in the control ranged 1–3 mg kg^{-1}. The Cd was higher at all depths in soils treated with sewage sludge than the control (**Figure 7**). The 10–20 cm depth of transect 3 had the highest Cd concentration. Transect 2 had the highest Cd at all other depths. There was no decline of Cd concentration with depth for all transects. The Cd concentrations were at or above the MPL (5 mg kg^{-1}) at all depths in all transects treated with sewage sludge, except the 0–10 cm depth of transect 3.

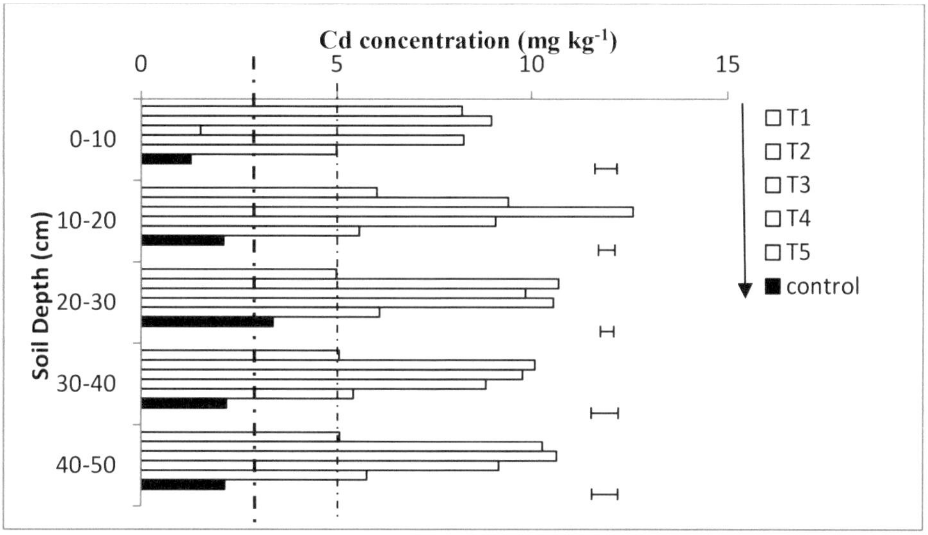

Figure 7.
Profile distribution of total Cd concentrations for all transects. Error bars represent least significant differences (LSD) at $p < 0.05$. TMT and MPL for Cd are 3 and 5 mg kg^{-1}, respectively [23].

All the layers of all transects had lower As than the MPL but exceeded the TMT limits in all transects. The control soil had a mean As of 5.6–7.4 mg kg^{-1}. The concentrations of As in transects 1, 2 and 3 at all depths (treated with sewage sludge) were higher than in the control. Transect 1 had the highest concentration in the 0–10 (11.3) and 10–20 (13.6) depth, whereas transect 2 had the highest in the 30–40 (13.9) and 40–50 cm (14.4 mg kg^{-1}) depth. While levels of As in transect 2 increased with depth, the concentrations remained constant in the other transects (**Figure 8**).

Correlation results suggested that soil pH and C, and not the available P, determined Cd mobility. At all depths, transect 1 had the least pH and Cd concentrations of transects treated with sludge, while the others had higher pH and Cd at all depths. The lower pH could be the result of sludge removal during the harvesting of turf grass at this transect. The mobility could have been related to sorption of the Cd^{2+} ion on the negative soil colloids at higher pH. Gao et al., [7] found that Cd mobility was found to decline with increase in pH from pH 4.5 to 6.5, in soils enriched with organic matter. In this case, the concentration of soil organic matter could therefore have affected the sorption. In the 0–10 cm depth, transects 1, 2 and 4 had the highest soil C and Cd concentrations, while at all other depths, transects 2, 3 and 4 had the highest soil C and Cd concentrations, which emphasized the importance of sorption of Cd to organic matter on mobility of Cd. Although Cd sorption is important, the weak sorption on organic matter, silicate clays and oxides at less than pH 7 [33] could have contributed the high total Cd (>MPL) beyond the 30 cm depth, which was in agreement with [34]. Formation of Cd-soluble organic matter complexes could also have resulted in the mobility and redistribution of the metal throughout the profile with greater risk for ground water pollution. The increased mobility also makes the element more available for plant uptake. The lack of reaction between Cd and P could explain the high concentrations of the metal at deeper layers.

The negative correlation between soil pH and As concentrations suggested that increasing pH increased the mobility of As with no effects of P and OM. Although As concentrations were below the MPL for all the transects, changes in soil pH affected

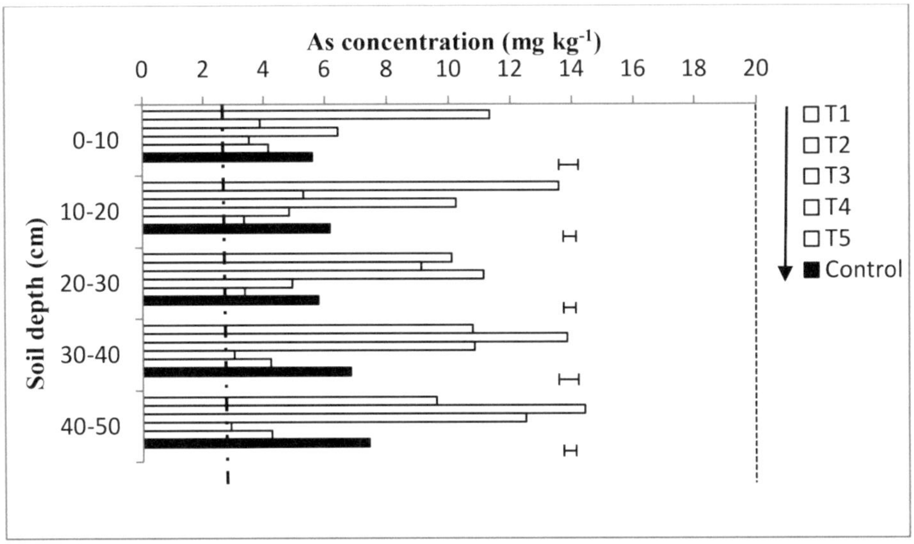

Figure 8.
Distribution of total arsenic concentrations for all transects. Error bars represent least significant differences (LSD) at $p < 0.05$. TMT and MPL for As are 2 and 20 mg kg^{-1} [23].

its mobility and accumulation in parts of the soil profile. The relationship with pH was particularly clear where transects treated with sludge had higher As concentrations than the control in part of the profile (transects 1, 2 and 3). For example, in the 0–10, 10–20 and 20–30 cm depths, transect 1 had the lowest pH and the highest As concentration, while transects 4 and 5 had the highest pH and the lowest As (lower than that of the control). Although transects 4 and 5 also had elevated pH in the surface layer, their As concentrations were lower than those of the control, and the effects of pH was therefore not evident. The reason for the lower As in these two transects than the control could be that the As had leached beyond the 40–50 cm depth, possibly into ground water, due to the elevated pH. On the other hand, in transects 2 and 3, As concentrations increased with depth whereas pH decreased. The elevation of pH due to sludge accumulation in transects 2 and 3 could have increased the negative charge on soil colloids, making As, in anionic form, to be more mobile and accumulating at deeper layers, where it gets sorbed on soil colloids with pH-dependent charge. The increased mobility of As could increase its availability to plants.

3.2.3 Correlation of soil pH, carbon and available P with heavy metal concentrations

Soil pH was strongly correlated with soil C and available P (**Table 4**). The soils treated with sewage sludge had total metal concentrations in the order Cr > Zn > C u > Pb > Ni > Cd = As. Soil Cr, Zn and Cu strongly correlated with soil pH, while Cd weakly correlated with soil pH. The correlation of pH with As was weak (significant) and negative. There was no correlation of pH with Pb and Ni. Total soil C was strongly correlated with the available P. Soil C was strongly correlated with all the metals except Pb, which was weakly correlated, and As, for which the correlation was not significant. The available P was correlated strongly with Cr, Zn, Cu and Ni, weakly with Pb and not correlated with As and Cd.

Parameter	pH	Total C	Avail P
Soil C	0.539*		
Avail P	0.471*	0.9928*	
Cr	0.578*	0.902*	0.879*
Zn	0.582*	0.891*	0.804*
Cu	0.540*	0.866*	0.812*
Pb	0.256	0.441**	0.416**
Ni	0.328	0.570*	0.500*
Cd	0.364**	0.482*	0.279
As	−0.386**	−0.184	−0.169

Soil C = total soil carbon, Avail P = available phosphorus. *Correlation is significant at p = 0.05 level.*
***Correlation is significant at p = 0.01 level.*

Table 4.
Correlation coefficients (r) of soil pH, total C and available P with soil-heavy metal concentrations.

4. Conclusions

Long-term land application of sewage sludge resulted in accumulation of most metals in the top 30 cm, with only Cr, Zn (only in the 0–30 cm depth for both), Cd and Pb exceeding maximum permissible limits, even in the deeper layers of the soil, with greater accumulation on lower slope positions. The accumulation of the metals in parts of the soil profile is affected by soil pH, available P, OM and relative concentrations of the metals.

Acknowledgements

The study was funded by the University of KwaZulu-Natal (UKZN), through a competitive grant, and the Department of Agriculture, Forestry and Fisheries (DAFF), through a research bursary. The views expressed in this article are not those of UKZN or DAFF. Staff of Umgeni Water provided assistance with site information and soil sampling, and Dr. C. Southway, of the Chemistry Discipline, UKZN, provided valuable inputs in sample analysis.

Conflict of interest

The authors declare no conflict of interest.

Author details

Thandile Mdlambuzi[1,2], Pardon Muchaonyerwa[2] and Awonke Mbangi[3*]

1 South African Sugarcane Research Institute, Blackburn, South Africa

2 School of Agricultural, Earth and Environmental Sciences, University of KwaZulu-Natal, Scottsville, South Africa

3 Department of Agriculture, Mangosuthu University of Technology, Umlazi, South Africa

*Address all correspondence to: mbangi.awonke@mut.ac.za

References

[1] Snyman HG, Van der Waals J. Laboratory and Field Scale Evaluation of Agricultural Use of Sewage Sludge. Gezina: Water Research Commission; 2004

[2] Röhrs LH, Fourie AB, Blight GE. Effects of sewage sludge and refuse composition on the rate of degradation and leachate quality of co-disposed waste in a water-deficient environment. Water. 1998;**24**:137-146

[3] Hue NV. Residual effects of sewage-sludge application on plant and soil-profile chemical composition. Communications in Soil Science and Plant Analysis. 1988;**19**(14):1633-1643

[4] Li C, Zhou K, Qin W, Tian C, Qi M, Yan X, et al. A review on heavy metals contamination in soil: Effects, sources, and remediation techniques. Soil and Sediment Contamination: An International Journal. 2019;**28**(4):380-394

[5] Tesfamariam EH, Annandale JG, Steyn JM, Stirzaker RJ, Mbakwe I. Municipal sludge as source of nitrogen and phosphorus in perennial pasture Eragrostis curvula production: Agronomic benefits and environmental impacts. Water. 2013;**39**(4):507-514

[6] Water Research Commission. Guidelines for the Utilisation and Disposal of Wastewater Sludge. Water Research Commission; 2006

[7] Gao S, Walker WJ, Dahlgren RA, Bold J. Simultaneous sorption of Cd, Cu, Ni, Zn, Pb, and Cr on soils treated with sewage sludge supernatant. Water, Air, and Soil Pollution. 1997;**93**(1):331-345

[8] McBride MB, Richards BK, Steenhuis T, Russo JJ, Sauvé S. Mobility and solubility of toxic metals and nutrients in soil fifteen years after sludge application. Soil Science. 1997;**162**(7):487-500

[9] Tu S, Ma LQ. Interactive effects of pH, arsenic and phosphorus on uptake of As and P and growth of the arsenic hyperaccumulator Pteris vittata L. under hydroponic conditions. Environmental and Experimental Botany. 2003;**50**(3):243-251

[10] Cao X, Wahbi A, Ma L, Li B, Yang Y. Immobilization of Zn, Cu, and Pb in contaminated soils using phosphate rock and phosphoric acid. Journal of Hazardous Materials. 2009;**164**(2-3):555-564

[11] Zhang M, Zhang H. Co-transport of dissolved organic matter and heavy metals in soils induced by excessive phosphorus applications. Journal of Environmental Sciences. 2010;**22**(4):598-606

[12] Sally B, Chancy R, Judith H. In situ treatment to reduce the phyto and bioavailability of lead, zinc and cadmium. Journal of Environmental Quality. 2004;**33**:522-531

[13] Ashworth DJ, Alloway BJ. Soil mobility of sewage sludge-derived dissolved organic matter, copper, nickel and zinc. Environmental Pollution. 2004;**127**(1):137-144

[14] Bergkvist B, Folkeson L, Berggren D. Fluxes of Cu, Zn, Pb, Cd, Cr, and Ni in temperate forest ecosystems. Water, Air, and Soil Pollution. 1989;**47**(3):217-286

[15] He ZL, Zhang M, Yang XE, Stoffella PJ. Release behavior of copper and zinc from sandy soils. Soil

Science Society of America Journal. 2006;**70**(5):1699-1707

[16] Kumpiene J, Lagerkvist A, Maurice C. Stabilization of As, Cr, Cu, Pb and Zn in soil using amendments—A review. Waste Management. 2008;**28**(1):215-225

[17] Richards BK, Steenhuis TS, Peverly JH, McBride MB. Metal mobility at an old, heavily loaded sludge application site. Environmental Pollution. 1998;**99**(3):365-377

[18] Richards NP, Botha GA, Schoeman P, Clarke BM, Kota MW, Ngcobo FN. Engineering Geological Mapping in Pietermaritzburg, South Africa: Constraints on Development. London: Engineering Geology Special Publication; 2009. p. 22

[19] Gee G, Bauder J. Particle-size analysis. In: Klute A, editor. Methods of Soil Analysis, Part 1. Madison, WI: American Society of Agronomy. Inc; 1986

[20] Non-affiliated soil analysis work committee. Handbook of standard soil testing methods for advisory purposes. Soil Science Society of South Africa, Pretoria. 2020;**1990**(160):10

[21] Murphy JA, Riley JP. A modified single solution method for the determination of phosphate in natural waters. Analytica Chimica Acta. 1962;**27**:31-36

[22] Chen M, Ma LQ. Comparison of three aqua regia digestion methods for twenty Florida soils. Soil Science Society of America Journal. 2001;**65**(2):491-499

[23] Herselman JE, Snyman HG. Guidelines for the utilisation and disposal of wastewater sludge. Water Research Commission. 2009

[24] Jaganyi D, Tesfai FK, Southway C, Mamabolo M. Determination of inorganic pollutants and assessment of the current south African guidelines on permissible utilisation of sewage sludges. Water. 2005;**31**(3):359-367

[25] Comber S, Gardner M, Georges K, Blackwood D, Gilmour D. Domestic source of phosphorus to sewage treatment works. Environmental Technology. 2013;**34**(10):1349-1358

[26] Kutu FR, Muchaonyerwa P, Mnkeni PN. Complementary nutrient effects of separately collected human faeces and urine on the yield and nutrient uptake of spinach (Spinacia oleracea). Waste Management & Research. 2011;**29**(5):532-539

[27] McBride MB. Environmental Chemistry of Soils. New York: Oxford University Press; 1994

[28] Hooda PS. Trace Elements in Soils. Chichester: Wiley; 2010;**618**

[29] Giller KE, Witter E, Mcgrath SP. Toxicity of heavy metals to microorganisms and microbial processes in agricultural soils: A review. Soil Biology and Biochemistry. 1998;**30**(10-11):1389-1414

[30] Griffiths BS, Diaz-Ravina M, Ritz K, McNicol JW, Ebblewhite N, Bååth E. Community DNA hybridisation and G+ C profiles of microbial communities from heavy metal polluted soils. FEMS Microbiology Ecology. 1997;**24**(2):103-112

[31] Camobreco VJ, Richards BK, Steenhuis TS, Peverly JH, McBride MB. Movement of heavy metals through undisturbed and homogenized soil columns. Soil Science. 1996;**161**(11):740-750

[32] Hettiarachchi GM, Pierzynski GM, Ransom MD. In situ stabilization of soil lead using phosphorus. Journal of Environmental Quality. 2001;**30**(4):1214-1221

[33] Hooda PS, Alloway BJ. The effect of liming on heavy metal concentrations in wheat, carrots and spinach grown on previously sludge-applied soils. The Journal of Agricultural Science. 1996;**127**(3):289-294

[34] Roy M, McDonald LM. Metal uptake in plants and health risk assessments in metal-contaminated smelter soils. Land Degradation and Development. 2015;**26**(8):785-792

Chapter 4

Some Metals Found in Stored Canned Fish Products Sold in Nigeria

Andress Edowaye Odiko

Abstract

Trace amount of some metals like zinc, iron, manganese is normally constituent of natural water bodies, canned products and fishes. Excessive levels of metals can become detrimental to consumers of fish and fish products especially those stored for prolonged period at varied temperatures. Toxicological and environmental concerns have therefore prompted interest into the investigation of some of these metals especially in canned fish stored in different media, sold in Nigeria. In order to curb preventable damage to consumers' vital organs as a result of unsuspecting elevated levels in consumed fish and fish products, this study will determine the presence of some metals in different storage media and compare their values with safe set-standard levels. Monthly mean concentration of metals was determined using Inductive Coupled Plasma-optical emission spectrometer (ICP-OES), and values compared with permissible set-values by different moderating organizations like the Codex, World Health Organization (WHO) among others when considering metal contamination. Based on all the different storage media, an average concentration of 2.88–29.45 mg/kg Fe, 7.04–72.09 mg/kg Zn, 0.09–0.67 mg/kg Mn, 0.55–5.61 mg/kg Ni, 0.02–0.27 mg/kg V were detected during the period of study. This study is therefore intended to encourage comprehensive periodic monitoring of canned products in Nigeria market.

Keywords: metals, canned fish, curry sauce, exposure, toxicity, Nigeria

1. Introduction

Many food-insecure nations are located in Africa, which happen to be the net importer of fish in terms of volume [1]. The Nigeria population which is estimated to be growing at 2.41% annually, depends massively on fish to meet the over 40% animal protein intake needs of its population. Demand for fish from industrial fishing countries around the world in the form of frozen and canned fish is valued at about $500 Million US Dollars or N250 Billion Naira per annum [2]. The state of World Fisheries and Aquaculture in 2022 reported that that global merchandise trade which increased at a rate of 6.8% per year in nominal terms between 1976 to 2020 and by 3.75% in real terms.

Fish consumption can supply over 50% high quality protein, low fat, essential poly unsaturated fatty acids, micro and macro nutrients [3]. World population growth has outweighed fish production due to increased fish and sea food consumption which have increased per capita across the world from as reported for as at 2017 [4].

Federal Department of Fisheries (FDF) and Food and Agriculture Organization (FAO) has previously reported Nigeria's self-sufficiency in fish production ratio of 98.8% in 1983, which dwindled to between 40% and 19.2% in 2005–2014. And with the present economic realities in Nigeria foreign exchange, weak institutional reforms and economic measures without the well withal to implement them; the challenge to increase protein consumption in Nigeria appears to be more urgent [5].

According to FAO statistics as reported [6], 6% of the 11% global population's animal protein intake is derived from fish. Presently, more than 30% of the world population is suffering from malnutrition, this is more evident in developing countries like Nigeria, where people experience just over one quarter of all possible deprivation as highlighted in the 2022 multidimensional poverty index survey which revealed that 63% (133 million people) of persons living within Nigeria are multidimensionally poor [7].

Carbohydrate which is the stable food in Nigeria is complemented richly with fish because of its provision of fatty amino acids, minerals and trace metals and it is also believed to be a healthier alternative to many other sources of animal protein. Until recently in Nigeria, fish was regarded as the cheapest and easily accessible animal protein source. And canned fish which has become very popular, makes marine fish available in different part of the world irrespective of season.

Humans are usually more exposed to toxic elements via food as they tend to absorb many of these pollutants from the natural environment, storage medium, storage conditions before sales among others [8]. Therefore, the determination of safety levels has become paramount due to the health challenges which hereto were not frequently diagnosed, but has become a recent trend in the poorly revitalized health sector of the country coupled with the dwindling value of the naira due to the high demand for dollars as a result of Nigeria's mono-economy.

Studies on the quality of frozen and canned fishes has attracted attention from relevant agencies and experts, as the beneficial and toxic level range of any product are very small [9]. Therefore, effects of exposure above certain threshold can be potentially life threatening especially if not properly diagnosed for immediate medical treatment.

Canned food are popular food sources all around the world [10], and it is a modern technological advancement in food processing, helps to increase shelf life and allow storage for several years. Operations in canning industries are manually and mechanically carried out, the final heat treatment gives protection from spoilage organisms. But this process of heat treatment is done in such a way as not to alter the flavor and texture of fish considerably [11].

Most of the marine fish are canned, thus, making it more available for consumption by humans living far away from sea sites [12]. Fish are constantly exposed to chemicals in polluted and contaminated waters; therefore, fish products need to be well established to be relatively free of excessive metals as some metals according to [13] can act as catalysts in the oxidative reactions of biological macromolecules resulting in induced alterations in almost all body functions and organ integrity. Metals can be classified as potentially toxic, probably essential and essential [14].

Many priority pollutants as classified by the United States Environmental Protection Agency (US EPA) have made their way into aquatic systems through anthropogenic sources and natural weathering. The major sources of pollution of surface waters include effluent discharges by industries, agricultural run offs, atmospheric depositions of pollutants and accidental spillage. Trace elements may be immobilized within the stream sediments and result in absorption, co-precipitate and form complexes by co-absorbing with other elements or form particulates as oxides hydroxides of Fe, Mn.

Toxicity and the resultant adverse effect based on its' concentration on humans has been of concern especially in developed countries which has resulted in the imposition of new and more restrictive regulations [15, 16]. Regulations in Nigeria are still not strictly followed due mostly to poor will-power and inadequate manpower. This has resulted in increased level of health concerns in the country, as more and more persons now embark on medical pilgrimages and increase pressure on the available medical facilities and personnel.

Canned fish and fish products considered by Nigerians as conventional, hygienic and tasty has resulted in their preference and patronage, but some of these products may be laden with deleterious contaminants whose effects may take a long time to manifest especially when consistently taken in large quantities.

During these last decades, environmental research on metal pollution on marine and freshwater environments has been conducted, focusing especially on the distribution of trace metals in freshwater as well as in seawater fish [17]. Since most of the marine fish are canned, it therefore becomes more available for consumption by humans living far away from sea sites [12].

Since metals present in seafood, at low concentration may be essential; however, they can become toxic having harmful effects when their intake exceeds the recommended quantities significantly [18–20].

Human exposure to heavy metals has risen dramatically in the last 50 years as a result of an exponential increase in the use of heavy metals in industrial processes and products [21]. Therefore, the lethal thresh of substances like zinc while working on the toxicity of zinc to *Clarias gariepinus* can be defined after a long period of exposure [22], which is characteristic of substances that may be slowly excreted or metabolized and at the same time a decreasing concentration may not appreciably increase the survival period of the fish. It is essential to identify the interaction between the foodstuff and its package, particularly when it is being purchased and consumed nationwide on a regular basis [23]. The most important heavy metals causing severe toxicity to fish are arsenic, lead, nickel, zinc, tin among others, as they are considered the most important form of pollutants of the aquatic environment because of their toxicity and accumulation by marine organisms [24]. Heavy metal pollution of the marine environment has long been recognized as a serious environmental concern as toxic elements can become very harmful even at low concentration when ingested over a long period of time [25, 26]. Heavy metals can be accumulated by marine organisms through a variety of pathways, including respiration, adsorption and ingestion [27, 28].

The aquatic environment is a major place for contamination by metals, as a result of bio-accumulation by fish during feeding, it may become bio-magnified in consumers of such contaminated fishes. The growing concern over the increasing level of heavy metals contamination, in agricultural, sea foods and industrially processed food products has therefore birthed this study. This study therefore aims to determine the levels of some metals (Fe, Zn, Mn and Ni) in six different brands of canned fish products stored in curry sauce, tomato sauce and soybean oil with a view to evaluating how safe these canned fish products in these storage media, sold in the Nigerian market are against set safe standards.

2. Materials and methods

2.1 Sample collection

Six different brands of canned fish products manufactured in three different Asian countries were purchased from open markets, wholesales shops and super markets (chain-stores) in Nigeria. Their storage media were tomato sauce, soyabean oil and curry sauce and the species canned were mackerel, sardine and skipjack tuna with an average weight of between 106 and 155 g/Can respectively. The expiration date of samples was between two to four years as at the time of study and bought samples were taken to the laboratory, kept in clean dry/cool place prior to digestion and analysis.

2.2 Digestion of samples

The canned fish preservative medium (tomato sauce, curry sauce and soybean oil) was carefully decanted in the laboratory, rinsed with deionized water and samples were oven dried to constant weight at a temperature of 80°C for 72 hours. Each sample in three replicates per brand/month was milled separately to powder form using a porcelain mortar and pestle. They were stored in labeled plastic packs, sealed and stored at −10°C prior to digestion and analysis during the period of study. Fish samples were digested using the organic extraction technique described by Sreedevi et al. [29].

2.3 Statistical analyses

Data obtained were analyzed using computer software (SPSS version 21). One way analysis of Variance (ANOVA) test was used in all cases to test for significant differences between means at 5% probability level. Significant treatment means were separated using the New Duncan's Multiple Range Test.

3. Results

The results of the mean concentration of study metals (Fe, Zn, Mn, Ni and V) detected in six brands of canned fish products and at different months of study stored in curry sauce, tomatoes sauce and soyabean oil are as shown in **Tables 1** and **2**.

In the months of study (February–October) the highest concentration of all the metals were recorded in October, while the least concentrations were observed at the beginning of the study as shown in **Table 1**.

Month	Fe	Zn	Mn	Ni	V
February	7.30 ± 1.70^{d}	19.86 ± 4.24^{c}	0.17 ± 0.04^{d}	1.39 ± 0.32^{d}	0.07 ± 0.02^{b}
April	8.22 ± 1.92^{bcd}	22.37 ± 5.21^{bc}	0.19 ± 0.04^{bcd}	1.57 ± 0.37^{bcd}	0.08 ± 0.02^{b}
June	9.87 ± 2.30^{bc}	26.84 ± 6.26^{ab}	0.23 ± 0.05^{bc}	1.88 ± 0.44^{bc}	0.10 ± 0.02^{b}
August	10.30 ± 2.40^{b}	22.37 ± 5.21^{bc}	0.24 ± 0.05^{b}	1.96 ± 0.46^{b}	0.10 ± 0.03^{b}
October	14.72 ± 3.43^{a}	31.98 ± 7.46^{a}	0.35 ± 0.08^{a}	2.80 ± 0.65^{a}	0.15 ± 0.04^{a}

Note: Different superscripts in the same column are significantly different.

Table 1.
Mean and standard deviation of concentrations of study metals (mg/kg) in different months.

Spp/storage media	Fe	Zn	Mn	Ni	V
Sardine/CurrySauce$_1$	29.45 ± 2.09^a	72.09 ± 3.64^a	0.67 ± 0.05^a	5.61 ± 0.51^a	0.22 ± 0.02^b
Mackerel/ TomatoSauce$_2$	6.15 ± 0.42^c	15.05 ± 0.72^c	0.14 ± 0.01^d	1.17 ± 0.08^c	0.05 ± 0.00^c
Mackerel/ TomatoSauce$_3$	3.20 ± 0.23^{de}	7.83 ± 0.41^{de}	0.09 ± 0.01^d	0.61 ± 0.04^d	0.02 ± 0.00^e
Mackerel/ TomatoSauce$_4$	2.88 ± 0.20^e	7.04 ± 0.34^e	0.66 ± 0.01^{ab}	0.55 ± 0.04^d	0.02 ± 0.00^e
Sardine/SoybeanOil$_5$	15.18 ± 1.05^b	37.17 ± 1.80^b	0.35 ± 0.24^c	2.89 ± 0.20^b	0.27 ± 0.02^a
Skipjack/SoybeanOil$_6$	3.64 ± 0.25^d	8.92 ± 0.43^d	0.11 ± 0.01^d	0.69 ± 0.05^d	0.04 ± 0.00^{cd}

Note: Different superscripts in the same column are significantly different.
Different subscript in spp./storage media column refers to different brands.

Table 2.
Mean and standard deviation of concentration of metals (mg/kg) in the different species storage media.

Concentration of the study metals in the six brands of canned fish analyzed varied within the months and between the storage medium, although some were not significantly different during the period of study. The highest concentration value of Fe (29.45 mg/kg), Zn (72.09 mg/kg), Mn (0.67 mg/kg), Ni (5.6 mg/kg) and V (0.27 mg/kg) in sardine all stored in curry sauce except V which was in sardine stored in soybean oil. The least value of Fe (2.88 mg/kg), Zn (7.04 mg/kg), Mn (0.09 mg/kg), Ni (0.55 mg/kg) and V (0.02 mg/kg) were all observed in mackerel stored in tomato sauces$_{(2–4)}$ as shown in **Table 2**. The results obtained were statistically tested.

The month of October had the highest mean concentration values across the individual heavy metals and the least concentrations at the beginning of study as shown in **Table 1** above. Using Analysis of variance (ANOVA) to statistically test the results obtained for significant difference and a further separation of means with the Duncan Multiple Range Test (DMRT) between brands storage media. It was observed that there was significant difference ($p > 0.05$) in the metal contents among them. The brand of canned sardine in curry sauce had higher concentration of all the study metals except for V, followed by sardine stored in Soybean oil within brands of canned fishes (**Table 2**); there was also in the different months significant difference ($p > 0.05$) in the metal concentration in the different months (**Table 1**). Mn, Ni and V were not significantly different ($p < 0.05$) during the months of study and in the brands irrespective of their storage media.

4. Discussion

Fish has been considered a good indicator for heavy metal contamination in aquatic systems because they occupy different trophic levels therefore the levels of toxic elements in fish are related to age, sex, season and habitat [30, 31]. Fish being widely consumed in many parts of the world as this has been made possible by advances in packaging technology especially the use of cans, such that any level of pollution in stored fish will endanger human health all over the world. Results from this study shows that Fe, Zn, Mn, Ni and V were present in the six brands of canned fish products in the different storage medium marketed and frequently consumed in the Nigeria market.

Iron is an essential metal and it is an important constituent of hemoglobin, myoglobin and enzymes such as catalases and peroxidases. WHO set iron permissible limit at 0.30 mg/day and in the UK, women during their reproductive age and post-menopausal estimated daily iron requirement are set at 11.4 mg and 6.7 mg respectively, while 6.7–8.8 mg was for young male-adults. Estimated average daily requirements for children of between 0 and 10 years varied between 1.3 mg and 6.7 mg [32]. Although, there are no set permissible limits to be used as a standard in Nigeria with regards to ferrous concentration, the results obtained during the study period and in the different storage media exceeded the WHO set limit in all the brands. Except for sardine stored in curry sauce and soybean oil, all other values were within the UK and COMA set permissible limit. However previous studies in Nigeria [33], revealed that 0.0379 mg/kg and 0.0302 mg/kg were obtained while working on two different brands of canned fish. Results from this study are much higher than their findings but similar to results obtained earlier [34] while working on canned fish stored in brine and sunflower.

Zinc has been implicated in most metabolic pathways in humans; its deficiency can lead to loss of appetite, growth retardation, skin changes and immunological abnormalities. Maximum zinc level permitted for fish is 50 mg/kg according to Food Codex, while the US recommended daily intakes of zinc are 15 mg and 12 mg for adult male and female respectively. Results obtained shows two brands of mackerel in tomato sauce$_{(2,3)}$ and skipjack in soybean oil$_6$ did not exceed these daily requirement Although, United Nations Environmental Protection Agency and the European Commission (US-EPA and EC) have not set any standard limits for zinc concentrations, Codex [35] maximum permissible limit of 40 mg/kg was exceeded in the mean zinc concentration of 72.09 mg/kg observed in sardine in curry sauce$_1$ but all other values were within the permissible limit during the different months. and the lowest average zinc concentration was observed in costa mackerel in tomato sauce$_3$ 7.04 mg/kg. This value (7.04 mg/kg) obtained in this study, compare favorably with the result of [36] who also observe a low Zn content in mackerel (6.9 mg/kg), binito (5.8 mg/kg) and trout (6.4 mg/kg) and a higher content in anchovy and sardine (27.3 mg/kg and 17.6 mg/kg respectively). In canned tuna flakes stored in brine a zinc concentration level of 28.95 mg/kg had earlier been reported [34]. While working on four imported fish species stored in commercial cold store in Nigeria [37] it was observed that no value exceeded the WHO/FAO maximum permissible limit as the highest mean value of 21.52 mg/kg was obtained.

According to USEPA [38], there is no information on the carcinogenicity of manganese; COMA and WHO have been unable to set a specific recommendation for manganese intake. The EU scientific committee for food (EU-SCF) considered an adequate intake to be 1–10 mg/person/day. Results obtained from this analysis shows that the maximum and minimum concentration range of Mn was 0.09–0.67 mg/kg with a mean monthly value which was basically below the recommended values by the US National Research Council (NRC) [39] specified estimated safe and adequate daily dietary intakes of 0.3–1, 1–3 and 2–5 mg/day for infants, children and adults respectively. This result is comparable to [34] reported range of 0.07–0.51 mg/kgMn and 0.002 mg/kgMn and 0.0016 mg/kgMn [33] while working on brands of Geisha and Founty canned fish respectively. But it was however lower than the corresponding maximum level of 15.77 mg/kg reported for canned sardines in Brazil [40]. Although small daily intake of manganese is needed for a total wellbeing, inability to remove excess amount of manganese from the system can result in nervous system problems.

Nickel which can act as an activator of some enzyme systems at high levels can also cause respiratory problems because it can accumulate in the lungs and it is carcinogenic [41]. In addition to environmental contamination of nickel in foods, it can get into foods through processing activities such as canning and cooking. The upper tolerable intake level of nickel for children and adult are 7 mg/day and 40 mg/day respectively [42] but Codex [35] set the permissible limit in foods at 3.0 mg/day. The mean concentration range of 5.61 mg/kg-0.55 mg/kg Ni observed in this study was higher than that earlier obtained [43], which was in the range of 0.0 to 0.78 mg/kg while working on canned fishes sold in Brazil. Except for sardine stored in curry sauce, all other values obtained during this study were below set permissible limit in foods [35] especially in all the brands stored in tomato sauces$_{(2–4)}$.

Vanadium is a transition metal, with an estimated ˃ 60 thousand tons of this element being emitted into the atmosphere yearly from human activities as oxides or sulphates. But it also occurs naturally in soil, waters and air [44, 45]. Obtained values from this study showed a V concentration range of 0.27 mg/kg to 0.02 mg/kg which were much less than the 1.8 mg/kg upper tolerance intake level of vanadium for adults of between 19 and 70 years as set by Institute of Medicine [42], and 2 mg/kg and 2.9 mg/kg set by Codex and EC respectively.

5. Conclusions

Metal poisoning though can be diagnosed and treated, the best option however is to prevent poisoning. As metal consumption beyond certain threshold can be potentially life threatening as a result of their bio-toxic effects,

Canned fish brands used for this study can be certified as relatively safe for consumption haven detected concentrations lower than most maximum tolerance concentration set by many foods and fish regulatory bodies. Storage media did not significantly affect the fishes stored in the different media but for their brands, as mackerel stored in tomato sauce in three different brands showed a wide variation especially in one of the brands as compared to the others which were not significantly different. Sardine and skipjack both stored in Soybean oil were significantly different as sardine stored in the soybean oil had much higher percentage metal concentration compared to skipjack. Same brand of sardine stored in soybean oil and curry sauce respectively, were not significantly different from each other but both were significantly different from all the other brands.

The trend observed during this study also shows that these study metals concentration showed some level of variation as the storage period increased. These variations with time may be due to sales and storage temperatures to which these products are exposed to. Temperatures in Nigeria can be extremely high, and no deliberate attempt is made to adhere to manufacturer instructions on storage and keeping details by marketers, quality managers and regulatory bodies. Good management of the harvesting waters, production process and adequate hygienic sales outlet, with controlled atmosphere where possible for handling canned fish products must be advocated in Nigeria. Food regulatory bodies in Nigeria should ensure that good quality canned fish products are not only imported into Nigeria, but that the quality integrity of these imported products are kept at optimum level by monitoring possibly the chain line where products are likely to loss quality. Because of their storage and keeping conditions during sales and before consumption, so as to ensure the safety of unsuspecting consumers. Therefore, continuous monitoring of the metal levels in imported products distributed within the country at point of arrival and periodic evaluation of products in circulation and where possible other variables, will help sanitize the market and safe guild the health of her people.

Author details

Andress Edowaye Odiko
Department of Aquaculture and Fisheries Management, University of Benin, Benin City, Nigeria

*Address all correspondence to: andress.odiko@uniben.edu

References

[1] FAO. The State of World Fishery and Aquaculture. 2021. Available from: http://www.fao.org/fishery/en/statistics

[2] Oyinbo O, Rekwot GZ. Fishery production and economic growth in Nigeria: Pathway for sustainable economic development. Journal of Sustainable Development in Africa. 2013;**15**(2):100-109

[3] Mohanty BP, Mahanty A, Ganguly S, Mitra T, Karunakaran D, Anandan R. Nutritional composition of food fishes and their importance in providing food and nutritional security. Food Chemistry. 2019;**293**:561-570

[4] World Data Statistics. Fish and Sea Food Consumption in the World. 2017. Available from: http://www.ourworldindata.org

[5] Mbanasor JA. Reserve use patterns among poultry enterprises in Abia State Nigeria. Nigeria Journal of Animal Production. 2002;**29**(1):64-70

[6] Chinma N, Kafeelah Y, Comfort A. Heavy metal contamination and human health risk from selected fish in Lagos, Nigeria. Journal of Research and Review in Science. 2017;**2**:231-236

[7] National Bureau of Statistics. Multidimensional Poverty Index Survey. 2022. Available from: https://www.nigstats.gov.ng/news/78

[8] Wangboje OM, Akhaehomen S, Erhayimwen M. Human health risk assessment in relation to heavy metals in Atlantic mackerel (*Scomber scrombrus*, L., 1758) sold in some major markets in Benin City, Nigeria. Tropical Journal of Natural Product Research. 2017;**1**(1):32-38

[9] Tchounwou PB, Yedjou CG, Patlolla AK, Sutton DJ. Heavy Metals Toxicity and the Environment. 2012. Available from: https://www.ncbi.nih.gov./pmc/ncbi. 10.1007/978-3-7643-8340-4_6

[10] Mol S. Determination of trace metals in canned anchovies and canned rainbow trouts. Food and Chemical Toxicology. 2011;**49**:348-351

[11] Tawari CC. Effectiveness of Agricultural Agencies in Fisheries Management and Production in the Niger Delta. Nigeria: Rivers State University of Science and Technology, Port Harcourt; 2006. p. 180

[12] Dabeka RW, Mckenzie AD, Bradley P. Survey of total mercury in total diet food composites and an estimation of the dietary intake of mercury by adults and children from two Canadian Cities. 1988-2000. Journal of Food Additives and Contaminants. 2003;**20**(7):629-638

[13] Ercal N, Gorer-Orhan H, Aykin-Burn N. Toxic metals and oxidative stress Part 1: Mechanisms involved in metal-induced oxidative damage. Current Topics Medical Chemistry. 2001;**2001**:529-539

[14] Munoz-Olivas R, Camara C. Speciation related to human health. In: Ebdon L, Pitts L, Cornelis R, Crews H, editors. Trace Element Speciation for Environment, Food and Health. United Kingdom: Royal Society of Chemistry; 2001. pp. 331-353

[15] European Commission Regulation (EC). Setting maximum levels for certain contaminants in foodstuffs. No. 165/2010 of February, 2010

[16] Figueroa E. Are more restrictive food cadmium standards justifiable health

safety measures or opportunistic barriers to trade? An answer from economics and public health. Science of the Total Environment. 2008;**389**:1-9

[17] Ashraf W, Tariq MJ, Jaffar M. Contents of trace metals in fish sediment and water from three water reservoirs on the Indur River. Pakistan Fish Research. 1991;**12**(4):355-364

[18] Ray S. In: Kiceniuk JW, Ray S, editors. Analysis of Contaminants in Edible Aquatic Resources. New York, U.S.A.: VCH Publishers, Inc.; 1994. pp. 91-113

[19] Oehlenschlager J. Identifying heavy metals in fish. In: Bremmer HA, editor. Safety and Quality Issues in Fish Processing. Cambridge, United Kingdom: Woodhead Publishing Limited; 2002. pp. 95-113

[20] Soylak M, Saracoglu S, Tüzen M, Mendil D. Determination of trace metals in mushroom samples from Kayseri, Turkey. Food Chemistry. 2004;**92**:649-652

[21] Singh VP. Metal Toxicity and Tolerance in Plants and Animals. first ed. New Delhi: Sarup & Sons; 2005

[22] Odiko AE. Toxicity of zinc to *Clarias gariepinus* in soft water. Journal of Agriculture, Forestry and Fisheries. 2016;**15**(2):79-82

[23] Fong SS, Kanakaraju D, Ling SC. Evaluation of the acid digestion method with different solvent combination for the determination of Iron, zinc and lead in canned sardines. Malaysian Journal of Chemistry. 2006;**8**:10-15

[24] Madhuri S, Pandey G, Bhandari R, Shrivastav AB. Fish cancer developed by environmental pollutants. International Research Journal of Pharmacy. 2012;**3**(10):17-19

[25] Blakas IT, Tugre L, Salhogln J. Trace metal levels in fish and crustaceans from Northeastern and Mediterranean Coastal waters. Marine Environmental Research. 1982;**6**:281-289

[26] Tariq J, Jaffar M, Moazzam M. Concentration correlations between major cations and heavy metals in fish from the Arabian Sea. Marine Pollution Bulletin. 1991;**22**(11):562-565

[27] Boran M, Koracam H, Celikkale MS, Kose S, Fexzioglu M, Kutlu S. Levels of heavy metal in blue whiting caught from the eastern sea black sea area of Turkey. Toxicological and Environmental Chemistry. 2000;**75**:67-73

[28] Zhou JL, Salvador SM, Liu YP, Sequeria M. Heavy metals in the tissues of common dolphins (*Delphinus delphis*) stranded on the Portuguese coast. Science Total Environment. 2001;**273**:61-76

[29] Sreedevi PA, Suresh A, Siraramkrishna B, Prebbavathi B, Radhadrishanaiab K. Bio-accumulation of nickel in the organs of the freshwater fish *Cyprinus carpio* and the water mussel *Lamethhodens marginals* under lethal and sub lethal nickel stress. Chemosphere. 1992;**24**(1):29-39

[30] Kagi JH, Schaffer A. Biochemistry of metallothionein. Biochemistry. 1998;**27**:8509-8515

[31] Burger J, Gochfeld M. Mercury in canned tuna; white versus light and temporal variation. Environmental Research. 2004;**96**(3):239-249

[32] Committee on Medical Aspects of Food and Nutrition Policy (COMA). Dietary Reference Values for Food Energy and Nutrients for the UK. London: HMSO; 1991

[33] Iwuoha GN, Uporo VB, Onwuachu UI. Variation of heavy metals

in canned Gaisha and Founty Mackerel fish brands obtained from Choba market Port Harcourt, Nigeria. Journal of Applied Science Environmental Management. 2013;**17**(4):577-580

[34] Odiko AE, Agbozuadu B, Ikhala B. Investigation of availability of metals in some brands of canned fish stored in brine and sun-flower oil, sold in Nigeria. International Journal of Fisheries and Aquatic Studies. 2017;**5**(3):530-533

[35] Codex Alimentarius. General standards for contaminants and toxins in food and feed. 2015 Amendment (CODEX STAN 193-11995). International Food Standards. 2015;**2015**:59

[36] Manthey-Karl M, Ostermeyer U, Altinelataman C, Celik U, Oehlenschlager J. Chemical composition, cholesterol, trace metals and amino acid composition of different canned fish products produced and sold in Turkey. Journal of Fisheries Science. 2014;**8**(1):17-26

[37] Odiko AE, Eribo GE, Akenbor M. Determination of metal content in *Ethmalosa fimbiata, Pseudotolilthus epipercus, Clupea harengus and Merluccious merluccious* stored in commercial cold stores. ADAN Journal of Agriculture. 2021;**2**(1):170-178

[38] United States Environmental Protection Agency (USEPA). National recommended water quality criteria. 4304T Office of Water, Science and Technology. 2006; USEPA

[39] National Research Council. Arsenic in Drinking Water. Washington, DC: National Academy Press; 2001. pp. 4-13

[40] Tarley CRT, Coltro WKT, Matsushita M, de Souza NE. Characteristics levels of some heavy metals from Brazilian canned sardines (*Sardinella brasiliensis*). Journal of Food Composition and Analysis. 2001;**14**:611-617

[41] Agency for Toxic Substances and Disease Registry. Agency for Toxic Substances and Disease Registry. Atlanta, GA: Division of Toxicology; 2004. Available from: http://www.atsdr.cdc.gov/toxprofiles/

[42] Institute of Medicine. Dietary Reference Intakes: Applications in Dietary Planning, Subcommittee on Interpretation and Uses of Dietary Reference Intakes and the Standing Committee on the Scientific Evaluation of Dietary Reference Intakes. Washington DC: Institute of Medicine of the National Academics, Press; 2003. p. 248

[43] Ikem A, Egiebor NO. Original article assessment of trace elements in canned fishes (mackerel, tuna, salmon, sardines and herrings) marketed in Georgia and Alabama (United States of America). Journal of Food Composition and Analysis. 2005;**18**:771-787

[44] Aragon AM, Althamirano-Lozano M. Sperm and testicular modifications induced by subchronic treatments, with vanadium (iv) in CD-1mice. Regiod Toxiolgy. 2001;**15**:145-151

[45] International Programme on Chemical Safety. Vanadium pentoxide and other inorganic vanadium compounds. In: Concise International Chemical Assessment Document 29, World Health Organization. Geneva, Switzerland; 2001. Available from: http://www.who.int/ipcs/publications/cicad/en/cica d29.pdf

Chapter 5

Birds as Intrinsic Bio-Indicators for Probing Heavy Metal Contamination Signatures in Polluted Environmental Matrices

Sanchari Biswas

Abstract

The purpose of this study is to provide a comprehensive summary of the major functions of avifauna as bioindicators of contamination and their impact on conservation. Birds have been utilised as bioindicators of contamination globally for a long time. When their functions and numbers are taken into account, bioindicators—species that are used to assess the health of the environment—are capable of evaluating the integrity of the ecosystem. Birds are excellent indicators because they are very noticeable and their existence is simple to spot in any setting. They are also equipped with the ability to fly, so they may flee from an environment if it does not meet their ecological needs. In addition, due to their widespread distribution, individual species are simple to recognise in classification. When their functions and numbers are considered, bioindicators can evaluate the integrity of the ecosystem. Birds are excellent indicators since they are simple to spot and observe in any environment. Although necessary for life, heavy metals can be harmful at high levels and disrupt behaviour and productive function. They can also be easily ingested and biomagnified through food or the food chain. Therefore, the use of birds as bioindicators depends on the features of interest and the resources available for ecological evaluation, which encourages the conservation of bird species for the next generation.

Keywords: birds, environment, heavy metals, bio-indicators, pesticides

1. Introduction

Birds reflect the quantities of xenobiotics in the environment and have been utilised as bioindicators to assess the extent of developing contamination present in any particular ecosystem. Due to the omnipresence of the environment in the modern world, heavy metals produced due to anthropogenic pressure persists in the environment Elemental poisoning in the environment has increased in recent years, raising concerns that have not only afflicted industrialised countries in Southern Asia but also India [1]. Heavy metal deposition in soils is caused by natural geological processes, whereas atmospheric deposition, flooding, and industrial discharge are caused by anthropogenic sources. Higher-trophic level organisms absorb these elements from bacteria, macro-invertebrates, flora, and other abiotic elements, which are subsequently transferred to them [2]. Due to exposure parameters that are over acceptable limits, vital metals that would otherwise be necessary for the growth and development of the species throughout this process can instead become hazardous to the organism [3]. The contribution of heavy metal contamination to the terrestrial environment comes from several man-made sources, including urban activities, paint, manures, industrialisation, discharge from mines, biomedical and hazardous waste disposal, obnoxious use of fertilisers and pesticides, coal burning, vehicle exhausts, and open incineration [4]. Long-term accumulation of persistent toxicants in sediments exposes wildlife living in coastal environments to the effects of such toxicants, which can continue to occur for years [5]. Heavy metals, a significant ecotoxicological pollutant, have an impact on the diet, feeding habits, and evolutionary traits of several animals [6].

The first instance of using Aves as a biomonitor to assess environmental health was recorded in history in the 1960s [7]. Consumption of contaminated food is a significant method through which birds obtain metals [8]. Acute poisoning can cause adult birds to die [9], but chronic exposure to metals at low levels can continue to occur in birds with extended life spans [10]. Heavy metal toxicosis can cause physiological and genetic diseases, growth impairments, impaired development, reproductive issues, disease resistance, and population decrease [11, 12].

2. Background

Diverse marine bird and animal species have recently perished from anthropogenic causes and rare diseases. One of the primary causes of such events is anthropogenic sources. Animals may exhibit endocrine physiological problems as a result of several man-made substances. Toxicants can be deadly to animals, including birds, and can mess with their endocrine systems. One of these toxicants is organochlorine since it can penetrate all topical layers. Endocrine disruptive substances, such as polychlorinated biphenyls (PCBs) and organochlorine pesticides, can accumulate in the tissues of wildlife over time. Due to exposure to excessive quantities of hazardous compounds from different pesticides, wildlife has suffered as a result of mass fatalities. The incapacity of open-sea creatures like Albatrosses and cetaceans to metabolise hazardous persistent pollutants has been linked to higher levels of organochlorines in these creatures [13]. As a result of their quick adaptation to climate changes, birds are regarded as pioneer indicators for changes connected to global warming.

3. Consequences of environmental pollution

Due to heavy metal contamination, the environment has lost its capacity to support life and transmit its fundamental values. Heavy metals are present in the environment naturally, but because of human activity, their occurrence is considerably more extreme than it is. This puts the lives of humans, animals, and plants in danger. But it presents significant obstacles to our knowledge and way of life. Environmental contamination is the introduction of anything new to our environment. The primary causes of pollution in our environment are anthropogenic, or man-made, sources. The presence of pollutants in a region has the potential to disrupt the overall ecological equilibrium. Contaminants are the main type of environmental contaminant that harms our ecosystem. Pests, condensates, and other materials can carry bacteria, chemicals, and particulate contaminants. While adulteration brought on by environmental microbes causes content to decay and disease, chemical pollutants frequently induce harmful reactions. Heavy metals are metallic elements that are denser than water in comparison [14–16] asserts that arsenic, a metalloid, is hazardous when exposed at low levels and draws the assumption that toxicity and mass are related. A few of the environmental sources of heavy metals are the atmosphere, industries agriculture, geological pollution, pharmaceuticals, and home effluents [17].

Mine tailings, emissions from industrial regions, deposition of elemental waste produced by the paint, fertiliser, and pesticide industries, sewage, thermal power plant residues, petrochemical spillage, and atmospheric deposition are the principal sources of heavy metals and metalloids in soils. Regardless of where they come from, heavy metals are mostly absorbed by the soil. While inorganic contaminants like metals do not go through the oxidation process described above and stay relatively persistent in soils even after introduction, several organic contaminants do so as a result of microbial activity [18, 19].

Industrial, agricultural, and other anthropogenic activities harm aquatic ecology [20]. Metals continue to biomagnify in aquatic environments through the food chain

Figure 1.
Illustration of heavy metals and other toxic compounds which impacts avifauna, revering them as "bio-indicators."

and cause biological harm that poses a concern to humankind [21]. According to Canli and Atli [22], aquatic species occasionally accumulate heavy metals inside of them that is higher than ambient levels. Industrial discharges cause freshwater habitats to become more contaminated than other environments over time [23]. Numerous aquatic ecosystems may be able to withstand large levels of pollution, according to studies, however, those ecosystems may also suffer harm from the resulting living populations.

4. Bio-indicators

Bioindicators (**Figure 1**) are species that are used to assess the health of an ecosystem or the environment. They can assess the integrity of the environment based on their populations and functions. Monitoring the state of several ecological unit constituents is essential for protecting and managing the natural environment. Ecologists have utilised wild birds as indicators of heavy metal contamination over the years because they can gauge the impact of environmental heavy metal concentrations. In ecology and conservation science, surrogate species have been employed over time to ascertain the relationship between plants and animals in a particular geographic location [24]. Because managing ecosystems is challenging and there are few scientific tools available, conservation biologists have also employed target species as desirable shortcuts.

5. Biomonitoring using birds through the invasive method

5.1 Bones

Animals are exposed to accumulative sources of metals through food consumption, which elevates over time with rising metal concentrations. Fish that are exposed to trace concentrations of heavy metals over an extended period experience damage to their skeletal, renal, and respiratory systems [25]. According to clinical investigations Engström et al. and Rodríguez and Mandalunis [26, 27], Cadmium exposure increases the risk of bone deterioration, osteoporosis, and kidney impairment. Brzóska and Moniuszko-Jakoniuk [28] also noted that exposure to Cd causes a decrease in mineralisation, affecting the biomechanical characteristics and making the bones more prone to deformation and fracture. Long-term exposure to cadmium reduces bone volume, according to Chen et al. [4], however, Rodríguez and Mandalunis [27] found that it increases the number of tartrate-resistant acid phosphatase (TRAP) positive cells in tibial bone. According to Papa et al. [29], cadmium causes osteoblast apoptosis, which causes the cytoskeleton to rupture, as well as DNA fragmentation, a rise in the frequency of micronuclei, and reactive oxygen species [27, 30].

According to Morais et al. [31], a high Ni concentration inhibits alkaline phosphatase activity, which has an impact on bone mineralisation [32]. However, it was concluded by Rodríguez and Mandalunis [27] that there is no literature on the effects of Nickel poisoning on human skeletal tissue or in vivo experimental models. Studies on Goldfish indicate that methylmercury directly affects and lowers the metabolism of scale bone cells, increases the production of metallothionein, and decreases the expression of calcaemic and oestrogen receptors [33]. According to

Yachiguchi et al. [34], there is a decrease in TRAP and alkaline phosphatase (ALP) expression along with an increase in the production of metallothioneins. Mercury would thus inhibit both osteoclast and osteoblast function. According to Abd El-Aziz [35], experimental models with methylmercury injection into laboratory animals slowed ossification and had detrimental effects on foetus development. According to Rodríguez and Mandalunis [27], there is a dearth of information on the impact of mercury on mammal skeletal tissue, with only a few studies having been conducted.

According to Bier et al. and Ma et al. [36, 37], lead (Pb) blocks the signalling pathway, which in turn limits osteoblastic activity. The findings of Lv et al. [38] mentions that Pb's autophagy can protect osteoblast apoptosis and it induces osteoblasts to apoptosis. Exposure to aluminium (Al) causes osteomalacia, osteoporosis, and renal osteodystrophy. Chappard et al. [39] found that in the hydroxyapatite crystals of exostotic patients, Ca^{2+} was replaced by Al^{3+} and Fe^{3+}.

5.2 Tissues

Due to industrial, agricultural, and urbanisation-related activities that injure organisms, the environment is polluted with both essential and non-essential metals including Zn, Cu, Cd, Cr, and Pb. The food chain helps to mitigate the impacts on the raptors, even when they are not deadly. Studying the risk effects on the species and the degrees of contamination in their habitat is aided by the degree and extent of metal concentration in tissues [40]. Because of their biological behaviour, several species are exposed to the contamination that can be found if abiogenic sampling is not done. According to Jagar et al. [41], data on the degree of contamination and the movement of contaminants through the food chain could be obtained from raptors during large-scale sampling. When compared to other species of a similar kind, some aerial species are more exposed to pollution [42]. Information on temporary and geographical exposure to pollutants was proposed by Burger et al. and Pérez-López et al. [43, 44]. According to Rothschild and Duffy [45], the habitat, distribution, feeding habits, and life span of wild birds can all be used as indicators of environmental contamination. Only a few wild bird species, according to Movalli [42], can serve as bioindicators on a larger scale and be used to predict future environmental changes [46]. Burger et al. [43] claimed that because birds are such sensitive species to specific pollutants, they can provide information regarding spatial and temporal exposure to certain pollutants. According to Kler et al. [47] soft tissues of birds, such as their muscle, kidney, spleen, liver, heart, lungs, blood, and brain, can be widely employed as bioindicators to identify metal contamination in studies. Since different organs contain different amounts of metal, it is required to estimate different tissues for evaluation at the population level.

According to Jin et al. [48], metals that are not needed by the living organism do not dissolve and occasionally accumulate in the body, causing negative effects. Additionally, the ionisation of these ions causes them to react with biological components like protein or nucleic acid, which has an impact on how enzymes are activated and how three-dimensional protein structures are formed. Additionally, sometimes essential metals that are centrally situated in metalloenzymes are replaced by heavy metals. Furness and Greenwood [49] concluded that birds can be utilised as biological indicator species for determining the degree of heavy metal contamination and its negative effects.

According to Fisher et al. [50], the main sources of lead discharge into the environment are mining, industries, and hunting activities. According to Snoeijs et al. [51], persistent exposure to high concentrations had an impact on the wild birds' reproductive rates, immunity, and physiological systems. Cadmium was listed as one of the harmful trace elements by Battaglia et al. [52] due to its toxicity and persistence in both food and the environment. According to references [53, 54], cadmium accumulation in birds' bodies caused harm to their renal tubular cells and a decline in their physical conditions. It was concluded by Carpenter et al. [55] that copper (Cu) and manganese (Mn) have positive effects on biological processes as well as negative effects at greater concentrations on the kidneys, reproductive health, and the potential to cause mortality. According to Kim and Oh [56], determining the extent of accumulation worldwide as well as excretion patterns allowed for the conclusion of contamination levels and unfavourable reactions to certain species.

6. Biomonitoring using birds through the non-invasive methods

6.1 Feathers

Raptors have been the most frequently used species in biomonitoring research on the environment because of their wide geographic spread for foraging purposes and also because of their higher position on the food chain, which can reflect the number of contaminants flowing through it [57]. However, because a number of raptor species are protected, non-invasive methods can be used. According to Dauwe et al. [58], sampling of feathers is simple, and feathers have already been used in numerous research [59]. According to Burger's research, heavy metals have the propensity to bind to protein molecules during the brief embryonic stage of a feather when tiny blood capillaries connect it to the bloodstream. References [58, 59] came to the conclusion that after the formation of the feathers, the blood vessels become weak and physically detach from the bird. Birds shed a sizable amount of heavy metals during the moulting process through their plumage [57]. Some interior tissues begin displaying decreased metal levels as soon as the metals are sewn into the feathers. When a moult is finished, the internal level of a few heavy metals rises until the next moult, which is a constant process [60]. According to research by [61], if the overall amount of heavy metals in a bird's body is reflected in the feathers as they form, the feathers that are moulted at the conclusion of the process should do so with lower concentrations than those that are moulted at the beginning. The concentration of various heavy metals inside the body of the bird during moulting may vary, being higher at the start and lower towards the end of the process. Birds also deposit toxins like Cd, Cr, and Pb into their developing feathers, which attach to the keratin in feathers and become biologically separated [62]. In addition, the accumulation of these metals into the calcareous tissues is a significant way for female birds to eliminate Pb [63], but Zn and Cu are favourably deposited in the feathers when concentrations surpass permitted limits [64]. Studies by [65, 66], revealed that during the brief period of egg hatching and while it matures, hatchling feathers can indicate local contamination that may result from their parents consuming food from the area [65].

6.2 Eggs

Over time, it has been discovered that the presence of metals in feathers, excrement, and eggs also causes harm to the bird population [65]. Hashmi [67] reported using feathers and eggs for biomonitoring soil, but due to their lengthy storage times and ease of sampling, this method is related to one's own. Additionally, it was determined that because eggs are formed at a specific point in a female bird's life cycle, they can act as a superb signal for local exposure in addition to belonging to a particular subset of egg-laying birds. According to Jayakumar and Muralidharan [68], anthropogenic inputs are the cause of the presence of unnecessary trace elements in both domestic and wild birds. The harmful effects of metal pollution might include everything from kidney and testicular damage to reduced egg production [69]. Bird eggs, according to reference [68] are among the most complex and distinctive types of reproductive cells. They have a protective coating (or shell) that encloses albumin and yolk contents that vary greatly in volume, shape, weight, and amount. From tropical to temperate zones, according to reference [57], female bird species tend to forage on their breeding grounds before to egg production after egg laying. According to reference [67], utilising eggs as a biomonitoring technique provides a number of benefits over using feathers since eggs are more stable and do not change with age, size, or body structure. Eggs have recently been employed in a number of biomonitoring studies, and it has thus been demonstrated that the collection of a single egg without harming the environment has a negligible impact on the population of the species [5]. A given metal's circulation concentration in birds, which is then reflected in egg quality, can be closely correlated with recent exposure [67]. According to reference [70], the egg shell and egg content are typically where metals collect throughout the female's metal deposition. In a similar vein, Burger et al. [43] came to the conclusion that heavy metal exposure from the parent's local environment is what puts eggs in the greatest danger of being harmed.

6.3 Excreta

According to reference [47], investigations on heavy metal contamination in birds are extremely rare in nations like India. Additionally, the Wildlife Protection Act of India (1972) forbids the capture and sacrifice of birds, which makes it challenging for scientists to conduct any analytical research on eggs and tissues. Excreta might therefore be chosen as a different source that can provide clear information on environmental contamination [47, 71]. Birds are exposed to heavy metals through contaminated water sources, excessive metal concentrations in the atmosphere, and food sources close to point sources [72]. The rate of heavy metal absorption varies according to the physiology of the species and is influenced by the metals' bioavailability and other factors. By storing them in the salt gland, uropygial gland, or excreta, birds can get rid of heavy metals [47]. Bird excretions are the best indicator of the environment's level of heavy metal contamination since they excrete more metals than other animals [73]. Excreta from birds serve as helpful non-destructive bioindicators and provide information on the bird's diet [74]. It also details the kinds of metals absorbed and removed [75].

A total of five types of pollution indicator species were identified by [76] which are represented in **Table 1**.

S. no	Type of species	Function
1	Sentinels	These extremely vulnerable animals were introduced into a specific area to provide early warning signs of ecosystem contamination
2	Detectors	These are species that are unique to a particular area and exhibit demonstrable responses to environmental changes, such as adjustments in behaviour, mortality, or age-class structure. In addition, it is anticipated that the species would react quickly to changes in the environment
3	Exploiters	These are the kinds of species that, by virtue of their very existence, can signal environmental damage. The ability of exploiter species to make use of a polluted environment determines whether or not they will survive. Because rivals cannot survive in the distorted environment, they frequently thrive in disturbed or contaminated places
4	Accumulators	These are species that accumulate pollution in significant amounts in their skin, feathers, lungs, liver, and kidney tissues

Table 1.
Type of pollution indicator species.

7. Advantages of Avifauna as indicator species in probing heavy metal pollution

Birds have historically served as reliable environmental indicators. Despite the fact that they still have a limited capacity for directly and quickly altering the features of ecosystems and for altering the behaviour of other taxa [77, 78]. The advantages of utilising bird species as indicators are:

i. They can be easily observed in any given environment.

ii. Due to their diurnal nature and ability to call, they are easily recognisable.

iii. Avifaunal classification and speciation make them easily identifiable across any field.

iv. Birds prefer different habitat niches and are widely distributed geographically over a wide range of areas. Thus, their foraging activities constantly expose them to heavy metals present in the atmosphere, or in any given environment according to their habitat preferences.

v. They are top-level predators which makes them susceptible to the changes going on in the food chain and food web at the producer, primary consumer and secondary consumer levels.

vi. Birds are great pollinators and help in seed dispersal, thereby maintaining the ecological balance in the ecosystem. In a polluted environment, plants sequester heavy metals into their fruits through their roots and upon consuming these fruits, birds are either exposed to heavy metals or continue to accumulate them in their system.

8. Disadvantages of using Avifauna as indicator species in probing heavy metal pollution

i. Because birds are mobile, it is possible to see them on a wide range. Making it simple for researchers to link their responses to particular alterations in the environment.

ii. Bird species are very mobile in comparison to other terrestrial animals, and they use resources in three dimensions as opposed to other animals, which leads to erroneous reactions to environmental change.

iii. Birds are similarly affected by secondary or tertiary changes in stressor components, which lessens the value of birds as bioindicators.

iv. In comparison to other animals, they also possess behavioural and physiological characteristics that make them less vulnerable to changes in the ecosystem. For instance, birds have significantly greater control over the amount of fat and metal present in their body tissues than do invertebrates.

9. Conclusions

High levels of toxicity in some heavy metals can disrupt behaviour and productive function. Lead poisoning can be triggered by the presence of heavy metals, impairing the immune system and harming the nervous system. Birds' ability to reproduce and grow can both be affected negatively by Cadmium. In addition, methylmercury can hinder bird reproduction, resulting in a decrease in egg size, fertility, and hatchability. Once a certain heavy metal's allowable limit is exceeded, the aforementioned impacts of heavy metals on bird species are immediately apparent. As a result, wild birds are good markers of environmental heavy metal pollution. A function of the features of interest and resources available for ecological evaluation, using birds as bioindicators encourages the conservation of bird species for future generations. It is crucial to investigate potential mechanisms through which heavy metals may be to blame for the development of illnesses in wild birds. This will most certainly support the protection of threatened and endangered bird species.

Acknowledgements

The author is thankful to the Department of Environment Sciences, GITAM Institute of Science for providing basic infrastructure, library, and technical support and to the University Grants Commission (UGC) Rajiv Gandhi National Fellowship for funding the current work.

Conflict of interest

The authors declare no conflict of interest.

Acronyms and abbreviations

PCB	polychlorinated biphenyls
Cd	cadmium
TRAP	tartrate-resistant acid phosphatase
ALP	alkaline phosphatase
Pb	lead
Zn	zinc
Cu	copper
Al	aluminium
Cr	chromium
Mn	manganese
Ni	nickel

Author details

Sanchari Biswas
Department of Environmental Science, GITAM (Deemed to be University), Visakhapatnam, Andhra Pradesh, India

*Address all correspondence to: biswasanchari@gmail.com

References

[1] Bhuiyan MN, Bhuiyan HR, Ahmed K, Dawlatana M, Haque KF, Rahim M, et al. Organochlorine insecticides (DDT and heptachlor) in dry fish: Traditional washing and cooking effect on dietary intake. Bangladesh Journal of Pharmacology. 2009;**4**(1):46-50

[2] Blakely JK, Neher DA, Spongberg AL. Soil invertebrate and microbial communities, and decomposition as indicators of polycyclic aromatic hydrocarbon contamination. Applied Soil Ecology. 2002;**21**(1):71-88

[3] Gupta S, Jena V, Jena S, Davić N, Matić N, Radojević D, et al. Assessment of heavy metal contents of green leafy vegetables. Croatian Journal of Food Science and Technology. 2013;**5**(2):53-60

[4] Chen X, Zhu G, Jin T, Gu S, Xiao H, Qiu J. Cadmium induces differentiation of RAW264. 7 cells into osteoclasts in the presence of RANKL. Food and Chemical Toxicology. 2011;**49**(9):2392-2397

[5] Furness Robert W. Birds as monitors of pollutants. In: Birds as Monitors of Environmental Change. Dordrecht: Springer; 1993. pp. 86-143

[6] Rajeev M, Sushmitha TJ, Aravindraja C, Toleti SR, Pandian SK. Exploring the impacts of heavy metals on spatial variations of sediment-associated bacterial communities. Ecotoxicology and Environmental Safety. 2021;**209**:111808

[7] Denneman WD, Douben PET. Trace metals in primary feathers of the barn owl (Tyto alba guttatus) in the Netherlands. Environmental Pollution. 1993;**82**(3):301-310

[8] Burger J, Gochfeld M. Mercury and other metals in feathers of common eider (*Somateria mollissima*) and tufted puffin (*Fratercula cirrhata*) from the Aleutian chain of Alaska. Archives of Environmental Contamination and Toxicology. 2009;**56**(3):596-606

[9] Weyers B, Glück E, Stoeppler M. Investigation of the significance of heavy metal contents of blackbird feathers. Science of the Total Environment. 1988;**77**(1):61-67

[10] Scheuhammer AM. The chronic toxicity of aluminium, cadmium, mercury, and lead in birds: A review. Environmental Pollution. 1987;**46**(4):263-295

[11] Snoejis P, Wänstrand I, Pinto E, de Barros MP, Colepicolo P, Pedersén M. Thiamine (vitamin B1) dynamics in phytoplankton and copepods is regulated by inorganic nutrient supply. 2004

[12] Tallon PP. Inside the adaptive enterprise: An information technology capabilities perspective on business process agility. Information Technology and Management. 2008;**9**(1):21-36

[13] Tanabe S. Contamination and toxic effects of persistent endocrine disrupters in marine mammals and birds. Marine Pollution Bulletin. 2002;**45**(1-12):69-77

[14] Dasari S, Tchounwou PB. Cisplatin in cancer therapy: Molecular mechanisms of action. European Journal of Pharmacology. 2014;**740**:364-378

[15] Fergusson JE. The Heavy Elements: Chemistry, Environmental Impact and Health Effects. Oxford (UK): Pergamon Press; 1990

[16] Duffus JH. Heavy metals—a meaningless term. Pure and Applied Chemistry. 2002;**74**(5):793-807

[17] He ZL, Yang XE, Stoffella PJ. Trace elements in agro ecosystems and impacts on the environment. Journal of Trace Elements in Medicine and Biology. 2005;**19**(2-3):125-140

[18] Kirpichtchikova TA, Manceau A, Spadini L, Panfili F, Marcus MA, Jacquet T. Speciation and solubility of heavy metals in contaminated soil using X-ray microfluorescence, EXAFS spectroscopy, chemical extraction, and thermodynamic modeling. Geochimica et Cosmochimica Acta. 2006;**70**(9):2163-2190

[19] Wuana RA, Okieimen FE. Heavy metals in contaminated soils: A review of sources, chemistry, risks and best available strategies for remediation. International Scholarly Research Notices. 2011;**2011**

[20] Unlu E, Akba O, Sevim S, Gumgum B. Heavy metal levels in mullet, *Liza abu* (Mugilidae) from the Tigris River, Turkey. Fresenius Environmental Bulletin. 1996;**5**:107-112

[21] Yilmaz AB, L. Yilmaz influences of sex and seasons on levels of heavy metals in tissues of green tiger shrimp (*Penaeus semisulcatus* de Hann, 1844). Food Chemistry. 2007;**101**:1664-1669

[22] Canli M, Atli G. The relationship between heavy metal (Cd, Cr, Cu, Fe, Pb, Zn) levels and size of six Mediterranean fish species. Environmental Pollution. 2003;**121**:129-136

[23] Yılmaz AB, Yılmaz L. Influences of sex and seasons on levels of heavy metals in tissues of green tiger shrimp (*Penaeus semisulcatus* de Hann, 1844). Food Chemistry. 2007;**101**(4):1664-1669

[24] Carignan V, Villard MA. Selecting indicator species to monitor ecological integrity: A review. Environmental Monitoring and Assessment. 2002;**78**(1):45-61

[25] Martin S, Griswold W. Human health effects of heavy metals. Environmental Science and Technology Briefs for Citizens. 2009;**15**:1-6

[26] Engström A, Michaëlsson K, Vahter M, Julin B, Wolk A, Åkesson A. Associations between dietary cadmium exposure and bone mineral density and risk of osteoporosis and fractures among women. Bone. 2012;**50**(6):1372-1378

[27] Rodríguez J, Mandalunis PM. Effect of cadmium on bone tissue in growing animals. Experimental and Toxicologic Pathology. 2016;**68**(7):391-397

[28] Brzóska MM, Moniuszko-Jakoniuk J. Low-level exposure to cadmium during the lifetime increases the risk of osteoporosis and fractures of the lumbar spine in the elderly: Studies on a rat model of human environmental exposure. Toxicological Sciences. 2004;**82**(2):468-477

[29] Papa V, Bimonte VM, Wannenes F, D'Abusco AS, Fittipaldi S, Scandurra R, et al. The endocrine disruptor cadmium alters human osteoblast-like Saos-2 cells homeostasis in vitro by alteration of Wnt/β-catenin pathway and activation of caspases. Journal of Endocrinological Investigation. 2015;**38**(12):1345-1356

[30] Hu KH, Li WX, Sun MY, Zhang SB, Fan CX, Wu Q, et al. Cadmium induced apoptosis in MG63 cells by increasing ROS, activation of p38 MAPK and inhibition of ERK 1/2 pathways. Cellular Physiology and Biochemistry. 2015;**36**(2):642-654

[31] Morais S, Sousa JP, Fernandes MH, Carvalho GS, De Bruijn JD, Van Blitterswijk CA. Effects of AISI 316L corrosion products in in

vitro bone formation. Biomaterials. 1998;**19**(11-12):999-1007

[32] Kanaji A, Orhue V, Caicedo MS, Virdi AS, Sumner DR, Hallab NJ, et al. Cytotoxic effects of cobalt and nickel ions on osteocytes in vitro. Journal of Orthopaedic Surgery and Research. 2014;**9**(1):1-8

[33] Suzuki N, Yamamoto M, Watanabe K, Kambegawa A, Hattori A. Both mercury and cadmium directly influence calcium homeostasis resulting from the suppression of scale bone cells: The scale is a good model for the evaluation of heavy metals in bone metabolism. Journal of Bone and Mineral Metabolism. 2004;**22**(5):439-446

[34] Yachiguchi K, Sekiguchi T, Nakano M, Hattori A, Yamamoto M, Kitamura KI, et al. Effects of inorganic mercury and methylmercury on osteoclasts and osteoblasts in the scales of the marine teleost as a model system of bone. Zoological Science. 2014;**31**(5):330-337

[35] Abd El-Aziz GS, El-Fark MM, Saleh HA. The prenatal toxic effect of methylmercury on the development of the appendicular skeleton of rat fetuses and the protective role of vitamin E. The Anatomical Record: Advances in Integrative Anatomy and Evolutionary Biology. 2012;**295**(6):939-949

[36] Bier RL, Voss KA, Bernhardt ES. Bacterial community responses to a gradient of alkaline mountaintop mine drainage in Central Appalachian streams. The ISME Journal. 2015;**9**(6):1378-1390

[37] Ma Y, Fu D, Liu Z. Effect of lead on apoptosis in cultured rat primary osteoblasts. Toxicology and Industrial Health. 2012;**28**(2):136-146

[38] Lv XH, Zhao DH, Cai SZ, Luo SY, You T, Xu BL, et al. Autophagy plays a protective role in cell death of osteoblasts exposure to lead chloride. Toxicology Letters. 2015;**239**(2):131-140

[39] Chappard D, Bizot P, Mabilleau G, Hubert L. Aluminum and bone: Review of new clinical circumstances associated with Al3+ deposition in the calcified matrix of bone. Morphologie. 2016;**100**(329):95-105

[40] Annalisa Z, Giulia A, Ferrante MC, Carpene E, Gloria I, Lucisano A. Metal concentrations in the liver and kidney of raptor species from the Calabria region, Italy. Acta Veterinaria. 2008;**58**(4):315-324

[41] Jager LP, Rijnierse FV, Esselink H, Baars AJ. Biomonitoring with the BuzzardButeo buteo in the Netherlands: Heavy metals and sources of variation. Journal für Ornithologie. 1996;**137**(3):295-318

[42] Movalli PA. Heavy metal and other residues in feathers of Laggar falcon Falco biarmicus jugger from six districts of Pakistan. Environmental Pollution. 2000;**109**(2):267-275

[43] Burger J, Laska M, Gochfeld M. Metal concentrations in feathers of birds from Papua New Guinea forests: Evidence of pollution. Environmental Toxicology and Chemistry: An International Journal. 1993;**12**(7):1291-1296

[44] Pérez-López M, de Mendoza MH, Beceiro AL, Rodríguez FS. Heavy metal (Cd, Pb, Zn) and metalloid (As) content in raptor species from Galicia (NW Spain). Ecotoxicology and Environmental Safety. 2008;**70**(1):154-162

[45] Rothschild RF, Duffy LK. Mercury concentrations in muscle, brain and bone of Western Alaskan waterfowl. Science of the Total Environment. 2005;**349**(1-3):277-283

[46] Moreno Grau MD. Toxicología Ambiental: Evaluación de Riesgo Para la Salud Humana. Madrid, ES: McGraw-Hill Interamericana; 2003

[47] Kler TK, Vashishat N, Kumar M. Heavy metal contamination in excreta of avian species from Ludhiana district of Punjab. International Journal of Advanced Research. 2014;**2**(7):873-879

[48] Jin SD, Seo SG, Shin YU, Bing KC, Kang TH, Paek WK, et al. Heavy metal accumulations of 4 species of Anseriformes in Korea. Journal of Korean Nature. 2012;**5**(4):345-349

[49] Furness RW, Greenwood JJ, editors. Birds as Monitors of Environmental Change. Springer Science & Business Media; 2013

[50] Fisher IJ, Pain DJ, Thomas VG. A review of lead poisoning from ammunition sources in terrestrial birds. Biological Conservation. 2006;**131**(3):421-432

[51] Snoeijs T, Dauwe T, Pinxten R, Vandesande F, Eens M. Heavy metal exposure affects the humoral immune response in a free-living small songbird, the great tit (Parus major). Archives of Environmental Contamination and Toxicology. 2004;**46**(3):399-404

[52] Battaglia A, Ghidini S, Campanini G, Spaggiari R. Heavy metal contamination in little owl (*Athene noctua*) and common buzzard (*Buteo buteo*) from northern Italy. Ecotoxicology and Environmental Safety. 2005;**60**(1):61-66

[53] Anteau MJ, Afton AD, Custer CM, Custer TW. Relationships of cadmium, mercury, and selenium with nutrient reserves of female lesser scaup (*Aythya affinis*) during winter and spring migration. Environmental Toxicology and Chemistry: An International Journal. 2007;**26**(3):515-520

[54] Wayland M, Gilchrist HG, Marchant T, Keating J, Smits JE. Immune function, stress response, and body condition in arctic-breeding common eiders in relation to cadmium, mercury, and selenium concentrations. Environmental Research. 2002;**90**(1):47-60

[55] Carpenter JW, Andrews GA, Beyer WN. Zinc toxicosis in a free-flying trumpeter swan (*Cygnus buccinator*). Journal of Wildlife Diseases. 2004;**40**(4):769-774

[56] Kim J, Oh JM. Metal levels in livers of waterfowl from Korea. Ecotoxicology and Environmental Safety. 2012;**78**:162-169

[57] Dauwe T, Janssens E, Pinxten R, Eens M. The reproductive success and quality of blue tits (*Parus caeruleus*) in a heavy metal pollution gradient. Environmental Pollution. 2005;**136**:243-251

[58] Dauwe T, Janssens E, Bervoets L, Blust R, Eens M. Heavy-metal concentrations in female laying great tits (*Parus major*) and their clutches. Archives of Environmental Contamination and Toxicology. 2005;**49**(2):249-256

[59] Denneman WD, Douben PET. Trace metals in primary feathers of the barn owl (*Tyto alba guttattus*) in the Netherlands. Environmental Pollution. 1993;**82**:301-310

[60] Braune BM, Gaskin DE. A mercury budget for the Bonaparte's gull during autumn moult. Ornis Scandinavica. 1987;**18**:244-250

[61] Altmeyer M, Dittmann J, Dmowski K, Wagner G, Müller P. Distribution of elements in flight feathers of a white-tailed eagle. Science of the Total Environment. 1991;**105**:157-164

[62] Malik RN, Zeb N. Assessment of environmental contamination using feathers of *Bubulcus ibis* L., as a biomonitor of heavy metal pollution, Pakistan. Ecotoxicology. 2009;**18**(5):522-536

[63] Lam JC, Tanabe S, Lam MH, Lam PK. Risk to breeding success of waterbirds by contaminants in Hong Kong: Evidence from trace elements in eggs. Environmental Pollution. 2005;**135**(3):481-490

[64] Boncompagni E, Muhammad A, Jabeen R, Orvini E, Gandini C, Sanpera C, et al. Egrets as monitors of trace-metal contamination in wetlands of Pakistan. Archives of Environmental Contamination and Toxicology. 2003;**45**(3):399-406

[65] Abdullah M, Fasola M, Muhammad A, Malik SA, Bostan N, Bokhari H, et al. Avian feathers as a non-destructive bio-monitoring tool of trace metals signatures: A case study from severely contaminated areas. Chemosphere. 2015;**119**:553-561

[66] Muralidharan S, Jayakumar R, Vishnu G. Heavy metals in feathers of six species of birds in the district Nilgiris, India. Bulletin of Environmental Contamination and Toxicology. 2004;**73**:285-291

[67] Hashmi MZ, Abbasi NA, Tang X, Malik RN. Egg as a biomonitor of heavy metals in soil. In: Heavy Metal Contamination of Soils. Cham: Springer; 2015. pp. 127-143

[68] Jayakumar R, Muralidharan S. Metal contamination in select species of birds in Nilgiris district, Tamil Nadu, India. Bulletin of Environmental Contamination and Toxicology. 2011;**87**(2):166-170

[69] Hui CA. Concentrations of chromium, manganese, and lead in air and in avian eggs. Environmental Pollution. 2002;**120**(2):201-206

[70] Mora MA. Heavy metals and metalloids in egg contents and eggshells of passerine birds from Arizona. Environmental Pollution. 2003;**125**(3):393-400

[71] Sharma C, Vashishat N. Assessment of heavy metals in excreta of house crow (*Corvus splendens*) from different agroecosystems of Ludhiana. Journal of Entomology and Zoology Studies. 2017;**5**(4):1891-1895

[72] Bala M, Sharma A, Sharma G. Assessment of heavy metal residues in excreta of blue rock pigeon in Jaipur, Rajasthan. International Journal of Advanced Research. 2017;**5**(9):983-988

[73] Berglund ÅM, Koivula MJ, Eeva T. Species-and age-related variation in metal exposure and accumulation of two passerine bird species. Environmental Pollution. 2011;**159**(10):2368-2374

[74] Chiang G, González-Acuña D, Espej W. Studying heavy metals on Antarctica by using non-invasive biotic samples of penguins. Oceanography and Fisheries. 2018;**7**(1):23-24. DOI: 10.19080/OFOAJ.2018.07.5557059

[75] Sánchez-Virosta P, Espín S, García-Fernández AJ, Eeva T. A review on exposure and effects of arsenic in passerine birds. Science of the Total Environment. 2015;**512**:506-525

[76] Chambers SA. Birds as environmental indicators: Review of literature. Environmentally Friendly Paper. 2008:1-25

[77] Mac Nally R, Ellis M, Barrett G. Avian biodiversity monitoring in

Australian rangelands. Austral Ecology. 2004;**29**:93-99

[78] Gregory RD, Van Strien A, Vorisek P, et al. Developing indicators for European birds. Philosophical Transactions of the Royal Society Biological Sciences. 2005;**360**:269-288

Chapter 6

Heavy Metals in Surface Soils and Crops

Mohammad Velayatzadeh

Abstract

In the era of industrialization and technological progress, pollution has reduced the quality of life for humans. Heavy metal pollution is one of the main causes of environmental degradation. The underlying causes are natural as well as human. Heavy metal contamination of soil has become a worldwide environmental issue that has attracted considerable public attention, mainly due to increased concern for the safety of agricultural products. Heavy metals refer to some metals and metals with biological toxicity such as cadmium, mercury, arsenic, lead and chromium. These elements enter the soil agricultural ecosystem through natural processes resulting from raw materials and through human activities. Heavy metal pollution is a great threat to the health and well-being of animals and humans due to the risk of potential accumulation through the food chain. The main sources of heavy metal pollution are air pollution, river sediments, sewage sludge and municipal waste compost, agricultural chemicals such as fertilizers and pesticides, and industrial wastes such as factories that release chemicals. Heavy metals can enter the water supply through industrial and consumer wastes or even from acid rain that decomposes soils and releases heavy metals into streams, lakes, rivers and groundwater.

Keywords: soil pollution, heavy metals, toxic elements, human health, crops

1. Introduction

The progress of industries and the growth of urban communities have caused an increase in man-made pollution caused by industrial and agricultural activities and many pollutants enter the environment [1]. One of the most important pollutants in the environment is soil pollution [2]. Soil pollution includes the entry of physical, biological, and chemical substances into this environment, which will eventually enter the life cycle of animals, plants, and as a result, humans, and will cause negative effects in the life of living organisms [3]. One of the effects of soil pollution is the reduction of plant growth and development, which causes the loss of vegetation and ultimately leads to soil erosion and desertification [4]. Most of the pollution created in the soil is caused by the discharge or leakage of organic substances. Petroleum substances and their derivatives cause soil pollution as a result of transportation or storage, while the deeper the petroleum substances penetrate into the soil, the more difficult it is to remove the pollution and the cost will be several times higher [5].

Soil pollution is very dangerous and due to the fact that this type of pollution, like air and water pollution is not directly related to human life, less attention has been paid to it. Soil is one of the valuable resources of nature, which provides about 96% of the food needed by humans [6]. Healthy and clean soil is very necessary and important for life on earth. Day by day, soil ecosystems become a place for harmful substances, scum, waste, and receiving harmful substances, and more than the weather, their pollution burden increases, and on the other hand, more and more Due to the construction of buildings, roads and urban and industrial facilities, a large amount of soil is taken out of the natural circulation and also from the agricultural area and becomes dead soil. Therefore, proper management to have a healthy soil is necessary for human survival [7]. Soils have a special advantage called self-purification and they are considered to be nature's purifiers, but the self-purification power of soil is less than the self-purification power of water and air due to its low exchange with other regions and areas, for this reason soil pollution, It is considered one of the most important types of environmental pollution [8].

Heavy metals are a group of metals and quasi-metals, whose amounts and concentrations are toxic and dangerous. Mercury, lead, cadmium and arsenic can be mentioned from the group of toxic metals [9]. Because heavy metals enter the soil through anthropogenic activities or exist naturally in the soil texture, they can easily and very quickly cause soil pollution [10]. In addition to directly affecting the physical and chemical properties of the soil, reducing biological activity and reducing the bioavailability of soil nutrients, heavy metal pollution is also a serious threat to human health through entering the food chain and environmental security through penetration into They are considered underground waters [11].

2. Sources and origins of heavy metals

One of the basic problems and challenges of recent years has been the gradual accumulation of chemical pollution in the environment [12]. The most important soil pollutants include heavy metals, chemical compounds resulting from acid rain and organic materials [13]. The pollution whose importance has been realized for many years is soil pollution with heavy metals [14].

Spatial changes of heavy metal contents in surface soil may be influenced by parent soil materials and human resources [15]. In other words, these metals exist naturally in the soil, but they are added to the soil as a result of human activities [16]. Heavy metals have cytotoxic, carcinogenic and mutagenic effects on humans and other living organisms [17]. Pollution caused by heavy metals can spread in the air, water or soil. Among the aforementioned pollutions, more attention has been directed towards air and water pollution, and soil pollution has been neglected [18].

The rapid development of the industry and the increase in the release of chemicals used in industries and agriculture into the environment have led to increased concerns about the potential for the accumulation of heavy metals in the soils of big cities [19]. Heavy metal contamination in soils may lead to irregularity in the soil structure, negative impact on the growth of plants and animal ecosystems, and even damage to human health by entering the food chain [20].

Soil plays an important role in the cycle of mineral and organic elements and as a dynamic ecosystem provides the life of small and large living organisms, therefore monitoring and evaluating its pollution is of immense importance [21]. Heavy metals can have fatal effects on humans and animals that use contaminated plants in the area. Therefore, determining the amount of these metals in soil environments has attracted the attention of many researchers [22].

Various industries such as metal mines to computers and electronics, chemical fertilizer production factories, dyeing, textile, weapons production and thermal power plants, oil and petrochemical industries, steel and pipe making industries, hospitals and livestock and poultry slaughterhouses cause pollution. The results of heavy metals play a role [23]. Heavy metals in the soil, such as lead, zinc, copper and cadmium, can originate from car parts, tire friction, grease and oily substances, the output of industrial factories (smoke) and furnaces [24]. Metallic elements enter the soil through the use of chemical fertilizers, fungicides, industries and sewage sludge [25]. In other words, human activities, especially agricultural effluents, industrial wastewaters, and pollution from the transportation industry, cause a significant amount of heavy metals to enter the surrounding environment [26].

Health risk assessment of surface soil in urban and industrial areas is widely used to quantify the carcinogenic and non-carcinogenic risks of potentially toxic elements for humans through three routes: ingestion, skin contact, and inhalation [27].

3. Heavy metal toxicity

Heavy metals in the soil can enter the human body directly through swallowing and breathing [28] or enter the body through water and food after contamination of water and soil sources and entering the structure of plants [29]. The harmful effects of heavy metals on human health have been proven from various aspects, and exposure to these pollutants causes acute and chronic poisoning, as well as various diseases, including nervous disorders, hormone imbalances, and respiratory disorders and heart disease, memory loss, various types of cancer and eventually death [30, 31]. Arsenic, lead and cadmium are toxic elements that do not play a role in biological interactions in the human body and cause hemoglobin biosynthesis disorders and anemia, increased blood pressure, kidney damage, miscarriage and premature birth, nervous system disorders, and brain damage, male infertility, reduced learning ability and behavioral disorders in children [32]. Vanadium and nickel are heavy metals that have irreparable effects on humans. Vanadium causes respiratory abnormalities and has negative effects on the liver and kidneys, and increasing the accumulation of nickel in the body can cause lung, nose, larynx, prostate cancer, reduced reproductive ability, watery lungs, itching and skin problems [33]. Zinc and copper metals play a role in biological processes based on their amounts [18] and are among the essential elements in biological reactions, but the increase in biological accumulation in body tissues causes the high toxicity of these two metals [34].

Studies show that human industrial and agricultural activities are the only cause of soil pollution with toxic and dangerous compounds by heavy metals and other pollutants [35, 36]. In Iran, in many areas, the creation of petrochemical industries, the construction of refineries and the drilling of oil and gas extraction wells, steel industries, food and packaging industries, and agriculture have caused an increase in soil pollution in these areas. Factors such as improper disposal of wastes and wastes in industrial centers, pollutant spreading by refineries and power plants, pollutant leakage from underground oil tanks and gas stations, accidents of tankers and tankers contribute to this problem [33]. The adverse effects of heavy metals in soil and plants are determined when their concentration increases beyond a certain amount, and this increase depends on the type of metal, the type of soil, various human activities, and the time of metal accumulation [37]. Research has shown that heavy metals are effective in reducing animal and plant populations [38]. In general, the accumulation of heavy metals in surface soils is more than in deep soils, which indicates the recent pollution in the region and indicates the impact of environmental pollutants such as industrial, urban activities and especially motor vehicles on the soils of the region [39].

Heavy metals are one of the environmental pollutants whose amounts have been widely and differently reported [33] and the toxicity of heavy metals in humans, animals and plants has been investigated and proven in numerous studies [34]. Also, in other reports, researchers found that heavy metals can be transferred to sediments in water, which increases the possibility of environmental risk and poisoning of aquatic organisms [35, 36]. A very small amount of some heavy metals such as copper, zinc, iron and magnesium are necessary for all living organisms, but a large amount of them can cause fatal poisoning of living organisms [37]. Some plant species are able to absorb a large amount of heavy metals from soil and water, as a result, by consuming contaminated plants, heavy metals enter the human body [38]. As much as possible, the body's immune system tries to remove heavy metals through sweat, urine and feces, but some types of these metals are deposited in the tissues before the body can remove them and the effects They leave their mark [39, 40]. In general, neurological disorders (Parkinson's, Alzheimer's, depression, schizophrenia), various types of cancer, nutrient deficiency, hormonal imbalance, obesity, abortion, respiratory and cardiovascular disorders, damage to the liver, kidneys and brain. Allergy and asthma, endocrine disorders, chronic viral infections, lowering the body's tolerance threshold, dysfunction of enzymes, changes in metabolism, infertility, anemia, fatigue, nausea and vomiting, headache and dizziness, irritability, weakening of the body's immune system, destruction Genes, premature aging, skin disorders, loss of memory, anorexia, joint inflammation, hair loss, osteoporosis and in severe cases death are the results of the effects of heavy metals entering the human body. On the other hand, the accumulation property of heavy metals in plants and their entry into the food chain doubles the risks caused by them. With the growth of industry and the increase in the consumption of chemicals, their entry into water, soil and air and the pollution of the environment, the possibility of humans facing the dangers caused by them has increased [41, 42].

4. Soil heavy metals and effects of corps

Increasing the concentration of heavy metals is very dangerous because heavy metals are toxic, stable and non-degradable [43]. Physical and chemical processes (saturation and oxidation) can release the heavy metals accumulated in the soil, which means that the metals can be transferred to the surrounding waters and be consumed by crops and as a result from water supply and food chain to affect public health [44]. As a result of various human activities such as agriculture, mines, industries, the passage of vehicles and urban ecology, the soil becomes polluted [45]. Some of these pollutions are caused by the exploration and extraction of petroleum materials from the ground. Accidents involving vehicles that transport pollutants are another example of soil contamination by human activities. Other polluters that cause soil pollution include cars, trucks, and airplanes that carry materials such as fuel, and soil pollution occurs due to their spilling and exiting the vehicle. Spilling toxic substances such as solvents, dyes and detergents spread the pollution of the earth and soil [21]. Also, heavy metals can enter the soil through chemical processes of minerals, mother rocks and natural activities. Also, through other ways such as urban sewage, agricultural runoff is also absorbed into the soil [46, 47]. It has been reported in many researches that heavy metals can also affect the soils of wetland areas and when heavy metals enter the soil environment, the self-remediation function of the soil alone in the short term cannot effectively prevent its impact. Reduce the impact on the environment [48–52]. High concentration of heavy metals in soil can be a threat to human health and wildlife and should be investigated in agricultural soil [53, 54].

Microbial activity and soil enzyme activity can reflect soil quality sensitively [55]. Aceves et al. (1999) believed that soil microbial biomass is an important indicator in determining the degree of soil contamination. Microbial activity is significantly inhibited in soil contaminated with heavy metals [56]. Kandeler et al. (1997) showed that microbial biomass is strongly inhibited in soil contaminated with Cu, Zn, Pb and other heavy metals. The soil microbial biomass near the mine was significantly lower than the mine, and the effect of different concentrations of different heavy metals on the soil microbial biomass was different [57]. Chander et al. (1995) by studying the effect of different concentrations of heavy metals on soil microbial biomass, found that only if the concentration of heavy metals in the soil is three times higher than the EU environmental standard, it can inhibit the microbial biomass [58]. Fliebbach et al. (1994) found that low concentrations of heavy metals can stimulate microbial growth and increase microbial biomass, while high concentrations can significantly reduce soil microbial biomass. In addition, the enzymes in the soil play an important role in the decomposition of organic matter and the cycle of nutrients. Studies have shown that the activity of enzymes in soil is related to heavy metal contamination [59]. Chander et al. (1995) found that with the increase in the concentration of heavy metals, the activity of enzymes in the soil decreases significantly and approximately 10 to 50 times [58].

A low concentration of heavy metals in the soil, regardless of whether they are essential or unnecessary for plants, will not affect the growth of plants in a certain range, but if the concentration is too high, the content of heavy metals enriched by the plant will exceed its tolerance threshold. And so the plant is poisoned and even leads to the death of the plant. In a research, it was found that if the amount of copper in the soil is more than 50 mg/kg, it affects citrus seedlings. If the amount of soil copper reaches 200 mg/kg, wheat wilts [60]. The research showed that the growth of cabbage plant and bean plant is inhibited under the concentration of 30 micromol/liter of cadmium metal, root length is reduced and plant height and leaf area are reduced [61]. Cadmium may disturb the photosynthesis of crops and protein synthesis and cause membrane damage etc. [62, 63]. However, through a series of environmental effects, nature returns bad results to humans. The most important issue is to control the excessive discharge of pollution sources from the main stream, and source monitoring should correct the technical path. The ultimate goal of preventing soil erosion and pollution is sustainable human development, while heavy metal pollution seriously threatens human health. Its biological effectiveness is not only an important basis for evaluating soil pollution status, but also a theoretical basis for pollution control and a key parameter for evaluating its treatment results. It is of great importance in guiding the production of agricultural products and protecting the environment. By analyzing the sources of heavy metals in soil, it is possible to address the main pollution process and then find a logical way to intervene as soon as possible from the source to reduce the toxicity damage to the soil. Heavy metal pollution is a complex process involving metal elements that changes over time and space with the interaction of biological system forces and environmental system forces. It is necessary to systematically study the dynamic process of interaction between the environmental system and the biological system contaminated with heavy metals in order to increase the value of its application and the importance of its guidance in environmental governance. At present, there are still some deficiencies in the methods of analyzing the source of heavy metal pollution in soil, so it is necessary to conduct more systematic and comprehensive research in the next stage. It can also be combined with various analysis methods to make the heavy metal pollution source analysis method more complete, comprehensive and scientific. In the future, according to the actual situation of the contaminated soil, along with environmental factors such as crop planting, biological community, and weather conditions in the contaminated area, it is possible to select a targeted and suitable extracting for investigation. The chemical form of heavy metals in the soil, the bioavailability data of heavy metals in the human body can be obtained by using it in the simulated human absorption test. It can make the assessment of pollution and heavy metal remediation effect more objective, fast, accurate and efficient [64].

Soil contamination with heavy metals has become a serious environmental problem in many parts of the world [2, 65]. The most important soil pollutants include heavy metals, acid precipitation, and organic substances, among them, heavy metals have been receiving a lot of attention in recent years due to their polluting characteristics in the soil [66]. Spatial changes of heavy metal contents in surface soil may be influenced by parent soil materials and human resources. In other words, these metals exist naturally in the soil, but they are added to the soil as a result of human activities. In fact, human activities may lead to more accumulation of heavy metals in the soil [67]. The background concentration of elements in the soil depends on the mineralogical composition of the parent material and weathering processes affecting the formation of the soil, as well as characteristics such as particle size, amounts of clay and soil organic matter [68]. As a result, the natural concentration of elements in soils is widely variable, and the use of the background levels of other countries and the global average to identify the extent and risks of heavy metal pollution in the soils of areas where environmental limits have not been defined is incorrect [69]. Therefore, although the natural background concentrations of heavy metals in the soil have been studied in many countries such as Poland and many other European countries, the United States and China, and the basis for understanding the natural changes of elements and identifying soil pollution has been laid. It is also necessary to estimate background concentrations in Iran [42].

Heavy metals are naturally present in the parent materials of the soil, but the main human sources of metals in the soil and environment are mining and smelting of metals, agricultural activities, sewage sludge and combustion of fossil fuels, factories and metal industries, waste disposal, use and disposal of metal and electrical materials, electricity production, chemical production industries [70, 71]. There are two sources for soil heavy metal contamination; Natural resources and human resources. Natural resources include the entry of heavy metals through the erosion of parent soil materials and are therefore related to the geology of the region. Iron and steel industries, mining, road transportation, waste burning, and especially the use of fertilizers and chemicals in agriculture are very important human sources of heavy metals entering the soil and water in surface ecosystems [72, 73]. The use of fertilizers and chemicals in agricultural lands leads to an increase in the concentration of heavy metals such as chromium, cobalt, chromium, cadmium, lead, copper and zinc and causes an increase in the concentration of heavy metals in these areas [74, 75]. Considering the concern about heavy metals as environmental pollutants and also because of their stability in the ecosystem, biological monitoring can be a suitable method to measure the concentration of these metals and their bioavailability. For example, large fish and birds are useful indicators for measuring heavy metal pollution due to being at the high levels of food chains in an ecosystem and also their sensitivity to toxic substances. Bioaccumulation refers to the ability to accumulate a chemical substance from the surrounding environment, which occurs directly from water, soil and sediments by different organs or indirectly through eaten food [76].

5. Lead (Pb)

Lead is one of the most common and useful metals known to humans and can be detected in all environments and biological systems. The amount of lead in the environment has increased more than 1000 times during the last three centuries, which is the result of human activities, and between 1950 and 2000, the greatest increase in lead levels was observed [77, 78]. Lead is from the fourth group of the periodic table and has an atomic weight of 207.2. Lead is a gray heavy metal and is usually combined with two or more other elements [79]. Lead reaches aquatic ecosystems due to surface soil erosion and atmospheric sediments. The concentration of lead in the deep oceans is about 0.01–0.02 g/L, but it is about 0.3 g/L in the surface waters of the oceans [80, 81].

In general, lead is absorbed from food and air. Workers in smelting, casting and steel industries, battery manufacturing, plastic factories and printing industries are exposed to lead [82]. Lead is the most abundant and widely used heavy metal and its dispersion in the environment is wider [83]. It is easy to work with lead due to its low melting point, it can be easily made into various shapes. Due to the applications that have caused the uncontrollable dispersion of this element, its environmental concentration is increasing in most countries. Lead metal in car battery, ceramics, inside cans, cigarette ash, car exhaust fumes, leaded gasoline, hair dyes, insecticides, mascara, snow, soldering alloy, putties, paints, city water and or there are wells, alcoholic drinks, pipes [76].

Children are sensitive to the effects of lead, which is considered a primary environmental hazard. Metal poisoning in children causes sensitivity in the developing nervous system, which is due to the sensitivity to lead metal toxicity [84, 85]. Lead is classified in group 2B of IARC carcinogenic compounds, and its toxic effects in the body include the occurrence of disorders in four places, i.e. digestive system, central nervous system, peripheral nerves and hematopoietic system [86]. Lead may enter the human body through the intestine. It is also absorbed through the lungs, inhalation and skin or by direct ingestion and drinking [87]. Lead accumulates in high concentrations in bones, teeth, liver, lungs, kidneys, brain and spleen and passes through the blood–brain barrier and the fetal placenta. The symptoms of diseases caused by lead are completely different and unrecognizable in the first place. In the early stages, impatience, anorexia and lack of movement can be mentioned. Weight loss and blood loss are other symptoms of lead poisoning. Among women, monthly periods do not occur or are postponed. General changes also occur in the form of blood. Red blood cells change shape. As a result of lead deposition in the mouth, gums and teeth turn gray. This gray color can remain for some time even after the poisoning is removed [85, 88]. The biological half-life of lead may be much longer in children than in adults. Lead has a half-life of 35 days in blood, 40 days in soft tissue, and 20–30 years in bones [89]. The main route of excretion of absorbed lead is the urinary tract, which is usually done by glomerular filtration in the kidney. It can also be excreted through the digestive system through bile [90].

6. Zinc (Zn)

Zinc metal has been used for centuries due to its low boiling point. Like other metals, zinc reacts slowly. Zinc is a blue-white or silvery metal found in the earth's crust. The abundance of zinc in nature changes depending on the place and season [91]. It combines with oxygen and other non-metals and reacts with dilute acid and releases hydrogen gas. Zinc is the fourth most common and used metal, after iron, aluminum and copper, it is the most produced metal. Features such as corrosion resistance, reactivity with iron and electrochemical properties of zinc metal have led to its use as a suitable coating against corrosion or galvanization. Galvanized steel is used in construction, power, construction of urban amenities (such as benches and tables), agriculture and transportation. This metal is used in the production of various alloys such as brass, aluminum alloys, and magnesium alloys, which are used in the construction industry and electric cars [92].

Also, this metal can enter the environment through textile and cotton, battery, rubber, paint, cosmetic, fertilizer and medical industries. In addition to metal smelting industries, the impact of acid rain on construction materials containing zinc are the main sources of this element entering the environment [76].

Zinc is one of the important components of some important biomolecules in the human body. There are more than three hundred important enzymes in the body, zinc is one of these enzymes [93]. These enzymes play an important role in maintaining body function and natural health, and some of these enzymes also play a role in the process of gene expression [94]. Zinc plays a role in regulating the synthesis of important biological molecules such as blood sugar balance, insulin hormone, glucose transport, body metabolism and its availability in the physiological system. The zinc element plays a very vital role in enzymes, so that if zinc is removed from their composition, the activity of the enzyme in question stops [95, 96].

Zinc is one of the rare elements of the body, poisoning with it leaves many effects in different organs of the body. Its acute toxic effect, which manifests itself in the form of fever, has been fully proven, but regarding the complications of chronic contact with this metal, various studies have put forward many opinions [97]. Some of the adverse effects of zinc accumulation in the body are: poisoning, fever, confusion, nausea, vomiting and diarrhea after consuming acidic drinks or foods that are prepared and stored in galvanized containers. Zinc is considered a low-risk element, but its toxicity increases in large quantities in the presence of arsenic, lead, cadmium, and antimony. Fever caused by zinc poisoning has symptoms of cold, fever and nausea. Zinc chloride vapor causes the lungs to dry [91].

7. Copper (Cu)

Copper is the first known element that is relatively red in color and has high electrical and thermal conductivity. Copper is one of the most widely used metals in industry. Copper in industries such as electronics (in wires, cathode ray lamps, in IC, vacuum lamps, switches and electronic amplifiers), military industries (production of weapons), metal industries (production of alloys and preparation of coins), tools Kitchen, water purification, is used as a reagent in chemistry and preparation of agricultural pesticides [98, 99]. Copper is one of the common elements in nature, which is found in abundance in the environment due to natural phenomena [100]. Many copper compounds are deposited in sediments or soil particles or stick to these particles. Soluble compounds of copper may be harmful to human health. Usually copper is released in the environment as water-soluble compounds after agricultural activities. Copper is generally found near mines, industrial sites and waste disposal sites. Copper does not decompose in the environment, and for this reason, when it is in the soil, it accumulates in plants and animals [101]. In copper-rich soils, a limited number of plants have a chance of survival. This is the reason why there is not much vegetation near the copper factories. Due to the effects of copper on plants, depending on the acidity of the soil and the amount of organic matter, this element is considered a serious threat to farms. When fields are contaminated with copper, animals absorb higher concentrations of copper, which harms their health [92, 102].

Copper is abundant in the human body and is important for several physiological functions. Copper is actually a mineral that is essential and important for maintaining natural health and for survival. Liver, brain, heart, kidneys and skeleton are important organs where copper is stored. Copper plays an important role in increasing the absorption of other metals such as iron. In addition, copper is also present in the collagen of the supporting tissue. Copper acts as a cofactor for several enzymes and thus are involved in the regulation of various physiological processes. Copper is part of hair and elastic tissues. This element is present in several important substances containing structural and functional proteins in the body [103]. Also, studies show that plants and animals need sufficient amounts of copper for normal growth and maintaining their health. The appropriateness of yield of crops and seeds is also related to copper concentration. Therefore, copper is an important mineral for producers (green plants) and therefore is very important for maintaining the food chain [96].

It is obvious that the element copper is one of the essential elements of the body of living organisms, but it should be noted that large amounts of copper can cause acute effects such as discomfort in the digestive system, damage to the circulatory system such as the liver and kidney systems, and anemia.. The most well-known metabolic disorder resulting from the accumulation of copper in the body is Wilson's disease, in which the blood ceruloplasmin concentration is severely reduced [76].

8. Arsenic (As)

Arsenic is one of the natural elements and the source of global pollution that is found in rocks, soil, water, air and food [104]. Arsenic has a complex chemical structure and can be found in common inorganic forms, arsenite, arsenate, and ternary organic forms. Organic arsenic combines with carbon and hydrogen ions and forms. Organic arsenic compounds are found in fish and shellfish [105]. Inorganic arsenic found in soil and water has been classified by the US Environmental Protection Agency as a human carcinogenic pollutant [106]. High doses of organic arsenic can produce the same toxic effects as a lower amount of inorganic arsenic [107]. Mineral arsenic is present in some man-made resource industries, including waste of industrial products, coal, copper, lead and glass products [85]. The use of arsenic compounds as herbicides, pesticides and fungicides is another source of arsenic pollution [79].

Absorption of arsenic through inhalation is strongly dependent on the solubility and size of the particles that enter the respiratory system, and arsenic is well absorbed from the digestive system, in other words, soluble arsenic compounds can be absorbed from the digestive system [108]. The biological half-life of absorbed inorganic arsenic is about 10 hours and about 50–80% is excreted in about 3 days, while the half-life of methyl arsenic is 30 hours [85, 88]. Absorbed arsenic can cross the placenta and lead to cord blood concentrations similar to maternal blood concentrations [109]. The mechanism of arsenic toxicity involves a number of sulfhydryl-containing proteins and enzyme systems that change with arsenic exposure [110]. The specialized food committee determined the acceptable weekly intake for inorganic arsenic at 0.015 mg/kg [111]. It seems that consumption of organic arsenic at about 0.05 mg/kg of body weight per day causes dangerous effects for humans [112]. This element has no role in biological interactions in the human body and affects the cardiovascular system and skin, central and peripheral nervous system, kidneys and hematopoietic system of the body and causes carcinogenesis. The health effects of ingested inorganic arsenic include skin cancer, internal cancer, and non-cancerous effects on the skin, vascular system, digestive system, and liver. In general, soluble inorganic arsenic compounds are more toxic than organic types, and trivalent forms of arsenic are more toxic than pentavalent forms of arsenic, and various body systems and organs, including the skin, respiratory, cardiovascular, immune, genetic, and reproductive systems., digestive and nervous systems are affected by this substance and combined side effects are observed.

9. Chromium (Cr)

Chromium is one of the elements of the periodic table with the symbol Cr and atomic number 24. Chrome or Chromium is a hard, shiny metallic gray metal with high polishability and high boiling point and remarkable resistance to rust and tarnishing. Chromium oxide was used to coat metal weapons in the Chinese Empire for more than 2000 years. Chromium was discovered as an element in 1761 and was first used as a pigment [113]. In 1797, chrome metal was separated from its ore for the first time. Since then, almost all the chromium in the world is obtained from the ore chromite. The value of this metal is mostly due to its high resistance to rust and erosion, especially when it was discovered that adding chromium to steel has a significant effect in preventing corrosion and tarnishing of steel. Today, about 85% of the world's chrome consumption is used to make stainless steel, of which at least 10% of its volume is chrome, and chrome plating is also used [91]. Chromium metal is used in metallurgy to resist corrosion and in final polishing, as a component in alloys, for example in stainless steel, in chrome plating, in anodized aluminum, as a catalyst. Chromite is used to make molds for baking bricks. Chromium salts cause the glass to turn green, and chromates and oxides are used in hair color and ordinary colors [114, 115]. Chromium is used to harden steel and this element is used to make stainless steel products, these compounds have useful applications. This element is used for covering hard surfaces and for decorating and preventing corrosion and rusting. Chrome is glass-shaped and emerald green in color and widely used [76].

Surveys show that among heavy metals in the past, chromium was less important in agriculture and environment. While it can have severe destructive effects on plants and the environment, soil and water contamination with chromium is one of the major environmental concerns in recent decades [116]. Chromium has harmful effects on plant physiological processes such as photosynthesis, water relations and mineral nutrition, germination, growth and development of roots and leaves. The metabolic changes made by chromium in plants are either directly on enzymes and plant metabolites or through the creation of active oxygen species that cause oxidative stress [117, 118].

Chromium metal and chromium III compounds are usually not hazardous to health, but chromium VI compounds are toxic if swallowed. The amount of almost half a teaspoon of toxic chromium VI compounds is lethal, and it has been proven that non-lethal amounts of chromium VI are carcinogenic. Most chromium VI compounds are harmful to eyes, skin and mucous tissues. Permanent contact with these compounds can cause permanent damage to the eyes, except for cases where complete treatment is done [119, 120]. In 1958, the World Health Organization suggested the maximum allowable consumption of chromium VI from the health aspect of 0.05 mg per liter of drinking water. This proposal was reviewed many times and the announced amount did not change during this time [121, 122].

The amount of chromium in drinking water is very low, but it is possible that contaminated water contains some chromium IV and chromium VI, which are considered dangerous types of chromium. If the amount of chromium III consumed by humans is higher than usual, it affects human health and, for example, causes skin itching. Chromium IV has many effects on human health. There is usually chrome in leather products. This combination causes severe allergy such as skin itching in people. By breathing chromium IV, the nose is stimulated and nosebleeds occur. Other diseases caused by Chromium IV include skin itching, stomach ulcers, respiratory system problems, weakening of the body's immune system, kidney and liver damage, genetic material changes, lung cancer and death.

10. Cadmium (Cd)

Cadmium is an industrial and environmental pollutant that affects a number of human body organs. Cadmium is a group IIB metal with an atomic weight of 112.41 [79, 105]. In general, exposure to cadmium occurs mainly through two sources. The first food route is through water and food contaminated with cadmium, especially vegetable leaves, seeds, grains, fruits and fish [123]. The second source is through the inhalation of cadmium particles in industrial or daily activities, among which the inhalation of cigarette smoke is considered as a very dangerous source, because cadmium is easily absorbed by the lungs [124–126].

Forest fires and volcanoes, human activities such as industrial waste leachates, production of synthetic phosphate fertilizers are important sources of cadmium emissions. The main use of this element is as a stabilizer and pigment in plastic and electrolysis industries, but its main part is used in soldering and as an alloy in nickel and cadmium batteries.. This element is used in the industry as an anti-friction, catalyst, anti-rust agent or in the composition of alloys. Cadmium is also used in rod protection semiconductors in nuclear reactors, metal plating, ceramic making, PVC factories and plastic industries, battery production, fungicide compounds, engine oil, rubber making and photography [76].

The amount of cadmium absorption in foods is caused by the way animals are fed, kidneys and liver are suitable places for cadmium accumulation. Sea shells also have a high concentration of cadmium. Absorption of cadmium through the skin is very limited [21]. The biological half-life of cadmium in humans, in soft tissues and bones, is 10 to 30 years. The speed of cadmium methylation is much lower compared to mercury, arsenic, and lead, and only two bacteria named Pseudomonas sp. and Staphylococcus aureus are able to methylate cadmium in water environments [121]. A serious disease caused by it in humans is a disease called itai itai (rheumatism disease or painful skeletal deformity). The main effects of cadmium toxicity are on the lungs, kidneys, and bones. Acute effects caused by its inhalation include bronchitis, pneumonia and liver poisoning. Chronic inhalation of cadmium compounds, in the form of vapors or dust, causes pulmonary edema, in which the small air sacs enlarge and are eventually destroyed due to the reduced lung volume. Both chronic inhalation and absorption of cadmium through the mouth affect kidney secretions, which is the first stage of protein excretion by the proximal tubules of the kidney. Acute poisoning with cadmium may cause the death of animals and birds and cause severe poisoning in aquatic animals. Absorption of cadmium from the lungs is more effective than the intestine, and 50% of the cadmium inhaled through cigarette smoke may be absorbed. On average, the concentration of cadmium in the blood of smokers is 4–5 times and in the kidneys 2–3 times more than that of non-smokers. It seems that cadmium reduces the body's defense resistance, especially the host's resistance against bacteria and viruses. Cadmium may cause demineralization of the skeleton and increase bone fragility and the risk of fracture [91].

11. Effects of heavy metals on humans

The harmful effects of heavy metals on human health have been proven from various aspects, and exposure to these pollutants causes acute and chronic poisoning, as well as various diseases, including nervous disorders, hormone imbalance, respiratory and cardiac disorders, decrease memory, types of cancer and eventually death [30, 127]. The lethality of most heavy metals for humans is estimated in the range of 500–350 mg per day. Heavy metals cause various diseases such as infertility, poisoning, nervous system disorders, breaking chromosomes, premature aging and various cancers in humans [28, 128]. Cancer is the main cause of death in developed and developing countries of the world. The increase in cancer may be caused by the increase in environmental pollution [65]. Heavy metals are one of the most important environmental pollutants [129]. These metals have the greatest impact on the health of citizens due to the occurrence of health risks such as reduced growth of children, kidney diseases, cancer and other adverse effects [130].

Heavy metals accumulate in the vital organs of the human body due to their indestructible and stable nature and lead to serious health disorders. Heavy metals should be considered due to their indestructibility, stability, accumulation in living organs and damage to the health of living organisms [32]. Contact with heavy metals occurs chronically (contact over a long period of time) due to movement in the food chain, but acute poisoning by ingestion or skin contact with heavy metals is rare, but possible. There is [131].

Heavy metals in soils are harmful to human health, especially children [132, 133]. Children have a high rate of heavy metal absorption due to their active digestive system, small body size, developing nervous system, swallowing dust, soil or suspended particles, weak immune system and excessive use of hands [134]. Heavy metals are very harmful to the human body because they do not have any effective elimination mechanism in the body [135]. These metals affect health indirectly by consuming plants that grow in contaminated soil and directly by inhaling and consuming contaminated water [136]. Heavy metals may become a problem for human health and have adverse environmental effects [137]. Heavy metals can enter the human body directly through swallowing and breathing, or reach the earth's surface through atmospheric fallout, and enter the body through water and food after polluting water and soil sources and entering the structure of plants [29].

Some heavy metals are necessary for the continuation of life and activities of animals and they play a significant role in the body. Unnecessary and toxic heavy metals without having a role in the physiological activities of animals, even in low concentrations, cause disturbances in the body system of animals [27]. Among the pollutants in the environment, toxic metals in high concentrations cause poisoning for living organisms. Some metals, especially heavy metals, are of high environmental importance due to their toxic properties and accumulation in living organisms, even in relatively low concentrations [138]. Unlike some organic substances, these toxic metals are not biodegradable and their accumulation in living tissues can lead to death or serious threats to health [139]. Due to the fact that these metals are not decomposed by conventional biological processes, as a result, by accumulating in the tissues of living organisms, they are easily moved through the food chain, hence, by increasing their amounts in the soil over time, to It significantly damages plants. For example, copper and zinc in very low concentrations are essential trace elements for the survival of plant and animal life [140]. Long-term biological durability and remaining in the soil causes the accumulation of these metals in food chains and as a result, potential negative effects for human health. The amount of access to these metals depends on the type of plant and their required amount as micronutrients and the ability of plants to efficiently regulate their metabolism through the secretion of organic acids or protons into the root environment. In addition, the soil properties are effective on the mobility of these metals and therefore regulate their release rate in the soil solution. The ability of plants to absorb metals from the soil, their internal use and detoxification mechanisms have met with increasing popularity [141].

Physiological stages in plants are affected by increasing the concentration of heavy metals around the target plant. These metals cause oxidative stress in the plant, which is one of the harmful effects of this stress in the production of free radicals. In high concentrations of metals, substitution with essential metals occurs and since essential metals play an important role in the formation of pigments and enzymes, therefore the formation of pigments is disturbed and hence the elements It makes the soil unsuitable for plant growth and destroys biodiversity [142, 143]. For example, cadmium metal is one of the most toxic elements for plants and has no biological role. Cadmium metal mainly enters the environment and food chain through industrial processes and phosphate fertilizers. This toxic metal is easily absorbed by the roots and by forming complex complexes with organic compounds such as proteins, it prevents the necessary activity of cells. Cadmium, by increasing the peroxidation of lipids and the production of reactive oxygen species, provides membrane deterioration [100]. Since this metal has two positive charges (bivalent) and competes with elements such as magnesium in chlorophyll and with iron ions which are divalent and replaces them and the chlorophyll molecule in the plant is thus destroys Therefore, photosynthesis is very sensitive to cadmium. The high concentration of essential metals such as copper and zinc also harms the plant. Copper metal reduces plant growth by preventing the absorption of other elements such as calcium, iron and potassium, which are essential plant elements [144, 145].

12. Conclusions

Dietary exposure to toxic metals is a public health concern. As a result, food safety is an issue that threatens human health and agricultural business. The principles, advantages, and disadvantages of immobilization, soil washing, and phytoremediation techniques, often cited as among the best available technologies for cleaning up heavy metal-contaminated sites, are presented. Remediation of soils contaminated with heavy metals is necessary to reduce related risks, make land resources available for agricultural production, increase food security, and reduce land ownership problems caused by changes in land use patterns. Also, washing rice reduces some toxic and essential elements in rice. The distribution of elements in cereals from different regions helps countries make informed decisions about importing cereals such as rice. Rice and other cereal producers can develop strategies to reduce significant metal uptake from soil. Identification of appropriate rice treatment processes such as washing will provide information on reducing metal exposure to the rice consuming population while preserving essential elements in the grain.

Author details

Mohammad Velayatzadeh
Department of Industrial Safety, Caspian Institute of Higher Education, Qazvin, Iran

*Address all correspondence to: mv.5908@gmail.com

References

[1] Kharkan J, Sayadi MH, Rezaei MR. Investigation of heavy metals accumulation in the soil and pine trees. Environmental Health Engineering and Management Journal. 2019;**6**(1):17-25

[2] Solgi E, Esmaili-Sari A, Riyahi-Bakhtiari A, Hadipour M. Soil contamination of metals in the three industrial estates, Arak, Iran. The Bulletin of Environmental Contamination and Toxicology. 2012;**88**(4):634-638

[3] Ihedioha JN, Ukoha PO, Ekere NR. Ecological and human health risk assessment of heavy metal contamination in soil of a municipal solid waste dump in Uyo, Nigeria. Environmental Geochemistry and Health. 2017;**39**(3):497-515

[4] Olawoyin R, Oyewole SA, Grayson RL. Potential risk effect from elevated levels of soil heavy metals on human health in the Niger delta. Ecotoxicology and Environmental Safety. 2012;**85**:120-130

[5] Wu S, Peng S, Zhang X, Wu D, Luo W, Zhang T, et al. Levels and health risk assessments of heavy metals in urban soils in Dongguan, China. Journal of Geochemical Exploration. 2015;**148**: 71-78

[6] Bhuyana S, Islam S. A critical review of heavy metal pollution and its effects in Bangladesh. Environmental and Energy Economics. 2017;**2**(1):12-25

[7] Payandeh K, Nazarpour A, Velayatzadeh M. Human health risk of some heavy metals accumulated in Tomatoe, cucumber, potato, and onion grown in Dezful and Shushtar. Arch Hyg Sci. 2021;**10**(4):299-314

[8] Egbe ER, Nsonwu-Anyanwu AC, Offor SJ, Usoro CAO, Etukudo MH. Heavy metal content of the soil in the vicinity of the united cement factory in southern Nigeria. Journal of Advances Environmental Health Research. 2019;**7**:122-130

[9] Zhang D, Pan X, Lee D-J. Potentially harmful metals and metalloids in the urban street dusts of Taipei City. Journal of the Taiwan Institute of Chemical Engineers. 2014;**45**(4):1727-1732

[10] Islam MS, Ahmed MK, Raknuzzaman M, Habibullah-Al-Mamun M, Islam MK. Heavy metal pollution in surface water and sediment: A preliminary assessment of an urban river in a developing country. Ecological Indicators. 2015;**48**:282-291

[11] Keshavarzi B, Tazarvi Z, Rajabzadeh MA, Najmeddin A. Chemical speciation, human health risk assessment and pollution level of selected heavy metals in urban street dust of shiraz, Iran. Atmospheric Environment. 2015;**119**:1-10

[12] Yalcin MG, Battaloglu R, Ilhan S. Heavy metal sources in sultan marsh and its neighborhood, Kayseri, Turkey. Environmental Geology. 2007;**53**:399-415

[13] Mico C, Recatala L, Peris M, Sanchez J. Assessing heavy metal sources in agricultural soils of an European Mediterranean area by multivariate analysis. Chemosphere. 2006;**65**:863-872

[14] Proshad R, Islam MS, Kormoker T, Bhuyan MS, Hanif MA, Hossain N, et al. Contamination of heavy metals in agricultural soils: Ecological and health risk assessment. SF Journal of Nanochemistry and Nanotechnology. 2019;**2**(1):1012

[15] Cao S, Duan X, Zhao X, Ma J, Dong T, Huang N, et al. Health risks from the exposure of children to As, Se, Pb and other heavy metals near the largest coking plant in China. Science of the Total Environment. 2014;**472**:1001-1009

[16] Li MS, Luo YP, Su ZY. Heavy metal concentrations in soils and plant accumulation in a restored manganese mine land in Guangxi, South China. Environmental Pollution. 2007;**147**:168-175

[17] Velayatzadeh M, Askary Sary A, Hoseinzadeh Sahafi H. Determination of mercury, cadmium, arsenic and lead in muscle and liver of Liza dussumieri from the Persian Gulf, Iran. Journal of Biodiversity and Environmental Sciences. 2014;**5**(3):227-234

[18] Velayatzadeh M, Biria M, Mohammadi E. Determination of heavy metals (Hg, Cd, Pb and Cu) in Carasobarbus luteus in Karun River, Iran. World J Fish Mar Sci. 2015;**7**(3):158-163

[19] Wong SC, Li XD, Zhang G, Qi SH, Min YS. Heavy metals in agricultural soils of the Pearl River Delta, South China. Environment Pollution. 2002;**119**:33-44

[20] Rachwal M, Kardel K, Magiera T, Bens O. Application of magnetic susceptibility in assessment of heavy metal contamination of Saxonian soil (Germany) caused by industrial dust deposition. Geoderma. 2017;**295**:10-21

[21] Maqbool A, Xiao X, Wang H, Bian Z, Akram MW. Bioassessment of heavy metals in wheat crop from soil and dust in a coal mining area. Pollution. 2019;**5**(2):323-337

[22] Li Y, Wang YB, Gou X, Su YB, Wang G. Risk assessment of heavy metals in soils and vegetables around non-ferrous metals mining and smelting sites, Baiyin, China. Journal of Environmental Sciences. 2006;**18**(6):1124-1134

[23] Gu JG, Lin QQ, Hu R, Zhuge YP, Zhou QX. Translation behavior of heavy metals in soil-plant system-a case study of Qingchengzi Pb-Zn mine in Liaoning province. Journal of Agro-Environment Science. 2005;**4**:634-637

[24] Payandeh K, Velayatzadeh M. Introduction of heavy metals in urban dust and their source. In: International Conference of Dust Storm in Southwestern Asia. Zabol, Iran: University of Zabol; 2019

[25] Aitta A, El-Ramady H, Alshaal T, El-Henawy A, Shams M, Talha N, et al. Seasonal and spatial distribution of soil trace elements around Kitchener drain in the northern Nile Delta. Egypt. Agriculture. 2019;**9**(7):152

[26] Chibuike GU, Obiora SC. Heavy metal polluted soils: Effect on plants and bioremediation methods. Applied and Environmental Soil Science. 2014;**2014**:1-12

[27] Du P, Xie YF, Wang SJ, Zhao H, Zhang Z, Wu B, et al. Potential sources of and ecological risks from heavy metals in agricultural soils, Daye City, China. Environmental Science and Pollution Research. 2014;**22**:3498-3507

[28] Dehghani S, Moore F, Keshavarzi B, Hale BA. Health risk implications of potentially toxic metals in street dust and surface soil of Tehran, Iran. Ecotoxicoloy and Environmental Safety. 2017;**136**:92-103

[29] Cao Q, Liu B, Ren Z, Xiao H, Cheng J, Xue W. Temporal distribution characteristic and risk analysis of heavy

metals in greenhouse vegetable soils. Polish Journal of Environmental Studies. 2020;**29**(3):2071-2079

[30] Pathak AK, Yadav S, Kumar P, Kumar R. Source apportionment and spatial-temporal variation in the metal content of surface dust collected from an industrial area adjoining Dehli, India. Science of the Total Environment. 2013;**443**:662-672

[31] Doabi SA, Karami M, Afyuni M, Yeganeh M. Pollution and health risk assessment of heavy metals in agricultural soil, atmospheric dust and major food crops in Kermanshah province, Iran. Ecotoxicoloy and Environmental Safety. 2018;**163**:153e164

[32] Cooper AM, Felix D, Alcantara F, et al. Monitoring and mitigation of toxic heavy metals and arsenic accumulation in food crops: A case study of an urbancommunity garden. Plant Direct. 2020;**4**:1-12

[33] Feszterova M, Porubcova L, Tirpakova A. The monitoring of selected heavy metals content and bioavailability in the soil-plant system and its impact on sustainability in agribusiness food chains. Sustainability. 2021;**13**:7021

[34] Huang WL, Chang WH, Cheng SF, Li HY, Chen HL. Potential risk of consuming vegetables planted in soil with Lead and cadmium and the influence on vegetable antioxidant activity. Applied Sciences. 2021;**11**(9):3761

[35] Anyakora C, Ehianeta T, Umukoro O. Heavy metal levels in soil samples from highly industrialized of Lagos environment. Environment Science Technology Journal. 2013;7:917-924

[36] Wu S, Yang H, Guo F, Han R. Spatial patterns and origins of heavy metals in Sheyang River catchment in Jiangsu, China based on geographically weighted regression. Science of the Total Environment. 2017;**580**:1518-1529

[37] Sharma RK, Agrawal M, Marshall F. Heavy metal contamination of soil and vegetables in suburban areas of Varanasi, India. Ecotoxicology and Environmental Safety. 2007;**66**:258-266

[38] Wang YP, Shi JY, Wang H, Lin Q, Chen XC, Chen YX. The influence of soil heavy metals pollution on soil microbial biomass, enzyme activity, and community composition near a copper smelter. Ecotoxicology and Environmental Safety. 2007;**67**:75-81

[39] Ghoreishy F, Salehi M, Fallahzade J. Cadmium and lead in rice grains and wheat breads in Isfahan (Iran) and human health risk assessment. Human and Ecological Risk Assessment: An International Journal. 2018;**25**:1-11

[40] Keller C, Rizwan M, Davidian JC, Pokrovsky OS, Bovet N, Chaurand P, et al. Effect of silicon on wheat seedlings (Triticum turgidum L.) grown in hydroponics and exposed to 0 to 30 mM Cu. Planta. 2015;**241**:847-860

[41] Hajar EWI, Sulaiman AZB, Sakinah AMM. Assessment of heavy metals tolerance in leaves, stems and flowers of stevia Rebaudiana plant. Procedia Environmental Sciences. 2014;**20**:386-393

[42] Ali S, Chaudhary A, Rizwan M, Anwar HT, Adrees M, Farid M, et al. Alleviation of chromium toxicity by glycinebetaine is related to elevated antioxidant enzymes and suppressed chromium uptake and oxidative stress in wheat (Triticum aestivum L.). Environmental Science and Pollution Research. 2015;**22**:10669-10678

[43] Sheng JJ, Wang XP, Gong P, Tian LD, Yao TD. Heavy metals of the Tibetan top soils level, source, spatial distribution, temporal variation and risk assessment. Environmental Science and Pollution Research. 2012;**19**:3362-3370

[44] Hossain MB, Jahiruddin M, Panaullah GM, Loeppert RH, Islam MR, Duxbury JM. Spatial variability of arsenic concentration in soils and plants, and its relationship with iron, manganese and phosphorus. Environmental Pollution. 2008;**156**:739-744

[45] Salman SA, Zeid SAM, El-Montser M, Seleem EM, Abdel-Hafiz ME. Soil characterization and heavy metal pollution assessment in Orabi farms, El Obour, Egypt. Bulletin of the National Research Centre. 2019;**43**:42

[46] Xiao R, Bai JH, Huang LB, Zhang HG, Cui BS, Liu XH. Distribution and pollution, toxicity and risk assessment of heavy metals in sediments from urban and rural rivers of the Pearl River delta in southern China. Ecotoxicology. 2013;**22**(10):1564-1575

[47] Zhan SF, Peng ST, Liu CG, Chang Q, Xu J. Spatial and temporal variations of heavy metals in surface sediments in Bohai Bay, North China. Bulletin of Environmental Contamination and Toxicology. 2010;**84**(4):482-487

[48] Li QS, Wu ZF, Chu B, Zhang N, Cai SS, Fang J. Heavy metals in coastal wetland sediments of the Pearl River estuary. China. Environmental Pollution. 2007;**149**(2):158-164

[49] Pan K, Wang WX. Trace metal contamination in estuarine and coastal environments in China. Science of Total Environment. 2012;**421**(April):3-16

[50] Tang CWY, Ip CC, Zhang G, Shin PKS, Qian PY, Li XD. The spatial and temporal distribution of heavy metals in sediments of Victoria harbour, Hong Kong. Marine Pollution Bulletin. 2008;**57**(6-12):816-825

[51] Zhang WG, Feng H, Chang JN, Qu JG, Xie HX, Yu LZ. Heavy metal contamination in surface sediments of Yangtze river intertidal zone: An assessment from different indexes. Environmental Pollution. 2009;**157**(5):1533-1543

[52] Zhang R, Zhou L, Zhang F, Ding YJ, Gao JR, Chen J, et al. Heavy metal pollution and assessment in the tidal flat sediments of Haizhou Bay, China. Marine Pollution Bulletin. 2013;**74**(7):403-412

[53] Krishna AK, Govil PK. Soil contamination due to heavy metals from an industrial area of Surat, Gujarat, Western India. Environmental Monitoring and Assessment. 2007;**124**:263-275

[54] Reimann C, Demetriades A, Eggen OA, Filzmoser P, The EurogGeoSurveys Geochemistry expert group. The Euro GeoSurveys Geochemical Mapping of Agricultural and Grazing Land Soils Project (GEMAS)–Evaluation of Quality Control Results of Total C and S, Total Organic Carbon (TOC), Cation Exchange Capacity (CEC), XRF, pH, and Particle Size Distribution (PSD) Analysis (NGU Report 11.043). Vol. 90. Trondheim, Norway: Geological Survey of Norway; 2011

[55] Lee DH, Zo YG, Kim SJ. Nonradioactive methods to study genetic Profies of natural bacterial communities by PCR-single-strand-conformation polymorphism. Applied and Environmental Microbiology. 1996;**62**(9):3112-3120

[56] Aceves MB, Grace C, Ansorena J, et al. Soil microbial biomass and organic C in a gradient of zinc concentration in

soils around a mine spoil tip. Soil Biology and Biochemisty. 1999;**31**(6):867-876

[57] Kandeler E, Lurienegger G, Schwarz S. Influence of heavy metals on the functional diversity of soil microbial communities. Biology and Fertility of Soils. 1997;**23**:299-306

[58] Chander K, Brookes PC, Harding SA. Microbial biomass dynamics following addition of metal-enriched sewage sludge to a sandy loam. Soil Biology and Biochemistry. 1995;**27**(11):1409-1421

[59] Fliepbach A, Martens R, Reber H. Soil microbial biomass and activity in soils treated with heavy metal contaminated sewage sludge. Soil Biology and Biochemistry. 1994;**26**:1201-1205

[60] Zhang ZJ, Lu QF, Fang F. Effect of mercury on the growth and physiological function of wheat seedlings. Chinese Journal of Environmental Science. 1989;**10**(4):l0-l13

[61] Qin TC, Wu YS, Wang HX. Effect of cadmium, lead and their interactions on the physiological and biochemical characteristics of Brassica chinensis. Acta Ecologica Sinica. 1994;**14**(1):46-49

[62] Kale H. Response of roots of trees to heavy metals. Environmental and Experimental Botany. 1993;**33**:99-119

[63] Acar YB, Alshawabkeh AN. Principles of electrokinetic remediation. Environmental Science and Technology. 1993;**27**(13):2638-2647

[64] Zhang Q, Wang C. Natural and human factors affect the distribution of soil heavy metal pollution: A review. Water, Air, & Soil Pollution. 2020;**231**(7):2638-2647

[65] Zhao Q, Wang Y, Cao Y, Chen A, Ren M, Ge Y, et al. Potential health risks of heavy metals in cultivated topsoil and grain, including correlations with human primary liver, lung and gastric cancer, in Anhui province, eastern China. Science of the Total Environment. 2014;**470-471**:340-347

[66] Yang J, Ma S, Zhou J, Song Y, Li F. Heavy metal contamination in soils and vegetables and health risk assessment of inhabitants in Daye, China. Journal of International Medical Research. 2018;**46**(8):3374-3387

[67] Liao J, Wen Z, Ru X, Chen J, Wu H, Wei C. Distribution and migration of heavy metals in soil and crops affected by acid mine drainage: Public health implications in Guangdong Province, China. Ecotoxicology and Environmental Safety. 2016;**124**:460-469

[68] Franco C, Soares A, Delgado J. Geostatistical modelling of heavy metal contamination in the top soil of Guadiamar river margins (S Spain) using a stochastic simulation technique. Geoderma. 2006;**136**:852-864

[69] Bhowmik AK, Alamdar A, Katsoyiannis I, Shen H, Ali N, Ali SM, et al. Mapping human health risks from exposure to trace metal contamination of drinking water sources in Pakistan. Science Total Environment. 2015;**538**:306-316

[70] Madrid L, Diaz-Barrientos E, Madrid F. Distribution of heavy metal contents of urban soils in parks of Seville. Chemosphere. 2002;**49**:1301-1308

[71] Wang L, Ma L, Yang Z. Spatial variation and risk assessment of heavy metals in paddy rice from Hunan Province, southern China. International Journal Environment Science and Technology. 2018;**15**(7):1561-1572

[72] Hutton M, de Meeus C. "Analysis and Conclusions from Member States' Assessment of the Risk to Health and

the Environment from Cadmium in Fertilizers", Final report European Commission - Enterprise DG. Environmental Resource Management. Springer; 2001

[73] Hansen E, Lassen C, Stuer-Lauridsen F, Kjolholt F. Heavy Metals in Waste. London: European Commission DGENV.E3, Project ENV.E.3,ETU/200/0058, Final Report; 2002

[74] Al-Khashman OA. Heavy metal distribution in dust, street dust and soils from the work place in Karak industrial estate, Jordan. Atmospheric Environment. 2004;**38**:6803-6812

[75] Lado, L.R, Hengl, T, Reuter, H.I, "Heavy metals in European soils: A geostatistical analysis of the FOREGS geochemical database", Geoderma, No. 148, pp. 189-199, 2008.

[76] Askary SA, Velayatzadeh M, Beheshti M. Determination of heavy metals in Liza abu from Karkheh and Bahmanshir Rivers in Khoozestan from Iran. Advances in Environmental Biology. 2012;**6**:578-583

[77] Agency for Toxic Substance and Disease Registry (ATSDR). Toxicological Profile for Lead. Atlanta, GA: U.S. Department of Health and Humans Services, Public Health Service, Centers for Diseases Control; 2005

[78] Green RE, Pain DJ. Risks to human health from ammunition-derived lead in Europe. Ambio. 2019;**48**:954-968

[79] Castro-Gonzalez MI, Mendez-Armenta M. Heavy metals: Implications associated to fish consumption. Environmental Toxicology and Pharmacology. 2008;**26**:263-271

[80] Sepe A, Ciaralli L, Ciprotti M, Giordano R, Fumari E, Costantini S. Determination of cadmium, chromium, lead and vanadium in six fish species from the Adriatic Sea. Food Additives and Contaminants. 2003;**20**:543-552

[81] Ericson B, Otieno VO, Nganga C, Fort JS, Taylor MP. Assessment of the presence of soil Lead contamination near a former Lead smelter in Mombasa, Kenya. Journal of Health and Pollution. 2019;**9**(21):190307

[82] Jarup L. Hazards of heavy metal contamination. British Medical Bulletin. 2003;**68**:167-182

[83] Askary Sary A, Velayatzadeh M. Determination of lead and zinc in Cyprinus carpio and Oncorhynchus mykiss from Iran in 2001 and 2011. International Journal of Biosciences (IJB). 2014;**4**(11):1-9

[84] Goyer RA. Lead toxicity: Current concerns. Environmental Health Perspectives. 1993;**100**:177-187

[85] Goyer RA, Clarsksom WT. Toxic effects of metals. In: Klaassen CD, editor. Casarett and Doull's Toxicology. The Basic Science of Poisons. New York: McGraw-Hill; 2001. pp. 811-867

[86] Askary Sary A, Velayatzadeh M. Lead and zinc levels in Scomberomorus guttatus, Scomberomorus commerson and Otolithes ruber from Hendijan, Iran. Advances in Environmental Biology. 2012;**6**(2):843-848

[87] Weiss D, Shotyk W, Kramers JD, Gloor M. Sphagnum mosses as archives of recent and past atmospheric lead deposition in Switzerland. Atmospheric Environment. 1999;**33**:3751-3763

[88] Gwalteney-Brant SM. Heavy metals. In: Haschek WM, Rosseaux CG, Wallig AM, editors. Handbook of Toxicologic Pathology. New York: Academic Press; 2002. pp. 701-732

[89] Papanikolaou CN, Hatzidaki GE, Belivanis S, Tzanakakis GN, Tsatsakis MA. Lead toxicity update. A brief review. Medical Science Monitor. 2005;**No. 11**:RA329-336

[90] Gerofke, A, Ulbig, E, Martin, A, Muller-Graf, C, Selhorst, T, Gremse, et al. "Lead content in wild game shot with lead or non-lead ammunition – Does "state of the art consumer health protection" require non-lead ammunition?". PLoS One. 2018;**13**(7):e0200792

[91] Hambidge KM, Casey CE, Krebs NF. Zinc. In: Mertz W, editor. Trace Elements in Human and Animal Nutrition. 5th ed. Vol. 2. Orlando, Fla: Academic Press; 1986. pp. 1-137

[92] Trumbo P, Yates AA, Schlicker S, Poos M. Dietary reference intakes: Vitamin a, vitamin K, arsenic, boron, chromium, copper, iodine, iron, manganese, molybdenum, nickel, silicon, vanadium, and zinc. Journal of the American Dietetic Association. 2001;**101**:294-301

[93] Manzeke MG, Mtambanengwe F, Watts MJ, et al. Fertilizer management and soil type influence grain zinc and iron concentration under contrasting smallholder cropping systems in Zimbabwe. Scientific Reports. 2019;**9**:6445

[94] Krolak E, Marciniuk J, Popijantus K, et al. Environmental factors determining the accumulation of metals: Cu, Zn, Mn and Fe in tissues of Taraxacum sp. sect. Taraxacum. Bulletin of Environmental Contamination and Toxicology. 2018;**101**:68-74

[95] McCall KA, Huang C, Fierke CA. Function and mechanism of zinc Metalloenzymes. Journal Nutrition. 2000;**130**(5):1437S-1446S

[96] Ghosh D, Singha PS, Firdaus SB, Parida P, Ghosh D. Biometals in health and disease: A review. World Journal of Pharmaceutical Research. 2016;**5**(12):390-399

[97] Hussain S, Khan M, Sheikh TMM, Mumtaz MZ, Chohan TA, Shamim S, et al. Zinc essentiality, toxicity, and its bacterial bioremediation: A comprehensive insight. Frontiers in Microbiology. 2022;**13**:1912

[98] Ullah A, Heng S, Munis MFH, Fahad S, Yang X. Phytoremediation of heavy metals assisted by plant growth promoting (PGP) bacteria: A review. Environmental and Experimental Botany. 2015;**117**:28-40

[99] Xu L, Xing X, Liang J, Penge J, Zhou J. In situ phytoremediation of copper and cadmium in a co-contaminated soil and its biological and physical effects. Royal Society of Chemistry. 2019;**9**:993-1003

[100] Usman K, Al-Ghouti MA, Abu-Dieyeh MH. The assessment of cadmium, chromium, copper, and nickel tolerance and bioaccumulation by shrub plant Tetraena qataranse. Scientific Reports. 2019;**9**(1):5658

[101] Thanh NH, Ha TTL, Ha CV, Hung ND, Hung PQ, Kurosawa K, et al. Uptake of Pb, Zn and Cu by roots and shoots of fast growing plants grown in contaminated soil in Vietnam. African Journal of Soil Science. 2013;**1**(2):016-022

[102] Rieuwerts JS. The solubility and bioavailability of trace metals in tropical soils: A review. Chemical Speciation and Bioavailability. 2007;**19**:75-85

[103] Ghosh D, Paul S, Chattopadhyay A, Bandyopadhyay D. Melatonin and aqueous curry leaf extract

in combination protects against lead induced oxidative stress in rat heart: A new approach. World Journal of Pharmaceutical Research. 2015;**9**(12):619-634

[104] Farooq SH, Chandrasekharam D, Dhanachandra W, et al. "relationship of arsenic accumulation with irrigation practices and crop type in agriculture soils of Bengal Delta, India", applied water science. No. 2019;**9**:119

[105] Agency for Toxic Substance and Disease Registry (ATSDR). Toxicological Profile for Arsenic. Atlanta, GA: U.S. Department of Health and Humans Services, Public Health Service, Centers for Diseases Control; 2003

[106] USA Environmental Protection Agency (USEPA). Integrated Risk Information Systems (IRIS) on Arsenic. Washington, DC: National Center for Environmental Assessment, Office of Research and Development; 1999

[107] Dutton MD, Vasiluk L, Ford F, Bellantino Perco M, Taylor SR, Lopez K, et al. Towards an exposure narrative for metals and arsenic in historically contaminated Ni refinery soils: Relationships between speciation, bioavailability, and bioaccessibility. Science of the Total Environment. 2019;**686**:805-818

[108] Sanjrani MA, Zhou B, Zhao H, Bhutto SA, Muneer AS, Xia SA. Arsenic contaminated groundwater in China and its treatment options, a review. Applied Ecology And Environmental Research. 2019;**17**(2):1655-1683

[109] Patrick L. Toxic metals and antioxidants. Part II. The role of antioxidants in arsenic and cadmium toxicity. Alternative Medicine Review. 2003;**8**:106128

[110] Thomas JD, Styblo M, Lin S. The cellular metabolism and systemic toxicity of arsenic. Toxicology and Applied Pharmacology. 2001;**176**:127-144

[111] FAO/WHO. Expert Committee on Food Additives, Arsenic. FAO/WHO. Science Direct, Elsevier; 2005

[112] Uneyama C, Toda M, Yamamoto M, Morikawa K. Arsenic in various foods: Cumulative data. Food Additive and Contamination. 2007;**24**:447-534

[113] Rosariastuti R, Maisyarah S, Sudadi S, Hartati S, Purwanto P. Remediation of chromium contaminated soil by Phyto-bio system (PBS) application. Journal of Soil Science and Agroclimatology. 2019;**16**(1):90-102

[114] Barceloux DG. Chromium. Journal of Clinical Toxicology. 1999;**37**(2):173-194

[115] Pramono A, Rosariastuti R, Ngadiman Irfan DP. Bacterial Cr (VI) reduction and its impact in bioremediation. Ilmu Lingkungan. 2013;**11**(2):120-131

[116] Pramono A, Rosariastuti R, Ngadiman Irfan DP. Peran Rhizobakteri dalam Fitoekstraksi Logam Berat Kromium pada Tanaman Jagung. Ecolab. 2012;**6**(1):1-60

[117] Rani P, Arya RC, Dwivedi S. Chromium pollution: Impact on plants and its mitigation. In: Innovations in Food Technology. Singapore: Springer; 2020. pp. 323-340

[118] Cervantes C. Interaction of chromium with microorganisms and plants. FEMS Microbiology Reviews. 2001;**25**(3):335-447

[119] Cervantes C, Campos-García J, Devars S, Gutiérrez-Corona F,

Loza-Tavera H, Torres-Guzmán JC. et al. Interactions of chromium with microorganisms and plants. FEMS microbiology reviews. 2001;**25**(3):335-347

[120] Iskal TA. Effects of chromium exposure from a cement factory. Environmental Research. 2003;**91**:113-118

[121] World Health Organization (WHO). Health Criteria and Other Supporting Information. Vol. 2. Geneva: World Health Organization; 1984. pp. 195-201

[122] Sinha V, Pakshirajan K, Chaturvedi R. Chromium tolerance, bioaccumulation and localization in plants: An overview. Journal of Environment and Management. 2018;**206**:715-730

[123] Chen H, Yang X, Wang P, Wang Z, Li M, Zhao FJ. Dietary cadmium intake from rice and vegetables and potential health risk: A case study in Xiangtan, southern China. Science of the Total Environment. 2018;**639**:271-277

[124] Goyer AR. Toxic metals and essential metal interactions. Annual Review of Nutrition. 1997;**17**:37-50

[125] Saldivar RL, Luna M, Reyes E, Soto R, Fortul T. Cadmium determination in Mexican-produced tobacco. Environment Research. 1991;**55**:91-96

[126] Stohs SJ, Bagchi D, Bagchi M. Toxicity of trace elements in tobacco smoke. Inhalation Toxicology. 1997;**9**:867-890

[127] Doabi SA, Afyuni M, Karami M. Multivariate statistical analysis of heavy metals contamination in atmospheric dust of Kermanshah province, western Iran, during the spring and summer 2013. Journal of Geochemistry Exploration. 2017;**180**:61-70

[128] Yuen JQ, Olin PH, Lim HS, Benner SG, Sutherland RA, Ziegler AD. Accumulation of potentially toxic elements in road deposited sediments in residential and light industrial neighborhoods of Singapore. Journal of Environmental Management. 2012;**101**:151-163

[129] Nyarko B, Dampare S, Serfor-Armah Y, Osae S, Adotey D, Adomako D. Bio monitoring in the forest zone of Ghana: The primary results obtained using neutron activation analysis and lichens. International Journal of Environment and Pollution. 2008;**32**(4):467-476

[130] Saeedi M, Li LY, Salmanzadeh M. Heavy metals and polycyclic aromatic hydrocarbons: Pollution and ecological risk assessment in street dust of Tehran. Journal Hazard Material. 2012;**227-228**:9-17

[131] Lu X, Wang L, Li L, Lei YK, Huang L, Kang D. Multivariate statistical analysis of heavy metals in street dust of Baoji, NW China. Journal Hazard Material. 2010;**173**:744-749

[132] De Miguel E, Llamas JF, Chacon E, Berg T, Larssen S, Royset O, et al. Origin and patterns of distribution of trace elements in street dust: Unleaded petrol and urban lead. Atmospheric Environment. 1997;**31**(17):2733-2740

[133] Moller A, Muller HW, Abdullah A, Abdelgawad G, Utermann J. Urban soil pollution in Damascus, Syria: Concentrations and patterns of heavy metals in the soils of the Damascus Ghouta. Geoderma. 2005;**124**(1-2): 63-71

[134] Yazdi M, Behzad N. Heavy metal contamination and distribution in the Parks City of Islam Shahr, SW Tehran, Iran. The Open Environmental Pollution and Toxicology Journal. 2009;**1**:49-53

[135] Ghosh AK, Bhatt MA, Agrawal HP. Effect of long term application of treated sewage water on heavy metal accumulation in vegetables grown in northern India. Environmental Monitoring and Assessment. 2012;**184**:1025-1036

[136] Apostoae L, Iancu OG. "heavy metal pollution in the soils of Iaşi city and the suburban areas (Romania)", Studia Universitatis babes-Bolyai. Geologia. 2009;**MAEGS – 16**(Special Issue):142-146

[137] Duong TTT, Lee BK. Determining contamination level of heavy metals in rod dust from busy traffic areas with different characteristics. Journal of Environment and Management. 2011;**92**:554-562

[138] Wei B, Yang L. A review of heavy metal contaminations in urban soils, urban road dusts and agricultural soils from China. Micro chemical Journal. 2010;**94**(2):99-107

[139] Naja GM, Volesky B. Toxicity and sources of Pb, Cd, Hg, Cr, As, and radionuclides in the environment. Heavy Metals in the Environment. 2009:13-61

[140] Iyaka YA, Kakulu SE. Copper and zinc contents in urban agricultural soils of Niger state, Nigeria. An International Multi-Disciplinary Journal, Ethiopia. 2009;**3**(3):23-33

[141] Ghosh M, Singh SP. A review on phytoremediation of heavy metals and utilization of It's by products. Asian Journal on Energy and Environment. 2005;**6**(4):214-231

[142] Kumar N, Kulsoom M, Shukla V, et al. Profiling of heavy metal and pesticide residues in medicinal plants. Environmental Science and Pollution Research. 2018;**25**:29505-29510

[143] Zafarzadeh A, Rahimzadeh H, Mahvi AH. Health risk assessment of heavy metals in vegetables in an endemic esophageal cancer region in Iran. Health Scope. 2018;**7**(3):e12340

[144] McGrath SP, Zhao FJ, Lombi E. Plant and rhizosphere processes involved in phytoremediation of metal-contaminated soils. Plant and Soil. 2001;**232**:207-214

[145] Cong TL, Ma Q, Bondada B. Arsenic accumulation in the Hyperaccumulator Chinese brake and its utilization potential for phytoremediation. Journal of Environmental Quality. 2002;**31**:1671-1675

Chapter 7

The Efficiency of Phytoremediation of the Big-Sage Plant in Accumulating Some Heavy Metals in Their Tissues *In Vitro*

Majid Ibrahim, Mahmood Hashim and Anfas Okash

Abstract

Concerning the controlled environment and media technique in these studies, *in vitro* phytoremediation analyses might provide more precise and reliable findings. Hence, this chapter pursued to estimate the efficacy of the shoot and root organs of big-sage (*Lantana camera (L.) Czern.*) plantlets in assembling heavy metals (cadmium, cobalt, and lead) *via* the plant tissue culture technique. Many examinations achieved on the phytoremediation of the *Lantana camara* seedlings to heavy metals *in vivo* demonstrated that they were assembled in the shoot organs at a higher concentration compared with the root organs of this plant. Thus, *L. camara* can be regarded as a higher accumulation potential plant for heavy metals such as lead, chromium, cadmium, nickel, and arsenic, and a favorable plant for phytoremediation. As for the examinations executed on the effect of different levels of the heavy metals cadmium, cobalt, and lead on their assemblage and some growth traits in the shoot and root organs of the *L. camera* plantlets beneath *in vitro* culture conditions, they discovered that the assemblage of these metals in the shoot and root organs increased with the increase in the treatment level, except for the heavy metal lead, which assemblage in the roots without the shoots.

Keywords: assemblage, BCF, cadmium, cobalt, heavy metal, TF

1. Introduction

Heavy metals are a special type of toxins that cannot be damaged into non-toxic shapes. The level of these toxic heavy elements has risen dramatically since industrial development [1]. These toxic metals can get into the soil directly *via* the usage of heavy elements and are a special kind of poison that cannot be degraded into non-toxic forms. The concentrations of these toxic heavy metals have advanced dramatically since industrial evolution [1]. These toxic heavy metals can reach into the soil presently by the use of pesticides and fertilizers or indirectly because of wastewater remains, factory emissions, and fossil fuel burning, which might make soils unsuited for cultivation if this trouble aggravates and rises by surpassing certain edges [2]. In complement to selecting out of the agricultural specialization, soils polluted with heavy elements such as chromium, arsenic, lead, cadmium, copper, zinc, mercury, and nickel assess a significant risk to resources of groundwater through heavy elements filtering. Pollution of harvests cultivated in those soils passively impacts human health and food [3, 4]. The major attraction of environmental contamination investigations is discovering creative methods to rescue the environment from pollutants' damaging impacts [5]. Phytoremediation is a term usually supplied to the mechanisms by which living higher plants can completely attribute to the chemical impacts of the soil they are grown in. In other terms, it is an environmentally suitable technique to collect heavy metals in plant tissues to recycle contaminated soils. The origin of word phytoremediation came from the Greek term "Phyto," which means the plant and the Latin term "remedium," which demonstrates cleaning or rehabilitation [6]. Phytoremediation is a low-cost and practical method for operating soil in evolving countries [7]. The plant species utilized for this purpose are also found in several plant families, such as Asteraceae, Brassicaceae, Fabaceae, Poaceae, Euphorbiaceae, Verbenaceae, and Violaceae [8].

The big-sage (*Lantana camara L.*) plant is an ornamental evergreen shrub grown as a fence plant and the attractiveness of its flowers [9]. The original home of the *L. camara* plant is the subtropical and tropical regions of the American continent and in the tropical areas of Africa and Asia. This species was spread widely almost the world through the eighteenth, nineteenth, and twentieth centuries and evolved into a select evergreen shrub [10]. Further, this shrub earlier revealed favorable findings as shrub phytoremediation [11–14].

In vitro culture techniques include being near utilized in phytoremediation investigation [15–20]. Regarding the controlled environment and media technique in these investigations, *in vitro* phytoremediation examinations might provide more precise and dependable findings. Thus, this chapter desired to estimate the effectiveness of the root and vegetative tissues of *L. camara* seedlings in assembling heavy elements (cadmium, cobalt, and lead) via *in vitro* plant tissue culture conditions.

2. Botanical description of big sage (*L. camara* L.)

The *Lantana* genus has been described as shrubs of different species as the difference is in flower size, leaf shape and color, stem thorns, growth rates, shade tolerance, toxicity to organisms, chromosome number, and DNA content [21]. The big sage plant belongs to the family Verbenaceae, which includes 100 genera and about 2000 species. The genus Lantana has about 150 species that fall into the group of ornamental plants. As for its flowers, they are small in size and grouped in the form of small bouquets, ranging from 20 to 30 flowers in one inflorescence (**Figure 1**), as well as its fruits are of small size, and its seeds are solid and stone [22]. The *Lantana* shrub is characterized by ribbed stems covered with hairs and curved and sharp spines. It has opposite leaves, aromatic with a strong smell when crushed. This plant has cluster flowers of different colors; they may be white, yellow, pink, red, or orange. Its fruits are cluster aggregate, and each fruit contains a single seed inside it, which at the beginning of its growth takes a bright green color, and then turns to a blackish-purple color when it ripens [23]. These shrubs have many uses in the fields of health, as the oil extracted from the leaves and flowers of this plant has the property of acting as an antimicrobial and as a fungicide or insecticide to combat nematodes [9, 24].

Figure 1.
*The big-sage (*Lantana camara (L.) Czern.*) flowers.*

3. *In vivo* phytoremediation of *L. camara* for heavy-metal-polluted soil

One of the studies on the plants of *L. camara* and *Datura inoxia* planted on polluted sites such as industrial landfill areas, waste dumping areas, and mining mines indicated that they play a major role in controlling the accumulation and disposal of heavy metal pollutants [25].

A study was conducted on the phytoremediation of *L. camara* L. for soils contaminated with heavy metals resulting from factory wastes in the city of Bhopal in India. It was found that the leaves of *Lantana camera* plant had accumulated the largest amount of heavy metals in them compared with its branches. Chromium, lead, cadmium, and nickel accumulated in leaves at a concentration of 242.7, 262.2, 49.4, and 34.8 mg kg^{-1}, respectively, while the contents of heavy metals above accumulated in the vegetative branches were 72.3, 88.4, 28.8, and 22.8 mg kg^{-1}, respectively [13].

The study was maintained by Deepa et al. [26] to research the possibility of *L. camara* for the phytoremediation and accumulation of arsenic and nickel in the root and vegetative parts. The soil and plant samples utilized in this investigation were obtained from areas nearby Koradi Lake, the Northern of Nagpur, and then examined for arsenic and nickel levels. The accumulated heavy metals were analyzed utilizing an inductively connected plasma atomic emission spectrometer device. The arsenic and nickel concentrations in the soil were 2.29 mg L^{-1} and 58.344 mg L^{-1}, respectively.

The capability of plants to accumulate heavy metal from the soil was estimated by the bioconcentration factor, whereas their capability to translocate heavy metal from roots to vegetative parts was estimated by the translocation factor. On the basis of bioconcentration factor and translocation factor data, *L. camara* was determined as the phytoremediator for arsenic and nickel in contaminated soil. Nickel accumulated higher than the arsenic concentration in *L. camara*. These heavy metals accumulated in vegetative parts with more concentration compared with roots in this plant. Therefore, *L. camara* may be considered as an accumulator plant to nickel higher than arsenic heavy metal and a promising plant for phytoremediation.

4. *In vitro* phytoremediation of *L. camara* for some heavy-metal-polluted media

4.1 The aim of study

The effect of different concentrations of some heavy metals (cadmium, cobalt, and lead) on the vegetative and root growth characteristics of big sage (*L. camara* L.) plants under *in vitro* conditions and their efficiency in accumulating these elements [27].

4.2 Materials and methods

The study was conducted in the Plant Tissue Culture Laboratory, College of Agriculture, University of Basrah, Basrah, Iraq. The seeds of the local cultivar of the big sage (*L. camara* L.) plant obtained from Basrah nurseries were used. The fruits were soaked in sterile distilled water for 30 minutes to facilitate the removal of the fruit pulp. Then, the seeds were placed in a sterilizing solution of sodium hypochlorite at a concentration of 1.05% with the addition of three drops of Tween-20 for 20 minutes. Then, it was washed with distilled and sterile water thrice [27].

4.2.1 Preparation of nutrient medium

The nutrient medium was prepared from ready-made MS salts [28] at a concentration of 4.43 g L^{-1} obtained from Cassion Lab, USA (**Table 1**). Other chemicals were added to the MS medium (**Table 2**). The pH was adjusted to 5.7–5.8 with a solution of sodium hydroxide (NaOH) or hydrochloric acid (HCl) 0.1 N. Then add the agar at a concentration of 6 g L^{-1}. Then complete the MS to 1000 ml with distilled water. Then, the medium was heated to 90°C. After the medium became homogeneous and clear, the nutrient medium was poured into culture tubes of dimensions 2.5 × 18 cm (Pyrex) with a volume of 20 ml for each culture tube. Then, the tube nozzles were blocked with medical cotton, and the nozzles were wrapped with aluminum foil [27].

4.2.2 The proliferation of L. camara *plantlets under* in vitro *culture conditions*

Sterilized seeds of the big sage plant were cultured in MS medium without the addition of hormones to obtain seedlings from which the shoot tips are taken as explants for subsequent experiments. The regenerated shoots of the *L. camara* were produced from branch proliferation on MS medium supplemented with 0.6 mg L^{-1} BA and 0.1 mg L^{-1} NAA after 8 weeks of culturing (**Figure 2A**). Then, these proliferated shoots were rooted by growing them on an MS medium supplemented with

No.	Inorganic salts	Concentration (mg L^{-1})
1	Calcium chloride	332.02
2	Ammonium nitrate (NH_4NO_3)	1650
3	Magnesium sulfate ($MgSO_4$)	80.70
4	Boric acid (H_3BO_3)	6.2
5	Cobalt chloride ($CoCl_2.6H_2O$)	0.025
6	Cupric sulfate ($CuSO_4.6H_2O$)	0.025
7	Manganese sulfate ($MnSO_4.H_2O$)	16.90
8	Potassium iodide (KI)	0.83
9	Potassium nitrate (KNO_3)	1900
10	Potassium phosphate (KH_2PO_4)	170
11	Sodium molybdate ($Na_2MoO_4.2H_2O$)	0.25
12	Zinc sulfate ($ZnSO_4.7H_2O$)	8.60
Iron source		
13	Sodium EDTA (Na_2-EDTA)	37.26
14	Ferric sulfate ($FeSO_4.7H_2O$)	27.80

Table 1.
Inorganic nutrient components of MS medium [28].

No.	Inorganic salts	Concentration (mg L^{-1})
1	Sucrose	30,000
2	Glycine	1
3	Thiamin-HCl	1
4	Pyridoxin-HCl	1
5	Nicotinic	1
6	Adenine sulfate	40
7	Sodium hydrogen orthophosphate	170
8	Poly vinyl pyrrolidone (PVP)	1000
9	Phyto-agar	6000

Table 2.
Inorganic nutrient components of MS medium [27].

1.0 mg L^{-1} naphthalene acetic acid (NAA) and 0.1 mg L^{-1} benzyl adenine (BA) (**Figure 2B**). The plantlets having three pairs of leaves per plantlet were utilized in the accumulation of heavy metal investigations [27].

4.2.3 Heavy metal accumulation in root and vegetative part experiment

Plantlets were cultivated on the MS media supplemented with 0.0, 0.2, 0.4, 0.6, and 0.8 mg L^{-1} of Co ($CoCl_2.6H_2O$), Cd ($CdCl_2.2H_2O$), or Pb (Pb $(NO_3)_2$) [27]. After 30 days of cultivating, the subsequent data were registered:

1. Estimation of heavy metals content in the root and vegetative parts, cobalt, cadmium, and lead, according to Ref. [29] utilizing an atomic absorption spectrophotometer device (Phoenix-986 model) at wavelengths of 228.8, 240.7, and 283.3 nm, for Co, Cd, and Pb, respectively.

2. The bioconcentration factor (BCF) was estimated by the subsequent equation: BCF = Heavy metal concentration in vegetative and root parts/heavy metal concentration in MS medium [30].

3. The translocation factor (TF) was estimated by the following equation: TF = Heavy metal concentration in vegetative part/Heavy metal concentration in root part [30].

The investigations were designed by utilizing a randomized complete design. Each treatment included 10 replications (10 plantlets). The data were analyzed by utilizing analysis of variance with the statistical program SPSS Version 22. The treatments were compared between them utilizing the revised least significant difference test (R-LSD) at a probability level of 5% [31].

4.3 The heavy metal accumulation in *L. camara*

4.3.1 The heavy metal accumulation in vegetative organs

The increase in cadmium concentration that was added to the MS medium caused a significant increase in cadmium accumulation in the vegetative parts of the *L. camara* plant after 4 weeks of cultivating (**Table 3**) [27]. The 0.8 mg L^{-1} cadmium was significantly greatest than other treatments (0.192 mg kg^{-1} cadmium). This finding is in accord with the findings acquired by Kališová-Špirochová et al. [16], who investigated cadmium assembly in *Helianthus annuus*, *Populus tremula* × *tremuloides*, and *Zea mays*, and Milusheva et al.'s [19] investigation on *Petunia* × *hybrida* and *Ageratum houstonianum via* tissue culture technique. The findings also agree with the findings acquired by [32] when investigating the phytoremediation possibilities of the *Brassica juncea (L.) Czern.*, where an accumulation in cadmium concentration in the vegetative parts was noticed when the added concentration enhanced.

Parallel to cadmium, the assemblage of cobalt in the vegetative parts raised significantly with the enhancement of its concentration in the MS medium after

Figure 2.
Micropropagation of Lantanta camara *shrub; A—shoot multiplication; B—rooting shoots [27].*

Treatment concentration of Cd, Co, or Pb ($mg\ L^{-1}$)	**Accumulated heavy metal concentration ($mg\ kg^{-1}$)**		
	Cd	**Co**	**Pb**
0.0	—	—	—
0.2	0.015	0.055	—
0.4	0.063	0.180	—
0.6	0.132	0.228	—
0.8	0.192	0.326	—
R-LSD $P \leq 0.05$	0.018	0.044	—

Table 3.
In vitro *accumulation of cadmium, cobalt, and lead in the vegetative organs of the* Lantana camara *shrub [27].*

4 weeks of cultivating (**Table 3**). The 0.8 $mg\ L^{-1}$ cobalt registered the most increased cobalt accumulation among the examined concentrations reaching 0.326 $mg\ kg^{-1}$. No indications of toxicity were detected in the plants, which show that the *L. camara* is a phytoremediator that accumulates heavy metals without impacting its growth. These findings were alike to Al-Wahaibi's [33] findings, which indicated that assembling heavy metals in these plants is a natural direction for them.

Regarding lead accumulation, there was no lead assemblage in the vegetative parts in any the examined concentrations (**Table 3**). Comparable findings were registered in other plant species where lead assemblies were in the root parts instead of the vegetative parts [34].

4.3.2 Heavy metal accumulation in the root organs

Cadmium and cobalt concentrations of roots significantly accumulated with each rising in cadmium and cobalt concentrations in the MS medium (**Table 4**). The treatment of 0.8 $mg\ L^{-1}$ concentration of cadmium or cobalt caused the highest metal accumulation reaching 0.318 $mg\ kg^{-1}$ cadmium and 0.312 $mg\ kg^{-1}$ cobalt.

Furthermore, lead assembled in root parts was noticed under 0.6 and 0.8 $mg\ L^{-1}$ lead only, with the last recording the highest lead amount reaching 0.627 $mg\ kg^{-1}$ lead (**Table 2**). The findings of the current investigation oppose previous results on different plant species, as they noticed the assemblage of lead in both vegetative and root parts [15, 16, 18, 20].

Treatment concentration of Cd, Co, or Pb ($mg\ L^{-1}$)	**Accumulated heavy metal concentration ($mg\ kg^{-1}$)**		
	Cd	**Co**	**Pb**
0.0	—	—	—
0.2	0.099	0.013	—
0.4	0.148	0.117	—
0.6	0.198	0.166	0.501
0.8	0.318	0.312	0.627
R-LSD $P \leq 0.05$	0.018	0.044	0.052

Table 4.
In vitro *accumulation of cadmium, cobalt, and lead in the root organs of the* Lantana camara *shrub [27].*

4.3.3 Bioconcentration factor (BCF)

Bioconcentration is the concentration of a specific heavy element in the tissues of a plant in comparison with the plant's enclosing concentration of that element [27]. Accordingly, BCF is a necessary indicator of the response of plants to the existence of heavy metals in their environment and a direct indicator of the phytoremediation possibilities. The highest bioconcentration factor values for the cadmium and cobalt examined elements were noticed under 0.8 mg L^{-1} concentration for both metals, with 0.32 and 0.4 in cadmium and cobalt investigations, respectively (**Table 5**). Cadmium BCF under 0.8 mg L^{-1} concentration was significantly more increased than that of 0.4 mg L^{-1} concentration of this metal. Nevertheless, no significant differences were registered between cd BCF values under 0.2, 0.6, and 0.8 mg L^{-1} concentrations of cadmium. Furthermore, there were no significant differences in BCF factor between 0.4, 0.6, and 0.8 mg L^{-1} concentrations of cobalt (**Table 5**). The present results are alike to those of findings in Ref. [32] for the BCF factor of cadmium in *B. juncea (L.) Czern*.

As for the lead treatments, 0.6 mg L^{-1} Pb concentration registered the highest bioconcentration factor data; regardless, there was no significant distinction between bioconcentration data under 0.6 and 0.8 mg L^{-1} concentrations of lead (**Table 5**).

4.3.4 Translocation factor (TF)

The translocation factor means the level of contaminants assembled in the shoot organs of a plant to those in the root organs [27]. The most increased translocation value in the cadmium investigation was noticed under 0.6 mg L^{-1} cadmium level (0.67), which was significantly more increased than further levels (**Table 6**). This finding indicates the efficacy of *L. camara* in the translocation of cadmium from the root organs to the shoot organs. Alike findings were acquired by [35] for the transport of cadmium in *Populus alba* and *Morus alba* trees.

About cobalt, the MS medium with a level of 0.2 mg L^{-1} cobalt was significantly excellent compared with the other treatments with a translocation factor value reaching 4.23.

Furthermore, the TF of lead for all treatments was equal to zero since no lead assemblage was noticed in the shoot organs (**Table 6**).

It was apprised that the perfect plant for phytoremediation should be capable to absorb and assemble heavy metals from contaminated soils and have specific

Treatment concentration of Cd, Co, or Pb (mg L^{-1})	Bioconcentration factor (BCF)		
	Cd	Co	Pb
0.0	—	—	—
0.2	0.29	0.17	—
0.4	0.26	0.37	—
0.6	0.28	0.33	0.42
0.8	0.32	0.40	0.39
R-LSD $P \leq 0.05$	0.06	0.20	Non-significant

Table 5.
In vitro *bioconcentration factor (BCF) of Cd, Co, and Pb in* Lantana camara *shrub [27].*

Treatment concentration of Cd, Co, or Pb (mg L^{-1})	Translocation factor (TF)		
	Cd	Co	Pb
0.0	—	—	—
0.2	0.15	4.23	—
0.4	0.43	1.54	—
0.6	0.67	1.37	—
0.8	0.60	1.05	—
R-LSD $P \leq 0.05$	0.05	2.80	—

Table 6.
In vitro *translocation factor (TF) of Cd, Co, and Pb in* Lantana camara *shrub.*

characteristics such as deep and dense roots, large biomass, and rapid growth [36]. This study findings revealed that big sage (*L. camara* L.) could be an active phytoremediator in soils contaminated with cadmium and cobalt because of the suitable translocation factors of these heavy metals.

4.4 The impact of different concentrations of some heavy metals on some growth indicators of *Lantana camera* under *in vitro* culture conditions

4.4.1 Cadmium (Cd)

The data in **Table 7** indicate that there is no significant effect of the heavy metal cadmium concentrations in plantlet height, compared with the control treatment. It is also noted from the same table that there is no significant effect in each of the characteristics of the leaf numbers and the shoot dry weights among all treatments. While the addition of the cadmium heavy metal to the MS medium had a significant effect, as the plantlets treated with a concentration of 0.8 mg L^{-1} were significantly superior in the total shoot fresh weights, reaching 0.461 g, compared with the other treatments [27].

The data of the phytoremediation in **Table 7** for the *Lantana camera* plant show that there is no significant effect of different cadmium concentrations among all treatments in each of the shoot numbers plantlet^{-1}, leaf area (cm^2), and total chlorophyll content of the leaves (mg.100 g^{-1} fresh weight) (**Figure 3**).

The reason may be attributed to the use of plants to absorb these heavy metals from the culture media and translocate them to the vegetative organs or convert them into volatile compounds using the phytovolatilization technique. This technique exploits the ability of some plants to convert some heavy elements into volatile compounds for disposal [38].

The reason for this may be because the plant is a natural phytoremediator, as it can accumulate the contaminant, break it down, or assemble it in its biomass, and it is characterized by being a fast-growing plant and having a large biomass and having a widespread root system [39].

The data in **Table 8** show the effect of cadmium on the root growth indicators of the *L. camara* plantlets. It is noted that there are no significant differences in the number of main roots per plantlet in all treatments. It is also observed from the same table that there were no significant differences among all cadmium treatments in each of the root length and dry weight characteristics, while the data are shown in the same table

Cd concentration (mg L^{-1})	Plantlet height (cm)	Leaf numbers per plantlet	Fresh weight of vegetative parts (g)	Dry weight of vegetative parts (g)	Shoot numbers per plantlet	Leaf area (cm^2)	Total chlorophyll content (mg 100 g^{-1} FW)
0	7.27	4.00	0.113	0.035	2.67	1.80	1.424
0.2	7.47	10.67	0.233	0.030	4.00	2.63	1.746
0.4	10.23	5.33	0.170	0.026	2.67	1.93	2.029
0.6	9.03	8.67	0.240	0.025	2.67	2.40	1.857
0.8	10.13	7.67	0.461	0.048	3.33	2.37	2.042
R-LSD ($p \leq 0.05$)	NS*	NS	0.150	NS	NS	NS	NS

NS: Non-significance.

Table 7.
Effect of different concentrations of cadmium on some vegetative growth of Lantana camara *shrub [37].*

Figure 3.
Effect of different concentration of cadmium on plantlet growth of Lantana camara *shrub [37].*

Cd concentration (mg L^{-1})	Root numbers per plantlet	Root length (cm)	Fresh weight of root parts (g)	Dry weight of root parts (g)
0	4.33	2.57	0.026	0.009
0.2	5.67	2.77	0.050	0.021
0.4	3.33	3.77	0.041	0.010
0.6	3.67	4.70	0.062	0.011
0.8	5.00	4.87	0.114	0.015
R-LSD ($p \leq 0.05$)	NS*	NS	0.018	NS

NS: Non-significance.

Table 8.
Effect of different concentrations of cadmium on some root growth of Lantana camara *shrub [37].*

that there was a significant effect in the fresh weight of the roots in the MS medium to which the cadmium heavy metal was added. The MS medium supplemented with 0.8 mg L^{-1} cadmium was significantly superior in the root fresh weight, reaching 0.114 g. This is explained as the ideal concentration of *L. camara*. Despite the toxicity of the lead element, the big-sage shrub showed a phytoremediator for this heavy metal.

This can be explained by our findings is the ability of the big-sage plant to accumulate and be tolerant to cadmium heavy metal. Al-Wahaibi [40] indicated the characteristics of the accumulating plants when they absorb heavy elements, they stimulate the form of chelating compounds that surround the atoms of the contaminating elements and keep them within the vacuoles found in the cells of plant tissues.

4.4.2 Cobalt (Co)

Table 9 shows the effect of different concentrations of the heavy element cobalt on the vegetative growth indicators (**Figure 4**). The data showed that there was no significant effect on the characteristics of each of the plant's height (cm), leaf numbers, and the fresh and dry weights of the shoots (g) among all treatments [27].

Co concentration (mg L^{-1})	Plantlet height (cm)	Leaf numbers per plantlet	Fresh weight of vegetative parts (g)	Dry weight of vegetative parts (g)	Shoot numbers per plantlet	Leaf area (cm^2)	Total chlorophyll content (mg 100 g^{-1} FW)
0	7.27	4.00	0.113	0.0260	2.67	1.80	1.424
0.2	7.50	8.33	0.213	0.0150	2.00	1.57	1.555
0.4	6.73	5.67	0.100	0.0203	2.67	1.80	1.819
0.6	7.57	4.00	0.142	0.0167	2.33	2.70	2.019
0.8	6.73	7.00	0.112	0.0193	2.67	3.00	2.097
R-LSD ($p \leq 0.05$)	NS*	NS	NS	NS	NS	1.043	NS

NS: Non-significance.

Table 9.
Effect of different concentrations of cobalt on some vegetative growth of Lantana camara *shrub [37].*

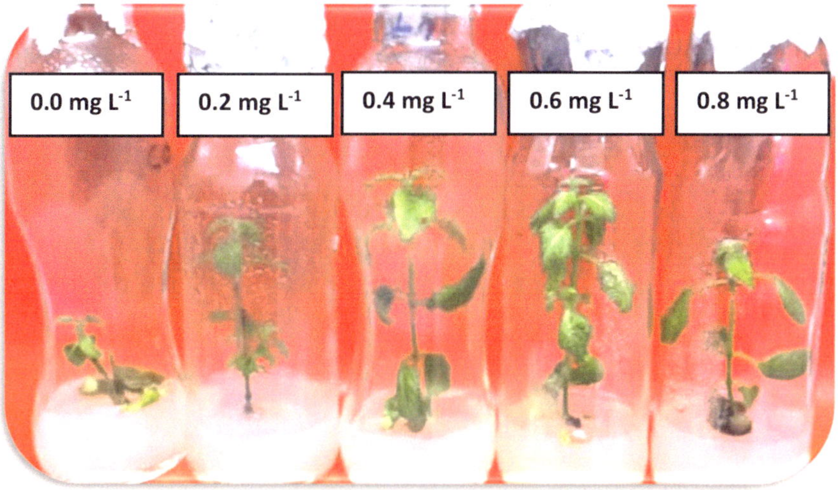

Figure 4.
Effect of different concentration of cobalt on plantlet growth of Lantana camara *shrub [37].*

Table 9 includes the effect of different cobalt heavy metal concentrations on the shoot numbers per plantlet. There was no significant effect of the element cobalt in this characteristic among all treatments. The addition of cobalt to the MS medium had no significant effect on the total chlorophyll content of leaves in all treatments.

The different concentrations of the heavy element cobalt had a significant effect on the leaf area. The treatment with 0.8 mg L^{-1} cobalt showed a significant effect on the leaf area, reaching 3.00 cm^2 compared with other treatments, except for the treatment with 0.6 mg L^{-1} cobalt, which did not differ significantly from it, reaching 2.70 cm^2.

The reason for this is that the heavy metal ions that enter the cell are associated with the chelators and companions. These chelating compounds remove the toxicity of metals by transporting minerals to the cytosol, while the companion transfer minerals to the organelles to reach the proteins that require metal. There are many chelating metal compounds and well-known chelators in plants, including phytochelatins, metallothioneins, organic acids, and amino acids [41].

The data of **Table 10** showed the effect of adding different cobalt concentrations of the MS medium on root growth indicators of *L. camara*. It was observed that there

Co concentration (mg L^{-1})	Root numbers per plantlet	Root length (cm)	Fresh weight of root parts (g)	Dry weight of root parts (g)
0	4.33	2.57	0.026	0.004
0.2	3.00	1.93	0.032	0.009
0.4	3.67	1.43	0.044	0.009
0.6	3.33	2.47	0.009	0.003
0.8	4.67	2.67	0.055	0.007
R-LSD ($p \leq 0.05$)	NS*	NS	0.028	NS

**NS: Non-significance.*

Table 10.
Effect of different concentrations of cobalt on some root growth of Lantana camara *shrub [37].*

Pb concentration ($mg\ L^{-1}$)	Plantlet height (cm)	Leaf numbers per plantlet	Fresh weight of vegetative parts (g)	Dry weight of vegetative parts (g)	Shoot numbers per plantlet	Leaf area (cm^2)	Total chlorophyll content ($mg\ 100\ g^{-1}$ FW)
0	7.27	4.00	0.113	0.026	2.67	1.80	1.424
0.2	5.53	6.00	0.347	0.046	2.67	2.20	1.600
0.4	5.47	6.33	0.237	0.032	2.33	2.50	1.966
0.6	6.07	7.33	0.376	0.047	2.67	2.43	1.642
0.8	6.33	5.67	0.165	0.034	3.67	2.93	1.737
R-LSD ($p \leq 0.05$)	NS*	NS	NS	NS	NS	NS	NS

**NS: Non-significance.*

Table 11.
Effect of different concentrations of lead on some vegetative growth of Lantana camara *shrub [37].*

was no significant effect on some root growth characteristics, including the main root numbers per the plantlet, root length, and root dry weights.

The contamination of the MS medium with cobalt had a significant effect on the total root fresh weights. The treatment at 0.8 mg L^{-1} cobalt was significantly superior in this characteristic, reaching 0.055 g compared with the treatments at 0.6 mg L^{-1} cobalt and the control, which reached 0.026 g.

The reason for this may be that plants exposed to high levels of cobalt, more than the permissible levels of heavy metals, show symptoms of toxicity due to excessive treatment of cobalt, which is more than what most species need. Moreover, cobalt toxicity rarely occurs when plants are exposed to low levels [42]. Therefore, the *L. camara* shrub showed tolerance for this heavy metal.

4.4.3 Lead (Pb)

The data in **Table 11** show that there are no significant differences in the characteristics of vegetative organs, plantlet height (cm), leaf numbers, and fresh and dry weights of the shoot (g) of *L. camara* shrub grown in MS media that supplemented with different concentrations of lead heavy metal (**Figure 5**) [27].

The data in **Table 11** show that there were no significant differences when the MS medium was contaminated with lead after 1 month of the experiment in each of the characteristics of the number of leaves per shoot, leaf area (cm^2), and total chlorophyll content of leaves (mg 100 g^{-1} fresh weight).

The data of **Table 12** indicate that there are no significant differences when adding lead at the different concentrations in the MS medium in each of the characteristics of the main root numbers per the plantlet, root length (cm), and the fresh and dry weights of the root parts (g).

This can be explained by the limited transport of lead through the root, as a result of the precluding caused by the Casparian strip in the root endodermis, which prevents the translocation of lead through the endodermis to the central vascular cylinder tissues.

Figure 5.
Effect of different concentration of lead on plantlet growth of Lantana camara *shrub [37].*

Pb concentration (mg L^{-1})	Root numbers per plantlet	Root length (cm)	Fresh weight of root parts (g)	Dry weight of root parts (g)
0	4.3	2.57	0.026	0.0090
0.2	9.3	3.73	0.151	0.0227
0.4	4.7	5.23	0.051	0.0137
0.6	5.7	5.23	0.214	0.0293
0.8	6.0	3.63	0.036	0.0237
R-LSD ($p \leq 0.05$)	NS*	NS	NS	NS

NS: Non-significance.

Table 12.
Effect of different concentrations of lead on some root growth of Lantana camara *shrub [37].*

Whereas the accumulation of lead depends on the species, variety, and plant organ, and then increases in the accumulation within the root organs compared with the vegetative organs, and then a decrease occurs in some characteristics of the vegetative organs such as total fresh weight of the shoots when the concentration of lead is increased, which causes a difference in the characteristics of the roots at the expense of the characteristics of the vegetative parts [43].

5. Conclusions

It is concluded from the studies conducted on testing the Lantana camera plant growing in soils and tissue cultures contaminated with heavy elements that it can be exploited as a promising ornamental plant in the phytoremediation of heavy metals such as lead, cadmium, cobalt, arsenic, and nickel. The accumulations of heavy elements in the vegetative organs were higher than the root organs. The accumulation of heavy metals in the tissues of this plant did not significantly affect some growth characteristics.

Acknowledgements

I extend my thanks and appreciation to the coauthors in preparing this chapter and to the editor of this book. I also extend my thanks and gratitude to the members of the publishing and printing committee at IntechOpen and the scientific evaluation of this product.

Conflict of interest

The authors declare no conflict of interest.

Author details

Majid Ibrahim*, Mahmood Hashim and Anfas Okash
University of Basrah, Basrah, Iraq

*Address all correspondence to: majid.abdulhameedl@uobasrah.edu.iq

References

[1] Jabeen R, Ahmad A, Iqbal M. Phytoremediation of heavy metals: Physiological and molecular mechanisms. The Botanical Review. 2009;**75**(4):339-364. DOI: 10.1007/s12229-009-9036-x

[2] Pulford ID, Watson C. Phytoremediation of heavy metal-contaminated land by trees - A review. Environment International. 2003;**29**(4):529-540. DOI: 10.1016/S0160-4120(02)00152-6

[3] Erakhrumen AA. Phytoremediation: An environmentally sound technology for pollution prevention, control and remediation in developing countries. Educational Research and Reviews. 2007;**2**(7):151-156

[4] Varsha M, Nidhi M, Anurag M. Heavy metals in plants: Phytoremediation: Plants used to remediate heavy metal pollution. Agriculture and Biology Journal of North America. 2010;**1**(1):40-46

[5] Gruca-Królikowska S, Wacławek W, Metale w środowisku CZ. Wpływ metali ciężkich na rośliny. Chemistry-Didactics-Ecology-Metrology. 2006;**11**:41-56

[6] Cunningham SD, Shann JR, Crowley DE, Anderson TA. Phytoremediation of contaminated water and soil. In: Kruger EL, Anderson TA, Coats JR, editors. Phytoremediation of Soil and Water Contaminants. Washington, DC: American Chemical Society; 1997. DOI: 10.1021/bk-1997-0664.ch001

[7] Shanker AK, Djanaguiraman M, Pathmanabhan G, Sudhagar R, Avudainayagam S. Uptake and phytoaccumulation of chromium by selected tree species. In: Proceedings of the international conference on water and environment held in Bhopal, India; 2003

[8] Baker AJ, McGrath SP, Reeves RD, Smith JA. Metal hyperaccumulator plants: A review of the ecology and physiology of a biological resource for phytoremediation of metal-polluted soils. Phytoremediation of Contaminated Soil and Water. 2020:85-107. DOI: 10.1201/9780367803148

[9] Reddy NM. *Lantana camara* Linn. Chemical constituents and medicinal properties: A review. Scholars Academic Journal of Pharmacy. 2013;**2**(6):445-448

[10] Choyal R, Sharma SK. Allelopathic effects of *Lantana camara* (Linn.) on regeneration in Funaria hygrometrica. Indian Journal of Fundamental and Applied Life Science. 2011;**1**:177-182

[11] Vara Prasad MN, de Oliveira Freitas HM. Metal hyperaccumulation in plants: Biodiversity prospecting for phytoremediation technology. Electronic Journal of Biotechnology. 2003;**6**(3):285-321

[12] Padmavathiamma PK, Li LY. Phytoremediation technology: Hyper-accumulation metals in plants. Water, Air, and Soil Pollution. 2007;**184**(1):105-126. DOI: 10.1007/s11270-007-9401-5

[13] Waoo AA, Khare S, Ganguly S. Evaluation of phytoremediation potential of *Lantana camara* for heavy metals in an industrially polluted area in Bhopal, India. International Journal of Engineering and Applied Sciences. 2014;**1**(2):1-3

[14] Veraplakorn V. Micropropagation and callus induction of *Lantana camara* L. – A medicinal plant. Agriculture and Natural Resources. 2016;**50**(5):338-344. DOI: 10.1016/j.anres.2016.12.002

[15] Abubacker MN, Sathya C. In vitro phytoremediation potential of heavy metals by duck weed *Lemna polyrrhiza* L.(Lemnaceae) and its combustion process as manure value. Journal of Environment and Biotechnology Research. 2017;**6**(1):82-87

[16] Kališová-Špirochová I, Punčochářová J, Kafka Z, Kubal M, Soudek P, Vaněk T. Accumulation of heavy metals by in vitro cultures of plants. Water, Air and Soil Pollution: Focus. 2003;**3**(3):269-276. DOI: 10.1023/A:1023933902452

[17] Fathy HM, Hashish KI, Taha LS. In vitro ability of paulownia hybrid as phytoremediator against some heavy metals pollution. Journal of Chemical and Pharmaceutical Research. 2016;**8**(7):525-532

[18] Santos-Díaz MD, Barrón-Cruz MD. Induction of in vitro roots cultures of *Thypha latifolia* and *Scirpus americanus* and study of their capacity to remove heavy metals. Electronic Journal of Biotechnology. 2007;**10**(3):417-424. DOI: 10.4067/S0717-34582007000300009

[19] Milusheva DI, Atanassova BY, Iakimova ET. Application of in vitro test system for evaluating the tolerance of *Ageratum houstonianum* and *petunia x hybrida* to cadmium toxicity. Bulgarian Journal of Agricultural Science. 2019;**25**(2):300-309

[20] Kovačević B, Miladinović D, Orlović S, Katanić M, Kebert M, Kovinčić J. Lead tolerance and accumulation in white poplar cultivated in vitro. South-East European Forestry: SEEFOR. 2013;**4**(1):3-12. DOI: 10.15177/seefor.13-01

[21] Binggeli P, Hall JB, Healey JR. An overview of invasive Woody plants in the tropics. In: School of Agricultural and Forest Science Publication No. 13. Bangor: University of Wales; 1998

[22] Mishra A. Allelopathic properties of *Lantana camara*. International Research Journal of Basic and Clinical Studies. 2015;**3**(1):13-28. DOI: 10.14303/irjbcs.2014.048

[23] Henderson L. Alien weeds and invasive plants. In: Plant Protection Research Institute Handbook no. 12. Pretoria, South Africa: Agricultural Research Coucil; 2001. p. 300

[24] Kalita S, Kumar G, Karthik L, Rao KV. A review on medicinal properties of *Lantana camara* Linn. Research Journal of Pharmacy and Technology. 2012;**5**(6):711-715. DOI: 10.5958/0974-360X

[25] Nagendran R, Selvam A, Joseph K, Chiemchaisri C. Phytoremediation and rehabilitation of municipal solid waste landfills and dumpsites: A brief review. Waste Management. 2006;**26**(12):1357-1369. DOI: 10.1016/j.wasman.2006.05.003

[26] Deepa P, Sheetal B, Ashoke K, Nisha D. Phytoremediation potential of *Lantana camara* and *Polygonum glabrum* for arsenic and nickel contaminated soil. Research Journal of Chemical Sciences. 2015;**5**(8):18-22

[27] Ibrahim MA, Sabti MZ, Mousa SH. In vitro accumulation potentials of heavy metals in big-sage (*Lantana camara* L.) plant. DYSONA-Life Science. 2021;**2**(2):12-17. DOI: 10.30493/DLS.2021.254481

[28] Murashige T, Skoog F. A revised medium for rapid growth and bio assays with tobacco tissue cultures. Physiologia Plantarum. 1962;**15**(3):473-497. DOI: 10.1111/j.1399-3054.1962.tb08052.x

[29] Jackson ML. Soil Chemical Analysis: Advanced Course. USA: UW-Madison Libraries Parallel Press; 2005

[30] Selvaraj K, Sevugaperumal R, Ramasubramanian V. Phytoremediation of arsenic chloride by indian mustard (*Brassica juncea*). Indian Journal of Fundamental and Applied Life Science. 2013;**3**(1):184-191

[31] Al-Rawi KM, Allah AA. Design and Analysis of Agricultural Experiments Ministry of Higher Education and Scientific Research. Dar Al-Kut for Publishing: University of Al Mosul; 2000

[32] Khudair ARM. Phytoremediation of Soil Contaminated with some Heavy Elements by Indian Mustard (*Brassica juncea* (L.) Czern.). M.Sc. Thesis,. Iraq: College of Agriculture, University of Basrah; 2014

[33] Al-Wahaibi MB. The phenomenon of accumulation of heavy elements in plants. Saudi Journal of Life Sciences. 2007;**14**(2):73-96

[34] Fahr M, Laplaze L, Bendaou N, Hocher V, Mzibri ME, Bogusz D, et al. Effect of lead on root growth. Frontiers in Plant Science. 2013;**4**:175. DOI: 10.3389/fpls.2013.00175

[35] Rafati M, Khorasani N, Moattar F, Shirvany A, Moraghebi F, Hosseinzadeh S. Phytoremediation potential of *Populus alba* and *Morus alba* for cadmium, chromuim and nickel absorption from polluted soil. International Journal of Environmental Research. 2011;**5**(4):961-970. DOI: 10.22059/IJER.2011.453

[36] Alkorta I, Hernández-Allica J, Becerril JM, Amezaga I, Albizu I, Garbisu C. Recent findings on the phytoremediation of soils contaminated with environmentally toxic heavy metals and metalloids such as zinc, cadmium, lead, and arsenic. Reviews in Environmental Science and Biotechnology. 2004;**3**(1):71-90. DOI: 10.1023/B:RESB.0000040059.70899.3d

[37] Mousa SH. The Phytoremediation of some Heavy Metals by Big-Sage (*Lantana camara* L.) Plant Propagated Via in Vitro Culture. Basrah, Iraq: Basrah University; 2013

[38] Flathman PE, Lanza GR. Phytoremediation: Current views on an emerging green technology. Journal of Soil Contamination. 1998;**7**(4):415-432. DOI: 10.1080/10588339891334438

[39] Ibrahim KM. Applications of Plant Biotechnology. Second volume ed. Iraq: Al-Nahrain University, Ministry of Higher Education and Scientific Research; 2017

[40] Al-Wahaibi MH. The phenomenon of accumulation of heavy elements in plants (brief review). Saudi Journal of Life Sciences. 2007;**14**(2):73-96

[41] Clemens S. Molecular mechanisms of plant metal tolerance and homeostasis. Planta. 2001;**212**(4):475-486. DOI: 10.1007/s004250000458

[42] Hammodi A, Ali F. Determination of some heavy metals quantities in some leguminous plants grown in polluted soil. Journal of Education and Science. 2008;**21**(3):53-65. DOI: 10.33899/edusj.2008.56039

[43] Gonçalves JF, Becker AG, Pereira LB, da Rocha JB, Cargnelutti D, Tabaldi LA, et al. Response of *Cucumis sativus* L. seedlings to Pb exposure. Brazilian Journal of Plant Physiology. 2009;**21**:175-186. DOI: 10.1590/S1677-04202009000300002

Chapter 8

Perspective Chapter: Environmental Assessment on the Effect of Chemical Waste from Dyeing Industries in Zaria

David Ebuka Arthur, Karimatu Abdullahi, Michael Abatyough and Chidinma Chinelo Arthur

Abstract

The research study was done on groundwater obtained from waste wells, well around the waste wells and wells about 5 km from the dyeing sites of Zaria, in Kaduna state of Nigeria. The objectives were to assess the status of pollution on local dyeing areas, the occupational hazard associated with this activity and the impact on the residents of the area. Parameters such as pH, coli form bacteria, conductivity, colour, DO, BOD, COD, chlorides, total available nitrogen, cadmium, chromium, lead, mercury and alkalinity were determined and statistical analysis carried out to evaluate the Significant difference of pollutants in the area studied.

Keywords: groundwater, dyeing industry, Zaria metropolis, heavy metals, pollution

1. Introduction

Water is essential to maintain and sustain human life, animals and plants [1], this is because it constitutes to a large extent, the major solvent in which many of the body's proteins and other substances are dissolved. It enables many metabolic activities of the body to take place [2]. Water is essential for growing food, for domestic uses and as a critical factor in industries, tourism and cultural purpose as it helps in sustaining the earth's ecosystem. According to Ajibade et al., [3] 90% of the population in Nigeria depends largely on hand dug wells and boreholes [4]. Rapid growth in urban populations, industrial activities, commercial and agricultural developments result in an increase in the search for potable water. The preference of groundwater as a source of drinking water in rural areas is because of its relatively better quality than that of river water [5]. Historically, the point of rural settlement was being determined by water source such as stream, river and spring [6]. The inhabitants of rural settlements relied on groundwater often within a few meters of the surface which they exploited by digging wells. Access to safe drinking water is a basic human need and a fundamental human right that is crucial for poverty reduction [7].

According to a report [8] this situation forces people to consume untreated water from rivers and ponds and represents a high risk to their health [9].

Kaduna State was famous for its traditional indigo dyeing pits, during the Trans Sahara trade in Northern part of the country. Local industries like the dyeing industry, proved vital to Nigeria's socio-economic development, apart from providing employment to a good number of people; it serves as source of tourist attraction. These local industries used a local technology of dyeing and discharge of waste. Ideally citing industries should strike a balance between socio- economic and environmental considerations. This was not the case with local dyeing industries in Zaria; which were mostly located within the city walls and surrounded by settlements. The mode of disposal was usually in pits dug for that purpose. Previous studies, when the dyeing activities were functional, have shown contamination of ground water in close proximity to the dye wastes well in Zaria [10]. The findings show environment-related ailments ranging from skin and eye problems to cancerous tumours and methaemoglobinaemia in children which were attributed to the disposal of dye wastes in these areas. Even though most of these dyeing pits have long been abandoned, their impacts on the settlements around them are still persisting.

2. Research methodology

2.1 Sample collection

Samples were collected from local dyeing areas located in northern Nigeria namely, Zaria a major city in Kaduna state (**Figure 1**). The dye wastes are normally disposed in waste wells dug around these local dyeing industries. Each well is approximately 3.5 m deep. Samples were collected in February 1998 and august 1998, which represents the peaks of dry and wet seasons in these areas respectively.

Samples of wastewater (about 2 litres) were collected using plastic containers. Samples were also collected from wells at a depth of approximately 15 m, 30 m, and 45 m radius from the waste wells. All the samples were stored in an airtight screw capped plastic bottles. Samples were also collected from wells located about 5 km away from local dyeing industries (**Table 1**). All the water samples collected were coded with reference to their site of collection as:

2.2 Measurements of physicochemical parameters

The physicochemical properties of water, plays a vital role in determining the extent to which heavy metal pollution of water occurs. Standard methods as recommended by relevant authorities such as World Health Organisation (WHO), United States Environmental Protection Agency (US-EPA), etc. were employed for the preparation of reagents and determination of all water quality parameters.

2.3 Determination of pH

The pH of the water samples was determined using a portable pH meter after being standardised with buffers of pH 4.0 and pH 9.2 [11].

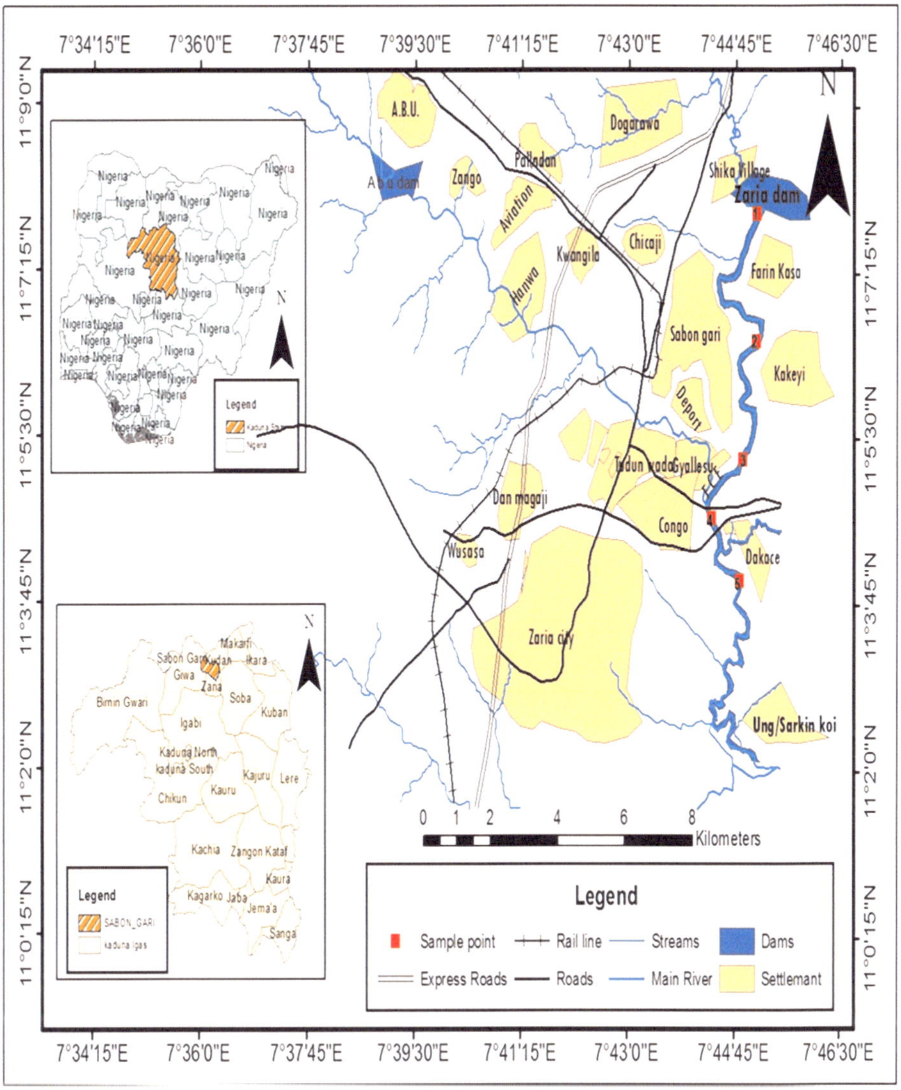

Figure 1.
Map of Zaria and its environment.

Code	City	Location of dyeing industry
ZA_1, ZA_2 and ZA_3	Zaria	Karofin Bakin Kasuwa
ZB_1, ZB_2 and ZB_3	Zaria	Karofin Mabuga
ZC_1, ZC_2 and ZC_3	Zaria	Karofin Karanka
ZN	Zaria	5 km away from dyeing industry

Table 1.
Location of dyeing industry and representative codes used in the study.

2.4 Determination of colour

Colour was determined by visual comparison using Lovibond colour disc (Pt-Co). The disc consists of different colours which is graduated. Sample was placed on disc and observed to see colour changes, and then the reading was taken for the colour which it corresponds with [12, 13].

2.5 Determination of conductivity

Conductivity of the surface water and the underground water samples was determined using the standard procedure approved by AOAC (1998). The conductivity meter (Hach model CO150) was used. The power key and the conductivity key of the conductivity meter was switched on, and the temperature of the meter adjusted; the instrument was calibrated with 0.001 M KCl to give a value of 14.7mS/m at 25°C. The probe was dipped below the surface of both samples. Time was allowed for the reading to be stabilised and the reading was recorded [14].

2.6 Determination of dissolved solids

Water sample (100 cm^3) was quantitatively transferred into an evaporating dish that has been previously weighed and dried in an oven for one hour and cooled in desiccators. The content of the dish was evaporated to dryness on a water-bath to a constant weight. The residue was dried in an oven between 103 and 105°C for two hours; then cooled in a desiccator and the difference in weight calculated using the following equation [12, 15].

$$\text{DS (mg/L)} == \text{difference in weight} \times 1000 \text{ ml of sample} \tag{1}$$

2.7 Determination of chemical oxygen demand (COD)

The COD of the surface water and the underground water samples was determined using the standard method described by Ademoroti [11]. 0.4 cm^3 of H_2SO_4 was placed in a refluxing flask. About 20 cm^3 of the samples was diluted with 20 cm^3 of distilled water. Exactly 10 cm^3 standard solution of $K_2Cr_2O_7$ was then added to glass leads already heated to 600°C for 1 hour. The flask was then attached to the reflux condenser and about 30 cm^3 of concentrated H_2SO_4 containing Ag_2SO_4 was added through the open end of the condenser. The resulting solution was thoroughly mixed by switching. The mixture was refluxed for 1 hour, cooled and the condenser was washed with about 25 cm^3 of distilled water. The mixture was diluted with 150 cm^3 of distilled water and cooled to room temperature. About 3 drops of (0.10–0.15 cm^3) ferroin indicator was added. The mixture was the titrated with $Fe(NH_4)_2(SO_4)_2$ taking as the end point the sharp colour change from blue-green to reddish brown. In the same manner a blank containing 20 cm^3 distilled water was refluxed together with the reagent.

$$\text{COD}\frac{\text{mg}}{\text{l}} = (\text{a} - \text{b}) \times \text{M} \times 8000 \text{ ml of sample} \tag{2}$$

Where a = cm^3 Fe $(NH_4)_2$ $(SO_4)_2$ used as blank,
b = cm^3 Fe $(NH_4)_2(SO_4)_2$ used for sample and M = Molarity of Fe $(NH_4)_2$ $(SO_4)_2$.

2.8 Determination of dissolved oxygen

The azide modification of the Winkler's method was used to determine dissolved oxygen (DO) and biological oxygen demand (BOD). 250 cm^3 of the sample was introduced into a stopped dark bottle and 2 cm^3 of manganese sulphate solution and 2 cm^3 alkali-iodide-azide reagent was added well below the surface of the liquid and mixed by inverting the bottle several times. Then 5 cm^3 of conc. H_2SO_4 was added immediately precipitate settled. The bottle was then shaken to ensure distribution of iodine, until titrant changed to pale-straw colour. 25 cm^3 of the mixture with 5 cm^3 of starch indicator was then titrated against 0.01 M sodium thiosulphate.

Titration continued until first disappearance of the blue colour. The titration was carried out three times and average titre value obtained was the equivalent value of dissolved oxygen (DO).

2.9 Determination of biological oxygen demand (BOD)

A fresh sample was incubated at 20°C for five days and the above procedure for the determination of dissolved oxygen was then repeated. The difference between DO for incubated sample and DO not incubated was determined [13].

BOD5 mg/L = DO (0)–DO (5) dilution factor.

Dilution factor = no of days ml of sample.

Where DO (5) = demand oxygen at day five and DO (0) = dissolved oxygen before incubation.

2.10 Determination of nitrate, phosphate and sulphate

The HANNA multi parameter logging spectrophotometer (HI83200) was used to digitally determine the nitrate, phosphate and sulphate in the surface water and ground water samples. The concentration of nitrate, sulphate and phosphate was determined using standard procedure. Sulphate was determined using Sulfa Ver methods 8051. Phosphate was determined using direct reading from HI 83200 HANNA multi parameter.

2.11 Determination of total alkalinity

Water sample (100 cm^3) was transferred into a conical flask, two drops of phenolphthalein indicator was added and the solution titrated with H_2SO_4 to the end point. Again, two drops of methyl orange was added to the titrated mixture and titration was continued to methyl orange end point [12].

$$\text{Total Alkalinity,mg CaCO3 A} \times \text{B} \times 1000 \text{ ml of sample} \tag{3}$$

Where A = Vol. of standard H_2SO_4 and B = Titre of standard acid.

2.12 Digestion of water sample

The determination of heavy metals in water is often regarded as the movement of total suspended and dissolved metals (soluble metals). In such cases consistent and dependable digestion procedures must be used so that data derived for total metal content is reliable. The water was immediately digested after sampling to prevent changes in composition of water samples according to standard procedures of the American Public Health Association [16].

2.13 Procedure for water digestion

Water samples (100 cm^3) were transferred quantitatively into beakers containing concentrated HNO_3 acid (10 cm^3) and concentrated HCl (5 cm^3) in ratio (2:1) and heated on a hot plate making sure the sample did not boil, until the volume was reduced to about 15 cm^3. The samples were then allowed to cool, filtered and quantitatively transferred into a 100 cm^3 standard volumetric flask and made up to mark with distilled water and further analysed using AAS.

3. Results and discussion

The chemical analysis of samples from the three local dyeing industries namely; Bakin Kasuwa, Mabuga and Karauka and surrounding wells was done and results can be discussed in the following order (**Tables 2–4**).

Except for the wastewater sample in Karofm Mabuga which has a pH value of 9.02 $\pm$ 0.16 during the dry season all pH values in the three dyeing industries; Bakin Kasuwa, Mabuga and Karauka fall within the accepted level of between 7.0 and 8.5. It could be inferred here that for the three places pollution is insignificant with regards pH values. The high pH value of ZB_3, during dry season could be as a result of high-alkalinity, the alkalinity of this waste well water is to the tune of 6600 $\pm$ 50 mg/L.

ZA_3 has alkalinity value of 300 $\pm$ 8.50, ZA_2, 200 $\pm$ 5.65 mg/L during dry and wet seasons respectively and a location where no dyeing activity took place has 300 $\pm$ 8.50 mg/L. Others are ZB_2, with 300 $\pm$ 5.00, ZB_3 with 300 $\pm$ 10.11 mg/L and also ZC_2, 300 $\pm$ 5.76, ZC_3 300 $\pm$ 4.87 mg/L. All these values are below the permissible level of 500 mg/L of alkalinity and are obtained during the wet season. All other values were found to be higher than the acceptable level. The lower values during the wet season are obtained due to dilution of underground water by penetration and increase in volume by ram water. The lower values can also be inferred from low values of conductivity and pH. This was found to be higher than those of WHO standard.

Considering the permissible level, the likelihood of pollution is imminent since the colour unit of more than 5 Hazen unit is unacceptable in drinking water and water for domestic purposes. All the waste water samples for the three areas (Bakin Kasuwa Mabuga and Karauka) have greatly exceeded this limit both during dry season and wet season, though higher values are obtained during wet season, ZC, well has 5500 $\pm$ 500 Hazen, ZB, has 12,500 $\pm$ 500. These are the highest values for waste wells recorded during wet season. The well water samples have values within acceptable levels with the exception ZA, and ZC_3 with 10 $\pm$ 0.0 Hazen respectively during the wet season. Colouration during the wet season may not be unconnected with presence of suspended particles as a result of movement of soil particles by penetration of rainwater. Comparing these values with that of the control site there is a certain degree of agreement in both dry and wet seasons.

Conductivity values of the wastewater samples in the three industries in Zaria (ZA_3) are high exceeding the levels permitted. This shows the presence of soluble, once and other dissolved solids. However, well water samples especially ZA_1, ZA_2, ZB_3 and ZC_1, all during dry season have concentrations slightly higher than the acceptable value [17, 18].

Parameter	Unit	Dry season samples				Wet season samples			
		ZA_1	ZA_2,	ZA_3	Z_N	ZA_1	ZA_2	ZA_3	Zn
pH		8.49 ± 0.11	8.28 ± 0.01	8.37 ± 0.01	7.65 ± 0.02	7.05 ± 0.02	7.03 ± 0.11	7.00 ± 0.07	7.05 ± 0.05
Coliform count	*MPN	7.00 ± 1.40	2	2	2	7.00 ± 1.60	5.00 ± 0.40	ND	2,00 ± 0.59
Conductivity x 10^2	mS/m	75 ± 5.00	20.00 ± 2.31	21.00 ± 1.29	10.2 ± 1.33	18.5 ± 2.3	17 ± 2.00	1.20 ± 0.2 1	8.50 ± 0.68
Colour	Hazen	1500 ± 250.00	5	5	5	1500 ± 50.50	10.00 ± 7.07	5	5
DO	mg/L	3.00 ± 0.01	8.00 ± 1.00	5.00 ± 1.00	10.00 ± 0.10	4.00 ± 1.00	9.00 ± 1.00	10.00 ± 1.00	12.00 ± 1.00
BOD	mg/L	2.40 ± 0.06	3.20 ± 0.07	2.28 ± 0.61	3.20 ± 1.25	4.80 ± 0.15	2.08 ± 0.10	3.68 ± 0.10	3.80 ± 0.25
COD	mg/L	144.00 ± 4.31	100.00 ± 0.39	92.00 ± 0.22	12–00 ± 0.03	120.00 ± 0.57	76.00 ± 1.97	64.00 ± 4.30	36.00 ± 2.01
NO_2^-N	mg/L	37.50 ± 1.11	43 00 ± 0 61	11.50 ± 0.67	17.50 ± 0.15	26.00 ± 0.12	37.00 ± 4.30	11.00 ± 2.40	15.00 ± 0.76
Chlorides	mg/L	351.00 ± 0.20	280.13 ± 0.01	209.21 ± 0.01	280.13 ± 38. 90	336.87 ± 32.90	3540.0 ± 0.02	3540 ± 0.01	531.10 ± 5.66
**TAN	mg/L	72.00 ± 0.05	74.00 ± 0 51	25.00 ± 0.23	30.00 ± 0.76	34.00 ± 5.67	46.00 ± 5.03	25.0 ± 1.4	31.00 ± 2.22
Cd	mg/L	2.10 ± 0.01	3.00 ± 0.01	3.50 ± 0.01	1.60 ± 0.26	1.90 ± 0.1 8	2.50 ± 0.01	2.20 ± 0.01	2.30 ± 0.21
Cr	mg/L	2.20 ± 0.21	1.9 ± 0.01	1.40 ± 0.01	2.40 ± 0.01	1.70 ± 0.21	15.00 ± 0.03	1.80 ± 0.11	2.20 ± 0.12
Pb	mg/L	2.60 ± 0.01	2.50 ± 0.00	2.60 ± 0.01	1.25 ± 0.12	3.20 ± 0.01	3.50 ± 0.01	2.00 ± 0.04	1.50 ± 0.11
Hg	mg/L	0.90 ± 0.12	1.10 ± 0.02	1.000 ± 0.001	1.20 ± 0.15	ND	1.00 ± 0.02	1.30 ± 0.07	1.10 ± 0.03
Alkalinity	mg/L	7600 ± 3610	1300 ± [illegible]	300.00 ± 8.50	700 ± 8.43	1300.00 ± 80.50	900.00 ± 6.78	200.00 ± 5.65	300.00 ± 8.50

*Most probable number.
**Total available nitrogen.
ND—not detected.

Table 2.
Results from Zaria (Karofin Bakin Kasuwa).

Parameter	Unit	Dry season samples				Wet season samples			
		ZC_1	ZC_2	ZC_3	Z_N	ZC_1	ZC_2	ZC_3	Z_N
pH		9.02 ± 0.16	8.04 ± 0.01	8.18 ± 0.66	7.65 ± 0.02	7.46 ± 0.15	5.73 ± 1.67	7.37 ± 0.71	7.05 ± 0.05
Coliform count	*MPN	9.00 ± 1.50	9.00 ± 1.00	2.00 ± 0.50	2	6.00 ± 1.67	4.00 ± 1.10	4.00 ± 0.80	2.00 ± 0.59
Conductivity x 10^2	mS/m	52.50 ± 0.90	19.00 ± 0.70	21.0 ± 0.90	10.00 ± 1.33	450.00 ± 3.45	6.00 ± 1.66	18.00 ± 0.59	8.50 ± 0.18
Colour	Hazen	1000 ± 115.00	5	5	5	12,500 ± 500.00	10.00 ± 5.00	5	5
DO	mg/L	4.00 ± 1.50	8.00 ± 1.11	5.00 ± 2.20	10	6.00 ± 1.80	8.00 ± 1.12	9.00 ± 1.30	12.00 ± 1.00
BOD	mg/L	3.60 ± 0.60	1.60 ± 1.21	3.20 ± 1.11	3.20 ± 1.25	6.00 ± 1.20	3.20 ± 0.87	1.80 ± 0.92	3.80 ± 0.25
COD	mg/L	156.00 ± 2.39	92 ± 8.09	84 ± 6.06	12.00 ± 0.03	180.00 ± 2.39	20.00 ± 1.17	40.2 ± 1.80	36 ± 2.00
$NO_2^{-}N$	mg/L	13.00 ± 5.50	130 ± 0.90	10.00 ± 1.19	17.50 ± 0.15	21.00 ± 0.40	20.00 ± 0.44	9.00 ± 1.26	15 ± 0.76
Chlorides	mg/L	138.29 ± 8.0 8	138.29 ± 0.50	138.29 ± 0.89	280.13 ± 3.89	141.84 ± 0.5 9	138.29 ± 0.89	124.11 ± 0.56	531.90 ± 5.66
**TAN	mg/L	35.0 ± 5.60	27.00 ± 2.42	19.00 ± 2.05	30.00 ± 0.76	39.00 ± 5.66	37.00 ± 0.72	21.00 ± 0.45	31.00 ± 2.22
Cd	mg/L	2.00 ± 0.16	ND	1.70 ± 0.09	1.60 ± 0.26	2.10 ± 0.04	1.60 ± 0.06	1.80 ± 0.02	2.30 ± 0.02
Cr	mg/L	1.80 ± 0.10	2.40 ± 0.03	2.50 ± 0.02	2.40 ± 0.01	1.80 ± 0.10	2.50 ± 0.03	2.60 ± 0.01	2.20 ± 0.21
Pb	mg/L	3.25 ± 0.02	2.50 ± 0.11	1.90 ± 0.11	1.25 ± 0.12	2.45 ± 0.05	1.75 ± 0.05	1.25 ± 0.04	1.50 ± 0.11
Hg	mg/L	1.20 ± 0.07	1.09 ± 0.01	1.00 ± 0.03	1.20 ± 0.15	1.10 ± 0.05	1.00 ± 0.09	0.95 ± 0.04	1.10 ± 0.03
Alkalinity	mg/L	6600 ± 50	800 ± 46.00	1700.00 ± 80.90	700.00 ± 8.43	8000 ± 100	300 ± 5.00	300.0 ± 10.1	300.00 ± 8.50

*Most probable number.
**Total available nitrogen.
ND—not detected.

Table 3.
Results from Zaria (Karofin Mabuga).

Parameter	Unit	Dry season samples				Wet season samples			
		ZB_1	ZB_2	ZB_3	Z_N	ZB_1	ZB_2	ZB_3	Z_N
pH		7.68 ± 0.29	7.65 ± 0.30	8.03 ± 0.67	7.65 ± 0.02	7.57 ± 0.66	7.63 ± 0.24	7.92 ± 0.11	7.05 ± 0.05
Coliform count	*MPN	6 ± 1.00	6.00 ± 0.50	ND	2	2.00 ± 0.60	6	2	2.00 ± 0.59
Conductivity x 10^2	mS/m	61.00 ± 17.14	20.00 ± 10.14	8.59 ± 0.49	10.00 ± 1.33	20.00 ± 0.80	17.00 ± 0.74	18.0 ± 1.7	8.50 ± 0.16
Colour	Hazen	1000	5	5	5	5500 ± 500.00	10	10	5
DO	mg/L	12 ± 2	8.00 ± 0.12	6.00 ± 2.10	10 ± 0	10 ± 0	9.00 ± 1.40	10.0 ± 1.0	12 ± 1.0
BOD	mg/L	3.20 ± 1.47	3.68 ± 1.20	2.28 ± 1.11	3.20 ± 1.25	7.00 ± 1.17	2.88 ± 0.03	2.68 ± 0.12	3.80 ± 0.25
COD	mg/L	208.00 ± 62.79	104.0 ± 31.4	92.00 ± 6.06	12.00 ± 0.03	156.00 ± 9.09	72.00 ± 1.02	84.00 ± 1.96	36 ± 2.00
NO_2^-N	mg/L	12.00 ± 3.70	9.00 ± 2.10	11.00 ± 1.18	17.50 ± 0.15	20.00 ± 0.06	13.00 ± 3.66	11.00 ± 3.3	15.00 ± 2.21
Chlorides	mg/L	138.29 ± 0.50	138.29 ± 1.90	138.29 ± 0.89	280.13 ± 38.9	3540.0 ± 13.8	177.30 ± 5.36	141.84 ± 0.04	531.90 ± 5.66
**TAN	mg/L	22.00 ± 6.74	16.00 ± 2.45	22.00 ± 0.76	30.00 ± 0.76	38.00 ± 2.06	25.00 ± 0.30	27.00 ± 0.66	31.00 ± 2.21
Cd	mg/L	3.00 ± 0.08	2.00 ± 0.06	2.10 ± 0.26	1.60 ± 0.26	3.70 ± 0.16	2.30 ± 0.02	2.90 ± 0.12	2.10 ± 0.02
Cr	mg/L	1.70 ± 0.06	1.80 ± 0.08	1.80 ± 0.01	2.40 ± 0.01	2.00 ± 0.11	2.30 ± 0.11	1.90 ± 0.10	2.20 ± 0.21
b	mg/L	2.40 ± 0.08	1.50 ± 0.06	1.75 ± 0.05	1.25 ± 0.12	3.25 ± 0.05	1.25 ± 0.05	1.90 ± 0.05	1.50 ± 0.11
Hg	mg/L	0.95 ± 0.01	1.50 ± 0.02	0.95 ± 0.15	1.20 ± 0.15	1.10 ± 0.09	1.06 ± 0.07	1.00 ± 0.02	1.10 ± 0.03
Alkalinity	mg/L	6900 ± 253	1200 ± 535	700.00 ± 7.67	700.00 ± 8.43	600.00 ± 5.64	300.00 ± 5.76	300.00 ± 4.87	300.00 ± 8.50

*Most probable number.
**Total available nitrogen.
ND—not detected.

Table 4.
Results from Zaria (Karofin Karauka).

The rest fall below the acceptable level, the fact that those that exceeded the acceptable value were only during dry season shows that ions are more concentrated during dry season because of the absence of dilution by rainwater.

The three respective areas studied in Zaria showed that the waste water wells, drinking water wells and wells in a location where no dyeing activity took place have all exceeded the possible level of 0.0 l mg/L cadmium concentration [19, 20]. The only exception is ZB_3, with non-determinable value during the dry season. These showed slight variations with those of WHO standard.

The maximum permissible level of chromium is 0.05 mg/L [21, 22]. The three respective areas have high chromium concentrations when compared to this standard.

Lead concentration of 0.1 mg/L is the concentration permissible in drinking water. Lead concentrations in the three respective areas of Zaria have all exceeded this level. The highest concentration was found in well ZA_2 during dry season with 3.50 ± 0.05 mg/L and the lowest is ZC_2 during wet season with 1.25 ± 0.05 mg/L. The lead concentration in wells from Kano and Katsina fall between the higher and the lower values here. This signifies little variation and high lead toxicity since the tolerance level has been exceeded [23, 24].

ZA, wastewater well is the only well with a non-determinable value of mercury concentration during the wet season in the three areas under study. All other wells here exceeded the permissible level of 0.000 l mg/L [25]. The environmental threat here associated with mercury is very significant for all the three areas. Concentrations of mercury as high as 1.50 ± 0.02 mg/L were obtained.

If the permissible level of chlorides concentration of 200 mg/L as reported by Ayoade is considered then, all the wells in Karofm Bakin Kasuwa have exceeded this level. However, in Karofin Mabuga only the well in the neutral location exceeded this level during dry and wet season with a concentration of 280.13 ± 3.89 and 531.90 ± 5.66 mg/L respectively. This may be due to the underlying soil structure as having less salt content as compared to Zaria Karofin Bakin Kasuwa. Apart from the well in the neutral location only one well ZC, during the wet season exceeded the level with a concentration of 3540 ± 13.83 mg/L. The reason for this variation may also be due to the soil texture and composition. The threat of chlorides can only be significant in the area of Karofin Bakin Kasuwa. Generally, values here are not as high as those found in Kano and Katsina [26–28]. This may also be due to variation in soil texture and composition.

The waste wells, the well in a location where no dyeing activity took place and the drinking water wells for both seasons in all the three areas (Bakin Kasuwa, Mabuga and Karauka) have all not exceeded the Nitrate permissible level of 45 mg/L [29].

Water quality can be indicated by DO values. DO values are between 12 ± 1.00 mg/L for the neutral well water wells and 3 mg/L for the ZA, waste water well during dry season. It could be seen that wastewater wells have lower DO values than the other wells. This may be due to chemical and biochemical demand because of the presence of organic and biological materials in the waste samples. However, it is noted that DO values are higher for all the three areas (Bakin Kasuwa, Mabuga and Karauka) during the wet season. This is because there may be more dissolved oxygen in rainwater, which eventually finds its way to the underground well waters through percolation.

BOD values for Bakin Kasauwa, Mabuga and Karauka waste wells during the wet season i.e. ZA_1, ZB_1, and ZC_1, are 4 .80 $\pm$ 0.15, 6.0 ± 1.2 and 7.00 ± 1.17 mg/L The ZC_1, exceeded the acceptable level of 6.0 mg/L BOD and ZB, slightly exceeding the level. However, not all other values for wastes and drinking water samples exceeded the permissible level. This shows in the wastewater wells outlined above, that biochemical activity is prominent during the wet season. This can be further enhanced by humid and other favourable conditions that are obtained during wet season. Therefore, coliform bacteria may be more prominent as the only surviving organism showing significant activity in all the samples here.

COD values should not exceed 10 mg/L standard. However, all the results for Bakin Kasuwa, Karauka and Mabuga have exceeded this level in all seasons. The only well close to this is the neutral well during the dry season with 12 ± 0.03 mg/L. There is no clear variation as to the content of the COD during wet and dry season for all the three areas. It is therefore noteworthy that many organic materials may be found herein.

4. Conclusion

The research agrees with what the medical examinations in the literature assert, which shows that local dye workers at Bakin Kasuwa, Mabuga and Karauka exhibited clear signs of lung and skin diseases notably, contact dermatitis and other fungal diseases like eczema. The skin diseases could have been caused by contact with organic dye materials that might have absorbed UV radiations from the sunlight. In addition, concentrations of heavy metals like chromium could have also contributed to causing dermatitis. Lungs diseases could be associated with chromium and high lead concentrations, which could have led to lung damage and ultimately cancer. Other diseases such as gastroenteritis, evidenced in some of the local dye workers as also reported in the literature could have been because of coliform bacteria, which were significantly present in the water of the well.

Funding

No funding was received.

Author details

David Ebuka Arthur[1]*, Karimatu Abdullahi[2], Michael Abatyough[3] and Chidinma Chinelo Arthur[4]

1 Department of Pure and Applied Chemistry, University of Maiduguri, Nigeria

2 Department of Chemistry, Baze University Abuja, Nigeria

3 Department of Chemistry, Bingham University Karu, Nigeria

4 Department of Physical and Health Education, University of Maiduguri, Nigeria

*Address all correspondence to: davidebukaarthur@gmail.com

References

[1] Pattanayak SK, Poulos C, Yang JC, Patil S. How valuable are environmental health interventions? Evaluation of water and sanitation programmes in India. Bulletin of the World Health Organization. 2010;**88**:535-542

[2] Davis J. Private-sector participation in the water and sanitation sector. Annual Review of Environment and Resources. 2005;**30**:145

[3] Ajibade I, McBean G, Bezner-Kerr R. Urban flooding in Lagos, Nigeria: Patterns of vulnerability and resilience among women. Global Environmental Change. 2013;**23**(6):1714-1725

[4] Adejide A, Ajibade LT. Quality of Well Water in Ede Area. Southwestern Nigeria Journal Human Ecology. 2005; **17**(3):223-228

[5] Obiri-Danso K, Adjei BKNS, Stanley KN, Jones K. Microbiological quality and metal levels in wells and boreholes water in some peri-urban communities in Kumasi, Ghana. African Journal of Environmental Science and Technology. 2009;**3**(3):059-066

[6] Okeola FO, Kolawole OD, Ameen OM. Comparative study of physic-chemical parameters of water from a river and its surrounding wells for possible interactive effect. Advances in Environmental Biology. 2010;**4**(3):336-344

[7] Laser M, Schulman D, Allen SG, Lichwa J, Antal MJ Jr, Lynd LR. A comparison of liquid hot water and steam pretreatments of sugar cane bagasse for bioconversion to ethanol. Bioresource Technology. 2002;**81**(1):33-44

[8] World Water Assessment Programme (United Nations). Water for People, Water for Life: The United Nations World Water Development Report: Executive Summary. Unesco Publisher; 2003

[9] Di Toro DM, Fitzpatrick JJ, Thomann RV. Water Quality Analysis Simulation Program (WASP) and Model Verification Program (MVP)-Documentation. Westwood, NY: Hydroscience, Inc; 1981

[10] Ajibola VO, Rilwanu R. Health hazard associated with waste dye wells in Zaria. Journal of Scientific and Industrial Research. 2000;**59**(2):132-135

[11] Ademoroti CMA. Standard Methods for Water and Effluents Analysis. Ibadan: Foludex Press Ltd; 1996. pp. 29-118

[12] Trivedy RK, Goel PK. Chemical and Biological methods for water pollution studies. Karad India: Environmental Publication; 1986;**1**:12-16

[13] U.S. EPA. Review of the International Surveillance Program for the Great Lakes as it relates to water quality. 905-R-79-106; 1979

[14] Feldsine P, Abeyta C, Andrews WH. AOAC international methods committee guidelines for validation of qualitative and quantitative food microbiological official methods of analysis. Journal of AOAC International. 2002;**85**(5): 1187-1200

[15] Trivedy RK, Goel PK. Chemical and biological methods for water quality studies. Karad: Environmental Publications; 1984. pp. 1-215

[16] American Public Health Association (APHA), American Water Works Association (AWWA), Water Pollution Federation (WPCF). Standard Methods for the Examination of Water and

Wastewater. 19th ed. Washington DC. 2005. pp. 47-65

[17] Leaper S, Abdel-Karim A, Gad-Allah TA, Gorgojo P. Air-gap membrane distillation as a one-step process for textile wastewater treatment. Chemical Engineering Journal. 2019;**360**:1330-1340

[18] Mashiatullah A, Qureshi R, Javed T, Ahmad N, Khalid F. Physico-chemical and biological water quality of Karachi coastal water. The Nucleus. 2020; **46**(1–2):53-59

[19] Cinperi NC, Ozturk E, Yigit NO, Kitis M. Treatment of woolen textile wastewater using membrane bioreactor, nanofiltration and reverse osmosis for reuse in production processes. Journal of Cleaner Production. 2019;**223**:837-848

[20] Lloyd RC, Clayton ED, Wilson RE, McBroom RC, Jones RR. Critical experiments using high-enriched uranyl nitrate with cadmium absorber. Nuclear Technology. 2020;**79**(1):82-91. DOI: 10.13182/NT87-A16006

[21] Balasubramanian S, Pugalenthi V. Determination of total chromium in tannery waste water by inductively coupled plasma-atomic emission spectrometry, flame atomic absorption spectrometry and UV–visible spectrophotometric methods. Talanta. 1999;**50**(3):457-467

[22] Razavi Mehr M, Fekri MH, Omidali F, Eftekhari N, Akbari-adergani B. Removal of chromium (VI) from wastewater by palm kernel shell-based on a green method. Journal of Chemical Health Risks. 2019;**9**(1):75-86

[23] Awual MR. An efficient composite material for selective lead (II) monitoring and removal from wastewater. Journal of Environmental Chemical Engineering. 2019;**7**(3):103087

[24] Awual MR, Islam A, Hasan MM, Rahman MM, Asiri AM, Khaleque MA, et al. Introducing an alternate conjugated material for enhanced lead (II) capturing from wastewater. Journal of Cleaner Production. 2019;**224**:920-929

[25] Goswami L, Kumar RV, Pakshirajan K, Pugazhenthi G. A novel integrated biodegradation—Microfiltration system for sustainable wastewater treatment and energy recovery. Journal of Hazardous Materials. 2019;**365**:707-715

[26] Mohammed M, Inuwa Y. Determination of serum chloride ion concentration in pregnant women from Minjibir local government area, Kano State Nigeria. ChemSearch Journal. 2012; **3**(1):19-23

[27] Wakawa R, Uzairu A, Kagbu J, Balarabe M. Impact assessment of effluent discharge on physico-chemical parameters and some heavy metal concentrations in surface water of river Challawa Kano, Nigeria. African Journal of Pure and Applied Chemistry. 2008;**2**(10):100-106

[28] Wakawa R, Uzairu A, Kagbu J, Balarabe M. Seasonal variation assessment of impact of industrial effluents on physicochemical parameters of surface water of river Challawa, Kano, Nigeria. Toxicological & Environmental Chemistry. 2010;**92**(1):27-38

[29] Adimalla N, Li P, Qian H. Evaluation of groundwater contamination for fluoride and nitrate in semi-arid region of Nirmal Province, South India: A special emphasis on human health risk assessment (HHRA). Human and Ecological Risk Assessment: An International Journal. 2019;**25**(5):1107-1124. DOI: 10.1080/10807039.2018.1460579

Chapter 9

Combined Effects of Earthworms and Plant Growth-Promoting Rhizobacteria (PGPR) on the Phytoremediation Efficiency of *Acacia mangium* in Polluted Dumpsite Soil in Bonoua, Côte d'Ivoire

Bongoua-Devisme Affi Jeanne,
Kouakou Sainte Adélaïde Ahya Edith, Hien Marie Paule,
Ndoye Fatou, Guety Thierry and Diouf Diégane

Abstract

The impact of earthworms and plant growth-promoting rhizobacteria (PGPR) on the remediation in polluted dumpsite soil was performed in a greenhouse pot culture with *Acacia mangium* inoculated or not (control: T_0) with *Pontoscolex corethrurus* (T_1) and with *Bradyrhizobium* (T_2); and inoculated with *Pontoscolex corethrurus* and *Bradyrhizobium* (T_3). Our results showed the presence of *Bradyrhizobium* and/or earthworms significantly increase ($P < 0.05$) in the height (2-fold), total dry biomass weight (7- to 15-fold) and metal uptake of the plant (2 to 10-fold), as compared with the non-inoculated plant. The presence of both inoculants (*Bradyrhizobium* and earthworm) enhanced soil Pb/Ni/Cr mobility and bioavailability in metal-contaminated soil, and increased 15-fold the total plant biomass and 10-fold metal accumulation in plant biomass, as compared with plant inoculated with earthworms or *Bradyrhizobium*. In addition, the presence of earthworms and/or *Bradyrhizobium* promoted the phytoimmobilization process of Ni, Cr and Pb preferentially in *Acacia mangium* roots than in shoot tissue. Our experiments highlight the importance of soil organisms on the phytoremediation efficiency. It appears that earthworms and/or *Bradyrhizobium* have the potential to enhance the phytoextraction efficiency of plants in metal-contaminated soil.

Keywords: acacia, phytoremediation, inoculation, phytoimmobilization, bioavailability

1. Introduction

Acacia mangium Wild. is a tropical plant, which has the capacity to improve soil fertility [1], to interact with soil bacteria and soil fauna, particularly in the rhizosphere [2], and to extract metal from polluted soils [3, 4]. In fact, *A.mangium* can accumulate 93.5 mg kg^{-1} of copper (Cu) and 79 mg kg^{-1} of zinc (Zn) in its biomass, was able to tolerate high concentration of cadmium (Cd) [3], and can require 5 and 17 years to remove 79.8 kg ha^{-1} of Zn and 47 kg ha^{-1} of Cu, respectively [4]. So, the phytoremediation efficiency of *A.mangium* has been reported in numerous studies [5, 6] in which it has been demonstrated that the success of phytoremediation may not solely depend on the plant itself but also on the interaction of plant roots with soil microorganism and soil fauna and the availability of heavy metals accumulated in soil [2, 7], because the interaction between plants and beneficial rhizosphere bacteria can enhance biomass production and the tolerance of plants to heavy metals.

It has recently been shown that rhizosphere bacteria may improve metal solubility and availability by decreasing the soil pH or by producing chelators and siderophores [8, 9]. Rhizosphere bacteria such as *Bradyrhizobium allorhizobium* stimulate plant growth either directly or indirectly and have been successfully used to reduce plant stress in metal-contaminated soils and to increase phytoremediation efficiency [6, 7, 10]. Moreover, plant growth-promoting rhizobacteria (PGPR) are known to affect heavy metal mobility and availability to the plant through the release of chelating agents, acidification, phosphate solubilization, and redox changes; therefore, they have the potential to enhance phytoremediation processes [2, 11, 12].

Hence, an alternative method to enhance phytoextraction efficiency and to improve plant growth is by using rhizosphere bacteria such as PGPR and rhizobia [7].

However, in recent studies, the action of earthworms, particularly *Pontoscolex corethrurus* to improve plant metal uptake during phytoremediation in contaminated soils, has been demonstrated [13–17]. Furthermore, the beneficial effects of *P. corethrurus* earthworms on *A. mangium* growth and its Pb, Ni, and Cr uptake have also been showed [5]. It was noted that in the presence of *P. corethrurus*, *A. mangium* promoted the phytoimmobilization process for Ni, Cr, and Pb, but its effectiveness depends on the nature of the plant, its behavior toward metals, rhizosphere function, and metal speciation in different soil compartments involved in the phytoremediation process [5].

Thus, it is of interest to study the conjugated actions of *P. corethrurus* earthworm, symbiotic bacteria (*Bradyrhizobium*), and of a metal tolerant plant such as A. *mangium* in the remediation of metal-contaminated dumpsite soil of M'Plouessoue Park at Bonoua, where the previous studies [18, 19] have demonstrated that the concentration of Cr (130 $mg.kg^{-1}$), Cd (81 $mg.kg^{-1}$), Pb (118 $mg.kg^{-1}$), and Ni (119 $mg.kg^{-1}$) are far above the permissible limits such as Canadian environmental quality criteria for contaminated sites (CEPA) recommendation [20] and World Health Organization limit (WHO-limit) recommendation [21].

The principal aims of this research were to evaluate the effects of *P. corethrurus* and/or *Bradyrhizobium* on lead, chromium, and nickel phytoextraction by *A. mangium* in polluted dumpsite soil metal-contaminated soil.

2. Materials and methods

2.1 Soil sampling and analysis

The metal-contaminated soils were sampled from the abandoned dumping site located in M'Ploussoue Park, Bonoua, Ivory Coast, at latitude 5°16′N and longitude 3° 36′W. Soil samples were collected at 18 different points from the surface horizon (0–30 cm) to cover the entire study area according to the random sampling technique.

The metal-contaminated soil samples were air-dried, sieved to 2 mm, and then were mixed and homogenized to obtain a composite sample. The composite sample was transferred to the laboratory for various analyses and their used for the pot experiment. Some properties of this metal-contaminated soil, which has been previously described by [18], are summarized in **Table 1**. The concentrations of Cd (81 mg kg^{-1}), Cr (130 mg.kg^{-1}), Pb (118 mg kg^{-1}), and Ni (119 mg kg^{-1}) in dumpsite soil are greater than limit values recommended for agricultural soil. However, previous studies performed in pot experiment with polluted soil and *A. mangium*[5] have revealed that only lead, chromium, and nickel concentrations in plant biomass were above the detection limit. Thus, cadmium was not detectable in plant biomass. This stipulates that in the soil, Cd is neither mobile nor exchangeable and is therefore not bioavailable for plants. While for Pb, Ni, and Cr, these metals are bioavailable for plants [5] justifying the choice of the three metals used in this study.

2.2 Biological material

Seeds of *A. mangium* were obtained from the CNRA (Centre National of Research Agronomy) at Oume, Ivory Coast). Seeds were treated with concentrated sulfuric acid (95%) to break hard seed dormancy before germination, as described by [22]. The treated seeds were pregerminated in a Petri dish containing 0.8% water-agar medium (w/v) and sterilized for 30 min at 110°C. Then, the Petri dish was stored at room temperature (30°C) in the dark for 72 hours, after packing with aluminum paper.

Soil properties		**values**
pH		6.9 ± 0.2
Particle Size (%)	Clay	21.3 ± 2
	Silt	1.6 ± 0.3
	Sand	77.1 ± 5
Organic matter (mg.kg^{-1} dry soil)	C	22,600 ± 90
	N	2400 ± 16
	MO	38,872 ± 155
Metal (mg.kg^{-1} dry soil)	Ni	119 ± 13
	Cd	81 ± 11
	Pb	118 ± 19
	Cr	130.1 ± 16

Table 1.
Physicochemical and chemical properties of the contaminated soil used in the pot experiment. ± standard errors.

Before germination in the Petri dish, three pregerminated seedlings were transplanted into polyethylene plastic nursery bags (15×40×150 cm) filled with polluted soil sieved at 2 mm. One month after transplantation in the plastic nursery bags, one seedling of uniform size was transferred into perforated pots filled with 5 kg of dry polluted soil.

Bradyrhizobium ORS strains were obtained from the collection of the Laboratory of Soil Microbiology by Institute Research Senegal Agricultural (IRSA), Dakar. *Bradyrhizobium* ORS strains immobilized in an alginate ball were suspended in sterile buffer solution (23 g K_2HPO_4, 14.6 g KH_2PO_4, and adjusted at 1liter of distilled water). Five milliliters of *Bradyrhizobium* solution was introduced around the roots of the seedlings after transplantation into plastic pots and then after 1 week. The uninoculated treatment (plant uninoculated with *Bradyrhizobium*) received a similar amount of buffer solution with sterilized inoculum to minimize any possible variation in soil properties.

Earthworms (*P. corethrurus*) were hand-collected from Felix Houphouet-Boigny University, Cocody, Abidjan, Ivory Coast, in not metal-polluted soil. The earthworms were then kept in plastic boxes filled with water for 1 week to monitor their health before starting the experiment. For the treatment with earthworms, five adult earthworms with 5 g as of weight biomass were placed in the perforated pot after transplantation of the seedling.

2.3 Experimental design

A greenhouse pot culture experiment was conducted at Felix Houphouet-Boigny University, Cocody, Abidjan, Ivory Coast, to study the effect of *Bradyrhizobium* ORS and earthworms (*P. corethrurus*) on the growth and phytoremediation capacity of *A. mangium*. The average temperatures in the greenhouse were 26, 38, and 32°C for morning, afternoon, and evening, respectively. The experiment was carried out using four treatments:

- Non-inoculated with *Bradyrhizobium* and/or earthworms (*P. corethrurus*), control treatment (T_0);
- Inoculated with *P. corethrurus* earthworms (T_1);
- Inoculated with *Bradyrhizobium* ORS (T_2);
- Co-inoculated with *Bradyrhizobium* ORS and *P. corethrurus* earthworms (T_3).

The experiment was conducted for 90 days, and each treatment was carried out in triplicate. Before filling the pot, it was perforated to allow aeration and then covered with a perforated filet to prevent the earthworms from escaping. The pots were placed in a factorial arrangement based on a completely randomized bloc design. The seedlings were watered daily with deionized water to maintain the moisture content at approximately 60% water-holding capacity of the soil.

2.4 Plant harvest and analysis

At the end of the experiment (90 days), three plants for each treatment were harvested. The rhizosphere soil (RS) and the drilosphere soil (DS) were collected. The soil that remained attached to the roots after gentle shaking was collected as rhizospheric soil (RS). Drilospheric soil is earthworm's structure (casts). The remaining bulk soil was the rest after collecting rhizospheric soils and drilospheric soils [23].

Growth parameters such as shoot length, fresh weight, and dry weight of the plants were measured. The height of acacia was measured for each treatment and each replicate. Shoots (leaves and stems) were harvested, and roots were carefully removed from the soil, rinsed with tap water, and washed three times with deionized water; nodules were detached and counted. The fresh weight of plant was determined for each plant part (shoots and roots) and then the plant part was dried at 60°C for 72 h, weighed, and stored for analysis. The total dry weight of biomass (shoots + roots) of each plant per pot was determined. Rhizobial infection was evaluated by counting the number of nodules per plant. All the different soil compartments were air-dried and stored prior to the analyses. The earthworms were hand-collected, counted, and weighed. Ni, Cr, and Pb concentrations in plant shoots (leaves and stems), roots, and the different soil compartments (RS and DS) were dosed by inductively coupled plasma atomic emission spectrometry (ICP-AES, Spectroblue) after total digestion of plant or soil samples.

The ability of the plant to accumulate metals from the soil and transfer metals from the roots to the shoots was estimated by the bioconcentration factor (BCF) and translocation factor (TF), respectively, as described by [3]. BCF is the ratio of the metal concentration in the total plants biomass to that in the soil used to fill into pot experiment. TF is the ratio of the metal concentration in the shoots to that in the roots of plants.

$$\text{Bioconcentration Factor (BCF)} : \text{BCF}_{\text{ETM}} = \frac{[ETM]totalPlantbiomass\ (\text{mg/kg dry } material)}{[ETM]soilusedtofilledintopot\ (mg/kgdrysoil)} \tag{1}$$

$$\text{Translocation Factor (TF)} : \text{FT}_{\text{ETM}} = \frac{[ETM]inShoots\ (\text{mg/kg dry } material)}{[ETM]inRoots\ (\text{mg/kg dry } material)} \tag{2}$$

According to [24], plants with both factors (TF and BCF) > 1 are suitable for phytoextraction while, plants with both factors <1 are suitable for phytoimmobilization. Plants with TF > 1 promote the phytoextraction process, while plants with TF < 1 are suitable for phytoimmobilization process [3]. Moreover, plants with BCF > 1 are qualified as a hyperaccumulator [3].

The phytoextraction efficiency (PEE) by *acacia* under different treatments was calculated as suggested in studies [25]:

$$(\text{PEE\%}) = \frac{[\text{ETM}]\text{in plant tissue (mgkg} - 1) \text{ x Wplant dry weight (g)}}{[\text{ ETM}]\text{in soil (mgkg} - 1) \text{ x Wsoil used to fill into pot(g)}} \text{x100.} \tag{3}$$

where:

[ETM]in plant tissue= metal (Pb, Ni or Cr) concentration in plant tissue (mg kg^{-1}).

Wplant dry weight= total plant dry biomass (g).

[ETM]insoil= metal (Pb, Ni or Cr) concentration in polluted soil for pot experiment (mg.kg^{-1}).

W soil used to fill into pot(g) = Weight of soil used to fill the pot (g).

2.5 Statistical analysis

The data were subjected to statistical analysis using 7.1 Statistica software. Significant differences between different treatments (non-inoculated, Control, (C); inoculated with *P. corethrurus* earthworms (IE); inoculated with *Bradyrhizobium* ORS (IB); co-inoculated with *Bradyrhizobium* ORS and *P. corethrurus* earthworms (Ci EB) in terms of height, biomass production, nodule numbers, and heavy metal contents in plant biomass, shoot tissue, root tissue, and different compartment RS and DS were performed using the Student–Newman–Keuls (SNK) test at 0.05 probability level.

3. Results and discussion

3.1 Earthworm mortality

After 90 days of exposure in the metal-polluted dumping soil, no mortality was observed throughout the experimental period (**Table 2**). The fresh weight of the earthworms remained stable, and no mortality was noted in each treatment throughout the experimental period. Active burrowing and surface casting were apparent in each treatment.

3.2 Plant growth performance under different treatments

Throughout the experimental period (90 days), regardless of the treatment applied, no visible heavy metal morphological toxicity symptoms, such as leaf chlorosis and root browning, appeared when A. mangium was planted in heavy metal-polluted dumping soil under greenhouse conditions (**Figure 1**). This result revealed that *A. mangium* is able to grow in metal-contaminated soils and is a metal-tolerant plant species, as suggested by [3, 4].

The significantly ($P < 0.05$) lowest height (**Figure 2a**) and total dry weight biomass (**Figure 2b**) were obtained under the non-inoculated (T_0) treatment, with 25.7 cm and 11 g, respectively (**Figure 2a** and **b**). The greatest height and total dry weight biomass were observed when *A. mangium* was co-inoculated with *P.*

Treatments	Times (days)	Earthworms parameters	
		Number	Weight
non-inoculated, control (T_0)	0	n.d	n.d
	90	n.d	n.d
inoculated with earthworm (T_1)	0	5	5
	90	5	5.3
inoculated with Bradyrhizobium (T_2)	0	n.d	n.d
	90	n.d	n.d
co-inoculated with Bradyrhizobium and earthworm (T_3)	0	5	5
	90	5	5.6

Table 2.
Evolution of number and weight of earthworm during 90 days. n.d: none determined.

Figure 1.
A. mangium *growth performance (number of leaves, length of stem, root system development) under different treatments: non-inoculated, control (T_0); inoculated with earthworms (T1); inoculated with Bradyrhizobium ORS (T2); co-inoculated with Bradyrhizobium ORS and earthworms (T3).*

corethrurus and *Bradyrhizobium* (T_3), at 54.5 cm and 141.7 g, followed by T_1 (*A. mangium* inoculated with earthworm) at 51.5 cm and 101 g, by T_2 (A. mangium inoculated with *Bradyrhizobium*) at 45.8 cm and 77 g (**Figure 2a** and **b**). Ours results indicated, for respective effect of *P. corethrurus* earthworms, *Bradyrhizobium* and of both inoculants, a growth stimulation of *A. mangium* by approximately twofold and 10-fold for the biomass under T_1 treatment, by approximately 1.5-fold and sevenfold for the biomass under T_2 treatment and by approximately twofold and 14-fold for the biomass under T_3 treatment. This phenomenon was probably due to the action of *P. corethrurus* earthworms, which have the potential to modify edaphic parameters such as soil structure, organic matter decomposition and indirectly improve soil microorganisms proliferation and activities, facilitate the uptake of many important nutrients by plant, and consequently promote plant growth [26, 27]. Our results are consistent with the well-known fact that earthworms enhance plant growth and biomass [28]. Because, by bioturbation, earthworms stabilize organic matter in soil, form soil aggregates, modify the structure and chemical composition of soil [29]. Such changes generally increase soil water holding capacity, soil nutrient content, and plant productivity [28–30]. Most previous studies justified this better enhancement of acacia growth performance in the presence of earthworm by the fact that in

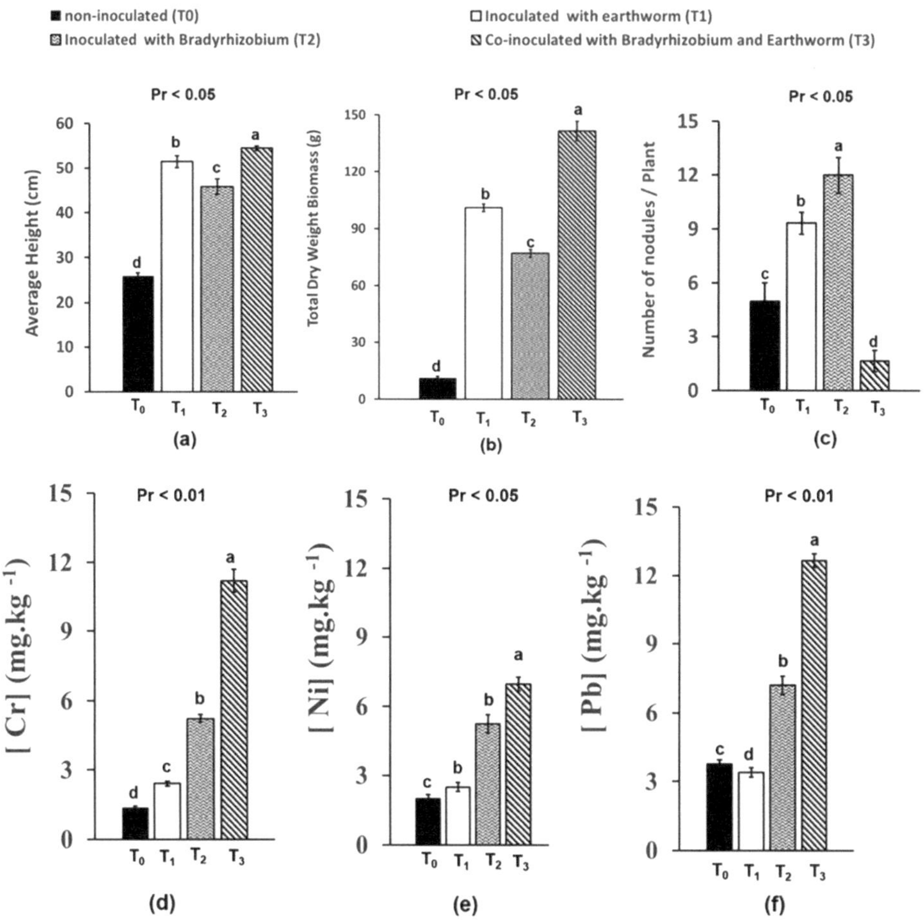

Figure 2.
Effect of different treatments (non-inoculated, control (T_0); inoculated with earthworms (T_1); inoculated with Bradyrhizobium *(T_2); co-inoculated with* Bradyrhizobium *and earthworms (T_3) on average plant height (a), total dry weight biomass (roots and shoots) (b), number of nodules (c), concentrations ($mg.kg^{-1}$ dry weight) of Chromium (Cr) (d), Nickel (Ni) (e), and Lead (Pb) (f), in* Acacia manguim *total biomass. Histograms with the same letters (a, b, c) indicated no significant differences between treatments at 0.05 probability level according to Student–Newman–Keuls test. **very highly significant at 0.01 probability level, *significant at 0.05 probability level according to Student–Newman–Keuls test.*

metal-contaminated soil, some earthworms species (*Eisenia fetida*, *Lumbricus terrestris*, *P. corethrurus*) can decrease the content of potential toxic elements (PTEs) in metal contaminated soil through the accumulation potential toxic elements (PTEs) in their tissues and consequently promote plant growth [31–33]. A similar finding has been documented by [16], who showed that the presence of *P. corethrurus* could enhance the biomass of *Lantana camara L.* by approximately 1.5–2-fold under Pb stress.

The lower increase of acacia growth and biomass under inoculated with *Bradyrhizobium* treatment, compared with inoculated with *P. corethrurus* treatment (T_1), may be due to the competitive effects that may occur between autochthonous soil microorganisms and exogenous strains (*Bradyrhizobium*). Because, several studies have demonstrated that inoculation of seedlings such as *A. mangium* with rhizobial strains results in the change of root morphology, that is, increases in nodules, lateral roots, root hairs, root surface area, and total root length [34] and thus improve plant growth [22, 35] in unpolluted soil and in metal-contaminated sites.

In comparison to the control treatment (T_0) or single inoculated (T_1 or T_2), the presence of *P. corethrurus* earthworms and *Bradyrhizobium* strain significantly ($P < 0.05$) increases better plant growth stimulation. This positive effect might be due to the additive action of the two bioinoculants, which are recognized to promote plant growth and biomass production in metal-contaminated soil [36, 37]. So, ours findings showed that *A. mangium* exhibited better growth and high biomass production when both bioinoculants were present.

However, the greatest number of nodules per plant was obtain when *A. mangium* was inoculated with *Bradyrhizobium* (T_2), at 12 nodules/plant, followed by T_1 (A. mangium inoculated with earthworm) at 9 nodules/plant, by T_0 (*A. mangium* non-inoculated) at 5 nodules/plant (**Figure 2c**). The lowest number of nodules (two nodules/plant) was observed when *A. mangium* was inoculated with the two bioinoculants. The presence of nodules in all the treatments, especially under non-inoculated control treatment, suggested that the soil contained autochthonous strains that were able to colonize the root system of *A. mangium* and to form symbiotic structures (nodules). Moreover, the lowest rate of nodules noted when *A. mangium* was co-inoculated with the two bioinoculants might be due to the interactions between the activities of *Bradyrhizobium* strain and *P. corethrurus* earthworms. In fact, by ingesting soil, *P. corethrurus* earthworms increased organic matter mineralization and nutrient availability, which indirectly stimulated the soil microorganisms. Therefore, the competitive action between autochthonous soil microorganisms and exogenous strains (*Bradyrhizobium*) could affect the capacity of exogenous symbionts (*Bradyrhizobium*) to colonize plant roots and to form symbiotic structures (nodules).

The presence of *P. corethrurus* appeared to reduce the positive effect of *Bradyrhizobium* on *A. mangium* nodulation. This result was in agreement with the findings of [38], who noted that the presence of earthworms (*Allolobophora chlorotica*) can reduce the positive effect of *Glomus intraradices* on the *Allium porrum* L roots biomass.

It was concluded that the interaction between *P. corethrurus* and *Bradyrhizobium* could promote growth and biomass production, but not nodulation, of *A. mangium*.

3.3 Effect of inoculation on metal uptake by *A. mangium*

In the control treatment (T_0), when *A. mangium* was non-inoculated, the concentrations of chromium, nickel, and lead were 1.33 $mg.kg^{-1}$; 1.98 $mg.kg^{-1}$, and 3.8 $mg.kg^{-1}$, respectively (**Table 3**). In addition, Cr and Ni contents were very highly significant ($P < 0.001$), three to fourfold greater in roots tissue, with 1.04 $mg.kg^{-1}$ for Cr and 1.6 $mg.kg^{-1}$ for Ni, than in shoot tissue, with 0.3 $mg.kg^{-1}$ for Cr and 0.44 $mg.kg^{-1}$ for Ni (**Figure 3**). Whereas, the concentration for Pb in roots tissue was lower (1.5 $mg.kg^{-1}$) than in shoots tissue (2.3 Cr $mg.kg^{-1}$) (**Figure 3**). Our results indicated that in the absence of inoculation, acacia preferentially uptake Cr and Ni in its roots and Pb in its shoots (**Figure 3**).

Moreover, the translocation factors ([metal]shoot/[metal]root), indicator of the effectiveness of the plant to translocate metals from roots to shoots of *Acacia* specie, were $TF < 1$ for Cr and Ni, and $TF > 1$ for Pb under non-inoculated treatment but under inoculated treatment, whatever metal dosed $TF < 1$ (**Table 4**). This emphasizes that acacia may possess metal exclusion strategy, which probably depended to the nature of the metal. The bioconcentration factors (BCF) ([metal]plant biomass/

Heavy metal	Different Compartments	Treatments			
		(T_0)	(T_1)	(T_2)	(T_3)
Chromium (Cr)	Rhizosphere soil (RS)	13.8	28.6	12.5	11.9
	Drilosphere soil (DS)	nd	17.8	nd	23.2
	Plant	1.33	2.40	5.23	11.2
Lead (Pb)	Rhizosphere soil (RS)	8.4	4.9	14.7	16.4
	Drilosphere soil (DS)	nd	4.7	nd	3.6
	Plant	3.75	3.4	7.2	12.7
Nickel (Ni)	Rhizosphere soil (RS)	2.6	4.7	13.4	12.7
	Drilosphere soil (DS)	nd	5.6	nd	2.9
	Plant	1.98	2.5	5.24	6.97

Table 3.
Content of chromium, nickel, and lead (mg.kg^{-1} dry material) in different compartments: Rhizosphere Soil (RS), Drilosphere Soil (DS), and in Acacia biomass after pot experiment under different treatments: non-inoculated, control (T_0); inoculated with earthworms (T_1); inoculated with Bradyrhizobium *(T_2); co-inoculated with* Bradyrhizobium *and earthworms (T_3). n.d (none determined).*

[metal]soil) were BCF <0.1 under non-inoculated and inoculated treatments (**Table 4**), which were indicated that acacia are not hyperaccumulator plant as demonstrated [39]. Our findings did not differ from those of various studies that observed a higher accumulation of Pb in the shoots of *A. mangium* compared with the roots, indicating that acacia is able to tolerate and uptake heavy metal in its tissues and therefore could be suitable for phytostabilization of metal-contaminated sites [3, 4, 40]. Furthermore, Cr, Ni, and Pb phytoextraction efficiency (PEE) of *A. mangium* non-inoculated was PEE <1 (**Table 4**) whatever the nature of metal, which could be attributed to the form of the metal in the soil rhizosphere. Thus, it appeared that, according to the nature of the metal in soil, acacia could have different phytoremediation processes (phytoimmobilization and phytoextraction) when it was non-inoculated. But, this phytoremediation process of acacia seems to depend on the nature and the mobile form of metal in the rhizosphere soil.

However, the inoculation of *A. mangium* with *P. corethrurus* earthworms, *Bradyrhizobium* or with both inoculants, significantly ($P < 0.05$) increased the concentrations of chromium, nickel, and lead taken up in the plant biomass, which ranged from 2.4 to 11.2 mg.kg^{-1} for Cr, 2.5 to 7 mg.kg^{-1} for Ni, and 3.4 to 12.7 mg.kg^{-1} for Pb compared to the control treatment with 1.98 mg.kg^{-1}, for Ni, 1.33 mg.kg^{-1} for Cr, and 3.8 mg.kg^{-1} for Pb (**Table 3**). The respective effect of *P. corethrurus* earthworms, *Bradyrhizobium* or of both inoculants, for Cr uptake by plant, was increased around twofold under T_1 treatment (2.41 mg.kg^{-1}), four-fold under T_2 treatment (5.23 mg.kg^{-1}) and 10-fold under T_3 treatment (11.2 mg.kg^{-1}) (**Figure 2d, e**, and **f**). For Ni uptake by plant, the effect of *P. corethrurus* earthworms, *Bradyrhizobium* or both inoculants was enhanced by 1.3-fold under T_1 (2.48 mg.kg^{-1}), threefold under T_2 treatment (5.24 mg.kg^{-1}), and fourfold under T_3 treatments (7 mg.kg^{-1}) (**Figure 2d, e**, and **f**). *P. corethrurus* earthworms decreased the Pb uptake by plant ranging from 3.8 to 3.4 mg.kg^{-1}. *Bradyrhizobium* individually or the combined *P. corethrurus* earthworms and *Bradyrhizobium* enhanced twofold (7.2 mg.kg^{-1}) and fourfold (12.7 mg. kg^{-1}), respectively. Pb uptake by *A. mangium* as compared with non-inoculated plants (**Figure 2d, e** and **f**). In addition, under inoculated with earthworm treatment (T_1),

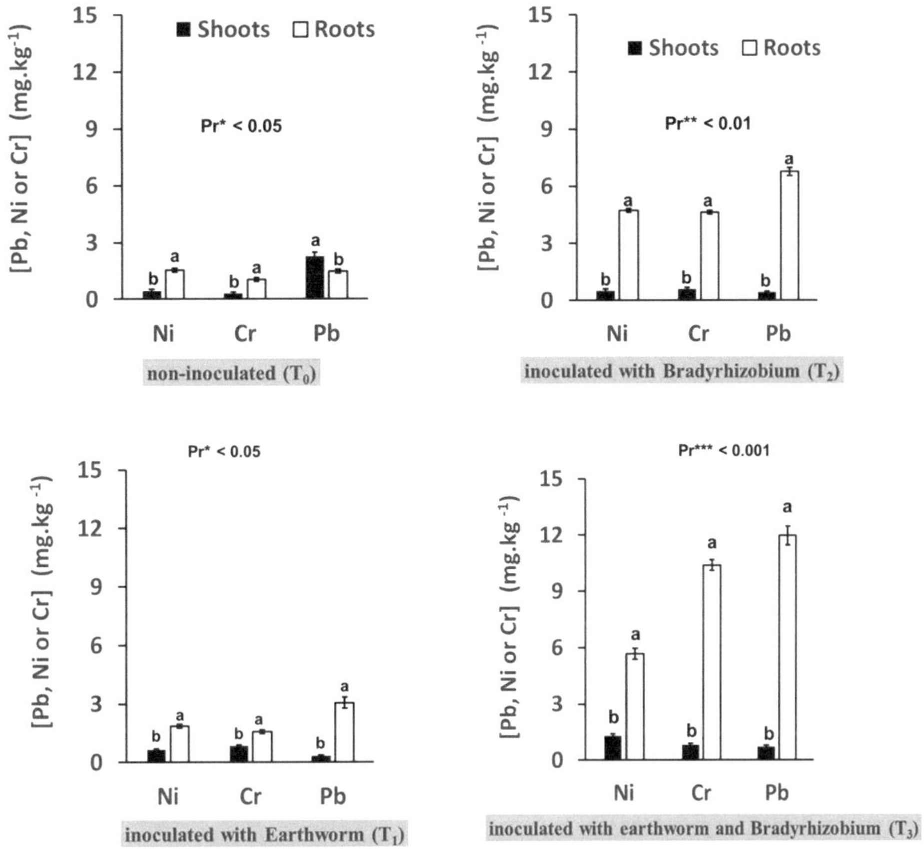

Figure 3.
Accumulation of Cr, Ni and Pb in Acacia mangium *shoot and root tissues under different treatments: non-inoculated, control (T_0); inoculated with earthworms (T_1); inoculated with* Bradyrhizobium *(T_2); co-inoculated with* Bradyrhizobium *and earthworms (T_3). Histogram with the same letters (a, b) indicated no significant differences between Cr, Ni, or Pb concentrations in shoot and root tissues under. *** very highly significant at 0.001 probability level, **highly significant at 0.01 probability level, * significant at 0.05 probability level according to Student–Newman–Keuls test.*

Cr, Ni, and Pb contents were very highly significant ($P < 0.001$), 2–10-fold greater in roots tissue, with 1.6 mg.kg^{-1} for Cr, 1.9 mg.kg^{-1} for Ni and 3.1 mg.kg^{-1} for Pb, than in shoot tissue, with 0.8 mg.kg^{-1}, 0.63 mg.kg^{-1}, and 0.3 mg.kg^{-1}, respectively (**Figure 3**). Furthermore, under inoculated with *Bradyrhizobium* treatment (T_2), Cr, Ni, and Pb contents were very highly significant ($P < 0.001$), 8–15-fold greater in roots tissue, with 4.6 mg.kg^{-1} for Cr, 4.7 mg.kg^{-1} for Ni, and 6.8 mg.kg^{-1} for Pb, than in shoot tissue, with 0.6 mg.kg^{-1}; 0.5 mg.kg^{-1}, and 0.4 mg.kg^{-1}, respectively (**Figure 3**). In the presence of *P. corethrurus* earthworms and *Bradyrhizobium*, Cr, Ni, and Pb contents were very highly significant ($P < 0.001$), 4–20-fold greater in roots tissue, with 10.4 mg.kg^{-1} for Cr, 5.7 mg.kg^{-1} for Ni, and 12 mg.kg^{-1} for Pb, than in shoot tissue, with 0.8 mg.kg^{-1}; 1.3 mg.kg^{-1}, and 0.7 mg.kg^{-1}, respectively (**Figure 3**). The phytoextraction efficiency (PEE) of *A. mangium* was much greater under inoculation treatments with 1–9% for Cr, 5–13% for Ni, and 7–18% for Pb (**Table 3**). Irrespective of the heavy metal dosed (**Table 4**), the significantly higher PEE ($P < 0.05$) was obtained when *A. mangium* was inoculated with *P. corethrurus* earthworms and *Bradyrhizobium* with PEE >9% (**Table 4**). This finding indicated that the

Factors	Heavy metal	Treatments				
		non-inoculated, control (T_0)	inoculated with earthworm (T_1)	inoculated with Bradyrhizobium (T_2)	co-inoculated with Bradyrhizobium and earthworm (T_3)	Probability (Pr)
BCF	Cr	0.01^d	0.02^c	0.04^b	0.08^a	0.05
	Pb	0.03^c	0.03^c	0.06^b	0.11^a	
	Ni	0.02^c	0.02^c	0.04^b	0.06^a	
TF	Cr	0.3^b	0.5^a	0.13^c	0.08^d	0.05
	Pb	1.6^a	0.1^b	0.07^c	0.06^c	
	Ni	0.3^a	0.3^a	0.1^c	0.2^b	
PEE (%)	Cr	0.2^d	1^c	6.4^b	9^a	0.01
	Pb	1^d	7^c	8^b	18^a	
	Ni	1.7^d	5^c	6^b	13^a	

Table 4.
Bioaccumulator (BCF), translocation factors (TF) and phytoextraction efficiency (PEE) of Cr, Ni and Pb in Acacia mangium biomass in metal-contaminated soil under different treatments: non-inoculated, control (T_0); inoculated with earthworms (T_1); inoculated with Bradyrhizobium *(T_2); co-inoculated with* Bradyrhizobium *and earthworms (T_3). Values with the same letters (a, b, c, d) indicated no significant differences between treatments according to Student–Newman–Keuls test.*

inoculation of *A. mangium* with *P. corethrurus* earthworms, *Bradyrhizobium,* or with both inoculants significantly increased the bioavailability of Cr, Pb and Ni in soil, then their uptake by *A. mangium* in biomass particularly in its roots tissue. In the presence of these organisms, the phytoextraction efficiency of *A.mangium* was significantly ($P < 0.05$) improved. The accumulation of potential toxic elements in acacia biomass may have been caused by the different soil organisms (*Bradyrhizobium* and/or earthworm), as demonstrated in previous studies in the presence of earthworms [5, 27, 31, 33] and of *Bradyrhizobium* [36, 41] only and also in presence of combined soil organisms such as earthworm and PGPR [42]. The metal uptake-stimulating effect of both inoculants was much greater than that of individual inoculated organism.

Moreover, under earthworm treatment, only the content of Cr at 29 $mg.kg^{-1}$ dry soil and Pb at 4.9 $mg.kg^{-1}$ dry soil was higher in the rhizosphere soil (RS) than in Drilosphere soil (DS) at 18 $mg.kg^{-1}$ Cr dry soil and 4.7 $mg.kg^{-1}$ Pb dry soil (**Table 3**). The lower content of Cr in the DS than in the RS compartment with earthworm treatment and the highest content of Cr (68 μg Cr/plant) in plant shoots suggested that Cr mobilized by earthworm in their structures (burrows and casts) was temporarily stored in these structure, which acted as sinks for the element [43], and transferred Cr to the RS compartment and subsequently to plant tissue. In contrast, despite the highest content of lead in the RS compartment, lead content was lowest in the plant shoots (25.2 μg Pb/plant). This phenomenon could be linked to the physiological behavior of *A. mangium,* which behaves as a Pb-excluder plant in the presence of bioinoculants. Likewise, despite the highest content of Ni in the DS (5.6 $mg.kg^{-1}$ Ni dry soil) under earthworm treatment, the content of Ni was higher in the shoot parts than in the root part, which suggested that Ni mobilized in the DS compartment was transferred to plants. Here, the DS compartment was used by the plant as a sink for the element [43].

3.4 *P. corethrurus* earthworm and *A. mangium* interaction on phytoremediation processes

Previous studies have been reported the interactive role of earthworms in improving plant growth in non-contaminated soils [26]. In addition to this, the benefit effect of earthworms in the remediation of metal contaminated soil has been very well demonstrated in numerous research studies [5, 27, 31, 33, 44], but only few studies have been conducted to assess their role in improving plant metal uptake during phytoremediation in contaminated soils [45].

Our findings have shown that the inoculation of acacia with *P. corethrurus* resulted a highly to very highly significant increase ($P < 0.01$) in plant height, total dry weight biomass, and metal concentration in plant biomass (**Figure 2**), as compared with the uninoculated treatment. Thus, acacia appeared to exhibit rapid growth and high biomass production when earthworms were present. For instance, the phytoextraction efficiency of the plant inoculated with *P. corethrurus* was enhanced by fivefold for Cr, twofold for Ni, and sevenfold for the Pb, as compared with non-inoculated plant (**Table 4**). This increase of growth-stimulating and of the amount of Cr, Ni in *A. mangium* biomass could be caused by the earthworms through their burrowing and casting activities, as suggested by [31], because the earthworms can facilitate metal conversion from the stable to the available form by changing physico-chemical and biological status of the soil such as soil pH decreases, production of organic acids, and stimulation of microbial activity, contributing to the increase of Ni, Pb, and Cr availability in soil and as a result increased their bioavailability for plants. Furthermore, the increase of metal accumulation in plant biomass could be due to the interactive action between *A. mangium* and *P. corethrurus.* In fact, some species such as *Acacia* secrete different types and quantities of organic acids into the rhizosphere [46], which were the main source of organic matter used by *P. corethrurus* earthworms in the rhizosphere, according to [13]. So, by decomposing different types of root exudates and organic acids secreted by *A. mangium* into the rhizosphere, P. *corethrurus* can probably reduce the stable form of metal while increasing its mobile form in the rhizosphere, enhancing Cr and Ni bioavailability for plant [31]. However, the higher significant ($P < 0.05$) content of Pb in plant non-inoculated than in plant inoculated with earthworm (**Figure 2**) could be attributed to the metal speciation in rhizosphere or drilosphere [44]. justified the decrease of Pb concentration in plant inoculated with earthworm, as compared with earthworm inoculation, by the fact that earthworm can also reduce the amount of Pb associated with the soluble and exchangeable fraction and subsequently plant uptake.

In addition, the higher content of Cr (29 $mg.kg^{-1}$) and Pb (4.9 $mg.kg^{-1}$) in RS and Ni in DS (5.6 $mg.kg^{-1}$) (**Table 3**) than in plant biomass could be related to the physiological character of *Acacia* species, which here seems to exclude a metal in its shoot tissue as demonstrated in previous studies [3, 4, 39, 46].

These results suggest that, although earthworms have the potential to improve the efficiency of plant phytoremediation in metal-contaminated soils, its effectiveness depends on the nature of the plant, its behavior toward metals, metal speciation in soil, rhizosphere function involved in the phytoremediation process.

3.5 Bradyrhizobium and *A. mangium* interaction on phytoremediation processes

Symbiosis between leguminous and rhizobacteria improves plant growth, nutrition and reduces the stress of plants, facilitating their development in metal-contaminated areas [47, 48]. Previous studies, specially studies from symbiotic microorganism, have demonstrated that rhizobia contribute to plant adaptation to multiple biotic and abiotic stresses, especially under metal-contaminated soils [41, 47, 49]. Among the rhizobacteria obtained from areas contaminated with different metals, there are strains of the genus *Rhizobium* sp., *Sinorhizobium, Mesorhizobium*, *Bradyrhizobium,* and *Azorhizobium* [48]. These strains are recognized as plant growth-promoting rhizobacteria [50].

Our findings have shown that the inoculation of acacia with *Bradyrhizobium* (T_2) resulted a highly to very highly significant increase ($P < 0.01$) in plant height, total dry weight biomass, and metal concentration in plant biomass (**Figure 2**), as compared with the non-inoculated plant. Thus, acacia appeared to exhibit rapid growth performance (seven-fold) when *Bradyrhizobium* was present. This growth stimulation could be attributed to the interactive action between *A. mangium* and with symbiotic rhizobia such as *Bradyrhizobium,* which have the capacity to form symbiotic association with *A. mangium.* and consequently influence positively plant P nutrition and growth and then soil microbial activities [51, 52].

Furthermore, positive effects from inoculation with *Bradyrhizobium* on metal uptake by *A. mangium* in metal-contaminated soil have been observed. The inoculation of *A. mangium* with *Bradyrhizobium* (T_2) increased very highly significant ($P < 0.001$) Cr, Ni, and Pb amounts by four-, three-, and twofold, respectively, as compared with the control treatment (T_0). In addition, the phytoextraction efficiency of the plant inoculated with *Bradyrhizobium* was enhanced by 32-fold for Cr, 4-fold for Ni, and 8-fold for the Pb, as compared with non-inoculated plant (**Table 4**). This positive effect may attribute to *Bradyrhizobium* Sp., which can increase the availability of soil metal through the production of metal chelating, agents siderophores, and organic acid [47, 53], and can also modify heavy metals speciation and metal/organic matter interaction by transformation of organic compounds [42], consequently increasing their bioavailability for plants. Ours findings showed that *Bradyrhizobium* effectively enhances *A.mangium* growth, its metal uptake, and also their accumulation in root than shoot tissues. Ours results indicated also that *Bradyrhizobium* improves metal bioavailability in soil and subsequently for plant. So, according to [47], this is possible because *Bradyrhizobium* can decrease the toxicity of metal contamination in plant by transforming pollutants into nontoxic or less toxic form and also by enhancing antioxidant defense in plants exposed to metal-contaminated soils. Similar reports also demonstrated that the inoculation with *Bradyrhizobium* higher increases Cu concentrations in *soybean* and especially in white *lupin* in inoculated plants [53, 54], showed that *Methylobacterium* sp. notably enhances the bioaccumulation of As in *Acacia farnesiana* biomass mainly in shoots. In contrast [55], showed that *Bradyrhizobium* Sp. reduced Ni and Zn uptake by Greengrass plants, which was probably due to the ability of *Bradyrhizobium* to protect plants against the inhibitory toxic effects of Ni and Zn.

Ours results demonstrated that under inoculated with *Bradyrhizobium* treatment (T_2), Cr, Ni, and Pb amounts were very highly significant ($P < 0.001$), 8, 10, to 15-fold greater in roots tissue than in shoot tissue (**Figure 3**). Likewise, the bioaccumulation factors (BCFs) and the translocation factors (TFs) of Cr, Ni, and Pb, which were < 1 respectively, revealed that the presence of *Bradyrhizobium* improved better the uptake of Cr and Ni mainly in roots. While for Pb, the presence of *Bradyrhizobium* improved Pb accumulating in roots. The presence of *Bradyrhizobium* modified the phytoextractor potential of non-inoculated plant to act as Pb phytoexcluders. Thus, in the presence of *Bradyrhizobium*, *A. mangium* is considered to have great potential for the phytoimmobilization of Cr, Ni, and Pb. A similar effect has been observed by [56], which after inoculation with *Cupriavidus taiwanensis*, *Mimosa pudica* showed higher capacity of metal uptake and improved Pb, Cu, and Cd accumulating mainly in roots.

These findings supported the ability of *Bradyrhizobium* to protect *A. mangium* plants against the inhibitory toxic effects of Ni, Cr, and Pb, as demonstrated by [57].

3.6 *P. corethrurus* earthworms and Bradyrhizobium interactions on Pb, Ni, and Cr uptake by *A. mangium*

Earthworms and rhizobacteria are essential for nutrient cycling and organic matter dynamics in terrestrial ecosystems. In soils, they tightly interact especially in the rhizosphere. In our experiment, we tested the effects of earthworm *P. corethrurus* and *Bradyrhizobium* on the growth performance of *A. mangium* and also its metal uptake in metal-contaminated soil. We observed a significant ($P < 0.05$) positive effect of both inoculants (earthworm *P. corethrurus* and *Bradyrhizobium*) on *A. mangium* growth and total biomass, as compared with plant non-inoculated and also with plant inoculated with earthworm *P. corethrurus* or *Bradyrhizobium* only (**Figure 2**), as demonstrated in numerous reports [26, 51, 52]. This growth stimulation in the presence of both inoculants could be related to [1] earthworm *P. corethrurus* activity, which by increasing the mineralization of soil organic matter enhances nutrient availability, stimulates microbial activities [2]; the production of plant growth regulator substances through the stimulation of microbial activity [3, 58]; the stimulation of plant symbionts in the soil rhizosphere [4]; the plant genotype *A.mangium,* a leguminous, which used as symbiont *Bradyrhizobium* recognized as a plant growth-promoting bacteria (PGPR) [50], and [4] to the bio-control of metal stress by earthworm [27, 31, 33, 44] and by *Bradyrhizobium* [41, 47, 48, 59] only and also by the combined action with both inoculants [45]. This synergistic interaction is probably due to the stimulation of plant growth-promoting rhizobacteria, such as *Bradyrhizobium*, population in the presence of earthworms [60].

Furthermore, a significant ($P < 0.05$) greater amounts of Cr, Ni, and Pb in total biomass of plant have been observed in the presence of both inoculants, which was increased by 2–10-fold for Cr, 2–3-fold for Ni, and 2–4-fold for the Pb, as compared with non-inoculated and individual inoculated plants (**Figure 2**). Likewise, the phytoextraction efficiency of the plant inoculated with both inoculants was enhanced by 2–9-fold for Cr, 2–3-fold for Ni, and 2-fold for the Pb, as compared with individual inoculation (**Table 3**). This increase could be attributed to the combined activities of the two inoculants that have the ability to enhance metal uptake in plant tissues and to protect plants against toxic effects have been demonstrated in previous studies [31, 42]. The improvement of metal uptake by *A. mangium* inoculated with both inoculants, as compared with its inoculation with *Bradyrhizobium* or earthworm individually, could be linked to the relationship between earthworm and rhizobacteria (*Bradyrhizobium*). This finding suggests that combined inoculants consisting of *Bradyrhizobium* and earthworms have potential for enhancing metal uptake by *A. mangium,* confirming our hypothesis. The results indicate that metal uptake by this tolerant plant species was greatly facilitated by the interactions among these organisms, most likely due to the concomitant stimulation of metal immobilization and biomass production, as demonstrated by [13, 42].

In the presence of both inoculants (*P. corethrurus* earthworms and *Bradyrhizobium*), *A. mangium* preferentially accumulated Pb, Cr, and Ni in the roots than in shoots tissue (**Figure 3**) with TF < 1 (**Table 3**) and BCF < 1, indicating that for Pb, Ni, and Cr, *A. mangium* promotes the phytoimmobilization process.

In addition, the content of Cr was higher in the DS compartment (23 $mg.kg^{-1}$ dry soil) than in the RS (12 $mg.kg^{-1}$ dry soil) in the presence of both inoculants (earthworms and *Bradyrhizobium*) (**Table 4**). Our results also showed that the concentration of Cr was higher in the root (12 $mg.kg^{-1}$) than in the shoot tissue (0.7 $mg.kg^{-1}$). This finding suggested that Cr mobilized in the DS compartment (23.2 $mg.kg^{-1}$ dry soil), was preferentially transferred to RS compartment and then to the plant root tissue. While, the content of Pb and Ni was significantly higher in the RS compartment, ranging from 16.4 and 12.7 $mg.kg^{-1}$ dry soil, respectively (**Table 4**), than in the DS compartment (range 3–4 $mg.kg^{-1}$ dry soil). Despite the highest content of Pb and Ni in the RS compartment, the concentrations of Pb and Ni were lower in the plant shoot biomass. The translocation of Pb and Ni from the root to the shoot tissue was weak. This phenomenon could be linked to the behavior of *A. mangium,* which, in the presence of both bioinoculants, behaved as Pb, Cr, and Ni-excluder plant and promoted the phytoimmobilzation process for Cr, Pb, and Ni.

Efficiency of the different phytoremediation treatments applied.

Our results showed that inoculation of *A. mangium* with *Bradyrhizobium* or earthworms only and with both inoculants significantly increased ($P < 0.05$) in the height (twofold), total dry biomass weight (7–15-fold), and metal uptake of the plant (2–10-fold), as compared with the non-inoculated plant. However, the presence of *Bradyrhizobium* and earthworms increases twofold the total plant biomass and two- to fivefold metal accumulation in plant biomass, as compared with inoculated with earthworms or *Bradyrhizobium*.

Furthermore, irrespective of the heavy doses, the phytoextraction efficiency (PEE) percentage rank was in the order $T_3 > T_2 > T_1 > T_0$ (**Table 3**). The PEE percentage of *A. mangium* increased significantly in the presence of earthworms and *Bradyrhizobium*, demonstrating values of 18% for Pb, 9% for Cr, and 12.6% for Ni, followed by T_2 (when *A. mangium* was inoculated with *Bradyrhizobium* only) with 8% for Pb, 6.4% for Cr, and 6% for Ni and by T1 with 7% for Pb, 1% for Cr, and 5% for Ni (**Table 3**). We found strong evidence that the inoculation of plant with PGPR and earthworm enhanced soil Pb/Ni/Cr mobility and bioavailability in metal-contaminated soil, facilitating their transfer and absorption by plant.

This result indicated that the phytoremediation capacity of *A. mangium* was improved in response to the inoculation and optimally improved in the presence of both inoculants. So, our finding reveled that it is possible to use the combination of metal-tolerant plant and soil organisms (*Bradyrhizobium* and earthworms) as a potential bioaugmentation tool to accelerate metal phytoremediation efficiency in metal-contaminated soils.

4. Conclusion and recommendation

Beneficial effects of combined inoculation with *P. corethrurus* earthworms and *Bradyrhizobium* and of individual inoculation with *P. corethrurus* earthworms or *Bradyrhizobium* on A. mangium growth and its Pb, Ni, and Cr uptake in metal-contaminated soil have been observed in this study. Ours results revealed that the concomitant stimulation of metal immobilization and biomass production in the presence of these organisms and also that the inoculation of plant with PGPR (*Bradyrhizobium*) and earthworm enhanced soil Pb/Ni/Cr mobility and bioavailability in metal-contaminated soil, facilitating their transfer and absorption by plant. However, the growth stimulation and the metal accumulation in plant were increase twofold for the total plant biomass and two- to fivefold for metal amount in plant biomass, as compared with inoculated with earthworms or *Bradyrhizobium*. In addition, the presence of these organisms promoted the phytoimmobilization process of Ni, Cr, and Pb preferentially in *A. mangium* roots than in shoot tissue. Our experiments highlight the importance of soil organisms on the phytoremediation efficiency. It appears that earthworms and/or PGPR (*Bradyrhizobium*) have the potential to enhance the phytoextraction efficiency of plants in metal-contaminated soil.

Acknowledgements

We sincerely thank the Education and Research Ministry of the Ivory Coast, as part of the Debt Reduction-Development Contracts (C2Ds) managed by IRD, for the financial support rendered to carry out the project on ReSiPol research. We also thank Bonoua City Hall for permission to carry out the studies at the dumpsites.

Notes

The authors declare no competing financial interest.

Author details

Bongoua-Devisme Affi Jeanne[1]*, Kouakou Sainte Adélaïde Ahya Edith[1], Hien Marie Paule[1], Ndoye Fatou[2], Guety Thierry[1] and Diouf Diégane[2]

1 UFR Sciences de la Terre et des Ressources Minières, Université Felix Houphouët-Boigny, Abidjan, Côte d'Ivoire

2 Fatou NDOYE - Centre de recherche de Bel Air, Laboratoire Commun de Microbiologie, Dakar, Sénégal

*Address all correspondence to: bongoua_jeanne@yahoo.fr

References

[1] Koutika L-S. Afforesting savannas with Acacia mangium and eucalyptus improves P availability in Arenosols of the Congolese coastal plains. Geoderma Regional. 2019;**15**:e00207. DOI: 10.1016/j.geodrs.2019.e00207

[2] Ma Y, Oliveira RS, Nai F, Rajkumar M, Luo Y, Rocha I, et al. The hyperaccumulator sedum plumbizincicola harbors metal-resistant endophytic bacteria that improve its phytoextraction capacity in multi-metal contaminated soil. Journal of Environmental Management. 2015;**156**:62-69. DOI: 10.1016/j.jenvman.2015.03.024

[3] Majid NM, Islam MM, Mathew L. Heavy metal uptake and translocation by mangium (*Acacia mangium*) from sewage sludge contaminated soil [online]. Australian Journal of Crop Science. 2012;**6**(8):1228-1123 Available from: http://www.cropj.com/islam_6_8_2012_1228_1235.pdf

[4] Mohd SN, Majid NM, Shazili NAM, Abdu A. Growth performance, biomass and phytoextraction efficiency of *Acacia mangium* and *Melaleuca cajuputi* in remediating heavy metal contaminated soil. American Journal of Environmental Sciences. 2013;**9**(4):310-316 Available from: http://www.thescipub.com/ajes.toc

[5] Bongoua-Devisme AJ, Akotto OF, Guety T, Kouakou SAAE, Ndoye F, Diouf D. Enhancement of phytoremediation efficiency of *Acacia mangium* using earthworms in metal-contaminated soil in Bonoua, Ivory Coast. African Journal of Biotechnology. 2019;**18**(27):622-631. DOI: 10.5897/AJB2019.16852

[6] Bongoua-Devisme AJ, Koffi AL, Kouadio KP, Diouf D, Ndoye F, Kouadio K-KH. Response of *Acacia mangium* to inoculation with symbiotic microorganisms in the remediation of metal-contaminated soil at Bonoua, Ivory Coast. Journal of Chemical, Biological and Physical Sciences. 2019;**9**(4):544-560. DOI: 10.5897/AJB2019.16852

[7] Mohammadzadeh A, Tavakoli M, Motesharezadeh B, Chaichi MR. Effects of plant growth-promoting bacteria on the phytoremediation of cadmium-contaminated soil by sunflower. Archives of Agronomy and Soil Science. 2016;**63**(6):807-816. DOI: 10.1080/03650340.2016.1235781

[8] Ojuederie O, Babalola O. Microbial and plant-assisted bioremediation of heavy metal polluted environments: A review. International Journal of Environmental Research and Public Health. 2017;**14**(12):1504. DOI: 10.3390/ijerph14121504

[9] Wood JL, Liu W, Tang C, Franks AE. Microorganisms in heavy metal bioremediation: Strategies for applying microbial-community engineering to remediate soils. AIMS Bioengineering. 2016;**3**(2):211-223. DOI: 10.3934/bioeng.2016.2.211

[10] Kamran MA, Eqani S, Bibi S, Xu R-K, Monis MFH, Katsoyiannis A, et al. Bioaccumulation of Nickel by E. Sativa and role of plan growth promoting Rhizobacteria (PGPRs) under Nickel stress. Ecotox Environ Safe. 2016;**126**:256-263. DOI: 10.1016/j.ecoenv.2016.01.002

[11] Fan M, Xiao X, Guo Y, Zhang J, Wang E, Chen W, et al. Enhanced phytoremediation of Robinia pseudoacacia in heavy metal-

contaminated soils with rhizobia and the associated bacterial community structure and function. Chemosphere. 2018;**197**:729-740. DOI: 10.1016/j.chemosphere.2018.01.102

[12] Khan N, Bano A. Role of PGPR in the Phytoremediation of Heavy Metals and Crop Growth Under Municipal Wastewater Irrigation. In: Ansari A, Gill S, Gill R, R. Lanza G, Newman L, editors. Phytoremediation. Cham: Springer; 2018. pp. 135-149. DOI: 10.1007/978-3-319-99651-6_5

[13] De Novais CB, De Oliveira JR, Siqueira JO, De Faria SM, Da Silva EMR, et al. Trophic relationships between the earthworm *Pontoscolex corethrurus* and three tropical arbuscular mycorrhizal fungal species. Applied Soil Ecology. 2019;**135**:9-15 Available from: https://www.sciencedirect.com/science/article/pii/S0929139318301410

[14] Taheri S, Pelosi C, Dupont L. Harmful or useful? A case study of the exotic peregrine earthworm morpho-species *Pontoscolex corethrurus*. Soil Biology and Biochemistry. 2018;**116**: 277-289. DOI: 10.1016/j.soilbio.2017.10.030

[15] Duarte AP, Melo VF, Brown GG, Pauletti V. Earthworm (*Pontoscolex corethrurus*) survival and impacts on properties of soils from a lead mining site in Southern Brazil. Biology and Fertility of Soils. 2014;**50**(5): 851-860. DOI: 10.1007/s00374-014-0906-y

[16] Jusselme MD, Poly F, Lebeau T, Rouland-lefèvre C, Miambi E. Effects of earthworms on the fungal community and microbial activity in root-adhering soil of *Lantana camara* during phytoextraction of lead. Applied Soil Ecology. 2015;**96**:151-158. DOI: 10.1016/j.apsoil.2015.07.011

[17] Buch AC, Brown GG, Niva CC, Sautter KD, Sousa JP. Toxicity of three pesticides commonly used in Brazil to *Pontoscolex corethrurus* (Müller, 1857) and *Eisenia andrei* (Bouché, 1972). Applied Soil Ecology. 2013;**69**:32-38. DOI: 10.1016/j.apsoil.2012.12.011

[18] Bongoua-Devisme AJ, Gueable YKD, Balland Boloun Bi C, Bolou Bi BE, Kassin KE, Adiaffi B, et al. Hazardous impacts of open dumpsite of municipal solid wastes on soil: Case of M'Ploussoue Park dump at Bonoua in Ivory Coast. International Journal of Sciences. 2018; 7(5):32-38. DOI: 10.18483/ijSci.1653

[19] Bongoua-Devisme AJ, Bolou Bi BE, Kassin KE, Bolou B, Bi C, Gueable YKD, et al. Assessment of heavy metal contamination degree of municipal open-air dumpsite on surrounding soils: Case of dumpsite of Bonoua, Ivory Coast. International Journal Of Engineering Research and General Science. 2008;**6**(5):28-42 Available from: https://hal.archives-ouvertes.fr/hal-01891039/file/Bongoua2_2018.pdf

[20] CEPA. Canadian soil quality guidelines for the protection of environmental and human health. Quebec, Canada: National Guidelines and Standards Office; 2007 Available from: https://ccme.ca/en/res/polychlorinated-dioxins-and-furans-pcdd_fs-canadian-soil-quality-guidelines-for-the-protection-of-environmental-and-human-health-en.pdf

[21] WHO. Guidelines for assessing quality for herbal medecines with references to contaminants and residus, WHO Library Cataloguing-in-Publication Data. 2007. Available from: https://apps.who.int/iris/bitstream/handle/10665/43510/?sequence=1

[22] Diouf D, Duponnois R, Tidiane BA, Neyra M, Lesueur D. Symbiosis of *Acacia*

auriculiformis and *Acacia mangium* with mycorrhizal fungi and *Bradyrhizobium* spp. improves salt tolerance in greenhouse conditions. Functional Plant Biology. 2005;**32**(12):1143-1152. DOI: 10.1071/fp04069

[23] Tan Y, Cui Y, Li H, Kuang A, Li X, Wei Y, et al. Rhizospheric soil and root endogenous fungal diversity and composition in response to continuous Panax notoginseng cropping practices. Microbiological Research. 2017;**194**: 10-19. DOI: 10.1016/j. micres.2016.09.009

[24] Buscaroli A. An overview of indexes to evaluate terrestrial plants for phytoremediation purposes (Review). Ecological Indicators. 2017;**82**:367-380. DOI: 10.1016/j.ecolind.2017.07.003

[25] Yang Y, Yichen G, Hongyuan Z, Xihong Z, Liang P, Qingru Z. Phytoextraction of cadmium-contaminated soil and potential of regenerated tobacco biomass for recovery of cadmium. Scientific Reports. 2017;**7**:7210. DOI: 10.1038/s41598-017-05834-8

[26] Braga LP, Yoshiura CA, Borges CD, Horn MA, Brown GG, Drake HL, et al. Disentangling the influence of earthworms in sugarcane rhizosphere. Scientific Reports. 2016;**6**:38923. DOI: 10.1038/srep38923

[27] Lv B, Xing M, Yang J. Speciation and transformation of heavy metals during vermicomposting of animal manure. Bioresource Technology. 2016;**209**: 397-401. DOI: 10.1016/j.biortech. 2016.03.015

[28] Mudrák O, Frouz J. Earthworms increase plant biomass more in soil with no earthworm legacy than in earthworm-mediated soil, and favour late successional species in competition. Functional Ecology. 2017;**32**(3):626-635. DOI: 10.1111/1365-2435.12999

[29] Bottinelli N, Jouquet P, Capowiez Y, Podwojewski P, Grimaldi M, Peng X. Why is the influence of soil macrofauna on soil structure only considered by soil ecologists? Soil and Tillage Research. 2015;**146**:118-124. DOI: 10.1016/ j.still.2014.01.007

[30] Wu J, Zhang W, Shao Y, Fu S. Plant-facilitated effects of exotic earthworm *Pontoscolex corethrurus* on the soil carbon and nitrogen dynamics and soil microbial community in a subtropical field ecosystem. Ecology and Evolution. 2017;**7**(21):8709-8718. DOI: 10.1002/ ece3.3399

[31] Lemtiri A, Liénard A, Alabi T, Brostaux Y, Cluzeau D, Francis F, et al. Earthworms Eisenia fetida affect the uptake of heavy metals by plants Vicia faba and Zea mays in metal-contaminated soils. Applied Soil Ecology. 2016;**104**:67-78. DOI: 10.1016/j.apsoil. 2015.11.021

[32] Sizmur T, Palumbo-Roe B, Watts MJ, Hodson ME. Impact of the earthworm *Lumbricus terrestris* (L.) on As, Cu, Pb and Zn mobility and speciation in contaminated soils. Environmental Pollution. 2011;**159**(3): 742-748. DOI: 10.1016/j.envpol.2010. 11.033

[33] Boughattas I, Hattab S, Boussetta H, Banni M, Navarro E. Impact of heavy metal contamination on oxidative stress of *Eisenia andrei* and bacterial community structure in Tunisian mine soil. Environmental Science and Pollution Research. 2017;**24**(22): 18083-18095. DOI: 10.1007/s11356-017-9449-8

[34] Lee J-T, Tsai S-M, Li C-H. The nitrogen-fixing *Bradyrhizobium elkanii*

significantly stimulates root development and pullout resistance of Acacia confuse. African Journal of Biotechnology. 2017;**16**(18):1067-1077. DOI: 10.5897/AJB2017.15971

[35] Tajini F, Mustapha T, Jean-Jacques D. Combined inoculation with *Glomus intraradices* and *Rhizobium tropici* CIAT899 increases phosphorus use efficiency for symbiotic nitrogen fixation in common bean (*Phaseolus vulgaris* L.). Saudi Journal of Biological Sciences. 2012;**19**(2):157-163. DOI: 10.1016/j.sjbs.2011.11.003

[36] Seneviratne M, Gunaratne S, Bandara T, Weerasundara L, Rajakaruna N, Seneviratne G, et al. Plant growth promotion by *Bradyrhizobium japonicum* under heavy metal stress. South African Journal of Botany. 2016; **105**:19-24. DOI: 10.1016/j.sajb.2016.02.206

[37] Reichman SM. Probing the plant growth-promoting and heavy metal tolerance characteristics of *Bradyrhizobium japonicum* CB1809. European Journal of Soil Biology. 2014; **63**:7-13. DOI: 10.1016/j.ejsobi.2014.04.001

[38] Milleret R, Le Bayon RC, Gobat JM. Root, mycorrhiza and earthworm interactions: Their effects on soil structuring processes, plant and soil nutrient concentration and plant biomass. Plant and Soil. 2009; **316**:1-12. DOI: 10.1007/s11104-008-9753-7

[39] Cipriani HN, Dias LE, Costa MD, Campos NV, Azevedo AA, Gomes RJ, et al. Arsenic toxicity in *Acacia mangium* willd and *mimosa Caesalpiniaefolia* benth. seedlings. Revista Brasileira de Ciência do Solo. 2013;**37**(5):1423-1430. DOI: 10.1590/s0100-06832013000500031

[40] Ang LH, Tang LK, Ho WM, Hui TF, Theseira GW. Phytoremediation of Cd and Pb by four tropical timber species grown on an Ex-tin mine in peninsular Malaysia. International Journal of Environmental, Chemical, Ecological, Geological and Geophysical Engineering. 2010;**4**(2):70-74

[41] Sobariu DL, Fertu DIT, Diaconu M, Pavel LV, Hlihor R-M, Drăgoi EN, et al. Rhizobacteria and plant symbiosis in heavy metal uptake and its implications for soil bioremediation. New Biotechnology. 2017;**39**:125-134. DOI: 10.1016/j.nbt.2016.09.002

[42] Mahohi A, Raiesi F. Functionally dissimilar soil organisms improve growth and Pb/Zn uptake by Stachys inflata grown in a calcareous soil highly polluted with mining activities. Journal of Environmental Management. 2019; **247**:780-789. DOI: 10.1016/j.jenvman.2019.06.130

[43] Brown G, Edwards CA, Brussaard L. How earthworm affect plant growth: burrowing into the mechanisms. In: Edwards CA, editor. Earthworm Ecology. Boca Raton, USA: CRC Press; 2004, ch2. pp. 13-49. DOI: 10.1201/9781420039719

[44] Boughattas I, Hattab S, Alphonse V, Livet A, Giusti-Miller S, Boussetta H, et al. Use of earthworms *Eisenia andrei* on the bioremediation of contaminated area in north of Tunisia and microbial soil enzymes as bioindicator of change on heavy metals speciation. Journal of Soils and Sediments. 2018;**19**(1):296-309. DOI: 10.1007/s11368-018-2038-8

[45] Kaur P, Bali S, Sharma A, Vig AP, Bhardwaj R. Effect of earthworms on growth, photosynthetic efficiency and metal uptake in *Brassica juncea L.* plants grown in cadmium-polluted soils.

Environmental Science and Pollution Research. 2017;**24**(15):13452-13465. DOI: 10.1007/s11356-017-8947-z

[46] Kabas S, Saavedra-Mella F, Huynh T, Kopittke PM, Carter S, Huang L. Metal uptake and organic acid exudation of native Acacia species in mine tailings. Australian Journal of Botany. 2017;**65**:357-367. DOI: 10.1071/BT16189

[47] Vilela LAF, Teixeira AFS, Lourenço FMO, Souza MD. Symbiotic Microorganisms Enhance Antioxidant Defense in Plants Exposed to Metal/Metalloid-Contaminated Soils. In: Hasanuzzaman M, Nahar K, Fujita M, editors. Plants Under Metal and Metalloid Stress. Singapore: Springer; 2018. pp. 337-366. DOI: 10.1007/978-981-13-2242-6_13

[48] Hao X, Taghavi S, Xie P, Orbach MJ, Alwathnani HA, Rensing C, et al. Phytoremediation of heavy and transition metals aided by legume-rhizobia symbiosis. International Journal of Phytoremediation. 2014;**16**(2):179-202. DOI: 10.1080/15226514.2013.773273

[49] Bruneel O, Mghazli N, Sbabou L, Héry M, Casiot C, Filali-Maltouf A. Role of microorganisms in rehabilitation of mining sites, focus on Sub Saharan African countries. Journal of Geochemical Exploration. 2019;**205**:106327. DOI: 10.1016/j.gexplo.2019.06.009

[50] Bharti A, Agnihotri R, Maheshwari HS, Prakash A, Sharma MP. *Bradyrhizobia*-Mediated Drought Tolerance in Soybean and Mechanisms Involved. In: Choudhary D, Kumar M, Prasad R, Kumar V, editors. In Silico Approach for Sustainable Agriculture. Singapore: Springer; 2018. pp. 121-139. DOI: 10.1007/978-981-13-0347-0_7

[51] Egamberdieva D, Jabborova D, Wirth SJ, Alam P, Alyemeni MN, Ahmad P. Interactive Effects of Nutrients and *Bradyrhizobium japonicum* on the Growth and Root Architecture of Soybean (*Glycine max* L.). Frontier in Microbiology. 2018;**9**:1000. DOI: 10.3389/fmicb.2018.01000. PMID: 29875740; PMCID: PMC5974200

[52] Masciarelli O, Llanes A, Luna V. A new PGPR co-inoculated with *Bradyrhizobium japonicum* enhances soybean nodulation. Microbiological Research. 2014;**169**(7–8):609-615. DOI: 10.1016/j.micres.2013.10.001

[53] Sánchez-Pardo B, Zornoza P. Mitigation of Cu stress by legume–Rhizobium symbiosis in white lupin and soybean plants. Ecotoxicology and Environmental Safety. 2014;**102**:1-5. DOI: 10.1016/j.ecoenv.2014.01.016

[54] Alcántara-Martínez N, Figueroa-Martínez F, Rivera-Cabrera F, Gutiérrez-Sánchez G, Volke-Sepúlveda T. An endophytic strain of *Methylobacterium* sp. increases arsenate tolerance in *Acacia farnesiana* (L.) Willd: A proteomic approach. Science of the Total Environment. 2018;**625**:762-774. DOI: 10.1016/j.scitotenv.2017.12.314

[55] Wani PA, Khan MS, Zaidi A. Effect of metal tolerant plant growth promoting *Bradyrhizobium* sp. (vigna) on growth, symbiosis, seed yield and metal uptake by greengram plants. Chemosphere. 2007;**70**(1):36-45. DOI: 10.1016/j.chemosphere.2007.07.028

[56] Brígido C, Glick BR, Oliveira S. Survey of plant growth-promoting mechanisms in native portuguese chickpea *Mesorhizobium* isolates. Microbial Ecology. 2017;**73**:900-915. DOI: 10.1007/s00248-016-0891-9

[57] Baharlouei J, Pazira E, Solhi M. Evaluation of inoculation of plant growth-promoting Rhizobacteria on cadmium uptake by canola and barley. African Journal of Microbiology Research. 2011;**5**(14):1747-1754. DOI: 10.5897/AJMR10.625

[58] Castellanos SDE, Gigon A, Puga-Freitas R, Lavelle P, Velasquez E, Blouin M. Combined effects of earthworms and IAA-producing rhizobacteria on plant growth and development. Applied Soil Ecology. 2014;**80**:100-107. DOI: 10.1016/j.apsoil.2014.04.004

[59] Rangel WM, Thijs S, Janssen J, Oliveira LSM, Bonaldi DS, Ribeiro PRA, et al. Native rhizobia from Zn mining soil promote the growth of *Leucaena leucocephalaon* contaminated soil. International Journal of Phytoremediation. 2016;**19**(2):142-156. DOI: 10.1080/15226514.2016.1207600

[60] Jusselme MD, Poly F, Miambi E, Mora P, Blouin M, Pando A, et al. Effect of earthworms on plant *Lantana camara* Pb-uptake and on bacterial communities in root-adhering soil. Science of the Total Environment. 2012;**416**:200-207. DOI: 10.1016/j.scitotenv.2011.10.070

Chapter 10

Perspective Chapter: Uptake Capacity of Metals (Al, Cu, Pb, Sn, Zn) in Contaminated Water Metal Production Trade Village Dong Xam, Thai Binh, Vietnam by *Vetiveria zizanioides*

Nguyen Trung Minh, Seong-Taek Yun, Jang-Soon Kwon and Doan Thu Tra

Abstract

This chapter describes experiments, carried out under controlled environment conditions to investigate the uptake capacity of metals (Al, Cu, Pb, Sn, and Zn) by *Vetiveria zizanioides* to treat contaminated water from "metal production trade village Dong Xam, Thai Binh, Vietnam." The roots have a high hyperaccumulation capacity of Al, and it is much more than "reference plant" about 17- up to 30-folds, and the upper parts of shoots S2, and S3 are higher 1.2-fold. In vetiver plant the Cu concentration can be obtained up to 660 mg/kg in root, and 46.2 mg/kg in shoot, and it can withstand and be alive at 46 mg/L of contaminated solution. The lead translocation from root to shoot reached to about 41%. The tin is absorbed in the leaf chop with ratio: Root varied from 82% up to ~27% in the leaf chop. The zinc may be moved from roots and accumulated by the shoots of vetiver. The ratio shoot: root gets up to 46%. The study shows that vetiver had the high tolerance to trace metals Al, Cu, Pb, Sn, and Zn than other species plants. This plant has potential for usage in the phytoremediation of metals contaminated soil and wastewater from trade villages of Vietnam and other countries.

Keywords: uptake, metals Al, Cu, Pb, Sn, Zn, *Vetiveria zizanioides*, metal production trade village Dong Xam, Thai Binh, Vietnam

1. Introduction

There are heavy metal contaminations in the soil erosion from agricultural lands, urban wastes, and the products from rural, industrial, and mining industries that attracts worldwide concern, especially in developing countries [1, 2].

IntechOpen

Nowadays, in Vietnam, lots of trade villages (about thousand villages) are developing by many kinds of professions, and they have problems with wastewater and solid waste. Among waste matter, there are many types of metal contaminations.

The vetiver grass was first developed for soil and water conservation in farmlands. Morphological, physiological, and ecological characteristics of vetiver have a key role in the environmental protection. The vetiver root system can be reached up to 3–4 m in the first year. Vetiver can be tolerant to extreme climatic variation flood, prolonged drought, submergence, and extreme temperature. Vetiver can live in very harsh environments where surface temperature from −13°C exceeds 55°C, soil pH, from 3.0 to 10.5, high soil salinity, sodicity, acidity [2–4].

It seems that vetiver as other Panicoideae plant subfamily follows the same conjugation detoxification pathway, and vetiver is close to sorghum [5]. The transformation known to be positive for the environment, due to major metabolism of atrazine in vetiver grown in hydroponics was conjugation, mainly in leaves [6].

The vetiver grass was selected for wastewater treatment purpose from metal production trade village Dong Xam, Thai Binh, because of many reasons as at firstly, it can tolerate in wide range of pollution conditions [7–9], second, low-cost alternative mean to vegetate the heavy metal-contaminated area [3]. Vetiver is fast growth, and has strong root system and a long-lived perennial and can survive up to 50 years or more [10]; and vetiver can be produced 99 tons/ha/year (average dry matter yield) [11].

Many previous studies [2, 3, 6, 12–18] had reported the uptake capacity of some heavy metals by vetiver, but metals such as Al, Cu, and Sn have not been investigated completely, especially the pollution likes in "metal production trade village Dong Xam, Thai Binh" with numberless of heavy metal contaminations.

2. Materials and methods

2.1 Vetiver growth conditions

The soil materials were collected from five points in the study area, then sieved through a 2-mm mesh, and well mixed to obtain composite homogeneous samples. Seedling of vetiver was wrapped with the composite soils and irrigated with different chemical pollution regimes (**Figure 1a**).

The contents of Al, Cu, Pb, Sn, and Zn elements at soil in two pots (TB10 and TB6) are the same for vetiver cultivation, respectively, at 2.5, 55.6, 0.15, 7.7, and 24.4 mg (take out from wastewater of metal production trade village Dong Xam (**Table 1**)) and one pot (control) in the clean tap water. No fertilizer was applied during the entire growing period. Temperature in the laboratory growth chamber was 25 ± 2°C.

Vetiver plants were harvested after 36 days of growth in laboratory chamber by contaminated water TB10, TB6, and control water. The plant's height was 0.7 m (**Figure 1b**). First, the plants were rinsed three times with tap water and then two times with deionized water to remove all soil and other materials; afterward, it was dried in shade at room temperature for 5 days, and then at 80°C for 2 days in oven to constant weight. The plants were partitioned into five parts: three parts of shoots (S1—10 cm of shoot is from the meristematic region, S2—next 10 cm of shoot, S3—remaining part (about 20–40 cm) in the chop of shoot, meristematic region (M), and root (R)). The samples were sieved through a 2-mm mesh and well mixed (**Figure 2**).

a b

Figure 1.
(a) Vetiver land, and (b) it was grown in laboratory chamber by contaminated water for 36 days.

Elements	TB10		TB6	
	Mean, mg/L	**SD**	**Mean, mg/L**	**SD**
Al	1.242	0.002	2.070	0.003
Cu	27.821	0.0009	46.369	0.0015
Pb	0.075	0.0005	0.125	0.0008
Sn	3.861	0.001	6.435	0.001
Zn	12.225	0.0003	20.375	0.0005

Table 1.
Analytical results of contaminated solutions from two wastewaters (metal production trade village Dong Xam) before treatment by vetiver (mean ± SD).

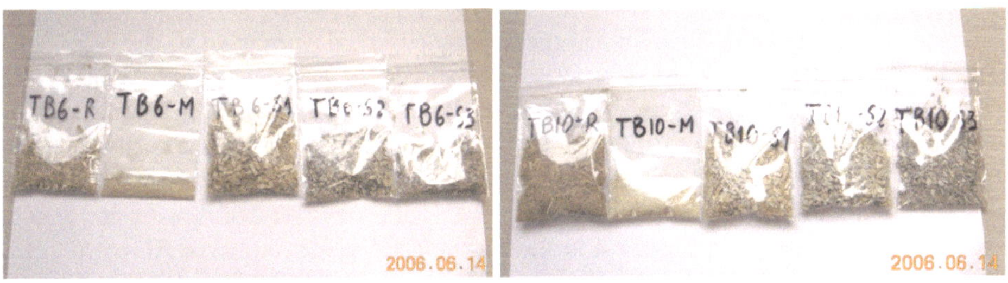

a b

Figure 2.
(a) Vetiver samples TB6 and (b) TB10 were sieved through a 2-mm mesh and mixed well.

2.2 Chemical analysis

Standard NIST 1568a (Rice Flour) and about 500 mg material from each part of vetiver were placed into 100-ml digesting Teflon bottles. The materials were digested at 5 ml 16 M HNO_3 and 1 ml 12 M $HClO_4$ (5:1, v/v) during 1 day in hotplate 180°C.

Element	Certificate, mg/kg		Found, mg/kg		Recovery (%)	
	Mean	SD	Mean	SD	Mean	SD
Cd	0.022	0.002	0.023	0.0006	104.8	2.6
Cu	2.400	0.3	2.176	0.087	90.7	3.6
Pb	<0.010		0.009	0.0005	91.5	5.0
Zn	19.400	0.5	20.301	0.819	104.6	4.2

Table 2.
Comparison of analytical results (mg/kg) for NIST 1568a (Rice flour).

After evaporation, the solutions were added 0.03 ml 18 M H_2SO_4 and kept at 180°C during 24 hours. The digested samples were brought to a volume 30 ml 2% HNO_3.

Table 1 shows the results of concentrations of Al, Cu, Pb, Sn, and Zn in the digesting solutions, and the standard deviation (SD) is calculated from three times analysis (n = 3). It was determined by ICP MS in Korea Basic Science Institute (KBSI).

A standard reference material NIST 1568a (Rice Flour) was used to verify the accuracy of metal determination by ICP-MS, and the recovery rates of Cu, Zn, Cd, and Pb elements were very high within 90.7 ÷ 104.8% ± 5.0% (**Table 2**). The analytical results are acceptable.

Chemical fingerprint: By the author [19], to overcome the problems of variety of data over scale, we use the type of data interpretation in the form of chemical fingerprints with normalization to "reference plant" for discussion of heavy metals Al, Cu, Pb, Sn, and Zn (**Figure 3**). The "reference plant" was set to zero (normalization), and the data of trace metals Al, Cu, Pb, Sn, and Zn concentrations of parts of vetiver will be given as deviations from the value of "reference plant."

3. Results and discussion

3.1 Aluminum (Al)

Follow [20]: The Al in the plants is controlling colloidal properties in the cell, possible activation of some dehydrogenases and oxydases. But the high availability of Al in nutrient soil is one of the limiting factors in the production of most field crops [21–23]. The physiological mechanisms of Al toxicity are still debate, but Al excess in plants is likely to interfere with cell division and with properties of protoplasm and cell walls [22]. The content of Al in plants varies greatly, depending on soil and plant factors.

Chemical fingerprint: In **Figure 4** is shown the relative deviation of Al from "reference plant." The concentration of Al in root materials is very high and much more than "reference plant" about 17- up to 30-folds (**Table 3**; **Figure 3**). The deviation in the lower parts (meristematic regions M and low parts of shoots S1) was less than zero, but upper parts of shoots S2 and S3 are higher and obtained at 120% (TB6-S2). It means that, in the shoots of vetiver, Al is concentrated in the leave top and the ratio of Al shoot: root is varied from 3 up to 8%.

The concentrations of Al in all parts of vetiver are increased by its increasing in contaminated water (**Tables 1** and **4**; **Figure 4**), and it was higher in the roots than in the shoots. The minimum concentrations are in the meristematic regions, because the amount of Al passively taken up by roots and then translocated to tops reflects the Al

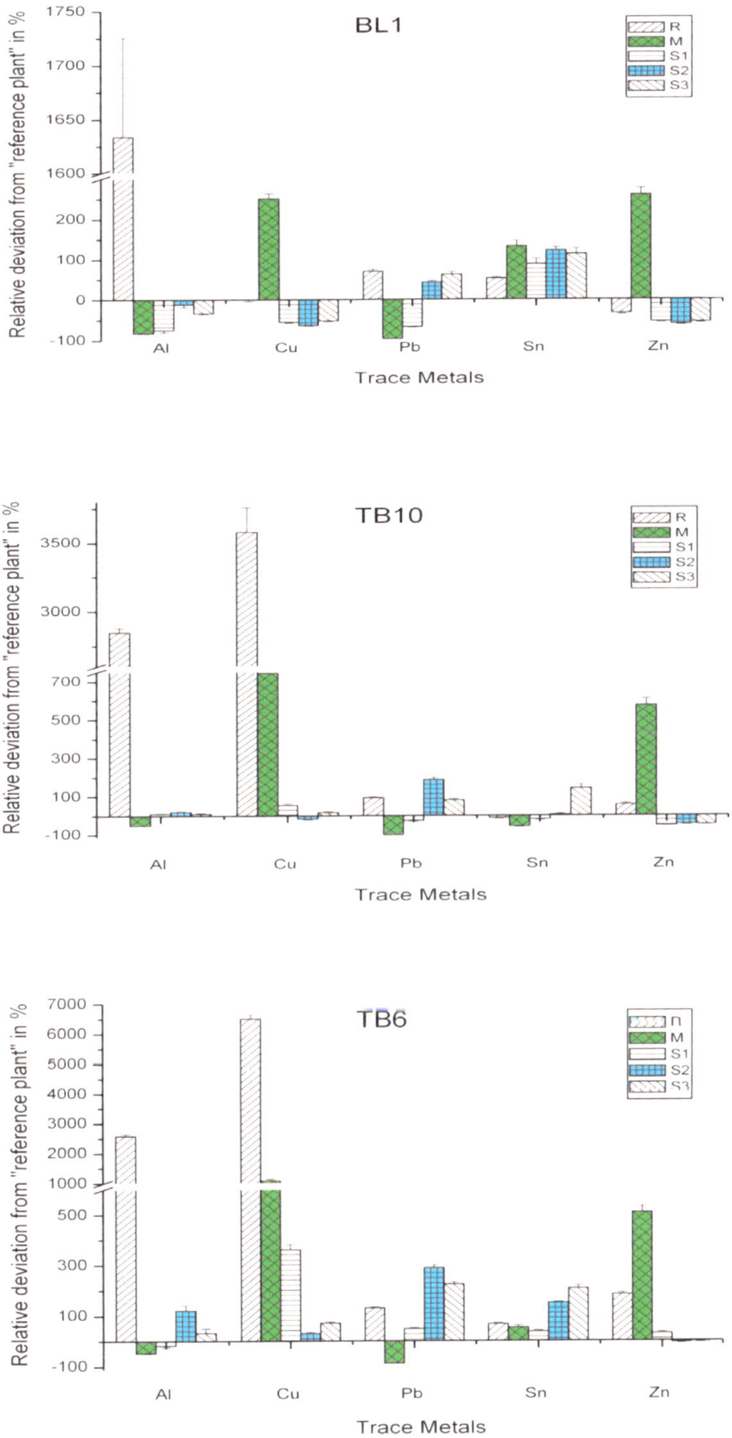

Figure 3.
Relative deviations of vetiver parts after normalization against "reference plant" [19].

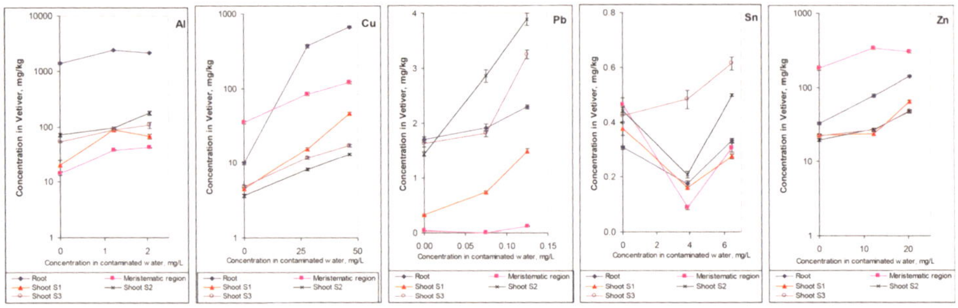

Figure 4.
Relationships between the concentrations of metals (Al, Cu, Pb, Sn, and Zn) in several parts of vetiver and those in contaminated water.

tolerance of plants, but the ability to accumulate Al in roots is not necessarily associated with Al tolerance [20].

3.2 Copper (Cu)

Copper had the major functions in plants as component of some enzymes role as catalyst [24], involved in oxidation, photosynthesis, protein, and carbohydrate metabolism, possibly in symbiotic N_2 fixation, and valence changes [20] (but it is toxic if concentration of Cu more than the plant needs). Cu is an essential element for the growth of most of aquatic organisms but is toxic at level as low as 10 mg/L [25]. In our experiment, vetiver plants were grown well in the solutions TB10 and TB6 with 27.821 and 46.369 mg/L Cu, respectively (**Table 1**).

In all parts of samples TB10 and TB6, copper concentration is higher in comparison with vetiver blank (BL1). In each vetiver sample, Cu is concentrate in root by the following order: R > M > S1 > S2, S3 (**Table 4**; **Figure 4b**) except blank BL1.

In the root tissue and the meristematic regions, Cu is almost entirely in complexed forms; it is most likely that the metal enters root cells in dissociated forms [20], and so it had strong capability to hold Cu, and Cu cannot be transported to shoots.

Chemical fingerprint: The Cu concentrations in "reference plant" are lower than in all vetiver parts, which were living in wastewater (except TB10-S2) (**Table 3**; **Figure 3**). The deviations with "reference plant" in the shoot oscillated from 16.7 (TB10-S3) to 361.5% (TB6-S1), in the meristematic region from 745 (TB10-M) to 1091% (TB6-M) and in the root from 3578 (TB10-R) up to 6507% (TB6-R). On contrary, in the root (−0.2%) and shoot (−52 ÷ −64%) of blank BL1, it is lower than zero (except meristematic region).

The trend of slope line is clearly in diagram "Cu concentration in Vetiver against Cu concentration in contaminated solution" (**Table 3**; **Figure 4b**): it is raised by increasing of Cu concentration in contaminated water. It seems that Cu concentration in vetiver is the function (in direct proportion) of its concentration in contaminated water. Cu concentrations in root (R), meristematic region, and shoots (S1, S2, S3) parts of vetiver are raised in proportion to its increasing in contaminated water. The Cu concentration increasing in root is faster than in meristematic region and in other parts M > S1 > S2, S3.

Cu has low mobility relative to other elements in vetiver, and most of this metal appears to remain in root and leaf tissues until it senesces [20].

Sample vetiver blank BL1										
Element	**R**		**M**		**S1**		**S2**		**S3**	
Al	1633.5	±91.9	−82.3	±1.3	−74.6	±6.1	−11.6	±6.5	−33.9	±2.2
Cu	−0.2	±4.5	250.9	±13.4	−55.4	±2.2	−63.9	±1.8	−52.3	±1.8
Pb	70.6	±4.8	−96.1	±0.2	−67.4	±1.2	43.4	±4.3	62.7	±5.9
Sn	53.2	±3.4	132.4	±13.1	88.4	±12.6	121.8	±8.4	112.9	±14.1
Zn	−33.6	±2.6	259.5	±16.4	−54.8	±2.2	−61.1	±1.7	−55.9	±1.6
Sample vetiver TB10										
Element	R		M		S1		S2		S3	
Al	2847.8	±32.9	−53.0	±1.5	10.4	±2.2	20.6	±3.0	10.2	±4.5
Cu	3578.3	±177.0	744.5	±44.9	53.9	±7.7	−18.1	±3.9	16.7	±4.7
Pb	91.9	±7.1	−100.0	±0.2	−26.4	±2.6	186.0	±11.0	80.9	±5.5
Sn	−12.5	±5.1	−56.8	±2.7	−20.2	±2.7	3.5	±5.9	142.2	±16.7
Zn	56.4	±8.0	573.9	±33.9	−52.7	±2.2	−46.0	±2.6	−46.7	±2.3
Sample vetiver TB6										
Element	R		M		S1		S2		S3	
Al	2585.4	±66.1	−47.9	±0.8	−16.7	±7.5	120.8	±21.0	32.7	±17.3
Cu	6506.7	±152.2	1091.0	±45.8	361.5	±21.8	30.5	±4.7	70.9	±5.8
Pb	130.3	±3.8	−88.3	±0.5	48.2	±4.2	288.5	±10.9	224.5	±8.1
Sn	66.5	±4.1	52.9	±7.8	37.1	±4.4	150.5	±2.6	207.0	±12.1
Zn	183.3	±7.6	507.6	±24.6	29.6	±6.2	−5.3	±3.9	−2.3	±3.3

Table 3.
Relative deviation concentration in parts of vetiver from "reference plant" (mean ± standard deviation) in %.

In the other plants, the excessive or toxic concentration of Cu is 20–100 mg/kg [20], but in vetiver plant it is much more, from 11 up to 660 mg/kg (**Table 4**).

The ratio of Cu in shoot: Root is low (4–7%) during living in the wastewater, and being higher (36–48%) in cleaning water that indicated the absorption capacity of vetiver root.

During the living in the difference concentrations of Cu in solution, the shoot of vetiver was uptake copper to the top. It seems to be raised by increasing of concentrations Cu in contaminated water (**Figure 4b**). For other plants, the level 10 mg/L of Cu in contaminated water is toxic, but vetiver can withstand and be alive at 46 mg/L.

The maximum Cu concentration in shoot of sample TB6 is 46.2, in meristematic region is 119.1, and in root is 660.7 mg/kg, which were much more than the previous results by the authors [3, 17, 26] (thresholds to shoot of Vetiver is 13–15, and root is 68 mg/kg).

In the contaminated water, there were both high Cu and Al contents, and its antagonism leads to reduction of Cu uptake by roots under high Al concentration [20].

3.3 Lead (Pb)

Pb is necessary for plant at the level of 2–6 μg/kg [27]. Pb received much attention as a major chemical pollutant of the environment and as the toxic element to plants [20].

Sample ID	Blank BL1 - Root		Blank BL1 - Meristematic region		Blank BL1 – Shoot S1		Blank BL1 – Shoot S2		Blank BL1 – Shoot S3	
Element	**Mean**	**SD**	**Mean**	**SD**	**Mean**	**SD**	**Mean**	**SD**	**Mean**	**SD**
Al	1386.78	73.52	14.142	1.029	20.289	4.843	70.735	5.222	52.912	1.724
Cu	9.978	0.448	35.089	1.337	4.460	0.220	3.614	0.180	4.770	0.183
Pb	1.706	0.048	0.039	0.002	0.326	0.012	1.434	0.043	1.627	0.059
Sn	0.306	0.007	0.465	0.026	0.377	0.025	0.444	0.017	0.426	0.028
Zn	33.188	1.301	179.735	8.191	22.612	1.077	19.463	0.842	22.060	0.801
Sample ID	TB10 – Root		TB10 - Meristematic region		TB10 - Shoot S1		TB10 - Shoot S2		TB10 - Shoot S3	
Element	Mean	SD	Mean	SD	Mean	SD	Mean	SD	Mean	SD
Al	2358.22	26.35	37.619	1.166	88.288	1.784	96.455	2.386	88.158	3.572
Cu	367.833	17.696	84.453	4.491	15.386	0.768	8.189	0.395	11.672	0.474
Pb	1.919	0.071	n.d.	n.d.	0.736	0.026	2.860	0.110	1.809	0.055
Sn	0.175	0.010	0.086	0.005	0.160	0.005	0.207	0.012	0.484	0.033
Zn	78.187	4.003	336.966	16.948	23.649	1.108	27.021	1.316	26.628	1.170
Sample ID	TB6 – Root		TB6 - Meristematic region		TB6 - Shoot S1		TB6 - Shoot S2		TB6 - Shoot S3	
Element	Mean	SD	Mean	SD	Mean	SD	Mean	SD	Mean	SD
Al	2148.32	52.91	41.668	0.604	66.628	6.035	176.675	16.775	106.164	13.811
Cu	660.674	15.220	119.105	4.578	46.151	2.177	13.053	0.471	17.095	0.583
Pb	2.303	0.038	0.117	0.005	1.482	0.042	3.885	0.109	3.245	0.081
Sn	0.333	0.008	0.306	0.016	0.274	0.009	0.501	0.005	0.614	0.024
Zn	141.641	3.777	303.817	12.303	64.808	3.086	47.334	1.971	48.860	1.669

n.d. = not detected.

Table 4.
Concentrations of trace metals in vetiver parts, (mean ± standard deviation), mg/kg.

Chemical fingerprint: Pb is concentrated in the roots of vetiver and deviation to compare with "reference plant" is 70.6 (BL1-R) up to 130% (TB6-R) (**Table 3**; **Figure 3**). But in the meristematic regions, the deviation is lower than zero and obtained −100% (TB10-R). Concentrate Pb in the shoots parts has the following order: (S2, S3) > S1, M, R, and it followed its concentrations in contaminated water and obtained fourfold more than "reference plant."

For other plants, the translocation of Pb from roots to tops is greatly limited, only 3% (Zimdahl R.L. 1975), but by our experiment for the vetiver, the translocation to shoot is obtained from 23 to 41%.

The trend of slope line is clearly in diagram "Pb concentration in Vetiver against Pb concentration in contaminated solution" (**Figure 4c**): It is raised very fast by increasing of concentration Pb in contaminated water.

The stimulating effect of Pb on Cd uptake by root may be an effect of the disturbance of the transmembrane transport of ions [20].

3.4 Tin (Sn)

Tin is very toxic to both higher plants and fungi [20].

Chemical fingerprint: The deviation of Sn to compare with "reference plant" in the low part of TB10 (R, M and S1) is lightly less than zero, but the upper parts (S2, S3) are higher and obtained 142%, and when the contaminated water raised (TB6), it is increased in all parts of vetiver and obtained to 207% (**Table 3**; **Figure 3**).

In the vetiver shoots TB10 and TB6: Concentrations of Sn are higher than in the root and meristematic region by the following order: S3, S2 > S1 > M, R (**Figure 4d**).

Not like to other plants, most of absorbed Sn remains in roots [28], the vetiver has the trend of uptake Sn, and it is accumulated in upper parts with ratio shoot: root varied from 82% (TB6-S1) to 277% (top of vetiver TB6-S3), and increased to the top by order S3/R > S2/R > S1/R.

3.5 Zinc (Zn)

The major functions of Zn in plants are: activates enzymes, regulates sugar consumption [24], and is involved in carbohydrate and protein metabolism [20].

As Kabata-Pendias Alina and Pendias Henryk suggest, soluble forms of Zn are available to vetiver and the uptake of Zn from soil to be linear with concentration in the contaminated water (**Figure 4e**).

Chemical fingerprint: The deviation of Zn concentration in meristematic regions is always higher than zero in comparison with the "reference plant," and it is obtained of 508 ÷ 574%, and root and shoot parts are obtained only lightly more than zero (**Table 3**; **Figure 3**).

Zn is concentrate much more in meristematic regions than in the roots. Roots and meristematic regions contain much more Zn than shoots, the ratio shoot: root obtains 30 up to 46%. It means Zn may be translocated from roots and accumulate by the shoots of vetiver. Vetiver has higher tolerance to Zn and Pb than other species [18]. The Zn-Pb antagonism adversely affects the translocation of each element from root to shoot [20].

4. Conclusions

In order to assess the uptake capacity of metals (Al, Cu, Pb, Sn, Zn) in contaminated water by *Vetiveria zizanioides* in laboratory condition, we have the conclusions as follows: Vetiver has higher tolerance to Al, Cu, Pb, Sn, and Zn than other species plants:

1. The roots are hyperaccumulated Al and much more than "reference plant" about 17- up to 30-folds, and the upper parts of shoots S2 and S3 are higher 1.2-folds.

2. In the other plants, the excessive or toxic concentration of Cu is 20–100 mg/kg, but in vetiver plant, it is much more and obtained up to 660 mg/kg in root, and 46.2 mg/kg in shoot, and it can withstand and be alive at 46 mg/L of contaminated water.

3. The translocation of Pb from root to shoot reached to 41%.

4. Sn is accumulated in upper parts with ratio shoot: In the root Sn, it varied from 82% up to 277% in the leave chop and increased to the leave chop by order S3/R > S2/R > S1/R.

5. Zn may be translocated from the roots and accumulated by the shoots of vetiver. The ratio shoot: the root obtains up to 46%.

The results of this study show that vetiver had the high tolerance to trace metals Al, Cu, Pb, Sn, and Zn in upper parts of shoot, and it can be used for wastewater treatment from "metal production trade village Dong Xam" and in many other trade villages of Vietnam and other countries.

Author details

Nguyen Trung Minh[1,2*], Seong-Taek Yun[2], Jang-Soon Kwon[2] and Doan Thu Tra[3]

1 Vietnam National Museum of Nature (VNMN), Vietnam Academy of Science and Technology, Hanoi, Vietnam

2 Department of Earth and Environmental Sciences, Korea University, Seoul, Republic of Korea

3 Institute of Geological Sciences, Vietnam Academy of Science and Technology, Hanoi, Vietnam

*Address all correspondence to: ntminh66@gmail.com

References

[1] Mejare M, Bulow L. Metal-binding proteins and peptides in bioremediation and phytoremediation of heavy metals. Trends in Biotechnology. 2001;**19**:67-73

[2] Tordoff GM, Baker AJM, Willis AJ. Current approaches to the revegetation and reclamation of metalliferous mine wastes. Chemosphere. 2000;**41**:219-228

[3] Truong PN, Baker D. Vetiver Grass System for Environmental Protection. Bangkok, Thailand: Office of the Royal Development Projects Board; 1998

[4] Truong PN. Vetiver grass for land rehabilitation. In: Proceedings of the First International Vetiver Conferences. Thailand; 1996. pp. 49-56

[5] Jensen K, Stephenson G, Hunt L. Detoxification of atrazine in three Gramineae subfamilies. Weed Science. 1977;**25**:212-220

[6] Sylvie M, Muriel R, Patrick R, Jean-Paul S. Conjugation of atrazine in vetiver (Chrysopogon zizanioides Nash) grown in hydroponics. Environmental and Experimental Botany. 2006;**56**:205-215

[7] Becker H. Hedging against erosion. Agricultural Research. 1992;**12**:8-10

[8] Grimshaw RG. A review of existing soil conservation technologies, and a proposed method of soil conservation using contour farming practices backed by vetiver grass hedge barriers. In: Proc. Vetiver Grass Seminar at the Int. Agric. Centre. Wageningen, the Netherlands; 1989

[9] Steven TS, Paul RA, Ricarda NK. Aquaculture sludge removal and stabilization within created wetlands. Aquacultural Engineering. 1999;**19**: 81-92

[10] Veldkamp JF. A revision of *Chrysopogon* Trin., including *Vetiveria* Bory (Poaceae) in Thailand and Malesia with notes on some other species from Africa and Australia. Austrobaileya. 1999;**5**:522-523

[11] Zhang J. Benefit and application future of sandy soils on windy Pingtan island. In: Vetiver Research and Development. China: Agricultural Science and Technology Press; 1998. pp. 179-191

[12] Chiu KK, Ye ZH, Wong MH. Enhanced uptake of As, Zn, and Cu by *Vetiveria zizanioides* and Zea mays using chelating agents. Chemosphere. 2005;**60**: 1365-1375

[13] Chiu KK, Ye ZH, Wong MH. Growth of *Vetiveria zizanioides* and *Phragmities australis* on Pb/Zn and Cu mine tailings amended with manure compost and sewage sludge: A greenhouse study. Bioresource Technology. 2006;**97**: 158-170

[14] Lai H-Y, Chen Z-S. Effects of EDTA on solubility of cadmium, zinc, and lead and their uptake by rainbow pink and vetiver grass. Chemosphere. 2004;**55**: 421-430

[15] Wilde EW, Brigmon RL, Dunn DL, Heitkamp MA, Dagnan DC. Phytoextraction of lead from firing range soil by vetiver grass. Chemosphere. 2005;**61**:1451-1457

[16] Xia HP. Ecological rehabilitation and phytoremediation with four grasses in oil shale mined land. Chemosphere. 2004; **54**:345-353

[17] Yahua C, Zhenguo S, Xiangdong L. The use of vetiver grass (*Vetiveria*

zizanioides) in the phytoremediation of soils contaminated with heavy metals. Applied Geochemistry. 2004;**19**: 1553-1565

[18] Yang B, Shu WS, Ye ZH, Lan CY, Wong MH. Growth and metal accumulation in vetiver and two Sesbania species on lead/zinc mine tailings. Chemosphere. 2003;**52**: 1593-1600

[19] Markert B. Establishing of "reference plant" for inorganic characterization of different plant species by chemical fingerprinting. Water, Air, and Soil Pollution. 1992;**64**:533-538

[20] Kabata-Pendias A, Pendias H. Trace Elements in Soils and Plants. 3rd ed. Boca Raton, Florida: CRC Press, Inc; 2001. p. 413

[21] Baker DE. Acid soils. In: Wright J, editor. Proceedings of Workshop on Plant Adaptation to Mineral Stress in Problem Soils. Vol. 4. Ithaca: Cornell University; 1976. p. 127

[22] Foy CD, Chaney RL, White MC. The physiology of metal toxicity in plants. Annual Review of Physiology. 1978;**29**: 511

[23] Frank R, Stonefield KI, Suda P. Metals in agricultural soils of Ontario. Canadian Journal of Soil Science. 1979; **59**:99

[24] Schlesinger WH. Treatise on geochemistry. In: Holland HD, Turekian KK, editors. Biogeochemistry. First ed. Oxford: Elsevier-Pergamon; 2004

[25] Leckie JO, Davis JA. Aqueous environmental chemistry of copper. In: Nriagu JO, editor. Copper in Environment. New York: Wiley; 1979. pp. 90-121

[26] Truong PN, Hart B. Vetiver System for Wastewater Treatment. Bangkok, Thailand: Office of the Royal Development Projects Board; 2001

[27] Broyer TC, Johnson CN, Paull RE. Some aspects of lead in plant nutrition. Plant and Soil. 1972;**36**:301

[28] Rommey EM, Wallace A, Alexander GV. Response of bush bean and barley to tin applied to soil and to solution culture. Plant and Soil. 1975; **42**:585

Chapter 11

Heavy Metals in Indonesian Paddy Soils

Dedik Budianta, Adipati Napoleon and Nanthi Bolan

Abstract

Long-term cultivation of paddy soils has resulted in Pb and Cd accumulation that exceeds the WHO tolerance levels of 2 mg kg^{-1} and 0.24 mg kg^{-1} in food. In Musi Rawas, South Sumatra, Indonesia, the paddy soils with the greatest levels of Pb and Cd were those that had been intensively farmed for 80 years, reaching the concentrations of 20.56 mg kg^{-1} Pb and 0.72 mg kg^{-1} Cd for soil, and 3.11 mg kg^{-1} Pb and 0.29 mg kg^{-1} Cd for rice. The lowest concentrations were obtained with 20 years of cultivation at 17.82 mg kg^{-1} and 0.26 mg kg^{-1}, for Pb and Cd in soils, respectively. The Pb content in the paddy fields in Pati, Central Java, ranged from 0.23 to 2.55 mg kg^{-1}, while the Pb content in the lowland watershed of Solo Hilir ranged from 0.20 to 2.94 mg kg^{-1}. The highest concentration of Pb and Cd in rice was found at 80 years old in paddy soils with the value of 3.11 mg kg^{-1} and 0.29 mg kg^{-1}, respectively. The lowest concentrations were found at 20 years old of soils with a value of 2.35 mg kg^{-1} Pb and 0.15 mg kg^{-1} Cd, respectively.

Keywords: cadmium, intensive farming, lead, paddy soil, P fertilizer, rice intensification

1. Introduction

Rice is a staple food that globally provides calories to more than 3.5 billion people. It has contributed almost 19% of global human per capita energy and 13% of per capita protein [1]. Paddy soils used for rice growth are contained by embankments, called galengan in Indonesia, or canal to hold water. Indonesian farmers have per capita paddy fields of only <0.5 ha, which decreases over time due to population growth and conversion to non-agricultural activities. The total area of rice fields in Indonesia is around 7,483,948 ha [2]. The average production is around 7–8 tons ha^{-1} when the soil is relatively fertile. However, when the soil is less fertile, rice production is very low, below 4 tons ha^{-1}. The fertility of paddy soils has decreased, as indicated by the decreasing availability of macro- and micro-nutrients, low organic matter content, and slightly low pH (**Table 1**) [3, 4].

The fertility of paddy fields continuously decreases with the time of land use due to harvest and irrigation.

According to **Table 2**, paddy soils in Musi Rawas, South Sumatra, have a pH ranging from 5.40 to 5.56 with a low organic C content of 1.75–1.85%, moderate to low soil CEC ranging from 15.31 to 19.58 cmol(+)kg^{-1}, moderately available P between 14.10 and 20.80 mg kg^{-1}, medium K-exchangeable of 0.58 cmol(+)kg^{-1}, exchangeable Na between 0.33 and 0.70 cmol(+)kg^{-1}, exchangeable Ca ranging from 2.10 to 6.48

No.	Soil pH	Org-C	N	C/N	CEC
		%			Cmol(+)kg^{-1}
1	6.14	2.58	0.13	19.85	5.41
2	6.36	2.37	0.13	18.23	4.97
3	6.72	1.19	0.12	9.9	6.07
4	5.71	0.96	0.11	8.7	6.61
5	5.85	1.52	0.15	10.1	5.42
6	6.35	0.89	0.10	8.9	7.26
Mean	6.19	1.58	0.12	12.61	5.96
Criteria	slightly acidic	low	low		low

Table 1.
Some characteristics of paddy soils in Sidoarjo (East Java) near the industrial area [3].

Soil properties	Age of soil cultivation (year)				Unit
	20	40	60	80	
pH (H_2O)	5.56 sa	5.52 sa	5.44 a	5.40 a	
C-organic	1.85 l	1.80 l	1.75 l	1.80 l	%
P-available	14.10 l	14.50 l	16.40 m	20.80 m	mg kg^{-1}
K-exchangeable	0.58 m	0.58 m	0.58 m	0.58 m	cmol(+)kg^{-1}
Na-exchangeable	0.44 m	0.44 m	0.33 l	0.70 m	cmol(+)kg^{-1}
Ca-exchangeable	2.10 l	2.50 l	6.08 m	6.48 m	cmol (+)kg^{-1}
Mg-exchangeable	0.57 l	0.35 vl	0.68 l	0.50 l	cmol (+)kg^{-1}
CEC	17.40 m	16.31 m	19.58 m	15.31 l	cmol (+)kg^{-1}

Note: a = acidic, m = moderate, sa = slightly acidic, vl = very low and l = low.

Table 2.
Characteristics of soils based on the age of use of paddy soils [5].

cmol(+)kg^{-1}, and low exchangeable Mg of 0.35–0.68 cmol(+)kg^{-1}. The low organic C content of 1.75 to 1.85% can increase the solubility of Pb and Cd in paddy soil while increasing the uptake by plant roots. Organic matter is vital as a regulating agent for heavy metal mobility in the soil [6]. Furthermore, Pb and Cd can form complex and chelate compounds with organic materials [7]. The complex form is a reaction between metal ions and ligands through electron pairs [8]. Paddy soils have low pH ranging from 5.56 to 5.4. The high soil acidity or low pH can increase the solubility of Pb and Cd in the soil with the uptake by plant roots [9]. Soil acidity is an essential factor that determines metal transformation and controls the chemical properties of Pb and Cd and other processes in the soil. The decrease in pH increases the availability of heavy metals except for Mo and Se. At low pH, the availability of Pb and Cd increases, and the more acidic the soils, the greater the heavy metal affects the rice [1, 10]. To increase the fertility of paddy soils, farmers intensively apply inorganic fertilizers such as urea, SP-36/TSP, and KCl. These inorganic fertilizers are essential to provide adequate nutrients for crop growth and ensure successful harvests [11]. This is supported by the data in **Table 3**, which indicate that the average lowland rice

No	Subdistrict	Fertilizer (kg ha^{-1})			
		Urea	SP36/TSP	KCl	NPK
1	Tugumulyo	166.37	144.68	67.50	123.70
2	Megang Sakti	189.65	156.25	74.37	141.66
3	Purwodadi	181.94	162.85	76.84	131.57
4	Muara Beliti	191.27	156.90	64.54	133.22
5	Sumber Harta	192.85	130.58	69.23	131.42
	Total	922.20	751.26	352.48	661.57
	Mean	184.44	150.26	70.49	132.31

Table 3.
Average fertilizer usage in five sub-districts of Musi Rawas Regency, Indonesia [5].

farmer uses around 150.26 kg ha^{-1} P fertilizer in each growing season, exceeding the recommended dose of 100 kg ha^{-1} [5].

Paddy soil is not a typical soil classification term but indicates how to manage various soil types for rice cultivation. There are four paddy soil ecosystems: (a) flood-prone rice ecosystem, characterized by a flat to slightly wavy or basin surface; it is flooded due to high tides for more than 10 consecutive days as deep as 50–300 cm during plant growth; (b) aerobic to anaerobic and rice cultivation is carried out by transferring or spreading seeds on dry plowed soil; (c) rainfed lowland rice ecosystem, characterized by a flat to the slightly wavy land surface, bordered by bunds, and inundated due to discontinuous tides with varying depths and periods; and (d) aerobic–anaerobic soil alternating with varying frequency and period, where rice planting is carried out by transferring seeds to silted soil [12]. Intensive management of paddy soils in the long term can reduce soil productivity and environmental quality. High inputs of agrochemicals can deplete nutrients in the soil and cause negative impacts in the form of increased residues of materials. Additionally, consumer demands for food or agricultural products that are safe and hygienic, have a high nutritional value, and are free of contamination are a public concern for the quality of the environment and human health [13]. Furthermore, [13] reported that around 21–40% of paddy soils in the Pantura of West Java were contaminated with these two types of heavy metals; even 4–7% of them were contaminated in the heavy metals category, which was Pb > 1.0 mg kg^{-1} and Cd > 0.24 mg kg^{-1}.

2. Heavy metals

Various sources and causes of contamination of paddy fields that can lead to soil degradation include agrochemicals, industrial waste, mining activities, and household waste. The two sources of heavy metals are natural and anthropogenic [1]. The use of synthetic fertilizers (inorganic fertilizers) and industrial activities play an important role as a source of pollution in rice fields [14]. There are many reports of contamination of rice fields, especially areas adjacent to factories [15–17]. The amount of waste generated from industrial processes causes water sources to be polluted. Furthermore, materials consisting of toxic compounds can settle in the rice soil. This process is repeated over time, accumulating these materials and heavy metals in the soil. Therefore, there will be undesirable changes in the physical, chemical, and biological properties of the soil. Productivity decreases with the ability to support plant growth [18]. Heavy metal contents in agricultural soils can directly affect human health by consuming crops grown in contaminated soils [17]. These metals are nonessential elements but can accumulate in plants and adversely affect human health [19]. Contaminated soil adversely affects the whole ecosystem when these toxic metals migrate into groundwater or are taken up by plants, which may threaten ecosystems [20]. In general, the metals are accumulated mostly in the root compared to the stem, leaf, and grain [1]. The occurrence of these metals in paddy field soils ranks in the order Mn > Zn > Pb > Cr > Cu > Cd [19]. Heavy metals are potentially toxic to crop plants, animals, and humans when contaminated soils are used for crop production [21]. Environmental contamination of the biosphere due to intensive agricultural and other anthropogenic activities poses severe problems for the safe use of agricultural land [22]. Heavy metals such as Cd and Pb are of primary concern in soil and food contamination because of their toxicity, particularly in the rice cropping system [23]. These toxic elements accumulate in the soils, contaminating the food chain, endangering the ecosystem's safety, and causing soil degradation.

Degraded soil will have properties that do not support rice growth. It will lose the topsoil or arable layer, lose nutrients needed by rice plants, and result in reduced levels of organic carbon. In addition to these observable characteristics of degraded soils, it can also be distinguished by plants that typically do not thrive in such conditions. The performance of plants is reduced when planted in soil with degraded physical, chemical, and biological qualities. The parameters used to evaluate the level of soil degradation are decreasing base saturation, available nutrients including N, P, K and trace elements, bulk density, soil permeability, and organic carbon [24].

Soil properties influence rice growth and development. The characteristics supporting plant growth should be maintained, one of which is soil conservation measures to prevent chemical damage/degradation. Degraded soil can also lose the top layer, impacting the loss of nutrients needed by plants, changes in soil structure, and reduced levels of organic carbon. The organic carbon has a major role in improving the physical, chemical, and biological properties of the soil [25]. It can also be identified by using plants with poor growth performance. In this regard, the plant can be used as an indicator of soil degradation. Many definitions of soil degradation have been reported, showing a decrease in soil chemical properties compared to non-degraded soil. Land degradation results from one or more processes that decrease the actual or potential ability to produce food and fiber and provide ecosystem services. This definition shows a general understanding of agriculture's broad scope [26]. Land or chemical degradation is often associated with a use that does not follow the aspects of the balance of inputs and outputs. Inputs are related to soil improvement or fertilization in cultivation activities. In contrast, the output is associated with plant nutrient uptake and the possibility of leaching through erosion mechanisms. The phenomenon of land degradation is found in areas of land that promote agricultural activities. Land degradation can be indicated by symptoms of poor plant growth or the growth of weeds on the soil. The marginalization will continue with low inputs for farming and dry land management technology, which ultimately causes physical and chemical degradation. On sloping land, land degradation will occur quickly due to erosion, which reduces the quality of the physical and chemical properties of the soil. Consequently, the soil will be damaged or degraded due to acidification, accumulation of salts (salinization), and contamination of heavy metals, organic compounds, and xenobiotics such as pesticides or oil spills.

3. Characteristics of Pb and Cd

3.1 Lead (Pb)

Lead (Pb) is accumulated in plant organs, namely, leaves, stems, roots, and tubers (shallots), and the transfer depends on the soil composition and pH. High Pb concentrations (100–1000 mg kg^{-1}) have a toxic effect on photosynthesis and growth [27]. Pb is one of the nonessential heavy metals that are toxic to living organisms. It causes stunted growth, irritates the eyes, and contributes to lung [28] and kidney [29] damage. The highest accumulation in roots was proven by [30] through a study of Pb in kale (*Brassica oleracea* var. *sabellica*). In the 6-week-old kale plant, Pb concentration in the roots reached about 3360 mg kg^{-1}, and in other parts of the plant, it reached 2090 mg kg^{-1}. In 3-week-old kale, the Pb content in the roots was 1.860 mg kg^{-1} in the sample but 1.130 mg kg^{-1} in other parts. These data indicate that most Pb in water spinach is accumulated in the roots.

The largest Pb pollution comes from burning gasoline, which produces PbBrCl and $PbBrCl_2PbO$. The pollution can come from Pb components in dissolved air or water, such as $PbCO_3$ [31]. According to [32], heavy metals in the media are rapidly absorbed by plants at very low concentrations. The mechanism of absorption and accumulation can be divided into three continuous processes: (a) Absorption by roots: metals should be brought into the solution around the roots (rhizosphere) in several ways to be absorbed. Water-soluble compounds are usually taken up by the roots with water, while the surface absorbs hydrophobic compounds. (b) Translocation of metals from roots to other plant parts: After penetrating the root endodermis, metal or other foreign compounds follow the transpiration flow through the transport tissue (xylem and phloem) to other parts. (c) Metal localization in cells and tissues: This aims to keep metals from inhibiting plant metabolism. Plants have detoxification mechanisms in certain organs, such as roots, to prevent metal poisoning of cells. Metals in the root cells are transported to other plant parts through the xylem and phloem network when translocation occurs in the plant body. At low concentrations, heavy metals do not affect plant growth but cause damage to the soil, water, and plant at high concentrations.

Satpathy et al. [33] argued that Pb originating from air/atmosphere pollution is in the form of dust particles, which will stay on the plant's surface. Clouds and rain can cause Pb to be dissolved and enter the plant through the stomata, which can cause damage and contaminate food. Air pollution by Pb mainly comes from exhaust fumes from motor vehicles, and this metal is the remnant of combustion between the fuel and the vehicle engine. The presence of Pb in motor vehicle fuel functions as an anti-knock agent. The Pb element is released into the air through the exhaust of the vehicle's gasoline. Some will form particulates in the free air with other elements, while others will stick and be absorbed by the leaves of plants along the way. Soil contamination by Pb is more extensive than other heavy metals. This is because the largest contribution is from anthropogenic sources. The research results [34] showed that the Mn, Co, Cr, and Ni on the soil surface come from lithogenic and anthropogenic sources. These results indicated a significant need for developing pollution prevention and reduction strategies for heavy metal pollution. Accumulation of heavy metals can degrade soil quality, reduce crop yields and agricultural product quality, and negatively impact humans, animals, and the ecosystem. The solution can be achieved by identifying the source and measuring the concentration of heavy metals and the spatial variability in the soil. The results revealed could be used to determine the increase in Cd and Pb concentrations [35].

3.2 Cadmium (Cd)

Soil Cd in igneous, metamorphic, and sedimentary rocks is 0.100–0.300, 0.100–1.00, and 0.300–11 mg kg^{-1}. In general, the Cd content in the soil from the weathering process of rocks is 1.00 mg kg^{-1} or lower. The elements Cd and Zn have almost similar chemical properties, and only their function in the plant body is different. Cd levels in plant tissues range from 0.100 to 1.00 mg kg^{-1}. Excessive Cd accumulation can occur from other materials, with a detrimental effect on plant growth. This is because it breaks down nitrate absorption and inhibits the activity of the enzyme nitrate reductase. The critical limit of Cd in plants is 5–30 mg kg^{-1} [36], and the content in the 0–20 cm layer, on average, is close to 0.5 mg kg^{-1}, which is the critical limit concentration of the metal [13]. Cadmium in the soil is an anthropogenic byproduct of fertilizer and garbage dumps. Most of the soil's Cd is affected by pH, organic materials, metallic oxides, clay, and organic and inorganic substances [28]. The average level of natural Cd in the earth's crust is 0.1–0.5 mg kg^{-1}.

The Cd content is influenced by the reaction of the soil and fractions capable of binding the ions. Due to the rise in the hydrolysis process, the adsorption complex, and the charge of the soil colloid, Cd concentration in soil solution reduces with increasing pH. Sarwar et al. [37] stated that there was a reduction in root and shoot length of about 45 and 35% in maize plants grown on media containing 28.1 and 11.2 mg kg^{-1} Cd(II) ions, respectively, at the age of 18 days. The contribution from atmospheric deposits occurs in industrial areas that use coal and oil as fuel. Cd is added to the soil through phosphate fertilizers, manure, incinerator waste (furnace), and sewage sludge [23, 36]. In addition, the increase in Cd can occur through phosphate fertilizers, whose levels vary greatly depending on the type of phosphorite as an industrial material for phosphate fertilizers [38]. Cadmium has chemical properties similar to those of Zn, especially in the process of absorption by plants and soil. However, Cd is more toxic, which can interfere with enzyme activity. Excessive levels of Cd in food can damage kidney function, interfere with Ca and P metabolism, and cause bone disease [39].

4. Sources and causes of Pb and Cd pollution in agriculture

4.1 Phosphate fertilizer

P fertilizer is regularly applied to the soil in intensive farming systems to increase plant growth. The compounds used contain heavy metals Pb and Cd [14, 40]. Triple super phosphate (TSP) fertilizer supplied excessively in rice fields in the long term will be accumulated and cause pollution because of heavy metals. These metals are also present in natural phosphate rock used as the raw material for the manufacture of fertilizer P [14, 41]. Pb and Cd are known to have no physiological activity/function, and applying certain phosphate fertilizers adds Cd and other potentially toxic elements to the soil, such as F, Hg, and Pb [42]. Phosphate fertilizers can significantly contribute to hazardous trace elements such as arsenic (As), Cd, and Pb in croplands. These trace elements have the potential to accumulate in soils and be transferred through the food chain [11]. Various inorganic fertilizers and those derived from phosphate rock contain heavy metals (**Table 4**). The results of different phosphate fertilizers showed the presence of P_2O_5; secondary Ca and Mg; microelements Fe, Mn, Cu, and Zn; and heavy metals Cd, Cr, Pb, Cu, and Hg in varying amounts, namely, Cd (0.1–170 mg kg^{-1}), Cr (66–245 mg kg^{-1}), Pb (40–2000 mg kg^{-1}), and Cu

Phosphate rock (PR)	Cd	Cr	Pb
		mg kg^{-1}	
PR Christmast	38	-	60
PR Tunisia	76	-	42
PR Marko	57	-	113
PR Jordan	5	344	im
PR China Huinan	3	-	im
PR Ciamis	28	20	im
PR Sukabumi	65	-	65
SP-36	11	4	im

im = immeasurable.

Table 4.
Heavy metals in various types of natural phosphate rock and SP-36 [43].

(1–300 mg kg$^{-1)}$ [44]. Long-term P fertilizer applications are likely sources of heavy metals (Pb and Cd) in agricultural soils and crops [45, 46].

From the analysis of P fertilizer, TSP fertilizer contains 120.60 Pb mg kg^{-1} and 4.90 mg kg^{-1} Cd, while single superphosphate (SP)-36 contains 5.3 mg kg^{-1} Pb and 10.43 mg kg^{-1} Cd. Rai et al. [47] explained that using P fertilizer could cause the soil accumulation of Pb and Cd. Furthermore, it contains heavy metals Pb and Cd from the raw material for making P fertilizer. Natural phosphate rock has various associated elements such as Pb and Cd in high enough quantities. Meanwhile, the elements can be dissolved in soil solution, adsorbed by organic and inorganic colloidal surfaces, firmly bound in soil minerals, deposited by compounds in the soil, and contained in living materials.

4.2 Pesticide

Generally, pesticides are widely used in agriculture and horticulture, containing heavy metals. For example, about 10% of pesticides in England contain compounds of Cu, Hg, Mn, or Zn. Fungicides are pesticides containing Cu, such as a mixture of Bordeaux (copper sulfate) and copper oxychloride. Lead arsenate has been used in orchards for many years to control several parasitic insects. Arsenic-containing compounds are also extensively used to control cow lice and banana pests in New Zealand and Australia. Wood preservation using formulations of Cu, Cr, and As pollute the air due to excessive concentrations of heavy metals [27, 48]. In Indonesia, the use of pesticides on vegetable crops is very intensive, especially on cash crops with high economic value. Based on research, 30–50% of the total production cost is used for pesticides [49]. Intensive use can increase soil and plant residues and even enter the bodies of animals, fish, or other aquatic biotas. Pesticides with a long half-life of degradation can harm the health of humans that consume products containing these residues.

Organic waste (biosolid) is a solid product produced through a wastewater treatment process and can be recycled. In the United States, it is estimated that more than 50% of the approximately 5.6 million tons of dry waste is used or distributed annually on land. Biosolids are applied to agricultural land in every region of the state. In European society, more than 30% of this waste is used as fertilizer. In Australia, 175,000 tons of dry biosolids are produced annually by local governments. Other biosolid materials are sawdust, rice/corn straw, or plant residues [50]. Heavy metal contamination of the soil may result from the continued application of biosolids. The most common heavy metals found in biosolids are Pb, Ni, Cd, Cr, Cu, and Zn, and the concentration depends on the intensity of industrial activity [51]. The application can be leached into the soil profile and potentially contaminate groundwater. It was shown that continued application of biosolid to several soils in New Zealand resulted in increased concentrations of Cd, Ni, and Zn in drainage water [52].

4.3 Industrial waste

It is estimated that 20 million hectares of agricultural land are irrigated with wastewater from industry or households. In several cities in Asia and Africa, studies show that agriculture based on wastewater irrigation accounts for 50% of the vegetable supply to urban areas. Farmers are less concerned about environmental benefits or harms, maximizing yields and profits. Even though the concentration of heavy metals in wastewater is relatively low, long-term irrigation can lead to its accumulation in the soil [53]. In Indonesia, heavy metal pollution has been identified in watersheds. According to [54], paddy soils in the Solo downstream watershed have been contaminated with Pb. Furthermore, paper mill effluent irrigation water has polluted the soil with Cu, Pb, and Cd. The Juwana sub-watershed indicates that heavy metal Cr contamination has the potential to contaminate agricultural land because the irrigation uses sugar industry waste. **Table 5** shows the industrial type producing some heavy metals.

Industry type	Hg	Pb	Cd	Cr	Cu	Zn	Ni	Al	Fe	Co	Mn
Plastic/resin	+	-	+	-	-	+	+	-	-	-	-
Pharmacy/cosmetic	+	+	-	-	-	+	-	-	-	-	-
Chlorine	+	-	-	-	-	-	-	-	-	-	-
Control/measurement tools	+	-	-	-	-	-	-	-	-	-	-
Electronics/electrical	+	-	-	-	-	-	-	-	-	-	-
Electroplating	-	-	-	+	+	+	+	+	-	-	-
Anti-rust paint	+	-	-	-	-	-	-	-	-	-	-
Textile	+	-	+	+	+	+	-	-	+	+	-
Ceramic	-	+	+	-	-	+	-	-	-	-	-
Skin tanning	+	-	-	+	-	-	+	-	-	-	-
Pulp and paper	+	+	-	+	-	-	-	-	-	-	-
Batteries	+	+	+	-	-	-	+	-	-	-	+
Soap/detergent	+	+	-	-	+	+	+	-	+	-	-
Metal/metal product	+	+	-	+	+	+	+	+	+	-	+
Pesticide	-	+	+	-	-	+	+	-	-	-	-

+ exist; - no exist.

Table 5.
Types of industries whose raw materials produce heavy metals [55].

5. Accumulation of Pb and Cd in soil and plant

5.1 Pb content in the soil

The results for Pb based on paddy soil use can be seen in **Tables 6** and 7. The data show that the highest and lowest soil Pb content was found in 80-year-old rice fields and control areas, namely, 20.56 and 1.20 mg kg^{-1}, respectively. Compared with 20-year-old paddy fields, Pb content in 80 years was increased by around 16-fold. The Pb content in Pati, Central Java, and the lowland watershed of Solo Hilir, Lamongan Regency, ranged from 0.23 to 2.55 mg kg^{-1} [54] and 0.20–2.94 mg kg^{-1} [59]. In the highland rice fields of Wonosobo, Central Java, Gowa, South Sulawesi, and India, Pb content was 9.32–14.82 mg kg^{-1} [60], 0.0151 mg kg^{-1} [61], and 5.3–19.8 mg kg^{-1} [19]. Meanwhile, [17] reported that Pb content in paddy soil in Tanzania ranged from 8.0 to 28.5 mg kg^{-1}. The heavy metal content of Pb in the paddy field of Semarang Regency of Central Java is still below the standard provisions of India (250–500 mg kg^{-1}) and Europe (300 mg kg^{-1}) [62]. Possible sources of Pb include pesticide spraying by farmers and car exhaust near the village road [63].

Soil cultivation (year)	Soil Pb (mg kg^{-1})	Percentage increase (%)
Control (0)	1.20	
20	17.82	1.385,00
40	19.48	1.523,33
60	20.46	1.605,00
80	20.56	1.613,33

Table 6.
Soil Pb content and percentage increase based on the age of use of paddy fields [5].

No	Heavy metals (mg kg^{-1})			
	Soil	Pb	Cd	Hg
1	1	1.48	0.48	0.93
2	2	1.44	0.42	1.02
3	3	1.3	0.24	0.7
4	4	1.57	0.35	1.04
5	5	1.57	0.17	0.66
6	6	1.26	0.14	0.58
	Threshold value	0.5[a]	3-8[b]	0.3[c]

[a][56]
[b][57]
[c][58].

Table 7.
Content of various heavy metals in rice fields in Sidoarjo, East Java [9].

The relationship between soil Pb and time follows a logarithmic pattern, where the content increases with land use. The mathematical equation for soil Pb content as a function of time is $Y = 11.88 + 2.02 \ln(x)$ and $R^2 = 0.956$, where Y = Pb content in soils (mg kg^{-1}) and x = age of land use (year).

5.2 Cd content in soil

The soil Cd analysis results based on paddy fields can be seen in **Table 8**. The table shows that the highest and lowest Cd content was found in 80- and 20-year-old rice fields, namely, 0.72 and 0.26 mg kg^{-1}. Compared with 20-year-old paddy fields, the highest soil Cd increase was in 80-year-old fields by 1340%. Pradika et al. [64] also reported that P fertilization could add Cd metal to agricultural land. This is because the raw material for making P fertilizer comes from phosphate rock, which naturally contains Cd metal. The concentrations in surface soils range from 0.06 to 1.10 mg kg^{-1} with an average of 0.41 mg kg^{-1} [65]. Satpathy et al. [19] reported that Cd content in Indian paddy soils ranged from 0.02 to 0.6 mg kg^{-1}.

The relationship between soil Cd content and time follows an exponential pattern, where the content increases with the age of land use. The mathematical equations as a function of time are $\ln(Y) = 0.17 + 0.02x$ and $R^2 = 0.913$, where Y = Cd content in soil (mg kg^{-1}) and x = age of land use (year). The presence of heavy metals Pb and Cd in the soil can be caused by the intensive use of P fertilizer and exceeding the recommended dose. Heavy metals Pb and Cd can increase in line with the age of paddy fields.

5.3 Pb content in rice

The results of the Pb analysis of rice based on the age of the paddy fields can be seen in **Table 9**. The data show that the highest and lowest Pb content was found in 80- and 20-year-old rice fields, namely, 3.11 and 2.35 mg kg^{-1}. The Pb content has exceeded the critical limit set by WHO, which is 2 mg kg^{-1}. The Pb content in grain from paddy fields in Pati, Central Java, and Wonosobo highlands ranged from 0.23 to 1.23 mg kg^{-1} [54] and 0.28–1.32 mg kg^{-1} [60]. In the present study, the concentration of Pb was found to be higher in roots than in shoots and grains [19].

The relationship between the Pb content of rice with time follows an exponential pattern. The mathematical equations as a function of time are $\ln(Y) = 2.06 + 0.005x$ and $R^2 = 0.928$, where Y = Pb in rice (mg kg^{-1}) and x = age of land cultivation (year).

Soil cultivation (year)	Soil Cd (mg kg^{-1})	Percentage increase (%)
Control (0)	0.05	
20	0.26	420
40	0.32	540
60	0.39	680
80	0.72	1,340

Table 8.
Soil Cd content and percentage increase based on the age of use of paddy fields [5].

Age of soil (year)	Pb in rice ($mg\ kg^{-1}$)	Percentage increase (%)
20	2.35	
40	2.40	2.12
60	2.86	21.70
80	3.11	32.34

Table 9.
Pb levels in rice based on the age of use of paddy fields [5].

5.4 Cd content in rice

Cadmium can affect enzyme activity in plants, leading to lower photosynthesis. Therefore, growth and development, including germination, root elongation, and leaf expansion, can be decreased [65–67]. The Cd analysis of rice based on the age of the paddy fields can be seen in **Table 10**. It shows that the highest and lowest Cd content is found in rice fields aged 80 and 20 years, namely, 0.29 and 0.15 mg kg^{-1}. The content of Cd in 80-year-old rice fields has exceeded the critical limit set by WHO, which is 0.24 mg kg^{-1}. In the shoots, the concentration was higher than that in roots and grains [19]. Jarvis et al. [68] reported that Cd was easily taken up by plants and transported to different parts, but it is nonessential and has no beneficial effects on plants.

The relationship between the Cd content of rice with time follows an exponential pattern. The mathematical equations as a function of time are Ln (Y) = 0.11 + 0.011 x and R^2 = 0.934, where Y = Cd in rice (mg kg^{-1}) and x = age of land use cultivation (year).

The high content of Pb and Cd was caused by the content in the soil and the low fertility of paddy fields. Lowland rice plants can absorb dissolved Pb and Cd in the soil and accumulate them in large quantities from their tissues. The elements contained in plant tissue in the vegetative phase will be translocated to fruit during vegetative growth. Aprilia and Purwani [69] added that heavy metals Pb and Cd are more easily absorbed by plant roots in the form of Pb^{2+} and Cd^{2+} ions. Munaf [70] explained that the accumulation of Pb and Cd in plant tissues derived from the absorption of roots could be influenced by several factors. These include the solubility properties of compounds in soil solution, pH, organic C content, cation exchange capacity (CEC), and clay content.

Age of soil (year)	Cd in rice ($mg\ kg^{-1}$)	Percentage increase (%)
20	0.15	
40	0.16	6.66
60	0.21	40.00
80	0.29	93.33

Table 10.
Cd content in rice based on the age of use of paddy fields [5].

6. Conclusion

Rice fields in Indonesia are used very intensively, and the fertility is negatively affected following the decrease in pH. Furthermore, the paddy fields in Indonesia have experienced heavy metal pollution, especially Pb and Cd, which is indicated by their accumulation in soil and rice. These heavy metals come from phosphate fertilizers and industrial waste. The proposed solution uses organic materials to immobilize the metal or form ligand bonds. Cadmium and lead input from phosphate fertilizers threaten the environment and human health due to soil contamination, crop absorption, and bioaccumulation in the food chain. A decrease in non-polluted recycled and mineral P fertilizer dependence could alleviate the Cd and Pb pollution of the paddy soils.

Author details

Dedik Budianta[1*], Adipati Napoleon[2] and Nanthi Bolan[3,4]

1 Faculty of Agriculture, Soil Chemistry and Fertility, Soil Science Department, Sriwijaya University, Indralaya, Indonesia

2 Faculty of Agriculture, Soil Biologi, Soil Science Department, Sriwijaya University, Indralaya, Indonesia

3 UWA School of Agriculture and Environment, The University of Western Australia, Perth, Australia

4 The UWA Institute of Agriculture, The University of Western Australia, Perth, Australia

*Address all correspondence to: dedik.budianta@unsri.ac.id; dedik_budianto@yahoo.com

References

[1] Zulkafflee NS, Redzuan NAM, Selamat J, Ismail MR, Praveena SM, Razis AFA. Evaluation of heavy metal contamination in paddy plants at the Northern Region of Malaysia using ICPMS and its risk assessment. Plants. 2021;**10**:3. DOI: 10.3390/plants10010003

[2] Susilowati SH, Maulana M. Farm business land size and farmer welfare: Smallholder's existence and Agrarian reform urgency. Analisis Kebijakan Pertanian. 2012;**10**(1):17-30 Indonesian edition

[3] Khasanah U, Mindari W, Suryaminarsih P. Assessment of heavy metals pollution on rice field in sidoarjo regency industrial area (in Indonesian). Jurnal Teknik Kimia. 2021;**15**(2):73-81

[4] Mutiara C, Soso J, Hutubessy JIB. Identifikasi Kesuburan Tanah Sawah Di Desa Lo'a Kecamatan So'a Kabupaten Ngada. Prosiding Seminar Nasional dalam Rangka Dies Natalis ke-45 UNS Tahun. Vol. 5. No. 1. 2021. pp. 544-552

[5] Merismon BD, Napoleon A, Hermansyah. The contamination of Pb and Cd in intensive paddy field at Musi Tawas Regency, South Sumatra, Indonesia. Pollution Research. 2017;**36**(4):1-6

[6] Bolan N, Kunhikrishnan A, Thangarajan R, Kumpiene J, Park J, Makino T, et al. Remediation of heavy metal(loid)s contaminated soils - to mobilize or to immobilize? Journal of Hazardous Materials. 2014;**266**:141-166. DOI: 10.1016/j.jhazmat.2013.12.018

[7] Budianta D, Napoleon A, Merismon HML. Save our soil from heavy metal (Pb and Cd) accumulation for rice growth. IOP Conf. Series: Earth and Environmental Science. 2022;**1005**:012001. DOI: 10.1088/1755-1315/1005/1/012001

[8] Strawn DG, Spark DL. Effects of soil organic matter on the kinetics of Pb (II) sorption and desorption in soil. Journal of Soil Science. 2000;**64**:144-146. DOI: 10.2136/sssaj2000.641144x

[9] Bolan N, Adriano DC, Curtin D. Soil acidification and liming interactions with nutrient and heavy metal transformation and bioavailability. Advances in Agronomy. 2001:78. DOI: 10.1016/S0065-2113(02)78006-1

[10] Taberima S. Peranan mikro organisme dalam mengurangi efek toksik pada tanah terkontaminasi logam berat. Program Pascasarjana IPB. 2004;**80**:35-61

[11] Jiao W, Chen W, Chang AC, Page AL. Environmental risks of trace elements associated with long-term phosphate fertilizers applications: A review. Environmental Pollution. 2012;**168**:44-53. DOI: 10.1016/j.envpol.2012.03.052

[12] Indrajati RP. Evaluasi Perubahan Kualitas Tanah Sawah Irigasi Teknis di Kawasan Industri Sub Das Bengawan Solo Daerah kabupaten Karanganyar. [Surakarta Skripsi]. Fakultas Pertanian. Universitas Sebelas Maret. 2008

[13] Wihardjaka A, Harsanti ES. Cadmium (cd) concentration in rice grains and soils in rainfed lowland affected by continuous fertilization. Ecolab. 2018;**12**(1):1-52

[14] Bolan NS, Adriano DC, Naidu R, de la Luz MM, Mahimairaja S. Phosphorus-trace element interactions in soil-plant systems. In: Sims JT, editor. Phosphorus: Agriculture and the Environment. American Society of Agronomy, Crop

Science Society of America, Soil Science Society of America, USA. 2005. pp. 317-352. DOI: 10.2134/agronmonogr46.c10

[15] Rogan N, Todor S, Dolenec M, Tasev G, Dolenec T. Heavy metal contamination of paddy soils and rice (Oryza sativa L.) from Kocani Field (Macedonia). Journal of Environmental Geochemical. 2009;**31**:439-541. DOI: 10.1007/s10653-008-9197-2

[16] Arao T, Ishikawa S, Murakami M, Abe K, Maejima Y, Makino T. Heavy metal contamination of agricultural soil and countermeasures in Japan. Paddy and Water Environment. 2010;**8**(3):247-257. DOI: 10.1007/s10333-010-0205-7

[17] Machiwa F. Heavy metal levels in paddy soils and rice (Oriza Sativa. L) from wetlands of lake Victoria Basin Tanzanian. Tanzanian Journal Science. 2010;**36**:140-148

[18] Suganda H, Setyorini D, Kusnadi H, Saripin I, Kurniawan U. Evaluasi pencemaran limbah industry untuk kelestarian sumberdaya lahan sawah. Prosiding seminar nasional multifungsi dan konversi lahan pertanian. Pusat Penelitian dan Pengembangan Tanah dan Agroklimat. Vol. 5. 2003. pp. 203-221

[19] Satpathy D, Reddy MV, Dhal SP. Risk assessment of heavy metals contamination in paddy soil, plants, and grains (Oryza sativa L.) at the East Coast of India. BioMed Research International. Article ID 545473. 2014:11. DOI: 10.1155/2014/545473

[20] Bhagure GR, Mirgane SR. Heavy metal concentrations in groundwaters and soils of Thane region of Maharashtra, India. Environmental Monitoring and Assessment. 2011;**173**(1-4):643-652. DOI: 10.1007/s10661-010-1412-96

[21] Wong SC, Li XD, Zhang G, Qi SH, Min YS. Heavy metals in agricultural soils of the Pearl River Delta. South China. Environmental Pollution. 2022;**119**(1):33-44. DOI: 10.1016/S0269-7491(01)00325-6

[22] FytianosK,KatsianisG,TriantafyllouP, Zachariadis G. Accumulation of heavy metals in vegetables grown in an industrial area in relation to soil. Bulletin of Environmental Contamination and Toxicology. 2011;**67**(3):423-430. DOI: 10.1007/s001280141

[23] Bolan NS, Makino T, Kunhikrishnan A, Kim PJ, Ishikawa S, Murakami M, et al. Cadmium contamination and its risk management in rice ecosystems. Advances in Agronomy. 2013;**119**:183-273. DOI: 10.1016/B978-0-12-407247-3.00004-4

[24] Kurnia U, Agus F, Adimihardja A, Dairah A. Sifat Fisik Tanah dan Metode Analisisnya. Badan Penelitian dan Pengembangan Pertanian Departemen Pertanian; Balai Besar Penelitian dan Pengembangan. Sumberdaya Lahan Pertanian. 2006. p. 282 (Indonesian)

[25] Ramesh T, Bolan NS, Kirkham MB, Wijesekara H, Kanchikerimath M, Srinivasa Rao C, et al. Soil organic carbon dynamics: Impact of land use changes and management practices: A review. In: Spark DL, editor. Advances in Agronomy. Vol. 156. Cambridge, MA: Elsevier; 2019. pp. 1-107. DOI: 10.1016/bs.agron.2019.02.001

[26] Merismon. Remediasi pencemaran timbal (pb) dan cadmium (cd) pada lahan sawah intensif di Kabupaten Musirawas dengan pemberian biochar dan kompos. Disertasi Program Doktor Ilmu Lingkungan. Program Passca Sarjana. Universitas Sriwijaya. 2017

[27] Ogunlade MO, Agbeniyi SO. Impact of pesticides use on heavy metal pollution in cocao soil of cross-river,

Nigeria. African Journal of Agriculture Research. 2011;**6**(16):3725-3728

[28] Sudarmaji J, Mukono CIP. Toksikologi logam berat B3 dan dampaknya terhadap Kesehatan. Jurnal Kesehatan Lingkungan. 2006;**2**(2):129-142

[29] Nakata H, Nakayama SMM, Oroszlany B, Ikenaka Y, Mizukawa H, Tanaka K, et al. Monitoring lead (Pb) pollution and identifying Pb pollution sources in Japan using stable Pb isotop analysis with kidneys of wild rats. International Journal of Environmental Research and Public Health. 2017;**14**:56

[30] Kohar I, Hardjo PH, Lika II. Studi kandungan logam Pb dalam tanaman kangkung umur 3 dan 6 minggu yang ditanam di media yang mengandung Pb. Makara, Sains. 2005;**9**(2):56-59

[31] Jiang J, Xua R, Jiang T, Li Z. Immobilization of Cu(II), Pb(II) and Cd(II) by the addition of rice straw derived biochar to a simulated polluted Ultisol. Journal of Hazardous Materials. 2012;**230**:145-150. DOI: 10.1016/j.jhazmat.2012.05.086

[32] Lu H, Zhang YY, Huang X, Wang S, Qiu R. Relative distribution of Pb^{2+} sorption mechanisms by sludge-derived biochar. Journal of Water Researches Development. 2012;**46**:854-862. DOI: 10.1016/j.watres.2011.11.058

[33] Satpathy KK, Sahu G, Mohanty AK, Prasad MVR, Panigrahy RC. Phytoplankton community structure and its variability during southwest to northeast monsoon transition in the coastal waters of Kalpakkam, east coast of India. International Journal of Oceans and Oceanography. 2009;**3**(1):43-74

[34] Wang QR, Cui YS, Liu XM, Dong YT, Christine P. Soil contamination and plant uptake of heavy metals at polluted sites in China. Journal of Environmental Science and Health. 2003;**38**(5):823-838. DOI: 10.1081/ESE-120018594

[35] Singh J, Upadhyaya SK, Pathaka RK, Guptab V. Accumulation of heavy metals in soil and paddy crop (Oryza sativa), irrigated with water of Ramgarh Lake, Gorakhpur, UP, India. Toxicological & Environmental Chemistry. 2011;**93**(3):462-473. DOI: 10.1080/02772248.2010.546559

[36] Alloway BJ. Heavy Metals in Soils. London: Blackie Academic Professional; 1995

[37] Sarwar N, Saifullah MSS, Zia MH, Naeem A, Bibi S, et al. Role of mineral nutrition in minimizing Cadmium accumulation by plants. Journal Science Food Agriculture. 2010;**90**:925-937. DOI: 10.1002/jsfa.3916

[38] Bolan N, Adriano DC, Naidu R. Role of phosphorus in (im)mobilization and bioavailability of heavy metals in the soil-plant system. Reviews of Environmental Contamination and Toxicology. 2003;**177**:1-44

[39] Mengel K, Kirkby EA. Principle of Plant Nutrition. 4th ed. Bern: International Potash Institute; 2011

[40] Gambuœ F, Wieczorek J. Pollution of fertilizers with heavy metals. Ecological Chemistry and Engineering A. 2012;**19**(4-5):353-360. DOI: 10.1007/s00128-001-0141-8

[41] Kurnia U, Suganda H, Saraswati R, Nurjaya. Teknologi pengendalian lahan sawah tercemar. Balai PenelitianTanah. 2022. pp. 249-283. Available from: http://balittanah.litbang.pertanian.go.id/eng/dokumentasi/buku/tanahsawah/tanahsawah9.pdf [Accessed: 1 December 2022]

[42] Rochayati S, Laing GD, Rinklebe J. Use of reactive Phosphate rocks

as fertilized on acid upland soil in Indonesia: Acumulation of Cadmium and Zinc in soil and shoot of maize plants. Journal of Plant Nutrition and Soil Science. 2011;**174**(2):186-194

[43] Setyorini D. Soeparto, Sulaeman. Kadar logam berat dalam pupuk. Dalam Prosiding Seminar Nasional Peningkatan Kualitas Lingkungan dan Produk Pertanian: Pertanian Produktif Ramah Lingkungan Mendukung Ketahanan dan Keamanan Pangan. Pusat Penelitian dan Pengembangan Tanah Agroklimat. 2003;**7**:114-125

[44] Sukarjo. Laporan Akhir Penelitian Deliniasi Sebaran Residu Pestisida dan Logam Berat di Lahan Pertanian. Balai Penelitian Lingkungan Pertanian. Balai Besar Penelitian dan Sumberdaya Lahan Pertanian. Badan Penelitian dan Pengembangan Pertanian. Kementrian Pertanian. 2015.

[45] Alkhader AMF. The impact of phosphorus fertilizers on heavy metals content of soils and vegetables grown on selected farms in Jordan. Agrotechnology. 2015;**5**(1):5.1. DOI: 10.4172/2168-9881.1000137

[46] Weissengruber L, Möller K, Puschenreiter M, Friedel JK. Long-term soil accumulation of potentially toxic elements and selected organic pollutants through application of recycled phosphorus fertilizers for organic farming conditions. Nutrient Cycling in Agroecosystems. 2018;**2018**(110):427-449. DOI: 10.1007/s10705-018-9907-9

[47] Rai MK, Kalia R, Singh R, Gangola MP, Dhawan AK. Developing stress tolerant plants through in vitro selection-an overview of the recent progress. Environmental and Experimental Botany. 2011;**71**:89-98. DOI: 10.1016/j.envexpbot.2010.10.021

[48] Choppala G, Bolan N, Park JH. Chromium contamination and its risk management in complex environmental settings. Advances in Agronomy. 2013;**2013**(120):129-172. DOI: 10.1016/B978-0-12-407686-0.00002-6

[49] Ardiwinata AN, Ginoga LN, Sulaeman E, Harsanti ES. Pesticide residue monitoring on agriculture in Indonesia. Jurnal Sumberdaya Lahan. 2018;**12**(2):133-144

[50] Notohadiprawiro T. Pengelolaan Kesuburan Tanah dan Peningkatan Efisiensi Pemupukan. 2006. Available from: http://soil.faperta.ugm.ac.id/tj/1981/1984%20penge.pdf. Diakses tanggal 16 Juni 2012

[51] Hoang SA, Bolan N, Madhubashini AMP, Vithanage M, Perera V, Wijesekara H, et al. Treatment processes to eliminate potential environmental hazards and restore agronomic value of sewage sludge: A review. Environmental Pollution. 2022;**293**:118564

[52] Grispen VMJ, Nelissen HJM, Verley JAC. Phytoextraction with *Brassia napus* L: A toll for sustainable management of heavy metal-contaminated soils. Environmental Pollution. 2006;**144**:77-83. DOI: 10.1016/j.envpol.2006.01.007

[53] Kunhikrishnan A, Bolan NS, Müller K, Laurenson S, Naidu R, Kim WI. The influence of wastewater irrigation on the transformation and bioavailability of heavy metal(loid)s in soil. Advances in Agronomy. 2012;**115**:215-297. DOI: 10.1016/B978-0-12-394276-0.00005-6

[54] Mulyadi M. Pb heavy metals in the soil and grain of rice field on Juwana sub-watersheds Central Java. Agrologia. 2013;**2**(2):95-101. DOI: 10.30598/a.v2i2.263

[55] Balittanah (Balai Penelitian Tanah). Laporan Tahunan. Bogor: Balai Penelitian Tanah; 2010

[56] Lacatusu R. Appraising levels of soil contamination and pollution with heavy metals. In: Land Information Systems for Planning the Sustainable Use of Land Resources. European Soil Bureau—Research Report No. 4. 2000; pp. 393-402. (9 pages)

[57] Alloway BJ. Heavy Metals in Soils: Trace Metals and Metalloids in Soils and their Bioavailability. 3rd. ed. Environmental Pollution 22. Springer; 2013

[58] Balittanah (Balai Penelitian Tanah). Petunjuk Teknis Kimia Tanah, Tanaman, Air, dan Pupuk [Technical Instruction: Chemical Analysis of Soil, Plant, Water, and Fertilizer]. 2nd ed. Bogor, Indonesia: Soil Research Center; 2009

[59] Purbalisa W, Mulyadi. Pb and Cu at body of water and paddy soil in downstream watershed of Solo Lamongan district. Agrologia. 2013;**2**(2):116-123. DOI: 10.30598/a.v2i2.266

[60] Mellyga D, Sukarjo, Hidayah A, Setyanto P. Identifikasi sebaran timbal (pb) pada lahan sawah dataran tinggi di Kabupaten Wonosobo dan serapannya pada tanaman padi. Report. Balai Penelitian Lingkungan Pertanian. Hal. 614-520. 2015

[61] Suriani S. Analisis kandungan logam berat timbal (Pb), cadmium (Cd) dan seng (Zn) pada tanah sawah Kelurahan Paccinongan Kecamatan Sombaopu Gowa. Skripsi. Universitas Islam Negeri Alaudin Makasar; 73 hal. 2016

[62] Hindarwati Y, Soeprobowati TR, Sudarno. Heavy metal content in terraced rice fields at SruwenTengaran Semarang – Indonesia. E3S Web of Conferences. 2018;**31**:03009. DOI: 10.1051/e3sconf/20183103009

[63] Notodarmojo S. Pencemaran Tanah dan Air Tanah. Bandung: Institut Teknologi Bandung; 2005

[64] Pradika V, Masykuri M, Supriyadi. Farmer awareness to the dangers of heavy metal Cadmium (Cd) pollution due to over-fertilization in sragen regency central Java. 76 Caraka Tani. Journal of Sustainable Agriculture. 2019;**34**(1):76-85. DOI: 10.20961/carakatani.v34i1.27222

[65] Kabata-Pendias A, Szteke B. Chapter 5: Trace Elements in Abiotic and Biotic Environments. Boca Raton: CRC Press; 2015. DOI: 10.1201/b18198

[66] Hameed A, Rasool S, Azooz MM, Hossain MA, Ahanger MA, Ahmad P. Heavy metal stress: responses and signaling. In: Ahmad P, editor. Plant Metal Interaction: Emerging Remediation Techniques. Amsterdam, Boston: Elsevier; 2016. pp. 557-583. DOI: 10.1016/B978-0-12-803158-2.00024-2

[67] He S, Yang X, He Z, Baligar VC. Morphological and physiological responses of plants to cadmium toxicity: A review. Pedosphere. 2017;**27**(3):421-438. DOI: 10.1016/S1002-0160(17)60339-4

[68] Jarvis SC, Jones LHP, Hopper MJ. Cadmium uptake from solution by plants and its transport from roots to shoots. Plant and Soil. 1976;**44**(1):179-191

[69] Aprilia DD, Purwani KI. Pengaruh pemberian mikoriza terhadap akumulasi logam Pb pada tanaman *Euphorbia milii*. Jurnal Sains dan Seni. 2013;**2**(2):79-87

[70] Munaf, Haryati, Suryani. Penyerapan logam berat dengan menggunakan enceng gondok dan studi regenerasinya. Jurnal Kimia Andalas. FMIPA. Unand. Padang. 2000;**5**:34-40

Chapter 12

Heavy Metal Bioaccumulation in Sediment and Benthic Biota

Sarah Gnanasekaran and S. Amal Raj

Abstract

Bioaccumulation can be used as a measurement tool for analyses of sediment and soil toxicity. Heavy metal toxicity in sediments can be measured with bioaccumulation tests. Metal bioaccumulation has recently achieved more concentration from researchers due to its feasibility to conduct both field and laboratory experiments with indicative organisms. Bioaccumulation can be measured directly or using models. For this study, the concentrations of trace metals (Zn, Pb and Cu) in earthworm tissues were analyzed and compared with the total contents of heavy metals in contaminated parts of soils of Pallikaranai marshland. Samples were taken from different parts of the marshland, which have been reported to have heavy metal presence decades ago. Mostly predominant species found in the marshland *L. mauritii* and *P. excavatus* were used for the experiment. Soil samples were collected at six points along a gradient of increasing pollution. A regression model was applied to the results, and the order of accumulation of heavy metals BAF in the present study is Zn > Cu > Pb, indicating that zinc is a potentially high accumulating metal compared to Cu and Pb.

Keywords: bioaccumulation, bioavailability, heavy metals, benthic biota, sediment

1. Introduction

Metals are naturally occurring non-biodegradable substances in the environment, existing in both natural ways and man-made pollution. Generally, they occur by the geo-chemical weathering of rocks in an environment.

Anthropogenic interventions such as mining, smelting industries, paints made of metals and batteries, and several other inventions have caused severe metal pollution. Metals that are distributed to other places through metal mobility and that get stored in the places where they are deposited pose a serious threat to humans and the overall ecosystem's health.

Though some heavy metals like iron, copper, and zinc are essential for both humans and animals to a certain extent for the proper functioning of the body, those that present in quantities exceeding their limits are extremely hazardous, causing physiological damages in both humans and animals. For example, excess copper, through contaminated vegetables and fruits and other sources, is linked with anemia and liver and kidney damage.

Mercury poisoning at Minamata, Japan, is one example of such contamination that reached humans through fish consumption from the affected Yatsushiro Sea and the Agano River. Lead poisoning of drinking water is another common problem. Arsenic contamination of groundwater has been reported in recent years.

As a result of scientific advancements, large quantities of raw materials using various chemical elements are produced, and the resulting wastes end up as pollutants in major aquatic and terrestrial ecosystems, mainly in rivers and streams as untreated industrial effluents and hazardous chemical wastes. Many freshwater ecosystems around the world have been studied for such pollution effects in the past years.

1.1 Heavy metal pollution

Among the inorganic pollutants, metal pollutants seems to have gained importance as they stay forever in the environment present and tend to bio-magnify when present in a food chain. Metals tend to bioaccumulate in sediments, bio-concentrate in water, and deposit as free radicals in the air and cause damage to the respective ecosystems. Accumulation happens mainly through the food source and prey–predator relationships. From one trophic level to the other, metals tend to bio-magnify, reaching the highest level in the organism at the highest trophic level.

In an aquatic ecosystem, metals generally are found as free metal ions in the water or the sediment mostly as a metal complex bound with soil constituents. They may be colloidal in nature or as suspended solids in an aquatic ecosystem. In colloidal and particulate phases, they exist as oxides, hydroxides, silicates or sulphides and as metal complexes combined with organic matter. The soluble forms are generally ions or non-ionized chelated metallic complexes. pH, ligand concentration, oxidation state, and redox potential are all solubility-controlling factors in an aquatic environment.

Heavy metal contamination of drinking water sources is extremely dangerous for human consumption. Metals in the dissolved phase are more toxic to aquatic organisms. Surface run-off attributes to the pollution of water sources.

Heavy metals also find their way into the agricultural soils in the form of heavily loaded pesticides and fertilizers which eventually end up in the soil, and crops grown in that soil are also affected. In sediments, metals have an affinity towards certain chemicals and, hence, are present as metal complexes.

In sediments, controlling factors include metal speciation, acid volatile sulphides, particulate organic carbon, and Fe and Mn oxyhydroxides. The bioavailability of heavy metals in soil highly depends on factors other than sediment texture and physio-chemical properties such as soil pH, temperature and other organic soil constituents like humus. Sand, silt, and clay also play a major part in the metal's behavior in each sediment.

Heavy metal toxicity in sediments can be measured with bioaccumulation tests. Metal bioaccumulation has recently achieved more concentration from researchers due to its feasibility to conduct both field and laboratory experiments with indicative organisms. Bioaccumulation can be measured directly or using models.

Direct measurement includes assessment of the biota, the organisms present, water, or sediment. Models include simple regression models to mechanistic and empirical models. With recent advances in bioaccumulation models, there is a huge scope for studying metal bioaccumulation at the pollution level and applying the model elsewhere to get results without disturbing the environment. For example, measuring bioaccumulation in a particular sediment can be compared with sediment elsewhere with a simple one-compartment bioaccumulation model.

Bioaccumulation endpoints can be considered for environmental impact assessments. Here, we will concentrate on the recent bioaccumulation models and the relative importance of one compartment model using earthworms as an indicator organism to measure bioaccumulation in a sediment biota.

1.2 Heavy metal toxicity in sediments

Sediments are considered a mixture of various sorbent phases such as organic matter, oxides, sulphides, carbonates, and clay or stilt minerals in which their abundance is based on pH, redox conditions, hydrological level, and the depositional environment [1].

Sediments can be a long-term source of metal pollution. Sediment contamination with metals poses a severe threat to the soil ecosystem, particularly the organisms that are in direct contact with the environment, affecting their survival, reproduction, and growth. Heavy metal toxicity in sediments can be present for a longer period since sediments act as sinks for all kinds of pollution, both organic and inorganic.

Sediments not only store potentially toxic metals but also enhance the transportation of the same. Especially, aquatic sediments are likely to attract heavy metals due to their finer texture and store metals through adsorption, complexation and precipitation processes. Most of the metals present as free metal ions remain in waters surrounding the sediment. The form of the metals present (speciation) impacts the toxicity level of the metals in sediments.

Thus, speciation and distribution of heavy metals in sediment are of major concern when indicating metal toxicity in sediments. Physio-chemical factors such as temperature, hydrodynamic conditions, redox state, the content of organic matter and microbes, salinity and particle size affect the chelation process at the sediment level [2].

However, the distribution of heavy metals in the sediment is highly based on the composition of the sediment, particle size and organic matter content. In finer sediments, organic carbon plays a vital role in the binding of the metals, and higher organic carbon reduces metal solubility and toxicity [3, 4]. Especially humus in soil binds metals more than any other organic matter. This is due to the affinity the humic substances have towards heavy metals, due to their chelating properties.

This may serve the purpose of protection as the toxicity is absorbed to a considerable amount. The aquatic sediments relate to surrounding waters and overlying interstitial water. Hence, any disturbance leads to changes in the sediment biota and therefore may alter its biodiversity. Under aerobic conditions, microorganisms may break down the periphytons and the other aquatic plants and release phosphorous, but in anaerobic conditions, the phosphorus may enter the water column resulting in eutrophication [5].

Any alteration in the sediment composition affects the benthic organisms, severely disrupting the habitat's normal functions such as the exchange of organic matter or nutrients. In terrestrial sediments, the toxicity depends largely on the sulphides that are present in the sediment layers.

These sulphides, termed acid volatile sulphides (AVS), are formed by sulfur-reducing bacteria from organic matter, which are more in anaerobic sediments. Other factors include soil pH, particle size, carbonates and Fe and Mn-oxy hydroxides [6].

2. Metal speciation and bioavailability

The form in which a metal exists in the sediment can be defined as metal speciation. It determines the level of toxicity in the environment in which the metal is present, whether aquatic or semi-aquatic or terrestrial. Most of the metals exist as free metal ions, metal complexes and metal species that occur in an undissolved state.

Speciation affects the transportation of metals from the sediment to the overlying water. Thus, the mobility, fate and transport of the metals stored in the sediment are highly influenced by speciation, particularly in an aquatic environment [7].

In an aquatic environment, sedimentation can occur if the composition of the above sediment is altered, and the normal functioning of the stream or river is severely affected [8]. In the hyporheic zone, where the surface water and groundwater mix, a unique set of microbes and macroinvertebrates is present. But due to severe toxicity, the oxygen supply may be suspended along with the organic matter, due to which the hyporheic organisms disappear.

Understanding metal speciation is crucial to know metal bioavailability and toxicity. Bioavailability is the amount of potentially available forms of metals for uptake by the organisms living in that environment. Hence, in recent studies, bioavailability has been used as a criterion for measuring metal pollution in an aquatic environment.

The bioavailability of free metal ions is the best predictor of metal uptake and toxicity. In sediments, this is further affected by the presence of soil-dwelling organisms through their nature of burrowing. Burrowing gives way to bioturbation, in which a considerable amount of water flows in and out of the burrows of the organisms.

Bioturbation leads to an upward transport of pollutants embedded in deeper soil layers. This may affect the measurement of the free metal ions transported through the course. But, in recent days, a substantial amount of bioavailability tools has been developed by a computer programmed to give more accurate results.

Differences in metal bioavailability are also affected by several other environmental factors such as pH and redox, thereby affecting metal solubility and metal complexation with organic matter present in the soil such as hummus. pH is the most important factor governing metal speciation. It affects the solubility of metal hydroxide minerals and the adsorption and desorption processes.

Metal hydroxides have very low solubilities under pH conditions in water. Other factors include water hardness, organic carbon content and dissolved oxygen content. Due to this, there exist differences in metal bioavailability, leading to metal toxicity. There are also substantial uncertainties in the data collected when bioavailability is considered one of the tools to measure soil bioaccumulation [9].

2.1 Key factors determining bioavailability in an aquatic sediment

Sorption: It is a process by which a solute becomes physically or chemically attached to a solid sorbent regardless of the mechanism (e.g., chemisorption, adsorption, and absorption).

Desorption: It involves the removal of a chemical from a solid to which it is attached or a liquid to which it is dissolved.

Adsorption: It occurs when dissolved metals are attached to surfaces of particulate matter such as iron, Mn and Al oxides, clay and organic matter.

Chelation: Chelation of metal ions happens with soil constituents, and chelated metal complexes are formed. It is the binding of molecules with metal ions. Chelation often reduces metal toxicity by reducing the concentration of free metal ions.

2.2 Sediment–water–column interactions

The water in and around the sediment plays an important role in the movement of free metal ions in and out of the sediment. Pore-water measurement should be considered when measuring bioavailability.

2.3 Organism behavior

The sediment-dwelling organisms mostly are burrowing in nature. Their burrows are constantly irrigated or immersed with the surrounding pore water. This may lead to the movement of metals from the sediment to the porewater and vice versa. Complex metal ions are dissipated in this process. Due to this process, the organism's exposure routes may also vary. The sediment biotic ligand model approach (sBLM) is used to predict the bioavailability of metals in overlying water in such cases [10].

3. Bioaccumulation in sediments

Bioaccumulation refers to an increased level of metal concentration in a living organism than present in its environment. Many such pollutants are taken up by the organisms and stored or metabolized and excreted. In the case of metals, although some are essential for the organism's survival, many are hazardous. They are equally hazardous to the environment in which they are prone to accumulate since they cannot be degraded. Such non-degradable metals accumulate in the sediments in soil and water and contaminate the ecosystem forever. Since aquatic sediments are finer, they tend to accumulate more metals than in land. But in a terrestrial environment, the agricultural soil is the most affected due to pesticides and chemical fertilizers applied over a period regularly stored up in the soil layers. This is taken up by the plants grown in that environment and by the benthic biota present in that place. Thus, sediment heavy metal contamination happens and cannot be redone, thereby affecting the entire flora and benthic fauna. Accumulated metals are stored in different soil fractions. Hence, measuring sediment-associated contaminants requires sequential extraction procedures. The various binding sites in an organism's tissue enhance metal binding and lead to improper function of the organism.

3.1 Bioaccumulation in benthic organisms

Benthic ecosystem often refers to the bottom of the ocean floor and in some cases the bottom sediment in any aquatic ecosystem, both fresh and salt water. The organisms that are present in such sediments are referred to as benthic organisms. These may include a variety of species from micro- to macro-organisms representing insects, polychaete worms, earthworms and snails. Benthic macroinvertebrates readily accumulate contaminants and have been suggested to be reliable indicators of metal bioavailability in metal-contaminated aquatic ecosystems. They are mostly sessile, have long life cycles and represent a range of ecological niches [11]. Variations in season, functional feeding group and size of the organisms should be considered while measuring bioaccumulation in these organisms. They are also an important part of the food web in an aquatic ecosystem, serving as a prey for many fish and birds and are potential candidates for biomagnification. The sediment-dwelling organisms can easily bioaccumulate as they are highly exposed to pollution through the ingestion of contaminated soil and food (**Figure 1**).

3.2 Biological receptors in benthic organisms

Chemicals present in the soil interact with the soil constituents in such a way that over a period, the absorbed contents are not easily available for uptake by benthic organisms. For example, soil pH modifies metal solubility by controlling metal dissolution and precipitation and influences the ionization of pH-dependent ion-exchange sites on organic matter and metal oxide clay minerals. The biological receptors present in the soil-dwelling organisms readily absorb the available fraction of metals and store them either for detoxification or accumulated them in the form of toxicological accumulation. The non-sequestered portion that is not modified by the soil constituents remains as bioavailable fractions for the organism's uptake. Metallotheniens and chlorogosomes are examples of biological receptors in earthworms. Cadmium exposure can induce the production of cysteine-rich metalloproteins called metallotheniens and can be stored in a distinct subtype of sulfur-rich granule termed as cadmosomes [12]. Metallotheniens are sulfur-rich proteins with a low molecular weight that bind metals. Chlorogosomes are phosphate-rich structures with significant cation exchange capacities. The organic matrix of chlorogosomes is a highly complex mixture of carbohydrates, amino acids and lipids as well as redox pigments such as riboflavin, thiamine, carotene and metalloporphyrins [13].

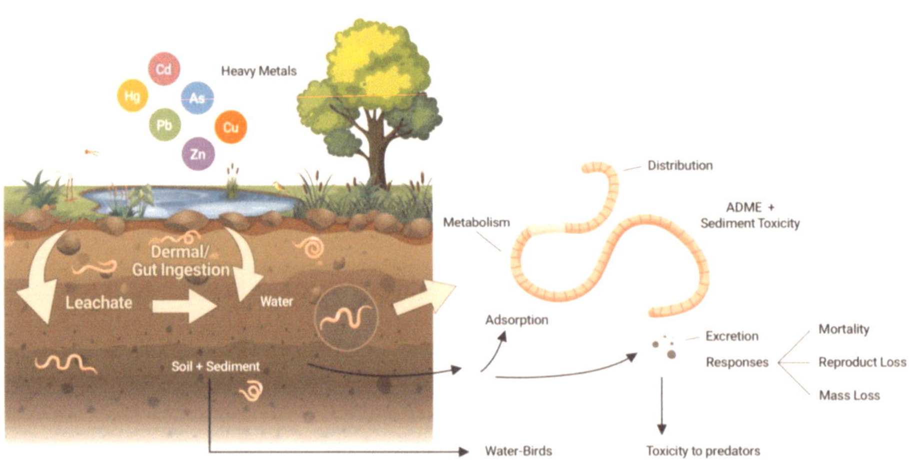

Figure 1.
Representation of bioaccumulation in earthworms through direct ingestion of soil and through dermal uptake.

4. Methodology

4.1 Measuring bioaccumulation

Bioaccumulation can be used as a measurement tool for analyses of soil toxicity. Bioaccumulation data involve field and laboratory analyses of test organisms exposed to spiked metal concentrations. Bioaccumulation data are metal and organism specific [14].

4.1.1 AVS/SEM

The sulphide minerals in sediments with more iron sulphide appear to be a controlling factor for certain divalent cationic metals affecting the metal activity and toxicity in sediments.

The sulphides termed acid volatile sulphides and metals that are combined with AVS are extracted through a process called fractionation and are termed sequentially extracted metals (SEM) and are used in measurements that can assess the potentially bioavailable metals in the sediment as the metals tend to bind well with the AVS content.

Hence, the SEM/AVS theory assumes that, if the AVS concentration is less in sediment than the concentration of SEM, toxicity will be observed.

In other words, if the SEM/AVS ratio is >1, sufficient AVS is not available to bind all the SEM, and therefore, benthic organisms might be exposed to toxic metals. If the ratio is <1, sufficient AVS exists to bind all SEM, and toxicity in benthic organisms is less expected.

4.1.2 Bioaccumulation tests

Bioaccumulation tests are conducted either in situ as field monitoring studies or using an indicator organism in a closed environment under controlled conditions with an artificially spiked substrate where the organisms are kept for certain days for uptake and elimination phases and are monitored during those days and the data are stored.

The degree to which bioaccumulation occurs can be expressed as follows:

BAF (bioaccumulation factor) or,

BSAF (biota sediment accumulation factor),

BAF/BSAF is the ratio of the chemical concentration in an organism to the concentration in the sediment.

$$BSAF = C_{biota}/C_{sediment} \tag{1}$$

Bioaccumulation endpoints include organisms' survival rate, mortality, growth or reproduction, or loss in growth and reproduction.

Sediment toxicity tests include the physicochemical characterization of sediment, toxicity-level assessments and benthic community surveys. This is called a sediment quality triad which is extensively used in decision-making frameworks for contaminated sediments.

4.2 Bioaccumulation models

Bioaccumulation endpoints in sediments can also be expressed using models such as those discussed in the following sections.

4.2.1 Equilibrium-partitioning model and kinetic model (regression model)

Equilibrium portioning models (EqP) assume a steady state concentration, which is achieved due to the thermodynamic equilibrium that exists among the organism and the sediment or the biota or substrate where it is present. Therefore, the fugacity of the compound (the particles' movement or their escape from their current phase) is assumed to be equal to the other compartments in the same environment [15].

And at equilibrium, bioaccumulation can be expressed as a simple partition coefficient or biota-sediment factor (BSAF). However, this model applies to organic contaminants since the lipid content in the organism is necessary to measure the hydrophobicity of the compound.

Based on the interconnections with the hydrophobicity of a compound and its lipid content, a portioning coefficient of 1.7 has been suggested for all compounds [16].

Kinetic models are mathematical models that need uptake and elimination data, the rates of which are modeled independently. The advantage of this model includes no assumption of equilibrium conditions, and hence, non-steady-state concentrations can be predicted.

The model also uses multiple exposure routes and different ways of bioaccumulation in the organism. Compartment-based models describe the movement of the chemicals through first-order equations.

4.2.2 One-compartment bioaccumulation model

A one-compartment bioaccumulation model assumes the organism as a single homogenous unit and follows a first-order reaction. It is expressed as:

$$\text{Rate of metal accumulation } dx/dt = \text{Rate in (uptake)–Rate out (excretion)} \quad (2)$$

Where the concentration of the metal in the body of the organism is directly proportional to the concentration of the metal in the soil.

The exchange of matter between the compartments is due to flux, which is given as K_a. The flux of the metal K_a in the earthworm is the product of an uptake rate constant K_{in} and the external metal conc C_e [17].

The parameters which affect the bioaccumulation of a substance include BAF, the uptake rate constant (K_{in}) and the elimination rate constant (K_{out}).

This model is gaining popularity and is used in environmental risk assessments (ERAs) extensively. It describes the organism as a single homogenous unit. This model is suitable for compounds that distribute rapidly throughout the body.

Two parameters govern the kinetics of the compounds in a one-compartment model [18].

The uptake or accumulation rate

$$K_{in} \text{ day}^{-1} \quad (3)$$

The elimination or excretion rate

$$K_{out} \text{ day}^{-1} \quad (4)$$

The uptake rate is proportional to the exposure concentration in the environment (C_{exp}, mg/kg), and the elimination rate is proportional to the concentration in the organism (C_{org}, mg/kg).

Therefore,

$$dC_{org}/d_T = K_{in}.C_{exp} - K_{out}.C_{org} \tag{5}$$

The rate of the contaminant is given as

$$C_{ew}/d_T = \text{Rate of uptake} - \text{Rate of elimination}$$

When the rate of absorption or intake is absent, the equation becomes.

$$dC_{org}/d_T = -\text{Rate out} \tag{6}$$

If the excretion rate follows first-order kinetics, then,

$$dC_{org}/dT = -K_{out}.x \tag{7}$$

Where K_{out} is the first-order elimination rate constant, and x is the amount of contaminant in the organism at a given time.

At initial time t, the concentration in the organism is 0 and the concentration in the substrate is constant. Eq. (1) has the following solution:

$$C_{org} = C_{exp}\ (K_1/K_2)\ (1 - e - k_{2t}) \tag{8}$$

Where

C_{org} = Concentration of contaminant in the organism.
C_{exp} = Concentration of contaminant in the substrate.
K_1 = uptake rate constant/day.
K_2 = elimination rate constant/day.
t = time/day.

And as the exposure time approaches infinity, the equation for the steady state condition becomes:

$$C_{org}/C_{exp} = K_1/K_2 = \text{BAF} \tag{9}$$

If uptake and elimination rate constants are determined, a BAF can be calculated using the above equations (**Figures 2** and **3**).

5. Results and discussion

Bioaccumulation by earthworms is non-linear, that is, decreases as the concentration increases. The biota to soil accumulation factor (BSAF) assumes that accumulation is linear and constant across all soil concentrations, and hence, the use of the log-linear regression model is used in this study to explain the bioaccumulation in the selected earthworm species. The log-log regression model explains that the bioaccumulation of metals or any other pollutants by earthworms decreases as soil concentration increases. As soil pollutant concentration and elimination rate increase,

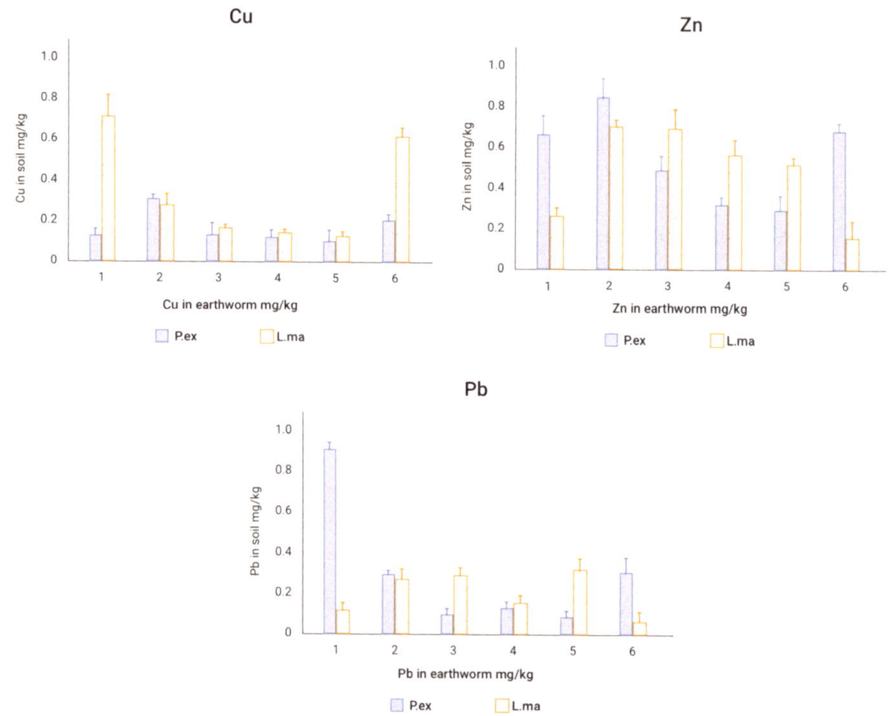

Figure 2.
Relationship between the concentrations of heavy metals (Zn, Cu and Pb) in soil and internal concentrations in earthworms of the earthworm species L. mauritii *and* P. excavatus.

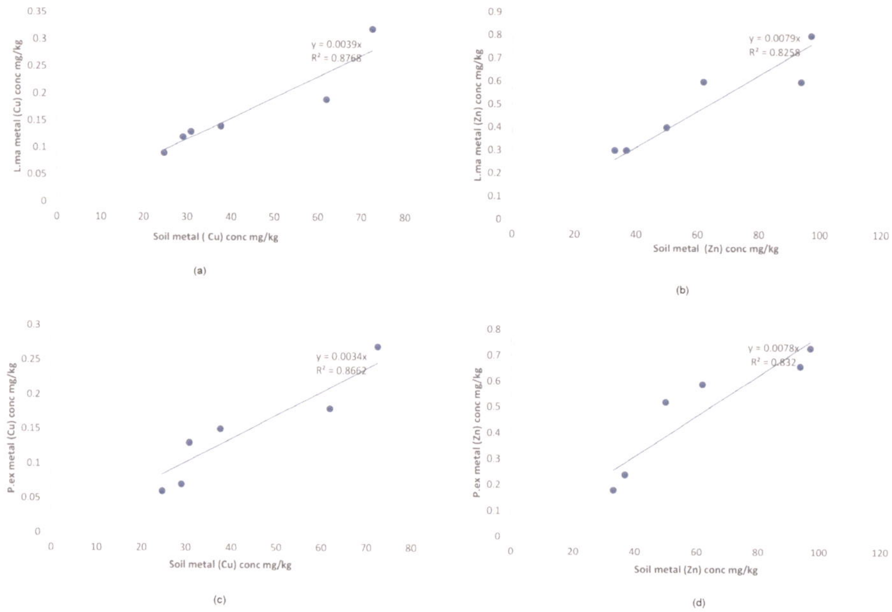

Figure 3.
Regression model applied on the study site between heavy metal concentrations in soil and those in earthworms L. mauritii *and* P. excavatus *(R^2 = 0.97). (a) and (c) Cu and Zn in soil vs. Cu and Zn in earthworm tissues in* L. mauritii; *(b) and (d) Cu and Zn in soil vs. Cu and Zn in earthworm tissues in* P. excavatus.

accumulation may decrease. In our study, heavy metal concentrations in the tissues of the earthworm species collected from the study area were generally not so high and were significantly different across the study sites ($p < 0.05$), especially for Pb and Cu. The earthworms showed increased accumulation pattern for Zinc. It is observed that Zn is the most bioaccumulated; Zn > Cu > Pb, with mean values in the range of Zn (0.13–0.7), Cu (0.28–0.82 Cu) and Pb (0.08–0.89) in *L. mauritii* and Zn (0.09–0.32), Cu (0.18–0.73) and Pb (0.05–0.3) in *P. excavatus*. The accumulation of zinc can be attributed to the readily available metal form in the soil. The order of accumulation of heavy metal BAF in the present study is Zn > Cu > Pb, indicating that zinc is a potentially high accumulating metal compared to Cu and Pb. Although zinc is an essential metal when present at higher levels, it causes cellular disruptions such as mitochondrial dysfunction and limits population growth by affecting reproduction in earthworm species. Copper and lead when present at higher levels are also known to cause higher mortality and reduction in the growth size of the organisms. *P. excavatus* being an epigeic species, which mostly feed on decomposed leaf, are comparatively less affected than the anecic *L. maurtii* that deeply burrow in the soil and are likely to be more exposed to soil metal pollution.

6. Conclusion

Heavy metals in sediments are the main cause of bioaccumulation in benthic organisms and in plants due to the uptake of water, minerals and other nutrients through direct body contact and via roots.

Soil-dwelling micro- and macro-organisms are in direct contact with metal pollution and are exposed to irreversible damage as sediments are very difficult to be remediated. Hence, sediment metal pollution should be considered a serious threat, and strict sediment quality guidelines should be applied.

Reducing the form of metal wastes that are generated through human activities can serve as a start. Going organic in food production can save the benthic biota and our future generation. Natural bioremediation techniques such as phytoremediation should be used to reduce the bioavailability to soil organisms.

Aerobic and anaerobic microorganisms can also be used to treat highly contaminated soils for biodegradation. However, organisms are adapted to high metal concentrations. In that case, the food web should be considered for biomagnification, and the organism that is much affected should be monitored closely.

Natural chelating substances can be used to bind metals that form organometallic complexes, which may be less hazardous for the soil-dwelling species. Also, some detoxifying mechanisms are present in benthic organisms, and hence, bioaccumulation measurements should include those as well.

The ADME process (adsorption, detoxification, metabolism and ejection) is common in all organisms. The level of absorbed pollutants should be considered only for toxicity tests. Identification of bio-accumulative metals may help in enhanced remediation processes.

Author details

Sarah Gnanasekaran* and S. Amal Raj
Department of Civil Engineering, Centre For Environmental Studies, Anna University, Guindy, Chennai, Tamil Nadu, India

*Address all correspondence to: sarahvasanth@gmail.com

References

[1] Burton ED, Phillips IR, Hawker DW. Factors controlling the geochemical partitioning of trace metals in estuarine sediments. Soil and Sediment Contamination. 2006;**15**:253-276. DOI: 10.1080/15320380600646290

[2] Zhao S, Shi X, Li C, Zhang H, Wu Y. Seasonal variation of heavy metals in sediment of Lake Ulansuhai, China. Chemistry and Ecology. 2014;**30**(1):1-14

[3] Besser JM, Brumbaugh WG, May TW, Ingersoll CG. Effect of organic amendments on the toxicity and bioavailability of cadmium and copper in spiked formulated sediments. Environmental Toxicology and Chemistry. 2003;**22**:805-815

[4] Di Toro DM, McGrath JA, Hansen DJ, Berry WJ, Paquin PR, Mathew R, et al. Predicting sediment metal toxicity using a sediment biotic ligand model: Methodology and initial application. Environmental Toxicology and Chemistry. 2005;**24**:2410-2427. DOI: 10.1897/04-413R.1

[5] Lampert W, Sommer U. Limnoecology: The Ecology of Lakes and Streams. Oxford: Oxford University Press; 1997

[6] Chapman PM, Wang F, Adams WJ, Green A. Appropriate applications of sediment quality values for metals and metalloids. Environmental Science & Technology. 1999;**33**:3937-3941. DOI: 10.1021/es990083n

[7] Leppard GG, editor. Trace Element Speciation in Surface Waters and its Ecological Implications. N.Y.: Plenum Press; 1983. p. 320

[8] Boulton AJ. The role of subsurface biological filters in gravel-bed river rehabilitation strategies. In: Rutherfurd I, Bartley R, editors. Second Australian Stream Management Conference: The Challenge of Rehabilitating Australia's Streams. Vol. 1. Clayton (Australia): CRC for Catchment Hydrology, Monash University; 1999. pp. 81-86

[9] OECD. Harmonized Integrated Hazard Classification System for Human Health and Environmental Effects of Chemical Substances and Mixtures. OECD Series on Testing and Assessment No. 23. Paris, France: Organization for Economic Co-operation and Development; 2001. p. 247

[10] Paquin PR, Gorsuch JW, Apte S, Batley GE, Bowles KC, Campbell PG, et al. The biotic ligand model: A historical overview. Comparative Biochemistry and Physiology Part C: Toxicology & Pharmacology. 2002;**133**:3-35

[11] Kiffney PM, Clements WH. Bioaccumulation of heavy metals by benthic invertebrates at the Arkansas River, Colorado. Environmental Toxicology and Chemistry. 1993;**12**: 1507-1517

[12] Morgan JE, Norey CG, Morgan AJ, Kay J. A comparison of the cadmium-binding proteins isolated from the posterior alimentary canal of the earthworms *Dendrodrilus rubidus* and *Lumbricus rubellus*. Comparative Biochemistry and Physiology. 1989;**92C**: 15-21

[13] Greville RW, Morgan AJ. Concentrations of metals (Cu, Pb, Cd, Zn, and Ca) in situ species of British terrestrial gastropods living near a disused Pb/Zn mine. Journal of Molluscan Studies. 1989;**56**:355-362

[14] Chapman PM, Wang F, Janssen C, Persoone G, Allen H. Ecotoxicology of metals in aquatic sediments: Binding and release, bioavailability, risk assessment, and remediation. Canadian Journal of Fisheries and Aquatic Sciences. 1998;**55**: 2221-2243

[15] Di Toro DM, Mahony JD, Hansen DJ, Scott KJ, Hicks MB, Mayr SM, et al. Toxicity of cadmium in sediments: The role of acid volatile sulfide. Environmental Toxicology and Chemistry. 1990;**9**:c1487-c1502

[16] McFarland VA, Clarke JU. Testing bioavailability of polychlorinated biphenyls from sediments using a two-level approach. In: Wiley RG, editor. Proceedings of the US Army Engineers Committee on Water Quality, 6th Seminar. Davis, CA: Hydraulic Engineering Research Center; 1986. pp. 220-229

[17] Nahmani J, Hodson ME, Devin S, Vijver MG. Uptake kinetics of metals by the earthworm Eisenia fetida exposed to field-contaminated soils. Environmental Pollution. 2009;**157**:2622-2628

[18] Van Straalen NM, Donker MH, Vijver MG, van Gestel CAM. Bioavailability of contaminants estimated from uptake rates into soil invertebrates. Environmental Pollution. 2005;**136**:409-417

Chapter 13

Use of Several Pollution Indices for Metal Contamination Assessment in Aquatic Ecosystems, A Case Study, Ebrié Lagoon-Côte d'Ivoire

Adama Diarrassouba Tuo, Issiaka Ben Chérif Traoré and Albert Trokourey

Abstract

In aquatic ecosystems, trace metals (TMs) are widely studied due to their harmful effects on living organisms and humans. The aim of the present study was to use different pollution indices to characterize the sediments contamination with six TMs (As, Cd, Fe, Hg, Mn and Pb). Sediments samples were collected in April 2006 with a Van Veen grab at five stations located in the Ebrie Lagoon (Côte d'Ivoire). TMs concentrations were determined using an ICP-MS Instrument for the calculation of the contamination index (CI), contamination factor (CF), pollution load index (PLI), enrichment factor (EF) and Muller's index of geoaccumulation (Igeo). The CI revealed the sediments contamination in As, Cd and Pb, while CF highlighted their contamination in Hg, As, Fe and Cd. Regarding the PLI, the sediments were uncontaminated with TMs. The EF showed the sediments enrichment with Hg, Pb and As, while the Igeo revealed their pollution with As, Fe, Pb and Cd. In conclusion, the PLI is a useful tool for different locations characterization, while the others (CI, CF, EF, and Igeo) allow individual characterization regarding each TM. Due to high contents in As, Cd and Pb, the studied area need a particular attention.

Keywords: trace metals, pollution indices, aquatic ecosystems, contamination, enrichment, Abobo-Doume fish market, Ebrié lagoon, Côte d'Ivoire

1. Introduction

Trace metals (TMs) are among pollutants assessed worldwide in environmental studies in general, and more particularly in aquatic ecosystems quality assessment [1, 2]. Therefore, due to their stability, bioaccumulative nature, persistence, and their various forms of toxicity in the environment [3, 4], TMs can affect the quality of the coastal ecosystems and present a considerable risk to the aquatic organisms on one hand, and on the other hand to human health [5, 6]. Natural major sources of TMs in coastal areas are the continental weathering of rocks and soil materials [7]. Natural concentrations of TMs in aquatic ecosystems are generally safe for marine organisms and also for human health. However, TMs from anthropogenic sources (domestic, mining, industrial, agriculture, transport activities, etc.) associated with those of natural origins can lead to TMs concentrations above the threshold levels in coastal areas such as lagoons. Such situation mostly increased when the effluents from anthropogenic activities are not properly treated prior to their introduction into the environment. In aquatic ecosystems, TMs are present in the three main matrices (waters, sediments, and living organisms) in various forms including both dissolved and particulate forms. While some TMs as Mn, Zn, and Fe are qualified as essential in relation to their biological benefits, several others like As, Cd, Hg, and Pb are considered to be toxic even at low concentrations [1]. The measurements of pollutants as metals in the water column only give an instant status of the ecosystem quality due to the fact of their low residence time [8]. Therefore, in coastal environments, the pollution status of marine sediments is widely used to understand the possible changes and impacts linked to the introduction of pollutants from anthropogenic activities [8–10]. Indeed, in aquatic environments, sediments act as an adsorptive sink for TMs and the metal concentrations found in sediments are higher than those observed in waters and organisms [8, 11–13]. For sediments, contamination/pollution with TMs in aquatic environments, several methods, including multivariate statistical methods, such as factor analysis, correlation analysis and cluster analysis, Sediment Quality Guidelines (SQGs), sediment contamination indices as enrichment factor, geoaccumulation index and contamination factor, and ecological risk assessment, such as ecological risk index and ecological risk factor, have been commonly used according to the aim of each of the studies undertaken [1, 3, 6, 8]. The Abobo-Doume Fish Market (ADFM), is well known by the population of Abidjan City (Côte d'Ivoire) due to the opportunities offered in terms of marine resources purchase. Several socioeconomics activities are also undertaken around the market, including domestic activities, sand extraction, restauration, artisanal, and SOTRA (a national transport company) boats navigation, industrial activities. This part, like the other ones of the Ebrie Lagoon, also receives significant sediment inputs from its banks and erosion that are generally introduced with pollutants adsorbed onto its. All of these human activities can introduce hazardous pollutants, including TMs into the waters and sediments of the part of Ebrie Lagoon located along the ADFM. The waters located along the ADFM are also used for fishing (fishes, mollusks, crustaceans, etc.). The aim of the present study was to use some contamination indices to assess the TMs spatial contamination in the area located along the ADFM. For this purpose, tools, such as the contamination index (CI), contamination factor (CF), pollution load index (PLI), enrichment factor (EF), and geoaccumulation index (Igeo), were performed to evaluate the contamination rank of the studied TMs regarding all of the sampling locations.

2. Materials and methods

2.1 Study area

The study area is the part of the Ebrié Lagoon located along the Abobo-Doume Fish market (ADFM), a well-known market of several marine organisms in Abidjan. Due to the differences regarding the contamination sources, five sampling stations were chosen and their main details are as follow. The main activities that take place in the study area, apart from those related to the trade in fishery products (S4), consist of restauration and domestic activities (S1), population transport with traditional boats (S2) and those of SOTRA, a state transport company (S3), and the presence of unused boats and also boats construction and reparation activities (S5). All of these activities produce both solid and liquid waste products (nutrients, trace metals, organic pollutants, etc.), which can have a negative impact on the ecosystem's quality.

2.2 Sampling and pretreatment

The sampling campaign of sediments was carried out in April 2016. Surface sediments were collected using a Van Veen grab, placed in polyethylene bags, stored below 4°C and transported to the laboratory for further treatments [14, 15].

2.3 Analytical procedures

For the determination of trace metal concentrations, dry sediment samples (0.3 g) were placed in a Teflon tube and underwent hot mineralization, using 1 mL of aqua regia (HNO_3: HCl; 1:3, v/v) and 6 mL of concentrated hydrofluoric acid (48% of purity). Heating is done at 120°C in a water bath for 2 hours 30 minutes. After cooling in ambient air, the residues are taken up in a solution of boric acid H_3BO_3 (2.70 g in 20 mL of bi-distilled water) for the neutralization of the hydrofluoric acid and the final volume is reduced to 50 mL. The resulting solution was left to stand overnight before analysis. The concentrations of the trace metals (As, Cd, Fe, Hg, Mn, and Pb) were then determined by Inductively Coupled Plasma Optical Emission Spectrometry (ICP-OES 720-ES Varian).

2.4 Pollution indices

To evaluate the trace metals (TMs) degree of contamination in sediments, six parameters were calculated as the contamination index (CI) and the Mean contamination index (MCI), contamination factor (CF), enrichment factor (EF), pollution load index (PLI), and geoaccumulation index (Igeo) [8, 15, 16].

2.4.1 Contamination index (CI) and mean contamination index (MCI)

The contamination index is defined according to the following formula:

$$CI_i = Cx/Mx \text{ and } MCI = (CI_1 + CI_2 + \ldots + CI_n)/n \tag{1}$$

With CI_i: Contamination index of the ith element for station x, Cx: the ith element concentration for station x, Mx: Mean concentration of the ith element for all of the studied stations, n: total number of trace metals analyzed. MCI is the contamination index of a selected station for all the studied trace metals. The mean contamination index (MCI) is used to classify many sampling sites or stations in consideration of the respective contamination index observed for all of the studied trace metals [8].

2.4.2 Contamination factor (CF)

The level of sediments contamination by trace metals is expressed in terms of a contamination factor (CF) calculated as:

$$CF = Cm\ Sample/Cm\ Background \quad (2)$$

where Cm Sample is the concentration of a given metal in lagoon sediment, and Cm Background is the value of the same metal equal to the world surface rock average given by [17]. CF values for describing the contamination level are shown in **Table 1**.

2.4.3 Pollution load index (PLI)

The pollution load index (PLI) is calculated for a selected site/station and determined according to the following method proposed by Tomlinson et al. [19]. The PLI is expressed as follows:

$$PLI = (CF_1 \times CF_2 \times CF_3 \times \ldots \times CF_n)^{1/n} \quad (3)$$

where n is the number of studied trace metals. The PLI provides simple but comparative means for assessing a site's quality. A value of $PLI < 1$ denotes perfection; $PLI = 1$ presents that only baseline levels of pollutants are presented and $PLI > 1$ would indicate a deterioration in the site quality [16].

2.4.4 Enrichment factor (EF)

The EF of metals is a useful indicator reflecting the status and degree of environmental contamination [20]. The EF calculations are used to compare each value with a given background level, either from the local site, using older deposits formed under similar conditions, but without anthropogenic impact, or from a regional or global average composition [21, 22]. The EF was calculated using the method proposed by [23] as follows:

$$EF = (Me/Fe)\ sample/(Me/Fe)\ background \quad (4)$$

where (Me/Fe)sample is the trace metal to Fe concentrations ratio in the selected sample of interest; (Me/Fe) background is the natural background value of the trace metal to Fe ratio. Due to the absence of trace metal background values for our study area, we used the values from surface world rocks [17]. Iron was chosen as the

Contamination factor (CF)	Contamination level
$CF < 1$	Low
$1 \leq CF < 3$	Moderate
$3 \leq CF < 6$	Considerable
$CF > 6$	Very high

Table 1.
Contamination factor (CF) and level of contamination [18].

Enrichment factor	Enrichment factor categories
EF < 2	Deficiency to minimal enrichment
2 ≤ EF < 5	Moderate enrichment
5 ≤ EF < 20	Significant enrichment
20 ≤ EF < 40	Very high enrichment
EF ≥ 40	Extremely high enrichment

Table 2.
Enrichment factor (EF) categories [25].

element of normalization because natural sources (1.5%) vastly dominate its input [24]. Enrichment factor categories are shown in **Table 2**.

2.4.5 Geoaccumulation index (Igeo)

Enrichment of metal concentration above baseline concentrations was calculated using the method proposed by Muller [17], termed the geoaccumulation index (Igeo), and expressed as follows:

$$\text{Igeo} = \text{Log}_2 \left[\text{Cm Sample}/(1.5\text{xCm Background})\right] \tag{5}$$

where Cm Sample is the measured concentration of element n in the sediment sample and Cm Background is the geochemical background value (world surface rock average given by [26]). Factor 1.5 is introduced to include the possible variation of the background values due to the lithogenic effect. Seven different grades or classes of the geoaccumulation index have been proposed by Muller [27]. These classes are given in **Table 3**. The overall total geoaccumulation index (Itot) is defined as the sum of Igeo for all trace elements obtain from the selected site [28].

3. Results

3.1 Trace metals concentrations in sediments

Arsenic, cadmium, iron, mercury, manganese, and lead concentrations were determined in surface sediment samples collected along the Abobo-Doumé Fish Market (ADFM). The results are presented in **Table 4**. Apart from station S4 with sediments free in As, the concentrations observed in the other samples ranged from 2.10^{-4} to 16668.43 mg/kg with an average value of 333.69 ± 746.14 mg/kg. Sediments collected at station S1 recorded the highest As concentration of 16668.43 mg/kg (**Table 4**). Cadmium concentrations ranged from 2.10^{-4} to 7.5 mg/kg with an average value of 1.50 ± 3.35 mg/kg (**Table 4**). The highest Cd content (7.5 mg/kg) was observed at station S4 located near the Abobo-Doume Fish Market. Fe concentrations varied from 1.95 to 4554.90 mg/kg, with a mean of 1444.70 ± 1868.45 mg/kg (**Table 4**). Hg concentrations were of 0.05 mg/kg at stations S1, S3, and S5. For stations S2 and S4, the mercury was below the detection limit (**Table 4**). The average

Class	Igeo value	Sediment quality
0	≤0	Unpolluted
1	0–1	From unpolluted to moderately polluted
2	1–2	Moderately polluted
3	2–3	From moderately to strongly polluted
4	3–4	Strongly polluted
5	4–5	From strongly to extremely polluted
6	>6	Extremely polluted

Table 3.
Muller's classification for geoaccumulation index (Igeo) [15].

Station	As	Cd	Fe	Hg	Mn	Pb
S1	1668.43	0.0002	1.95	0.05	0.850	253.5
S2	0.0002	0.0002	114.41	ND	0.001	0.001
S3	0.0002	0.0002	838.50	0.05	0.220	0.001
S4	ND	7.5	4554.90	ND	2.070	0.001
S5	0.0002	0.0002	1713.76	0.05	1.280	69.35
Average	333.69	1.50	1444.70	0.03	0.880	64.57
SD	746.14	3.35	1868.45	0.03	0.830	109.80
Minimum	0.0002	0.0002	1.95	ND	0.001	0.001
Maximum	1668.43	7.50	4554.90	0.05	2.070	253.50
UCC [29]	2.0	0.102	4.409	0.056	527.0	17.0

Concentrations are given in mg/kg dry weight; ND: Not detected; UCC: Upper Continental Crust.

Table 4.
Trace metals concentrations in Abobo-Doumé fish market sediments.

concentration of Hg was 0.03 ± 0.03 mg/kg. Mn concentrations ranged from 0.001 to 2.070 mk/kg. All of Mn contents observed in the studied area were below the Upper Continental Crust (UCC) value of 527 mg/kg (**Table 4**). For Pb, the determined concentrations ranged from 0.001 to 253.5 mg/kg with an average value of 64.57 mg/kg, more than three times higher than 17.0 mg/kg, the UCC value (**Table 4**). The highest contents of Pb observed in sediments were found at stations S1 and S5, the two extremities of the study area, with respective concentrations of 253.5 mg/kg and 69.35 mg/kg. Pb concentration (0.001 mg/kg) observed at the other three stations (S2, S3, and S4) was largely below the UCC value (**Table 4**).

3.2 Pollution indices

3.2.1 Contamination index and mean contamination index

Contamination indices determined for each of the six TMs in the five sampling stations and the mean contamination index are presented in **Table 6**. Sediments from stations S1 and S4 were respectively contaminated with As and Cd and recorded 5.00 as the contamination index (**Table 6**). The other stations were free in As and Cd.

Regarding Hg, a CI value of 1.67 was observed at stations S1, S3, and S5. For Mn, the lowest CI (0.00) was observed at station S2, while the highest (2.35) occurred at station S4. Pb exhibited CI values of 1.07 and 3.93 for stations S5 and S1, respectively. The mean contamination indices (MCI) varied from 0.01 to 1.93. The highest MCI value was observed in S1 sediments, while le lowest one occurred in S2 ones (**Table 6**).

3.2.2 Contamination factor (CF)

The spatial variation of CF values for the studied TMs is shown in **Figure 1**. The CF values for As ranged from 0.0 to 834.22 with a mean value of 166.84. The highest CF value for As was observed in sediments from station S1 (**Figure 1a**). For cadmium, only S4 presented a CF (73.53) greater than zero (**Figure 1b**). For the whole study area, the average CF value for Cd was 14.71. The CF values for Fe varied from 0.44 to 1033.09, with a mean value of 327.67 (**Figure 1c**).

The highest CF values for Fe were observed at S4 and S5 close to the ADFM and the boats' cemetery respectively (**Figure 1c**). Hg was detected in sediments from S1, S3, and S5 only and exhibited a CF value of 0.89 for the same three sampling points (**Figure 1d**). The lowest CF values calculated for the present study were observed for

Trace metal	ADFM area	Ebrié Lagoon	Ebrié Lagoon	Ebrié Lagoon	World Average
As	0.00–1668.43	NA	NA	NA	NA
Cd	2.10–4-7.50	5.57–372.5	NA	0.05–8.36	1.4
Fe	1.95–4554.90	0.0–4900	18.01–104.14		57405.9
Hg	ND-0.05	NA	NA	NA	NA
Mn	0.001–2.070	20.7–1284.03	NA	18.28–281	975.3
Pb	0.001–253.50	NA	63.95–188.63	1.45–536.09	230.75
Reference	Present study	[30]	[31]	[11]	[15]

ADFM: Abobo-Doume Fish Market; ND: Not detected, NA: Not available.

Table 5.
Concentrations of trace metals in Abobo-Doume in comparison to other local studies and the world averages.

Station	As	Cd	Fe	Hg	Mn	Pb	MCI
S1	5.00	0.00	0.00	1.67	0.97	3.93	1.93
S2	0.00	0.00	0.08	NC	0.00	0.00	0.01
S3	0.00	0.00	0.58	1.67	0.25	0.00	0.42
S4	NC	5.00	3.15	NC	2.35	0.00	1.75
S5	0.00	0.00	1.19	1.67	1.45	1.07	0.90

NC: Not calculated.

Table 6.
Contamination index and mean contamination index in Abobo-Doume fish market sediments.

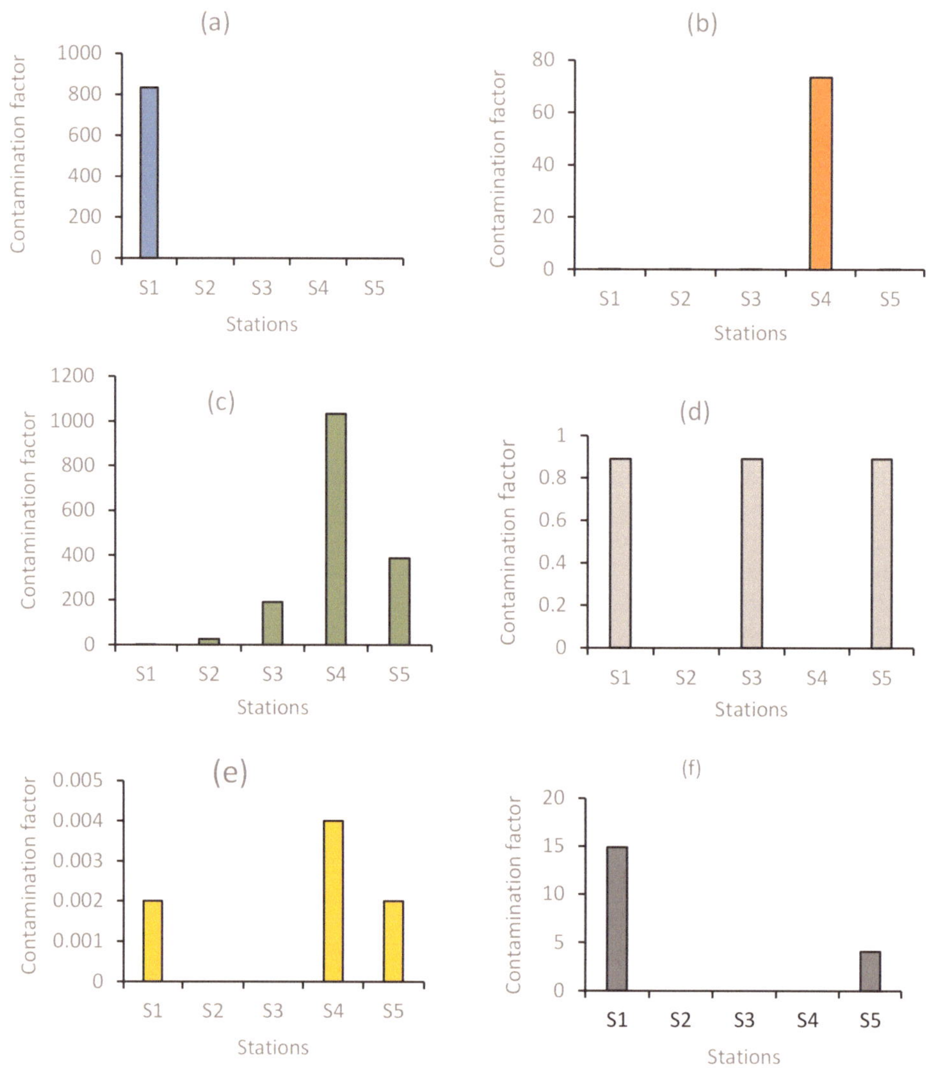

Figure 1.
Contamination factors of trace metals in Abobo-Doume fish market area. (a): arsenic, (b): cadmium, (c): iron, (d): mercury, (e): manganese, and (f): lead.

Mn with CF values ranging from 0.000 to 0.004, and a mean of 0.002 (**Figure 1e**). The CF values for Pb varied from 0.00 to 14.91. The highest CF for Pb was observed in sediments from S1 and S5 (**Figure 1f**). The average CF value for Pb was 3.80.

3.2.3 Pollution load index (PLI)

The pollution load index (PLI) was calculated for the different stations and the results are presented in **Figure 2**. The PLI values varied from 0.00 (S2) to 0.50 (S1). Regarding the PLI values, the sampling stations were in the following ascending rank in TMs contamination: S2 < S3 < S5 < S4 < S1.

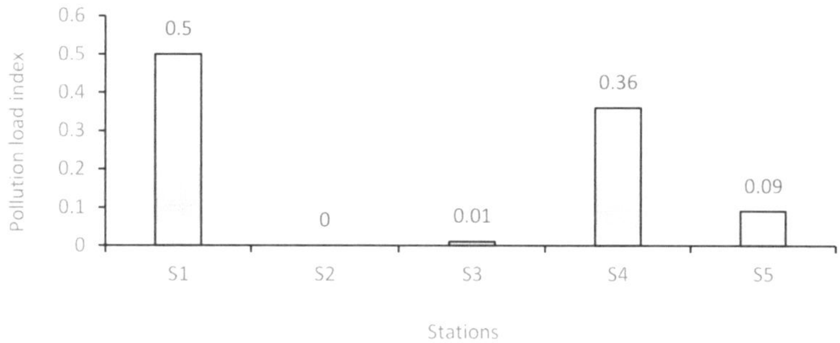

Figure 2.
Spatial variation of the pollution load index (PLI) in the sampling stations.

3.2.4 Enrichment factor (EF)

The spatial variation of enrichment factors for the TMs is presented in **Figure 3**. Arsenic exhibited a high EF value of 1886.182 in station S1. On the opposite, no enrichment effect was observed elsewhere (**Figure 3a**). EF values for Cd were low with values ranging from 0.0 to 0.071 in S4 sediments (**Figure 3b**). No enrichment in Cd was observed for S2, S3, and S5 sediments. For Hg, the calculated EFs ranged from 0.002 to 2.019. The highest EF value for Hg was determined for sediments in S1 (**Figure 3c**). The EF values observed for Mn were low compared to the other TMs ones with values that varied from 0.0 to 0.004 (**Figure 3d**). EFs for Pb ranged from 0.0 to 33.716 (**Figure 3e**). The highest EF value for Pb was observed in S1 sediments, while those in S2, S3, and S4 exhibited no enrichment status (**Figure 3e**).

3.2.5 Geoaccumulation index (Igeo)

Spatial variations of Igeo calculated for each of the TMs are presented in **Figure 4**. For arsenic, the Igeo values were of (−13.87) for stations S2, S3, and S5, and 9.12 for station S1, respectively. The Igeo value was not determined at station S4 due to its concentration that was below the detection limit (**Figure 4a**).

The Igeo values for Cd were all negative (−9.58), except S4 with a value of 5.62 (**Figure 4b**). Regarding Fe, the Igeo values were all positive and varied from 4.11 to 9.43, except S1 with a negative Igeo value of (−1.76) (**Figure 4c**). For Hg, the Igeo-obtained values were all negative with a common value of −0.75 (**Figure 4d**). The Igeo values for Mn were all negative and ranged from (−19.59) to (−8.58) (**Figure 4e**). For Pb, the Igeo values were 3.31 and 1.44 in S1 and S5 sediments, respectively, and a common and negative value (−14.64) in S2, S3, and S4 sediments (**Figure 4f**).

4. Discussion

4.1 Trace metals distribution in Abobo-Doume sediments

The arsenic concentration found in S1 sediments (1668.43 mg/kg, was more than 834 times higher than the Upper Continental Crust (UCC) value of 2.0 (**Table 4**). This high As content highlighted the presence of a potential source of arsenic located in the watersheds. For cadmium, only station S4 exhibited a concentration (7.5 mg/kg) higher than the UCC reference value of 0.102 mg/kg, so found

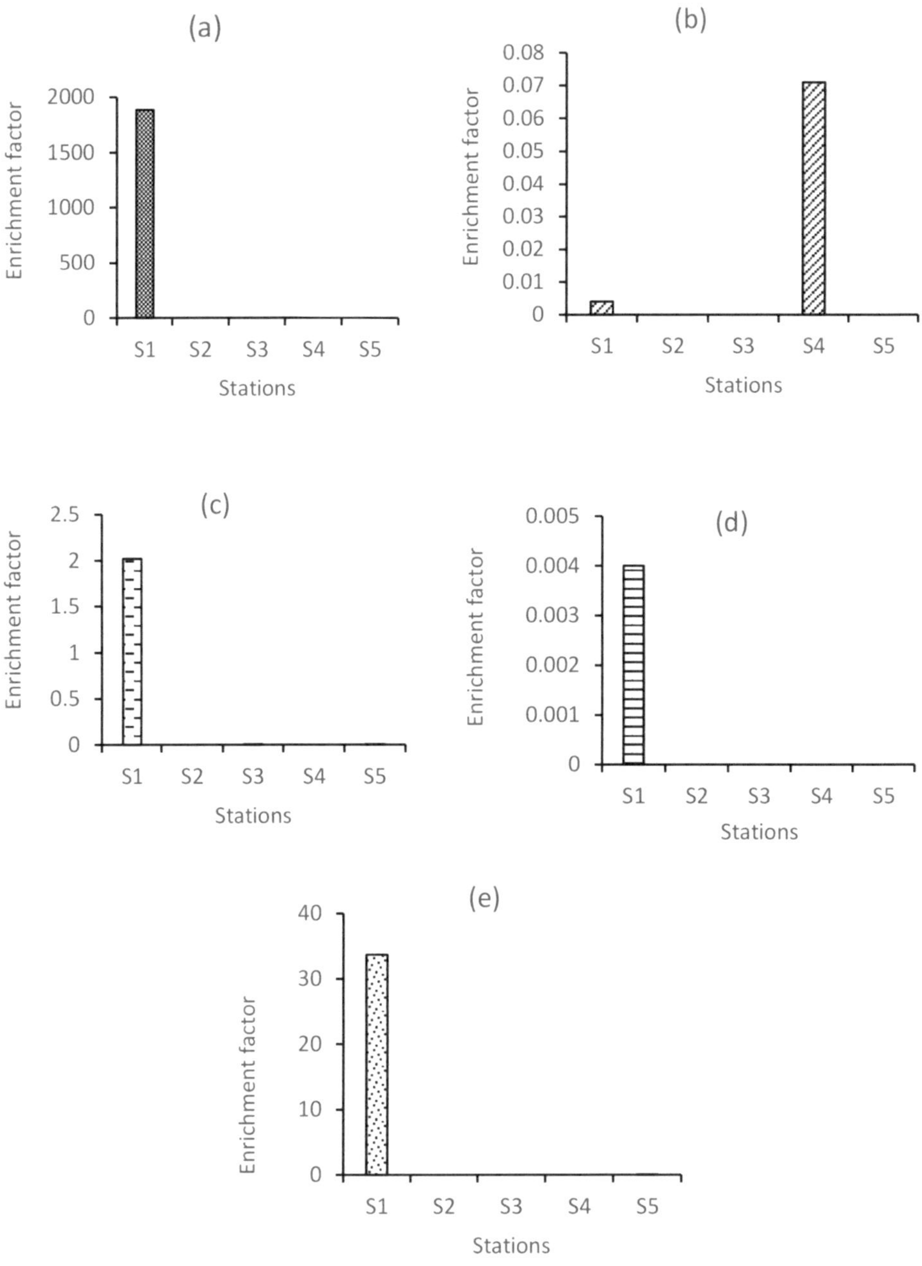

Figure 3.
Spatial variation of the enrichment factor (EF) in sediments. (a): arsenic, (b): cadmium, (c): mercury, (d): manganese, and (e): lead.

contaminated (**Tables 4** and 5). Station S4 is located in front of the Abobo-Doume Fish Market. Therefore, socioeconomic activities linked to fisheries and domestic activities seem to be the major sources of cadmium in this area. Concentration ranges of 5.57–372.5 mg/kg [30], and 0.05–8.36 mg/kg [11] were reported in sediments collected in urban bays in the Ebrié Lagoon. These ranges of Cd concentrations were above the ones in ADFM area (**Table 5**). In addition, the average value of 1.5 mg/kg observed for the whole study area was fairly above the reported World Average value of 1.4 mg/kg (**Table 5**) [16]. Iron concentrations values observed in the present study were in the range of 0.0–4900 mg/kg [30] and below those (18.1–104.14 mg/kg) reported for sediments collected elsewhere in the Ebrie Lagoon (**Table 5**) [31]. The maximum value of Fe observed in the present study was 12 times lower than the world average value of 57405.9 mg/kg [18]. The highest concentration of Fe observed in S4 sediments is linked to the corrosion of iron materials used by people working at the ADFM, such as fish-smoking materials. Mercury is one of the chemical that can have hazardous effects on marine organisms and finally on humans as these organisms' consumers. The observed Hg concentration was in the range of the Upper Continental Crust value of 0.056 mg/kg [29]. Thus, regarding the obtained results for this study, sediments along ADFMF were free of mercury (**Tables 4** and **5**). Mn is an essential element due to its biological benefits for both human and marine organisms [32]. As for iron and cadmium contents, the highest concentration of Mn in surface sediments was recorded at S4, the closest to the Abobo-Doume Fish Market (**Table 4**). Mn concentrations observed in sediments along the ADFM were below those of the range (18.28 to 281.0 mg/kg) reported by [11] on one hand, and, on the other hand than the world average value of 975.3 mg/kg and the Upper Continental Crust [29] one of 527.0 mg/kg (**Tables 4** and **5**).

Pb concentrations observed in S5 and S1 were found contaminated in Pb with contents four and fourteen times higher than the Upper Continental Crust value [29]. At S2, S3, and S4, Pb concentration was below the UCC value. Pb concentrations observed at S1 and S5 were higher than the concentrations reported by Ref. [31] in sediments collected in urban bays that are considered to be highly polluted with several organic and metallic compounds [31, 32]. The observed Pb contents were below the maximum value reported by Tuo et al. [11], and the average value for the study area (64.57 mg/kg) was lower than the world average concentration of Pb (**Tables 4** and 5).

4.2 Pollution indices

According to the contamination index (CI) value of 5.0 (**Table 6**), sediments collected in S1 (Restaurant) were contaminated in As, suggesting a source of As in the area. Regarding Cd, only sediments in S4 (Abobo-Doume Fish Market) were found to be contaminated with a CI value of 5.0. For Fe, the stations were in the following ascending rank of contamination: S1 < S2 < S3 < S5 < S4 (**Table 6**). The CI indices values highlighted the contamination in sediments from stations S3, S4, and S5. Sediments from stations S2 and S4 were safe regarding Hg. For Mn contents, sediments in S4 were the most contaminated, followed by those of S5S1 and S3, while S2 was found to be safe (**Table 6**). For Pb, sediments in S1 (Restaurant) were contaminated, followed by those in S5. Sediments from S2, S3, and S4 were safe regarding Pb contamination. In consideration of all of the six studied TMs through the mean contamination index (MCI), sediments in S1 (Restaurant) and S4 (Abobo-Doume Fish Market) were the most affected, followed by S5, S3, and S2 (**Table 6**). According to the MCI values, the Abobo-Doume Fish Market sediments were in the following descending rank of contamination: S1 > S4 > S5 > S3 > S2.

Considering the highest observed contamination factor (CF) value (834.22 > 40), sediments in S1 (Restaurant) were of an extremely high enrichment status (**Table 1** and **Figure 1a**). For S2 to S5, a deficiency to minimal enrichment occurred (**Table 1** and **Figure 1a**). According to the determined CFs, sediments collected near the ADFM were extremely enriched with Cd, a toxic trace metal even at low concentrations in the environment (**Table 1** and **Figure 1b**), and the other stations were found to be safe for Cd contamination. CF values for Fe revealed an extremely high enrichment in S4, S5 and S3, sediments, a very high enrichment in S2 sediments and a deficiency to minimal enrichment for S1 (**Table 1** and **Figure 1c**). For Hg, all of the determined CF values were below 2.0, which suggests a deficiency to minimal enrichment of sediments for the present study, with a particular concern for stations S1, S3, and S5 (**Table 1** and **Figure 1d**). CF values observed for Mn were very low and close to zero. These CF values highlighted a minimal enrichment of the studied sediments with Mn (**Table 1** and **Figure 1e**). Considering the CFs for Pb, significant and moderate enrichments were respectively observed in sediments from S1 and S5 (**Table 1** and **Figure 1f**). The observed pollution load indices (PLI) observed in sediments for all of the samples were less than one. Therefore, according to the PLI criteria, no deterioration of sediment quality occurred during the present study (**Figure 2**).

The enrichment (EF) observed for As in S1 sediments suggests an extremely high enrichment, so potential source of As in this part of the lagoon banks (**Table 2** and **Figure 3a**). EF values for Cd observed were less than 2.0 and indicate that the sediments were of minimal enrichment status in the study area (**Table 2** and **Figure 3b**). For Hg, the EF values denote a deficiency to minimal enrichment, particularly for sediments in S1 (**Table 2** and **Figure 3c**). EF values for Mn were close to zero, indicating the absence of any excessive enrichment of sediments in Mn (**Table 2** and **Figure 3d**). For Pb, a high EF value was observed in sediments from S1 in relation with an extremely high enrichment (**Table 2** and **Figure 3e**). Due to the fact that Pb is a very toxic TM, there's a need regarding all of the potential sources of sediment contamination in the lagoon's banks around this sampling point (Restaurant).

Sediments in S1 exhibited a high Igeo value for As, which indicates extreme pollution of these sediments (**Table 3** and **Figure 4a**). According to Muller's [16] classification for geoaccumulation index, sediments collected at station S4 were in the extremely contaminated class, while those of S1, S2, S3, and S5 were unpolluted with the Cd (**Table 3** and **Figure 4b**). According to these observed Igeo values, the sediments were unpolluted with Fe at S1, from strongly to extremely polluted quality at S2, and extremely contaminated at S3 to S5 (**Table 3** and **Figure 4c**). For Hg, the exhibited Igeo values in the range of NC to ($-0.75 < 0$), indicate that the sediments along the ADFM were unpolluted with Hg (**Figure 4d**). The negative Igeo values for Mn indicated that the sediments were in the unpolluted range for all of the studied areas (**Table 3** and **Figure 4e**). The Igeo values of Igeo values observed for Pb indicate that sediments from S2, S3, and S4 were unpolluted regarding Pb (**Figure 4f**). However, sediments from station S1 (Igeo = 3.31) and S5 (Igeo = 1.44) were strongly and moderately polluted, respectively (**Figure 4f**).

Several pollution indices were used to assess the contamination level of surface sediments collected in Ebrié Lagoon along with the Abobo-Doume Fish Market (ADFM) (**Table 6** and **Figures 1–4**). The metallic contamination levels in sediments observed at the different sampling stations are summarized in **Table 7**. According to the Contamination Factor (CF) values, very high contamination of sediments was observed at all of the studied stations with arsenic, cadmium, iron, lead, and low contamination for the mercury contents (**Table 7**). All of the observed PLI values were below 1. Therefore, for the present study and considering the PLI observed values, the sediments collected at the five stations were found safe

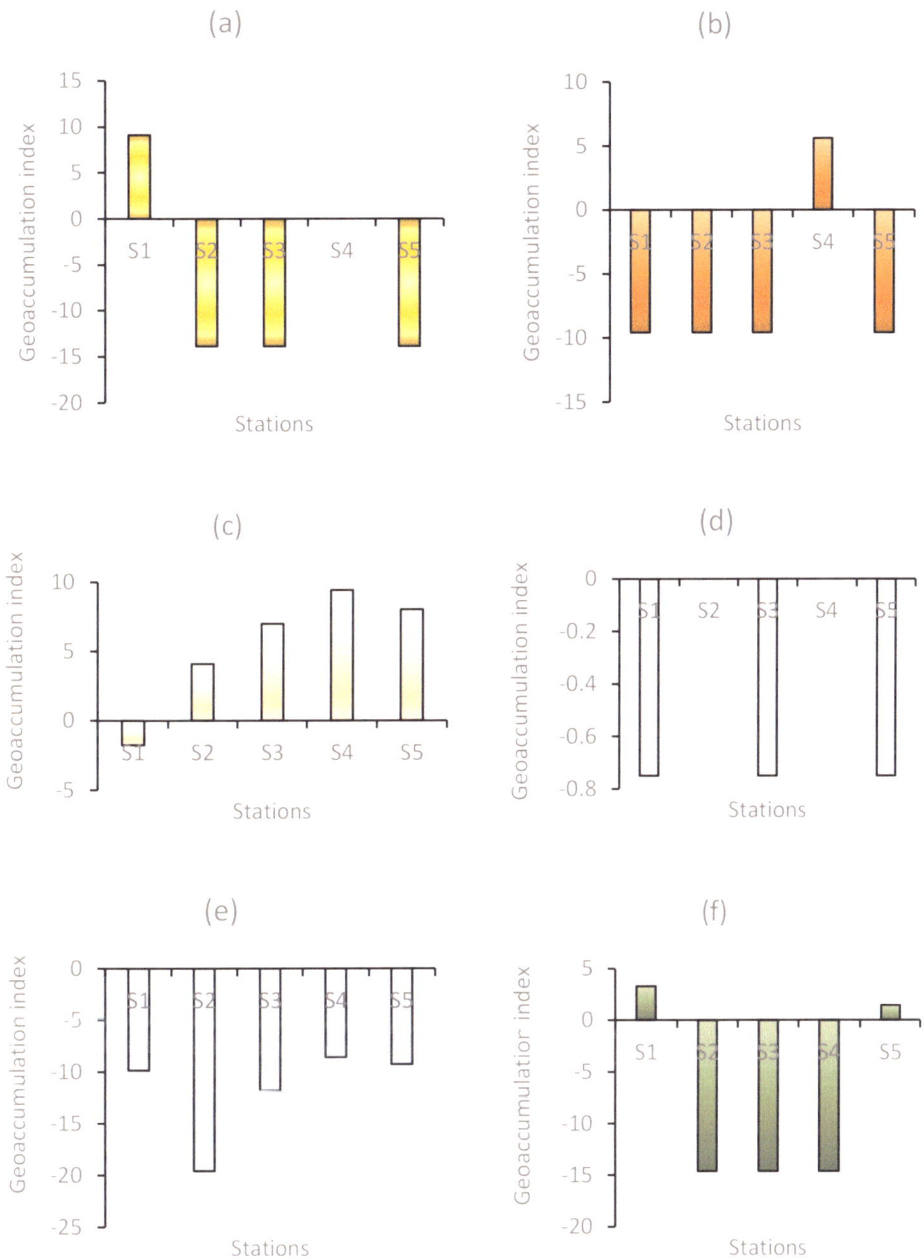

Figure 4.
Spatial variation of Igeo values in Abobo-Doume fish market sediments. (a): arsenic, (b): cadmium, (c): iron, (d): mercury, (e): manganese, and (f): lead.

regarding metallic contamination (**Figure 2** and **Table 7**). At station S1, the EF values highlighted extremely high and very high enrichments of sediments in As and Pb, respectively. A moderate enrichment in Hg was observed in sediments from station S1 (**Table 7**). For station S2, the sediments were safe in trace metals regarding the observed EF values. A deficiency to minimal enrichment levels in Hg

Station	Contamination levels according to the pollution indices			
	CF	PLI	EF	Igeo
S1	• Very high contamination (As, Pb), • Low contamination (Hg)	Safe	• Extremely high enrichment (As), • Very high enrichment (Pb), • Moderate enrichment (Hg)	• Extremely polluted (As), • Strongly polluted (Pb).
S2	• Very high contamination (Fe)	Safe	Safe	• From strongly to extremely polluted (Fe).
S3	• Very high contamination (Fe), – Low contamination (Hg)	Safe	• Deficiency to minimal enrichment (Hg)	• Extremely polluted (Fe)
S4	• Very high contamination (Cd, Fe)	Safe	• Deficiency to minimal enrichment (Cd)	• From strongly to extremely polluted (Cd), • Extremely polluted (Fe)
S5	• Very high contamination (Fe), • Considerable contamination (Cd), • Low contamination (Hg)	Safe	• Deficiency to minimal enrichment (Hg, Pb)	• Extremely polluted (Fe), • Moderately polluted (Pb)

Table 7.
Contamination levels according to the selected contamination indices in sediments from the different sampling stations.

and Cd were observed at stations S3 and S4, respectively. Sediments from station S5 were of deficiency to minimal enrichment levels in mercury and lead (**Table 7**). According to the Igeo values, sediments from station S1 were of bad quality due to their respective extremely and strongly polluted levels in As and Pb (**Table 7**). Sediments from stations S2 to S5 were of strong to extreme pollution levels regarding the Igeo values for Fe. Otherwise, extreme pollution of sediments from station S4 was observed, while moderate pollution in Pb occurred in sediments collected at station S5. The PLI values do not take into account the real state of sediment contamination with each of the studied trace metals taken individually. This parameter is rather useful when it is used for the classification of several sampling stations. Indeed, by considering this PLI parameter alone, one would be tempted to believe that the sediments studied were safe despite the high contents of As, Pb, and Cd observed at several stations.

With regard to Hg, a very toxic element and without any biological positive effect, its presence in sediments is already alarming. Indeed, the contamination of sediments by mercury as well as arsenic, lead, and cadmium could have serious consequences on the survival of benthic organisms on one hand, and on the quality of the fishery resources of this part of the Ebrie Lagoon, on the other hand, the other pollution indices seem to better reflect the levels of contamination of the chemical elements taken individually. For the present study, the contamination factor (CF) seems to be the one that takes into account the low levels of concentrations for the determination of the sediments quality. Indeed, among the calculated pollution indices, it was the one that took into account the presence of mercury, even at a low concentration.

5. Conclusion

The present chapter presents a case study regarding the use of pollution indices for trace metals (TMs) contamination in sediments in Abobo-Doume Fish Market (ADFM) area in Côte d'Ivoire. According to the obtained data, the sediments collected along with the ADFM were particularly found contaminated in As, Cd, and Pb. The contamination index (CI), the mean contamination index (MCI), the contamination factor (CF), the pollution load index (PLI), the enrichment factor (EF), and the geoaccumulation index (Igeo) were applied for the assessment of the studied sediments. The CI values revealed a general contamination status of all of the sampling stations with TMs. In consideration with the MCI, the sampling stations were in the following contamination descending rank: S1 > S4 > S5 > S3 > S2. The CF suggests that the ADFM sediments were highly contaminated by As, Pb, Fe, and Cd, lowly contaminated by Hg. In consideration with the PLI values, sediments in the Abobo-Doume Fish Market were not polluted with TMs. The EF values have revealed an extremely high contamination for As, a very high enrichment for Pb, a deficiency to minimal enrichment for Cd, and a moderate enrichment for Hg. The Igeo values have highlighted that sediments in the ADFM area were extremely polluted for As, strongly and moderately polluted for Pb, and strongly to extremely polluted for Fe and Cd. The pollution indices used in the present study are useful tools for sediment contamination by TMs assessment in aquatic environments such as coastal areas. In conclusion, the CF was the most sensitive index as it revealed the contamination of the sediment by Hg despite its observed low concentrations and the PLI should not highlight the risk link to toxic TMs as As, Cd, Pb, Hg, etc. even at low concentrations in the environment.

Author details

Adama Diarrassouba Tuo[1*], Issiaka Ben Chérif Traoré[2] and Albert Trokourey[2]

1 Centre de Recherches Océanologiques (CRO), Abidjan, Côte d'Ivoire

2 Laboratoire de Constitution et Réaction de la Matière de l'Université Félix Houphouët-Boigny, Abidjan, Côte d'Ivoire

*Address all correspondence to: adamatuo76@yahoo.com

References

[1] Tuo AD, Soro MB, Trokourey A, Bokra Y. Assessment of waters contamination by nutrients and heavy metals in the Ebrié lagoon (Abidjan, Ivory Coast). Research Journal of Environmental Toxicology. 2012;**6**(5): 198-209. DOI: 10.3923/rjet.2012.198.209

[2] Aroua M, Boulahdid M, Radakovitch O, Benhalima M, Guendouzi Y, Fowler SW, Ambrosi JP. Sediment pollution: An assessment of anthropogenic and geogenic trace element contributions along the central Algerian coast. Marine Pollution Bulletin. 2023;**186**:114398. DOI: 10.1016/j.marpolbul.2022.114398

[3] Inal A, Boulahid M, Angeletti B, Radakovitch O. Levels and ecological risk assessment of heavy metals in surface sediments of fishing grounds along Algerian coast. Marine Pollution Bulletin. 2018;**136**:322-333. DOI: 10.1016/j. marpolbul.2018.09.029

[4] Gao L, Chen J, Tang C, Ke Z, Wang J, Shimizu Y, et al. Distribution, migration and potential risk of heavy metals in the Shima River catchment areaSouth China. Environmental Science. Processes & Impacts. 2015;**17**:1769-1782. DOI: 10.1039/c5em00156k

[5] Abderrahmani K, Boulahdid M, Bendou N, Guenachi B, Hacene OR, Masino F, et al. Partitioning of trace elements in the tissues of Mediterranean mussels (Mytilus galloprovincialis) sampled from industrial sites along the Algerian coast. Marine Pollution Bulletin. 2021;**173**:113006-113016. DOI: 10.1016/j.marpolbul.2021.113006

[6] Guendouzi Y, Soualili DL, Fowler SW, Boulahdid M. Environmental and human health risk assessment of trace metals in the mussel ecosystem from the southwestern Mediterranean. Marine Pollution Bulletin. 2020;**151**:110820. DOI: 10.1016/j. marpolbul.2019.110820

[7] Viers J, Dupre B, Gaillardet J. Chemical composition of suspended sediments in world Rivers: New insights from a new database. Science Total Environment. 2009;**407**:853-868. DOI: 10.1016/j.scitotenv.2008.09.053

[8] Ouattara Y, Trokourey A, Tuo AD. Interrelationship between Ecological Risk Assessment with Trace Metals Levels in Surface Sediments from Fresco Lagoon, Côte d'Ivoire, West Africa. Current Advances in Geography, Environment and Earth Sciences. 2022;**8**: 80-101. DOI: 10.9734/bpi/cagees/v8/3624A

[9] Antizar-Ladislao B, Mondal P, Mitra S, Sarkar SK. Assessment of trace metal contamination level and toxicity in sediments from coastal regions of West Bengal, eastern part of India. Marine Pollution Bulletin. 2015;**101**(2):886-894

[10] Mondal P, de Alcântara MR, Jonathan MP, Biswas JK, Murugan K, Sarkar SK. Seasonal assessment of trace element contamination in intertidal sediments of the meso-macrotidal Hooghly (Ganges) river estuary with a note on mercury speciation. Marine Pollution Bulletin. 2018;**127**:117-130

[11] Tuo AD, Yeo KM, Soro MB, Trokourey A, Bokra Y. Contamination of surface sediments by heavy metals in Ebrié lagoon (Abidjan, Ivory Coast). International Journal of Chemical Technology. 2013;**5**(1):10-21

[12] Liang X, Song J, Duan L, Yuan H, Li X, Li N, et al. Source identification and risk assessment based on fractionation of heavy metals in surface sediments of

Jiaozhou Bay, China. Marine Pollution Bulletin. 2018;**128**:548-556. DOI: 10.1016/j.marpolbul.2018.02.008

[13] Varol M. Environmental, ecological and health risks of trace metals in sediments of a large reservoir on the Euphrates River (Turkey). Environmental Research. 2020;**187**: 109664. DOI: 10.1016/j. envres.2020.109664

[14] American Public Health Association (APHA). Standard Methods for the Examination of Water and Wastewater. 21st ed. Washington: APHA, AWWA, WPCF; 2005. p. 535

[15] Muller G. Index of geo-accumulation in sediments of the Rhine River. geojournal. 1969;**2**(3):108-118

[16] Salah MEA, Zaidan TA, Al-Rawi AS. Assessment of heavy metals pollution in the sediments of Euphrates River, Iraq. Journal of Water Resource and Protection. 2012;**4**:1009-1023. DOI: 10.4236/jwarp.2012.412117

[17] Martin J, Meybeck M. Elemental mass- balance of material carried by major world rivers. Marine Chemistry. 1979;7(3):178-206. DOI: 10.1016/0304-4203(79)90039-2

[18] Hakanson L. An ecological risk index for aquatic pollution control a sedimentological approaches. Water Research. 1980;**14**(8):975-1001. DOI: 10.1016/0043-1354(80)90143-8

[19] Tomlinson D, Wilson J, Harris C, Jeffrey D. Problems in the assessment of heavy- metal levels in estuaries and the formation of a pollution index. Helgoland Marine Research. 1980;**33**(1–4):566-575

[20] Feng H, Han X, Zhang WG, Yu LZ. A preliminary study of heavy metal contamination in Yangtze River intertidal zone due to urbanization. Marine Pollution Bulletin. 2004;**49**(11–12):910-915. DOI: 10.1016/j. marpolbul.2004.06.014

[21] Cato I. Recent sedimentological and geochemical conditions and pollution problems in two marine areas in Southwestern Sweden. Striae. 1977;**6**:1-150

[22] Choi K, Kim S, Hong G, Chon H. Distribution of heavy metals in the sediments of South Korean harbors. Environmental Geochemical Health. 2012;**34**(1):71-82. DOI: 10.1007/s10653-011-9413-3

[23] Sinex S, Helz G. Regional geochemistry of trace elements in Chesapeak Bay sediments. Environmental Geology. 1981;**3**(6):315-323. DOI: 10.1007/BF02473521

[24] Tippie V. An environmental characterization of Chesapeak Bay and a framework for action. In: Kennedy V, editor. The Estuary as a Filter. Academic Press: New York; 1984. pp. 467-487

[25] Mmolawa K, Likuku A, Gaboutloeloe G. Assessment of heavy metal pollution in soils along roadside areas in Botswana, Afric. Journal of Environmental Science and Technology. 2011;**5**(3):186-196

[26] Fikret U, Tepe Y. Water quality and sediment contamination assessment of Pazarsuyu Stream, Turkey using multivariate statistical methods and pollution indicators. International Soil and Water Conservation Research. 2018; 7(1):47-56. DOI: 10.1016/j. iswcr.2018.09.001

[27] Muller G. The heavy metal pollution of the sediments of Neckars and its tributary. A Stocktaking Chemische Zeit. 1981;**150**:157-164

[28] Ya ZG, Zhou LF, Bao ZY, Gao P, Sun XW. High efficiency of heavy metal removal in mine water by limestone. Chinese Journal of Geochemistry. 2007;**8**(3):293-298. DOI: 10.1007/s11631-009-0293-5

[29] Wedepohl KH. The composition of the continental crust. Geochimica et Cosmochimica Acta. 1995;**59**(7):1217-1232. DOI: 10.1016/0016-7037(95)00038-2

[30] Affian K, Robin M, Maanan M, Digbehi B, Djagoua EV, Kouamé F. Heavy metal and polycyclic aromatic hydrocarbons in Ebrié lagoon sediments, Côte d'Ivoire. Environmental Monitoring and Assessment. 2009;**159**:531-541. DOI: 10.1007/s10661-008-0649-z

[31] Yao KM, Soro MB, Trokourey A, Bokra Y. Assessment of sediments contamination by heavy metals in a tropical lagoon area (Ebrié lagoon, Côte d'Ivoire). European Journal of Scientific Research. 2009;**34**(2):280-289

[32] Tuo AD, Soro MB, Trokourey A, Bokra Y. Evidence of organic pollution observed in Ebrié lagoon around Abidjan City (Côte d'Ivoire). American International Journal of Research in Formal, Applied & Natural Sciences. 2015;**11**(1):40-45

Chapter 14

Heavy Metal Pollution Resulting from Informal E-Waste Recycling in the Greater Accra Region of Ghana

Albert Kwame Teye and Isaac Kow Tetteh

Abstract

This study investigated concentrations and spatial distributions of four heavy metals: Cadmium (Cd), Chromium (Cr), Copper (Cu), and Lead (Pb) in the soil and drainage systems resulting from informal e-waste recycling at Ashaiman, a town in the Greater Accra Region of Ghana. Twenty-four soil samples were randomly taken from two open burning sites, and three water samples from a drainage that flows through the scrapyard were digested using standard wet digestion methods. An atomic absorption spectrophotometer (AAS) was used to analyze three replicates per sampling location for the heavy metals. The results revealed that the soil and drainage samples were polluted, with the metallic levels exceeding the World Health Organization (WHO), the Food and Agricultural Organization (FAO) of the United Nations, and the Environmental Protection Agency (EPA) of Ghana limits. Geoaccumulation index (I_{geo}), pollution load index (PLI), and contamination factor (CF) further confirmed the contamination of the scrapyard by the heavy metals. Spatial distribution maps showed elevated levels of the heavy metals at portions designated for open burning and disposal of e-waste materials. The research corroborates studies on pollution of the environment by informal e-waste activities and underscores the urgent need for policy implementation and law enforcement to halt further pollution.

Keywords: e-waste recycling, heavy metals, pollution indices, spatial distribution, Ashaiman scrapyard-Ghana

1. Introduction

E-waste has other names, such as waste electrical and electronic equipment (WEEE) and e-scrap. There is no universally-agreed e-waste definition in both legislation and daily usage. This has generated countless definitions in e-waste regulations, policies, and guidelines. In this paper, we adopt the non-legal definition provided by Solving the E-Waste Problem (StEP) Initiative White Paper [1], which is as follows: "E-Waste is a term used to cover items of all types of electrical and electronic equipment (EEE) and its parts that the owner has discarded as waste without the intention of re-use." By this definition, selecting the Ashaiman scrapyard in the Greater Region of Ghana as the study area is seamlessly connected and justified. Due to their different lifespan profiles, different e-waste materials generate different volumes, potential environmental and health impacts, and economic values [2].

Owing to rapid changes in technological updates and upgrades of EEE, industrialization and modernization, an increase in disposable income, and the popularized increase in the use of EEE, there is an upsurge in the acquisition and utilization of electrical and electronic products. Consequently, e-waste generation has the world's largest and fastest growth rate. Asia contributed most to the generation of e-waste in 2019, generating close to 24.9 million Metric tons (Mt), followed by Europe (12.0 Mt), Americas (13.1 Mt), Africa (2.9 Mt), and Oceania (0.7 Mt) [2, 3].

As a result of free and illegal trading activities and the lack of implementation of environmental policies, Africa receives high quantities of potential e-waste materials from these continents. Liberia, Nigeria, Ghana, Benin, and Ivory Coast are major destinations for these "slightly used" materials [4]. Around 600,000 used EEE were imported into Nigeria in 2010. Also, close to 30% of second-hand imports into the country were considered non-functioning and thus regarded as e-waste. Ghana's e-waste quantities rose from 63,000 tons per year in 2003 to 169,000 tons per year in 2008, with a further increment to 215,000 tons in 2009. Only 30% of the total electrical appliances that arrived in Ghana in 2009 were determined to be new, with the rest regarded as used products, 15% of which were either faulty or outmoded and thus could not be sold, eventually ending up in the informal recycling sector [5].

However, e-waste materials in Ghana or Africa need to be better managed due to ignorance on the part of the public on the dangers of poor disposal systems, lack of safe systems of disposal, and absence of government policy and legislation or the enforcement of same. E-waste management in Ghana, just like in most African countries, is managed by the unhindered and poorly equipped informal sector [1]. Manual dismantling, acid leaching, open burning, and indiscriminate disposal of e-waste material are usually the methods of choice in the informal sector to recover valuable metals such as Cu, Au, and Ag that can be resold. These actions release toxic substances, including Polycyclic Aromatic Hydrocarbons, Organochlorine compounds, Phthalates, and heavy metals [2, 6]. The release of these compounds results in atmospheric pollution, and a reduction in the physicochemical characteristics of water quality, including pH, phosphate, oxygen, and chloride levels. Soil composition and viability to support plant life are compromised severely following e-waste contamination [7–9]. Essentially, poor e-waste management has a negative outlook on the realization of the Sustainable Development Goals (SDGs), particularly Goals 3 (Good health and wellbeing), 6 (Clean water and sanitation), 8 (Decent work and economic growth), 11 (Sustainable cities and communities), 12 (Responsible consumption and production), and 14 (Life below water) [1].

Heavy metals are significant components of e-waste materials. The application of these metals in electrical gadgets is influenced by good electrical conductivity to minimize power losses, an inert environment in operations to ensure reliable functioning, and using metals compatible with manufacturing processes [10]. Heavy metals make up about 60.2% of significant constituents of e-waste, including elements such as Tin (Sn), Mercury (Hg), Antimony (Sb), and Arsenic (As) [11]. The parent circuit board of many electrical gadgets houses heavy metals like As, Cd, Pb, and Hg [2]. For example, Pb constitutes nearly 0.4–1.0 kg of the total mass of cathode ray tubes found in computer monitors and television sets, respectively. Also, personal desktop computers (which weighed approximately 32 kg) contain Pb (6.3%), Cu (6.9%), Cobalt (Co) (0.02%), and Iron (Fe) (20.5%) [2, 11].

Heavy metals released during informal e-waste recycling are absorbed into living tissues, usually through inhaling toxic fumes and particulate matter and ingesting contaminated food and water [6]. Cadmium (Cd) is a known carcinogen of the lungs, kidneys, and prostate. Exposure to Cr causes cardiovascular diseases, hematological and neurological effects, and sometimes even death. Pollution resulting from Pb induces memory loss, dullness, anemia, convulsions, tremors, headache, and irritability, while respiratory irritation such as coughing and sneezing, gastrointestinal effects, including nausea, anorexia, diarrhea, and hematological effect result from exposure to Cu [12–15].

The informal sector dominates e-waste recycling activities in Ghana. This is mainly due to the need for more implementation of environmental-related laws and poverty. It is estimated that between 121,800 and 201,600 individuals are involved in the informal e-waste sector in Ghana. The formal recycling sector in Ghana receives only about 0.2% of e-waste for treatment [16]. Unsurprisingly, Ghana is noted for having one of the most significant e-waste recycling in Africa, at Agbogbloshie in Accra, Ghana.

This research assessed the pollution levels and spatial distributions of four heavy metals (Cd, Cr, Pb, and Cu) at two burning sites within the Ashaiman scrapyard in the Greater Accra Region of Ghana, where informal e-waste recycling occurs. Pollution levels of the heavy metals were investigated using selected pollution and contamination indices. The spatial distributions were also investigated using inverse distance weighted (IDW) method. The rest of the chapter is structured as follows: Section 2 describes Materials and Methods; Section 3 focuses on the Results. One of the key findings is that soil and drainage systems of the Ashaiman scrapyard were polluted with Cu, Pb, Cr, and Cd, mainly due to open burning and dumping of e-waste materials; Section 4 is devoted to Discussion; and Section 5 presents the Conclusions of the study.

2. Materials and methods

The Ashaiman scrapyard is located at the entry into the township from the Tema metropolis, about 0.12 km from the Accra-Tema Motorway. Covering a land size of about 0.07 km^2, it is located on latitude 5° 41′4.99″ N and longitude 0° 01′37.28″ W. The region is generally flat, with savannah grasses and shrubs being the dominant vegetation. The topsoil is primarily sandy clay, with the subsoil predominantly clay [17, 18].

The scrapyard houses large metal containers, which store e-waste materials until they are ready to be worked on. Dismantling and sorting activities were performed in sheds and wooden structures at sections of the scrapyard. Burning of e-waste to isolate valuable metals was done on the open field, though few burning activities were observed at the dismantling and sorting areas. At the time of research, two main sites were identified where open burning occurred. E-waste materials in the scrapyard included refrigerators, television sets, computers, cables, radios, and incandescent light bulbs.

Located very close and separating the scrapyard from a dumpsite is a drain that flows from the northern end of the scrapyard and serves as irrigation for farming crops and drinking water for herds of cattle.

2.1 Sample collection

2.1.1 Soil samples

Two burning sites (F and H) were chosen for soil sampling. Site F is located at the central portion of the scrapyard. Major parts of this site were used for the open burning of e-waste, though a few sections served as dumping grounds for e-waste material after dismantling, sorting, and burning. Site H lay closer to the drain running through the scrapyard. This site was used both for open burning and dumping of e-waste materials. **Figure 1** shows a map of the scrapyard area.

Samples were taken in the early hours of 16 July 2019. At each burning site, five topsoil samples (marked 1A, 2A, 3A, 4A, and 5A) within a soil depth of 0–10 cm and five subsoil samples (marked as 1B, 2B, 3B, 4B, and 5B) within a depth profile of 10–20 cm, were randomly collected from different sections. Thus, 10 soil samples were taken from each burning site, and 20 samples were obtained from the two burning areas. Four other topsoil samples were taken at distances of 25 m, 50 m, 75 m, and 100 m from the scrapyard (marked as HV 20, HV 50, HV 75, and HV 100, respectively) to test the detection of heavy metals as one moved away from the scrapyard. Sampling was done with a newly purchased stainless-steel garden shovel and a standard measuring rule to determine the vertical depth of the soil profile. The coordinates at sampling points were recorded using GPS software. A map showing the sampling points of the soil samples is shown in **Figure 2**.

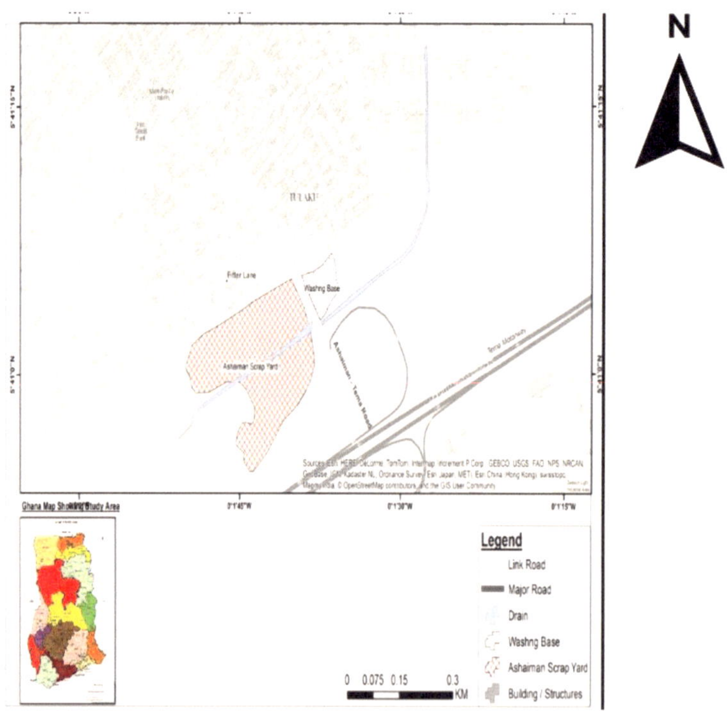

Figure 1.
Map showing scrapyard at Ashaiman.

FIgure 2.
Aerial view of the sampling points at the scrapyard.

2.1.2 Water samples

Three water sediment samples were collected about 140 m north of the scrapyard and mixed to form the control sample (WS C). Within the scrapyard, a water sediment sample was collected (WS 1), about 370 m from the control sample. In contrast, a second sample (WS 2), purely water without sediment from the drain, was also obtained, about 30 m from the second water sediment sample. Coordinates were taken using GPS software. pH of these samples was taken on-site using a Hanna pH meter calibrated with buffer solutions of pH 4, 7, and 10.

Samples were collected into plastic bowls with tightly fitting lids pre-cleaned with nitric acid and sent to the Ghana Standards Authority for treatment and analysis.

2.2 Soil sample preparation and determination of pH

Soil samples were air dried at around 105°C to eliminate wetness and obtain only constant weights representing the soil. They were then passed through a 2 mm non-metallic mesh to separate and remove rocks exceeding 0.25 inches (6.35 mm). Manual milling with mortar and pestle thoroughly homogenized the soil particles passing the mesh. These preparations were necessary for good dissolution during chemical treatments to increase the accuracy of the analysis [19]. To 3 g of each of the dried and sieved soil samples in a 25 ml beaker (which had been pre-cleaned and thoroughly washed with distilled water), 15 ml of aqua regia was added, and the resulting solution digested in a fume chamber for about 30 minutes to remove foreign materials that might interfere with the analytical test. Following cooling, distilled water was added to the digested sample and filtered into a 100 ml volumetric flask using the Johnson test paper filter paper with a diameter of 125 mm. Distilled water was added to the solution to the 100 ml mark.

Soil samples were prepared for pH analysis by dissolving 2 g of each sample in distilled water in a 1:1 ratio and stirring to a uniform suspended mixture using a clean glass rod. The samples were then allowed to settle for about 10 minutes. The samples were continually stirred for about 15 minutes using a magnetic stirrer on a magnetic sitter plate. The samples were allowed to settle, and their pH was determined by a handheld Hanna pH meter calibrated with pH buffer solutions 4, 7, and 10 [20, 21].

2.3 Water sample preparation

Water and sediment samples collected were filtered using the Johnson test paper filter paper with a diameter of 125 mm. In total, 10 ml of each filtrate was drawn into a 250 ml beaker, to which a 25 ml mixture was made up of 15 ml conc. HNO_3 and 10 ml conc. HCL was added and digested in a fume chamber for about 20 minutes. For efficient digestion of the water samples, an additional 10 ml of conc. HCL was added and heated in the fume chamber for about 15 minutes. Upon cooling, further filtration was carried out. Distilled water was added to the filtrate and made up to the 100 ml mark.

2.4 Determination of heavy metal concentrations

For each soil/water sample, calibration curves were prepared using heavy metal standards: 0.50, 1.00, 2.00, and 4.00 mg L^{-1} standards were each prepared in the spectrophotometric quantification of Cr, Cu, and Pb while calibration curves of concentrations 0.10, 0.20, 0.40, and 0.80 mg L^{-1} were ready in the case of Cd. For each soil/water sample, duplicate analyses were performed using Perkin Elmer 400 atomic absorption spectrophotometer with air-acetylene gas serving as fuel for the flame. Serial dilutions were performed on the samples where concentrations were very high, and after that, their dilution factors were factored in determining their concentrations.

2.5 Permissible limits of heavy metals in soil and water

As a guide, levels of heavy metals from the study will be compared with local and international standards of specification of heavy metals in soil and water bodies, as depicted in **Table 1**. Ghana's EPA utilizes the standards Romania set for heavy metal levels in the soil. This and Ghana EPA permissible limits for heavy metals were used as national standards to compare the heavy metals investigated in the samples. The WHO/FAO standards, available at https://www.researchgate.net/publication/345903051 and http://www.fao.org/3/t0234e/T0234E06.htm#ch5.5, were used as heavy metal limits in soil and water. These standards have also been used by other authors elsewhere.

2.6 Indices for determination of soil pollution

Three pollution indices were employed to evaluate how much the four heavy metals had polluted the scrapyard and its environment. The geoaccumulation index

Heavy metal	EPA Ghana limit in soil (ppm)	EPA Ghana limit in water (ppm)	WHO/FAO limit in soil (ppm)	WHO/FAO limit in water (ppm)
Cd	1	—	3	0.01
Cr	30	0.1	100	0.1
Cu	20	—	100	0.2
Pb	20	0.1	50	5

Table 1.
Ghana EPA and WHO/FAO permissible levels of heavy metals in soil and water.

(I_{geo}) determines the contamination of heavy metals by assessing their concentrations in sampled soils relative to background concentrations during pre-industrial periods [22, 23]. I_{geo} is computed using the mathematical formula:

$$I_{geo} = \log_2 (C_n / 1.5B_n) \quad (1)$$

C_n measures the heavy metal levels in the sediment under investigation, while B_n represents the geochemical background level [24]. The constant of 1.5 is to lessen the possible variations in the background data. Based on the results obtained, soils can be categorized into seven quality grades of pollution, as follows: practically unpolluted, where $I_{geo} < 0$; unpolluted to moderately polluted, where $I_{ge,o} = 0–1$; moderately polluted, where $I_{geo} = 1–2$; moderately to strongly polluted, where $I_{geo} = 2–3$; strongly polluted, where $I_{geo} = 3–4$; strongly to extremely polluted, where $I_{geo} = 4–5$ and extremely polluted, where $I_{geo} > 5$ [25].

The contamination factor (CF) evaluates quantities of an element in a sample normalized over the pre-industrial baseline value of the component [26]. Mathematically, CF is expressed as

$$CF - C_e / C_1 \quad (2)$$

Where C_e and C_i are, respectively, the heavy metal concentration levels in the sample of interest and the background value of the heavy metal of interest, based on values obtained, soil or sediments can be classified as no or low contamination, where CF < 1; moderate contamination, where 1 < CF < 3; considerable contamination, where 3 < CF < 6; very high contamination, where CF > 6 [27].

The pollution load index (PLI) [28] examines the mutual contribution of groups of metals to the pollution of a site. Mathematically,

$$PLI = (CF_1 \times CF_2 \times CF_3 \times CF_4 \times CF_5 \times ...CF_n)^{1/n} \quad (3)$$

where CF represents the contamination factor of each heavy metal element in a sampled soil and n is the number of heavy metals under consideration. The PLI indicates whether the site under consideration is lightly polluted, where PLI $\leq$ 1; moderately polluted; where 1 < PLI $\leq$3; highly polluted, where PLI > 3 [27, 28].

For instance, the metallic pollution levels in a study conducted by Fosu-Mensah et al. [28] were assessed using the WHO/FAO standards, whose links have been provided already.

2.7 Statistical and data analysis

Descriptive statistical variables, such as the mean, maximum, minimum, and standard deviation of heavy metal concentrations computed using Microsoft Excel software 2016 version. Pearson correlation, t-test, and coefficient of variation (CV) of the heavy metal concentrations were calculated by SPSS, version 21.0.

3. Results

3.1 Soil pH

Tables 2 and **3** show levels of heavy metals with their respective pH. pH values ranging from a mildly acidic pH of 5.88 to high alkaline pH of 8.03 was recorded for site F with an average of 7.13. At site H, samples had pH values between 6.07 and 7.78 and a mean of 6.94. pH values recorded were within the WHO benchmark of 6.5–8.5, except for three samples (5.88 at site F and 6.07, 6.38 at site H), which recorded pH values below the 6.5 minimum threshold.

Site /sample	Latitude	Longitude	Concentration of selected heavy metals (ppm)				pH
F			Cd	Cr	Cu	Pb	
1A	5.682713N	0.029065W	0.92	73.02	82.60	99.63	7.36
1B	5.682713N	0.029065W	n.d.	123.07	70.74	13.58	7.14
2A	5.682441N	0.029085W	0.05	49.08	129.37	276.78	6.86
2B	5.682441N	0.029085W	0.03	13.97	29.97	38.73	5.88
3A	5.682558N	0.028997W	0.17	162.50	122.30	83.48	6.80
3B	5.682558N	0.028997W	0.12	117.27	92.22	14.10	7.35
4A	5.682700N	0.028879W	0.11	60.81	219.82	32.97	7.38
4B	5.682700N	0.028879W	0.02	36.37	73.64	40.25	7.46
5A	5.682575N	0.029130W	0.52	67.49	74.42	38.75	7.04
5B	5.682575N	0.029130W	1.57	69.74	253.42	132.45	8.03
MEAN			0.39	77.33	114.85	77.07	7.13
MIN			0.02	13.97	29.97	13.58	5.88
MAX			1.57	162.50	253.42	276.78	8.03
STD.V			0.53	44.57	70.34	80.10	0.56

n.d., not detected.

Table 2.
Concentration and pH of soil samples at site F.

Site /sample	Latitude	Longitude	Concentration of heavy metals (ppm)				pH
H			Cd	Cr	Cu	Pb	
1A	5.683217N	0.029051W	0.29	17.11	108.76	84.29	7.78
1B	5.683217N	0.029051W	0.34	15.40	7.07	59.43	6.58
2A	5.683190N	0.028980W	1.67	30.11	5.24	300.25	6.91
2B	5.683190N	0.028980W	1.05	15.95	10.76	514.80	7.09
3A	5.683153N	0.028938W	13.56	25.25	12.40	17.81	7.38
3B	5.683153N	0.028938W	2.34	22.32	14.05	46.61	7.44
4A	5.683145N	0.029027W	6.87	21.11	48.00	572.79	6.50
4B	5.683145N	0.029027W	0.53	17.31	6.85	261.26	6.07
5A	5.683328N	0.028788W	7.96	23.10	185.77	1000.85	7.24
5B	5.683328N	0.028788W	6.79	22.32	84.76	556.20	6.38
MEAN			4.14	21.00	48.37	341.41	6.94
MIN			0.29	15.40	5.24	17.81	6.07
MAX			13.56	30.11	185.77	1000.85	7.78
STD.V			4.46	4.65	60.58	317.96	0.54

Table 3.
Concentration and pH of soil samples at site H.

3.2 Heavy metal concentrations at the burning sites of the scrapyard

The concentrations of the four heavy metals obtained are detailed in **Tables 2** and **3**. At site F, Cd ranged from a non-detection level to a maximum concentration of 1.57 ppm and an average of 0.48 ppm. Except for a subsoil sample with a concentration of 1.57 ppm, all Cd concentrations at site F were below the Ghana EPA permissible limit of 1.0 ppm and the WHO/FAO standard of 3 ppm. At site H, Cd concentrations were comparatively higher, with a minimum of 0.29 ppm, a maximum of 13.56 ppm, and an average concentration of 4.14 ppm, which exceeded Ghana EPA and the WHO/FAO standards.

A minimum concentration of 13.97 ppm and a maximum concentration of 162.50 ppm were recorded for Cr, with an average concentration of 77.33 ppm at site F. Most Cr concentrations exceeded the Ghana EPA threshold value of 30 ppm. Three samples also had Cr concentrations (123.07 ppm, 162.50 ppm, and 117.27 ppm) above the WHO/FAO standard of 100 ppm. However, Cr levels from site H were below the permissible limit of WHO/FAO and Ghana EPA with minimum and maximum concentrations of 15.95 ppm and 30.11 ppm, respectively, and an average of 21.00 ppm.

At site F, minimum and maximum concentrations of 29.97 and 253.42 ppm were recorded for Cu, with an average concentration of 114.85 ppm. They exceeded the permissible levels of Ghana EPA (20 ppm) and the WHO/FAO standards of 100 ppm. Also, at site H, Cu recorded minimum and maximum concentrations of 5.24 and 108.76 ppm, respectively, and an average concentration of 48.37 ppm, above the national and international standard limits.

At site F, a minimum concentration of 13.58 ppm and a maximum concentration of 276.78 ppm were recorded for Pb with an average concentration of 77.07 ppm, which

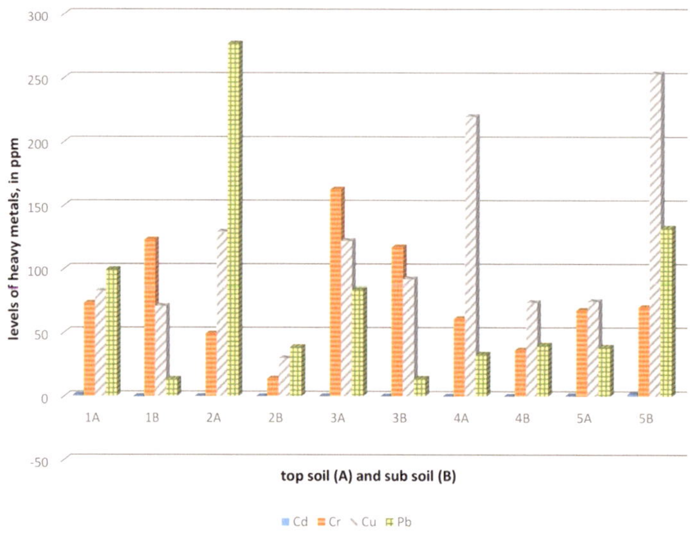

Figure 3.
Heavy metal levels in samples from site F.

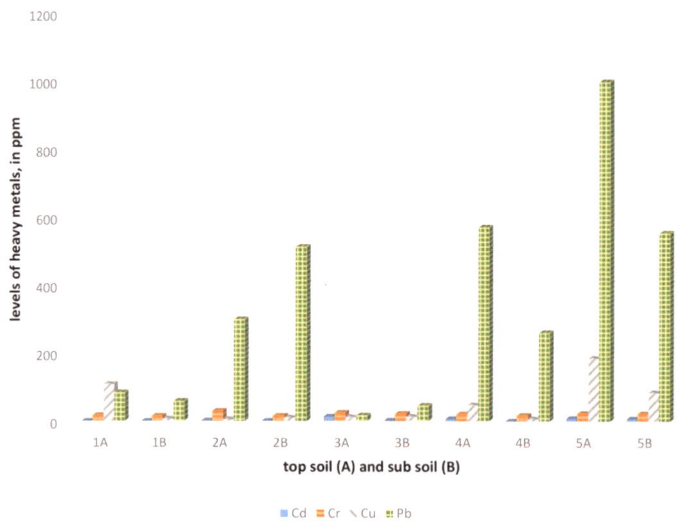

Figure 4.
Heavy metal levels in samples from site H.

was above the WHO/FAO and Ghana EPA standard of 50 ppm and 20 ppm, respectively. Pb in the sampled soil had a minimum concentration of 17.81 ppm, a maximum concentration of 1000.85 ppm, and an average concentration of 341.43 ppm at site H, exceeding the Ghana EPA and WHO/FAO standard limits.

The extent of pollution at sites F and H can be respectively expressed as Cu > Cr > Pb > Cd and Pb > Cu > Cr > Cd. This is also illustrated graphically in **Figures 3** and **4**.

Cities	Cd	Cr	Cu	Pb	Reference
Koforidua, Ghana	3	47	14,300	3530	[6]
Ibadan, Nigeria	2.50 ± 0.08	42.4 ± 2.6	3483 ± 980	5650 ± 750	[29]
Bangalore, India	0.478	54	429	126	[30]
Wenling, China	3	101.29	180.66	187.30	[31]
Ashaiman, Ghana	7.96	23.10	185.77	1000.85	Current research

Table 4.
A review of research studies on heavy metals concentrations (ppm) in e-waste soil.

Heavy metal concentrations in this study were similar to other research works in e-waste research, as shown in **Table 4**. Generally, Cu and Pb were in high concentrations in most of the research studies. Also, Cd concentrations are lower in most e-waste soils in other research works.

3.3 Statistical analysis

Pearson correlation revealed a positive relationship ($p < 0.05$ and $p < 0.01$) between Cd and Cu, Cd and Pb, and Cu and Pb in the subsoil at both investigated sites. These relationships are depicted in **Tables 5** and **6**.

Positive correlations were established between concentrations of heavy metals within the subsoil at both sites. This is shown in **Table** 7.

Coefficient of variation (CV) was computed for each heavy metal at both e-waste burning sites. Cr showed the least variation with CV values of 57.63% and 22.14%, respectively, at sites F and H. At sites F and H, CV values for Cd were high (136.77% and 107.76%, respectively), whiles Pb had CV values of 103.92% at site F and 93.13% at site H. A CV value of 61.24% was obtained for Cu at site F. Comparatively, a higher measurement of Cu was recorded at site H, with a CV value of 125.25%.

An independent t-test was conducted at $p < 0.05$ to determine the statistical significance of the mean concentration of the heavy metal. Results showed the following:

i. A significant difference between Cd concentrations at site F (Mean (M) = 0.39, standard deviation (SD) = 0.53) and Cd concentrations at site H (M = 4.14, SD = 4.46), with a t-value of −2.50 and p-value =0.02 (data are not normally distributed; skewed);

ii. A significant difference between Cr concentrations at site F (M = 77.33, SD = 44.57) and Cr concentrations at site H (M = 21.00, SD = 4.65), with a t-value of −3.98 and p-value = 0.001 (data are normally distributed);

iii. A significant difference between Cu concentrations at site F (M = 114.85, SD = 70.33) and Cu concentrations at site H (M = 48.37, SD = 6.58), with a t-value of 2.27 and p-value = 0.04 (data are normally distributed); and

iv. A significant difference between Pb concentrations at site F (M = 77.07, SD = 80.10) and Pb concentrations at site H (M = 341.43, SD = 317.96), with a t-value of −2.55 and p-value = 0.02 (data are not normally distributed; skewed).

	[Cd]	[Cr]	[Cu]	[Pb]
[Cd]	1.00	−.02	**.97b**	**.96a**
[Cr]	−.02	1.00	.17	−.28
[Cu]	**.97b**	.17	1.00	**.90a**
[Pb]	**.96a**	.28	**.90a**	1.00

[a]*significant correlation at the 0.05 level (2-tailed).*
[b]*significant correlation at the 0.01 level (2-tailed).*

Table 5.
Pearson correlation between heavy metal levels in the subsoil at site F.

	[Cd]	[Cr]	[Cu]	[Pb]
[Cd]	1	.79	**.98b**	.54
[Cr]	.79	1.00	.65	.08
[Cu]	**.98b**	.65	1.00	.61
[Pb]	.54	.08	.61	1.00

[b]*significant correlation at the 0.01 level (2-tailed).*

Table 6.
Pearson correlation between heavy metal levels in the subsoil at site H.

	[Cd]	[Cr]	[Cu]	[Pb]	**Site H**
[Cd]	**.97b**	.65	**.99b**	.59	
[Cr]	.05	.30	−.01	−.76	
[Cu]	**.95a**	.72	**97b**	.41	
[Pb]	**.90a**	.50	**.95a**	.76	
Site F					

[a]*significant correlation at the 0.05 level (2-tailed).*
[b]*significant correlation at the 0.01 level (2-tailed).*

Table 7.
Pearson correlation between heavy metals in the subsoil at site F and site H.

3.4 Indices of pollution

3.4.1 Index of geoaccumulation

Table 8 shows the I_{geo} of sampled soil of the two sites. The I_{geo} showed site F was practically uncontaminated with Cd (average I_{geo} = −1.58) and Cr (average I_{geo} = −1.07) but moderately polluted with Cu (average I_{geo} = 0.53) and Pb (average I_{geo} = 0.76). At site H, Igeo showed moderate to strong pollution with Cd (average I_{geo} = 2.16) and Pb (average I_{geo} = 2.64), a practically unpolluted soil with Cr (average I_{geo} = −2.72), and Cu (average I_{geo} = −1.63). Site H appears more contaminated than site F, probably due to its use as a burning and dumping site for e-waste materials.

Soil sample	Igeo of heavy metals							
	Site F				Site H			
	Pb	Cu	Cr	Cd	Pb	Cd	Cu	Cr
1A	1.73	0.29	−0.89	1.03	1.49	−0.64	0.69	−2.98
1B	−1.14	0.07	−0.13	n.d.	0.99	−0.39	−3.26	−3.13
2A	3.21	0.94	−1.46	−3.24	3.32	1.89	−3.69	−2.17
2B	0.37	−1.17	−3.27	−3.99	4.10	1.22	−2.65	−3.08
3A	1.48	0.86	0.27	−1.39	−0.75	4.91	−2.45	−2.42
3B	−1.09	0.45	−0.20	−1.86	0.64	2.38	−2.26	−2.60
4A	0.14	1.70	−1.15	−2.01	4.26	3.93	−0.49	−2.68
4B	0.42	0.13	−1.89	−4.76	3.12	0.23	−3.30	−2.96
5A	0.37	0.14	−1.00	0.22	5.06	4.15	1.46	−2.55
5B	2.14	1.91	−0.95	1.81	4.21	3.92	−0.33	−2.60
MEAN	0.76	0.53	−1.07	−1.58	2.64	2.16	−1.63	−2.72

n.d., not detected.

Table 8.
Igeo of selected heavy metals at sampled sites F and H.

Soil sample	Site F					Site H				
	CF				PLI	CF				PLI
	Pb	Cu	Cd	Cr		Pb	Cd	Cu	Cr	
1A	4.98	1.84	3.06	0.81	2.18	4.22	0.96	2.42	0.19	1.17
1B	0.68	1.57	-	1.37	1.36	2.97	1.14	0.16	0.17	0.55
2A	13.8	2.88	0.16	0.55	1.36	15.01	5.55	0.12	0.34	1.34
2B	1.94	0.67	0.09	0.16	0.37	25.74	3.50	0.24	0.18	1.40
3A	4.17	2.72	0.57	1.81	1.85	0.89	45.22	0.28	0.28	1.33
3B	0.71	2.05	0.42	1.30	0.94	2.33	7.80	0.31	0.25	1.09
4A	1.65	4.89	0.37	0.68	1.19	28.64	22.91	1.07	0.24	3.58
4B	2.01	1.64	0.06	0.40	0.52	13.06	1.75	0.15	0.19	0.91
5A	1.94	1.65	1.75	0.75	1.43	50.04	26.54	4.13	0.26	6.13
5B	6.62	5.63	5.25	0.78	3.51	27.81	22.64	1.88	0.25	4.14
MEAN	3.85	2.55	1.30	0.86	1.47	17.07	13.80	1.08	0.24	2.16

Table 9.
CF and PLI of selected heavy metals at sampled sites F and H.

3.4.2 Contamination factor and pollution load index

Table 9 provides information on the two sites' CFs and PLIs. The CF values showed site F was moderately contaminated with Cd (average CF = 1.30) and Cu (average CF = 2.55), less contaminated with Cr (average CF = 0.86), and considerably contaminated by Pb (average CF = 3.85). At site H, CF showed very high contamination of the soil with Cd (average CF = 13.80) and Pb (average CF = 17.07), no or low contamination of the soil with Cr (average CF = 0.24), moderately contaminated soil with Cu (average CF = 1.08). Again, results show more contamination at site H than at site F, possibly because of the dumping activities and the open burning.

The PLIs of both sites were found to be moderately polluted, with respective average values of 1.47 and 2.16 at sites F and H.

3.5 Spatial distributions of the heavy metals

Figures 5–8 show spatial distribution patterns of thc hcavy metals at the two sites, which were analyzed using the inverse distance weighted (IDW) interpolation method. The analysis revealed elevated levels of heavy metals in subsoil (Cd, Pb at site H and Cr, Cu at site F). Spatial maps also showed that site H was more polluted with Pb and Cd, while site F was mainly Cr and Cu.

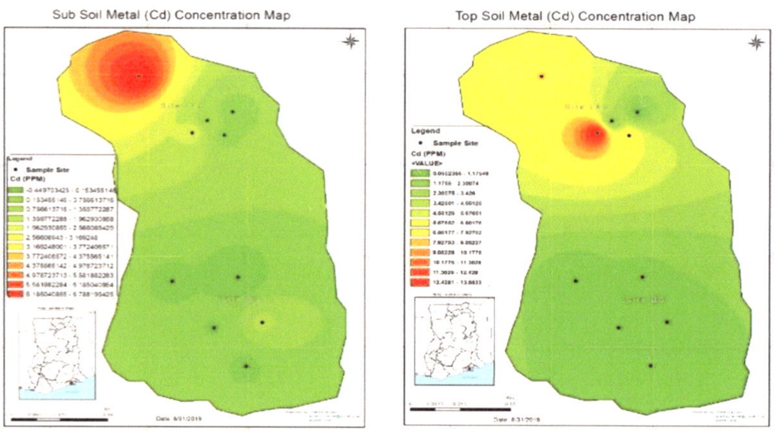

Figure 5.
Spatial distributions of Cd in subsoil (L) and topsoil (R) of sites F and H.

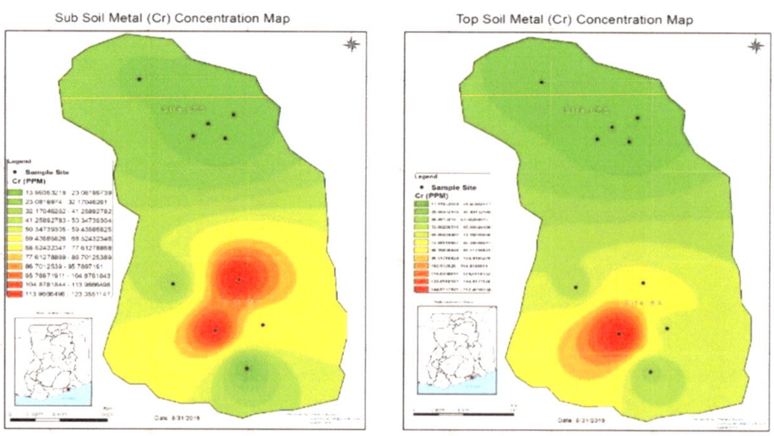

Figure 6.
Spatial distributions of Cr in sites F and H's subsoil (L) and topsoil (R).

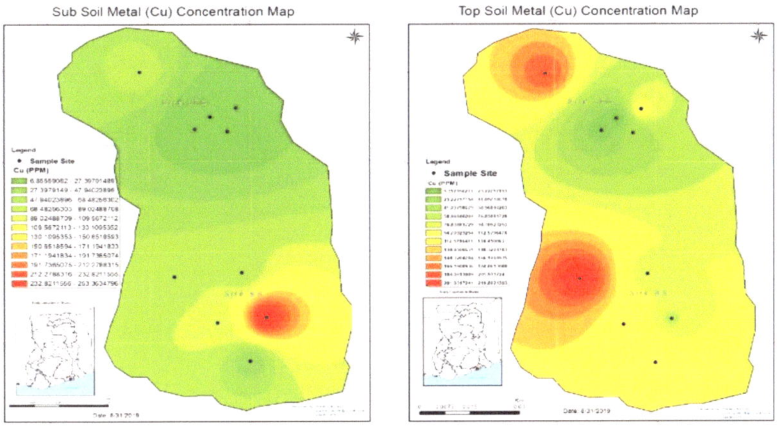

Figure 7.
Spatial distributions of Cu in subsoil (L) and topsoil (R) of sites F and H.

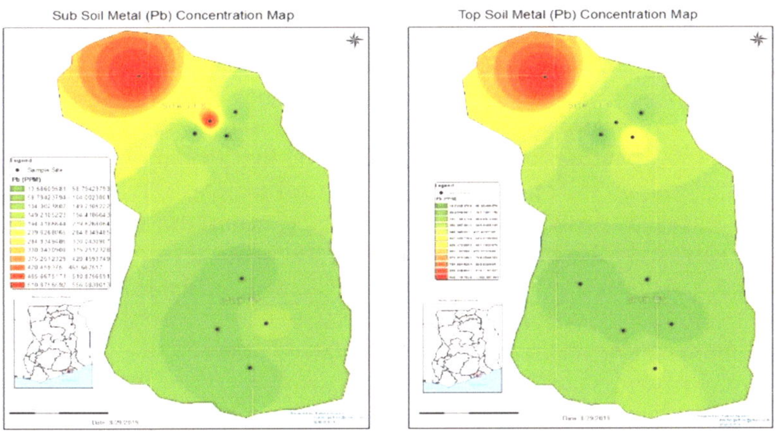

Figure 8.
Spatial distributions of Pb in subsoil (L) and topsoil (R) of sites F and H.

3.6 Concentration differences at increasing distance from the scrapyard

The study also sought to determine levels of heavy metals in the soil components at different distances from the scrapyard. This was important to evaluate the extent to which informal e-waste activities affected nearby communities. The result is provided in **Table 10**. Soil samples taken 25, 50, 75, and 100 m from the scrapyard were mainly sandy. pH values were mildly acidic and were within the 6.5–8.5 WHO thresholds. This indicates a decreasing pH as one moves away from the scrapyard.

Results revealed no level of Cd in these soil samples. Samples within the 25 m distance recorded respective concentrations of 20.73 and 24.94 ppm for Cr and Cu and were within safe levels set by WHO/FAO but slightly above permissible levels of Ghana EPA concerning Cu. However, Pb recorded concentrations of 155.17 ppm, which exceed the safe levels of Pb as determined by both WHO/FAO and Ghana EPA. The pH of the soil sample at 25 m was almost neutral at 6.97.

Sample	Latitude	Longitude	Concentration of selected heavy metals (ppm)				pH
			Cd	Cr	Cu	Pb	
HV 25	5.682392N	0.027973W	n.d	20.73	24.94	155.17	6.97
HV 50	5.682438N	0.027473W	n.d	25.18	96.73	74.72	6.58
HV 75	5.682677N	0.026773W	n.d	4.12	4.60	5.97	6.71
HV 100	5.6826190N	0.026582W	n.d	n.d	1.26	8.97	6.72
MEAN				16.68	31.88	61.21	6.75
MIN				4.12	1.26	5.97	6.58
MAX				25.18	96.73	155.17	6.97
STD.V				11.10	44.48	70.22	0.16

n.d., not detected.

Table 10.
Heavy metal levels at distances from the scrapyard.

Levels of Cr and Cu in samples within the 50 m boundary were within the safe limits set by WHO/FAO, but above permissible levels of Ghana EPA, with respective concentrations of 25.18 and 96.73 ppm. Pb quantified in these samples shows levels were above the 50-ppm threshold of WHO/FAO and the 20-ppm threshold of Ghana EPA, reaching levels of 74.72 ppm. The pH of the soil sample at 50 m was mildly acidic (6.58). Samples taken 75 m from the scrapyard had concentrations of 4.122 ppm (Cr), 4.600 ppm (Cu), and 5.965 ppm (Pb), while at a 100 m distance variation samples analyzed revealed no levels of Cr, 1.260 ppm of Cu, and 8.970 ppm of Pb.

3.7 Heavy metal concentration in water and water sediment

Table 11 compares levels of heavy metals at different sections and depths of a drain near the scrapyard. Water sediments outside the scrapyard showed lower concentrations of heavy metals than those obtained within the scrapyard. In contrast, none of the four heavy metals was detected in the water samples. Water sediment outside the scrapyard contained levels of Cd at 0.03 ppm, Cr at 11.95 ppm, and Cu and Pb concentrations, respectively, at 5.84 and 5.89 ppm. Water sediment within the scrapyard contained 0.49 ppm Cd and a concentration of 217.98 ppm for Cu. Cr had a concentration of 12.28 ppm, while Pb had a concentration of 44.77 ppm.

4. Discussion

Uncontrolled levels and spatial variabilities of e-waste have serious environmental repercussions. E-waste-laden environments have significant amounts of heavy metals. Such metals interfere with ecosystem integrity and health. Bioaccumulation and biomagnification of the metals remain persistent in the food webs. They pose severe hazards and risks to the biota. More significantly to humans, chronic exposure to these metals in uncontrolled scrap settings and using substandard resource recovery methods put them at high risk of several health damages, which include carcinogenicity, teratogenicity, mutagenicity, genotoxicity, immunosuppression, and physiological

Sample	Latitude	Longitude	Heavy metal concentration (ppm)				pH
			Cd	Cr	Cu	Pb	
WS C	5.684598N	0.026144W	0.03	11.95	5.84	5.89	7.55
WS 1	5.683145N	0.029027W	0.49	12.28	217.98	44.77	7.49
WS 2	5.683328N	0.028788W	n.d	n.d	n. d	n. d	7.61
MEAN			0.26	12.12	111.91	25.33	7.55
MIN			0.03	11.95	5.84	5.89	7.49
MAX			0.49	12.28	217.98	44.77	7.61
STD.V			0.33	0.23	150.00	27.49	0.06

n.d., not detected.

Table 11.
Heavy metals concentrations in the water sample.

and biochemical disorders [32–34]. While the toxicological analysis of the above effects on scrap workers was beyond the scope of the study, the environmental profile analysis has discovered some heavy metals in the study area, which hitherto was unknown. This sets a baseline upon which future research dimensions can evolve.

The concentrations and spatial variations (using IDW) of the heavy metals and the pollution levels using Igeo, CF, and PLI have been quantified and presented in the previous section, together with pH variations at the two e-waste burning sites. Generally, the study revealed that topsoil concentrations of heavy metals were higher than those of the subsoil, with few exceptions. This general trend could be due to the strong affinity of the heavy metals, mostly Pb and Cu, with the abundance of organic matter and minerals found in the topsoil, preventing the percolation of the heavy metals into the subsoil [35–37]. Additionally, it can be inferred from the results that anthropogenic pollution of heavy metals has more effect on the topsoil than subsoil. However, heavy metals found in the subsoil are particularly worrying, as the absorption of nutrients and water by plants takes place through the root system in the subsoil. The subsoil is also home to diverse microorganisms, and toxic metals can destabilize their niche. The high concentrations of heavy metals in some subsoils than in topsoil can be attributed to the leaching capability of the topsoil. Due to the high porosity of the top, sandy soil, heavy metals such as Cd and Cr are retained less in the topsoil and are percolated towards the subsoil [38]. Furthermore, Cd and Cr are less bonded to organic matter and minerals in the soil [38–39]. The above trends are comparable to other studies [39–43]. Pb and Cu were the heavy metals with the highest concentration, possibly because they find more applications in EEE, such as printed circuit boards, cathode ray tubes, bare/insulated wires, refrigeration units, fluorescent bulbs, batteries, and fuses. Furthermore, since Pb is not biodegradable, concentrations of Pb could build up for all the operational years of the scrapyard, resulting in the high concentrations measured. Sources of chromium in the scrapyard include steel alloy, and colored plastics, which are used as combustible materials for the burning of e-waste materials. Comparatively, Cr concentrations at site F were higher than at site H, possibly because the metal containers housing e-waste materials were closer to site F. These are typically composed of steel and chromium, so any wear and tear on the metal add Cr concentration to the soil. Cd was the heavy metal with the least concentration. In addition to the leaching and percolation effect of the soil structure, the mild to acidic pH of soils has also been shown to be a factor in the high mobility of Cd, resulting in its lower concentrations in the soil component [29]. Site H had higher concentrations of Cd than site F because Cd-containing e-waste materials, including printed circuit boards, batteries, accumulators, cathode ray tubes, and ultraviolet lights, were located more at the former site than the latter. However, the low concentrations of Cd should not be underestimated, as Cd is one of the most toxic heavy metals, especially to aquatic organisms. Cd pollution is related to an increased mortality rate from obstructive lung disease. Cadmium absorption also causes shortness of breath and emphysema. All heavy metals under consideration exceeded national and international standards, suggesting that the open burning of e-waste materials to extract valuable metals leads to excessive pollution of the environment. Indeed, other studies and research reach similar conclusions, and in some cases, other pollutants, such as Poly Aromatic Hydrocarbons (PAHs), are further identified. Igeo, CF, and PLI metrics have substantiated the current study's heavy metal pollution variations.

Heavy metal adsorption and retention by soil increases generally within a pH range of 4–7 [38, 44], and therefore the pH ranges from the study could account for the elevated levels of heavy metals found in the samples. According to a study by [30], dumpsite samples could retain heavy metals within a pH range of 2–8. The high pH value recorded in sample 5B (8.03) could be due to alkaline batteries, steel mills, and ashes from the incineration processes at the e-waste site. The range of pH values for this study is comparable to other e-waste research [9, 28].

The general decline of pH at increasing distance from the scrapyard was expected as increasing distance from the scrapyard meant decreasing heavy metal concentrations, most alkaline. This result is comparable to a study by Tang et al. [45], where the pH at a dumpsite decreased from 5.9 to 4.7 at 18 m from the dump site. The current research indicates that activities at the scrapyard had an effect 25–50 m away from it. However, since soil samples taken at 25 and 50 m were close to the Accra-Tema motorway, contamination from road dust is still possible since heavy metals are found in tires and brake abrasion, combustion exhaust, and pavement wear [46]. Further research will be needed to evaluate this assertion. With a general decline in the concentrations of heavy metals from the 75 and 100 m distance, the high levels of heavy metals within the scrapyard can be attributed mainly to the e-waste activities. Comparably, Cr, Cu, and Pb concentrations were several times higher within the scrapyard than outside. This decreasing concentration of heavy metals with increasing distances from the scrapyard agrees with other studies, which explored the effect of increasing distance from the source on concentration levels of heavy metals [47–49].

Analysis of water sediments showed that levels of the toxic metals in the water sediments increased significantly within the scrapyard compared to the control sample, which was taken outside the scrapyard area. With the drainage lying at a lower plain to the two burning sites, and with the movement of air current across the drainage from the two burning sites, it can be fairly postulated that the e-waste activities are a possible source of heavy metals in the water-sediment, through the actions of wind drift, wet and dry depositions. Another possibility is the presence of e-waste materials near or inside the drainage, causing heavy metals to leach into it [50]. Compared to no detection levels within the water itself, the relatively concentrated amounts of heavy metals in sediments affirm studies associating the high affinity of heavy metals with the suspended matter in water environments [51–52]. Heavy metals in the wastewater were above the standard permissible levels of Ghana EPA and WHO/FAO. This is of major concern as it serves as an irrigation source for farming crops and as drinking water for herds of cattle near the scrapyard. Studies conducted on vegetation and animals near the e-waste scrapyard revealed high levels of toxic metals in plants' root, stem, and leaves [53].

Sediment and water from the drain generally had neutral pH for both control samples and those taken within the scrapyard area. This observation differs from other studies [54–56] where the pH of water samples was in the acidic range (3.78–6.53). At lower pH, metals tend to have higher solubilities, leading to higher metal levels. This could be one of the primary reasons for detecting higher heavy metal concentrations in the water samples reported in the other studies [54–56] than in the current research. The highly positive correlation coefficients observed between pairs of heavy metals (0.90, 0.96, 0.97 in **Table 5** and 0.98 in **Table 6**) may be due to their dual complementary usage in certain EEE products. For instance, Cd and Pb find close applications in cathode ray tubes where Cd is used as the fluorescent powder coatings to produce color, while Pb is employed to absorb the UV lights and X-rays built. Cd-Cu alloy wires are more resistant to softening at higher temperatures, hence their co-occurrence in the waste. Pb alloyed to Cu acts as a lubricant and assists in chip breakup, increasing the machinability of the Cu metal. Since site H is a burning site and dumping grounds for e-waste materials, heavy metals can be carried from site F to site H. This could explain the high positive correlation (0.90, 0.95, 0.97, and 0.99) between heavy metals at different sites in **Table 7**. The weak correlations between heavy metals at the two sites could also indicate different and unrelated sources of contamination of the heavy metals.

Coefficient of variation (CV) results suggest that most heavy metals are widely dispersed rather than contained at locations. Wind effects, dry and wet deposition, and migration through water and soil are the primary sources of heavy metal dispersion or transport. High levels due to transport are a worry as it indicates e-waste recycling pollution is not limited to its immediate surroundings but can extend to other parts of the environment. Additionally, CV values provide insight into the sources of contamination. According to a study by [57], a CV of less than 20% indicates natural sources, while values greater than 50% imply anthropogenic sources. By inference, the heavy metal pollution was primarily due to anthropogenic sources, specifically e-waste activities, amplified by environmental factors.

5. Conclusions

The research showed that the heavy metal levels exceeded the permissible limits of the WHO/FAO and Ghana EPA standards. Pollution indices suggest the e-waste scrapyard were polluted with the four heavy metals investigated in varying degree. The CV results indicate that metal pollution is primarily anthropogenic-given and widely dispersed. Spatial distribution maps suggest contamination of the scrapyard, especially at the western north of site H and the central portion of site F. Consequently, environmental laws and regulations on the management and recycling of e-waste should be enforced by local authorities to prevent further pollution of the scrapyard and its environment. Public awareness and education on the adverse effect of informal recycling practices should be intensified. The study has further shown that Cd and Pb levels in the scrapyard suggest skewed distributions relative to Cr and Cu, which are normally distributed. This outcome provides insight into modeling the behavior of these metals in the future. Finally, future studies can also focus on investigating heavy metal contamination in workers at the scrapyard and herds of cattle around the environment.

Acknowledgements

The authors express appreciation to the workers of the Ashaiman scrapyard in the Greater Accra Region of Ghana for their cooperation in carrying out this study. Our appreciation also goes to the laboratory staff of Ghana Standard Authority, where the soil and water samples were analyzed.

Funding

The present study was self-funded.

Competing interests

The authors declare there is no conflicting interest regarding this study.

Availability of data and material

All data generated or analyzed during this study are included in this published article.

Authors' contributions

The second and corresponding author supervised the entire research— conceptualization, synthesis of ideas, scope, experimental design, analysis framework, data interpretation, review, and manuscript editing. The first author performed the field and lab work. He also contributed to data analysis, manuscript writing, and editing.

Author details

Albert Kwame Teye and Isaac Kow Tetteh*
Department of Environmental Science, Kwame Nkrumah University of Science and Technology, Kumasi, Ghana

*Address all correspondence to: iktetteh.cos@knust.edu.gh

References

[1] Solving the E-Waste Problem (Step) White Paper. One Global Definition of E-Waste. Tokyo Japan: United Nations University Press; 2014. p. 13

[2] Baldé CP, Forti V, Gray V, Kuehr R, Stegmann P. The Global E-Waste Monitor 2017: Quantities, Flows and Resources2017. pp. 1-117. Available from: https://www.itu.int/en/ITU-D/Climate-Change/Documents/GEM2017/Global-E-wasteMonitor2017.pdf [Accessed: February 19, 2021]

[3] Mmereki D, Li B, Baldwin A, Hong L. The Generation, Composition, Collection, Treatment and Disposal System, and Impact of E-Waste. E-Waste in Transition - From Pollution to Resource. InTech; 2016. pp. 1-31. DOI: 10.5772/61332

[4] World Economic Forum. A New Circular Vision for Electronics: Time for a Global Reboot2019. pp. 1-24. Available from: http://www3.weforum.org/doc/WEF_A_New_Circular_Vision_for_Electronics.pdf [Accessed: February 19, 2021]

[5] Malakata M. West Africa Turns into a Dumping Ground for E-Waste. 2015. Available from: https://www.cio.com/article/2878493/west-africa-turns-into-dumping-ground-for-ewaste.html [Accessed February 19, 2021]

[6] Amoyaw-Osei Y, Agyekum OA, Pwamang JA, Mueller E, Fasko R, Schluep M. Ghana E-Waste Country Assessment. SBC E-waste Africa Project; Dubendorf, Switzerland: EMPA; 2011. pp. 1-123

[7] Brigden K, Labunska I, Santillo D, Johnston P. Chemical Contamination at E-Waste Recycling and Disposal Sites in Accra and Koforidua, Ghana. Greenpeace Research Laboratories Technical note 2008; 2008. pp. 1-23. Available from: www.greenpeace.to/publication/chemical-contamination-at-e-wa.pdf [Accessed February 19, 2021]

[8] Luther L. Managing Electronic Waste: Issues with Exporting E-Waste. Congressional Research Service; 2010. pp. 5-12

[9] Khadijah AI. Environmental risk assessment for an informal e-waste recycling site in Lagos State, Nigeria. Ph.D thesis Middlesex University Research Repository; 2016. Available from: https://eprints.mdxac.uk/22233/

[10] Dharini K, Bernadette JC, Kamalambikai B, Arul SCJP, Muthu D. Hazardous e-waste, and its impact on soil structure. IOP Conf Series: Earth and Environmental Sciences. 2017;**80**(1):2-5

[11] Holloway PH. Gold/chromium metallizations for electronic device. Gold Bull. 1979;**12**(3):99-106

[12] Vats MC, Singh SK. E-waste characteristic and its disposal. International Journal of Ecological Science and Environmental Engineering. 2014;**1**(2):49-61

[13] Plachy J. Cadmium recycling in the United States in 2000. U.S; Geological Survey Circular. 2003;**1**:6-12

[14] Agency for Toxic Substances and Disease Registry U.S. Department of Health and Human Services. Toxicological Profile for Chromium. Atlanta, GA: Public Health Service; 2008. pp. 1-20. Available from: www.atsdr.cdc.gov/toxprofiledocs/index.html#tabs-2-3 [Accessed: February 19, 2021]

[15] Tchounwou PB, Yedjou CG, Patlolla AK, Sutton DJ. Heavy metal toxicity and the environment. Molecular, Clinical and Environmental Toxicology. 2012;**101**:133-164. DOI: 10.1007/978-3-7643-8340-4_6

[16] Agency for Toxic Substances and Disease Registry, US Department of Health and Human Services. Toxicological Profile for Copper2004. pp. 1-15. Available from: www.atsdr.cdc.gov/toxprofiledocs/index.html#tabs-2-3 [Accessed February 19, 2021]

[17] Pwamang JA. Government Policy and Initiatives on E-Waste in Ghana. 2011. Available from: https://www.epa.gov/sites/production/files/2014-05/documents/ghana_2.pdf [Accessed February 19, 2021]

[18] Ashaiman Municipal Assembly. ASHMA Municipal Profile. National Development Planning Commission. 2018. pp. 1-3

[19] Okine HA. E-waste imports and management practices in Ghana: A case study of Accra-Tema metropolitan area [Mphil thesis]. Ghana: University of Ghana; 2014. Available from: https://197.255.68.203/handle/123456789/7480 [Accessed February 19, 2021]

[20] Wong C, Xiang-Dong L. Analysis of heavy metal contaminated soils. In: Practice Periodical of Hazardous Toxic Radioactivation Waste Management. 2003. pp. 2-15

[21] US EPA. Method 9045D soil and waste pH. In: Test Methods for Evaluating Solid Waste. 2004. pp. 1-4. Available from: www.epa.gov

[22] United States Department of Agriculture. Soil Survey Field and Laboratory Methods Manual. Soil Survey Investigations Report No. 51. Version 2.0. 2014. pp. 28-32

[23] Atiemo S, Faabeluon L, Manhart A, Nyaaba L, Schleicher T. Baseline Assessment on E-Waste Management in Ghana. Sustainable Recycling Industries. 2016. pp. 11-24

[24] Muller G. Index of gcoaccumulation in sediments of the Rhine River. GeoJournal. 1969;**2**:108-118

[25] Taylor SR, McLennan SM. The Continental Crust: Its Composition and Evolution. Journal of Geology. 1985;**94**:632-633. DOI: 10.1017/S0016756800032167

[26] Lu X, Li Y, Wang L, Lei K, Huang J, Zhai Y. Contamination assessment of mercury and arsenic in roadway dust from Baoji, China. Atmos. Environ. 2009;**43**(15):2489-2493

[27] Özkan EY. A new assessment of heavy metal contaminations in an Eutrophicated Bay (Inner Izmir Bay, Turkey). Turkish Journal of Fisheries and Aquatic Sciences. 2012;**12**:135-147

[28] Fosu-Mensah BY, Addae E, Yirenya-Tawiah D, Nyame F. Heavy metals concentration and distribution in soils and vegetation at Korle Lagoon area in Accra, Ghana. Cogent Environmental Science. 2017;**3**:1-14. DOI: 10.1080/23311843.2017.1405887

[29] Adelekan BA, Abegunde KD. Heavy metals contamination of soil and groundwater at automobile mechanic villages in Ibadan, Nigeria. International Journal of Physical Sciences. 2011;**6**(5):1047-1055

[30] Ha NN, Agusa T, Ramu K, Tu NCT, Murata S, Bulbule Ka, et al. Contamination by trace elements at e-waste recycling sites in Bangalore, India. Chemosphere. 2009;**76**(1):9-15

[31] Chen TB, Zheng YM, Lei M, Huang ZC, Wu HT, Chen H. Assessment of heavy metal pollution in surface soils of urban parks in Beijing, China. Chemosphere. 2005;**60**:543-550. DOI: 10.1016/j.chemosphere.2004.12.072

[32] Huo X, Peng L, Xu X, Zheng L, Qiu B, Qi Z, et al. Elevated blood lead levels of children in Guiyu, an electronic waste recycling town in China. Environmental Health Perspectives. 2007;**115**:1113-1117

[33] Feldt T, Fobil JN, Wittsiepe J, Wilhelm M, Till H, Zoufaly A, et al. High levels of PAH-metabolites in urine of e-waste recycling workers from Agbogbloshie, Ghana. Science of the Total Environment. 2014;**466-467**:369-376

[34] Li Y, Xu X, Wu K, Chen G, Liu J, Chen S, et al. Monitoring of lead load and its effect on neonatal behavioral neurological assessment scores in Guiyu, an electronic waste recycling town in China. Journal of Environmental Monitoring. 2008b;**10**:1233-1238

[35] Kerndorff H, Schnitzer M. Sorption of metals on humic acid. Geochima et Cosmochimica Acta. 1980;**44**(11):1701-1708

[36] Sherene T. Mobility and transport of heavy metals in polluted soil environment. Biol. Forum-Int. J. 2010;**2**:112-121

[37] Elinder CG. Cadmium as an environmental hazard. IARC Scientific Publications. 1992;**118**:123-132

[38] Alloway BJ, Steinnes E. Anthropogenic additions of cadmium to soils. In: McLaughlin MJ, Singh BR, editors. Cadmium in Soils and Plants. Developments in Plant and Soil Sciences. 1999;**85**:98-115. DOI: 10.1007/978-94-011-4473-5_5

[39] Kuo S, Heilman PE, Baker AS. Distribution and forms of copper, zinc, cadmium, iron, and manganese in soils near a copper smelter. Soil Science. 1983;**135**(2):101-109

[40] Wieczorek J, Baran A, Urbanski K, Mazurek R, Klimowicz-Pawlas A. Assessment of the pollution and ecological risk of lead and cadmium in soils. Environmental Geochemistry and Health. 2018;**40**(6):2325-2342

[41] Alloway BJ. Soil Processes and Behavior of Heavy Metals. In: Alloway BJ, editor. Heavy Metals in Soils. New York: Blackie Academic and Professional Publication; 1995. p. 368. DOI: 10.1007/978-94-011-1344-1

[42] Lenntech WT. Chemical Properties, Health, and Environmental Effects of Copper. 2009. Available from: www.lenntech.com/periodic/elements/cu.htm.

[43] Kabata-Pendias A. Trace elements in soils and plants. Boca Raton, FL, USA: CRS Press, Taylor and Francis Group; 2011;**4**:505-545

[44] Adesokan MD, Osibanjo O, Adie G. Soil pollution by toxic metals near E-waste recycling operations in Ibadan, Nigeria. Journal of Health and Pollution. 2016;**6**(11):26-33. DOI: 10.5696/2156-9614-6-11.26

[45] Tang X, Chaofeng S, Dezhi S, Sardar AC, Muhammad IK, Congkai Z, et al. Heavy metal, and persistent organic compound contamination in soil from Wenling: An emerging e-waste recycling city in Taizhou area, China. Journal of Hazardous Materials. 2009;**173**(1-3):653-660. DOI: 10.1016/j.jhazmat.2009.08.134

[46] Guo G, Wu F, Xie F, Zhang R. Spatial distribution and pollution assessment of heavy metals in urban soils from Southwest China.

2012;**24**(3):411-416. DOI: 10.1016/S1001-0742(11)60762-6

[47] Aisien FA, Okoduwa IG, Aisien ET. Levels of heavy metals in and around scrap car dumpsite at Uwelu, Nigeria. Current Journal of Applied Science and Technology. 2013;**3**(4):1519-1522. DOI: 10.9734/BJAST/2014/5052

[48] Radziemska M, Fronczyk J. Level and contamination assessment of soil along an expressway in an ecologically valuable area in Central Poland. International Journal of Environmental Research and Public Health. 2015;**12**(10):13372-13387. DOI: 10.3390/ijerph121013372

[49] Svendsen ML, Blom HA, Steinnes E. Vertical and horizontal distributions of Zn, Cd, Pb, Cu, and Hg in uncultivated soil in the vicinity of a zinc smelter at Odda, Norway. Soil and Sediment Contamination. 2007;**16**(6):585-603. DOI: 10.1080/15320380701623644

[50] Fujimori T, Takigami H. Pollution distribution of heavy metals in surface soil at an informal electronic-waste recycling site. Environmental Geochemistry and Health. 2014;**36**(1):159-166. DOI: 10.1007/s10653-013-9526-y

[51] Adewumi B, Akingunsola E, Femi-Oloye OP, Oloye FF. Evaluation of the heavy metals composition of soil at e-waste dumping sites. Asian Journal of Environment & Ecology. 2018;**5**(4):1-8. DOI: 10.9734/AJEE/2017/39500

[52] Uchida N, Matsukami H, Someya M, Tue NM, Tuyen TH, Viet PH, et al. Hazardous metals emissions from e-waste-processing sites in a village in northern Vietnam. Emerging Contaminants. 2018. DOI: 10.1016/j.emcon.2018.10.001

[53] Radulescu C, Stihi C, Dulama LD, Chelarescu ED, Bretcan P, Tanislav D. Assessment of heavy metals content in water and mud of several salt lakes from Romania by atomic absorption spectrometry. Romanian Journal of Physics. 2015;**60**(1-2):249-251

[54] Ansah E, Nukpezah D, Hogarh JN. Levels and distribution of heavy metals in Weija reservoir, Accra, Ghana. West African Journal of Applied Ecology. 2019;**26**(1):1-87

[55] Barrow NJ. A brief discussion on the effect of temperature on the reaction of inorganic ions with soil. European Journal of Soil Science. 1992;**43**(1):37-45. DOI: 10.1111/j.1365-2389.1992.tb00118.x

[56] Paripurnanda Loganathan, Saravanamuth Vigneswaran, Ravi Naidu and Jaya Kandasamy. Cadium Sorption and Desorption in Soils: A review. Critical Reviews in Environmental Science and Technology. 2012;**42**(5):489-533. DOI: 10.1080/10643389.2010.520234

[57] Xiao Q, Zong Y, Lu S. Assessment of heavy metal pollution and human health risk in urban soils of steel industrial city (Anshan), Liaoning, Northeast China. Ecotoxicity and Environmental Safety. 2015;**120**:377-385. DOI: 10.1016/j.ecoenv.2015.06.019

Section 2

Removal of Heavy Metals

Chapter 15

Biological Treatment of Heavy Metals with Algae

Ahmad Mohammadi and Fahimeh Mahmoudnia

Abstract

The development of industrial activities has caused an increase in the production of various water pollutants, of which heavy metals are among the most important due to their toxicity and harmful environmental effects. Bioabsorption is a promising and environmentally friendly technology, which has been widely used in various wastewater treatment applications in recent years. Among the bioabsorbents, algae are particularly important due to their high absorption efficiency, availability, and cost-effectiveness. In this chapter, the advantages of using algae and their use as biosorbents for removing heavy metals such as copper, aluminum, cadmium, zinc, mercury, chromium, nickel, and lead from aqueous solutions have been investigated. The effect of various factors, including factors related to biomass and process conditions (solution pH, adsorbent dosage, contact time, temperature, and initial concentration of heavy metal ions) has been evaluated. Also, the mechanisms of biological absorption of heavy metal ions in algae have been analyzed. Numerous studies show that algae are effective and economic bioabsorbents for the removal of heavy metals from industrial wastewater, and due to their predictability with simple equilibrium and kinetic mathematical equations, they are suitable for large-scale applications in continuous processes.

Keywords: wastewater treatment, removal of heavy metals, biological treatment, microalgae, absorbing heavy metals

1. Introduction

Nowadays, the development of various industries (such as mining stone refining, battery making, and the creation of pesticides) and the increase of industrial wastewater from factories are considered serious problems for the environment and humans. The lack of usable water resources on the planet and water pollution with various contaminants such as metals, semi-metals, pesticides, drugs, and other persistent organic elements has become a major concern worldwide. Among the various types of water pollutants, heavy metals are among the most important due to their toxicity, stability, resistance to environmental degradation, and long-term accumulation in the food chain. The most dangerous ions for human health and other living organisms are Cr, Fe, Se, V, Cu, Co, Ni, Cd, Hg, As, Pb, and Zn [1–3].

Despite the negative environmental effects, heavy metals are still used as main and important materials in various industries such as mining, coating, smelting, plastic, fabric, painting, etc. Therefore, it is necessary to remove heavy metals from industrial wastewater to reduce their impact on the environment. Also, the separation of heavy metals from wastewater can be economically important due to the high price of these metals. All kinds of physical and chemical methods have been investigated and used to treat industrial wastewater. Methods such as oxidation/reduction, sedimentation, ion exchange, reverse osmosis, membrane filtration, coagulation and flotation and chemical precipitation, electrochemical techniques, and adsorption on activated carbon and plant residues [2–7]. Choosing a specific purification method depends on various factors such as the type and concentration of heavy metals, whether the wastewater is homogeneous or heterogeneous, the required removal percentage, and the cost of the treatment process. But limitations such as cost, time, and efficiency have limited the use of these methods. Activated carbon and Nano adsorbents have a very high efficiency in this regard, but the cost of activation and synthesis of nanoparticles is high [8].

In recent years, biological adsorption has become a promising alternative method for wastewater treatment. Especially in research, it has been determined that surface adsorption is the most effective method for separating heavy ions and dyes from wastewater, and biological adsorbents are very promising to achieve this goal. The most important advantages of this method include high absorption capacity, low cost and economic productivity (especially when the bio absorbent can be recycled and heavy metals can be reused), high efficiency and productivity, minimal consumption of chemicals and creating sludge, the possibility of recovering metals, the renewable nature of biological adsorbents, the ability to be used in a wide range of changes in operating conditions and environmental compatibility pointed out [9].

In this chapter, researches carried out in the field of biological absorption and the use of algae to remove heavy metals from wastewater since 1977 have been studied. The main characteristics of algae that cause the high capacity to absorb metal ions heavy in them are investigated and the mechanisms of biological absorption of metals in algae have been reviewed. Also, the effect of operating parameters on the absorption efficiency of heavy metals by algae and the process conditions reported for optimal absorption of heavy metal ions have been discussed and investigated.

2. Biological removal of heavy metals (mechanism of biosorption)

The process of biological absorption includes two phases: the solid phase and the liquid phase (including particles that will be absorbed). Due to the high affinity of the adsorbent with metal ions, complex processes with mechanisms such as; there are chemical absorption, surface absorption, ion exchange, absorption by physical forces, entrapment inside the fibrillar capillaries and the space between the polysaccharide network in passing through the cell wall and membrane. The amount of absorption is obtained from the following formula, where C0 is the initial concentration and C is the final concentration after contact with the adsorbent, and in the following, the absorption power of all types of adsorbents will be expressed with the help of this formula [9–11]:

$$\%adsorbed = \left(\frac{C_0 - C}{C_0} \right) \times 100 \tag{1}$$

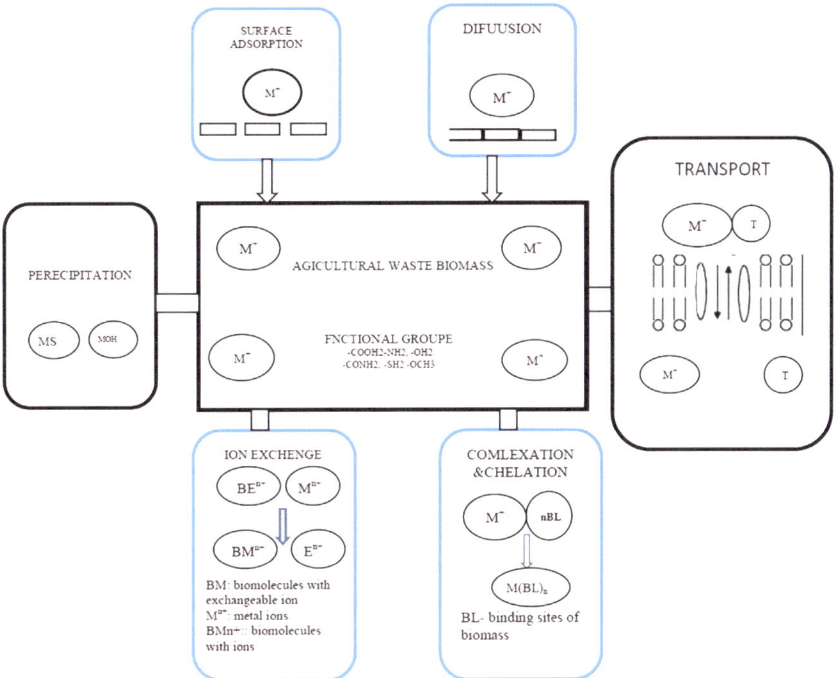

Figure 1.
A plausible mechanism of biosorption.

Accumulation of heavy metals in microorganisms mainly takes place in two phases; the first phase occurs at the cell surface. Absorption is fast and passive and is completely dependent on cell metabolism, the second phase of active absorption of ions into the algal cell cytoplasm, which is dependent on cell metabolism, and its other name is called intracellular absorption (**Figure 1**) [12].

3. The role of algae in the biological removal of heavy metals

The use of algae to remove heavy metals from wastewater has been noticed for more than 40 years, and many studies have been carried out to use algae as bio-absorbents, especially for the removal of heavy metals from aquatic environments.

In recent years, the cell structure of algae, modification and genetic change, and effective factors in their biological absorption, such as biomass concentration, primary ion concentration in the environment, size of present ions, temperature, and pH have been studied and investigated and it has been confirmed that the use of algae in The process of biological absorption of toxic and radioactive ions is a cheap, safe and efficient method; Also, valuable elements such as gold and silver can be extracted by them [12, 13].

In terms of mechanism, heavy metals are absorbed by groups such as hydroxyl, phosphoryl, carboxyl, sulfuryl, amino, sulfate, phosphate, and carbohydrate on the surface of algae. The availability of active sites of algae is determined through the FTIR test. Also, the amount of absorption depends on the number of effective groups in algae cells, their availability, the orientation of metal ions, and the chemical state of

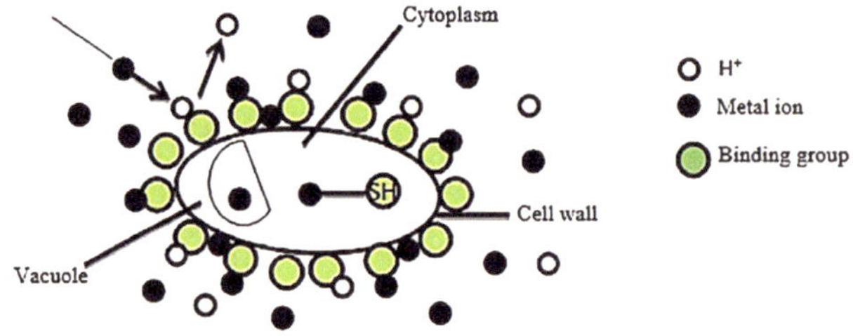

Figure 2.
Adsorption of metal ions on the algal cell surface.

Ligands class	ligands	Metal classes
1. Ligands preferred to class A	F-, O2-, OH, H2O, CO32-, SO4-, ROSO3-, NO3-, HPO42-, PO43-, ROH, RCOO-,C=O, ROR	A: Li, Be, Na, Mg, K, Ca, Sc, Rb, Sr., Y, Cs, Ba, La, Fr, Ra, Ac, Al, lanthanides, actinides
2. Other important ligands	Cl-, Br-, N3-, NO2-, SO32-, NH3, N2, RNH, R3N, =N-, -CO-N-, R, O2, O2-,O22-	Borderline ions: Ti, V, Cr, Mn, Fe, Co, Ni, Cu, Zn, Ga, Cd, In, SN, Sb, As
3. Ligands preferred to class B	H-, i-, R-, CN-, CO, S2-, RS-, R2S, R3AS	Class B: Rh, Pd, Ag, Lr, Pt, Au, Hg, Ti, Pb, Bi

Table 1.
Specific ligands for the absorption of each ion.

the active sites. Usually, functional groups (OH-, PO4 2-, RS-, RO, SH-, COO-, NO3-, and RNH2-) create a negative charge on the surface. These bonds are located in the cytoplasm of the cell, especially the vacuole, and improve absorption. This is shown in **Figure 2** [14].

Cytosolic proteins transport ions into the cell. Therefore, the vacuole is an organelle for accumulating metal ions. According to **Table 1**, specific ligands for the adsorption of each ion are listed.

Algal cell walls are the first barrier against absorption, and according to the abundance of cell wall compounds in different algal strains, the capacity to absorb metals will be different. According to research, brown algae is a very good absorbent in this area [12, 15].

4. The potential of macro algae in absorbing heavy metals

Macromolecular seaweeds are among the living and renewable resources of the seas and oceans, which are classified into three groups: brown, red, and green algae. Green and brown algae and their derivatives have a high absorption capacity for most metals. The adsorption capacity of brown algae is directly related to the algae content, its availability, and its specific macromolecular structure. It has been determined in the studies that the absorption power of *Sargassum* algae biomass for Cd+2, Pb2+, Zn2+, and Cu2+ ions is 79, 78, 227, and 51 mg/g, respectively. Two filamentous algae, Spirogyra and Cladophora, were also investigated for lead and copper absorption, and the results showed that Spirogyra's absorption power was higher. Also, the capacity of some species of macroalgae to absorb copper has been reported according to the following order [8, 16].

Fucus spiralis > *Ascophyllum nodosum* > *Chondrus crispus* > *Asparagopsis armata* > *Spirogyra insignis* > *Codium vermilara.*

5. The potential of microalgae in absorbing heavy metals

Microalgae are considered microscopic photosynthetic organisms that are found in all aquatic environments (freshwater and saltwater) and their cultivation is possible in closed and open environments. The size of microalgae can be from several micrometers to several hundred micrometers. Criteria and various methods are used to classify microalgae, including their pigment, life cycle, or their primary cell structure. Considering their abundance, the most important types of microalgae are diatoms (Bacillariophyceae), green microalgae (Chlorophyceae), and golden microalgae (Chrysophyceae). The difference between these types of seaweed is mainly in the structure of the cell walls, where the absorption of heavy metal ions occurs. The cell wall of microalgae generally contains significant amounts of starch and glycogen, as well as cellulose, hemicellulose, and polysaccharides. These compounds contain numerous reactive active groups (e.g., amino, hydroxyl, carboxyl, sulfate, etc.) that can It is involved in chemical bonding with metal ions and are known as the main factor of very good biological absorption potential of microalgae [17–19].

Investigations have shown that microalgae are at the forefront of wastewater treatment because, in addition to the biological absorption of heavy elements, they also do nitrogen removal, phosphorus removal, and COD reduction well [20–22].

Dirbaz and Rosta conducted a study on the kinetics and thermodynamics of cadmium biosorption by Parachlorella microalgae. They observed that the absorption capacity of this microalgae at a temperature of 30 degrees Celsius and a pH of 7 is 90.72 mg/g, which is between 3 and 5.5 times that of other studied absorbents [23].

PhongVo et al. presented a review of different designs and applications of microalgae-based photobioreactors for pollutant treatment. In their review article, in addition to summarizing the progress made in the field of removing pollutants with the help of microalgae, they provided a vision of the future of using photobioreactors in this field [24].

Moreira et al. investigated the biological removal of copper metal using the microalgae *Chlorella pyrenoidosa* with experiments designed by the factorial Box–Behnken method. They reported the removal of 83.14% under optimal conditions of pH 3.6, metal ion concentration of 5 mg/L, and adsorbent dose of 1.28 g/L [25].

Saavedra et al. conducted a comparative study on the removal of toxic elements arsenic, boron, copper, manganese, and zinc from solutions containing single metal ions and mixed metal ions by four different species of green microalgae [26].

Their results, in addition to confirming microalgae as efficient and economic adsorbents for removing heavy metals, showed that the presence of other metal ions strongly affects the removal rate of metal ions by microalgae.

Areco et al. studied the effect of zinc metal ions on the growth and photosynthetic metabolism of microalgae *Botryococcus braunii* and the ability of this microalgae to remove metal ions from aqueous solutions. They studied concentrations of 0 to 80 mg/L of metal ions and observed that increasing the concentration of zinc metal ions in the solution significantly reduces the growth of microalgae. Also, the metal absorption capacity in the period of 200 days was 3.4 grams per gram of absorbent [27].

In another study, Pradhan et al. studied the removal of hexavalent chromium using the Scenedesmus microalgae. They investigated the factors affecting this process, including initial pH, contact time, initial metal ion concentration, adsorbent dose, particle size, and temperature, and reported the effective removal of hexavalent chromium with a maximum of 92.89%. They also found the presence of aldehyde, amide, carboxylic acid, phosphate, and halide functional groups to be effective in this process by examining the FTIR spectra taken from the microalgae used. Regarding the adsorption mechanism, they concluded that the removal is done by anionic surface adsorption [28].

5.1 The mechanism of removing heavy metals by microalgae

Adsorption processes are usually very complex, and the mechanism of metal adsorption includes a combination of various elementary mechanisms such as electrostatic collisions, ion exchange, complex formation, adsorption with chelate formation, micro-deposition, etc., which occur simultaneously or sequentially. The basic mechanism of the biological absorption process can be divided into two categories: chemical biological absorption and physical biological absorption. As their names suggest, the first category includes chemical reactions and the second category involves the absorption of metal ions by van der Waals forces or electrostatic attraction forces. Ion exchange, complex formation, and microscopic sedimentation are the main mechanisms of heavy metal absorption by microalgae [29].

The primary interactions in the ion exchange mechanism can be from electrostatic or van der Waals forces to chemical bonds (ionic or covalent). In general, microalgae have mobile metal ions in their structure such as K+, Na+, Ca2+, Mg2+, etc. attached to the functional groups of microalgae. In the biosorption process, these mobile metal ions are exchanged with heavy metal ions according to the following reaction:

$$R^{-}-X^{+}+M^{+}\leftrightarrow R^{-}-M^{+}+X^{+} \tag{2}$$

where R- is the functional group of the microalgae surface, X+ is the mobile metal ion and M+ is the heavy metal ion in the aqueous solution. The mechanism of complex formation includes the formation of a complex on the cell surface, between heavy metal ions in the solution and a functional group of microalgae. For example, it has been shown in research that the absorption of Cu(II) ions on *Chlorella vulgaris* is done by a complex formation mechanism in which dative bonds are formed between metal ions and the amino and carboxyl groups of the microalgae cell wall polysaccharide. Microscopic sedimentation occurs when the pH of the biosorption solution increases sharply and/or the concentration of metal ions in the aqueous solution increases to the saturation level. In this case, heavy metals in an aqueous solution can be precipitated and the resulting microscopic sediments settle on the surface of the bioabsorbent [30].

5.2 Effective factors on the biological absorption of heavy metals by microalgae

In the case of microalgae, the most important factors affecting their biological performance in the process of removing heavy metals can be divided into two categories:

1. Biomass factors such as the growth environment, specific surface characteristics of microalgae, and pretreatment of cells.

2. Process factors such as the initial pH of the aqueous solution, bioabsorbent concentration, contact time, temperature, test method, bed height, solution flow intensity, and heavy metal concentration.

Growing conditions can affect the biological performance of microalgae. Although the data reported in the articles show that microalgae grown in saline environments contain higher amounts of polysaccharides than freshwater microalgae, their efficiency in the biological absorption process varies widely.

It was also shown that microalgae that have a large number of functional groups available on their surface show better biosorption characteristics. Of course, this depends on the nature of the microalgae and the pretreatment of the biomass cells before being used as a biosorbent. Normally, to obtain raw bioabsorbent, microalgae biomass is separated by centrifugation at different speeds and at different time intervals. This biomass is then pretreated. In most cases, pretreatment involves drying the biomass so that it can be stored more easily and for a longer period of time. The main process factors that affect the biological performance of microalgae and should be optimized for the discontinuous system are solution pH, adsorbent dose, contact time, and temperature. These factors for the continuous process are the pH of the solution, height of the bioabsorbent bed, flow rate of wastewater containing heavy metals, and initial concentration of heavy metal ions. A summary of the optimal process conditions for the removal of heavy metals from wastewater with microalgae is presented in **Table 2**.

The pH of the solution is one of the most important experimental parameters that not only affects the characteristics and solubility of heavy metal ions but also the degree of separation and dissociation of functional groups that are considered adsorption sites (such as hydroxyl, carboxyl, carbonyl, amino, etc.) from the bio absorbent surfaces [31].

According to research and review articles, the highest absorption rate of algae occurs at pH between 3 and 5 (because the acidity of the environment affects the surface bands of ions and biomass and the chemical structure of ions), and the dried biomass of algae has a much higher capacity and in the first 120 minutes, most of the absorption process will take place.

Biosorbent dosage is another parameter that should be optimized in order to ensure the economic and environmental efficiency of the bioremediation process. Using large amounts of bioabsorbent not only increases the cost of the bioabsorption process but also leaves a large amount of waste contaminated with heavy metals, which has a negative impact on the environment. On the other hand, the use of very small amounts of microalgae will significantly affect the efficiency of biological absorption, and the biological treatment process with low efficiency will not be used for industrial applications.

Contact time also plays an important role in ensuring the efficiency of the biological absorption process. The inappropriate value of this parameter can significantly limit the practical and industrial use of a biological adsorption process, even if its efficiency in removing heavy metal ions is high. The absorption rate of heavy metal ions on microalgae increases with increasing contact time, and the absorption process usually reaches equilibrium in about 180 min [32].

The investigated microalgae	heavy metal investigated	pH	Contact time (minutes)	Absorbent dose (gr/L)	Temperature	Reference
Chlamydomonas reinhardtii	Au(III)	2–3	70	0.1	26°C	[25]
Chlorella vulgaris	Cr(VI)	2	240	1	25°C	[26]
Cyanophyta & Chlorophyta	Ra226 U238	6–7 4–5	60	0.2	Environment temperature	[27]
Spirulina platensis	Cd2+	8	90	2	26°C	[28]
Scenedesmus obliquus	Cu(II)	5–7	60	0.03	Environment temperature	[29]
Scenedesmus quadricauda	Cd(II) & Pb(II)	5	60	0.2	Environment temperature	[30]

Table 2.
Optimal conditions for removing heavy metals using microalgae.

The important of the effect of temperature in the case of microalgae bioabsorbents is more important for the thermodynamic description of the absorption process than increasing the efficiency of heavy metal absorption. Many studies have shown that increasing or decreasing the temperature (even up to 40°C) has a small effect on the absorption of microalgae.

Discontinuous systems are usually only suitable for the biological treatment of small volumes of wastewater, and for larger scales, it is necessary to use continuous systems in which biological absorbents are used in several cycles of absorption and desorption (recovery). However, it should be noted that the use of microalgae in continuous systems has an important drawback, which is column clogging due to the small size of the bioabsorbent particles. Therefore, in order to ensure the sufficient intensity of wastewater flow through the column, in many types of research, the immobilization of microalgae in different matrices has been proposed, which increases the mechanical strength, particle size, and resistance to chemicals in wastewater [30].

According to the results reported in the articles in continuous systems, the metal absorption capacity and breakthrough time increase with the increase in bed height, which means an increase in the total surface area of the surface absorber. The metal adsorption capacity during adsorption decreases with an increasing initial metal concentration in the solution because the biosorbent becomes saturated faster at high concentrations. For this reason, optimal values for each of these parameters should be determined for practical applications.

Generally, the process of biological absorption in non-living microalgae follows a chemical mechanism, and the main factors that determine the nature of the primary processes are the type of functional groups on the microalgae surface, the nature of heavy metals in the aqueous solution, and the characteristics of the aqueous solution (pH, ionic strength, presence of competing ions, etc.). The absorption of various heavy metals such as Pb(II), Cd(II), Cu(II), Zn(II), etc. using different types of microalgae is mainly done by ion exchange. In confirmation of this point, laboratory studies showed that the concentration of light metal ions increases at the end of the biological absorption process [33].

In the complex formation mechanism, both electrostatic interactions and covalent and/or dative bonds are involved and compared to the ion exchange mechanism, the formed surface complexes are more stable. For this reason, the recovery of such biological attractants requires the use of strong agents. Nevertheless, the complex formation mechanism has been proven as the primary interaction in many adsorption processes on different types of microalgae, especially in high initial concentrations of heavy metal ions [34].

Microscopic sedimentation can occur depending on the nature of the microalgae or independent of it and can distort the results of biological absorption and prevent the determination of the absorption rate of metal ions. Although processes such as liming and turning into activated carbon are also used to increase the absorption capacity of microalgae. If the pretreatment of microalgae is done only by drying at 50–60°C (usually for 12–24 h), biomass is not decomposed and the functional groups of its surface are not changed [35–37].

Many studies have shown that pH values in the range of 2–8 lead to an increase in the absorption capacity of most heavy metals by microalgae. In this pH range, heavy metals have high solubility and are in solution as simple ions with the most toxic effect and the highest biological absorption. At lower pH values, the adsorption capacity of microalgae is lower due to the competition between protons and heavy metal ions to bind to biosorbent sites. At higher pH, heavy metal ions are precipitated as hydroxides and only a small amount of heavy metals remain in solutions to be absorbed by interacting with surface groups of microalgae. Due to the insignificant effect of temperature on the absorption capacity of heavy metals by microalgae, it is recommended that, for large-scale applications, the absorption of heavy metals from aqueous solutions in microalgae is carried out at an ambient temperature, because operating costs will be lower in this case. The use of microalgae to absorb heavy metals in continuous systems facilitates the treatment of a large volume of aqueous wastewater, however, research in the fields related to biological absorption in continuous systems is still ongoing [38, 39].

5.3 The difference between the performance of the living and non-living microalgae in the absorption of heavy metals

Although living microalgae have shown promising capabilities in the process of removing heavy metals from various types of wastewater, the ability to treat wastewater depends on their growth rate, biomass concentration, absorption capacity, and their use by various factors that affect their growth. The pH of wastewater, the concentration of heavy metals in wastewater, etc., is limited and can severely affect the efficiency of the treatment process. Therefore, non-living microalgae are more economical for industrial applications for the following reasons:

- Dead biomass can be stored at room temperature for a long time.
- The toxicity of heavy metals in wastewater does not affect them.
- Non-living microalgae have an absorption capacity comparable to or even higher than living microalgae, and their absorption capacity can be significantly improved by different chemical methods.

In general, there are few reports of the use of live algae because the life of algae is very much affected by environmental conditions. Absorption of heavy ions in living species is more complicated because absorption occurs in the growth phase and is intracellular. On the other hand, non-living algae absorb ions on their surface and it is considered an extracellular process that is easier to control and optimize.

A great advantage of non-living algae is the possibility of reusing biomass, which can be used in deionized water while living algae have little resistance to regeneration. Another advantage of non-living algae is the ease of use in physical and chemical modification and no need to add nutrients to the environment. Due to the growth of live algae, substances resulting from its metabolism may interfere with the absorption process. The only advantage of living algae is higher absorption power and absorption of a wide range of elements [12, 16].

5.4 The functional potential of stabilized microalgae in the absorption of heavy metals

The methods of flocculation, surface absorption, creating covalent bonds with carriers, transverse bands of algae, and trapping algae in a polymer network are among the methods of immobilization or stabilization of algae. Natural biopolymers such as agar and alginate or synthesized materials such as silica gel and polyacrylamide can be used as base materials. Natural polymers such as calcium alginate are widely used to create a stationary substrate for algae due to their non-toxicity. Among the synthetic polymers, polyacrylamide is the most used because it has good resistance and does not create a toxic environment, and is relatively cheaper [40].

Ashfaq Ahmad et al. studied the biosorption of Fe2+, Mg2+, and Zn2+ ions from aqueous solutions by free and stabilized *C. vulgaris* microalgae biomass on calcium alginate and the factors affecting this process in a laboratory manner. The results of their investigations showed that the biological absorption of all tested metal ions by cells fixed on calcium alginate is economical and with higher efficiency [41].

5.5 Use of modified algae (metal ion sorption by pretreated algae biomass)

With a variety of physical and chemical methods, the strength and number of active sites of algae can be increased. Physical improvement methods such as heating, boiling, crushing, freezing, and drying usually increase biological absorption because they provide more surface area for connecting bonds, and removing cell contents increases the possibility of forming metal bonds. The most famous chemical

Chemical modification	Its effect on the biological absorption process
$CaCl_2$	By binding calcium to algae, ion exchange becomes stronger
Formaldehyde	It strengthens the crosslink between effective groups (especially hydroxyl and amino)
NaOH	It increases electrostatic interactions and creates better conditions for ion exchange.
HCl	It replaces light metal ions with a proton and dissolves cell wall polysaccharides

Table 3.
The effect of different types of chemical modifications on algae absorption.

modifications are with formaldehyde, $CaCl_2$, HCl, NaOH, and glutaraldehyde, whose effects can be seen in **Table 3** [12].

6. Conclusions

The presence of heavy elements from the effluents of factories and industries has a severely destructive effect on the structure of living organisms and their performance. Because these elements cause various severe kidney, nervous, genetic, and cancer diseases in humans, therefore their absorption from polluted wastewater is very important. Absorption of heavy elements with the help of algae is a suitable, productive, and useful method in this field due to its availability, very low cost, and significant effect on the removal of pollutants.

Biological treatment of wastewater containing various heavy metals with biological absorption by microalgae is a simple method that has received much attention in recent years due to its high efficiency, leaving minimal secondary waste and using cheap materials. The research shows that these bioabsorbents have a very good performance in removing heavy metals from large amounts of wastewater with low concentrations of heavy metals. Microalgae can be grown in large quantities in different climates and in fresh and salty waters. The performance of microalgae in absorbing heavy metals is mainly due to the presence of various functional groups on the surface of microalgae and the tendency to replace the alkaline earth metal ions present in them with heavy metal ions. The efficiency of biological absorption of metals by microalgae depends on the characteristics of microalgae (particle size, growth conditions, biomass pretreatment, etc.) and process operating conditions (solution pH, adsorbent dose, contact time, temperature, work method, adsorbent bed height, the wastewater flow rate through the absorption column, heavy metal concentration, etc.) that should be optimized.

The biochemical sorption process can be easily modeled using several well-known equilibrium and kinetic models, which provide useful information about the mechanism of heavy metal sorption onto microalgae. Preliminary absorption experiments should be performed in batch systems to obtain preliminary information. Then the design and sizing of industrial-scale absorption system equipment and the necessary economic estimates are examined by conducting tests in continuous and dynamic systems. In general, the studies conducted in this research show that microalgae have a very high potential to be used as a cost-effective and efficient bio-absorbent in the removal of all types of heavy metals from industrial wastewater.

Acknowledgements

The Authors wish to thank the Research Center of Applied Plant Science Islamic Azad University, Arak Branch, for support in this research.

https://arak.iau.ir/rcaps/fa.

Author details

Ahmad Mohammadi[1*] and Fahimeh Mahmoudnia[2]

1 Department of Biosystem Engineering, Islamic Azad University, Arak Branch, Arak, Iran

2 Faculty of Science, Department of Biology, Farhangian University, Tehran, Iran

*Address all correspondence to: a-mohamadi@iau-arak.ac.ir

References

[1] Gonçalves AL, Pires JCM, Simões M. A review on the use of microalgal consortia for wastewater treatment. Algal Research. 2017;**24**:403-415. DOI: 10.1016/j.algal.2016.11.008

[2] Abdel-Ghani NT. Biosorption for metal ions removal from aqueous solutions: A review of recent studies. IJLRST. 2014;**3**(1):24-37

[3] Liu R, Liu B, Zhu L, et al. Effects of fluoride on the removal of cadmium and phosphate by aluminum coagulation. Journal of Environmental Sciences. 2015;**32**:118-125. DOI: 10.1016/j.jes.2014.10.024

[4] Llanos J, Williams PM, Cheng S, et al. Characterization of a ceramic ultrafiltration membrane in different operational states after its use in a heavy-metal ion removal process. Water Research. 2010;**44**(11):3522-3530

[5] Huang C-H, Chen L, Yang C-L. Effect of anions on electrochemical coagulation for cadmium removal. Separation and Purification Technology. 2009;**65**(2):137-146. DOI: 10.1016/j.seppur. 2008.10.029

[6] Gautam RK, Mudhoo A, Lofrano G, Chattopadhyaya MC. Biomass-derived biosorbents for metal ions sequestration: Adsorbent modification and activation methods and adsorbent regeneration. Journal of Environmental Chemical Engineering. 2014;**2**(1):239-259. DOI: 10.1016/j.jece.2013.12.019

[7] Wan Ngah WS, Hanafiah MA. Removal of heavy metal ions from wastewater by chemically modified plant wastes as adsorbents: A review. Bioresource Technology. 2008;**99**(10):3935-3948. DOI: 10.1016/j.biortech.2007.06.011

[8] Lakshmipathy R, Sarada NC, Jayaprakash N. Agricultural wastes as low cost adsorbents for sequestration of heavy metal ions and synthetic dyes from aqueous solution: A mini review. International Journal of ChemTech Research. 2015;**8**(5):25-31

[9] Sud D, Mahajan G, Kaur MP. Agricultural waste material as potential adsorbent for sequestering heavy metal ions from aqueous solutions- a review. Bioresource Technology. 2008;**99**(14):6017-6027. DOI: 10.1016/j.biortech.2007.11.064

[10] Agency for Toxic Substances and Disease Registry. Toxicological profile for 1,4-dioxane, 2012. Available from: https://www.atsdr.cdc.gov/toxprofiles/tp187.pdf. [Accessed: April 16, 2021]

[11] S. H & S. B. Masumeh Farasati. Cd removal from aqueous solution using agricultural wastes. In: Desalination and Water Treatment. Kermanshah, Iran: Razi University. DOI: 10.1080/19443994.2015.1043588

[12] Zeraatkar AK, Ahmadzadeh H, Talebi AF, Moheimani NR, McHenry MP. Potential use of algae for heavy metal bioremediation, a critical review. Journal of Environmental Management. 2016;**181**:817-831. DOI: 10.1016/j.jenvman.2016.06.059

[13] Horikoshi T, Nakajima A, Sakaguchi T. Uptake of uranium by *Chlorella regularis*. Agricultural and Biological Chemistry. 1979;**43**(3):617-623. DOI: 10.1080/00021369.1979.10863502

[14] Davis TA, Volesky B, Mucci A. A review of the biochemistry of heavy metal biosorption by brown algae.

Water Research. 2003;**37**(18):4311-4330. DOI: 10.1016/S0043-1354(03)00293-8

[15] Gardea-Torresdey JL, Gonzalez JH, Tiemann KJ, Rodriguez O, Gamez G. Phytofiltration of hazardous cadmium, chromium, lead and zinc ions by biomass of *Medicago sativa* (Alfalfa). Journal of Hazardous Materials. 1998;**57**:29-39. DOI: 10.1016/S0304-3894(97)000721

[16] Zeyu Lin WD, Li J, Luan Y. Application of algae for heavy metal adsorption: A 20-year meta analysis. In: Ecotoxicology and Environmental Safety. Beijing, China: College of Forestry, Beijing Forestry University; 2020. DOI: 10.1016/j.ecoenv.2019.110089

[17] Demirbas A. Use of algae as biofuel sources. Energy Conversion and Management. 2010;**51**(12):2738-2749. DOI: 10.1016/j.enconman.2010.06.010

[18] Wang J, Chen C. Biosorbents for heavy metals removal and their future. Biotechnology Advances. 2009;**27**(2):195-226. DOI: 10.1016/j.biotechadv.2008.11.002

[19] Arief VO, Trilestari K, Sunarso J, et al. Recent Progress on biosorption of heavy metals from liquids using low cost biosorbents: Characterization. Biosorption Parameters and Mechanism Studies. Clean. 2008;**36**(12):937-962

[20] Khan NA, Ibrahim S, Subramaniam P. Elimination of heavy metals from wastewater using agricultural wastes as adsorbents. Malaysian Journal of Science. 2004;**2004**:43-48

[21] Ho YS, McKay G. Kinetic model for lead (II) sorption on to peat. Adsorption Science and Technology. 1998;**1998**:243-255. DOI: 1783.1/25059

[22] Khattri SD, Singh MK. Colour removal from synthetic dye wastewater using a bioadsorbent. Water, Air, Soil Pollution. 2000;**2000**:283-294. DOI: 10.1023/A:1005207803041

[23] Dirbaz M, Roosta A. Adsorption ,kinetic and thermodynamic studies for the biosorption of cadmium onto microalgae Parachlorella sp. Journal of Environmental Chemical Engineering. 2018;**6**(2):2302-2309. DOI: 10.1016/j.jece.2018.03.039

[24] VoHNP NHH, Guo W, et al. A critical review on designs and applications of microalgae-based photobioreactors for pollutants treatment. Science Total Environment. 2019;**651**(Pt 1):1549-1568. DOI: 10.1016/j.scitotenv.2018.09.282

[25] Moreira VR, Lebron YAR, Freire SJ, et al. Biosorption of copper ions from aqueous solution using *Chlorella pyrenoidosa*: Optimization, equilibrium and kinetics studies. Microchemical Journal. 2019;**145**:119-129. DOI: 10.1016/j.microc.2018.10.027

[26] Saavedra R, Munoz R, Taboada ME, et al. Comparative uptake study of arsenic, boron, copper, manganese and zinc from water by different green microalgae. Bioresource Technology. 2018;**263**:49-57. DOI: 10.1016/j.biortech.2018.04.101

[27] Areco MM, Haug E, Curutchet G. Studies on bioremediation of Zn and acid waters using *Botryococcus braunii*. Journal of Environmental Chemical Engineering. 2018;**6**(4):3849-3859. DOI: 10.1016/j.jece.2018.05.041

[28] Pradhan D, Sukla LB, Mishra BB, Devi N. Biosorption for removal of hexavalent chromium using microalgae *Scenedesmus* sp. Journal of Cleaner Production. 2019;**209**:617-629. DOI: 10.1016/j.jclepro.2018.10.288

[29] Volesky B. Biosorbents for metal recovery. Trends in Biotechnology. 1987;**5**(4):96-101. DOI: 10.1016/0167-7799(87)90027-8

[30] Valdman E, Leite SGF. Biosorption of Cd, Zn and Cu by *Sargassum* sp. waste biomass. Bioprocess Engineering. 2000;**22**(2):171-173. DOI: 10.1007/PL00009109

[31] Han X, Wong YS, Wong MH, Tam NF. Biosorption and bioreduction of Cr (VI) by a microalgal isolate, *Chlorella miniata*. Journal of Hazardous Materials. 2007;**146**(1-2):65-72. DOI: 10.1016/j.jhazmat.2006.11.053

[32] Gao R, Wang J. Effects of pH and temperature on isotherm parameters of chlorophenols biosorption to anaerobic granular sludge. Journal of Hazardous Materials. 2007;**145**(3):398-403. DOI: 10.1016/j.jhazmat.2006.11.036

[33] Bulgariu L, Gavrilescu M. Chapter 30 - Bioremediation of heavy metals by microalgae. In: Kim S-K, editor. Handbook of Marine Microalgae. Academic Press; 2015. pp. 457-469. DOI: 10.1016/B978-0-12-800776-1.00030-3

[34] Aksu Z, Sag Y, Kutsal T. The biosorpnon of copperod by *C. vulgaris* and *Z. ramigera*. Environmental Technology. 1992;**13**(6):579-586. DOI: 10.1080/09593339209385186

[35] Han X, Wong YS, Tam NF. Surface complexation mechanism and modeling in Cr (III) biosorption by a microalgal isolate, Chlorella miniata. Journal of Colloidal Interface Science. 2006;**303**(2):365-371. DOI: 10.1016/j.jcis.2006.08.028

[36] Van Den Hende S, Vervaeren H, Boon N. Flue gas compounds and microalgae: (bio-) chemical interactions leading to biotechnological opportunities. Biotechnology Advances. 2012;**30**(6):1405-1424. DOI: 10.1016/j.biotechadv.2012.02.015

[37] Durán I, Rubiera F, Pevida C. Microalgae: Potential precursors of CO2 adsorbents. Journal of Carbon dioxide Utilization. 2018;**26**:454-464. DOI: 10.1016/j.jcou.2018.06.001

[38] Zhao Y, Wang B, Liu C, Wu Y. Biosorption of trace metals from aqueous multimetal solutions by green microalgae. Chinese Journal of Geochemistry. 2013;**32**(4):385-391. DOI: 10.1007/s11631-013-0646-y

[39] Priyadarshani I, Sahu D, Rath B. Microalgal bioremediation: Current practices and perspectives. Journal of Biochemical Technology. 2011;**3**(3):299-304

[40] Jinsong He JPC. A comprehensive review on biosorption of heavy metals by algal biomass: Materials, performances, chemistry, and modeling simulation tools. Bioresource Technology Journal. 2014;**2014**:67-78. DOI: 10.1016/j.biortech.2014.01.068

[41] Ahmad A, Bhat AH, Buang A. Biosorption of transition metals by freely suspended and Ca-alginate immobilised with Chlorella vulgaris: Kinetic and equilibrium modeling. Journal of Cleaner Production. 2018;**171**:1361-1375. DOI: 10.1016/j.jclepro.2017.09.252

Chapter 16

Metals as Catalysts for Ozonation

Jacqueline A. Malvestiti, Rodrigo P. Cavalcante, Valdemar Luiz Tornisielo and Renato Falcão Dantas

Abstract

Ozonation is an efficient process for water and wastewater treatment, widely used for the disinfection and oxidation of organic pollutants. This process is effective, however, some pollutants are ozone-resistant. For better oxidation, enhanced production of hydroxyl radicals ($HO^•$) can be obtained through the transition metals insertion in solution, known as homogeneous catalytic ozonation. These metals may react directly with O_3 to produce $HO^•$ or interact with organics such as humic substances in the water matrix to promote O_3 transformation to $HO^•$. In this chapter, a short review of the homogeneous catalytic ozonation, including key aspects, such as pH effect, metals concentration, catalytic mechanisms, drawbacks of the homogeneous catalytic ozonation application, and the possible solution for it was provided.

Keywords: transition metals, pH, catalytic mechanism, $HO^•$, drawback

1. Introduction

Access to clean and safe drinking water has become an emergency concern and requires immediate action. The population growth with consequent city development, especially in developing and emerging countries, has increased the volume of municipal wastewater produced every year and, this is the major contributor to a variety of water pollution problems [1].

Innovations in water and wastewater technologies are needed to solve challenges of climate change, resource shortages, emerging contaminants, urbanization, sustainable development, and demographic changes [2]. About 47% of the world's population has no access to clean and reliable drinking water supply and, according to the WWDR [3], this ratio is expected to increase 57% by 2050.

The removal of the new and wide range of pollutants, especially those of emerging concern in secondary effluents started to be included in several legislations around the world. However, most of the current wastewater treatment plants (WWTP) were not designed to remove these types of pollutants, thus an additional advanced tertiary treatment is necessary to achieve this goal.

Ozonation is considered one of the most effective methods for disinfection and removal of organic pollutants, even in low concentrations [4–8]. Ozone has a two-polar resonance structure, which makes ozone behave as both electrophilic and nucleophilic dipoles [4]. The organic pollutants' reactivity is selective, occurring mainly by specific reaction pathways, such as electrophilic, nucleophilic, or dipolar addition reactions [9] and the reactions predominate at low pH levels [10]. These reactions are known as direct reactions.

Direct reactions can be divided into four categories. The first one is the oxidation-reduction reaction, which occurs mostly due to the electron transfer process, such as the reactions between O_3 and HO_2^- (or $O_2^{-\bullet}$) (Eqs. 1 and 2) [11, 12].

$$O_3 + HO_2^- \rightarrow O_3^-\bullet + HO_2\bullet \tag{1}$$

$$O_3 + O_2^-\bullet \rightarrow O_3^-\bullet + O_2 \tag{2}$$

The second one is the cycloaddition reaction, which generally occurs between an unsaturated compound (with a carbon double bond or π electrons) and an electrophilic compound, forming a new compound. The cycloaddition reaction mechanism between O_3 and olefinic substance was proposed by Criegee (**Figure 1**): (1) formation of primary ozonide (or five-member ring); (2) generation of the zwitterion; (3) different reaction pathways of zwitterion and formation the final products, such as ketones, aldehydes or acids (in aqueous solution) [13].

The third one is the electrophilic substitution reaction, in which the ozone, as an electrophilic agent, attacks the nucleophilic position of the organic substances and substitute one part of the organic molecule. The last one is the nucleophilic reaction, in which the ozone molecule can react with molecules at their electrophilic positions, especially, when the compound contains carbonyl or double and triple nitrogen carbon bonds [14].

The indirect reaction occurs when the hydroxyl radical ($HO^\bullet$) and other reactive oxygen species (ROS), are formed by O_3 decomposition, it's a nonselective oxidant and highly reactive with almost all types of organic moieties at diffusion-controlled rates ($\sim 10^8$–10^9 M^{-1} s^{-1}), which may promote the complete degradation of organic pollutants [10, 15], prevailing at high pH levels.

Therefore, ozone-resistant pollutants are abated almost exclusively by ROS, mainly $HO^\bullet$ oxidation during ozonation [15, 16], and are usually less effectively abated due to the low $HO^\bullet$ yield from O_3 decomposition in real water matrices. The HO yield $\left(\frac{\text{moles of OH produced}}{\text{moles of} O_3 \text{ consumed}}\right)$ varies between 10% and 40% during conventional ozonation of water and municipal wastewater [16–19].

Ozone decomposition is the result of chain reactions with initiation, propagation, and ending phases [4]. The reaction between ozone and OH^- ions form hyperoxides radicals $HO_2^\bullet$ (initiation phase). $HO_2^\bullet$ is in equilibrium with the superoxide radical ($O_2^{\bullet -}$), and the reaction between ozone and superoxide radical produces ozonide ($O_3^{\bullet -}$) which reacts with H^+ to form $HO_3^\bullet$. Then $HO_3^\bullet$ is dissociated into $HO^\bullet$ and $O_2^\bullet$, and the reaction between O_3 and $HO^\bullet$ forms $HO_4^\bullet$ (propagation phase). The ending phase occurs with the dissociation of $HO_4^\bullet$ into $HO_2^\bullet$ and O_2. However, the presence

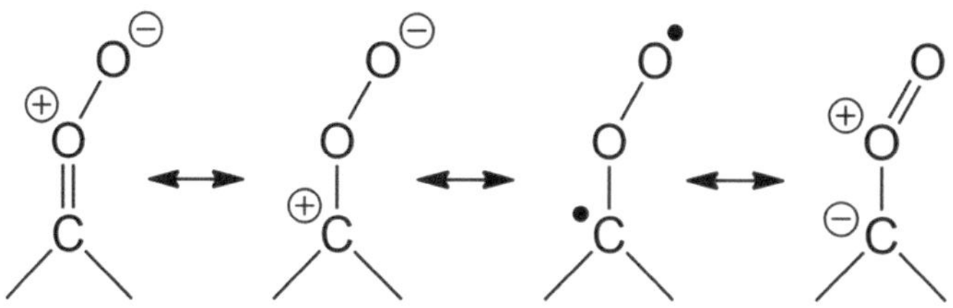

Figure 1.
Ozone reaction by the Criegge mechanism.

of inorganic and organic matter could initiate promote and prohibit the radical chain reaction [20]. In fact, a wide variety of compounds are able to initiate (i.e. hydrogen peroxide, humics, reduced metals, formate), to promote (i.e. primary and secondary alcohols, humics, ozone itself) or to inhibit (i.e. tertiary alcohols, HCO_3^-, CO_3^{2-}, HPO_4^{2-} and $H_2PO_4^-$) (Eqs. (3)–(6)) [20, 21] the radical chain reaction [21].

$$HO^{\bullet} + HCO_3^- \rightarrow OH^- + HCO_3^{\bullet} \tag{3}$$

$$HO^{\bullet} + CO_3^{2-} \rightarrow OH^- + CO_3^{-\bullet} \tag{4}$$

$$HO^{\bullet} + H_2PO_4^- \rightarrow OH^- + H_2PO_4^{\bullet} \tag{5}$$

$$HO^{\bullet} + HPO_4^{2-} \rightarrow OH^- + H_2PO_4^- \tag{6}$$

Ozone decomposition in water is strongly pH-dependent and occurs faster with an increase in pH [4].

In order to increase the production of hydroxyls radicals ($HO^{\bullet}$), and at the same time increase the oxidation capacity, ozonation can be performed in the presence of catalysts, namely catalytic ozonation.

The catalytic decomposition of O_3 in the presence of catalysts can lead to various ROS, such as ozonide radical ($O_3^{\bullet-}$), hydroxyl radical ($HO^{\bullet}$), superoxide radical ($O_2^{\bullet-}$), hydrogen peroxide (H_2O_2), and singlet oxygen ($^1O_2$) [18–22], these ROS and O_3 can react with pollutants simultaneously, thus bringing about their abatement. Due to the enhanced transformation of O_3 to ROS, higher abatement efficiencies can often be obtained for ozone-resistant pollutants during catalytic ozonation compared to conventional ozonation [9, 18–23].

The catalytic ozonation through the transition metals insertion in solution is known as homogeneous catalytic ozonation. They may react directly with O_3 to produce $HO^{\bullet}$ or interact with organics such as humic substances in the water matrix to promote O_3 transformation to $HO^{\bullet}$ [19–24].

The reaction mechanism follows two main pathways. The first one is based on the acceleration of ozone decomposition by the generation of the $^{\bullet}O_2^-$ and $^{\bullet}O_3^-$ radicals and subsequently $HO^{\bullet}$ formation [24, 25]. The other one is based on the formation of complexes between the catalyst and the organic compound, followed by a final oxidation reaction [25, 26]. Therefore, metal ions are able to enhance the efficiency of single ozonation for the removal of different organic compounds in aqueous solution, particularly those recalcitrant to direct ozone oxidation [27, 28]. In this catalytic process, the pH of the solution and the concentration of the transition ion can influence both the efficiency of the process and its mechanism [25]. The most widely metal ions used as catalysts of the ozonation process are Mn(II), Fe(III), Fe(II), Co(II), Cu (II), Zn(II), and Cr(III) [29–40].

The heterogeneous catalytic ozonation uses solid catalysts (e.g., metal oxides and metals on supports). There are two ways for O_3 to be decomposed in this system: at the catalyst surface and/or react with organics adsorbed on the catalyst surface to produce $HO^{\bullet}$ [41–43]. Many noble metals and metal oxides, immobilized or not on supports, have been used for heterogeneous catalytic ozonation [44, 45], being the order of catalytic activity for the decomposition of ozone following one [46]: Pt > Pd > Ag > Ru, Rh, Ir > Ni > Cd > Mn > Fe > Cu > Zn, Zr ≫ Co, Y, Mo, Ti, Au.

In this chapter, we only discuss homogeneous catalytic ozonation. Nevertheless, there are many homogeneous catalytic ozonation systems described in the literature. Some of them are collected in **Table 1**.

Catalyst	Pollutant	Operational conditions	Ozonation efficiency (%)	Reference
Co(II) and Fe (II)	p-chlorobenzoic acid (p-CBA) and benzotriazole	p-CBA and benzotriazole: 500 μg/L Ozone concentration: 2 mg/L Catalyst dose: 1 mg/L pH 7.8, T: 23 ± 2°C.	Degradation efficiency O_3/Fe(II): p-CBA: 57% benzotriazole: 98%	[29]
Fe(II)	Acid black Azo Dye	Ozone concentration: 70 mg/L, Dye concentration: 200 mg/L pH 6, Reaction time: 20 min	Degradation efficiency: 95.5%	[30]
Mn(II),Fe(II), Fe(III), Zn (II), Co(II), Ni (II)	C.I. Reactive Red 2 (RR2)	Catalyst dose: 0.6 mM Ozone concentration::200 ml/min Dye concentration: 100 mg/L pH: 2 and 5	Degradation efficiency: pH 2: Mn(II) > Fe (II) ≥ Fe(III) > Zn (II) > Ni(II) > Co(II) pH 5: Fe(III) > Fe (II) > Ni(II) ≥ Zn (II) ≥ Co(II) ≥ Mn (II)	[31]
Mn(II), Pb(I), Cu(II), Zn(II), Fe(II), Ti(II)	2-chlorophenol	Catalyst dose: 0–2 ppm 2-chlorophenol: 100 ppm Ozone concentration: 18 mg/min Reaction time: 20 min, pH = 3.	Degradation efficiency: Pb^{+}, Cu^{2+}, Zn^{2+}: 65% Ti^{2+} or Fe^{2+}: 80% Mn^{2+}: 90.5%	[32]
Fe(III) and Fe_2O_3/Al_2O_3	Oxalic acid	Ozone concentration: 30 mg/L, Gas flow rate: 24 L/h, Catalyst concentration: 1 mg/L, Al2O3 or Fe2O3/Al2O3: 1.25 g/L (part. Size: 1.6–2 mm), Reaction time: 3 h. T: 20°C. pH 2.5.	Degradation efficiency: 7% homogeneous (Fe (III)), 30% heterogeneous (Fe2O3/Al2O3)	[33]
Fe(II), Co(II), Cr(III), and Mn(II)	Di-(2-ethyl hexyl) phthalate (DEHP)	Ozone concentration: 96 mg/min. Flow rate of 2 L/min, DEHP concentration: 300 μg/L, Reaction time: 120 min. T:20 ± 0.5°C.	Degradation efficiency: Cr(III): the most active catalyst: 75% removal. Co(II): the least active catalyst: 45%	[34]
Co(II)	p-chlorobenzoic acid (pCBA)	Ozone concentration: 0.1 mM, Catalyst concentration:10-5 M Co (II), pCBA concentration: 3 Mm, Reaction time: 60 min, near-neutral pH, T24 °C.	Degradation efficiency: 75%	[35]
Mn(II)	2,4-dichlorophenol (DCP)	Ozone concentration: 8.2 mg/L, Ozone gas-flow rate: 200 mL/min, Catalyst concentration: 200 mg/L, DCP concentration:10 mg/ L. Reaction time: 30 min	Degradation efficiency: 100%, TOC removal: > 80%	[36]

Catalyst	Pollutant	Operational conditions	Ozonation efficiency (%)	Reference
Ni(II), Fe(II), Mn(II), Zn (II), Sr.(II), Cr (III), Cd(II), Hg(II), and Cu (II)	1,3,6-naphthalenetrisulfonic acid (NTS)	Ozone concentration: 1.04 × 10–4 mol/dm3 NTS concentration:1.23 × 10–4 mol/dm^3. Catalyst concentration: 0.1 millieq/dm^3 Reaction time: 30 min, pH: 2 and 7	Degradation efficiency: Fe(II): 79% and Mn(II): 72%	[37]
Electro-permanganate	Nitrobenzene (NB), Atrazine (ATZ), Sulfamethoxazole (SMX), Diclofenac (DCF), Phenol, and Carbamazepine (CBZ)	Ozone concentration: 1 mg/L, 60 mL/ min initial organic concentration = 0.060 mM pH = 5, T:293 K, Reaction time: 20 min Catalyst concentration = 0.100 mM	Degradation efficiency: CBZ and phenol: 100%. DCF: 87.08%, SMX: 84.23%, ATZ: 38.43%, NB: 21.56%.	[38]
Ce(III)	Phenol	Ozone concentration: 0.13 g/L, Flow rate: 100 ml/min, Reaction time: 120 min, Catalyst concentration: 50 mM, Phenol concentration: 2.7 mM	TOC removal: 71%, Degradation efficiency: 99%	[39]
Fe(II)	Reactive Red 2 (RR2)	Catalyst concentration: 0.9 Mm Ozone concentration: 2 mg/L/min RR2 dyeconcentration: 0.45 mM, pH 7 Reaction time: 70 min	k2: 2248 M/s	[40]

Table 1.
Catalytic ozonation processes of pollutants in water and wastewater.

1.1 Mechanism of homogeneous catalysis

The metal ions (Fe^{2+}, Cu^{2+}, Cr^{2+}, Mn^{2+}, Ni^{2+}, Co^{2+}, Cd^{2+}, Ag^{+}, Zn^{2+}, etc) influence the rate of reaction, the selectivity of ozone oxidation, and the efficiency of ozone utilization. A variety of different mechanisms has been proposed to explain the metal ions effects on ozonation, but there are two major mechanisms of homogeneous catalytic ozonation [22, 31, 35].

1.1.1 Mechanism 1: Decomposition of ozone by metal ions leading to the generation of free radicals

The oxidation mechanism of organic compounds via ozonation is dependent on the pH of the reaction medium, (i) at basic pH ozone decomposes producing $HO^{\bullet}$ radicals and other radical species in solution (Eqs. (7)–(9) and, (ii) at acidic pH, ozone is stable and reacts directly with organic substrates [31]. As it is well described in the literature, the generation of free radicals can subsequently oxidize the organic compounds more efficiently [23].

$$O_3 + OH^- \rightarrow O_2^{\bullet -} + HO_2^{\bullet} \quad (7)$$

$$O_3 + O_2^{\bullet -} \rightarrow O_2 + O_3^{\bullet -} \quad (8)$$

$$O_3^{\bullet -} + H^+ \rightarrow HO_3^{\bullet} \rightarrow O_2 + HO^{\bullet} \quad (9)$$

The homogeneous catalytic ozonation occurs mostly at acidic pH values because at the real pH range for waters/wastewaters (6–8), the effect of metal ions is almost diminished [32].

In general, the mechanism of metal-catalyzed ozone decomposition with the generation of $HO^{\bullet}$ radicals can be briefly expressed according to Eqs. (10)–(12) [31], being very similar to the Fenton process. The metal ions react with ozone or enhance its decomposition to generate $HO^{\bullet}$ radicals and their regeneration occurs via the oxidation by $HO_2^{-\bullet}$ radicals [9, 10].

$$M^{n+} + O_3 + H^+ \rightarrow M^{(n+1)+} + HO^{\bullet} + O_2 \quad (10)$$

$$O_3 + HO^{\bullet} \rightarrow O_2 + HO_2^{-\bullet} \quad (11)$$

$$M^{(n+1)+} + HO_2^{-\bullet} + OH^- \rightarrow M^{n+} + H_2O + O_2 \quad (12)$$

The formation of $HO^{\bullet}$ would be scavenged in the presence of excess metals (Eq. 13) [20, 21], so the optimization of catalyst dosage is also vital for catalytic ozonation process [23, 40].

$$M^{n+} + HO^{\bullet} \rightarrow M^{(n+1)+} + HO^- \quad (13)$$

One of the biggest challenges of this review was to find publications that represented the conditions found in the real waters (aquatic environments and wastewaters), for both the contaminants concentrations and pH. Since that, in the real waters, the contaminants concentrations range from ng/L and at the pH near neutral and most works present high values for them.

We gathered some works using metal ions for ozone catalysis and most of them report that Mn(II) and Fe(II) were the metals that showed the best results to increase ROS production. Sánchez-Polo and Rivera-Utrilla [37] tested the Mn(II) and Fe(II) ions as catalysts for the removal of 1,3,6-naphthalenetrisulfonic acid at acidic pH values; Xiao et al. [36] used the Mn(II) for the removal of 2,4-dichlorophenol; Ni et al. [32] used various metals for the removal of 2-chlorophenol at the acidic pH value and they found that Mn(II) was the most efficient catalyst tested. Okawa et al., [47] found that the presence of Fe(III) and Mn(II) enhanced the degradation of 2,4-dichlorophenol by ozone in acetic acid. However, Trapido et al. [48] observed no catalytic activity of Mn(II) for ozonation of dinitrobenzene. Wu et al. [31], Li et al. [49] and Ma & Graham [50] identified the optimal concentrations for metals ions to act as a catalyst for the decomposition of ozone into $HO^{\bullet}$ radicals, in order to remove emerging contaminants, with an emphasis here on C.I. reactive red 2, alachlor and atrazine.

1.1.2 Mechanism 2: Complexes formation between organic molecule and the catalyst

In this mechanism, the metal ions combine with organic molecules to form complexes, which are then oxidized by O_3 and other oxidizing species [23].

Pines and Reckhow [35] reported that high mineralization of oxalic acid takes place via ozonation in the presence of Co(II) ions. This process was determined to have a high reaction rate, which increase with a decrease in pH. To prove that the oxidation reaction did not depend on the formation of $HO^{\bullet}$ radicals, the researchers tested the reaction rate in the presence of tert-butanol, which is known as a $HO^{\bullet}$ radical scavenger. Based on the results obtained, the authors confirmed that it is not changed in mineralization rates, proving that the $HO^{\bullet}$ radicals was not responsible for mineralization of oxalate in the $Co(II)/O_3$ system (**Figure 2**).

Soon after the work of Pines and Reckhow [35], Beltrán et al. [51] tested the same oxalic acid mineralization in the $Co(II)/O_3$ system with the presence of tert-butanol and also confirmed that the $HO^{\bullet}$ radicals were not responsible for mineralization of oxalic acid. Continuing their work with the mineralization of oxalic acid, Beltrán et al. [33] also found that Fe(III) ions act as a catalyst for the mineralization of oxalic acid in the same way as Co(II). In both cases (Co and Fe) ozone reacts with both negatively and positively charged complex moieties and $HO^{\bullet}$ radicals are formed as secondary by-products. They proposed a sequence of reactions to explain the process (Eqs. (14)–(18)):

$$Fe^{3+} + C_2O_4^{-} \rightarrow FeC_2O_4^{+} \tag{14}$$

$$FeC_2O_4^{+} + C_2O_4^{-} \rightarrow Fe(C_2O_4)_2^{-} \tag{15}$$

$$Fe(C_2O_4)_2^{-} + C_2O_4^{-} \rightarrow Fe(C_2O_4)_3^{3-} \tag{16}$$

$$FeC_2O_4^{+} + O_3 \rightarrow 2CO_2 + Fe^{3+} + 2O_3^{\bullet-} \tag{17}$$

$$Fe(C_2O_4)_2^{-} + 2O_3 \rightarrow 2CO_2 + Fe(C_2O_4)^{+} + 2O_3^{\bullet-} \tag{18}$$

$$Fe(C_2O_4)_3^{3-} + 2O_3 \rightarrow 2CO_2 + Fe(C_2O_4)_2^{-} + 2O_3^{\bullet-} \tag{19}$$

Andreozzi et al. [52] explained the mechanism of the catalytic effect observed in Mn^{2+}/O_3 systems, as related to the formation of Mn(III) complex with oxalate ions (Ox). The molecular ozone attacked the oxalate ion radical and that leads to the formation of $HO^{\bullet}$ radicals [53] (Eqs. (20)–(25)):

$$Mn^{2+} + O_3 + 2H^{+} \rightarrow Mn^{4+} + O_2 + H_2O \tag{20}$$

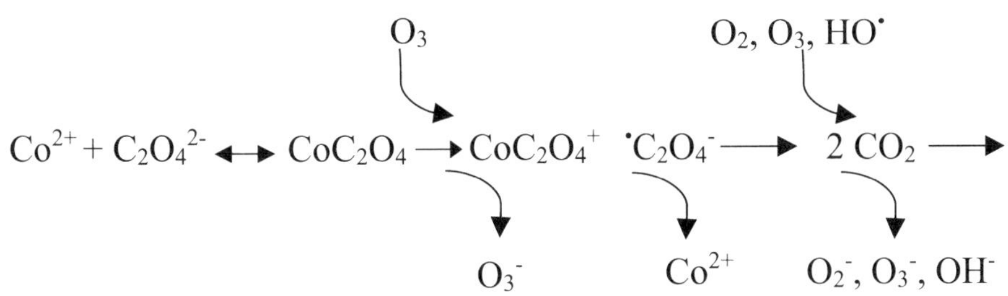

Figure 2.
Oxalic acid catalytic ozonation mechanism by means of the $Co(II)/O_3$ system.

$$Mn^{2+} + Mn^{4+} \rightarrow Mn^{3+} \quad (21)$$

$$Mn^{4+} + nOx^{2-} \rightarrow Mn^{3+}\left(Ox^{2-}\right)_n \quad (22)$$

$$Mn^{3+}\left(Ox^{2-}\right)_n \rightarrow Mn^{2+} + Ox^{-} + (n-1)Ox^{2-} \quad (23)$$

$$Ox^{\bullet -} + O_3 + H^{+} \rightarrow 2CO_2 + O_2 + OH^{\bullet} \quad (24)$$

$$OH^{\bullet} + Ox^{2-} \rightarrow \ldots \quad (25)$$

The transition metals are very important catalysts due to their characteristics [54]:

a. Bonding ability: Transition-metal catalysis is founded on the principle that electron donation from a metal to a ligand is accepted by an antibonding orbital of the ligand, thereby weakening one of the bonds in the ligand. Without this, the initial step of bond activation in many catalytic processes would simply not occur [55].

b. Wide choice of ligands: Transition elements readily form alloys and lose electrons to form stable cations, forming a wide variety of stable coordination compounds, in which the central metal atom or ion acts as a Lewis acid and accepts one or more pairs of electrons [54].

c. Ligand effects: the variation of the steric or electronic environment at the active site of a ligand can influence the behavior of a transition metal catalyst;

d. Oxidation state variability and the co-ordination number;

e. Ability to readily interchanges between oxidation states during a catalytic reaction: transition metals can be readily involved in redox processes.

In summary, in mechanism 2, the ozone may equally efficiently attack neutral, positively and negatively charged metal-complex species, which could be a major reaction pathway for catalytic ozonation, especially for some low-molecular-weight acids, such as oxalic acid [35, 50].

2. Concluding remarks

It is well described in the literature that catalytic ozonation in homogeneous phase is effective in removing a wide range of industrial effluents, products from the pharmaceutical industry, pesticides, and recalcitrant organics. However, homogeneous catalytic ozonation has the disadvantage of producing secondary contaminants from the addition of metallic ions [56]. In addition to the possibility of the residual concentration of metals exceeding regulatory limits for drinking water. Therefore, one more step in this system must be considered, to remove the metal ions from the treated matrices. This is the major drawback of applying homogeneous catalytic ozonation, especially when this process is applied for drinking water treatment. As a promising alternative to this inconvenience, heterogeneous catalysis appears which uses metals in the solid state (metallic oxides and metals on supports).

Acknowledgements

The authors thank the financial support of the São Paulo Research Foundation (FAPESP) [grant numbers #2022/00454-0; 2022/04015-1; #2019/26210-8], project CAPES/COFECUB [grant number 88881.191742/2018-2100], and National Council for Scientific and Technological Development (CNPq) [grant number 311674/2021-6].

Conflict of interest

The authors declare no conflict of interest.

Author details

Jacqueline A. Malvestiti[1,2]*, Rodrigo P. Cavalcante[1], Valdemar Luiz Tornisielo[2] and Renato Falcão Dantas[1]

1 School of Technology, University of Campinas, Limeira, São Paulo, Brazil

2 University of São Paulo, Piracicaba, São Paulo, Brazil

*Address all correspondence to: jacque.edf@gmail.com

References

[1] Edokpayi J N, Odiyo J O, Durowoju O S. Impact of wastewater on surface water quality in developing countries: A case study of South Africa. Chapter 18, in: Water Quality. London, United Kingdom: IntechOpen, 2017 [Online]. Available from: https://www.intechopen.com/chapters/53194. [Accessed: October 10, 2022]

[2] Hussain S, Aneggi L, Goi D. Catalytic activity of metals in heterogeneous Fenton-like oxidation of wastewater contaminants: A review. Environmental Chemistry Letters. 2021;**19**:2405-2424

[3] World Water Assessment Programme (Nations Unies), The United Nations World Water Development Report 2018 (United Nations Educational, Scientific and Cultural Organization, New York). 2018 [Online]. Available from: www.unwater.org/publications/world-water-development-repor. [Accessed: October 10, 2022]

[4] von Gunten U. Ozonation of drinking water: Part I. disinfection and by-product formation in presence of bromide, iodide or chlorine. Water Research. 2003;**37**:1469-1487

[5] Boczkaj G, Fernandes A. Wastewater treatment by means of advanced oxidation processes at basic pH conditions: A review. Chemical Engineering Journal. 2017;**320**:608-633

[6] Malvestiti JA, Dantas RF. Modelling secondary effluents disinfection by UV/H2O2 in presence of HO* scavengers using the ROH concept. Journal of Environmental Chemical Engineering. 2022;**10**(3):107879

[7] Miklos DB, Remy C, Jekel M, Linden KG, Hubner U. Evaluation of advanced oxidation processes for water and wastewater treatment—A critical review. Water Research. 2018;**139**:118-131

[8] Zhao R, Deng Y. Advanced oxidation processes (AOPs) in wastewater treatment. Water Pollution. 2015;**1**:167-176

[9] von Gunten U. Ozonation of drinking water: Part II. Disinfection and by-product formation in presence of bromide, iodide or chlorine. Water Research. 2003;**37**:1469-1487

[10] Lee Y, von Gunten U. Oxidative transformation of micropollutants during municipal wastewater treatment: Comparison of kinetic aspects of selective (chlorine, chlorine dioxide, ferrate VI, and ozone) and non-selective oxidants (hydroxyl radical). Water Research. 2009;**44**(2):555-566

[11] Masschelein WJ. Unit Processes in Drinking Water Treatment. Boca Raton: CRC Press; 1992

[12] Hoigné J. Chemistry of aqueous ozone and transformation of pollutants by ozonation and advanced oxidation processes. In: Quality and Treatment of Drinking Water II. Berlin, Heidelberg: Springer; 1998. pp. 83-141

[13] Bailey PS. The reactions of ozone with organic compounds. Chemical Reviews. 1958;**58**:925-1010

[14] Beltrán FJ. Ozone Reaction Kinetics for Water and Wastewater Systems. Boca Raton: CRC Press; 2003

[15] Yao W, Qu Q, von Gunten U, Chen C, Yu G, Wang Y. Comparison of methylisoborneol and geosmin abatement in surface water by

conventional ozonation and an electro-peroxone process. Water Research. 2017; **108**:373-382

[16] Zhu H, Ma W, Han H, Han Y, Ma W. Catalytic ozonation of quinoline using nano-MgO: Effcacy, pathways, mechanisms and its application to real biologically pretreated coal gasifcation wastewater. Chemical Engineering Journal. 2017;**327**:91-99

[17] von Sonntag C, von Gunten U. Chemistry of Ozone in Water and Wastewater Treatment: From Basic Principles to Applications. London, UK: IWA Publishing; 2012. p. 312

[18] Elovitz MS, von Gunten U, Kaiser H-P. Hydroxyl radical/ozone ratios during ozonation processes. II. The effect of temperature, pH, alkalinity, and DOM properties. Ozone: Science & Engineering. 2000;**22**(2):123-150

[19] Yu G, Wang Y, Cao H, Zhao H, Xie Y. Reactive oxygen species and catalytic active sites in heterogeneous catalytic ozonation for water purification. Environmental Science & Technology. 2020;**54**:5931-5946

[20] Staehelin J, Hoigne J. Decomposition of ozone in water in the presence of organic solutes acting as promoters and inhibitors of radical chain reactions. Environmental Science & Technology. 1985;**19**(12):1206-1213

[21] Legube B, Leitner NKV. Catalytic ozonation: A promising advanced oxidation technology for water treatment. Catalysis Today. 1999;**53**(1): 61-72

[22] Nawrocki J, Hordernb KB. The efficiency and mechanisms of catalytic ozonation. Applied Catalysis B: Environmental. 2010;**99**:27-42

[23] Wang J, Chen H. Catalytic ozonation for water and wastewater treatment: Recent advances and perspective. Science of the Total Environment. 2020; **704**:135249

[24] Martini J, Orge CA, Faria JL, Pereira MFR, Soares OSGP. Catalytic advanced oxidation processes for sulfamethoxazole degradation. Applied Sciences. 2019;**9**(13):2652

[25] Agulló-Barceló MI, Polo-López F, Lucena J, Jofre P, Fernández-Ibáñez. Solar advanced oxidation processes as disinfection tertiary treatments for real wastewater: Implications for water reclamation. Applied Catalysis B: Environmental. 2013;**136-137**:341-350

[26] Gracia R, Aragües JL, Cortés S, Ovelleiro JL. Study of the catalytic ozonation of humic substances in water and their ozonation byproducts. Ozone: Science & Engineering. 1996;**18**(3): 195-208

[27] Wang Y, Yu G. Challenges and pitfalls in the investigation of the catalytic ozonation mechanism: A critical review. Journal of Hazardous Materials. 2022;**436**:129157

[28] Khuntia S, Majudar SK, Ghosh P. Catalytic ozonation of dye in a microbubble system: Hydroxyl radical contribution and effect of salt. Journal of Environmental Chemical Engineering. 2016;**4**:2250-2258

[29] Psaltou S, Sioumpoura K, Kaprara E, Mitrakas M, Zouboulis A. Transition metal ions as ozonation catalysts: An alternative process of heterogeneous catalytic ozonation. Catalysts. 2021; **11**(9):1091

[30] Rekhate CV, Shrivastava JK. Decolorization of azo dye solution by ozone based advanced oxidation

processes: Optimization using response surface methodology and neural network. Ozone: Science & Engineering. 2020;**42**(6):492-506

[31] Wu C-H, Kuo C-Y, Chang C-L. Homogeneous catalytic ozonation of C.I. reactive red 2 by metallic ions in a bubble column reactor. Journal of Hazardous Materials. 2008;**154**(1–3): 748-755

[32] Ni CH, Chen JN, Yang PY. Catalytic ozonation of 2-dichlorophenol by metallic ions. Water Science and Technology. 2003;**47**:77-82

[33] Beltrán FJ, Rivas FJ, Montero-de-Espinosa R. Iron type catalysts for the ozonation of oxalic acid in water. Water Research. 2005;**39**(15): 3553-3564

[34] Khan HM, Jung JY. Ozonation catalyzed by homogeneous and heterogeneous catalysts for degradation of DEHP in aqueous phase. Chemosphere. 2008;**72**:690-696

[35] Pines DS, Reckhow DA. Effect of dissolved cobalt(II) on the ozonation of oxalic acid. Environmental Science & Technology. 2002;**36**(19): 4046-4051

[36] Xiao H, Liu R, Zhao X, Qu J. Effect of manganese ion on the mineralization of 2,4-dichlorophenol by ozone. Chemosphere. 2008;**72**(7):1006-1012

[37] Sanchez-Polo MS, Rivera-Utrilla J. Ozonation of 1,3,6-naphthalenetri-sulfonic acid in presence of heavy metals. Journal of Chemical Technology and Biotechnology. 2004;**79**(902):909

[38] Song Y, Zhao C, Wang T, Kong Z, Zheng L, Ding H, et al. Simultaneously promoted reactive manganese species and hydroxyl radical generation by electro-permanganate with low additive ozone. Water Research. 2021; **189**:116623

[39] Matheswaran M, Balaji S, Chung SJ, Moon IS. Studies on cerium oxidation in catalytic ozonation process: A novel approach for organic mineralization. Catalysis Communications. 2007;**8**: 1497-1501

[40] Zhang X, Dong W, Yang W. Decolorization efficiency and kinetics of typical reactive azo dye RR2 in the homogeneous Fe(II) catalyzed ozonation process. Chemical Engineering Journal. 2013;**233**:14-23

[41] Teng Y, Yao K, Song W, Sun Y, Liu H, Liu Z, et al. Preparation and characterization of Cu-Mn-Ce@γ-Al2O3 to catalyze ozonation in coal chemical wastewater-biotreated effluent. International Journal of Environmental Research and Public Health. 2019;**16**(8): 1439

[42] Xu Y, Lin Z, Zheng Y, Dacquin JP, Royer S, Zhang H. Mechanism and kinetics of catalytic ozonation for elimination of organic compounds with spinel-type $CuAl_2O_4$ and its precursor. Science of the Total Environment. 2019; **651**:2585-2596

[43] Roshani B, McMaster I, Rezaei E, Soltan J. Catalytic ozonation of benzotriazole over alumina supported transition metal oxide catalysts in water. Separation and Purification Technology. 2014;**135**:158-164

[44] Tang S, Yuan D, Zhang Q, Liu Y, Liu Z, Huang H. Fe-Mn bi-metallic oxides loaded on granular activated carbon to enhance dye removal by catalytic ozonation. Environmental Science and Pollution Research. 2016;**23**: 18800-18808

[45] Lu J, Wei X, Chang Y, Tian S, Xiong Y. Role of Mg in mesoporous $MgFe_2O_4$ for efficient catalytic ozonation of acid orange II. Journal of Chemistry Technology & Biotechnology. 2015;**91**(4):985-993

[46] Lin J, Kawai A, Nakajima T. Effective catalysts for decomposition of aqueous ozone. Applied Catalysis B: Environmental. 2002;**39**(2):157-165

[47] Okawa K, Tsai T-Y, Nakano Y, Nishijima W, Okada M. Effect of metal ions on decomposition of chlorinated organic substances by ozonation in acetic acid. Chemosphere. 2005;**58**(4):523-527

[48] Trapido M, Veressinina Y, Munter R, Kalla J. Catalytic ozonation of m-dinitrobenzene. Ozone Science and Engineering. 2005;**27**:359-363

[49] Li HY, Qu JH, Zhao X, Liu HJ. Removal of alachlor from water by catalyzed ozonation in the presence of Fe^{2+}, Mn^{2+}, and humic substances. Journal of Environmental Science Health B: Pesticides, Food Contaminants, Agricultural Wastes. 2004;**39**:791-803

[50] Ma J, Graham NJ. Degradation of atrazine by manganese-catalysed ozonation: Influence of humic substances. Water Research. 1999;**33**: 785-793

[51] Beltrán FJ, Rivas FJ, Montero-de-Espinosa R. Industrial & Engineering Chemistry Research. 2003;**42**:3210-3217

[52] Andreozzi R, Caprio V, Insola A, Marotta R, Tufano V. The ozonation of pyruvic acid in aqueous solutions catalyzed by suspended and dissolved manganese. Water Research. 1998;**32**: 1492-1496

[53] Nawrocki J. Catalytic ozonation in water: Controversies and questions. Discussion paper. Applied Catalysis B: Environmental. 2013;**142–143**:465-471

[54] Gusevskaya EV. Organometallic catalysis: Some contributions to organic synthesis. Química Nova [online]. 2003; **26**(2):242-248

[55] Braunschweig H, Damme A, Dewhurst R, et al. Bond-strengthening π backdonation in a transition-metal π-diborene complex. Nature Chemistry. 2013;**5**:115-121

[56] Guo Y, Yang L, Cheng X, Wang X. The application and reaction mechanism of catalytic ozonation in water treatment. Journal of Environmental & Analytical Toxicology. 2012;**2**:150

Chapter 17

Advances in the Adsorption Capacity, Rupture Time and Saturation Curve of Natural Zeolites

Carlos Montaño and Javier Montaño

Abstract

Reviewing the bibliography, it is found that the amount of heavy metals that natural zeolites are capable of adsorbing under normal conditions is 30%, $^{+}/_{-}$ 10, 20, with respect to the weight of the zeolite used as an adsorbent material in the best cases, highlighting the family of clipnoptilolite, it has been proven that with physical/chemical modifications, as well as, in non-normal conditions of pressure and temperature, superior adsorption results can be achieved. The present study analyzes the capacity of a certain family of natural zeolites that, by presenting a different chemical configuration, that is, instead of having 1,2,3 interchangeable bases such as Ca, K, Mg, has a compound such as (O Mg) and therefore a reorganized unit cell with the capacity to adsorb heavy metals up to 80% with respect to the total weight that is used as adsorbent material, this would be a new parameter to be considered in the adsorption of heavy metals by natural zeolites. According to the scientific literature, it is precisely the amount of exchangeable bases, diameter, and the weight and size of the zeolite pore that largely determines the adsorption of heavy metals.

Keywords: cation exchange, rupture curve, rupture time, saturation time, adsorption, chemical configuration

1. Introduction

According to the FAO 2018. There are 110 million mines and explosives scattered in approximately 64 nations in the world that contaminate water resources with heavy metals that could seriously affect the food chain, ecosystems, therefore, public health could be seriously threatened [1]. On the Ecuador-Colombia border is one of the red points of this contamination of water resources by heavy metals in Latin America, until 2013 more than 200 illegal open-pit gold extraction mines were established, according to E. Rebolledo et al.'s [2] studies carried out using the rapid index of water quality B.M.W.P [3] found the presence of heavy metals in water, sediments and aquatic species and the concentrations of heavy metals found are above the permissible limits of environmental regulations, [4], even in very superficial studies there are already cases that affect the health of the population due to heavy metals. A simple and quick method to evaluate biological quality of running freshwater based on Hellawell [3]: A new approach to the original B.M.W.P. [3] have been performed. Most of the macroinvertebrate families living in the Iberian Peninsula have been added to the original index and some of the scores have been changed. By comparison with some others biotic and diversity indexes, the different values of the new approach (B.M.W.P') have been correlated with quality classes and rate of pollution. The situation worsens when more than half of the population, in rural areas, on this border do not have drinking water and sewage services. The UN "in resolution 64/ 292 approved in 2010 establishes that water and sanitation are basic rights for life and for the dignity of all people." In addition, it is established that industry, mining and agriculture are the main polluters of water resources [5].

Natural zeolites are presented as an alternative with experimental scientific support as means of environmental remediation in this and other fields [6]. Natural zeolites are chemically hydrated aluminum silicates, and structurally they belong to the group of technosilicates. This chapter corresponds to a review of the adsorption capacity of natural zeolites considering the chemical formula of the zeolite. The efficiency of adsorption of heavy metals, making use of natural Zeolites, depends on several indicators such as pH, ionic strength, conductivity, initial concentration of cations and anions, the mass of the zeolite used, the particle size of zeolite, the rate of adsorption. This work intends to have experimental information if the chemical structure of the zeolites has any influence on the adsorption capacity. With the exception of the chemical formula of the zeolite, the other indicators can be potentiated by pre-treating the zeolite: ZNSP, ZNAT, ZNAA or ZNATA [7], the following samples were used: Clinoptilolite, Haulandite and a mixture between Haulandite and Clinoptilolite, Morante F. [8], and a fourth sample that differs in its chemical structure and unit cell (X), using fixed-bed concentration models with zeolites (ZNAA), that is, natural zeolites activated in an acid medium, with granulometries of 0.25 mm–1 mm, solutions with known concentrations (0.032 N of ZnSO4 H_2O) are prepared and the fractions are collected in 100 ml volumetric flasks that are analyzed by atomic absorption to determine the concentration in ppm of cation zn2+, the analysis finished when the concentration of the Zn2+ cation in the zeolite column effluent is close to or similar to the initial concentration of the Zn2+ cation. The columns have the same conditions (sample mass in grams g of zeolites, height in cm, volume in cm^3, diameter in cm, density in g/cm3, flow rate in cm^3/h, and T°).

Zn2+ adsorption results: Clinoptilolite 6.3 mg of Zn2+/g, Haulandite-Clinoptilolite 22.26 mg of Zn2+/g, Haulandite 5.5 mg of Zn2+/g, Morante [8] and sample X 45.1 mg of Zn2+/g. When applying X-ray diffraction (XRD), significant differences were observed in the chemical and structural structure of natural zeolites and sample X [6]. Regarding the rupture time tb and saturation time tsat, the following results were obtained in the respective order 0.7 h–25 h, 2.0 h–44 h, 1.6 h–11.67 h and for sample X 5.0 h–21 h, it is suggested that if there is an influence by the chemical structure of the zeolites with respect to the adsorption capacity and the amplitude of the breaking time.

2. The adsorption process

Experimentally now zeolites are "clartrates" or also called inclusion group due to the ability of zeolites to adhere or attract various substances in their structure as a guest. According to Armbruster and M Gunter they explain: "A zeolite mineral is a crystalline substance with a structure characterized by a tetrahedral molecular structure, which consists of four oxygen atoms surrounding a cation. This molecular structure contains open cavities in the form of channels and portals. These are generally occupied by water molecules and structural cations that are commonly exchangeable. These channels are large enough to allow invited species to pass through. In stages of hydration and dehydration they occur at temperatures almost always below 4000 C and vice versa it is higher. The atomic structure can be interrupted by groups (OH, F), these are placed externally on the tetrahedron that is not shared with an adjacent tetrahedron.

Figure 1 establishes that the basic structure of the natural zeolite is made up of Silicon^{4+} atoms, surrounded by four Oxygen atoms; Al^{3+} is replacing Si, creating a deficiency of positive charges or an increase in negative charges, which is compensated by the castions Ca^{2+}, Mg^{2+}, Mg^{2+} and Na^{+}, thus maintaining the balance of the zeolite.

The figure with a yellow background shows the basic structure of a natural zeolite and the location of the elements that form it together with their functions. The figure to the right with a white background shows the cation exchange process (CEC).

Adsorption is a process by which atoms, ions, and molecules are trapped or included on the surface or interior of an adsorbate, unlike absorption, which represents a volume. In chemistry, the adsorption of a species; cation, ion, molecule, among others, is its inclusion in a part of the interfacial adsorbate between two phases [9]

In this sense, zeolites are also called molecular sieves for their ability not only to trap cations, but also molecules. The purpose of this work is to investigate if there is a difference in adsorption capacity when zeolites have interchangeable bases such as K, Ca, Mg or when they have a compound such as O Mg.

Figure 2 establishes the way in which physical adsorption and chemical adsorption develop.

Figure 1.
FONT: F. Morante. Natural zeolites of Ecuador: Geology, Characterization and Applications. ISBN:978-9978-310-90-8. ESPOL 2014. Basic structure of a zeolite [8].

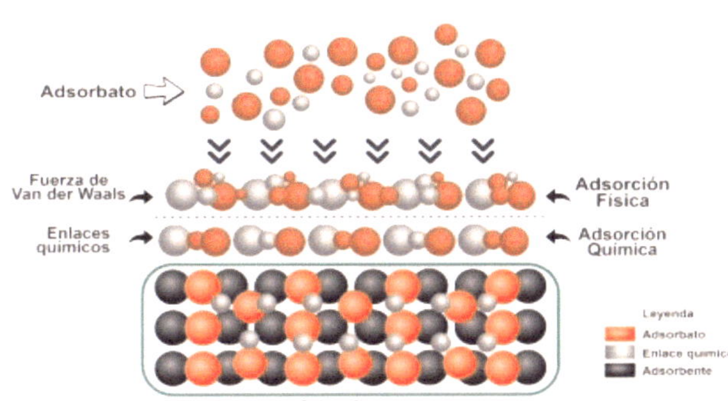

Figure 2.
FUENTE: Molina Vergaray . Physicochemical adsorption.

Physical adsorption is based on Van der Waals forces and chemical adsorption is established through chemical bonds. Ortega [10], Corona [11].

3. Chemical composition of natural zeolites

In natural zeolites, the presence of the trivalent aluminum element instead of the tetravalent silicon element detracts from the positive charges, increasing the negative charges. The remaining negative charges are compensated by monovalent and divalent cations, such as Na+, K+, Ca2+ elsewhere in the structure, so the empirical formula of zeolite is established as follows.

$$\mathbf{M_2/n\,O - Al_2O_3 - x\,SiO_2 - y\,H_2O} \qquad (1)$$

Where M corresponds to any alkali/alkaline earth cation, n is the valence of that cation, x is a number from 2 to 10 and y is a number from 2 to 7.

The unit cell of the clinoptilolite zeolite would look like this:

$$\mathbf{(Na_4K_4)O{-}(Al_8Si_{40}){-}O_{96\_}24\,H_2O} \qquad (2)$$

In the **Figure 3** the empirical formula of the zeolites is observed, the formation of the geometric figure of the tetahedr is observed, where the Silicate is the central cation of the figure, the structure is electrically neutral as occurs in quartz (SiO_2), however in some zeolites, some tetravalent silicons are replaced by trivalent aluminum, decreasing the positive charges Chiappim [12].

According to Morante F. [13] The ions located within the first parenthesis of the unit cell formula are called exchangeable cations; those in the second parentheses are called structural cations, because they build the tetrahedral structure with oxygen. In natural zeolites water is in free form and represents 10–20% of the hydrated phase, the entire volume of this water can be extracted continuously or reversibly through temperature around 3500 C. According to Mumpton and Ormsby. The dehydration or activation of a zeolite is an endothermic process; conversely, rehydration is exothermic. The following figure shows the chemical structure of the most widely used natural zeolites Georgiev [14].

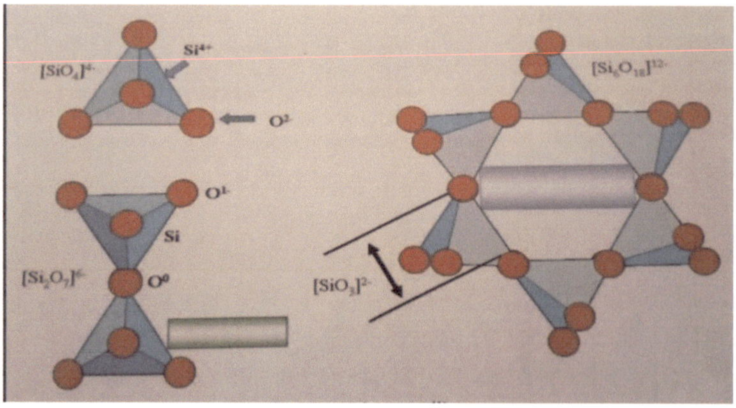

Figure 3.
FONT: Terkimitda. Substructural units of zeolites.

4. Types of adsorption

The adsorption of cations using natural zeolites is possible by the exchange of the zeolite's own cations or also called exchangeable bases.

The adsorption surface area of zeolites is considered to be approximately 10 m^2/g, unlike sand which is 0.01 m^2/g, this allows a larger contact area for the adsorption of suspended solids, microorganisms, and other materials in solution. Considering these characteristics of zeolites, the following types of adsorption are proposed: physical and chemical, in active as well as passive treatments, an active treatment is considered when continuous energy and reactive consumption methods are used, passive when neither energy nor reactives are used. In a broader sense, the adsorption of pollutants is also present in environmental remediation alternatives such as: Geological, Microbiological, Reactive barriers. According to [15], zeolites can be modified in order to increase their adsorption efficiency through the use of acids, bases, inorganic salts or hydrothermal treatment, however, the most appropriate modification is the one that agrees to specific pollutants [16, 17].

5. Significant advances in adsorption with natural zeolites

According to Wang, Xu and Sheng [18], Hawash et al. [19] they experienced advances in the purification of polluted water resources, as well as the adsorption of total phosphorus through CW (constructed wetlands).

According to Rahimi and Mahmoudi [20], Choi [21], Mahmood Ibrahimi [7] they obtained significant advances by modifying zeolites with (Sodium Hydroxide and Magnesium Oxide) to remove (Lead, Cobalt, Chromium and Zinc) from an aqueous solution, obtaining an adsorption result of 98.38%, 89.51 %, 81.07% and 78.24%, of Pb2+, Co2+, Cr2+, Zn2+ respectively, concluding that the capacity of the zeolite modified with MgO, in the adsorption of lead was more significant than that modified with NaOH, under similar conditions. Bezerra et al. [22].

Other significant advances with modified zeolites are; [23]; De-La-Vega et al. [24] Experimented hydrothermally modified zeolites from residual quartz sand and cal cium fluoride, doped with copper and synthetic faujasite type to remove heavy metals present in aqueous solutions, the results with respect to the percentage of lead removal was 93%, 95.9% and 70% respectively, they concluded that the removal process with modified zeolite is a viable, profitable and efficient alternative for adsorption processes Goñi [25].

Other significant tests were carried out with modified zeolite with high and low calcium fly ash for the removal of heavy metals in aqueous media, lead removal results were around 80%, they concluded that they are very significant advances [26, 27]. Zheng [28] in this same order of modifications, zeolites enriched with another cation in their chemical structure increase heavy metal removal by 40% [29].

On natural zeolites mixed with hexadecyltrimethylammonium bromide (HTAB), it was found that the HTAB-modified zeolite showed significant advances with respect to the natural zeolite, and that the zeolite is effective for the removal of colorants in aqueous solutions. In more significant advances, according to Yurekli, [30]; The removal and filtration processes were analyzed from a nanoparticle (zeolite membrane) combined with polysulfone (Psf) in the removal of lead and nickel in laboratory solutions [31].

Acid mine drainage (AMD) is one of the environmental challenges that has an urgent character for radiation, according to C Ayora et al., [32]. Caustic magnesia (MgO) maintains the pH between 8.5 and 10, allowing complete elimination of divalent metals. And according to Zendelska in her research, she used natural zeolite as an effective sieve for the removal of heavy metals from acid mine drainage (AMD) focusing on zinc ions, obtaining a removal percentage of 74%.

6. Advances in adsorption of heavy metals between natural zeolites with different chemical formulas

The following figure shows the main formulas of the unit cell of the main natural zeolites used in the world. It is observed that all these structural characteristics have an incidence or directly influence the adsorption of contaminants. But it is observed that the ion exchange capacity is related in a special way to the number of exchangeable cations of the first parenthesis (LINDEA, NATROLITE, ANALCIMA, LAUMONTITE).

6.1 Physicochemical characteristic of the main zeolites

Figure 4 shows the physical and specific characteristics of most zeolites, especially the ionic exchange capacity for those zeolites that have a single exchangeable base, which would be in accordance with the present work in which the study is presented. where greater adsorption of heavy metals was obtained with a zeolite that did not have interchangeable bases, but rather a binary compound (O Mg).

Observing that **Figure 5** of natural zeolites had a different adsorption activity, a characterization was carried out by means of X-Ray Diffraction (XRD).

It is observed that **Figure 5** under the same adsorption conditions as **Figure 6**, but with different chemical formulas in its unit cell, have a different adsorption capacity, favorable for **Figure 5**. It is also observed that depending on the chemical formula it is possible that the spatial configuration or shape of the unit cell has changes that favor adsorption.

In this work, a sample with the compound formula (O Mg) corresponding to **Figure 5** and another sample in which the chemical formula has interchangeable bases (K, Na, Mg) which corresponds to **Figure 6**.

In the **Figure 5** (Sample 14-0078), shows a chemical structure with an amorphous material that corresponds to 86%, this indicates that this sample space does not behave as a crystalline structure, on the other hand, unlike the chemical structure of other zeolites, the cations in the first bracket of sample X is a compound: MgO, in the other zeolites it is a cation, in this sense, it is considered that this structure favors the adsorption process due to the presence of oxygen as well as the large space amorphous (86%). In addition, this property of this sample (**Figure 5**), allows to have a longer rupture time and saturation time with respect to sample 2 (**Figure 6**).

In the **Figure 6** (Sample 14-0079), Meets the standard of a chemical structure of zeolites, that is, it follows a crystalline model; the amorphous space is only 3.3%. and its exchangeable cations that corresponding to the first bracket is not a compound, and the other natural zeolites behave like this. It is suggested that this is the reason for the decrease in its adsorption process when compared to **Figure 5.**

ZEOLITAS	FORMULA TÍPICA DE LA CELDA UNITARIA[1]	VOL. DE PORO	GRAVEDAD ESPECIFICA[4]	DIMENSIÓN DE LOS CANALES[1]	ESTABILIDAD TÉRMICA	CAPACIDAD DE INTERCAMBIO IONICO[2]
ANALCIMA	$Na_{16}(Al_{16}Si_{32}O_{96})*16H_2O$	18	2.24-2 29	2.6 Å	Alto	4.54
CHABAZITA	$(Na_2,Ca_6)(Al_{24}Si_{24}O_{72})*40\ H_20$	47	2.05-2.10	3.7x4.2	Alto	3.81
CLINOPTILOLITA	$(Na_4K_4)(Al_8Si_{40}O_{96})*24\ H_20$	39?	2.16	3.9x5.4	Alto	2.54
ERIONITA	$(Na,Ca_{0.5},K)_9(Al_9Si_{27}O_{72})*27H_2O$	35	2.02-2.08	3.6x5.2	Alto	3.12
FAUJASITA	$Na_{58}(Al_{58}Si_{134}O_{384})*240H_2O$	47	1.91-1.92	7.4	Alto	3.39
FERRIERITA	$(Na_2Mg_2)(Al_6\ Si_{30}O_{72})*18H_2O$	-	2.14 - 2.21	4.3 X 5.5 3.4 x 4.8	Alto	2.33
HEULANDITA	$Ca_4(Al_8Si_{28}O_{72})*24\ H_20$	39	2.10-2.20	4.0x5.5 4.4x7.2 4.1 X4.7	Bajo	2.91
LAUMONTITA	$Ca_4(Al_8Si_{16}O_{48})\ *16H_20$	34	2.20-2.30	4.6 x 6.3	Bajo	4.25
MORDENITA	$Na_8(Al_8Si_{40}O_{96})\ *24\ H_2O$	28	2.12-2.15	2.9 x 5.7	Alto	2.29
NATROLITA	$Na_{16}(Al_{16}Si_{24}O_{80})*16H_2O$	23	2.20-2.26	6-7 x 7.0 2.6 x 3.9	Bajo	5.26
PHILLIPSITA	$(Na,K)_{10}(Al_{10}Si_{22}O_{64})\ *\ 20H_2O$	31	2.15-2.20	4.2x4.4 2.8x4.8 3.3	Bajo	3.87
WAIRAKITA	$Ca_8(Al_{16}Si_{32}O_{96})\ *\ 16H_20$	20?	2.26	-	Alto	4.61
LINDE A[3]	$Na12(Al_{12}Si_{12}O_{48})\ *27H_20$	47	1.99	4.2	Alto	5.48
LINDE X[3]	$N_{86}(Al_{86}Si_{106}O_{384})\ *264\ H_20$	50	1.93	7.4	Alto	4.73

[1]Taken mainly from Breck (1974), Meier and Olson (1971). Void volume is determined from water content
[2]Calculated from unil-cell formula (meq/g). [3]Linde A and Linde X are synthclic phases.
[4]Taken mainly from Breck (1974), Deer el al. (1963); mostly on crystals from amygdules in basalt.

ZEOLITES	TYPICAL FORMULA OF THE UNIT CELL	VOL DE PORO	SPECIFIC GRAVITY	DIMENSION OF THE CHANNELS	ION EXCHANGE CAPACITY	THERMAL STABILITY
ANALCIMA	$Na_{16}(Al_{16}Si_{32}O_{96}).16H_2O$	18	2.24-2 29	2.6 A°	4.54	Alto
CHABAZITE	$(Na_2Ca_6)(Al_{24}Si_{24}O_{72})40H_2O$	47	2.05-2.10	3.7 X 4.2	3.81	Alto
CLINOPTILOLITE	$(Na_4k_4)(Al_8Si_{40}O_{96}).24H_2O$	39	2.6	3.9 X 5.4	2.54	Alto
ERIONITE	$(Na,Ca_{0.5},K)_9(Al_9Si_{27}O_{72})\ 27H_2O$	35	2.02-2.08	3.6 X 5.2	3.12	Alto
FAUJASITE	$Na_{58}(Al_{58}Si_{134}O_{384}).240H_2O$	47	1.91-1.92	7.4	3.39	Alto
FERRIERITE	$(Na_2Mg_2)(Al_6Si_{30}O_{72}).18H_2O$	----	2.14-2.21	4.3 X 5.5 3.4 X 4.8	3.33	Alto
HEULANDITE	$Ca_4(Al_8Si_{28}O_{72}).\ 24H_2O$	39	2.10-2.20	4.4 X 7.2	2.91	Bajo
LAUMONTITE	$Ca_4(Al_8Si_{16}O_{48}).\ 16\ H_2O$	34	2.20-2.30	4.6 X 6.3	4.25	Bajo
MORDENITE	$Na_8(Al_8Si_{40}O_{96}).\ 24H_2O$	28	2.12-2.26	2.9 X 5.7	2.29	Alto
NATROLITE	$Na_{16}(Al_{16}Si_{24}O_{80}).16H_2O$	23	2.15-2.20	2.6 X 3.9	5.26	Bajo
PHILLIPSITE	$(Na\ k)_{10}(Al_{10}Si_{22}O_{64}).20H_2O$	31	2.15-2.20	2.8 X 4.8	3.87	Bajo
WAIRAKITE	$Ca_8(Al_{16}Si_{32}O_{96}).16H_2O$	20	2.26	------------	3.87	Alto
LINDE A	$Na\ 12(Al_{12}Si_{12}O_{48}).27H_2O$	47	1.99	4.2	4.61	Alto
LINDE B.	$Na_{86}(Al_{86}Si_{106}O_{384}).264H_2O$	50	1.93	7.4	4.73	Alto

Taken mainly from Breck (1974). Meier and Olson (1971). Void volumen is determined from wáter content calculated from unil-cell (meq/g). Linde A and Linde x are synthclic phases. Taken mainly from Brek (1974). Deer el al. (1963); mostly on crystals from amygdules in basalt.

Figure 4.
FONT: F. Morante. Zeolitas naturales del Ecuador: Geología, Caracterización y Aplicaciones. ESPOL. 2014 ISBN:978-9978-310-90-8.

7. Fixed bed concentration models

According to Morante, F., [13], Hartini et al. [33], Erdem [34] in the fixed bed concentration models, the concentrations in the fluid phase and the solid phase vary with time and the position of the bed, the greatest mass transfer takes place near the inlet. of the bed, where the fluid comes into contact with the adsorbent. If the solid initially has no adsorbate, the concentration in the fluid decreases exponentially with distance to almost zero before reaching the far end of the bed. After a few minutes, the solid near the inlet is nearly saturated, and most of the mass transfer takes place away from the inlet. The concentration gradient assumes an S shape.

Figure 7 represents the physical process of a fixed bed concentration model based on natural zeolites or also called molecular sieves. Using fixed-bed concentration models with zeolites (ZNAA), that is, natural zeolites activated in an acid medium,

ANEXOS

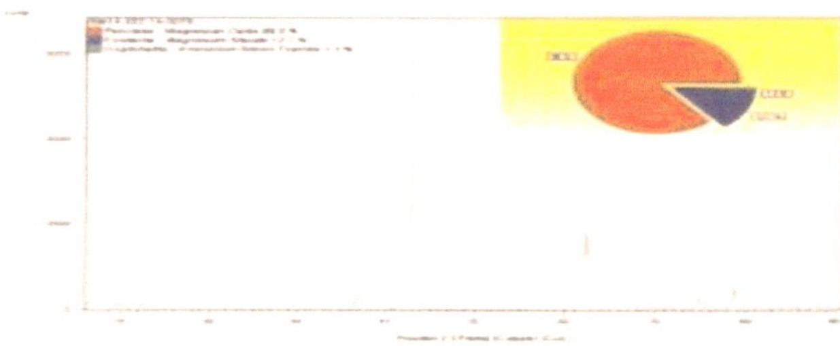

RESULTADOS DE DIFRACTOMETRÍA POR RAYOS X

MATERIAL:	Rocas absorbente color oro 1
CÓDIGO DE LA MUESTRA:	14-0078
EQUIPO UTILIZADO:	Difractómetro Panalytical XPertPro

Ref. Code	Nombre del compuesto	Fórmula química	SemiQuant(%)
00-045-0946	Periclase, syn	Mg O	86
00-004-0768	Forsterite, syn	Mg2 Si O4	12,7
00-044-1484	Cryptohalite, syn	(N H4)2 Si F6	1,3
	Amorfo		0
			100

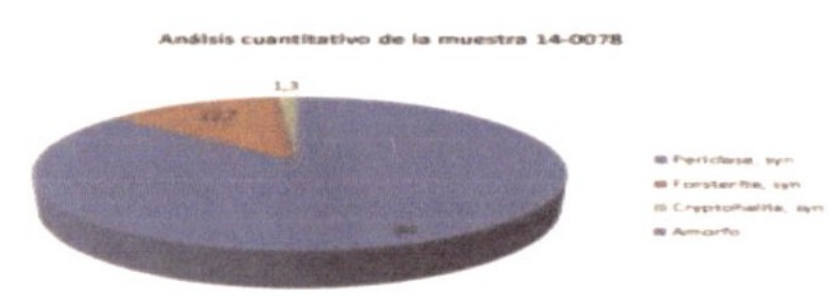

Figure 5.
FONT: C. Montaño 2022.

with granulometries of 0.25 mm–1 mm, solutions with known concentrations (0.032 N of $ZnSO_4$ H_2O) are prepared and the fractions are collected in 100 ml volumetric flasks that are analyzed by atomic absorption to determine the concentration in ppm of cation Zn^{2+}, the analysis finished when the concentration of the Zn^{2+} cation in the zeolite column effluent is close to or similar to the initial concentration of the Zn^{2+} cation. The columns have the same conditions (sample mass in grams g of zeolites, height in cm, volume in cm^3, diameter in cm, density in g/cm^3, flow rate in cm^3/h, and T°).

Figure 7 also shows the operation of the liquid phase on the solid phase, the liquid phase is represented by the solution of 0.01 N Zn SO4. The solution passes through the zeolite column, the dark shading represents the adsorption of the Zn2 + cation, as it moves downward, initially the concentration in the effluent is zero (C 1), until the adsorption zone reaches the base of the column, then the breaking point (C2) is reached. The rupture time is established when the concentration of the Zn2+ cation in the effluent reaches 5% of the initial concentration (C0), from the rupture time the concentration of the Zn2+ cation grows rapidly (C3) until reaching the initial concentration (C 0), at this moment the zeolite column is totally saturated (C 4)

At the beginning of the test in these columns with time both in the liquid phase as well as in the solid phase and its limits are considered between C/C0 (concentration ratio corresponding to the fluid and the feed) is from 0.95 to 0.05.

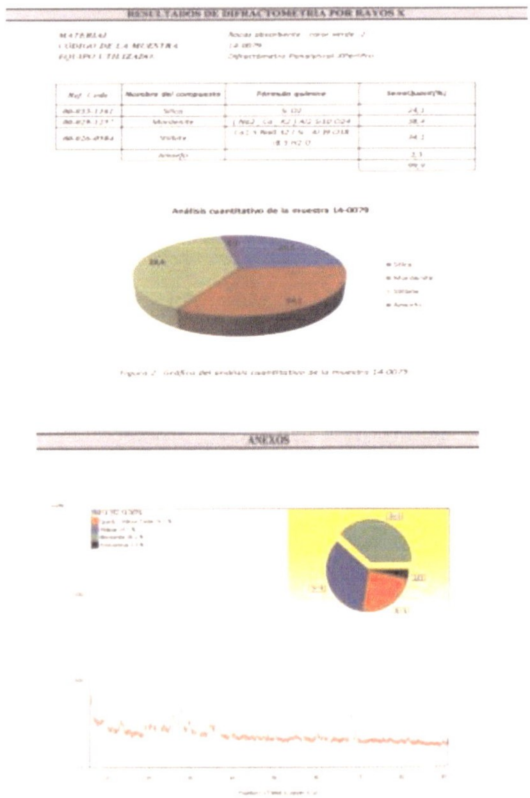

Figure 6.
FONT: C. Montaño 2022.

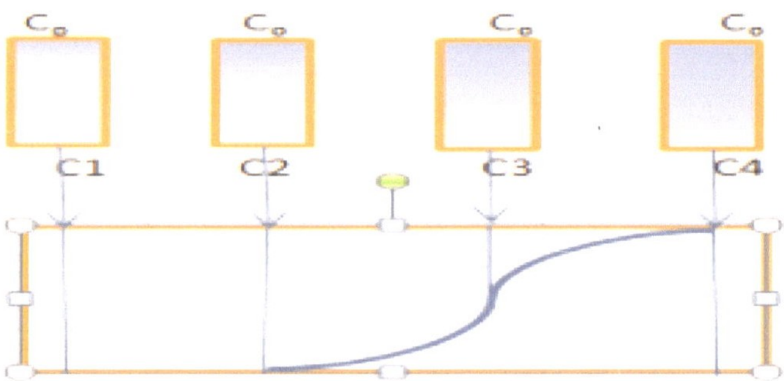

Figure 7.
Font : C. Montaño, autor.

8. Rupture curve/rupture time

Breakdown curve and breakup time are understood as the concentration curve or amount of the cation to be adsorbed in a certain time for the fluid or solution that comes out through the natural zeolite columns that act as adsorbent material.

The rupture time tb is always less than t, and the amount of adsorbed cations at the rupture point is established by integrating the rupture curve at the time tb

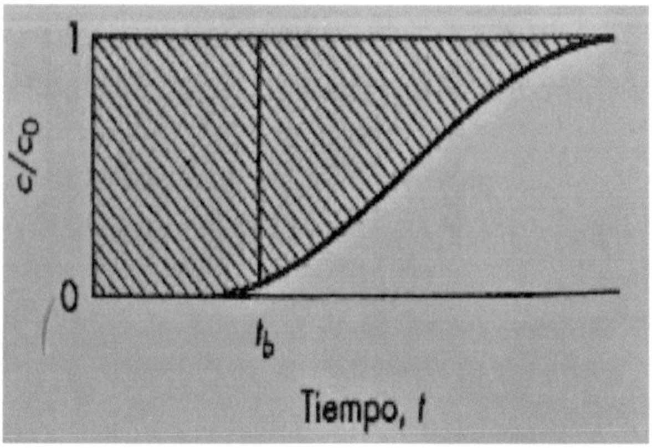

Figure 8.
FONT: F. Morante. ZEOLITAS naturales del Ecuador: Geologìa, CaracterizaciònY Aplicaciones ESPOL 2014 ISBN:978-9978-310-90-8.

The graph below shows how the adsorption breakup curve is formed and when it starts. The size of the curve is useful to determine the amount of adsorbed material.

Figure 8 represents how the natural zeolites adsorb the heavy metal (Zn^{2+}) over time until the saturation time, that is, when the natural zeolites do not have the capacity to adsorb more heavy metals. The rupture time tb is always less than t, and the amount of adsorbed cations at the rupture point is established by integrating the rupture curve at the time tb.

9. Experimental data

9.1 Materials

1.2 80 cm BURETTES

2.2 VOLUMETABLE FLAKS OF 100 ML

3.2 SAMPLES OF ZEOLITES (**Figure 9**)

In the adsorption test with fixed filters, the concentrations in the fluid phase and in the solid phase suffer variations over the time that the test lasts. As can be seen in **Table 3**, (**Figure 6**), which represents the zeolites with a chemical formula that have interchangeable bases (K, Mg, Na), had a rupture time of 1.4 h and a saturation time of 4 h in the adsorption process achieving a total Zn^{2+} adsorption of 161 Mg. Under the same conditions, (**Figure 5**), represents the zeolites where the chemical configuration is different and instead of interchangeable bases, they have a binary compound such as (O Mg), in the adsorption process it had a rupture time of 10 h and a saturation time of 14 h achieving a Zn^{2+} adsorption of 813 Mg (**Table 1**).

10. Results and discussion

The results obtained in the test are:

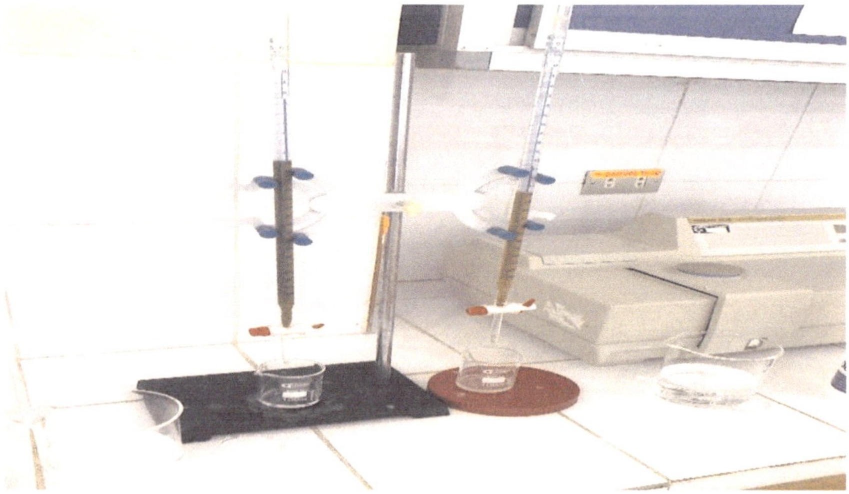

Figure 9.
FONT: C. Montaño 2022.

Zeolite (g)	Diámeter (cm)	Height (cm)	Volume (cm^3)	Density (g/cm^3)	Flow (cm^3/h)
18,0545	1	10,02	15,85	1,13	60

Table 1.
SAMPLE #1: Adsorption Column Data, Zeolite sample 1.

- Breakdown time: **tb**
- Mass of zinc adsorbed per gram of samples until saturation **Wsat**
- Equivalent length of unused bed: **LUB**

With the data in SAMPLE # 1 and with the feed flow rate, the superficial velocity of the fluid (dissolution) is obtained, expressed in cm/h.

$$u_o \frac{caudal\ cm3 \over h}{section(cm2)} \tag{3}$$

$$u_o\ 76,40\,cm/h$$

The feed rate of the cation (Zn^{2+}) per cm^2 of cross section, expressed in $g/cm^2/h$, is obtained:

$$\mathbf{F_A}\ \frac{u * C}{1000} \tag{4}$$

Where:
$u = \frac{cm}{h}$ = Surface speed in centimeters per hour
$\mathbf{C_o}$ Initial cation concentration, expressed in mg/cm^3

$$F_A 0,763 \frac{gZn}{cm^3/h} \quad (5)$$

The breaking time of the ordinate C/CO 0.5 1.40 h.
The saturation time of the ordinate C/CO 1 10 h.
In such a way that the mass of Zn^{2+} adsorbed per gram of zeolite until saturation is.

$$\mathbf{W_{sat.}} \frac{FA * . \int_0^{tsat} 1 - \frac{C}{C}}{d * h} * d \quad (6)$$

From where:
$\mathbf{F_A}$ Cation feed rate per cm^2 of cross section in g/m^2 h
d Zeolite density in g/cm^3
h Height in cm of the filter bed
$\int(1- C/C_O)$ι * dt. Area bounded by the break curve and the ordinate C/C_O 1, expressed in hours. The upper limit of integration is the saturation time (**t_{sat}**), which corresponds to the ordinate of the curve where C/C_O

$$\mathbf{W_{sat.} 161} \frac{mgZn}{gzeolita} \text{ is equal to } 1 \quad (7)$$

To calculate the fraction of the unused bed, we have.

$$\mathbf{LUB}h^* \left(1 - \frac{Wb}{Wsat}\right) \quad (8)$$

h length (height) of the filter bed, expressed in cm
Wb adsorbed mass (Zn^{2+}) per gram of zeolite to rupture time
Wsat adsorber mass (Zn^{2+}) per gram of zeolite at saturation point
LUB **4,8 cm (Table 2)**
The results with the two samples of zeolites are as follows (**Table 3**):

Sample 2 (g)	Diámeter (cm)	Height (cm)	Volume (cm^3)	Density (g/cm^3)	Flow (cm^3)
18,0606	1	9,85	15,03	1,20	60

Table 2.
SAMPLE #2: Adsorption Column Data, Zeolite sample 1.

Zeolites	Break time	Saturation time	Mass of zeolite adsorbed
MUESTRA 1	1.40 H	4 H	161 MG / Zn2+
MUESTRA 2	10. 0 H	14 H	813 MG / Zn2+

Table 3.
Final results of the adsorption process of the zeolite samples.

11. Discussion

It is observed in this work that there is an influence that favors adsorption when natural zeolites have a chemical composition where there are no interchangeable bases (Na, K, Mg), if not, a binary compound (O Mg) and an upper amorphous space. at 80%. However, it is necessary to carry out other adsorption tests with other metals, especially those with a higher molecular weight than Zn^{2+}, and also, it would be important to observe when there is more than one heavy metal in the solution or liquid phase.

The other scenario would be if the natural zeolites are enhanced with additives mentioned in this work: 1.1.4 (Significant advances with natural zeolites), these additives would surely improve the adsorption capacity of natural zeolites.

12. Conclusions

1. The adsorption capacity of natural zeolites is more optimal when the chemical formula has as its central axis a binary compound, (**Figure 5**), (O Mg) and no interchangeable bases,

   (**Figure 6**) (Na, K, Mg) and in the liquid phase there is only a heavy metal (Zn^{2+}).

2. The second physical characteristic of natural zeolites that favors adsorption is the large amorphous space in this case, according to the RX diffraction analysis, it showed 86% (**Figure 5**).

3. Of course, the rupture and saturation times are observed increased in favor of (**Figure 5**), which is very obvious.

4. The saturation time in the test with (**Figure 5**) was a bit slow after the 5th hour.

5. According to **Figure 4** where the physical characteristics of the main natural zeolites are, such as pore size, cation exchange capacity, among others, it agrees with the results obtained in this work, in (**Figure 4**) it is observed that the zeolites with 1 exchangeable base (Na, k, Mg) have higher cation exchange capacity.

6. Reviewing the literature regarding the adsorption of heavy metals, it is observed that in general all the families of natural zeolites have an average of 5–6 Mg for each gram of natural zeolite [13]. This would mean that (**Figure 5**) in the test achieved 44.7 Mg for each gram of natural zeolite.

Author details

Carlos Montaño[1*] and Javier Montaño[2]

1 Catholic University, Esmeraldas, Ecuador

2 Brigham Young University, Quito, Ecuador

*Address all correspondence to: cm96478@gmail.com

References

[1] Reyes YC, Vergara I, Torres OE, Díaz M, y González EE. Contaminación por metales pesados: Implicaciones en salud, ambiente y seguridad alimentaria. Ingeniería Investigación y Desarrollo. 2016;**16**(2):66-77. DOI: 10.19053/1900771X.V16.N2.2016.5447. ISSN 1900-771X

[2] Rebolledo E, Jiménez P. Affectations to water quality in the north of the esmeraldas province as a product of illegal gold mining. In: Conference: First International Scientific Seminar, Environment, Economy and Development. 2011. DOI: 10.1314O/2.1.4153.0563

[3] Hellawell JM. Biological Surveillance of Rivers. Stevenage: Water Research Center; 1978. p. 332

[4] Londoño Franco LF, Londoño Muñoz PT, Muñoz Garcia FG. The risks of heavy metals on human and animal health. Biotechnology in the Agricultural and Agroindustrial Sector. 2016;**14**(2):145. DOI: 10.18684/bsaa(14)145-153. ISSN 1909-9959

[5] Jara E, Gómez J, Montoya H, Sanchez T, Tapia L, Cano N, et al. Acumulación de metales pesados en Calamagrostis rigida (Kunth) Trin. ex Steud. (Poaceae) y Myriophyllum quitense Kunth (Haloragaceae) evaluadas en cuatro humedales altoandinos del Perú Accumulation of heavy metals in Calamagrostis rigida quitense. 2017;**24**(2):583-598

[6] Pérez L, Osorio P. Removal of Lead and Zinc in Polluted Water using Zeolites. Thesis. Lima Peru: Cesar Vallejo University; 2020

[7] Mahmood Ibrahimi M, Salih Sayyadi A. Application of natural and modified zeolites in removing heavy metal cations from aqueous media: An overview of including parameters affecting the process. International Journal of Geology, Agriculture and Environmental Sciences. 2015;**3**

[8] Morante F. Natural zeolites from Ecuador: Geology, Characterization and Applications. Edition Espol; 2014. pp. 181-192

[9] Gworek B, Kondzielski I. The Simultaneous Removal of Zinc and Cadmium from Natural Multicomponent Aqueous Solutions by Their Sorption onto Selected Natural and Synthetic Zeolites. 2020

[10] Ortega K, Angel Hernandezb M, Portilloc R, Ayala E, Romerob O, Rojas F, et al. Adsorption of Ar and N2 on dealuminated mordenite tuffs. Procedia Chemistry. 2015;**15**:65-71

[11] Corona OL, Hernández MA, Rojas F, Portillo R, Lara VH, Carlos FM. Propiedades de Adsorción en Zeolitas con Anillos de 8 Miembros I. Microporosidad y Superficie Externa. Revista Materia. 2009;**14**(3):918-931

[12] Chiappim W, Testoni GE, Moraes RS, Pessoa RS, Sag JC, Origo FD, et al. Structural, morphological, and optical properties of TiO_2 thin films grown by atomic layer deposition on fluorine doped tin oxide conductive glass. Vacuum. 2016;**123**:91-10

[13] Morante F. Natural zeolites from ecuador. Geology, Characterization, and Applications. ESPOL. 2014. ISBN: 978-9978-310-90-8

[14] Georgiev D, Bogdanov B, Markovska I, Hristov Y. A study on the synthesis and structure of zeolite Nax. Journal of

Chemical Technology and Metallurgy. 2013;**48**(2):168-173

[15] Kovacova Z, y Balintova M. Using of natural zeolites modification to heavy metals. 2018:1027-1035

[16] Shi J, Yang Z, Dai H, Lu X, Peng L, Tan X, et al. Preparation and application of modified zeolites as adsorbents in wastewater treatment. Water Science and Technology. 2017;**2017**(3):621-635. DOI: 10.2166/wst.2018.249. ISSN 02731223

[17] Gorimbo J, Taenzana B, Muleja AA, Kuvarega AT, y Jewell LL. Adsorption of cadmium, nickel and lead ions: Equilibrium, kinetic and selectivity studies on modified clinoptilolites from the USA and RSA. Environmental Science and Pollution Research. 2018;**25**(31):30962-30978. DOI: 10.1007/s11356-018-2992-0. ISSN 16147499

[18] Wang H, Xu J, y Sheng L. Purification mechanism of sewage from constructed wetlands with zeolite substrates: A review. Journal of Cleaner Production. 2020;**258**. DOI: 10.1016/j.jclepro.2020.120760. ISSN 09596526

[19] Hawash HBI, CHimielewska E, Netriova Z, Mazjlam J, Palcova H, Hudec P, et al. Innovative comparable study for application of iron oxyhydroxide and manganese dioxide modified clinoptilolite in removal of Zn(II) from aqueous medium. Journal of Environmental Chemical Engineering [en línea]. 2018;**6**(5):6489-6503. DOI: 10.1016/j.jece.2018.09.024

[20] Rahimi M, y Mahmoudi J. Heavy metals removal from aqueous solution by modified natural zeolites using central composite design. Periodica Polytechnica Chemical Engineering. 2020;**64**(1):106-115. DOI: 10.3311/PPch.13093. ISSN 15873765

[21] Choi HJ, Yu SW, Kim KH. Efficient use of Mg-modified zeolite in the treatment of aqueous solution contaminated with heavy metal toxic ions. Journal of the Taiwan Institute of Chemical Engineers. 2016;**63**:482-489. DOI: 10.1016/j.jtice.2016.03.005

[22] Bezerra BGP, Parodia A, da Silva DR, Pergher SBC. Cleaning produced water: A study of cation and anion removal using different adsorbents. Journal of Environmental Chemical Engineering. 2019;**7**:2. DOI: 10.1016/j.jece.2019.103006

[23] Fanta FT, Dubale AA, Bebizhu DF, Atlabachew M. Copper doped zeolite composite for antimicrobial activity and heavy metal removal from waste water. BMC Chemistry. 2019;**13**(1):1-13. DOI: 10.1186/s13065-019-0563-1

[24] de la Vega DP, Gonzalez C, Escalante CA, Gallego J, Salamanca M, Manrique-Losada L. Use of faujasite-type zeolite for ion adsorption in municipal wastewater. Tecnologia y Ciencias del Agua. 2018;**9**:4. DOI: 10.24850/j-tyca-2018-04-08

[25] Goñi S, Peña R, Guerrero A. Síntesis Hidrotermal de Zeolita a Partir de Ceniza Volante Tipo F: Influencia de la Temperatura. Materiales de Construcción. 2008;**60**(298):51-60. DOI: 10.18502/keg.v3i1.1476 Page 73

[26] He K, Chen Y, Tang Z, y Hu Y. Removal of heavy metal ions from aqueous solution by zeolite synthesized from fly ash. 2016:2778-2788. DOI: 10.1007/s11356-015-5422-6

[27] Ji XD, Ma YY, Peng SH, Gong YY, y Zhang F. Simultaneous removal of aqueous Zn2+, Cu2+, Cd2+, and Pb2+ by zeolites synthesized from low-calcium and high-calcium fly ash. Water Science and Technology. 2017;**76**(8):2106-2119.

DOI: 10.2166/wst.2017.361. ISSN 02731223

[28] Zheng F. Synthesis of NaY zeolite from coal gangue and its characterization for lead removal from aqueous solution. Advanced Powder Technology [en línea]. 2020;**2020**. DOI: 10.1016/j.apt.2020.04.035

[29] Limlamhong M, Yip ACK. Recent advances in zeolite-encapsulated metal catalysts: A suitable catalyst design for catalytic biomass conversion. Bioresource Technology. 2020;**297**: 122488. DOI: 10.1016/j.biortech. 2019.122488

[30] Yurekli Y. Removal of heavy metals in wastewater by using zeolite nano-particles impregnated polysulfone membranes. Journal of Hazardous Materials. 2016;**309**:53-64. DOI: 10.1016/ j.jhazmat.2016.01.064. ISSN 18733336

[31] Uman LS. Information management for the busy practitioner: Systematic reviews and meta-analyses. Journal of the American Academy of Child and Adolescent Psychiatry. 2011;**20**(1):57-59. DOI: 10.1016/j.revmed.2014.05.011. ISSN 1719-8429

[32] Ayora C. Los sistemas terrestres y sus implicaciones medioambientales [en línea]. Madrid: Ministerio de Educación y Formación Profesional. 2004. ISBN 8436939247. Disponible en: shorturl.at/ eftMO

[33] Hartini S, Pio C, Mater S, Ing S, Hartini S, Lakahina O, Kristijanto AI, Efectividad de los adsorbentes naturales para reducir el contenido de Cu y Pb del tratamiento de aguas residuales del laboratorio de química Efectividad de los adsorbentes naturales para reducir el contenido de Cu y Pb del tratamiento de aguas residuales. 2020

[34] Erdem E, Karapinar N, Donat R. The removal of heavy metal cations by natural zeolites. Journal of Colloid and Interface Science. 2004;**280**:309-314

Chapter 18

Biorefinery for Rehabilitation of Heavy Metals Polluted Areas

Teodor Vintila, Eniko Gaspar, Maria Mihaela Antofie, Luca Magagnin, Adina Berbecea and Isidora Radulov

Abstract

Biorefinery applied in heavy metals polluted lands proposed here describes a process starting from soil (polluted and unfit for food and feed production) and solar energy stored in carbohydrates (regarded here as a solar energy carrier) to deliver liquid and gaseous biofuels, green building block chemicals for the market and return the rest of the matter (not delivered to the market) as fertilizer and soil improver, extracting the heavy metals from the polluted soil for safe reuse and remediating the land to sustainably deliver resources in a circular bioeconomy. The circular economy proposed in this chapter offers a novel approach to land rehabilitation by investigating the opportunity for economic value creation as an integral part of a rehabilitation strategy and production of biomaterials and biofuels as renewable energy carriers. The case study approached here can be developed in a complete circular biorefinery process and value chain enabling the use of heavy metals polluted lands for production of renewable energy and biomaterials and at the same time serve as a means of rehabilitation of contaminated lands. This biotechnology can be transferred and adapted in other areas improper for food/feed production due to contamination human industrial activity.

Keywords: biofuels, biorefinery, circular bioeconomy, Copsa mica, heavy metals, pollution, soil remediation

1. Introduction

In recent years, agricultural use of the anaerobic digestate as organic fertilizer has aroused extensive public criticism due to its increased heavy metal (HM) contents, [1, 2]. Consequently, some novel approaches are developed to treat organic materials and avoid the negative effect of HMs to environment [3, 4]. Comparing with the fresh crops, their biogas residues after anaerobic digestion retained main fibrous texture which is essential for soil texture and fertility. On the other hand, organic materials such as digestate, manure, compost, have been reported to effectively reduce the availability of heavy metals in contaminated soils. This effect can be explained by the enhanced contents of organic matter in amended soil associated with the improvement of the biological, microbiological and biochemical properties of contaminated soils amended with organic materials [5, 6]. Organic amendments decreased significantly metal availability in soil, due to the binding of metals to organic matter as metal-organic complexes. The addition of 45 t ha^{-1} of organic ammendaments dry matter, led to an increase biomass yield in a mix of perennial grasses and straw cereals, belonging to the Leguminosae (*Trifolium pratense*) and Gramineae family (*Dactylis glomerata, Lolium perenne, Agropyron repens*), cultivated in polluted area of Copsa Mica (Romania). The contents of Cd and Pb in plants treated with organic amendments were significantly decreased [7]. The Pb concentrations in plants from field treatments with organic amendments were below the threshold for green fodder (40 mg kg^{-1}) according with Directive 2002/32/EC on undesirable substances in animal feed [8]. Administration of organic materials as amendments to contaminated soils is used in many cases with in order to improve soil fertility, improve vegetation on polluted lands and decrease availability of toxic substances for the plants cultivated on marginal lands [9, 10]. The addition of organic amendments to contaminated soils can affect bioavailability of heavy metals by forming metal oxides or carbonates associated with organic matter, which reduce the bioavailable forms to more stable fractions [11]. In addition to these effects, organic amendments are known to improve other soil characteristics such as water and nutrient holding capacities or aeration in the soil particles.

Sorghum crop is selected in this work as a tool with multiple purpose. It is applied as a green cover for un-used, or not-properly used polluted soil; being a robust, drought tolerant and low-demanding for nutrients plant, can deliver important yields of biomass and sugars for industrial purpose; intensive cultivation can be a tool to extract pollutants from soil; residues generated along cascade biorefining of sorghum biomass are regarded as heavy metals carriers, which can be delivered by the biorefinery in concentrated form.

The proposed circular approach proposed in this work consists in processing sorghum crop in cascade. Biorefinery will primarily process sugars to liquid biofuels-biochemicals and sorghum crops harvested from the HM polluted area (high or low polluted) is considered as feedstock for an industrial scale biorefinery. The harvested biomass (highly polluted and low polluted) from the envisaged area is used as feedstock in biorefinery for liquid biofuels, anaerobic digestion and thermal decomposition (pyrolysis or combustion) applied in cascade:

Biorefinery – Anaerobic digestion – Thermal decomposition

Thermal decomposition (combustion-pyrolysis) is performed to concentrate HM from digestate obtained by processing biomass from very polluted areas, primarily envisaged to carry HM in high concentration.

The aim of this work is to design a truly circular production process for bioenergy production that simultaneous serves as a contaminated land rehabilitation strategy. Thus, our approach will provide a means to address past wrong doings, yield a new opportunity for already polluted lands to be used sustainably and offer an example of how value chains can be made self-sustainable.

In our approach, we start from the paradigm "*considering the plants as the greenest battery able to accumulate and store solar energy, simultaneously delivering organic matter as feedstock for bio-based economy and cure the environment*". It all starts with unique and wonderful ability of the photo-autotrophic living organisms (plants) to convert CO_2, H_2O and sun energy (by simultaneously emitting free O_2) into $C_6H_{12}O_6$ (simple sugars)– the molecule used as basic energy source in all living organisms and consequently the main energy carrier in bio-based economy. Plants have also the ability to grow in a wide range of environments, including those affected by humans, and can therefore re-introduce "lost" molecules in natural cycles. We just need to understand how to harness these organisms, their metabolism and to find them the right place in a sustainable bioeconomy.

Circular bioeconomy is about integration and inter-connection. We propose to connect several processes and technologies in a comprehensive, sustainable and circular bioeconomic value chain with the objective to offer a method of biological remediation of environment affected by humans that integrates a complete biorefinery of plants to deliver products and eco-service at the same time by smart management of bioresources.

2. Concept of remediating biorefinery

We propose a study case using a sugary plant – *sweet sorghum* as feedstock for a biorefinery and remediating tool delivering sugars for bioethanol, lignocellulose for second generation biofuels and for biogas to capture residual energy. This crop is considered here as a tool to extract heavy metals (HM) from an industrially polluted area, where the main pollutants are Cd, Pb, Cu and Zn. The area heavily polluted in the proximity of the heavy metals' smelter stretches on around 22,000 ha (the surface polluted by lead according to previous research [12, 13]) and pollution radially decreases. From this point of view, we divided the studied area in two categories: "heavily polluted" and "lightly polluted". Crop from both categories will be used as feedstock in biorefinery where it is converted to liquid biofuels and biochemicals. The by-products resulting from this process carry organic matter, energy and pollutants (HM). These by-products can contain HM in a wide range of concentrations depending on the soil used to grow the crop.

Both lightly polluted and heavily polluted biomass (by-products) are sent to anaerobic digestion (AD) to continue this way the biorefinery of the feedstock and to deliver more energy (gaseous biofuel - biogas) and to continue decomposition of the vegetal organic complex structures. After AD process, the obtained digestate is separated in liquid and solid fraction, which will contain HM in varying concentrations. The solid fraction containing HM above legal limits, is sent to combustion/pyrolysis. The ash/char is used to extract metals in concentrated form and re-delivered to metals industry. The solid fraction containing HM within legal limits, is used as soil improver and immobilizer for lightly polluted soil, this way gradually returning polluted soil back to produce edible crops. From the liquid fraction, nitrogen is extracted to obtain two types of products: nitrogen organic fertilizers and liquid digestate reach in other elements (P, K, S). This way eutrophication by using N-reach digestate to improve soil quality is avoided. Liquid digestate containing HM above legal limits, is sent to extraction of metals and the resulted digestate is returned back to soil. **Figure 1** displays the overall concept of the circular remediating biorefinery described here.

By this concept of circular remediating biorefinery we intend to develop a new approach for biorefining agricultural feedstock, addressing as case study/main feedstock the biomass obtained in marginal lands with reference to industrially polluted areas that take into consideration not only the economy related to the biorefinery products but also to create a complete value and social chain that will allow agriculture to bloom again in polluted areas.

By implementing our concept, we intend to develop a complete circular biorefinery including remediation technologies (more specific: phytoremediation, immobilization and phytoextraction, applied in two main scenarios) of heavy metals contaminated land using as feedstock carbohydrates (regarded here as solar

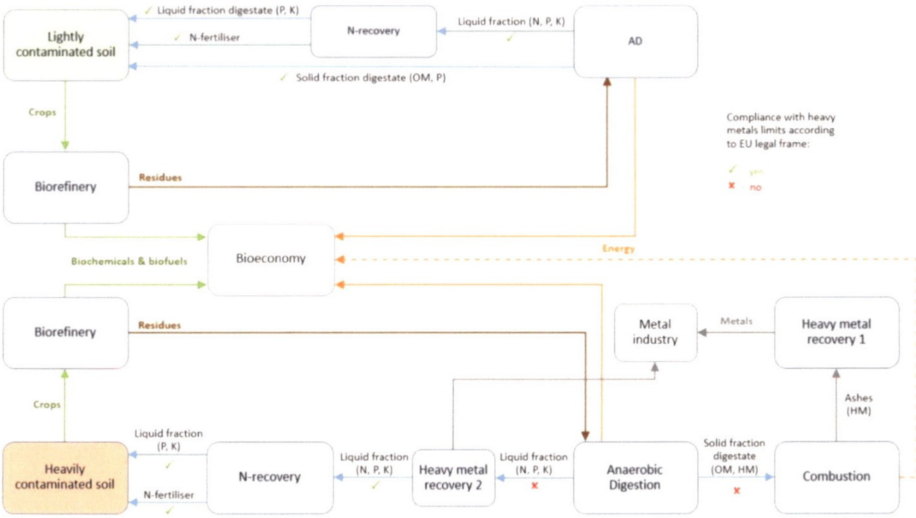

Figure 1.
Scheme of the circular remediating biorefinery concept.

energy carriers). The biorefinery will convert feedstock to green energy and a range of biochemicals, (ethanol, lactic acid for PLA etc) regarded as biofuels and green building block chemicals for the market. Other part of the feedstock is returned as fertilizer and soil improver, while residues containing metals in concentrated form are extracted for safe use.

"*Circular Remediating biorefinery*" developed here will bridge the gap between phytoremediation strategies and clean biofuel production in a sustainable and optimum manner, using and simultaneously remediating the contaminated land to sustainably deliver resources for bioeconomy - as case study- to be transferred and adapted in other areas with land improper for food/feed production.

3. Polluted area of Copsa Mica – Romania, as case study

Copsa Mica (see **Figure 2**) has for decades been known as" *the most polluted town in Europe*". The main pollutants identified in this area were cadmium, copper, lead and zinc. Moreover, this city was presented in Blacksmith Institute and Green Cross Switzerland Report 2012 - "*World's Worst Polluted Places*" [15] as examples of high cadmium pollution. In Romania there are some critical areas in terms of heavy metal pollution (Baia Mare, Zlatna, Moldova Noua, Copșa Mică). Of these, Copșa Mică area presents the highest risk of interception of heavy metals through locally produced local food, due to the large abundance of agrosystems in the structure of local socio-ecological systems. The Copsa Mica polluted area can be defined as the surface of land where the pollutant content in the top level of the soil (upper 20 cm) exceeds the alert thresholds defined by Romanian legislation. According to Vrinceanu and Lacatusu [12, 13], this polluted area covers 7040 ha where zinc content in soil is over 300 mg/kg; 10,320 ha of land where cadmium content in soil is over 3 mg/kg, or 22,565 ha where the lead content in soil exceed 50 mg/kg.

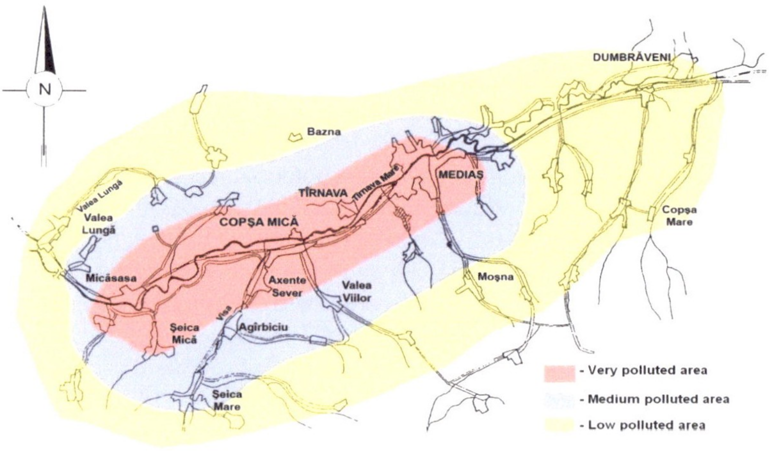

Figure 2.
Polluted area of Copsa mica (adaptation after: Barbu Horia, Lucian Blaga University of Sibiu [14]).

In Copsa Mica, the amount of Pb and Cd in vegetable samples exceeded the maximum permissible limits in carrots (median concentration 0.32 mg/kg for Pb and Cd) and in yellow onions (median concentration 0.24 mg/kg for Cd).

The European Commission has recently set new maximum levels for Cd and Pb in a range of food products to improve public health protection, with these measures entering into force from Aug. 30th 2021. Such *actions aim to further reduce the presence of carcinogenic contaminants in food and make healthy food more accessible* — a key aim of Europe's Beating Cancer Plan. Examples of these thresholds include 0.030 milligram per kilogram wet weight (ppm WT) for Cd and 0.10 ppm WT for Pb in stem vegetables; 0.10 ppm WT for Cd and 0.10 ppm WT for Pb in root and tuber vegetables (EU Regulations/2021 [16, 17]). In the case study area, the daily TM intake rates via local vegetable consumption are well above these values, more precisely 2 to 4 times higher for Pb and 5 to 10 times higher for Cd; yielding potential adverse public health effects [18]. Even after 10 years of ceasing production of the nonferrous smelter (production was ceased in 2009), the HM contents in the soil and plants are high inside the polluted area [7]. Such soils, or those with various other components that can be absorbed by crop plants and endanger public health, are recommended for use in the production of non-food crops, ideal for obtaining biomass to be used as a raw material in biorefinery.

Chronic effects in human health mainly result from exposure to low levels of cadmium and are represented by chronic obstructive diseases of the lungs and the renal system. There may also be effects on the cardiovascular and bone system. The fetus, young children, and pregnant women are unanimously recognized as sensitive populations with increased risk of developing adverse effects in chronic exposure to lead, including relatively low concentrations. Another indicator influenced by chronic lead exposure is somatic development, the height and weight can be changed, in the sense of growth delay, of about 1–1.5 years [19].

Over the decades, many options to clean the soils polluted with heavy metals have been considered. Taking into account the problems posed by top layer removal and replacement, chemical washing and many other "hard" methods, "gentle remediation" options have been explored.

Phytoextraction, using plants that can accumulate large amounts of potentially toxic metals in their above ground parts was proposed by (Gosh and Singh, 2005) as a feasible option, but the proposed plants (*Thlaspi caerulescens, Brassica juncea, Ipomoea carnea, Datura innoxia, Phragmytes karka* etc), either were exotic in Europe, raise cultivation problems, deliver smalls yields in terms of feedstock (carbohydrates, biomass) for biorefinery and/or concerns about their invasive potential. On the other hand, common European plants with high productivity have a small metal accumulation capacity [20]. Even several remediation attempts in Copsa Mica area have applied, none of them have been proven particularly successful [21–24] and the local community and farmers have adopted coping strategies, looking for other crops with a low uptake of heavy metals (i.e. excluders, as defined by Baker [25]) (**Figures 3** and **4**).

An important research focused on bioremediation of polluted soil in Copsa Mica area is *The RECARE project funded by the European Commission FP7 Programme, ENV.2013.6.2–4 'Sustainable land care in Europe'. EU grant agreement: 603498*. The research in this project regarding Copsa Mica case was focused on immobilization of these pollutants using different types of inorganic additives, such as zeolitic tuff, bentonite, volcanic tuff, and organic materials such as biosolids, cattle manure [26]. Although the situation is well documented, and there have been some attempts to remediate these soils, until now, no feasible solution for the decontamination of this large region has been found.

When phytoextraction is approached, important attention is dedicated to plants able to accumulate a particular metal from soil with higher efficiency compared with other plants. These plants are defined as hyperaccumulator plants and can accumulate metals in high concentration in some of their tissues - 100-fold or 1000-fold when compared to other plants growing in that soil. The bad news is that such plants are relatively rare, are endemic only in scattered areas around the world, are less adaptive

Figure 3.
Area in vicinity of heavy metals smelter.

Figure 4.
Miscanthus giganteus (left) and sweet sorghum (right) cultivated on polluted soil in Copsa mica area.

to the polluted areas where they need to be cultivated and the number of known species are low - less than four hundred species identified for as little as eight polluting metals [27]. The good news is that heavy metals can be extracted by very common plants, even they do not have ability to accumulate metal in high concentrations in their tissues as hyperaccumulator plants. If we take into account plants that produce high amounts of biomass per hectare of land, such as corn (*Zea mays*), sorghum (*Sorghum bicolor*), alfalfa (*Medicago sativa L.*) cup plant (*Silphium perfoliuatum*) [27] or sunflower (*Helianthus annus L.*) [28], the quantity of metals extracted from soil by plants producing high yields even containing lower concentrations of metals, can be more important than quantity of metals extracted by low yielding hyperaccumulator plants. Large quantity of biomass produced by such crops can remove higher amount of metals from the polluted soil even at lower concentrations of metal within the plants compared with low quantity of biomass with high concentrations of metal produced by hyperaccumulator plants [27].

Even high levels of heavy metals have been found in edible plants in the polluted area, people continue to farming and to produce food. There are farms producing different types of crops, cattle, chicken and other animal farms. There is no constraining regarding farming in the polluted area. The people are not informed regarding the risks. Still, in the polluted area there are land owners avoiding cultivation of edible plants. There are several hectares of *Miscanthus giganteus* as energy crops. *Sweet sorghum* has been cultivated as well, as trials in a former research projects, obtaining high yields: 60–100 to/ha of fresh mater, or up to 35 to/ha of biomass dry matter. **Figure 4** is an example of *Miscanthus giganteus* cultures adopted by local land owners to use the polluted soil for energy purpose.

Although is generally accepted that phytoremediation of metals from contaminated soil is possible and several plants have the ability to extract the pollutants from soil [29], the utilization of the biomass containing heavy metals raise important questions. Several approaches for disposal of metals containing biomass include decaying, thermal decomposition including burning or pyrolysis, chemical extraction, or even recovery of precious and semiprecious metals - called phytomining [30]. These technologies applied for processing polluted biomass generates wastes. For example, in the process of thermal decomposition by pyrolysis of heavy metals polluted biomass, almost all metals accumulate in the char, while by burning, metals can be found both in flue gasses and ash. Consequently, questions are raising regarding disposal of the heavy metals contaminated byproducts/residues such as char or ash. An attractive alternative approach to use heavy metal-polluted soils is to produce sugars crops and high-yielding biomass crops such as sweet sorghum. Sugars extracted from such crops can be converted in biorefineries to a wide range of biochemical; the organic residues resulted in biorefinery to be digested for production of biogas and digestate and to return this digestate in the same polluted soil as a fertilizer. By this circular bioeconomy approach, heavy metals can be confined in the polluted area and the risks of disposal of contaminated residues and further disperse the pollutants is eliminated.

Plant biomass can be used for different energy-recovery techniques, such as anaerobic digestion, incineration, gasification and liquid biofuels production. It is important to be sure that the metal burden, toxic metals such as Cd, in plant biomass will not affect biofuels production [31]. So, it is essential to assess these effects of metals concentrations on biofuels production systems and the design of biotechnological processes (ethanol fermentation, anaerobic digestion for biogas production etc). Moreover, selecting suitable plants is essential, as species accumulating high concentrations of pollution may raise difficulties in conversion processes [31].

Sweet sorghum has been chosen by our team for bioremediation/biorefinery application. This type of plant biomass could be used to supplement the metals needed in fermentation systems and anaerobic digester, contributing to the implementation of circular economic strategies and closing the loop in resource utilization chain.

Why sweet sorghum has been chosen for this study case? Sweet sorghum can be an important source of fermentable sugars for industrial biotechnology. A wide range of bioproducts for industrial, pharmaceutical or agricultural use can be obtained by microbial fermentation processes (alcohols, organic acids, amino acids, proteins, antibiotics, etc.). Sweet sorghum is an annual plant with a short production cycle and can be harvested after about 140–150 days of cultivation. The optimal seeding season is at the end of April - beginning of May and the harvesting can be done in September-October. It can be cultivated as secondary or precursor crops in combination with other short-cycle plants (for example triticale from early spring to June, followed by sweet sorghum for biofuels production). In classical biogas production technology, sugars in sorghum stems are converted in lactic acid during ensiling and by this the biomass is preserved until anaerobic digestion for biogas production.

Several studies indicates a variety of advantages of sorghum bicolor (L.) Moench over other sugar and starchy crops, which make this plant a highly studied energy crop [32]. Exhibiting high tolerance to draught, both sweet and grain sorghum varieties produce high yields even under a wide range of environmental conditions. More than that, studies demonstrates that sorghum can be cultivated on marginal lands and require low inputs [33–35]. In the context of climate change adaptation, temperatures and higher atmospheric level of CO_2 may beneficially affect sorghum crops in terms of a higher biomass yield. Sorghum cultures have been previously applied in *phytoextraction* experiments in polluted areas [36–38] and our own research results [39] indicated sweet sorghum as a good alternative for utilizing land polluted by heavy metals for bioethanol production, which would not only avoid food-fuel competition issues, but also provide way forward for land that is uncultivated due to human pollution.

Compared to other energy crops, sorghum has a global potential, being one of the most variable plants in terms of genetic resources, making breeding and development of new cultivars, adapted to different climate zones around the globe, easier [40].

Crops	**Corn (for grains)**		**Sugar beet**		**Sweet sorghum**	
Fresh mass (ton/ha)	20 (≈ 30–50% moisture)		50 (≈ 75–80% moisture)		60 (≈ 60–65% moisture)	
Products	Grains	Stalks	Sugar	Press cake	Sugar	Bagasse
(ton/ha) ≈	9	9	8	20	6	30
Ethanol (ton/ha) ≈	2.5	—	5	—	3	—
Methane (m^3/ha)	—	750	—	1200	—	2500
(MWh/ha)*	20.6	7.5	41.2	12	24.8	25
1st + 2nd gen (MWh/ha)	28		53		50	
Climate change, (drought)	Sensitive to drought, production dramatically affected		Very sensitive to drought, production dramatically affected		Resilient to drought, high production in hot summer 40–60 MWh/ha	
Technological advantages	Low cost, long term and easy storage of main energy carrier (starch)		Direct conversion of main energy carrier - sugar (no hydrolysis phase)		Direct conversion of main energy carrier - sugar (no hydrolysis phase); availability of high energy content lignocellulose (bagasse)	
Technological disadvantages	Hydrolysis of starch (added costs); difficult storage of high moisture stalks (too dry for ensilage, too moist for baling); large surface to cover for harvest and transport of stalks		Seasonal production, loss of sugar during storage of beets, sorghum stalks; difficult to storage beets cake and sorghum bagasse; high costs, energy intensive technology for sugar concentration and refining. Problems solved by patented innovation 131,499 /2021 [41]			

**1 kg EtOH = 8.25 kWh; 1 m^3 methane = 10 kWh.*

Table 1.
Relevant features of application of three main crops cultivable in European climate conditions.

Calculations show that sorghum can be comparable and, in some cases, competitive to sugar cane and corn in terms of sugar and bioethanol output per hectare (**Table 1**), while requiring much less water [42–45].

In this work we propose an integrated process in cascade using as feedstock sorghum biomass produced on heavy metals polluted soil (15 km distance from the smelter – the core of polluted area) having as objective soil remediation. Preliminary results obtained in lab scale [39] recommended sorghum as a crop able to grow on polluted soil and to provide readily fermentable sugars by juice extraction. In this research, metals concentration in sorghum juice are between 0.5 and 1.0 mg·kg^{-1} for Pb and between 22.7 and 86.2 mg·kg^{-1} for Zn, while Cd and Cu are not detected. When bagasse resulted after juice extraction is analyzed, the concentration of heavy metals increase as bagasses are pretreated - the average metals levels found in the thermo-chemically pretreated biomass were higher than those in the unpretreated biomass. Concentrations of all four analyzed metals increased after pretreatment: Cd from 3.60 to 4.03 mg·kg^{-1}, Cu from 15.57 to 25.56 mg·kg^{-1}, Pb from 11.24 to

19.38 mg·kg^{-1} and Zn from 123.50 to 134.19 mg·kg^{-1}. These increased values after pretreatments indicates higher availability of free metals after decomposition of lignocellulosic complex. More than that, the fate of metals was tracked in the biorefinery of bagasse to produce second generation bioethanol and after distillation, portions of Cu and Pb were found in the distilled ethanol, while Cd and Zn remained in vinasse.

In order to define "highly polluted biomass" and "low polluted biomass", we refer to the European legal frame [46, 47]. In this EU legal frame are defined the highest levels of heavy metals accepted in products used as fertilizers and soil improvers (**Table 2**). Assuming that the concentration of metals increases in the solid part of the biomass along cascade processing (regarded as by-products in biorefinery) during treatments (juice pressing, hydrolysis, anaerobic digestion), we expect higher concentration of heavy metals to be found in digestate than in the raw material (sorghum biomass). According to European legal frame above mentioned, contaminants must not exceed the following limit values:

In this respect and according to previous research [39], in average Cd is the main pollutant found in sorghum in the area selected as study case – average concentration of 5 mg/kg dry matter of sorghum biomass, while lead is found in average concentration of 20 mg/kg (under the limit of 120 mg/kg in EU legislation). It is expected to harvest sorghum biomass containing higher concentrations of cadmium and lead closer to smelter and lower in low polluted areas (blue and yellow on the map, **Figure 2**). When the limit of Cd < 1,5 mg/kg DM (EU threshold for fertilizer) is regarded here to define low polluted biomass, the surface delivering very polluted biomass in the map from **Figure 2** includes red and blue zones.

Respecting the above-mentioned legal frame, we propose to developed a new approach for biorefining agricultural feedstock, addressing as case study/main feedstock the biomass obtained from an energetic crop cultivated in marginal lands with reference to industrially polluted areas that take into consideration not only the economy related to the biorefinery products but also to deliver an integrated soil remediation system and create a complete value and social environment that will allow agriculture to bloom again in the selected area.

Regarding extraction of metals from biorefinery by-products, ashes/chars obtained through combustion/pyrolysis of solid digestate fraction can be considered as renewable secondary sources for the recovery of heavy metals. These ashes are usually classified as hazardous material due to their high content of toxic metals and soluble components. The most widespread leaching method is acidic leaching using strong mineral acids as many metal compounds have high solubility at low pH.

Metal	In soil improver, mg/kg dry matter	In organic fertilizer, mg/kg dry matter
cadmium (Cd)	2	1.5
hexavalent chromium (Cr VI)	2	2
mercury (Hg)	1	1
nickel (Ni)	50	50
lead (Pb)	120	120
inorganic arsenic (As)	40	40

Table 2.
Highest levels of heavy metals accepted in products used as fertilizers and soil improvers according to EU legal framework.

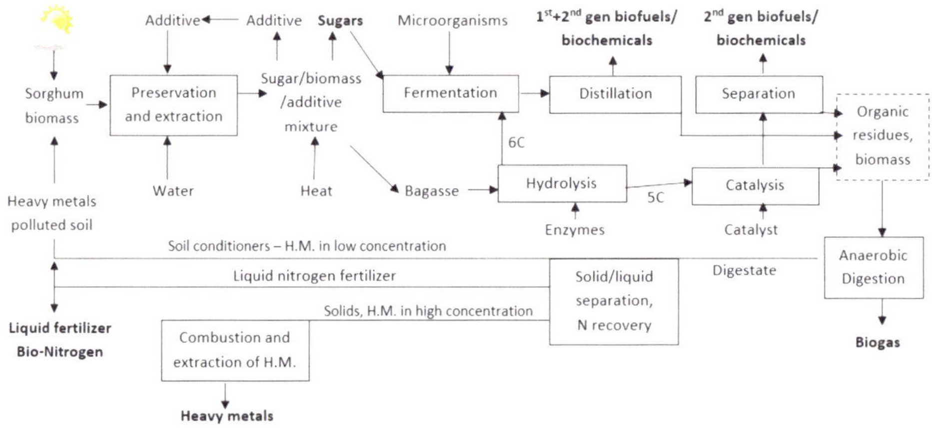

Figure 5.
Draft of the proposed circular bioeconomy.

However, due to the alkalinity of the ash large amounts of acid are needed. This problem can be overcome by alternative leaching media (especially organic acids obtained in biorefinery) in order to favor the achievement of high efficiency in the dissolution and in the electrowinning of heavy metals. In this respect, the biorefinery can be an important provider for metals industry. Firstly, redistributing back "lost metals" extracted by plants and delivered through biorefinery by-products and secondly, providing catalysts for metals extraction. Organic acids can be produced in biorefinery by fermentation of sugars and delivered to industry to prepare leaching media for dissolution and in the electrowinning of heavy metals. This can be another aspect of circular (bio)economy proposed here. The schematic approach of the proposed circular economy is presented in **Figure 5**.

4. Preservation and extraction of sugars from biomass for biorefinery

There are numerous research results indicating sweet sorghum as one of the best crops for biofuels industry. Still, industrial exploitation of sorghum cultures as an energy carrier is inhibited by a short harvesting period in temperate regions (1–2 months in Romania) and storage challenges leading either to high costs (in either processing or storage facilities) or high losses of fermentable sugar [48, 49]. Apart from tropical climates where there is a minimum of two harvests a year of sugar cane, the production of biofuels from sugary plants in other part of the world (and in particular in Europe) does not allow processing cycles on an annual basis. This means that the entire production capacity of the plants is not exploited, with the consequent problems of scale economy that penalize these types of feedstock in Europe for bioethanol production. In North America market corn is widely used since this type of feedstock (corn grains) can be stored for long periods and the full capacity of the plant is fulfilled on annul basis. In Europe this approach is not considered environmentally friendly due to the large soil and water usage for non-food application. On the other hand, pretreatments of lignocellulose matrices have not yet exploited their full potential in order to be competitive with bioethanol produced in Brazil and North America.

The main disadvantage is the impossibility of preserving the sugar-containing sorghum crop throughout the year without losing the sugar content accumulated in sorghum stems. Having a considerable moisture content (30–35%), sugar sorghum biomass is easily colonized by microorganisms, which consume sugars and cause the degradation of biomass. Classical ensiling (by lactic bacteria) has the same effect of sugar consumption, and drying of the biomass and preservation in dried form until processing is an alternative that cannot be applied on an industrial scale due to high energy inputs and storage costs. Therefore, the main bottleneck in using sweet sorghum as feedstock in biorefinery is preservation of sugar content in the biomass to be available around the year as feedstock for biorefinery.

Without preservation of sugars, the processing of sweet sorghum biomass would resemble sugar-beet processing: high capacity processing within a short period of time to minimize loss of sugars contained in the biomass. Processing facilities have large capacity and are only operated for a short period. This classical approach makes the process expensive due to high cost of capital per outputs unit. Furthermore, current sugar extraction technologies involve pressing the sorghum stalks (sometimes hydrated by the addition of water) and the harvesting of the sugar-containing juice. This process extracts only part of the sugar content stored in the sorghum biomass; the rest of the sugars remain in the bagasse. Sugars left in the bagasse are lost shortly after pressing by microbial proliferation. Incomplete use of sugars from sugar-containing biomass results in a much lower energy balance, reflected in a low degree of sustainability, low economic efficiency and higher carbon footprint.

Laboratory studies made by BUAS team have validated analytical predictions regarding sugars preservation and extraction from sorghum biomass and sugar beet. The analysis was carried out in small scale batches preserved during periods of over 12 months and the whole biorefinery of biomass was carried out in laboratory equipment. This technology, patented by BUAS Timisoara, is registered to Romanian Office for Inventions and Marks (OSIM) under the title "*PROCESS FOR BIOREFINING OF SUGAR YIELDING PLANTS WITH CONSERVATION AND EXTRACTION OF SUGARS FOR PRODUCTION OF BIOFUELS AND OTHER BIOPRODUCTS*", Patent no. 131499/2021 [41]. Recently, at the global "Climate Launchpad" event - the world's largest green business ideas competition, the invention reached the world semifinal and won the 1st place at national Climate Launchpad final in Romania.

The innovation of our approach lies in:

The use of a sorghum pre-treatment approach, proved at laboratory-scale (TRL3), consisting of preservation of sorghum biomass containing sugars for several months, more than one year, to make possible the use of sweet sorghum crops as feedstock for biofuels and biochemicals. The preservation of the main energy carrier (sugar) in the plants is the main criterion in considering a certain crop as feedstock for biorefinery. The patent developed a process of preservation of sugars in sorghum biomass by additivated ensilage, using a cost effective, recoverable, produced in situ additive. In traditional ensiling techniques, the sugars are lost by fermentation and transformed into organic acids such as lactic acid.

The invention has triple effect: (1) preservation of sugars, (2) extraction of sugars and (3) release of cellulose from lignocellulosic complex and access of cellulolytic enzymes to hydrolyse cellulose to glucose.

The advantages of the pre-treatment method can be found on three levels:

- Firstly, this method preserves sugars in the biomass for year around, improve extraction and releasing the sugars from biomass resulting in higher sugar yields and energy yields on the surface of land and the possibility to produce sweet juice for fermentation on an ongoing basis; reducing the required pressing capacity and hence investment costs.

- Secondly and perhaps more importantly, during preservation period of the sugar-containing biomass by this original process, the chemical structure of lignocellulosic biomass is affected, facilitating the access of enzymes for hydrolysis of cellulose, this way excluding conventional methods of thermal-chemical pretreatment (alkaline/steam).

- And last, but not least, the economy of the complete biorefinery is much improved by lower capital expenditures, since smaller factories can be installed to process the same amount of biomass which currently have to be processed by large capacity refineries in short time, exclusively in the cold season, to prevent sugar lost by natural decaying. The economically-improved complete biorefinery is made possible by the original preserving process since first generation technologies can be directly linked to second generation biofuels and biochemicals and to anaerobic digestion which closes the loop of the circular economy.

In order to protect patent priority, data regarding sugars extraction and preservation have not yet been published.

By chaining the three main processes: (1) alcoholic fermentation of sugars extracted from sweet sorghum (2) enzymatic hydrolysis and fermentation of bagasse results after extraction of sugars and (3) anaerobic digestion of waste resulted after hydrolysis and fermentation for biogas production, total energy production is maximized (see **Table 3**), and the digestate returns to soil as organic fertilizer.

In order to increase technology readiness level (TRL), the pre-treatment process will be tested at pilot scale, and coupled with down-stream pressing, hydrolysis and fermentation to produce amounts of bioethanol and biochemicals in sufficient amounts to prove higher TRL. More than that, modeling of environmental impact and energy balance, profitability and costs need to be assessed to support transferability of project results from research to industrial application. Equipment located in University of Life Science from Timisoara, part of the pilot scale biorefinery, will be used to fulfill these objectives. **Figure 6** summarizes the innovative processes of preservation and biorefinery of sugary plants.

Circular bioeconomy proposed here integrate several technologies such as energy crops production, extraction of sugars, biorefinery, anaerobic digestion and remediation of heavy metals polluted soil. The concept can be summarized as follows.

Sweet sorghum harvested from highly and low polluted area is sent to preservation phase. In order to make sugars available for biorefinery around the year, the innovative technology for preservation of sugar containing biomass is proposed. In the next phase of the integrated circular bioeconomy, the extracted sugars are converted to first generation biofuels & biochemicals and the resulted lignocellulosic portion of the biomass is converted to second generation biofuels & biochemicals. The by-products generated in biofuels technology (bagasse, thin stillage, vinasse etc) are converted by anaerobic digestion to biogas (for bio-electricity or bio-methane) and digestate. From this phase, the bioeconomy will follow two paths:

a. Digestate containing high concentration of pollutants (over the limits according to EU legal frame) is sent to combustion and to extraction of heavy metals. In

	Process	1st gen. Ethanol, (kJ·kg^{-1} biomass D.M.)	2nd gen. Ethanol, (kJ·kg^{-1} D.M.)	Biogas (kJ·kg^{-1} biomass D.M.)	Integrated process, cumulated energy production (kJ·kg^{-1} biomass D.M.)
A	Solid State Fermentation of sugar containing ensiled biomass	6522	0	7236	13,758
B	Hydrolysis and fermentation of bagasse resulted from (A)	6522	913	5256	12,691
C	Pretreatment, hydrolysis and fermentation of bagasse resulted from (A)	6522	1696	2149	10,367
D	Fermentation of freshly harvested juice, without ensilage	804	0	9000	9804
E	Fermentation of juice harvested from ensiled bagasse resulted from (D)	3996	0	14,400	18,396
F	Fermentation of juice (single extraction) from ensiled biomass	9343	0	5940	15,283
G	Hydrolysis and fermentation of bagasse resulted from (F)	9343	7500	3679	20,522
H	Fermentation of juice (multi-stage extraction) from ensiled biomass	16,620	0	4932	21,552
I	Hydrolysis and fermentation of bagasse resulted from (H)	16,620	2857	5904	25,381

Table 3.
Energy production by several processes applied in laboratory scale for biorefinery of sweet sorghum preserved by original preservation method.

industrial scale application, it is envisaged to deliver the heavy metals containing ash back to the smelter responsible for dispersion of pollutants in the area.

b. The digestate containing low concentration of pollutants (between the limits according to EU legal frame) is used as soil improver to remediate, improve soil characteristics and to immobilize heavy metals inside the polluted area.

According with results from previous studies carried out in experimental field from Copsa Mica [26] the addition of organic amendments enhances the formation of stable complexes with organic compounds, improving the heavy metals immobilization processes in soil and leads to a decreasing of metals available amounts in soil. By addition of organic solids such as digestate resulted from AD of biorefinery by-products, it is foreseen to immobilize important part of pollutants (heavy metals), this way allowing the cultivation of edible crops at least in the low polluted area. Previous studies [7] indicated important rates of immobilization in polluted area due to the addition of organic fertilizer (digested plants) in the ranges of 38–40% of extractable Cd, 77–83% of extractable Pb and 43–47% of extractable Zn. Addition of organic residues will have significant effects on metal accumulation in biomass

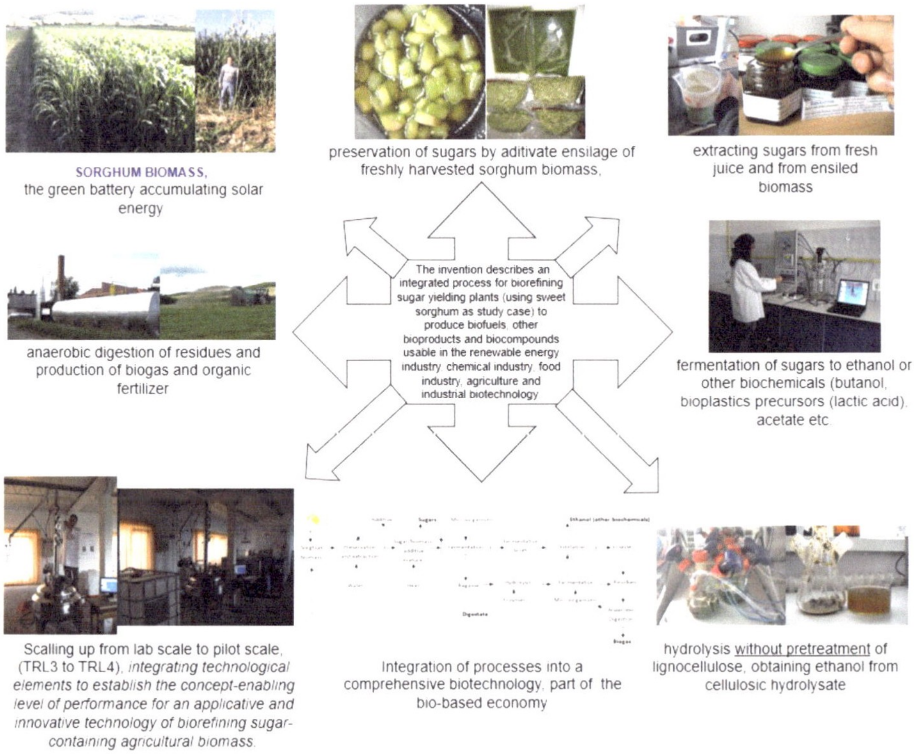

Figure 6.
Summary of innovative process of preservation and biorefinery of sugary plants.

and the contents of pollutants in plants will significantly decrease. The target is to reach below the threshold for green fodder (40 mg kg^{-1}) according with EU legislation on undesirable substances in animal feed [8]. Also, by reducing metal toxicity and improving soil fertility, the application of organic amendments, will result in the development of a permanent vegetation cover in non-arable land of polluted area. Development of a dense plant cover can halt erosion and thus prevent pollutants from spreading to other areas.

Particular outcome is envisaged in low polluted area, where application of digestate for immobilization of metals can restore the agriculture lands for food and feed production in shorter time frame, improving social attractiveness of the area and consolidating social acceptance of the biorefinery-bioenergy-remediation circular economy in polluted areas.

Summarizing, the overall methodology proposed here consists of the following main aspects to be approached:

- Production of sugar crops (sweet sorghum) while assessing the effects of organic fertilizers and improvers (liquid nitrogen and digestate) on soil quality, crops production and quality and degree of pollution in HM polluted area;
- Application of the innovative process consisting of preservation of sugars by ensiling the freshly harvested biomass in the presence of a preservation additive, which simultaneously preserves the sugars, pretreat biomass and enhances the yields regarding sugars extraction;

- Fermentation of extracted sugars to produce ethanol/lactate as biofuel and bio-based building block chemicals, hydrolysis of cellulose (approaching additional pretreatments in relation to additivated ensiling regarded as pretreatment here), fermentation of hexoses and pentoses from hydrolysate to produce second generation biofuels/biochemicals;

- Treatment of by-products generated in previous stages by anaerobic digestion to obtain biogas and digestate to be used as soil conditioner and organic fertilizer, pyrolysis of highly polluted by-products (solid digestate) to provide bioenergy and treatment of ash by electro-chemical techniques to extract heavy metals in concentrated form.

5. Sustainability of the proposed biorefinery

The sustainability of biofuels production has been widely addressed in former projects and actions and shared among the chain actors which accepted the technologies through the technical, logistic, economic, financial, energetic, environmental and administrative aspects. Consequently, the main market players have been encouraged to start up new entrepreneurships to increase the economic competitiveness and at the same time the environmental sustainability of biofuels. The challenges in the biofuels market are to enhance raw material diversification, decentralization of the production and sustainability of biofuels (mainly as GHGs saving). The circular economy proposed here regarding the production of biofuels and other products in a biorefinery approach using sweet sorghum (as case study crop) and in the same time to cure the environment (polluted areas) contributes to address the current debates on land use and sustainability and to facilitate and promote a well-informed and balanced attitude among decision makers and the general public.

In order to better evaluate the impact of our circular biorefinery approach in the case study area, let us have an insight look on two possible scenarios to use local bioresources for biofuels production, considering *50,000 tons commercial scale ethanol biorefinery.* This is not approached as part of this research in this project, it can be considered a possible future scenario. This impact scenario is not restricted exclusively to the industrial polluted area, it can be extrapolated in general as circular bioeconomy approach.

Scenario 1: consider recuperation of lignocellulosic agricultural by-products and their conversion to second generation biofuels (ethanol).

Previous own laboratory results, in concordance with other published data indicate an average production potential of 200–250 ml ethanol per 1 kg of agricultural residual biomass (wheat straw and corn stover). *The question raised is: what are the potentials in different regions to provide agricultural residues for a 50,000 tons commercial scale lignocellulosic ethanol biorefinery?* We assume that half of an entire region straw production can be harvested from the fields and transported to the biorefinery.

Table 4 shows the comparison of the average production of grains (main products) and straw (by-products) of the most used energy crops as well as the proportion of main products/by-products.

Sustainable harvesting of straw from the field, without affecting the humus content of the soil, generally depends on local climate and soil conditions. As a general rule, according to the scientific analysis of the above cited authors, up to 40% of the available straw can be harvested from the field for energy production, without

Crop	Average grains production, t/ha	Average straw production, t/ha	Grain – straw rate
Wheat	6.5	5.2	1: 0.8
Corn	6.8	8.9	1: 1.3
Sunflower	2.5	10.2	1: 4.1
Soy	3.5	2.3	1: 0.6
Rape	3.5	10.1	1: 2.9

(Source: Energie aus Biomasse [50])

Table 4.
Proportion of main products/by-products in energy crops.

damaging the quality of the soil. This important fact is respected in all related projects and business plans.

Assessing regions with large agricultural areas and regions with lower agricultural productions, the availability in space of the by-products to be delivered to a 50,000 tones capacity second-generation biorefinery differs. In our calculations, the surface needed to provide the feedstock necessary for the biorefinery is around 5000 km^2, more specific ≥70 x 70 km is need to be covered to harvest the agricultural residues for biofuels production in agricultural areas with high crops productions (such as western Romanian planes). As for forestry and hilly regions, such as center Romania (case study polluted area), in average 34,000 km^2 can provide similar quantity of feedstock as in intensive agricultural areas.

When calculating ethanol yields per surface of cultivated land, data in **Table 5** are obtained, as the average ethanol potential for two of the main crops in Romania.

According to this approach, to provide feedstock for a 50,000 tons commercial scale lignocellulosic ethanol biorefinery, needs harvesting of wheat straw and corn stover from ≈ 88,000 hectares. The resulted numbers are generated from a theoretical potential analysis approach. Still, even if more criteria are introduced in the potential study, the main conclusions does not change, namely: if the approach is to produce ethanol exclusively from agricultural residues, large surfaces of land are needed. The biomass need to be harvested from large surfaces and transported long distances.

Scenario 2: consider using marginal lands, or areas improper for edible crops for biorefinery converting sweet sorghum as feedstock.

In our research, in laboratory and pilot scale trials, we obtained 3–4 tons of ethanol / hectare of sweet sorghum by fermentation of sweet juice harvested from sorghum stems and around 3 tons of lignocellulosic ethanol from sorghum bagasse resulted after juice extraction. This is a total amount of 6–7 tons of ethanol/hectare of sweet sorghum.

Biomass	Grains average production, t/ha	Proportion grains-straw	Straw average production, t/ha	40% available for biofuels	Ethanol, L/ha
Wheat	6.5	1:0.8	5,2	2.08	416
Corn	6.8	1:1.3	8,9	3.56	712
				Average:	564

Table 5.
Average ethanol potential for two of the main crops.

Another scenario, proved by our team in laboratory scale is fermentation of sweet sorghum juice to lactic acid. Yields obtained in lactic acid fermentation indicate that 5 tons of lactate can be produced by lactic fermentation of sweet juice obtained from one hectare of sweet sorghum. This can be converted in biodegradable bio-plastic PLA, replacing plastic waste generated by 360 Romanians/year. If bagasse is anaerobically digested, the BMP preliminary assays indicate an average of 6000 m^3 of methane potential from bagasse resulted from one hectare of sorghum.

If we consider that one hectare of sweet sorghum crop yields 6 tons of ethanol (from sugars + cellulosic ethanol), 50,000 tons ethanol biorefinery can be operated using as feedstock the sweet sorghum crops cultivated on ≈ 8000–9000 hectares. This is around 10% from the surface to be covered to transport the by-products for second-generation biorefinery in the first scenario.

In our case study, the total area considered as polluted is 50 x 20 km, meaning around 100,000 hectares. Approximately 30% is agricultural land, which totalize around 30,000 ha. The total surface of the farmers associated in GAL Podişul Mediaşului (association in Copsa Mica polluted area) is approximately 25,000 ha arable land. If 30% of the total agricultural area is cultivated with energy crops, a biorefinery can count on 10,000 ha of total crops production as feedstock to produce biofuels and other chemical building blocks (a biorefinery 50,000 L ethanol capacity relying on sweet sorghum needs around 8000–9000 ha). If the biorefinery is located in the centre of the polluted area, the maximum distance to transport the feedstock is 20 km.

Consequently, comparing the efforts to gather feedstock in the two above scenarios, the balance indicates the scenario 2 as the most attractive.

The system proposed to be developed in the polluted area consists of:

- One" flexible" lignocellulose/sugar to ethanol biorefinery. The main feedstock: sugar crops from polluted area, such as non-food crops (to not interfere with food sector). Example of non-food crop: sweet sorghum. Problem regarding the preservation of sugars in the biomass around the year is solved by "*PROCESS FOR BIOREFINING OF SUGAR YIELDING PLANTS WITH CONSERVATION AND EXTRACTION OF SUGARS FOR PRODUCTION OF BIOFUELS AND OTHER BIOPRODUCTS*" [41].

- Several biogas plants in the proximity of the biorefinery. The total capacity of biogas plants to collaborate with ethanol biorefinery is flexible, but ideally is 5 MWe installed power to serve 50,000 tones/year ethanol biorefinery with sugar and part of the lignocellulosic biomass. More clearly is a picture of the biorefinery collaborating with five one-MWe biogas plants (**Figure 7**).

Biorefinery uses as feedstock: (a) syrup extracted from sweet sorghum grown in polluted area and (b) lignocellulose (sorghum bagasse, other biomass from polluted area). The process can be multiple: production of first and second generation biofuels or building block chemicals. The residual lignin resulted after second generation biorefinery, originating from biomass grown in the highly polluted area (red zone), containing high concentration of heavy metals is pyrolysed. The char containing heavy metals is sent to the local smelter to extract metals simultaneously with the extraction process from mineral ores.

Biorefinery can have installed on-site biogas plant. The large storage capacities for preservation of sugar plants, the large volume of digesters for A.D. does not allow

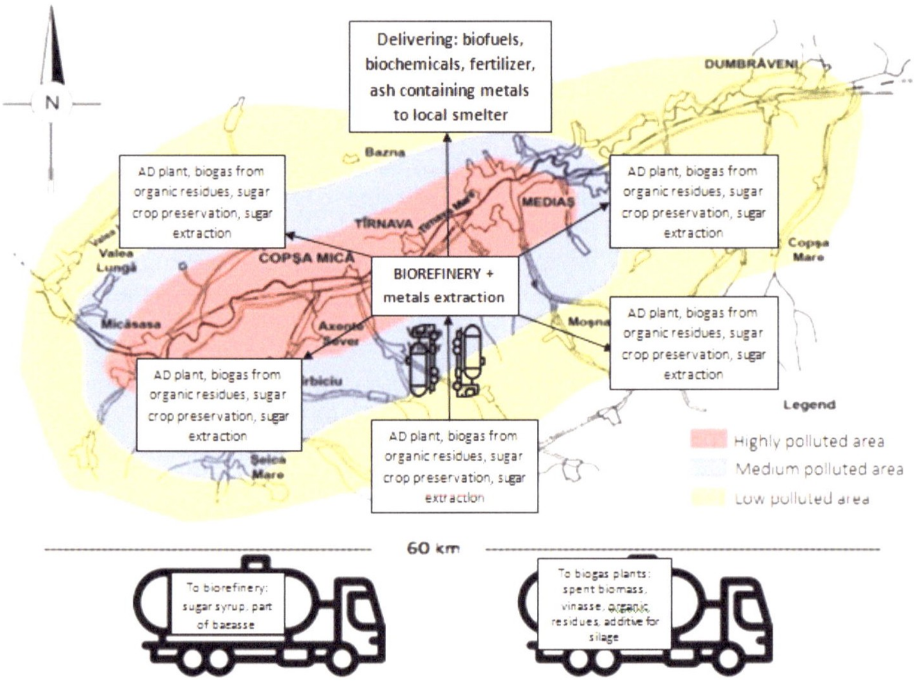

Figure 7.
System proposed to be developed in polluted area.

construction of such a large surface industrial facility [51, 52]. Therefore, it is recommended to outsource the storage of sugar plants connected to biogas technology in agricultural area (on-farm biogas plants).

The on-farm biogas plants are upgraded with the followings:

- Storage capacities for preservation of sugar containing biomass (sweet sorghum, sugar beet).

- Roller press to harvest sweet juice from preserved sugar plants.

- Evaporation equipment (operated by waste heat from biogas CHP unit plus part of the biogas) to concentrate the sweet juice from Brix 16–18 (obtainable in local climate conditions of sweet sorghum crops) to Brix 50 or higher; concomitantly, the preservation additive is recovered by distillation and stored on-site for preservation of the next harvest.

Description of the system operation, in general lines.

A trading system will be developed between the biorefinery and biogas plants (**Figure 7**):

- Biorefinery provide equipment needed to upgrade the biogas plants to be able to deliver sugar syrup (storage facilities, preservative chemical, press for harvesting sweet juice, evaporation equipment). Biorefinery provide as well a part of the spent biomass resulted after ethanol production process (containing low concentration of metals); vinasse resulted after distillation and other organic residues that can be used as feedstock for anaerobic digestion in biogas production.

- Biorefinery will adjust the production process according to a yearly plan agreed with biogas plants. For example, in cold season, when the electricity needs are higher, the energy from biogas will be dedicated to CHP unit. In this period, the ethanol production will be based mainly on residual lignocellulosic biomass. In warm season, part of the energy of biogas is used to produce syrup. The syrup will be added in biorefinery to the cellulosic hydrolysate, this way the ethanol production is much increased.

- Biogas plants provide to biorefinery the sugar syrup when situation on biofuels and biochemicals is favorable and when energy needs in the grid are covered by other sources. The sweet juice need to be concentrated to lower the transport costs and carbon emissions associated to transport. For production of syrup, part of the energy of the biogas produced on-site is dedicated to this process.

- Biogas plants can use the bagasse resulted after extraction of sweet juice to produce biogas (case of biomass containing low concentration of heavy metals), or can sell it to biorefinery for production of lignocellulosic ethanol, to be incinerated and sent for extraction of heavy metals (the highly polluted biomass obtained in the core of the polluted area - red zone).

- In case of deficit in energy grid, or other criteria recommending production of higher amount of electric energy, the biogas plant is operated to ensure electricity production of CHP unit at full capacity.

- Another scenario regarding flexibility consists of biogas plants equipped with biogas upgrading technology and delivering of biomethane. In this case, the biomethane can be compressed and liquefied to be used as second-generation biofuel. Biomethane can be also injected into the grid, can be stored in national storage system and used in thermo-power plants in case of deficit in electricity (instead of importing natural gas). The trading system with biorefinery is similar as in biogas plants + CHP technology.

6. Conclusions

To summarize, in our scenario, biomass containing sugars is stored around-the-year and sugar is extracted on daily basis. Sugar is converted in biorefinery to biofuels/biochemicals. Bagasse is converted to second generation biofuels/biochemicals and the by-products are sent to AD. The highly polluted solid digestate is pyrolysed to obtain process heat and the char is sent to local smelter for metal extraction. The highly polluted liquid digestate is sent to smelter for metals extraction. By-products containing low concentrations of pollutants are digested to obtain biomethane. The digestate containing low concentration of metals (within legal limits) is returned in the soil to maintain fertility and to immobilize heavy metals.

The whole biorefinery approach is an innovative aspect and can be adapted to other cases of polluted areas, to be remediated for production of edible crops for food and feed. To note that crops containing pollutants within legal limits are subject to be used for food sector, while polluted bioresources are subject to biorefinery for non-food products. The main impacts of the circular bioeconomy developed here are foreseen to be materialized in the following directions:

- **Environment,** by remediation of polluted areas and contribution to climate neutrality through the proposed innovative bioremediating farming system, decreasing Greenhouse Gases (GHG) emissions through cascading use of bioresources;

- **Human health,** by redirecting polluted lands from producing contaminated food responsible for degradation of public health to the production of materials which contributes to the improvement of public health (biofuels and biochemical that replace fossil based materials and fuels currently produced by a polluting industry);

- **Economy**, developing an interconnected agro-industrial system based on smart management of main resources (sun, water and soil) by optimization of agricultural feedstock production in the emerging circular bioeconomy and delivering bio-based goods and services for a healthy and sustainable human society;

- **Social dimension**, improving quality of life in communities from polluted areas as social distribution of environmental quality is unequal, and often biased against poorer or socially excluded groups which are more likely to live in areas of poorer environmental quality than other groups.

Regarding the impact of the biorefinery on soil remediation (environment), the addition of immobilizing amendments is a promising and suitable technique for remediation of contaminated soils even if the total content of contaminants is not decreased. Organic amendments like, manure, compost, bio-solids and bio-solids compost may effectively reduce the availability of HM due to its high content of organic matter and improve the biochemical properties of contaminated soils. Immobilized in the soil, the pollution poses much less of a threat to e.g. groundwater, organisms in the soil or uptake by crops.

All the previous impacts converge into one important impact: food security (integrating health, environment, economy and society). Circular bioeconomy is meant to be one of the tools needed by today's society to improve food security and decrease pollution and impact of human activities on the environment.

Author details

Teodor Vintila[1*], Eniko Gaspar[2], Maria Mihaela Antofie[2], Luca Magagnin[3], Adina Berbecea[1] and Isidora Radulov[1]

1 University of Life Science "King Mihai I" from Timișoara, Romania

2 "Lucian Blaga" University of Sibiu, Romania

3 Polytechnic University of Milan, Italy

*Address all correspondence to: tvintila@animalsci-tm.ro

References

[1] Demirel B, Göl NP, Onay TT. Evaluation of heavy metal content in digestate from batch anaerobic co-digestion of sunflower hulls and poultry manure. Journal of Material Cycles Waste Management. 2013;**15**:242-246. DOI: 10.1007/s10163-012-0107-4

[2] Huang Y, Tian Y, Xie L, Liu Y, Dai B, Guo X, et al. The application of two-dimensional correlation spectroscopy for the binding properties of heavy metals onto digestate-derived DOM from anaerobic digestion of chicken manure. Ecotoxicology and Environmental Safety. 2020;**204**:111129. DOI: 10.1016/j.ecoenv.2020.111129

[3] Guo X, Xie X, Liu Y, Wang C, Yang M, Huang Y. Effects of digestate DOM on chemical behavior of soil heavy metals in an abandoned copper mining areas. Journal of Hazardous Materials. 2020;**393**:122436. DOI: 10.1016/j.jhazmat.2020.122436

[4] Jiang B, Lin Y, Mbog JC. Biochar derived from swine manure digestate and applied on the removals of heavy metals and antibiotics. Bioresource Technology. 2018;**270**:603-611, ISSN 0960-8524,. DOI: 10.1016/j.biortech.2018.08.022

[5] Marques APGC, Oliveira RS, Rangel AOSS, Castro PML. Application of manure and compost to contaminated soils and its effect on zinc accumulation by Solanum nigrum inoculated with arbuscular mycorrhizal fungi. Environmental Pollution. 2008;**151**(3):608-620

[6] Katoh M, Kitahara W, Sato T. Role of inorganic and organic fractions in animal manure compost in Lead immobilization and microbial activity in soil. Applied and Environmental Soil Science. 2016;**2016**, Article ID 7872947:9. DOI: 10.1155/2016/7872947

[7] Vrînceanu NO, Motelică DM, Dumitru M, Calciu I, Tănase V, Preda M. Assessment of using bentonite, dolomite, natural zeolite and manure for the immobilization of heavy metals in a contaminated soil: The Copșa Mică case study (Romania). Catena. 2019;**176**:336-342. DOI: 10.1016/j.catena.2019.01.015

[8] Directive 2002/32/EC of the European Parliament and of the Council of 7 May 2002 on Undesirable Substances in Animal Feed (OJ L 140). 2002. p. 10

[9] Williamson A, Johnson MS. Reclamation of metalliferous mine wastes. In: Lepp NW, editor. Effect of Heavy Metal Pollution on Plants. Pollution Monitoring Series. Vol. 2. Dordrecht: Springer; 1981. DOI: 10.1007/978-94-009-8099-0_6

[10] Abbott D, Essington M, Mullen M, Ammons J. Fly ash and lime-stabilized biosolid mixtures in mine spoil reclamation: Simulated weathering. Journal of Environmental Quality. 2001;**30**:608-616. DOI: 10.2134/jeq2001.302608x

[11] Walker DJ, Rafael Clemente M, Bernal P. Contrasting effects of manure and compost on soil pH, heavy metal availability and growth of Chenopodium album L. in a soil contaminated by pyritic mine waste. Chemosphere. 2004;**57**(3):215-224. DOI: 10.1016/j.chemosphere.2004.05.020

[12] Vrinceanu N, Motelica D, Dumitru M, Gamen E, Toti M, Veronica T, et al. Estimation of some heavy metals accumulation in plants and soils from Copsa mica area. Scientific

Papers, USAMV Bucharest, Series A. 2009;**LII**:76-81

[13] Lacatusu R, Lacatusu AR. Evolution of heavy metals pollution from Copsa mica. Scientific Papers, UASVM Bucharest, Series A. 2010;**LIII**:85-92

[14] Barbu CH, Pavel PB, Sand C, Pop MR. Reduced uptake of Cd and Pb by Miscanthus sinensis x giganteus cultivated on polluted soil and its use as biofuel. Environmental Engineering and Management Journal. 2013, 2013;**12**(2):1535-1538

[15] http://www.worstpolluted.org/projects/display/113

[16] Commission Regulation (EU) 2021/1323 of 10 August 2021 Amending Regulation (EC) No 1881/2006 as regards Maximum Levels of Cadmium in Certain Foodstuffs

[17] Commission Regulation (EU) 2021/1317 of 9 August 2021 Amending Regulation (EC) No 1881/2006 as regards Maximum Levels of Lead in Certain Foodstuffs.

[18] Nedelescu M, Baconi D, Tsatsakis AM, et al. Environmental metal contamination and health impact assessment in two industrial regions of Romania. Science of the Total Environment. 2017;**580**:984-995

[19] Bellinger DC. Very low lead exposures and children's neurodevelopment. Therapeutics and Toxicology. 2008;**20**(2):172-177

[20] Dickinson NM, Baker AJM, Doronila A, Laidlaw S, Reeves RD. Phytoremediation of inorganics: Realism and synergies. International Journal of Phytoremediation. 2009;**11**(2):97-114

[21] Barbu HC, Sand C. Teoria și practica modernă a remedierii solurilor poluate cu metale grele. Sibiu: Alma Mater; 2004

[22] Muntean E, Muntean N, Mihăiescu T. Cadmium and lead soil pollution in Copsa mica area in relation with the food chain. Research Journal of Agricultural Science. 2010;**42**(3):731-734

[23] Iordache V, Dumitru M, Neagoe A, Sandu R, Bodescu F, Marinoiu L. Decision support for the management of agrosystems contaminated with metals: Research program and preliminary results. In: Barbu C-H, Sand C, editors. 2nd International Conference on Modern Technologies and Biotechnologies for Environmental Protection. "Lucian Blaga" University of Sibiu; 2010. pp. 83-93. ISSN 2068-5610

[24] Szanto Prodan M, Micle V, Gamenț E, Prodan VC. Investigații privind starea de calitate a solului din zona Copșa Mică. Ecoterra – Journal of Environmental Research and Protection. 2012;**31**:110-116

[25] Baker AJM. Accumulators and excluders strategies in response of plants to heavy metals. Journal of Plant Nutrition. 1981;**3**(1-4):643-654

[26] https://www.recare-hub.eu/case-studies/copsa-mica-romania

[27] Nescu V, Ciulca S, Sumalan RM, Berbecea A, Velicevici G, Negrea P, et al. Physiological aspects of absorption, translocation, and accumulation of heavy metals in *Silphium perfoliatum* L. Plants Grown in a Mining-Contaminated Soil. Minerals. 2022;**12**:334. DOI: 10.3390/min12030334

[28] Rizwan M, Ali S, Rizvi H, Rinklebe J, Tsang DCW, Meers E, et al. Phytomanagement of heavy metals in contaminated soils using sunflower: A review. Critical Reviews in Environmental Science and Technology. 2016;**46**:1498-1528

[29] Tordoff G, Baker A, Willis AJ. Current approaches to the revegetation

and reclamation of metalliferous mine wastes. Chemosphere. 2000;**41**:219-228. DOI: 10.1016/S0045-6535(99)00414-2

[30] Cao Z, Wang S, Wang T, Chang Z, Shen Z, Chen Y. Using contaminated plants involved in phytoremediation for anaerobic digestion. International Journal of Phytoremediation. 2015;**17**:201-207

[31] Tian Y, Zhang H. Producing biogas from agricultural residues generated during phytoremediation process: Possibility, threshold, and challenges. International Journal of Green Energy. 2016;**13**:1556-1563

[32] Zhao YL, Dolat A, Steinberger Y, Wang X, Osman A, Xie GH. Biomass yield and changes in chemical composition of sweet sorghum cultivars grown for biofuel. Field Crops Research. 2009;**111**(1-2):55-64

[33] Bennett S, Anex RP. Production, transportation and milling costs of sweet sorghum as a feedstock for centralized bioethanol production in the upper Midwest. Bioresource Technology. 2009;**100**(4):1595-1607

[34] Chohnan S, Nakane M, Rahman MH, Nitta Y, Yoshiura T, Ohta H, et al. Fuel ethanol production from sweet sorghum using repeated-batch fermentation. Journal of Bioscience and Bioengineering. 2011;**111**(4):433-436

[35] Marx S, Ndaba B, Chiyanzu I, Schabort C. Fuel ethanol production from sweet sorghum bagasse using microwave irradiation. Biomass and Bioenergy. 2014;**65**:145-150

[36] Epelde L, Mijangos I, Becerril JM, Garbisu C. Soil microbial community as bioindicator of the recovery of soil functioning derived from metal phytoextraction with sorghum. Soil Biology and Biochemistry. 2009;**41**(9):1788-1794

[37] Zhuang P, Shu W, Li Z, Liao B, Li J, and Shao J, "Removal of metals by sorghum plants from contaminated land," Journal of Environmental Sciences, vol. 21, no. 10, pp. 1432-1437, Jan. 2009.

[38] Székely Á, Poór P, Bagi I, Csiszár J, Gémes K, Horváth F, et al. Effect of EDTA on the growth and copper accumulation of sweet sorghum and sudangrass seedlings. Acta Biologica Szegediensis. 2011;**55**(1):159-164

[39] Vintila T, Negrea A, Barbu H, Sumalan RS, Kovacs K. Metal distribution in the process of lignocellulosic ethanol production from heavy metal contaminated sorghum biomass. Journal of Chemical Technology and Biotechnology. 2016;**91**(6):1607-1614

[40] Davila-Gomez FJ, Chuck-Hernandez C, Perez-Carrillo E, Rooney WL, Serna-Saldivar SO. Evaluation of bioethanol production from five different varieties of sweet and forage sorghums (Sorghum bicolor (L) Moench). Industrial Crops and Products. 2011;**33**(3):611-616

[41] Teodor Vintilă. Process for biorefining sugar plants with preservation and extraction of sugars to obtain biofuels and other biochemicals. brevet de invenție nr. 131499/26.02.2021. publ. in BOPI nr. 11/2016, OSIM Romania.

[42] Misook K, Donal FD. Composition of sugar cane, energy cane and sweet sorghum suitable for ethanol production at Louisiana sugar mills. Journal of Industrial Microbiology & Biotechnology. 2011;**38**:803-807

[43] Mumm R, Goldsmith P, Rausch K, Stein H. Land usage attributed to corn

ethanol production in the United States: Sensitivity to technological advances in corn grain yield, ethanol conversion, and co-product utilization. Biotechnology for Biofuels. 2014;**7**:61

[44] Tumbalam P, Thelen KD, Adkins A, Dale B, Balan V, Gunawan C, et al. Corn Stover ethanol yield as affected by grain yield, Bt trait and environment. Biomass and Bioenergy. 2016;**85**:119-125

[45] Supriya M, Umakanth AV, Tonapi VA, Sharma R, Manoj Sharma K. Sweet sorghum as biofuel feedstock: Recent advances and available resources. Biotechnology for Biofuels. 2017;**10**:146

[46] Regulation (EU) 2019/1009 of the European Parliament and of the Council of 5 June 2019 laying down rules on the making available on the market of EU fertilising products and amending Regulations (EC) No 1069/2009 and (EC) No 1107/2009 and repealing Regulation (EC) No 2003/2003. Official Journal of the European Union, L 170. Vol. 62, 2019

[47] Regulation (EU) 2019/1010 of the European Parliament and of the Council of 5 June 2019 on the alignment of reporting obligations in the field of legislation related to the environment, and amending Regulations (EC) No 166/2006 and (EU) No 995/2010 of the European Parliament and of the Council, Directives 2002/49/EC, 2004/35/EC, 2007/2/EC, 2009/147/EC and 2010/63/EU of the European Parliament and of the Council, Council Regulations (EC) No 338/97 and (EC) No 2173/2005, and Council Directive 86/278/EEC

[48] Wu X, Staggenborg S, Propheter JL, Rooney WL, Yu J, Wang D. Features of sweet sorghum juice and their performance in ethanol fermentation. Industrial Crops and Products. 2010;**31**(1):164-170

[49] Lingle S, Tew T, Rukavina H, Boykin D. Post-harvest changes in sweet sorghum I: Brix and sugars. Bioenergy Research. 2012;**5**(1):158-167

[50] Kaltschmitt M, Hartman H, Hofbauer H. Energie aus Biomasse, E-book. Elsevier; 2009

[51] Dj D, Effenberger M, Martinov M. Method for assessing and improving the efficiency of agricultural biogas plants based on fuzzy logic and expert systems. Applied Energy. 2014;**134**:163-175. DOI: 10.1016/j.apenergy.2014.08.021

[52] Dj D, Effenberger M, Lehner A, Martinov M, Tesic M, Gronauer A. New method for assessing the performance of agricultural biogas plants. Renewable Energy. 2012;**40**(1):104-112. DOI: 10.1016/j.renene.2011.09.026

Chapter 19

Perspective Chapter: Removal of Heavy Metals and Salmonella Pathogens from Sewage Sludge Using a Novel Chelating Agent and Its Reuse as a Fertilizer

Sara Mohamed Yonues and Abdou Saad El-Tabl

Abstract

The direct use of sewage as fertilizers in agriculture without proper treatment has led to substantial economic environmental and healthy ramifications. Proper treatment as well as adequate environmental management of sewage sludge is a necessity in order to eliminate the negative sequences of its utilization in the agriculture field. In this chapter, a novel organic Schiff base chelator derived from hydroxybenzylidene succinohydrazide (HBSH) has been successfully synthesized and characterized by elemental analysis, ^{1}H-NMR as well as infrared spectroscopy. The effect of sewage treated with varying concentration of the Schiff base chelator (0.8, 1.6 and 2.4 g/L) as well as the untreated sewage on the sludge solid reduction, removal of heavy metals and salmonella pathogens has been investigated. The implementation of raw as well as treated sludge on the growth as well as the heavy metal content of radish plant has been also investigated. It was observed that the treated sample showed a reduction in the total content of Zn, Ni, Cr and Cu and enhancements in the yield, stem length, leaf number and flourishing.

Keywords: sewage sludge, succino-hydrazide, fertilizer, heavy metals, wastewater treatment, radish plant

1. Introduction

Sewage sludge is a non-homogeneous material constituting of a combination of various compounds including organic and inorganic materials as well as microorganisms, and moisture [1, 2]. It is counted up as the major by-product resulted from treatment of wastewater. The sludge undesirable content of heavy metals, synthetic organic compounds and pathogenic bacterial and other microorganisms represents a major harmful environmental risk. Therefore, disposal process of this by-product surely lead to unwelcomed environmental impacts including human beings health threats and the possibility of atmospheric polluting, as well as water and soil resources contaminating [1, 3]. The remarkable high phosphorus, nitrogen and potassium nutrients content imparts sewage sludge a property of being used as agricultural fertilizer however a series harmful substances content oppose this beneficial application [4–6]. Hence the main aim of sewage sludge treatment is to eliminate sludge unfavorable contents while to retain sludge nutrients. The primary two steps in the treatment processes of sewage sludge are thickening and dewatering. The first one is aiming to thickened sludge to lower solid percent, while the other one reduces the water content by (centrifugation, filtration, and/or evaporation) in order to reduce transportation costs of disposal, or to improve suitability for advanced processing [7, 8]. On the other hand, digestion (anaerobic and aerobic), incineration, and composting aim to diminish the organic matter content and the amount of harmful microorganisms existing in the residue matter [9–15].

The high heavy metal content in the sewage sludge represents another major obstacle against sludge utilization. The non-biodegradability nature, unlike organic contaminants, leads to accumulation of heavy metals in the biota, which involves a health risk and an environmental worry. Although metabolism of living organism needs metal ions in order to carry out many metabolic pathways, higher concentrations can cause expected acute as well as chronic toxicity. Therefore, rigorous parameters have been approved for release of various metal ions in wastewaters to evade health risk and environmental contamination. Various chemical, physical, and biological treatment methods such as chemical precipitation, adsorption, membrane filtration, ion-exchange, electrochemical treatment and microorganisms have been established for removing metal ions from water and wastewaters [16–29]. Among these wide scope of heavy metal treatment methods, utilizing of chelating agents has demonstrated a pronounced impact in eliminating of harmful metals [30, 31]. Challenges of metal ions removal can be represented in that metal ions are adsorbed on soil, so it shows resistance to be removed upon washing by surface and ground water moreover the actual low water solubility of some transition metals hydroxides. The remarkable metal binding capabilities enables chelating agents to overcome these challenges [32–36]. The most commonly ligand, Ethylenediamine-tetraacetic acid (EDTA), a hexadentate ligand, has the ability to bind most of heavy metals forming very stable complexes [37–41]. However the problem of non-biodegradability is the main drawback of EDTA utilization, as the degradation of EDTA results in formation of a stable organic pollutant (3- ketopiperazine-N,N-diacetate) [42]. The bio-degradable isomer S,S-ethylenediamine disuccinic acid (S,S-EDDS) has been proposed as a likely alternative chelating agent [43, 44], however its ability to metal ions is inferior and special pH conditions should be taken into consideration as the influential pH range is narrower [45].

In this study a novel organic Schiff base chelator derived from hydroxybenzylidene succinohydrazide (HBSH) has been successfully synthesized and characterized by elemental analysis, ^{1}H-NMR as well as infrared spectroscopy. The ability of this novel Schiff base to decontaminate semi-solid sewage has been investigated. The utilization of the treated sludge has been also tested as a plant fertilizers. The influence of raw as well as treated sludge on the growth and heavy metal content in radish plant have been also investigated.

2. Material and methods

2.1 Instrumentation and measurement

The C, H and N content in the obtained compounds was analyzed at the Microanalytical Laboratory, Cairo University, Egypt. Metal ion content was determined using Standard analytical methods [46–48]. Jasco FT/IR 300E Fourier transform infrared spectrophotometer covering the range 400–4000 cm^{-1} was used to record FT-IR spectra of the ligand and its metal complexes using KBr discs. ^{1}HNMR spectrum was obtained on a JEOL EX-270 MHz FT-NMR spectrometer in d6-DMSO as solvent. Where the chemical shifts were determined relative to the solvent peaks. All metal concentrations were detected using Perkin Elmer ICP (ICP-MS-1).

2.2 Preparation of the Schiff base (HBSH)

The Schiff base, (HBSH) was prepared by refluxing (4 gm, 0.022 mole) of 2,3 dihydroxy succcinohydrazide in ethanol with (4.8 ml) of salicylaldehyde (1: 2 molar ratio), for 5 hours at 80°C (**Figures 1** and **2**). The formed yellow precipitate was left to cool to room temperature, then filtered off and dried under vacuum over anhydrous CaCl2. [C18H26N4O10] Yield: 75%, Color: yellow. Elemental Anal. Calc.: C, 47.16; H, 5.72; N, 12.22. Found: C, 46.98; H, 5.39; N, 12.10. IR, (KBr, cm-1): 3650,3665,3620,3180 ʋ(OH/H2O), 3320-2730 ʋ(H-bonding) 1705, 1690 ʋ(C=O), 1318,1272ʋ(COH), 1630,1622 ʋ(C=N).

2.3 Settled sludge volume (SV30)

The sludge used in this study was collected from the sewage outcome of the aeration tank of Al kharry waste-water plant, El-Behira Governorate, Egypt. Settled Sludge Volume was estimated using reported standard method [49, 50]. In briefly, 1 L of the sludge sample (raw or treated) was places in settling column and the solid content was uniformly distributed by inverting the covered cylinder for three times,

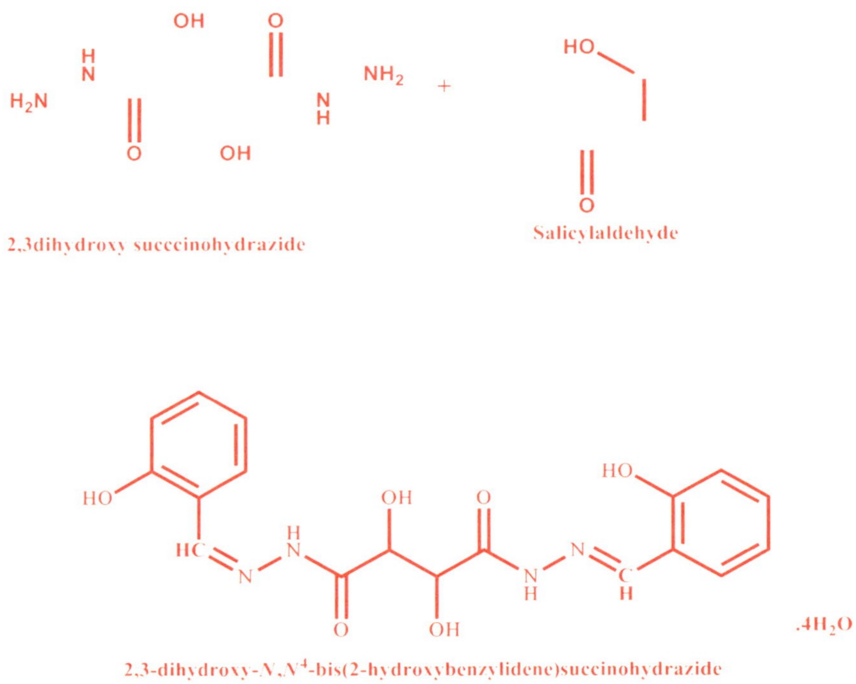

Figure 1.
Preparation of the Schiff base 2,3-dihydroxy-N,N4-bis(2-hydroxybenzylidene) succinohydrazide (HBSH).

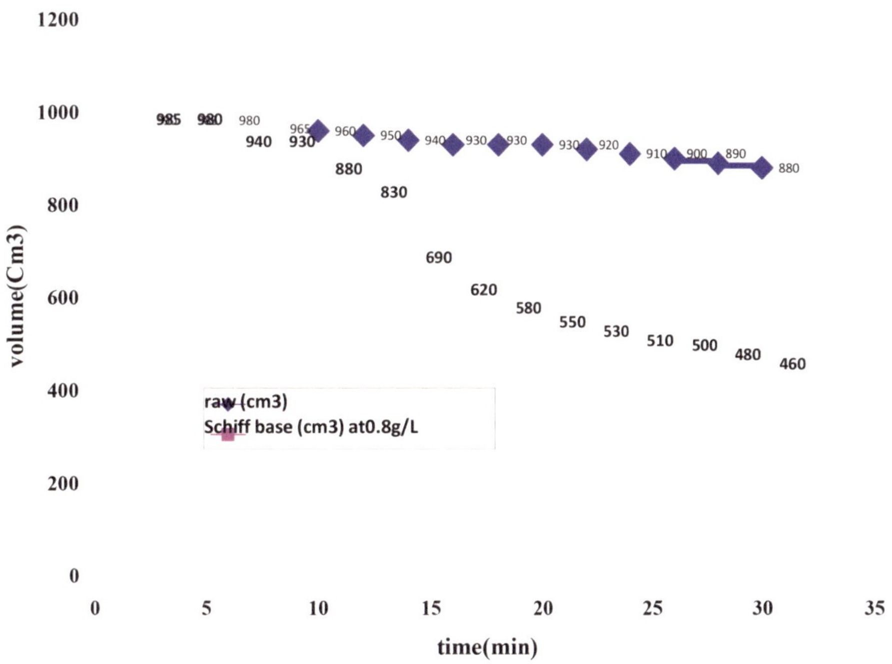

Figure 2.
Variation in TSS concentrations along raw and treated sewage at 0.8gL^{-1}.

then stirred using stirring rod. The suspension is kept under stirring throughout the experiment. The volume occupied by the suspension was determined for 30 minutes at 2 minutes intervals. The same procedures were carried out in presences of different concentration of the succinohydrazide Schiff base (0.8, 1.6, 2.4) g/L.

2.4 Sludge and radish digestion

Digestion of raw sludge, treated sludge as well as radish plant was carried out by well suspension of 1gm of the dry sample in 100 ml of distilled water. Three milliliter of conc. HNO3, and the mixture was evaporated cautiously to 4 ml, then 5 ml of conc. HNO3 (15.8 M) was added and refluxed for 1 hour. The mixture was cooled then solution of (15 ml of HCl (11.65 M) + 15 ml H_2O) was added heated again for 15 minutes then cooled. Finally 100 ml of distilled water was added, the mixture was filtered and the heavy metal was estimated on ICP (ICP-MS-1) - Germany [50].

2.5 Salmonella detection

One gram of the sample was diluted in 9 ml 1% NaCl to a dilution up to 5-10 times. One ml from each dilution was transferred into another five tubes containing buffered peptone water (9 ml). The inoculated tubes were incubated at 37° C for 24 h, then 0.1 ml from each tube showing bacterial growth (turbidity) was transferred into 10 ml of Rappaport-Vassiliadis (RV) broth. The inoculated tubes were incubated at 43.5 ± 1°C for 24 h, then three loops from each tube were taken, the first was streaked out onto bismuth sulfite agar according to [51]. The plates incubated up to about 48 h at 37 ± 0.5°C. Growth of Salmonella on bismuth sulfite agar plates showed are black center, light edges surrounded by a black precipitate with metallic sheen (so-called rabbits or fish- eye). A typical colony was collected and streaked on slants of tryptic soya agar (TSA) (contains 10% glycerol) and stored at 4°C not more than 1 year. Confirmation of Salmonella was carried out using the API E20 Enterobacteriaceae test system and RISA molecular profiling [52].

3. Results and discussion

3.1 Characterization of sludge

Table 1 displays the general information regarding heavy metals (HMs), parasite and solid volume contents of the sludge sample accompanied by the Egyptian Code and Environmental Protection Agency allowed standard limits. The content values of lead and cadmium are two- folds the allowed Egyptian code standard limit, in the time that nickel and zinc recorded 801.25, 8275 mg/kg respectively which are about four times more than the allowed Egyptian code standard limit. Iron and chromium are greatly pronounced with 468,750 and 1637.5 mg/kg respectively. The values of the main metal species present in the sludge under experiment are all beyond the Environmental Protection Agency limit values. Sludge volume after 30 min (SV30) recorded 650 cm^3/L an average value. The parasitic content evaluation showed a higher presence of salmonella (2.6×10^4 unit/mg). Disposal of such untreated sludge on nearby unused lands would inevitability lead to serious parasitic and heavy metal contaminations accompanied with a dangerous chain of negative effects on human and environment [52].

3.2 ^{1}H-NMR spectrum

The ^{1}H- NMR spectrum of the ligand in deuterated DMSO showed the absence of protons of the amine and aldehyde groups belonging to the 2,3 dihydroxy

Parameters (unit)	Raw sludge	Raw dry Sludge (mg/kg)	limit EG*	EPA limit**
Heavy metals (mgL^{-1})				
Pb	0.5 ± .01	637 ± 10	300	840
Cd	0.064 ± .01	80 ± 10	39	85
Cu	1.39 ± .01	1737.5 ± 10	1500	4300
Fe	375 ± .01	468,750 ± 10	—	—
Ni	0.641 ± .01	801.25 ± 10	240	420
Zn	6.62 ± .01	8275 ± 10	2800	7500
Cr	1.31 ± .01	1637.5 ± 10	1200	—
Parasites (unit/100mL)				
Salmonella	2.6×10^4	2.6×10^4		
Solid Volume (cm^3/L)				
***Sv30		625		

**EG referred to (Egyptian Code).*
***EPA referred to (Environmental Protection Agency).*
****Sludge volume after 30 min.*

Table 1.
Physicochemical characteristics of the sewage sludge.

succcinohydrazide and Salicylaldehyde starting material and the appearance of new peaks at 4.29 ppm corresponding to protons of the two azomethine groups (CH=N-). The chemical shift appeared in the 6.2–7.1 ppm range were assigned to the aromatic protons of the benzene moiety which appearing as multiplets, whereas the chemical shifts observed at 5.35 ppm was assigned to the protons of aromatic C-OH groups. The two protons of -NH recorded a chemical shift at 8.9 ppm, whereas the chemical shift observed at 2.8 ppm corresponded to two protons of the alcoholic OH groups (OHC-COH) [53].

3.3 Infrared spectrum

The infrared spectrum of the ligand showed broad bands at 3650, 3665, 3620 and 3180 cm^{-1} are assigned to the ʋ (OH)/H2O groups. The two bands located at 1705 and 1690 cm^{-1} are assigned to the two (C=O) groups, whereas the other two bands observed at 1630 and 1622 cm^{-1} are attributed to ʋ(C=N). Presence of the two (C-OH) groups was further supported by the two vibrational bands observed at 1318, 1272, whereas the presence (N-N) group was supported by the vibrational band observed at 1115 cm^{-1}. The IR spectrum showed a band at 1569 cm^{-1} which is related to ʋ (C=C)Ar. [54].

3.4 Effect of different concentration of HBSH on sludge solid reduction

The settleability i.e. settling rate expressed as TSS was measured by collecting samples with time in 1000 cm^3 (sampling point) of the settling column. The same amount of raw and the treated sludge with (0.8, 1.6 and 2.4 g/L) HBSH was used to conduct to evaluate the settling performance as compared to untreated sludge. The settling rate as TSS concentration was measured with settling time is illustrated in **Figures 2–4**. The observed results showed that the 0.8 treated sample, at the first

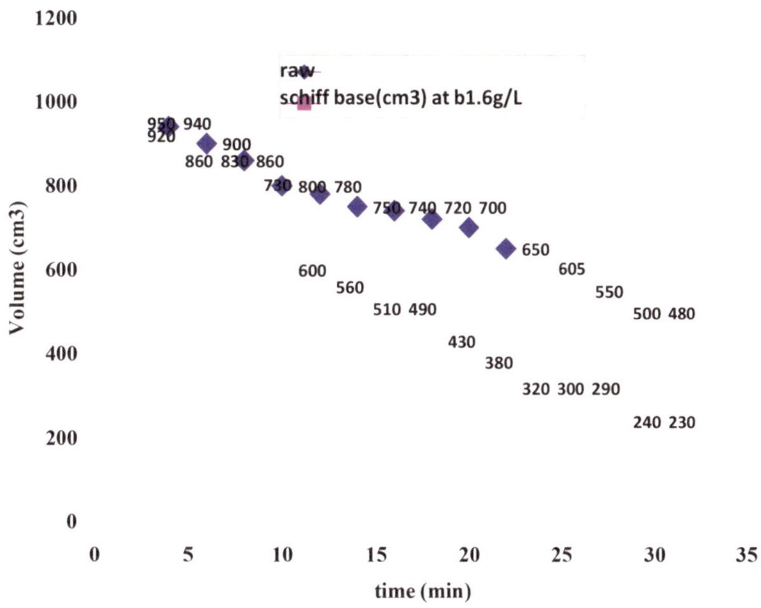

Figure 3.
Variation in TSS concentrations along raw and treated sewage at 1.6gL⁻¹.

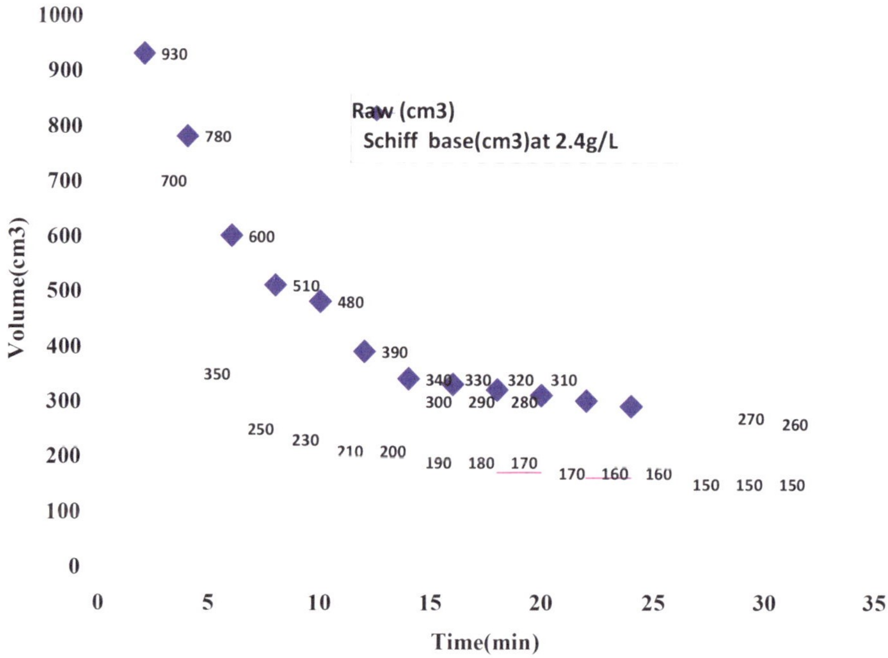

Figure 4.
Variation in TSS concentrations along raw and treated sewage at 2.4gL^{-1}.

part of the curve (up to 12 min) displayed a slightly higher settlement of TSS than that recorded by the raw sludge. However the settleability was doubled after that time and the overall TSS decreased to about half that in the untreated sludge at the end of 30 min period. The settlement rate was enhanced upon further addition of HBSH (1.6 g/L), the TSS decreased quickly and exceeded the TSS value recorded at 30 min by the untreated sludge within only about 17 min. The figure also shows a sharp reduction of TSS recorded after 10 min of settling then gradual decrease in TSS takes places to reach 230 g/L at 30 min. The faster settling rate was recorded by the 2.4 g/L treated sample. It was noticed that the TSS concentration decreased quickly and within less than 5 min reached the TSS concentration recorded by the raw sludge in 30 min. After such very sharp a decrease in the TSS concentration gradually continued to reach the lowest TSS value (150 g/L) recorded by the samples under experiment. The highest percentage of solids settlement (58.38%) was recorded by the 2.4 g/L HSBH treated sample after about 6 min of settling operation [55].

3.5 HMs binding capacity of HBSH at different concentration

Figures 5–7 shows the heavy metals (HMs) distribution in the liquid sludge phase before and after treating with the HBSH ligand at three different concentrations (0.8, 1.6 and 2.4 g/L). It was found that heavy metal concentrations were very high in the raw sludge. On the other hand, the heavy metal concentration showed a remarked decrease upon treating with the ligand (HBSH); this finding could be assigned to the ability of the HBSH to chelate the heavy metals ions. The results show that among all the HBSH concentrations used, Cd extraction efficiencies are noticeably the highest. The concentration of the chelator is inversely correlated with the general trend of HMs removal. As the concentration of HBSH raised from 0.8 to 2.4 g/L, removal

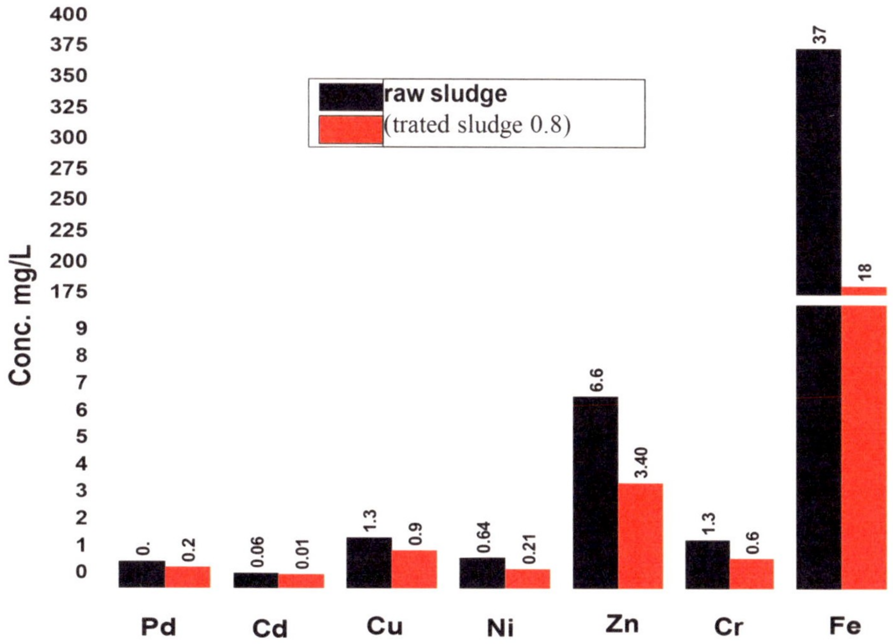

Figure 5.
Variation in heavy metals concentrations along raw and treated sewage at 0.8 gL⁻¹ dose.

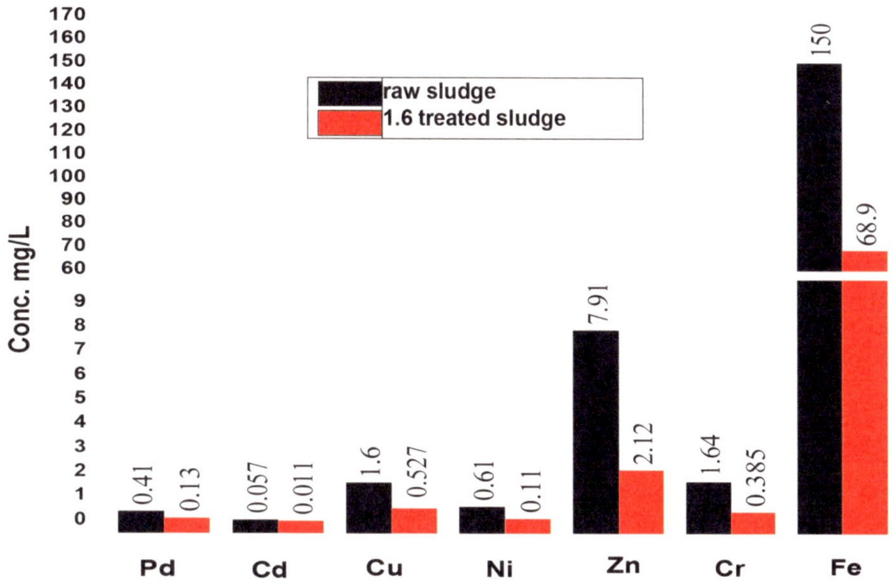

Figure 6.
Variation in heavy metals concentrations along raw and treated sewage at 1.6 gL⁻¹ dose.

of lead, cadmium, copper, zinc and chromium showed a continuous progress. This behavior is not the same in case of iron and nickel removal which firstly improved significantly by increasing the addition of HBSH from 0.8 to 1.6 g/L, while there is almost no change occurs upon raising the concentration from 1.6 to 2.4 g/L [56].

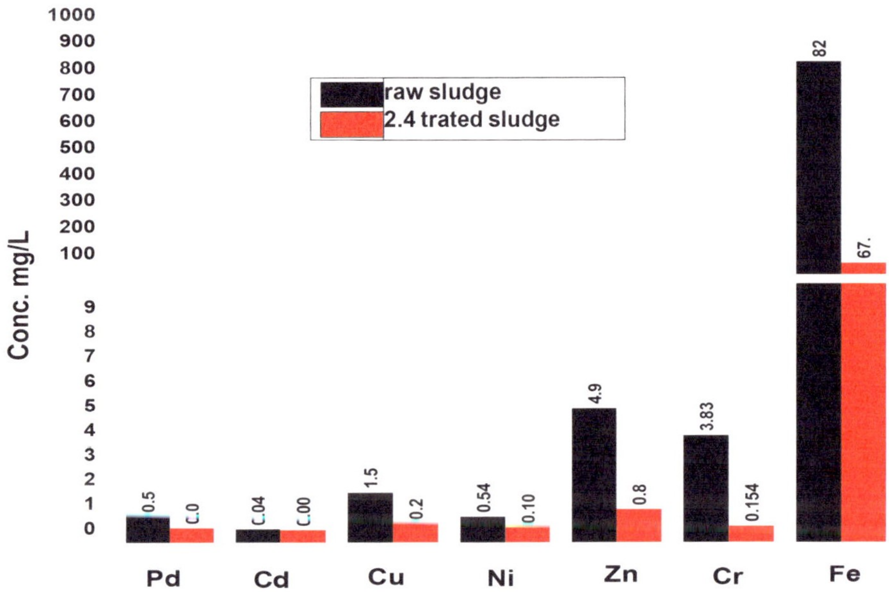

Figure 7.
Variation in heavy metals concentrations along raw and treated sewage at 2.4 gL^{-1} dose.

Figure 8 depicted the percentage of HMs removal upon addition of varying concentration of the HBSH. At 0.8 g/L, the ligand showed the highest removal potential toward cadmium with percentage 81.25% followed by nickel (66.3%). On the other hand, the lowest removal percentage was recorded for cupper. For other metal species (Pb, Fe, Zn and Cr) the removal percentages was around 50%. Upon further addition of HBSH (1.6 g/L), the removal percentages of all metal species have been enhanced to reach over than: 80% for iron and nickel, 70% for lead and chromium, and 60% for copper and zinc. However there is almost no change in removal percentage of cadmium (1.5%). Finally, further addition of HBSH (2.4 g/L) resumes its ability

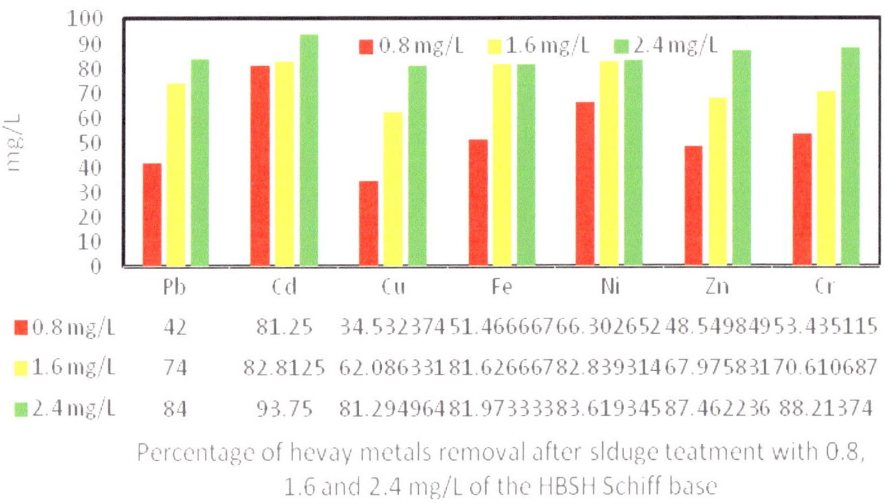

Figure 8.
Variation in percentage of heavy metal concentrations along raw and treated sewage.

to remove cadmium content of the sludge to reach 93.75%. Removal of other metal species was elevated to: over than 80% of cupper, iron, nickel and lead and more than 85% for zinc and chromium [57].

The results suggest also a higher HBSH lead to more effective extraction of heavy metals, this could be explained on basis that increasing dose of chelating ligand would facilitate the complexing reaction between metal ion and the chelating ligand leading to the formation of a chelate. Enhancing of the removal efficiency could be also related to the reason that there are many substances presented in the sludge beside the studied metal species such as: Ca, and Mg which consequently compete with the targeted metal species to bind the ligand and hence participate in the consumption of HBSH and according a large excess of ligand is required to solubilize the target metal due to the co-solubilization of Ca and Fe. Nowack et al. [29].

3.6 Removal of salmonella pathogens

Specific tests for the presence of Salmonella sp. were carried out, in both raw and dry sludge, we identified the presence of Salmonella spp. Confirmation of Salmonella was carried out using the API E20 Enterobacteriaceae test system and RISA molecular profiling. The data demonstrated that the total count of Salmonella sp. have been markedly lowered upon contacting with Schiff base for 30 min (**Figure 9**). The total counts for salmonella in the raw sludge recorded an averages of 2.8×10^3 and 1.7×10^4 MPN index/100 ml in the liquid and dray raw sludge respectively. These values have drastically decreased to only (25–28) MPN index/100 ml with successful removing of salmonella by about 99% upon treating with different concentration of Schiff base. From the results, it is clear the Salmonella sp. present in the effluent were in fact successfully being removed or inactivated upon treating with the Schiff base. The removal

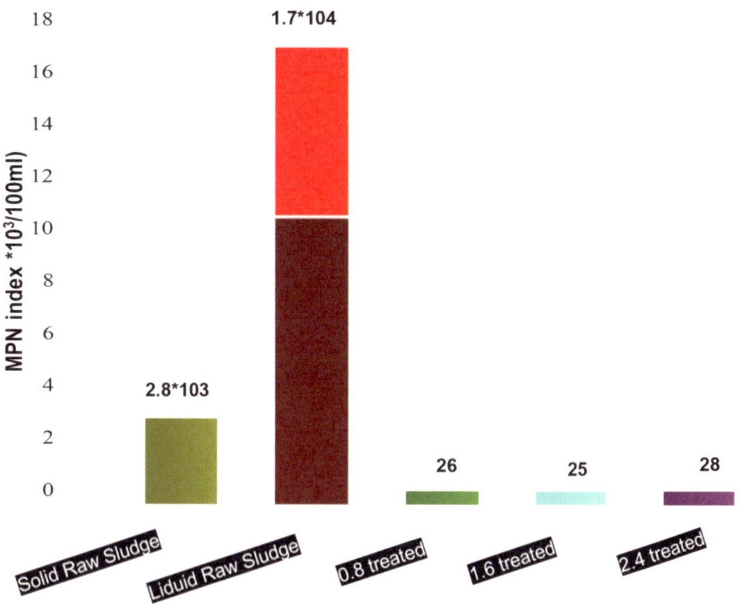

Figure 9.
MPN index/100 ml of salmonella count in solid, liquid raw sludge as well as treated with varying concentrations (0.8, 1.6, 2.4 g/L) of Schiff base (HBSH).

of Salmonella was clearly observed, however no pronounced effect on the salmonella surviving was shown upon changing the Schiff base concentration from 0.8 to 2.4 g/L.

3.7 Raw and treated Schiff base-sludge as fertilizers for radish plant

Organic fertilizers are used to encounter the requirements of vegetable and plant production. Sewage sludge is an important factor that can be utilized for vegetable and plant production in view of its high organic matter content and rich macro and micro nutrients. However extreme care is to be taken to avoid the entry of the heavy metals and other organic compounds in the food chain. In our study, we investigated

Figure 10.
Effect of (a) raw and (b) treated sludge with 2.4 g/L Schiff base on the growth and flourishing of radish plant.

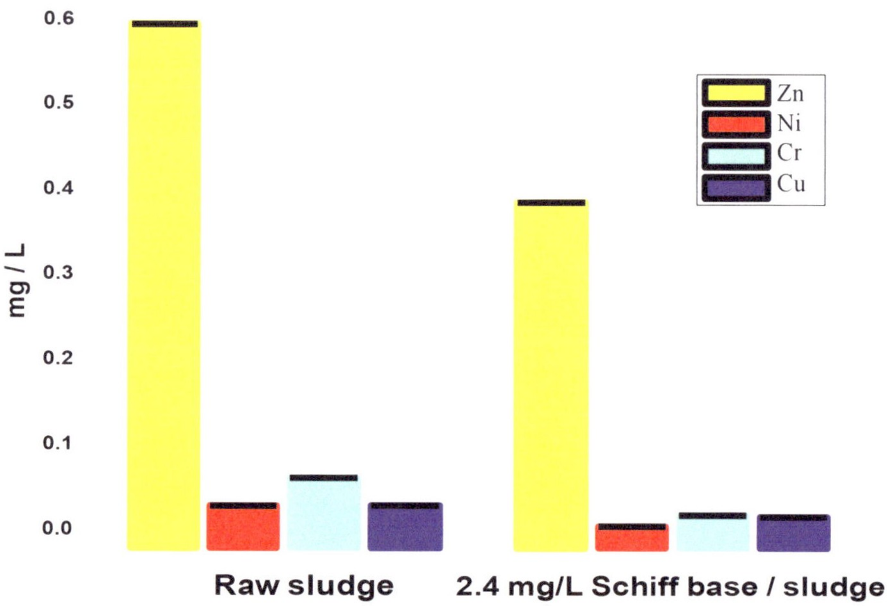

Figure 11.
Variation in heavy metals concentrations in radish plant fertilized by raw and treated sewage at 2.4 gL^{-1} of the Schiff base.

the implementation of raw as well as treated sludge with Schiff base at 2.4 g/L on the growth as well as the heavy metal content of radish plant. It was observed that the application of sewage sludge enhanced soil fertility and that crop yield in the treated sludge was higher than in the raw one. The treated sample showed enhancement in the radish yield, stem length, leaves number and flourishing (**Figure 10**). The total heavy metal content in radish have been also estimated in the radish treated with the raw and Schiff base- sludge. The results showed the application of 2.4 g/L of Schiff base showed a reduction in the total content of zinc (Zn), nickel (Ni), chromium (Cr) and cobber (Cu) in the plant. Concentrations of nickel and chromium ions showed a significantly reduction upon treating the plant with the Schiff base rather than that treated with the raw sludge with 82.75 and 70.96% respectively (**Figure 11**). Schiff base treating sludge showed also promising results regarding copper and zinc ions [58, 59].

4. Conclusion

A novel organic Schiff base chelator (HBSH) has been successfully synthesized and characterized by elemental analysis. The effect of sewage treated with varying concentration of the Schiff base chelator (0.8, 1.6 and 2.4 g/L) as well as the untreated sewage on the sludge solid reduction, removal of heavy metals and salmonella pathogens has been investigated. The settlement rate was enhanced upon addition of HBSH Schiff base, the faster settling rate as well as the highest percentage of solids settlement (58.38%) was recorded by the 2.4 g/L treated sample. The highest removal potential was recorded toward cadmium with percentage 81.25% followed by nickel (66.3%). The total counts for salmonella in the raw sludge have drastically decreased to only (25–28) MPN index/100 ml referring to successful removing of salmonella by about 99% upon treating with different concentration of Schiff base. The implementation of raw as well as treated sludge on the growth as well as the heavy metal content of radish plant have been also investigated. It was observed that, the treated sample showed a reduction in the total content of Zn, Ni, Cr and Cu and enhancements in the yield, stem length, leaves number and flourishing.

Author details

Sara Mohamed Yonues[1*] and Abdou Saad El-Tabl[2]

1 Chemical Engineering Department, Borg El Arab Higher Institute Engineering and Technology, Alexandria, Egypt

2 Chemistry Department, Faculty of Science, Menoufia University, Shebin ElKom, Egypt

*Address all correspondence to: ch_sara2011@yahoo.com

References

[1] Huang H-J, Yuan X-Z. The migration and transformation behaviors of heavy metals during the hydrothermal treatment of sewage sludge. Bioresource Technology. 2015. DOI: 10.1016/j.biortech.2015.10.099

[2] Fonts I, Gea G, Azuara M, Ábrego J, Arauzo J. Sewage sludge pyrolysis for liquid production: A review. Renewable and Sustainable Energy Reviews. 2012;**16**:2781-2805. DOI: 10.1016/J.RSER.2012.02.070

[3] Manara P, Zabaniotou A. Towards sewage sludge based biofuels via thermochemical conversion—A review. Renewable and Sustainable Energy Reviews. 2012;**16**:2566-2582. DOI: 10.1016/j.rser.2012.01.074

[4] Chen Y, Yu F, Liang S, Wang Z, Liu Z, Xiong Y. Utilization of solar energy in sewage sludge composting: Fertilizer effect and application. Waste Management. 2014;**34**:2014-2021. DOI: 10.1103/PhysRevLett.121.062003

[5] Dassanayake KB, Jayasinghe GY, Surapaneni A, Hetherington C. A review on alum sludge reuse with special reference to agricultural applications and future challenges. Waste Management. 2015;**38**:321-335. DOI: 10.1016/j.wasman.2014.11.025

[6] Vogel C, Adam C, Kappen P, Schiller T, Lipiec E, McNaughton D. Chemical state of chromium in sewage sludge ash based phosphorus-fertilisers. Chemosphere. 2014;**103**:250-255. DOI: 10.1016/j.chemosphere.2013.12.012

[7] Yukawa H, Yoshida H, Kobayashi K, Hakoda M. Fundamental study on electroosmotic dewatering of sludge at constant electric current. Journal of Chemical Engineering of Japan. 1976;**9**:402-407. DOI: 10.1252/JCEJ.9.402

[8] Pepping R. Method for Dewatering Sludge. Google Patents; 1978

[9] Asztalos JR, Kim Y. Enhanced digestion of waste activated sludge using microbial electrolysis cells at ambient temperature. Water Research. 2015;**87**:503-512. DOI: 10.1016/j.watres.2015.05.045

[10] Suanon F, Sun Q, Mama D, Li J, Dimon B, Yu C-P. Effect of nanoscale zero-valent iron and magnetite (Fe_3O_4) on the fate of metals during anaerobic digestion of sludge. Water Research. 2016;**88**:897-903. DOI: 10.1161/CIRCOUTCOMES.119.005616

[11] Tandukar M, Pavlostathis SG. Co-digestion of municipal sludge and external organic wastes for enhanced biogas production under realistic plant constraints. Water Research. 2015;**87**:432-445. DOI: 10.1016/j.watres.2015.05.045

[12] Kavouras P, Pantazopoulou E, Varitis S, Vourlias G, Chrissafis K, Dimitrakopulos GP, et al. Incineration of tannery sludge under oxic and anoxic conditions: Study of chromium speciation. Journal of Hazardous Materials. 2015;**283**:672-679. DOI: 10.1016/j.jhazmat.2014.09.066

[13] Gunaseelan VN. Anaerobic digestion of biomass for methane production: A review. Biomass and Bioenergy. 1997;**13**:83-114. DOI: 10.1016/S0961-9534(97)00020-2

[14] Siripong S, Rittmann BE. Diversity study of nitrifying bacteria in full-scale municipal wastewater treatment plants.

Water Research. 2007;**41**:1110-1120. DOI: 10.1016/j.watíes.2006.11.050

[15] D. EPA. Washington, Primer for Municipal Waste water Treatment Systems, (2004). Document No. EPA 832-R-804-001

[16] Al-Jlil SA, Alsewailem FD. Saudi Arabian clays for lead removal in wastewater. Applied Clay Science. 2009;**42**:671-674. DOI: 10.1016/j.clay.2008.03.012

[17] Alvarez MT, Crespo C, Mattiasson B. Precipitation of Zn (II), Cu (II) and Pb (II) at bench-scale using biogenic hydrogen sulfide from the utilization of volatile fatty acids. Chemosphere. 2007;**66**:1677-1683. DOI: 10.1016/J.CHEMOSPHERE.2006.07.065

[18] Alyüz B, Veli S. Kinetics and equilibrium studies for the removal of nickel and zinc from aqueous solutions by ion exchange resins. Journal of Hazardous Materials. 2009;**167**:482-488. DOI: 10.1016/j.jhazmat.2009.01.006

[19] Aman T, Kazi AA, Sabri MU, Bano Q. Potato peels as solid waste for the removal of heavy metal copper (II) from waste water/industrial effluent. Colloids and Surfaces B: Biointerfaces. 2008;**63**:116-121. DOI: :10.1016/j.colsuífb.2007.11.013

[20] Amini M, Younesi H, Bahramifar N. Statistical modeling and optimization of the cadmium biosorption process in an aqueous solution using Aspergillus Niger. Colloids and Surfaces A: Physicochemical and Engineering Aspects. 2009;**337**:67-73. DOI: 10.1016/J.COLSURFA.2008.11.053

[21] Nędzarek A, Drost A, Harasimiuk FB, Tórz A. The influence of pH and BSA on the retention of selected heavy metals in the nanofiltration process using ceramic membrane. Desalination. 2015;**369**:62-67. DOI: 10.1016/J.DESAL.2015.04.019

[22] Zhu W-P, Gao J, Sun S-P, Zhang S, Chung T-S. Poly(amidoamine) dendrimer (PAMAM) grafted on thin film composite (TFC) nanofiltration (NF) hollow fiber membranes for heavy metal removal. Journal of Membrane Science. 2015;**487**:117-126. Available from: https://pubmed.ncbi.nlm.nih.gov/25016298

[23] Al-Qahtani KM. Water purification using different waste fruit cortex for heavy metals removal. Journal of Taibah University for Science. DOI: 10.1016/j.jtusci.2015.09.001

[24] Basha CA, Somasundaram M, Kannadasan T, Lee CW. Heavy metals removal from copper smelting effluent using electrochemical filter press cells. Chemical Engineering Journal. 2011;**171**:563-571. DOI: 10.1016/J.CEJ.2011.04.031

[25] Huang Z, Lu L, Cai Z, Ren ZJ. Individual and competitive removal of heavy metals using capacitive deionization. Journal of Hazardous Materials. 2016;**302**:323-331. DOI: 10.1016/j.jhazmat.2015.09.064

[26] Ihsanullah A, Abbas AM, Al-Amer T, Laoui MJ, Al-Marri MS, Nasser M, et al. Atieh, heavy metal removal from aqueous solution by advanced carbon nanotubes: Critical review of adsorption applications. Separation and Purification Technology. 2016;**157**:141-161. DOI: 10.1016/j.seppur.2015.11.039

[27] Kim G, Igunnu ET, Chen GZ. A sunlight assisted dual purpose photoelectrochemical cell for low voltage removal of heavy metals and organic pollutants in wastewater. Chemical Engineering Journal. 2014;**244**:411-421. DOI: 10.1016/J.CEJ.2014.01.090

[28] Colin VL, Villegas LB, Abate CM. Indigenous microorganisms as potential bioremediators for environments contaminated with heavy metals. International Biodeterioration & Biodegradation. 2012;**69**:28-37. DOI: 10.1016/j.ibiod.2011.12.001

[29] Muñoz AJ, Ruiz E, Abriouel H, Gálvez A, Ezzouhri L, Lairini K, et al. Heavy metal tolerance of microorganisms isolated from wastewaters: Identification and evaluation of its potential for biosorption. Chemical Engineering Journal. 2012;**210**:325-332. DOI: 10.1016/j.cej.2012.09.007

[30] Flora SJ, Pachauri V. Chelation in metal intoxication. International Journal of Environmental Research and Public Health. 2010;7:2745-2788. Available from: https://pubmed.ncbi.nlm.nih.gov/20717537

[31] Chang F-C, Lo S-L, Ko C-H. Recovery of copper and chelating agents from sludge extracting solutions. Separation and Purification Technology. 2007;**53**:49-56. DOI: 10.1016/J.SEPPUR.2006.06.011

[32] Jing J, Logan TJ. Chelating resin method for estimation of sludge-cadmium bioavailability. Communications in Soil Science & Plant Analysis. 1991;**22**:2029-2035. DOI: doi/abs/10.1080/00103629109368555

[33] Lake D, Kirk P, Lester J. Fractionation, characterization, and speciation of heavy metals in sewage sludge and sludge-amended soils: A review. Journal of Environmental Quality. 1984;**13**:175-183. DOI: 10.2134/JEQ1984.00472425001300020001X

[34] Chaiyaraksa C, Sriwiriyanuphap N. Batch washing of cadmium from soil and sludge by a mixture of Na 2 S 2 O 5 and Na 2 EDTA. Chemosphere. 2004;**56**:1129-1135. DOI: 10.1016/j.chemosphere.2004.03.016

[35] Ludmer Z, Golan T, Ermolenko E, Brauner N, Ullmann A. Simultaneous removal of heavy metals and organic pollutants from contaminated sediments and sludges by a novel technology, sediments remediation phase transition extraction. Environmental Engineering Science. 2009;**26**:419-430. DOI: 10.1089/EES.2007.0198

[36] Zhang L, Zhiliang Z, Zhang R, Zheng C, Zhang H, Yanling Q, et al. Extraction of copper from sewage sludge using biodegradable chelant EDDS. Journal of Environmental Sciences. 2008;**20**:970-974. Available from: https://www.ntnu.edu/employees/zhiliang.zhang

[37] Dede G, Ozdemir S. Effects of elemental Sulphur on heavy metal uptake by plants growing on municipal sewage sludge. Journal of Environmental Management. 2016;**166**:103-108. DOI: 10.1016/j.jenvman.2015.10.015

[38] Hanay O, Hasar H, Kocer NN. Effect of EDTA as washing solution on removing of heavy metals from sewage sludge by electrokinetic. Journal of Hazardous Materials. 2009;**169**:703-710. DOI: 10.1016/j.jhazmat.2009.04.008

[39] Hau NT, Chen S-S, Nguyen NC, Huang KZ, Ngo HH, Guo W. Exploration of EDTA sodium salt as novel draw solution in forward osmosis process for dewatering of high nutrient sludge. Journal of Membrane Science. 2014;**455**:305-311. DOI: 10.1016/j.biortech.2013.12.037

[40] Liu C-C, Lin Y-C. Reclamation of copper-contaminated soil using EDTA or citric acid coupled with dissolved organic matter solution extracted from distillery sludge. Environmental Pollution.

2013;**178**:97-101. DOI: 10.1016/j.envpol.2013.02.034

[41] Zaier H, Ghnaya T, Ben Rejeb K, Lakhdar A, Rejeb S, Jemal F. Effects of EDTA on phytoextraction of heavy metals (Zn, Mn and Pb) from sludge-amended soil with Brassica napus. Bioresource Technology. 2010;**101**:3978-3983. Available from: https://www.sciíp.oíg/íefeíence/íefeíencespapeís.aspx?íefeíenceid=81567

[42] Kneller LA, Uhl HS, Brem J. Successful calcium disodium ethylene diamine tetra- acetate treatment of lead poisoning in an infant. New England Journal of Medicine. 1955;**252**:338-340. DOI: 10.1056/NEJM195503032520902

[43] Subramanian B, Christou S, Efstathiou A, Namboodiri V, Dionysiou D. Regeneration of three-way automobile catalysts using biodegradable metal chelating agent—S, S-ethylenediamine disuccinic acid (S, S-EDDS). Journal of Hazardous Materials. 2011;**186**:999-1006. DOI: 10.1016/j.jhazmat.2010.11.089

[44] Tandy S, Schulin R, Suter MJ-F, Nowack B. Determination of [S, S′]-ethylenediamine disuccinic acid (EDDS) by high performance liquid chromatography after derivatization with FMOC. Journal of Chromatography A. 2005;**1077**:37-43. DOI: 10.1016/J.CHROMA.2005.04.064

[45] Ullmann A, Brauner N, Vazana S, Katz Z, Goikhman R, Seemann B, et al. New biodegradable organic-soluble chelating agents for simultaneous removal of heavy metals and organic pollutants from contaminated media. Journal of Hazardous Materials. 2013;**260**:676-688. DOI: 10.1016/j.jhazmat.2013.06.027

[46] Svehla G. Vogel's Textbook of Macro and Semimicro Qualitative Inorganic Analysis. Longman London; 1979. Available from: https://www.cademia.edu/37549627/Vogels_ľextbook_Of_M

[47] Welcher FJ. Analytical uses of ethylenediaminetetraacetic acid. 1958. DOI: 10.1111/j.2042-7158.1959.tb12543.x

[48] Vogel A. A Text Book of Quantitative Inorganic Analysis. London: ELBS; 1978. p. 474. Available from: https://aíchive.oíg/details/vogelstextbookof0000voge

[49] Zhou S, Mak BW, Leong E, Hon-Shing K, Shu-kuen I, Tze-Kwan L. Optimized Centre-feed clarifier design for the Tai Po Sewage Treatment Works, Hong Kong. Water and Environment Journal. 2010;**24**:140-146. DOI: 10.1111/j.1747-6593.2009.00167.x

[50] Association APH, Association AWW, Federation WPC, Federation WE. Standard Methods for the Examination of Water and Wastewater. American Public Health Association; 1915. Available from: https://www.sciíp.oíg/(S(351jmbntvnsjt1aadkposzje))/íefeíence

[51] Apha A. WEF, 2005. Standard methods for the examination of water and wastewater. 2005;**21**:258-259. Available from: https://www.wef.oíg/íesouíces/publications/books/StandaídMethods

[52] Yonues SM, El-Tabl AS, Wahba MA. Removal of heavy metals and salmonella pathogens from sewage sludge using a novel chelating agent derived from succino-hydrazide and its reuse as a fertilizer. Journal of Chemistry and Chemical Sciences. 2016;**6**(4):277-292. Available from: https://www.sciíp.oíg/íefeíence/íefeíencespapeís.aspx?íefeíenceid=332416

[53] Ali AE, Salem WM, Younes SM, Elabdeen AZ. Ferrite nanocomposite (Rice Straw-CoFe2O4) as new

chemical modified for treatment of heavy metal from waste water. Hydrology Current Research. 2018:10. DOI: 10.4172/2157-7587.100031

[54] El-Tabl AS, Wahaab RA, Younes SM, Journal LS. Downflow hanging sponge (DHS) reactor as a novel post treatment system for municipal wastewater. 2013;**10**(3). Coípus ID: 13837870

[55] Younes SM, sayed G e. Use chlorine doses as disinfection of filamentous bacteria by eliminating physio-chemical parameters in activated sludge wastewater treatment. Journal of Chemical Science and Chemical Engineering. 2021:2. DOI: 10.4236/ajac.2022.139023

[56] El-Tabl AS, Wahaab RA, Younes SM. Downflow hanging sponge (DHS) reactor as a novel post treatment system for municipal wastewater. Life Science Journal. 2013;**10**(3):60. Available from: http://www.lifesciencesite.com

[57] Younes SM. Static study of activated sludge treatment plant and its ideal adjustment. Life Science Journal. 2018;**15**(12)

[58] Younes SM, Fahim SM. Synthesis and characterization of novel oxime derivatives and its application in oncology. European Journal of Biotechnology and Bioscience. 2021;**9**(3):63-69. Available from: http://www.biosciencejournals.com

[59] Ali AE, Salem WM, Younes SM, Kaid M. Modeling climatic effect on physiochemical parameters and microorganisms of stabilization pond performance. Heliyon. 2020;**6**:e04005. DOI: 10.1016/j.heliyon.2020.e04005

Chapter 20

New Advancements in the Field of Pollution Treatment, Including Contamination of the Soil and Water

Ahmad Akhavan

Abstract

The food security of human societies has become a major source of worry due to heavy metal contamination in soils and water supplies. Water and soil sources are becoming more and more contaminated with heavy metals every day as a result of the development of several mining techniques and technologies as well as the expansion of numerous enterprises. A career assessment predicts a 7–10% increase in employment for soil and plant scientists between 2018 and 2028. Because the production of wholesome food and the safety of food are very important issues. Therefore, some of the innovative techniques for eliminating organic and mineral contamination from water and soil sources are addressed in this book chapter.

Keywords: pollution, remediation, heavy metals, toxicity, comparison

1. Introduction

Heavy metals refer to a group of toxic elements that are very important both biologically and industrially. According to definition, heavy metals are naturally occurring metals having an atomic number greater than 20 and an elemental density greater than 5 g/cm [1]. The rapid developments of industrialization and unplanned urbanization have introduced heavy metals into the environment through improper dumping of industrial wastes directly on land and near water sources [2]. Pollution of soil and water sources with heavy metals is one of the most severe environmental problems that can seriously affect the quality of the environment and human health [3]. Today, the entry of heavy metals into water and soil sources from various natural and anthropogenic sources has been confirmed. Although the formation process of environmental pollution has a long history, the growth of this abnormality after the industrial revolution grew increasingly due to the very significant use of heavy metals in various industrial technologies [4]. Today, the amount of global production of heavy elements in various industries is very high. Heavy metals found in soils and water resources include nickel, chromium, lead, cadmium, arsenic, copper, cobalt, zinc, manganese, aluminum, mercury and antimony. Among the mentioned heavy metals, arsenic, cadmium, lead, and mercury are among the 20 most dangerous substances that have been determined by the Agency for Toxic Substances and Disease Registration and the US Environmental Protection Agency (USEPA). Among the effects of environmental pollution with heavy metals is the occurrence of bio-toxicity and its effects on the biological degradation process [5]. One of the ways of entering heavy elements into the environment, especially agricultural soils, is the use of different fertilizers that contain heavy elements as impurities. Excessive accumulation of heavy metals in agricultural soils causes more uptake of these metals by food crops and vegetables, which in turn can cause serious risks to human health [6]. The entry of, including cardiovascular diseases, cancer, Alzheimer's, chronic anemia, cognitive impairment, kidney damage, skin problems, memory loss, Aplastic anemia, infertility and nervous system weakness [7]. With the continuous entry of heavy metals into water and soil sources, there has been a concern that the concentration of these metals will exceed the permissible limits and disrupt the majority of biological activities. In addition, with the growth of public awareness, people have become sensitive to the contamination of soil and water sources with toxic elements and have understood that these compounds can have very important effects on the quality and quantity of their lives [3]. As a result of growing public awareness and sensitizing societies to the threats ahead, innovations and technologies have been formed that can be effective in cleaning and reducing the risk of sites contaminated with heavy metals. The distinguishing feature of pollution related to heavy metals is that, unlike organic pollution, these pollutants are not degradable and are resistant to biological and chemical processes [8]. Therefore, due to the fact that these compounds are resistant to decomposition, the extent of contaminated areas increases every year. There are about 100,000 contaminated areas in the United States, while the extent of contaminated agricultural land in China reaches more than 3.5 million hectares [9]. Of course, it is estimated that there are 2.5 million other contaminated areas in this country. For the decontamination of these areas, more than 6 billion euros should be spent annually [10]. If proper information of contaminated sites in other countries of the world are prepared, the cost of their decontamination will probably exceed thousands of billions of euros. The purification of soils, sediments, and polluted water has been the subject of a lot of research, and scientists have suggested a variety of technologies. The effectiveness and efficiency of methods used to treatment the contaminant from different sources is a critical aspect. Here, we will review some of the various purification methods that have been suggested for soil and water pollution, as well as an emphasis on more modern methods.

The technologies used for the treatment of contaminated areas can be divided into two main category: In-situ technologies and ex-site technologies. In the in-situ technology, the process of remediation and treatment of pollution is carried out at the place of its origin. The purpose of this type of treatment is to remove pollutants from soil, water and sediments without moving the soil and sediment. In ex-site technologies, drilling, refining, and treatment of contaminated materials are done outside the contaminated sites [10]. In the in-situ remediation process, the cost-benefit ratio is generally higher than the ex-situ remediation method. While removal or extraction the pollutant from soil and water is much better than immobilization or containment the pollutant. In addition to the above, in the in-situ remediation process, the contact of workers and people with the polluted environment is less and the possibility of contamination spreading to other areas is reduced [11]. To become more efficient and economical, these techniques can sometimes be used simultaneously [3].

2. Soil remediation

2.1 Physical remediation

2.1.1 Soil replacement

Replacement of contaminated soil is called complete or partial replacement of contaminated soil with non-contaminated and clean soil. In this method, the concentration of heavy metals in the soil decreases (dilutes) and leads to an increase in soil fertility and functionality [3]. Earlier to 1984, excavation, remove the contaminated soil and off-site disposal in specific places and replacement it with clean soil were the most common method for cleaning-up in contaminated areas. In soil spading, the contaminated soil is spaded deeply with special devices so that the surface contaminated soil is mixed with the uncontaminated under layer clean soil and the concentration of heavy metals is reduced (diluted) [12]. Another method related to soil replacement is adding clean and non-contaminated soil to the surface of contaminated soil. In this method, we can mix the imported non-contaminated soil with the contaminated soil so that the concentration of heavy metals per unit weight of the soil is reduced and a suitable environment is provided for the growth of plants [13]. The soil replacement method can isolate the contaminated soil and the ecosystem and reduce its harmful effects on the ecosystem [14]. But this method is very expensive because it requires a lot of labor and physical work, and it is suitable for small and highly polluted areas. The cost of doing this method is about 270 to 460 dollars for each ton of moving and adding clean soil. It is natural that the longer the distance, the higher the costs.

2.1.2 Soil isolation

Isolation means separating the soil contaminated with heavy metals from non-contaminated soil [15] or preventing the movement and transmission of contamination from one point to another [16], but for the complete purification of contamination in this method, other engineering methods are also needed. Contaminated soil isolation measures are based on engineered barriers and include hydrological barriers and stabilization approaches [17].

In general, isolation technologies are designed to prevent the off-site movement of heavy metals and other contaminants by confining them to a specific area [3] Engineering barriers, which may be on the surface or below the surface, are generally used to limit the contact of surface water or groundwater with waste materials and transfer to the surrounding environment. An underground barrier restricts the flow of ground and/or surface water at a contaminated site, allowing contaminated water and soil to be separated [16]. By far, the most common engineering barrier is a surface barrier called a cap, which is usually placed on top of waste piles. Vertical subsurface engineering barriers limit the lateral movement of groundwater and dissolved pollutants. These vertical barriers are installed downstream, upstream or generally surrounding a site and are generally used in combination with the cap system.

2.1.3 Vitrification

In the vitrification process, contaminated soil is transformed into a crystal and glass product due to heating and melting with electric energy. In this method, the mobility of heavy metals in the soil decreases, which is due to the formation of vitreous materials [18]. The Pacific Northwest Laboratory, which is working on the development of vitrification, is conducting research that can make this technology operational for buried waste and underground tanks of the United States Department of Energy [19]. A vertical array of electrodes is inserted into the contaminated soil during in situ vitrification in order to pass electrical current through it. Of course, it should be noted that in dry soils, due to low conductivity, the vitrification process is not performed well. Temperature is a key factor in immobilization of heavy metals in vitrification method [20]. Vitrification can be performed both in-situ and ex-situ. But preference is given to the in-situ method because it is easier and its energy supply is more accessible. In situ vitrification is limited by the possibility of melting soil and allowing current to pass through it. Furthermore, soils with a high alkali content (1.4 wt%) are unlikely to conduct current efficiently [21]. As a result, vitrification can only take place under wet soils with low alkali levels.

2.1.4 Electrokinetic remediation

Electrokinetic remediation is an in-situ process in which an electrical field is created in a soil matrix by applying a low-voltage direct current (DC) to electrodes placed in the soil. As a result of the application of this electric field, heavy metal contaminants may be mobilized, concentrated at the electrodes, and extracted from the soil [22]. In this method the separation of heavy metals (loids) in soil is accomplished via lectrophoresis, electric seepage, or electromigration and thus decrease the contamination [12]. Other techniques and processes, such as electrokinetic microbe joint remediation, electrokinetic-chemical joint remediation, electrokinetic-oxidation/reduction joint remediation, coupled electrokinetic phytoremediation, electrokinetics coupled with electrospun polyacrylonitrile nanofiber membrane, and electrokinetic remediation conjugated with permeable reactive barrier, are also used in conjunction with electrokinetic remediation methods [23, 24]. Soils with low permeability respond well to electrokinetic remediation. Because electrokinetic remediation is simple to set up and utilize, it is cost-effective [25]. Pollutant concentrations in soil are reduced when electrochemical adsorption is combined with extraction using low-molecular-weight organic acids [26–28]. Fluctuations in soil pH are the key limiting factor for direct electrokinetic remediation since it cannot sustain soil pH value [3].

2.2 Chemical remediation

2.2.1 Immobilization techniques

This technique, also known as Solidification and Stabilization. Immobilization is the process of adding immobilizing chemicals to polluted soils to reduce the mobility, bioavailability, and bioaccessibility of heavy metal(loid)s in the soil [29]. The immobilization of heavy metals in soil can be achieved through complexation, precipitation, and adsorption. By redistributing heavy metal(loid)s from soil solution to solid particles, these processes limit heavy metal(loid) transport and bioavailability in soil [30]. Binders, cement, clay, zeolites, phosphates, alkaline materials, termitaria, industrial eggshell, red-mud, chemical compounds, and more recently nanomaterials are a few of the mixing ingredients employed in the imobilization procedure [31].

2.2.2 Encapsulation

Encapsulation of contaminated soil stops the pollutants from spreading by covering the contaminant source with layers of concrete, lime, clay caps, or synthetic textiles, to limit the leaching and migration of contaminants away from thc isolatcd zone [12]. By becoming immobile, the polluted soil avoids contaminating the nearby materials [32]. Several binding materials are used in the production of solid blocks, but cement is chosen due to its accessibility, adaptability, and affordability [33]. Encapsulated soil can never be used to grow anything, hence this method of soil cleanup is only used as a last resort. Various immobilization agents are utilized during encapsulation, including polyvinyl alcohol, chitosan, alginate, agar, polyacrylamide, and polyurethanes [3]. The leaching of organic materials may be prevented effectively by encapsulation. Various immobilization agents, such as polyvinyl alcohol, chitosan, alginate, agar, polyacrylamide, and polyurethanes, are employed during encapsulation. While asphalt encapsulation is utilized for soils contaminated with hydrocarbons, encapsulation by lime and concrete has been used concurrently in the efficient treatment of soil contaminated with heavy metals and oil [34].

2.2.3 Soil washing

A technique known as soil washing uses two processes to remove pollutants from soil: physical separation and chemical leaching by aqueous solutions. This method starts with a homogenization step in which the coarse particles are divided based on their densities [35]. Depending on the type of metal and soil, the contaminated soil is dug up and combined with an appropriate extractant solution during soil washing [36]. For a predetermined amount of time, the extractant solution and dirt are fully blended. The heavy metals in soil are transported from soil to liquid phase and then removed from the leachate through precipitation, ions exchange, chelation, or adsorption [37]. If the contaminated soil passes regulatory tests for heavy metals after the washing process is complete, it will be returned to its original location. It is very common to use soil washing to purify heavy metals from contaminated soils, because it completely removes heavy metals. In addition, soil washing is a rapid method which can easily meet the researchers' criteria [38, 39]. A variety of chemicals, such as synthetic chelating agents (EDTA, EDDS), organic acids, humic compounds, surfactants, and cyclodextrins, have been employed to mobilize and remove heavy metals from soil [40, 41]. The capacity of the extractant to dissolve the heavy metal in soils determines the effectiveness of soil cleaning [42].

2.3 Biological remediation

2.3.1 Phytoremediation

Phytoremediation is a recently developed technology that offers a cost-effective solution by using plants, and associated soil microbes, to reduce the content, or toxic effects, of contaminants in the environment [43]. Botano-remediation, vegetative remediation, green remediation, and agro-remediation are all synonyms for phytoremediation [44]. Recently, there has been a lot of interest in and usage of phytoremediation, a natural, solar-powered, and environmentally benign method, especially in combination with other methods like biological, physical, and chemical methods for the treatment of hazardous pollutants [45]. A phytoremediation system can effectively clean-up sites with low-to-moderate levels of heavy metals while being environmentally friendly, appealing, esthetically pleasing, non-invasive, energy efficient, and cost-effective. Through mental accumulation, precipitation, or root surface absorption, heavy metals in the soil are strengthened during the solidification process [46]. Phytoremediation is usually divided into phytoextraction, phytostabilization, phytotransformation, and phytovolatilization. In practice, the selection of phytoremediation technology should be based on the types of soil and plants, the structure of rhizosphere microorganisms, and the complex coupling between the geochemical forms of pollutants.

2.3.2 Phytovolatilization

Phytovolatilization involves the uptake of contaminants by plant roots and its conversion to a gaseous state, and release into the atmosphere. This process is driven by the evapotranspiration of plants [47]. Metals are absorbed into volatile organic compounds during phytovolatilization, and these compounds are then released as biomolecules into the environment [48]. Succulent plants are regarded as a choice of plants in mining areas in arid and semi-arid environments.

2.3.3 Phytostabilization

Phytostabilization aims to contain contaminants within the vadose zone through accumulation by roots or precipitation within the rhizosphere. As a result of phytostabilization, heavy metals concentrations in contaminated soil are not reduced, but their movement is prevented [49]. When phytoextraction is not feasible or desirable, phytostabilization is used. In addition, phytostabilization can be used at sites with technical or regulatory limitations that make the selection and implementation of more appropriate remediation techniques difficult [50]. In abandoned contaminated sites such as mine wastelands, urban landfills, and sewage treatment plants, phytostabilization is commonly used. In tailings areas, pioneer plants are typically used to enhance physicochemical properties, provide cover, and establish a vegetation cap for long-term stability [51]. To maintain optimal stabilizing conditions, the site must be monitored regularly since heavy metals are stabilized within soil. A hyperaccumulator plant with the best phytostabilization properties (a) reduces heavy metals leaching by reducing water percolation through the soil matrix, (b) inhibits soil erosion and moves heavy metals to other areas, and (c) prevents direct contact with soil contaminated with heavy metals [52]. The most commonly used plant species for phytostabilization of Pb, Zn, and Cu polluted soils in Europe are Festuca spp. and Agrostis spp. [53].

2.3.4 Phytoextraction

In phytoextraction, heavy metals are removed from contaminated materials (soil and water) by uptake into harvestable plant parts [54]. As a result, phytoextraction reduces soil contamination. Most plant species cannot sustain in heavily polluted environments, so phytoextraction is suitable for sites with low-moderate levels of metal pollution [55]. There are four characteristics of plant species that can be used effectively for phytoextraction: (a) high metal-accumulation capability in their aboveground parts, (b) tolerance to high metal concentrations, (c) ability to grow rapidly with high biomass, and (d) profuse root systems. A number of common hyperaccumulator plants have been discovered and used to treat heavy metal-contaminated soil, including *Pteris vittata* L, *Sedum plumbizincicola*, *Solanum nigrum* L, *Polygnoum hyiper* L, *Thlaspi rave service* L, *Calendula officinalis*, and many others [56]. There are some limitations to heavy metal extraction by hyperaccumulating plants, including poor extraction efficiency, low biomass, easy environmental impact, heavy metal poisoning, and long repair times. However, they can be avoided by combining them with other technologies. Heavy metals in the soil can be effectively activated by adding chelators, creating a water-soluble metal chelator complex, which could change the occurrence form of heavy metals in the soil and then encourage the enrichment of heavy metals by plants, given the limitations of hyperaccumulation plants in the extraction of heavy metals [57]. The synthetic chelating agents EDTA, DTPA, EGTA, and EDDS are the most often utilized ones. Lead has the greatest ability to be activated by EDTA when compared to other heavy metal ions like Cu, Zn, and Cd.

2.4 Nanoremediation

The use of nanoparticles in nanoremediation enables the removal of heavy metals contaminants from soils and other environments in a cost-effective and eco-friendly manner [58]. Through the use of this novel remediation method, heavy metals can be absorbed, reduced to a stable metallic state, and catalyzed to leave a site [59, 60]. Different technical processes are employed in nanoremediation, including adsorption, heterogeneous catalysis, electrical field deployment (electronanoremediation), photodegradation, and the use of microorganisms (nanobioremediation) to remove or immobilize heavy metals from contaminated soils [61]. Metal nanoparticles, metallic oxides nanoparticles, carbonaceous nanoparticles, polymeric nanoparticles, and nanocomposites have all been successfully used and applied to remove heavy metals. Through pore spaces, nanoparticles can also reach inaccessible areas, such as crevices and aquifers, eliminating the need for traditional methods. These remediation materials have three modes of action: (1) A physical process that involves the adsorption and immobilization of contaminants on the surface of the particles. In one study, iron oxide Fe_3O_4 particles (12 nm in diameter) were used to remove arsenic from water after sorption and magnetic separation. (2) Toxic compounds are transformed into less harmful products through the process of detoxification, which induces and/or catalyzes the initial chemical breakdown. The dominant mechanisms are oxidation-reduction reactions, as in the photocatalytic oxidation of organics by titanium oxide TiO_2 nanoparticles [62], or the reduction of organics by nanoscale zero-valent iron (nZVI). (3) In bio-cooperative degradation, the particles increase bioavailability while degrading pollutants into more bioremediable species [63]. As an illustration, Fe_3O_4 NPs were employed as ion suppliers to enhance the production of biogas during anaerobic digestion procedures [64]. The tendency of nanoparticles to biostimulate bacterial cells was highlighted in a recent review by Abdelsalam and Samer, which also highlighted how this increased bacterial activity and growth kinetics [65].

3. Water treatment technologies

Reclaiming freshwater for use in agriculture and human activities requires wastewater treatment. Every year, as global water demands rise, many pollution schemes have threatened water sources [66]. Proper treatment and permanent removal of heavy metals are of immediate necessity. Many effective ways to remove pollutants such as heavy metals from wastewater are currently available [67]. Conventional techniques include ion exchange, membrane filtration membrane filtration, and chemical precipitation. Due of its simplicity, the chemical precipitation process is extensively utilized. Other alternative treatment techniques like photocatalysis, electrochemical, flotation, coagulation, and adsorptions have garnered a lot of attention in recent years. In order to remove heavy metals from wastewater, this study analyzes numerous treatment systems, their mechanisms, and the most recent developments.

3.1 Photocatalysis

Photocatalysis is a photo-activated chemical reaction occurring when free radical mechanisms are initiated as contact is made between the compound and photons that have sufficiently high energy levels. The words photo, which has to do with photons, and catalyst, which is a chemical that affects the rate of a process when it is present, are combined to form the term "photocatalyst." As a result, photocatalysts are substances that, when exposed to light, alter the rate of a chemical reaction. The term "photocatalysis" refers to this occurrence [68]. This technique was created as a result of research to emulate photosynthesis and the evolution of hydrogen for use in environmental applications. Semiconductors known for their photocatalytic properties, such as TiO_2, ZnO, CeO_2, CdS, and ZnS, was used in photocatalytic processes [69]. Strong oxidizing power, the ability to destroy heavy metal complexes and release them from the metal ions, and the capacity to oxidize and degrade organic complexes simultaneously are the characteristics that define photocatalysis.

Three processes make up the basic mechanism of photocatalysis. The first step is the production of charge carriers, which happens when a semiconductor is exposed to light that has a high energy or is equal to its bandgap. Second, the produced electron-hole pair moves onto the semiconductor's surface as electrons transition from the photocatalyst's valence band (vb) to conduction band (cb). Thirdly, electrons decrease the O_2 molecule to make superoxide radical anion (O_2) in the conduction band while photogenerated holes oxidize the H_2O molecule to yield OH in the valance band [70]. Various metal oxide-based photocatalytic materials such as TiO_2, ZnO, CuO, CdS, etc. have been used to remove organic and inorganic pollutants present in wastewater.

3.2 Coagulation/flocculation

Coagulation flocculation is a highly efficient physicochemical method for removing heavy metals [71]. In this process, fine particles and colloids agglomerate into larger particles, reducing turbidity, NOM and other wastewater pollutants. In the first stage, a coagulant added to the water stimulates the coalescence of colloidal material into small aggregates known as flocs [72]. The most commonly used coagulants include aluminum sulfate, ferrous sulfate, polyaluminum chloride (PACl), polymeric ferrous sulfate (PFS), and polyacrylamide (PAM) [73]. In the second stage, with gentle agitation, the flocs agglomerate, settle and are then disposed of as sludge. This process is used as a pre-treatment, post-treatment or main wastewater treatment due to its versatility [74]. This process is relatively economical and simple in operation, but limitations are incomplete removal of heavy metals, generation of sludge, and high operating costs due to chemical consumption.

3.3 Chemical precipitation

Chemical precipitation is an effective technique for removing heavy metals, mainly from effluents from the papermaking and electroplating industries. In this process, chemical precipitants such as alum, lime, iron salts and some polymers react with heavy metals present in the wastewater, resulting in insoluble precipitates [75]. This reaction allows metals to be removed more easily. Removal capacity and efficiency can be improved by optimizing parameters such as pH, temperature, initial concentration and ionic charge [76]. The mechanism of heavy metal removal by chemical precipitation is given by Eq:

$$M^{2+} + 2(OH)^{-} \rightarrow M(OH)_2 \quad (1)$$

where M^{2+} and OH are the metal ions and the precipitant, respectively, and $M(OH)_2$ is the metal hydroxide. The pH is adjusted to basic conditions (pH 9–11), which has the greatest impact in this treatment. Chemical precipitation is divided into hydroxide and sulfide precipitation. The use of coagulants in hydroxide precipitation can improve heavy metal removal by filtration or sedimentation. On the other hand, the sludge generated in the metal sulfide precipitation is removed by gravity separation or filtration. This process requires pre- and post-treatment as well as precise control over the addition of reagents due to the toxicity of sulfide ions and H_2S. Although this method has the following advantages: low capital investment, simple operation, and easily automated treatment method but it also brings problems that can be produce a large amount of sludge containing toxic compounds that require further treatment, requires a large number of chemicals to reduce metals to an acceptable level for discharge, slow metal precipitation, poor settling, and the long-term environmental impacts [74].

3.4 Ion exchange

A reversible ion exchange takes place between the solid and liquid phase. In particular, an insoluble substance removes the ions from an electrolyte solution and releases other ions of similar charge in chemically equivalent amounts. The most common ion exchange materials are synthetic organic resins [77], inorganic three-dimensional matrix and new generation hybrid materials [78]. Using an adequate replacement resin can provide an effective and economical solution to contamination control requirements. In the case of heavy metals, more highly concentrated metals are obtained by elution with suitable reagents after separating the loaded resin. The acid functional resin contains sulfonic acid in its structure. Therefore, the physicochemical interactions occurring during the removal of metal ions. Various optimization goals can be investigated for ion exchange. For example, use less resin to achieve a greater removal rate and optimize contact time with a smaller device size [79]. Anionic resins are generally used at a lower pollutant concentration, while cationic resins contain strong and weak acidic resins with more extensive use [80]. Weakly acidic resins with (COOH), while acidic resins with ($-SO_3H$) group are among the most popular cation exchangers [81]. However, ion exchange has some disadvantages, such as B. the need for a pre-treatment process, for example to remove fat or oil, as well as the need for chemical reagents to recover resins, which also cause secondary pollution [82].

3.5 Electrochemical technologies

Heavy metal ions from water sources can be effectively removed using electrochemical treatment techniques. These techniques involve recovering metals in their elemental metal state by employing cathodic and anodic processes in an electrochemical cell. Electrochemical treatments include electrocoagulation, electroflotation, and electrodeposition [83]. Traditional chemical coagulation is where electrocoagulation gets its start [84]. In this procedure, anode and cathode electrode sets serve as the sites for the oxidation and reduction reactions, respectively. An appropriate anode material is electrolytically oxidized to produce the coagulant as the charged ionic metals react with the anion in the effluent. By depositing pollutants on the cathode or removing them via flotation, the simultaneous cathodic reaction enables the removal of contaminants [74]. This method produces less sludge, is simple to use, and does not require any chemicals. The recovery of harmful metal ions from industrial wastewaters, such as Pb, Cd, Cu, Ni, Zn, or Cr, or the recovery of valuable metals from solutions, such as Ag, Pt, Au, etc., both involve considerable use of electrodeposition. The cost of treating water electrochemically has been reduced through a number of initiatives. In this regard, a comparison between platinum plate and stainless steel AISL904L was described. When treating Cu (II) from industrial contaminants, these plates are employed in place of three-dimensional electrodes. Cu foam can be used as an alternative because it has a wide surface area and performs better for the removal of effluents, but it makes the process more expensive. It was discovered that treating industrial water with tin dioxide anodes during the electrochemical process reduced water and electrolyte consumption by up to 70% [85].

3.6 Membrane technologies

A membrane acts as a barrier, allowing some substances to pass through while obstructing others. This technology is controlled by the Donnan exclusion effect (charge-charge repulsion), the size exclusion or steric hindrance mechanism, and the adsorption capacity of particular pollutants [86]. This form of treatment can be used to get rid of organic and inorganic pollutants, suspended solids, and other things. Membranes are categorized as either organic (made of synthetic organic polymers like polyethylene or cellulose acetate) or inorganic (made of ceramics, metals, zeolites, silica, among other materials) depending on the substance used to make them [87]. Microfiltration, ultrafiltration, and distillation are examples of low-pressure membrane processes. Nanofiltration, reverse osmosis, and electrodialysis are examples of high-pressure membrane processes. Direct osmosis, electrodialysis, and liquid membrane processes are examples of osmotic pressure-driven membrane processes. The removal performance of a membrane is greatly influenced by a number of variables, including the size and distribution of the pores, surface charge, degree of hydrophilicity, solution flow, and the presence of functional groups. These variables must be taken into account.

3.7 Adsorption

One of the finest ways to remove heavy metals and other impurities from water is adsorption. Its benefits include the potential to prevent significant secondary pollutants, a high removal capacity, relatively low energy consumption, and technical requirements for operation [88]. Adsorbents should possess a number of desirable qualities, including a sizable specific surface area, high mechanical strength, strong thermal stability, predictable morphology, and processing that is ecologically benign. Given the high adsorption capacity and efficiency, selectivity, low cost, and reusability, this should result in a high performance. Some of the most popular adsorbents are activated carbon (AC), polymer-based materials, biomaterials, magnetic materials, and industrial and agricultural wastes. Agricultural waste (fruit peels, bagasse, coir pith, cobs of corn, sawdust, and bark); Activated carbon (wood peat, coconut shells, coals); polymeric substances (Lignin, Chitosan, Cellulose, Alginate, Silk, and Cyclodextrin); sludge, metal hydroxide, red mud, fly ash, and other industrial byproducts; Ilmenite, Hematite, Magnetite, Spinel ferrite, and other magnetic adsorbents, Metal oxide particles/graphene composites, polymer matrix composites, and lignocellulosic residues/magnetic particles are all examples of composite adsorbents that are utilized for the removal of metals from wastewater [74].

3.8 Nanotechnology

Treatments based on nanotechnology make use of nanomaterials, which have drawn interest in recent years due to their high surface-to-volume ratios and distinctive electrical, optical, and magnetic capabilities [89, 90].

One of the most popular nanotechnology technologies for heavy metal removal is nanofiltration. Chemisorption is a highly effective method for eliminating dissolved heavy metals in systems made of alumina nanofibers. Additionally, low dimensional structures like nanoclays, magnetic nanoparticles, single or multi metal oxides, non-metal oxides, and nanocarbon are the most frequently used for the purification, disinfection, and removal of heavy metals from water [91, 92]. All of these nanostructures have huge, highly reactive surfaces, and many of them may be produced synthetically or using abundant natural resources. Similar technologies for wastewater treatment include nano-assemblies, nanoplates, microspheres with nanosheets, and hierarchical ZnO nano-rods. However, the dearth of knowledge regarding the toxicity, effects on the environment, and health of nanomaterials is significant and prevents their full utilization [93]. Nanocarbon (carbon nanotubes, graphene and other carbon derivatives), 2D materials, also known as single-layer materials, include graphene and borophene, germanene, silicene, 2D metal carbides (MXenes), MoS_2 nanosheets, silicon, iron, titanium, zinc, magnesium, and manganese oxides are commonly used for heavy metal remediation in the form of nanoparticles [94, 95].

3.9 Cold plasma technology

Cold plasma is characterized by a range of temperatures that correlate to various types of particles. A considerable number of energetic and chemically reactive species, such as free radicals, excited atoms, ions, and molecules, are produced in cold plasmas thanks to the high energy of the electrons (up to 1–10 eV), which serves as the catalyst for the start and spread of plasma chemical reactions. One of the method's most significant benefits is that it does not require high temperatures, which lowers energy use [96, 97].

Due to the possibility of using various operating gases (air, Ar, O_2, N_2, etc.) or types of plasma discharges (such as glow discharge, corona discharge, radio frequency discharge, gliding arc discharge, and dielectric barrier discharge), various plasma properties may develop, leading to the emergence of a number of applications. Due to the special characteristics of cold plasma, it is widely used in various fields [98]. Different settings have been studied for the efficient degradation of contaminants depending on the type of electrical discharges and reactor layouts. Three steps could be used to summarize the cold plasma process: Highly energetic electrons, OH radicals, ozone, O- and N-contained excited species, as well as other reactive species, are produced during the first step, contributing to the initiation and progression of the plasma chemical reactions; the second step entails the intrusion of the reacted species on the soil surface or soil pores or the dissolution or diffusion of the reacted species [99]. The capacity of easy mass transfer has a significant impact on the efficiency of remediation in both soil and water treatment as well as the effectiveness of the contact between the reactive species and the soil/water. When making electrical discharges during water cleanup, whether in a liquid or at a gas-liquid interface, the transport is changed since it is carried out by the slow aquatic ions, which is significantly impacted by the liquid conductivity. While other influencing factors (such as the impact of ionic charges present in water on the RONS/pollutant interaction) that also affect the process are not fully understood, it is discovered that the dissolved gases that create plasma micro-bubbles inside the liquid play a significant role in the process. The final step in the cleanup of contaminated sites is the chemical reaction of the reactive species with the organic pollutants. The process is impacted by the pollutants' type. For example, during soil treatment, highly volatile molecules are broken down via a two-path decomposition (evaporation of the contaminants into the gas phase where gas phase reactions are occurring or/and direct oxidation in soil due to the presence of the active species), whereas in the case of less volatile compounds, the oxidation processes are primarily occurring on the soil granules (the reactive species get in touch with the soil through diffusion or adsorption) [96].

4. Conclusion

In this research, modern and conventional technologies for remediation of soils and waste waters contaminated with heavy and toxic elements were briefly presented. Among these approaches, we can consider physical, chemical, bioremediation and combined methods. Effective factors, advantages, disadvantages and cost comparison in these methods were mentioned. Today, the methods of purifying water and soil pollution have grown almost adequately and progress has been made in this field, but it does not meet the real needs of the environment. With the continuous progress of science and the emergence of new technologies, newer methods are proposed.

At the same time, all the proposed methods have limitations. For example, in the treatment of contaminated soils through replacement, the method of storage and subsequent leaching of heavy toxic elements from the transferred soil is problematic and is still a matter of controversy among scientists. Also, phytoremediation is a long and time-consuming process. In comparison to green methods such as phytoremediation, chemical remediation methods have better advantages, including faster application-response and larger scales.

However, pollution purification methods, especially chemical methods, need a long way to reach full maturity. Among these cases, reducing secondary pollution due to leaching of heavy elements and identifying newer economic chemicals for chemical oxidation-reduction and optimal immobilization of pollution. To achieve the mentioned goal, it is necessary to turn to group and interdisciplinary researches. Group and interdisciplinary researches can lead to the formation of new technologies that overcome the weaknesses of existing methods. For example, extensive research should be done to make selective separation of pollution from water environments.

Author details

Ahmad Akhavan
Faculty of Agriculture, Soil Science Department, University of Zanjan, Zanjan, Iran

Address all correspondence to: ahmad.akhavan@znu.ac.ir

References

[1] Ali H, Khan E. What are heavy metals? Long-standing controversy over the scientific use of the term 'heavy metals'—Proposal of a comprehensive definition. Toxicological & Environmental Chemistry. 2018;**100**(1):6-19

[2] Dixit R et al. Bioremediation of heavy metals from soil and aquatic environment: An overview of principles and criteria of fundamental processes. Sustainability. 2015;7(2):2189-2212

[3] Khalid S et al. A comparison of technologies for remediation of heavy metal contaminated soils. Journal of Geochemical Exploration. 2017;**182**:247-268

[4] Akhavan A, Golchin A. Estimation of arsenic leaching from Zn–Pb mine tailings under environmental conditions. Journal of Cleaner Production. 2021;**295**:126477

[5] Masindi V, Muedi KL. Environmental contamination by heavy metals. Heavy Metals. 2018;**10**:115-132

[6] Xiong T et al. Kinetic study of phytotoxicity induced by foliar lead uptake for vegetables exposed to fine particles and implications for sustainable urban agriculture. Journal of Environmental Sciences. 2016;**46**:16-27

[7] Bhargava P et al. Health issues and heavy metals. Austin Journal of Environmental Toxicology. 2017;**3**(1):3018

[8] Anderson A et al. Treatment of heavy metals containing wastewater using biodegradable adsorbents: A review of mechanism and future trends. Chemosphere. May 2022;**295**:133724. DOI: 10.1016/j.chemosphere.2022.133724-

[9] Song B et al. Evaluation methods for assessing effectiveness of in situ remediation of soil and sediment contaminated with organic pollutants and heavy metals. Environment International. 2017;**105**:43-55

[10] Thomé A et al. Remediation technologies applied in polluted soils: New perspectives in this field. In: The International Congress on Environmental Geotechnics. Singapore: Springer; 11 Oct 2018. ISBN: 978-981-13-2220-4

[11] Kuppusamy S et al. In-situ remediation approaches for the management of contaminated sites: A comprehensive overview. Reviews of Environmental Contamination and Toxicology. Cham: Springer. 2016;**236**: 1-115. ISBN: 978-3-319-20012-5, ISBN: 978-3-319-20013-2

[12] Yao Z et al. Review on remediation technologies of soil contaminated by heavy metals. Procedia Environmental Sciences. 2012;**16**:722-729

[13] Zhou D et al. Advances in remediation technologies of contaminated soils. Ecology and Environmental Sciences. 2004;**13**(2):234-242

[14] Douay F et al. Contamination of urban soils in an area of Northern France polluted by dust emissions of two smelters. Water, Air, and Soil Pollution. 2008;**188**(1):247-260

[15] Zheng C, Wang PP. A field demonstration of the simulation optimization approach for remediation system design. Groundwater. 2002;**40**(3):258-266

[16] National Research Council, Commission on Geosciences, Environment, and Resources, Board on Radioactive Waste Management, Committee on the Remediation of Buried and Tank Wastes. Long-Term Institutional Management of U.S. Department of Energy Legacy Waste. 2000

[17] Zhu L et al. Isolation of aerobic denitrifiers and characterization for their potential application in the bioremediation of oligotrophic ecosystem. Bioresource Technology. 2012;**108**:1-7

[18] Mallampati SR et al. Dynamic immobilization of simulated radionuclide 133Cs in soil by thermal treatment/vitrification with nanometallic Ca/CaO composites. Journal of Environmental Radioactivity. 2015;**139**:118-124

[19] Buelt JL, Farnsworth RK. In situ vitrification of soils containing various metals. Nuclear Technology. 1991;**96**(2):178-184

[20] Dellisanti F, Rossi PL, Valdrè G. In-field remediation of tons of heavy metal-rich waste by joule heating vitrification. International Journal of Mineral Processing. 2009;**93**(3-4):239-245

[21] Buelt J, Thompson L. The in Situ Vitrification Integrated Program: Focusing an Innovative Solution on Environmental Restoration Needs. Richland, WA (United States): Pacific Northwest Lab; 1992

[22] AEC U. In-situ electrokinetic remediation of metal contaminated soils-technology status report. US Army Environmental Center, Report Number: SFIM-AEC-ET-CR-99022. 2000

[23] Yang Z et al. Bioleaching remediation of heavy metal-contaminated soils using Burkholderia sp. Z-90. Journal of Hazardous Materials. 2016;**301**:145-152

[24] Mao X et al. Electro-kinetic remediation coupled with phytoremediation to remove lead, arsenic and cesium from contaminated paddy soil. Ecotoxicology and Environmental Safety. 2016;**125**:16-24

[25] Virkutyte J, Sillanpää M, Latostenmaa P. Electrokinetic soil remediation—Critical overview. Science of the Total Environment. 2002;**289**(1-3):97-121

[26] Yang X et al. Remediation of heavy metal contaminated soils by organic acid extraction and electrochemical adsorption. Environmental Pollution. 2020;**264**:114745

[27] Xu J et al. Remediation of heavy metal contaminated soil by asymmetrical alternating current electrochemistry. Nature Communications. 2019;**10**(1):1-8

[28] Han J-G et al. Enhanced electrokinetic (E/K) remediation on copper contaminated soil by CFW (carbonized foods waste). Journal of Hazardous Materials. 2010;**177**(1-3):530-538

[29] Tica D, Udovic M, Lestan D. Immobilization of potentially toxic metals using different soil amendments. Chemosphere. 2011;**85**(4):577-583

[30] Wuana R et al. Evaluation of copper and lead immobilization in contaminated soil by single, sequential and kinetic leaching tests. African Journal of Environmental Science and Technology. 2013;**7**(5):249-258

[31] Soares MA, Quina MJ, Quinta-Ferreira RM. Immobilisation of lead and zinc in contaminated soil using compost derived from industrial eggshell. Journal of Environmental Management. 2015;**164**:137-145

[32] Ucaroglu S, Talinli I. Recovery and safer disposal of phosphate coating sludge by solidification/stabilization. Journal of Environmental Management. 2012;**105**:131-137

[33] Pandey B, Kinrade SD, Catalan LJ. Effects of carbonation on the leachability and compressive strength of cement-solidified and geopolymer-solidified synthetic metal wastes. Journal of Environmental Management. 2012;**101**:59-67

[34] He J, Chen JP. A comprehensive review on biosorption of heavy metals by algal biomass: Materials, performances, chemistry, and modeling simulation tools. Bioresource Technology. 2014;**160**:67-78

[35] Rodríguez MF et al. Soil Pollution Remediation. Reference Module in Biomedical Sciences. Encyclopedia of Toxicology. 3rd ed. Elsevier; 2014:344-355. DOI: 10.1016/B978-0-12-386454-3.00579-0

[36] Hillel D, Hatfield JL. Encyclopedia of Soils in the Environment. Vol. 3. Amsterdam: Elsevier; 2005

[37] Ferraro A et al. Application of an electrochemical treatment for EDDS soil washing solution regeneration and reuse in a multi-step soil washing process: Case of a Cu contaminated soil. Journal of Environmental Management. 2015;**163**:62-69

[38] Bolan NS et al. Cadmium contamination and its risk management in rice ecosystems. Advances in Agronomy. 2013;**119**:183-273

[39] Park B, Son Y. Ultrasonic and mechanical soil washing processes for the removal of heavy metals from soils. Ultrasonics Sonochemistry. 2017;**35**:640-645

[40] Shahid M et al. Influence of plant species and phosphorus amendments on metal speciation and bioavailability in a smelter impacted soil: A case study of food-chain contamination. Journal of Soils and Sediments. 2014;**14**(4):655-665

[41] Kulikowska D et al. Feasibility of using humic substances from compost to remove heavy metals (Cd, Cu, Ni, Pb, Zn) from contaminated soil aged for different periods of time. Journal of Hazardous Materials. 2015;**300**:882-891

[42] Fabbri D et al. Removal and degradation of aromatic compounds from a highly polluted site by coupling soil washing with photocatalysis. Chemosphere. 2008;**71**(1):59-65

[43] Patra DK, Pradhan C, Patra HK. Toxic metal decontamination by phytoremediation approach: Concept, challenges, opportunities and future perspectives. Environmental Technology & Innovation. 2020;**18**:100672

[44] Sarwar N et al. Phytoremediation strategies for soils contaminated with heavy metals: Modifications and future perspectives. Chemosphere. 2017;**171**:710-721

[45] Yadav KK et al. Mechanistic understanding and holistic approach of phytoremediation: A review on application and future prospects. Ecological Engineering. 2018;**120**:274-298

[46] Wuana RA, Okieimen FE. Heavy metals in contaminated soils: A review of sources, chemistry, risks and best available strategies for remediation. International Scholarly Research Network. 2011;**2011**. Article ID 402647, 20 pages. DOI: 10.5402/2011/402647

[47] Chandra R, Kumar V. Phytoremediation: A green sustainable

technology for industrial waste management. In: Phytoremediation of Environmental Pollutants. 1st ed. CRC Press; 2017. pp. 1-42. eBook ISBN: 9781315161549

[48] Kumar V, Shahi S, Singh S. Bioremediation: An eco-sustainable approach for restoration of contaminated sites. In: Microbial Bioprospecting for Sustainable Development. Singapore: Springer; 19 Sep 2018. pp. 115-136. ISBN: 978-981-13-0052-3, ISBN: 978-981-13-0053-0

[49] Bolan NS et al. Phytostabilization: A green approach to contaminant containment. Advances in Agronomy. 2011;**112**:145-204

[50] Chan SS et al. Recent advances biodegradation and biosorption of organic compounds from wastewater: Microalgae-bacteria consortium-a review. Bioresource Technology. 2022;**344**:126159

[51] Zine H et al. Guidelines for a phytomanagement plan by the phytostabilization of mining wastes. Scientific African. 2020;**10**:e00654

[52] Pachaiappan R et al. A review on biofiltration techniques: Recent advancements in the removal of volatile organic compounds and heavy metals in the treatment of polluted water. Bioengineered. 2022;**13**(4):8432-8477

[53] Galende MA et al. Field assessment of the effectiveness of organic amendments for aided phytostabilization of a Pb–Zn contaminated mine soil. Journal of Geochemical Exploration. 2014;**145**:181-189

[54] Suman J et al. Phytoextraction of heavy metals: A promising tool for clean-up of polluted environment? Frontiers in Plant Science. 2018;**9**:1476

[55] Sabir M et al. Phytoremediation, Soil Remediation and Plants. Elsevier Inc.; 2015

[56] Reeves RD et al. Global distribution and ecology of hyperaccumulator plants. In: Agromining: Farming for Metals. Cham: Springer; 08 Dec 2020. pp. 133-154. DOI: 10.1007/978-3-030-58904-2_7. ISBN: 978-3-030-58903-5, ISBN: 978-3-030-58904-2

[57] Shahid M et al. Water extraction kinetics of metals, arsenic and dissolved organic carbon from industrial contaminated poplar leaves. Journal of Environmental Sciences. 2013;**25**(12):2451-2459

[58] Baragaño D et al. Nanoremediation of As and metals polluted soils by means of graphene oxide nanoparticles. Scientific Reports. 2020;**10**(1):1-10

[59] Gil-Díaz M et al. A nanoremediation strategy for the recovery of an As-polluted soil. Chemosphere. 2016;**149**:137-145

[60] Bradford SA et al. Environmental Applications and Risks of Nanomaterials: An Introduction to CREST Publications during 2018-2021. Critical Reviews in Environmental Science and Technology. Taylor & Francis; 2022. pp. 3753-3762. DOI: 10.1080/10643389.2021.2020425

[61] Hua M et al. Heavy metal removal from water/wastewater by nanosized metal oxides: A review. Journal of Hazardous Materials. 2012;**211**:317-331

[62] Abdelraheem WH, Nadagouda MN, Dionysiou DD. Solar light-assisted remediation of domestic wastewater by NB-TiO2 nanoparticles for potable reuse. Applied Catalysis B: Environmental, 2020;**269**:118807

[63] Marcon L, Oliveras J, Puntes VF. In situ nanoremediation of soils and

groundwaters from the nanoparticle's standpoint: A review. Science of the Total Environment. 2021;**791**:148324

[64] Casals E et al. Programmed iron oxide nanoparticles disintegration in anaerobic digesters boosts biogas production. Small. 2014;**10**(14):2801-2808

[65] Abdelsalam EM, Samer M. Biostimulation of anaerobic digestion using nanomaterials for increasing biogas production. Reviews in Environmental Science and Bio/Technology. 2019;**18**(3):525-541

[66] United nations. UN-Water coordinates the United Nations' work on water and sanitation. Geneva, Switzerland; 2018

[67] Saleh NB et al. Removal of poly- and per-fluoroalkyl substances from aqueous systems by nano-enabled water treatment strategies. Environmental Science: Water Research & Technology. 2019;**5**(2):198-208

[68] Braslavsky SE et al. Glossary of terms used in photocatalysis and radiation catalysis (IUPAC recommendations 2011). Pure and Applied Chemistry. 2011;**83**(4):931-1014

[69] Daghrir R, Drogui P, Robert D. Modified TiO2 for environmental photocatalytic applications: A review. Industrial & Engineering Chemistry Research. 2013;**52**(10):3581-3599

[70] Collivignarelli MC et al. Overview of the main disinfection processes for wastewater and drinking water treatment plants. Sustainability. 2017;**10**(1):86

[71] Zhu Y et al. Removal of chelated heavy metals from aqueous solution: A review of current methods and mechanisms. Science of the Total Environment. 2019;**678**:253-266

[72] Teh CY et al. Recent advancement of coagulation–flocculation and its application in wastewater treatment. Industrial & Engineering Chemistry Research. 2016;**55**(16):4363-4389

[73] Xu D, Zhou B, Yuan R. Optimization of coagulation-flocculation treatment of wastewater containing Zn (II) and Cr (VI). In: IOP Conference Series: Earth and Environmental Science. IOP Publishing; 2019;**227**(5):052049. DOI: 10.1088/1755-1315/227/5/052049

[74] Zamora-Ledezma C et al. Heavy metal water pollution: A fresh look about hazards, novel and conventional remediation methods. Environmental Technology & Innovation. 2021;**22**:101504

[75] Verma B, Balomajumder C. Hexavalent chromium reduction from real electroplating wastewater by chemical precipitation. Bulletin of the Chemical Society of Ethiopia. 2020;**34**(1):67-74

[76] Gunatilake S. Methods of removing heavy metals from industrial wastewater. Methods. 2015;**1**(1):14

[77] Chaemiso TD, Nefo T. Removal methods of heavy metals from laboratory wastewater. Journal of Natural Sciences Research. 2019;**9**(2):36-42

[78] Bashir A et al. Removal of heavy metal ions from aqueous system by ion-exchange and biosorption methods. Environmental Chemistry Letters. 2019;**17**(2):729-754

[79] Torres-Perez J et al. Two-stage optimization of Allura direct red dye

removal by treated peanut hull waste. SN Applied Sciences. 2020;**2**(3):1-12

[80] Carolin CF et al. Efficient techniques for the removal of toxic heavy metals from aquatic environment: A review. Journal of Environmental Chemical Engineering. 2017;**5**(3):2782-2799

[81] Chandrasekara NGN, Pashley R. Study of a new process for the efficient regeneration of ion exchange resins. Desalination. 2015;**357**:131-139

[82] Pan L et al. Oil removal in tight-emulsified petroleum waste water by flocculation. In: IOP Conference Series: Materials Science and Engineering. IOP Publishing; 2018;**392**(4):042005. DOI: 10.1088/1757-899X/392/4/042005

[83] Maarof HI, Daud WMAW, Aroua MK. Recent trends in removal and recovery of heavy metals from wastewater by electrochemical technologies. Reviews in Chemical Engineering. 2017;**33**(4):359-386

[84] Collivignarelli MC et al. Electrolytic recovery of nickel and copper from acid pickling solutions used to treat metal surfaces. Water, Air, & Soil Pollution. 2019;**230**(5):1-13

[85] Orts F et al. Effect of chloride on the one step electrochemical treatment of an industrial textile wastewater with tin dioxide anodes. The case of trichromy procion HEXL. Chemosphere. 2020;**245**:125396

[86] Abdullah N et al. Recent trends of heavy metal removal from water/wastewater by membrane technologies. Journal of Industrial and Engineering Chemistry. 2019;**76**:17-38

[87] Obotey Ezugbe E, Rathilal S. Membrane technologies in wastewater treatment: A review. Membranes. 2020;**10**(5):89

[88] Chai WS et al. A review on conventional and novel materials towards heavy metal adsorption in wastewater treatment application. Journal of Cleaner Production. 2021;**296**:126589

[89] Gehrke I, Geiser A, Somborn-Schulz A. Innovations in nanotechnology for water treatment. Nanotechnology, Science and Applications. 2015;**8**:1

[90] Upadhyay AK, Singh R, Singh D. Restoration of Wetland Ecosystem: A Trajectory Towards a Sustainable Environment. Springer; 2019. ISBN: 978-981-13-7665-8

[91] Borji H et al. How effective are nanomaterials for the removal of heavy metals from water and wastewater? Water, Air, & Soil Pollution. 2020;**231**(7):1-35

[92] Vázquez-Núñez E et al. Use of nanotechnology for the bioremediation of contaminants: A review. PRO. 2020;**8**(7):826

[93] Wołowiec M et al. Removal of heavy metals and metalloids from water using drinking water treatment residuals as adsorbents: A review. Minerals. 2019;**9**(8):487

[94] Singh N, Shukla SK. Properties of two-dimensional nanomaterials. In: Two-Dimensional Nanostructures for Biomedical Technology. A Bridge Between Material Science and Bioengineering. Elsevier; 2020. pp. 73-100. DOI: 10.1016/B978-0-12-817650-4.00003-6

[95] Feng X et al. Self-assembling 2D/2D (MXene/LDH) materials achieve ultra-high adsorption of heavy metals Ni2+ through terminal group modification.

Separation and Purification Technology. 2020;**253**:117525

[96] Aggelopoulos CA. Recent advances of cold plasma technology for water and soil remediation: A critical review. Chemical Engineering Journal. 2022;**428**:131657

[97] Daletou M, Aggelopoulos C. Highly-energy efficient oxidation of MWCNT with nanosecond pulsed dielectric barrier discharge plasma. Applied Surface Science. 2021;**563**:150139

[98] Pankaj SK, Keener KM. Cold plasma: Background, applications and current trends. Current Opinion in Food Science. 2017;**16**:49-52

[99] Dimitrakellis P, Gogolides E. Hydrophobic and superhydrophobic surfaces fabricated using atmospheric pressure cold plasma technology: A review. Advances in Colloid and Interface Science. 2018;**254**:1-21

Chapter 21

Heavy Metals Removal from Water and Wastewater

Lavinia Lupa and Laura Cocheci

Abstract

The chapter summarizes the main treatment processes used for the removal of heavy metals from water and wastewater. Heavy metals present a recalcitrant and persistent character, a bioaccumulating ability in the environment, therefore their removal from water and wastewater represents a worldwide concern. This chapter reviews the recent advances and technical applicability of the methods that have been used to treat heavy metal-containing water. The discussed technologies include chemical precipitation, coagulation-flocculation, flotation, ion exchange, adsorption, membrane filtration, and electrochemical methods. The main advantages/disadvantages and limits of each method are evaluated. It was observed that even if are made a lot of research on each mentioned method in practice is still necessary for intensive work for each case. In all methods, there are things that could be improved; therefore, future studies must be focused on the development of cost-effective materials and methods that involve low treatment costs, high efficiency, and minimal impact on the environment.

Keywords: heavy metals, water, wastewater, environmental issues, removal processes

1. Introduction

Metals present numerous benefits for everyday life. They have contributed to the development of civilization, to the modernization and development of industries. In some cases, they are essential for maintaining the metabolism of the human or animal body, or they are indispensable in the growth of plants as microelements [1, 2]. However, if metals exceed certain concentration levels (even trace amounts), they can contribute to environmental pollution and can lead to devastating effects on living organisms [3, 4]. **Figure 1** shows the main sources of metal pollution and their transport through an environment. Water is the most responsible environmental factor, with the help of which the heavy metals present at improper concentrations can reach living organisms. Furthermore, heavy metals have the property of bioaccumulation since they could not be biodegradable, leading to critical health issues [5–7]. Therefore, water pollution with heavy metals represents a global concern and the World Health Organization (WHO) established the maximum admitted level of heavy metal concentrations in drinking water [8].

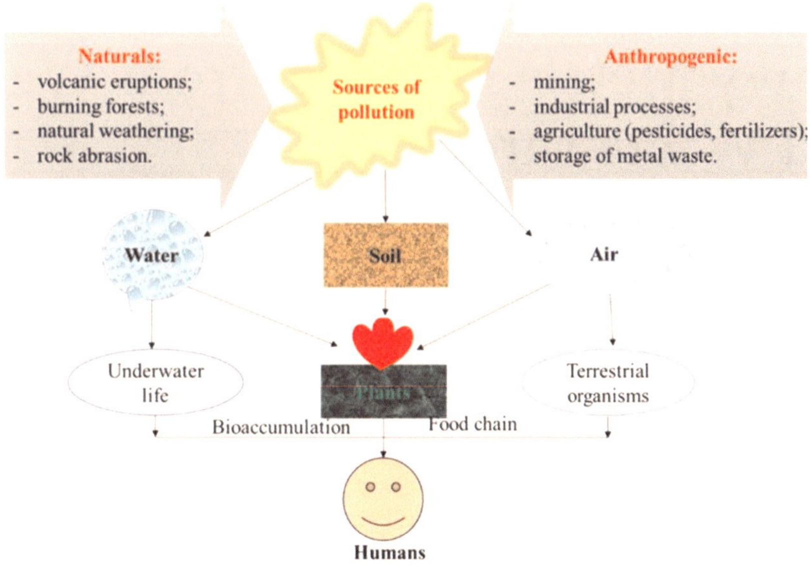

Figure 1.
Sources and transport of heavy metals through an environment.

The removal of heavy metals from water sources or wastewater before discharge is an important problem that must be carried out to minimize pollution, reduce their interference with beneficial uses, and eliminate their negative effects on the environment. In some cases, in addition to eliminating the toxic effect of the metals present in waters upon the environment, these water treatment methods also have the purpose of recovering and revaluing the metals, especially if we are talking about rare and precious metals (such as gold, silver, platinum, palladium, rhodium, ruthenium, iridium, and osmium). The present chapter summarizes the treatment methods employed for heavy metals removal, describing new advanced developments, and highlighting the advantages and disadvantages of each in terms of efficiency, accuracy, feasibility, and kinetic.

2. Chemical precipitation

Chemical precipitation (**Figure 2**) is the process of transforming soluble metal ions into insoluble metal compounds using various precipitating agents, such as hydroxides, carbonates, sulfides, sulfates, phosphates, chlorides, and sodium borohydride, usually followed by a separation step (sedimentation, filtration, settling, and centrifugation) [9–11].

The most frequently used method in industry is precipitation under hydroxide form. The method is based on the low solubility of metal hydroxides (reaction 1) at alkaline values of the mass reaction pH [1, 3].

$$M^{n+}{}_{aq} + n(OH)^{-}{}_{aq} = M(OH)_{n\ pp} \tag{1}$$

The solubility of hydroxides is dependent on pH. For each metal, there are solubility diagrams that represent the graphical representation of the solubility function of

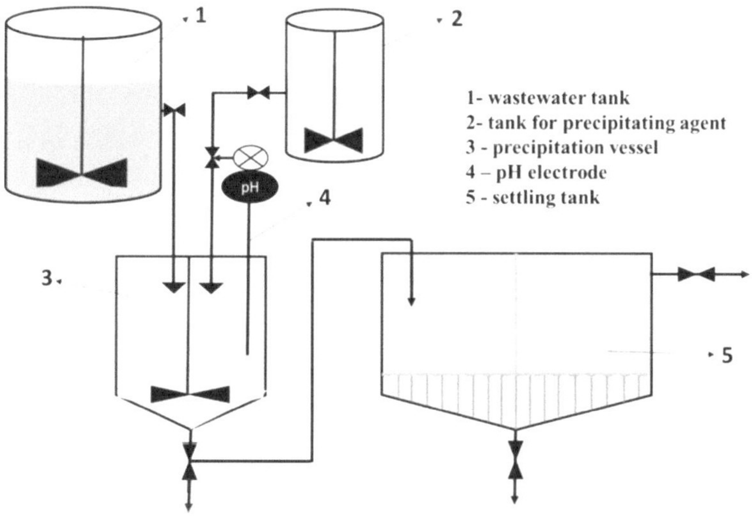

Figure 2.
Scheme of the precipitation system.

pH, which present importance in the establishment of the pH value at which the desired metal ions present the lowest solubility [12]. Different metals have different values of the optimal precipitation pH, so maximizing the removal efficiency of a certain metal can lead to a significant decrease in the removal degree of another metal.

Sometimes for this reason it is necessary to use two stages of treatment to obtain a high removal degree of all the metals present in the wastewater. On the other hand, the fact that metals present in water precipitate at different pH values can lead to a selective separation of them. The main advantages of the method are: (1) easy of operation, could be applied without requiring a pretreatment; (2) could remove several parameters at once; (3) could be performed at ambient temperature; (4) could be automated; and (5) presents low costs if lime is used as a precipitating agent. At the same time, the method presents also some disadvantages, such as requires a high amount of precipitating agent, consequently generating a large amount of sludge (especially if lime is used as precipitating agent); the pH must be strictly controlled; does not remove complex metals; the efficiency of the separation step influence the treatment performance [13]. Among the metals that can be removed using this method, could be mentioned: iron, copper, zinc, cadmium, beryllium, cobalt, mercury, manganese, and aluminum. Chromium can only be removed if it is found in the trivalent form.

Besides pH, other parameters and factors must be considered (such as temperature, ionic strength, formation of complexes, formation of other solid phases, alkalinity, and formation of buffer solutions) because they affect the solubility of metal hydroxides and consequently the efficiency of metal ions removal through precipitation. The metal ions' solubility increases with the temperature increasing, with ionic strength increasing, or with the formation of complexes. Around the optimum pH value of metal ions precipitation under hydroxide form takes place in the formation of another solid phase, such as carbonates form. Certain anions present in wastewater from metal processing (chlorines, sulfates) can also precipitate the metals present. The carbonates and sulfates present in the water, as well as some organic anions can precipitate with the calcium ions added in the form of lime to adjust the pH. The metal ions' solubility decreases as the formatted precipitate is subject to aging and is transformed into ideal crystals [12].

Instead of hydroxide precipitation or together with it is used precipitation under carbonates forms (reaction 2).

$$M^{n+}{}_{aq} + n(CO_3)^{2-}{}_{aq} = M_2(CO_3)_{n\ pp} \tag{2}$$

In contrast to metal hydroxides, metal carbonates are formed at lower values of the reaction mass pH, and present higher density, which leads to the improvement of the removal process performance. Lower metal solubilities can be obtained by increasing the carbonate dose at pH values lower than the optimal pH for the precipitation of hydroxides. The main disadvantages of this method are represented by its reduced kinetic and by the possible foaming of the reaction mass due to CO_2 release [14]. This method is frequently used for lead removal due to the high solubility of lead hydroxide [15, 16].

Metal sulfides are compounds that also present a much lower solubility than metal hydroxides. Thus, metal ions can be removed from aqueous solutions by precipitating them in the form of sulfides according to reactions 3–6 [1, 11, 17, 18].

$$H_2S = HS^- + H^+ \tag{3}$$

$$HS^- = S^{2-} + H^+ \tag{4}$$

$$M^{2+} + S^{2-} = MS_{pp} \tag{5}$$

$$M^{2+} + HS^- = MS_{pp} + H^+ \tag{6}$$

For the precipitation of metals in the form of sulfides, soluble sulfides can be used, for example, sodium sulfide (Na_2S), calcium polysulfide (CaS), or sodium hydrosulfide (NaHS), or insoluble sulfides, such as ferrous sulfide (FeS). The partially soluble ferrous sulfide is added as a suspension. Most metals have lower solubility than FeS, so heavy metals precipitate as sulfides, while FeS is solubilized (reaction 7). The reaction takes place around pH = 8, and Fe precipitates in the form of hydroxides [1, 11, 17].

$$M^{2+} + FeS = MS + Fe^{2+} \tag{7}$$

Due to its reducing character, an important advantage of this method is the fact that FeS can be used directly in the removal process of Cr(VI) from water (reaction 8). The sulfur is oxidized to its basic state, and chromium precipitates in the form of hydroxide [19–21].

$$H_2CrO_4 + FeS + 2H_2O = Cr(OH)_3 + Fe(OH)_3 + S \tag{8}$$

Another advantage of the precipitation under sulfide ions is the ability to direct the precipitation of complex metals. In this case, it does not require any pretreatment steps or multiple stages to remove different metals. Unfortunately, the sulfide precipitates are in the form of small particles that present poor sedimentation properties, sometimes requiring the use of coagulation agents. The main disadvantage of this method consists of the strict conditions for operation and handling of the resulting sludge due to the possibility of toxic hydrogen sulfide formation. The advantages and disadvantages of the main precipitation agents used for the removal of metal ions from water and wastewater are summarized in **Table 1**.

Precipitation agent	Advantages	Disadvantages
NaOH	• Good kinetic • It is suitable for removing many parameters. • The process takes place under ambient conditions. • It could be automated.	• Expensive • Require high values of pH to obtain precipitates with lower solubility. • The pH must be strictly controlled. • Could not remove complex metals. • Treatment performance is determined by the separation stage.
$Ca(OH)_2$	• Inexpensive • It is suitable for removing many parameters. • The process takes place under ambient conditions. • It could be automated.	• Require a higher amount of precipitation agent dose to achieve the optimum pH. • Formation of a higher quantity of sludge. • Could not remove complex metals. • Treatment performance is determined by the separation stage.
Carbonates	• Operation at lower pH values. • Easy sedimentation	• Slow kinetic • Foaming of the reaction mass. • Could not remove complex metals.
Sulfides	• Ability to directly eliminate Cr (VI) without other pretreatment steps. • Precipitates complex metals. • It does not require a final pH adjustment.	• Poor sedimentation characteristics. • It requires increased attention in the handling and storage stage of the sludge due to its hazardous nature. • Excess sulfides present in the effluent can lead to the development of H_2S.

Table 1.
The advantages and disadvantages of the main precipitation agents used for the removal of metal ions from water and wastewater.

Precipitation in the form of sulfates is generally applied to remove barium from aqueous solutions [22–24]. Precipitation in the form of phosphates is generally applied for the elimination of trivalent metals (iron, aluminum, and chromium). The precipitation in the form of chlorides is particularly applied in combination with the oxidation of cyanides for the removal and recovery of silver [25]. Sodium borohydride is an effective reducing agent and is used to remove Pb, Hg, Ni, Cd, Co, Cu, and some precious metals [26, 27].

All these mentioned factors and variables that influence the precipitation raise difficulties in the evaluation of the treatment process. So that predicting the efficiency of the process requires the evaluation of a theoretical and experimental study for each individual case. For example, Serrano and his coworkers studied the removal of Fe (III), Cd(II), and Zn(II) using hydroxides precipitation combined with flotation. The precipitation efficiency was determined function of pH, metal ions initial concentration, treatment time, and dosage of the precipitating agent. The simultaneous removal of 99% for all three studied metal ions was obtained using a pH value of 10.3, an initial concentration of metal ions between 1 mol/L to 15 mol/L, and a treatment time of 15 minutes [9]. Zhang and Duan studied the removal of heavy metals by precipitation using magnesium hydroxy carbonate as a precipitating agent. They obtained a residual concentration under the maximum admitted value using a dosage of the precipitation agent of 0.3 g/50 mL of residual solution at a pH value of 7.1 [12]. Sadeghi et al. studied the removal of lead through precipitation using sodium sulfide and sodium carbonate as precipitation agents. In each case was obtained a removal efficiency

>95% at pH = 11 [15]. The lead ions removal was also studied by precipitation with sodium carbonate by Hu et al. In this case, the researchers used a ball milling process to increase the reaction between lead salts and the precipitation agent, achieving a removal degree of 99% [16]. Regarding precipitation in the form of sulfides, studies have been focused on minimizing the production of toxic hydrogen sulfide. In this regard, Phol studied the metal ions precipitation using other sulfur-containing precipitation agents, such as potassium/sodium thiocarbonate (STC), 2,4,6-trimercaptotiazine (TMT), sodium dimethyldithiocarbamate (SDTC), 1,3-benzenediamidoethanethiol (BDETH 2), 2,6-pyridinediamidoethanethiol (PyDET), or pyridine-based thiol ligand (DTPY) [17]. In this way, the metal ions are binding by the precipitation agent and form metal complexes and are avoided the H_2S formation but to obtain an efficient removal of metal ions is necessary to use a higher dose of precipitation agents. Prokkola et al. studied the metal ions precipitation from acidic mine drainage (AMD) by using HS, resulting from sulfate reduction reaction. The resulted H_2S gas and ionic HS^- during anaerobic treatment were recycled in the precipitation process. The optimum pH value of the precipitation process was 5.5, when is achieved a residual concentration of metal ions <30 μg/L [18]. It is observed that regardless of the precipitation agent used, the degree of metal ions removal from aqueous solutions is influenced by the pH value, treatment time, precipitation agent dose, and initial concentration of metal ions in the solution. By adjusting these influencing factors, the removal capacity of metal ions from wastewater can be significantly improved, which provides the theoretical basis for the practical application of these technologies from the laboratory scale, updated to pilot scale, and further to the industrial level.

3. Coagulation – flocculation

Coagulation – flocculation is a physical-chemical process of metal ions removal from water and wastewater, which consist of three steps (**Figure 3**). In the first step, a coagulation agent is introduced under vigorous stirring having the role of colloidal

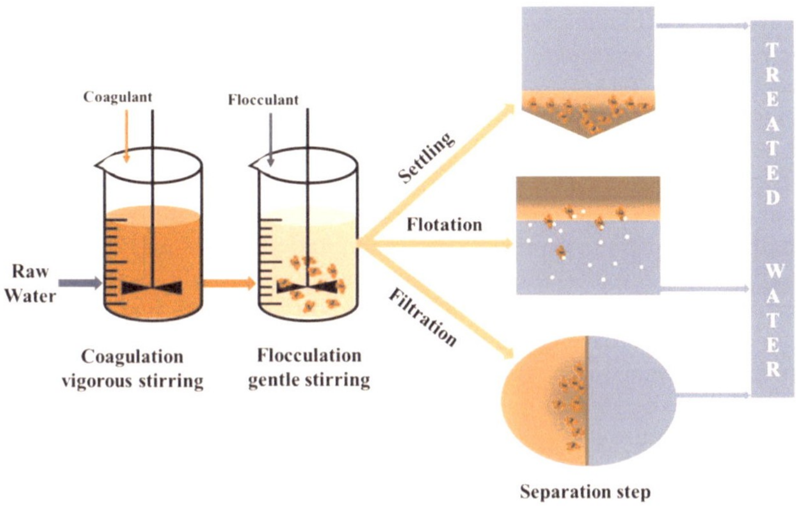

Figure 3.
Scheme of the coagulation-flocculation system.

Coagulant/ flocculant agent	Water type	Methodology	Obtained results	Ref.
Alum	Landfill leachate	Coagulation/flocculation process combined with continuous adsorption using eggshell waste materials (CES). Optimum dose of alum is 3.0 g/L; CES adsorbent mass 25.5 g. Hydraulic loading rate HLR = 6.37 L/m^2• min.	Removal efficiencies of Fe, Pb, Cu, Zn, Ni, and Cr of 92%, 93%, 87%, 76%, 65%, and 60%, respectively.	[28]
Ferric chloride: polymer	Industrial waste waters	Ferric chloride: polymer = 200 mg/ L:20 mg/L	Removal efficiencies of Cr, Fe, and Zn of 97%, 92%, and 0%, respectively. Resulted volume sludge of 120 mL/L.	[29]
Ferric chloride	Industrial waste waters	Jar test, C_{iW} = 321 ppm, Fe/W ratio = 4, or 8, pH = 4, 6, 8 and 10.	Tungsten removal efficiency of 99%, at pH < 6. Residual concentration of tungsten was <10 ppm.	[32]
Micro-alloyed aluminum composite (MAlC)	Wastewater containing Zn^{2+} and Cu^{2+}	Spontaneous reduction coagulation process in a laboratory semi-flow system varying several parameters (treatment time, pH, initial metal concentration, and flow rate).	Residual concentration of Zn and Cu ions under admissible level was obtained at 20 minutes of treatment, at neutral pH, and increased initial concentration of metal ions. The presence of copper favor zinc removal efficiency and copper removal is not affected by zinc presence.	[33]
Cationic polymers	Tannery wastewater	Jarr test methodology using cationic polymers with different weights and charge densities.	A removal efficiency of Cr > 96% was obtained using cationic polymers with molecular weight between 4 and 6 and charge density between 40 and 55% and an optimum dose of 20%.	[34]
Cactus Opuntia ficus indica bioflocculant	River water with Pb^{2+}, Zn^{2+}, Cd^{2+}, and Cu^{2+} content.	The coagulation-flocculation process was optimized for pH, initial concentration of Pb(II) ions, particle size, dosage, ionic strength, contact time, and temperature.	Removal efficiency of Pb, Zn, Cd, and Cu of 100%, 85.74%, 84.16%, and 93.02%, respectively, were achieved at pH 5, natural ionic strength, dosage of 8 mg/L, corresponding to zero zeta potential, with particle size<75 μm, at 35°C for a floc settling time of 180 min.	[36]
Commercial tannin-based flocculant	Raw surface water with Cu^{2+}, Zn^{2+} and Ni^{2+} content.	Jar-test procedure	Removal efficiency of Cu, Zn, and Ni of 90%, &5% and 70%, respectively, was achieved using pH value of 6, 7, and 8, respectively, and at a flocculant dose between 100 and 150 ppm.	[38]
Chitosan/ montmorillonite	Synthetic water with Co^{2+}, Ni^{2+}, and Cu^{2+} content	Synergic coagulation-flocculation process.	Chitosan: montmorillonite, weight ratio = 5% gave the highest cation removal yields at pH 6.8, and a concentration of 20–100 ppm of Co^{2+}, Ni^{2+}, and Cu^{2+}.	[39]

Table 2.
The efficiency of various coagulation/flocculation processes applied to different wastewater treatment.

substance destabilization. Vigorous stirring facilitates the contact between the metal ions and the coagulant. In the second step, under gentle stirring, a flocculant is introduced, which has the role of inducing the union of destabilized particles with the aim of forming large flocs, easily separable. The collision between particles and the growth of flocs is facilitated by the gentle stirring of the reaction mass. In the third stage, the resulting sludge and clarified effluents are separated. Separation can be done by settling, flotation, or filtration [2, 3].

The first and most intensive coagulants used in water treatment are those based on Fe^{3+} and Al^{3+}, such as ferric sulfate—$Fe_2(SO_4)_3\bullet 8H_2O$; ferrous sulfate—$FeSO_4\bullet 7H_2O$; ferric chloride—$FeCl_3$; aluminum sulfate—$(Al_2SO_4)_3\bullet 14H_2O$; aluminum chloride—$AlCl_3\bullet 6H_2O$; poly aluminum chloride—$(Al(OH)_{1.5}(SO_4)_{0.125}Cl_{1.25})n$; and sodium aluminate—$NaAlO_2$ [28–33]. Among the two trivalent metal ions, coagulants based on Al^{3+} are preferred because red coloring is avoided, as can happen when using coagulants based on Fe^{3+}. In both cases, a subsequent pH adjustment is needed, and a great dose of coagulant is required for efficient flocculation. The main disadvantage of using inorganic coagulants is a large amount of resulted sludge and its toxic nature, which leads to serious health problems. For this reason, researchers have turned their attention to the use of organic coagulants, especially polymers [34–37]. Even if the polymers are user-friendly, they do not require pH adjustment, and the obtained flocs are large and easily separated, their high price minimizes their use on a large scale. Thus, an alternative to the drawbacks of synthetic polymers is the use of natural polymers, which are found in abundance and are environmental-friendly, due to their biodegradable property [38–40]. But this characteristic, biodegradability, limits their applications, because the formed flocs present instability and loss of strength in time, the natural polymers presenting a shorter lifetime of the active compound. Coagulation–flocculation cannot achieve a 100% heavy metals removal efficiency; therefore, these techniques must be followed, or combined with other treatment processes. **Table 2** summarizes the methodology and the results of various studies of single or combined coagulation/flocculation processes applied on different waste waters with their metal ions removal efficiencies.

Coagulation/flocculation processes could be applied as primary or tertiary treatment of different types of water. The operating parameters (such as coagulant/flocculant dosage, pH, stirring speed, stirring time, settling time, and temperature) present a significant influence on the metal ions removal efficiency. In order to improve the obtained performance, in some cases the coagulation/flocculation process is combined with different treatment methods.

4. Ion-flotation

The process which uses different surfactants for the destabilization of the metal ions present in the wastewaters, and their transformation into hydrophobic compounds, followed by removal through flotation using air bubbles is called ion-flotation [41, 42]. The principle of the ionic flotation process consists on the one hand of attaching the metal ion to the polar end of the surfactant and on the other hand of the interaction between the nonpolar end of the surfactant and the air bubbles, which leads to the concentration and separation of the metal ions on the surface of the residual water by froth forming. In addition to surfactants, foaming agents/frothers (such as ethanol, polypropylene glycol, and methyl isobutyl carbinol), can also be used, which have the role of obtaining air bubbles of the smallest possible size to increase the surface of interaction between them and the target metal ions [42]. Ionic flotation can be used both for the treatment of wastewater and for the recovery of precious or rare metal ions. As surfactants/collectors were studied sodium dodecyl sulfate [42], polyethylenimine [43], hexadecyltrimethyl ammonium bromide [44], 4-thiazolidinone derivatives [45], etc. This treatment method does not require high energy consumption and sophisticated and large equipment, it is easy to operate, and can be used for selective separations presenting high efficiency and obtaining low amounts of sludge. Ion-flotation could be conducted at different work conditions (pH, flow rate, treatment time, and molar ratio of surfactant/metal ions), function of the metal ions which need to be removed, and function of the used surfactant/collector. The results presented in the specialty literature describe this treatment method as a simple, rapid, and economic one, but also underline the necessity of developing new low-cost collectors, with greater selectivity, environmentally friendly, and present a higher efficiency using a lower dosage, all these leading to a decreasing of the treatment costs.

5. Ion exchange

Chemical precipitation and coagulation-flocculation are used to treat water with a high content of metal ions. Their efficiency decreases if the metal ions are found in trace concentrations. In this situation, it is necessary to use advanced water treatment methods. Ion exchange is one such method that can be successfully used to remove metal ions found in low concentrations from water. Ion exchange treatment of waters with metal ions content requires the use of ion exchangers (IX), a water-insoluble compound, that release in water some harmless ions, such as H^+, Na^+, or OH^-, retaining in return the target metal ions, which present a higher affinity for the used IX, according to reaction (9). This process is reversible and takes place until the exhaustion of IX. Usually, the treatment is carried out with two columns filled with IX (**Figure 4**), one is used for operation, while the other is regenerated. After

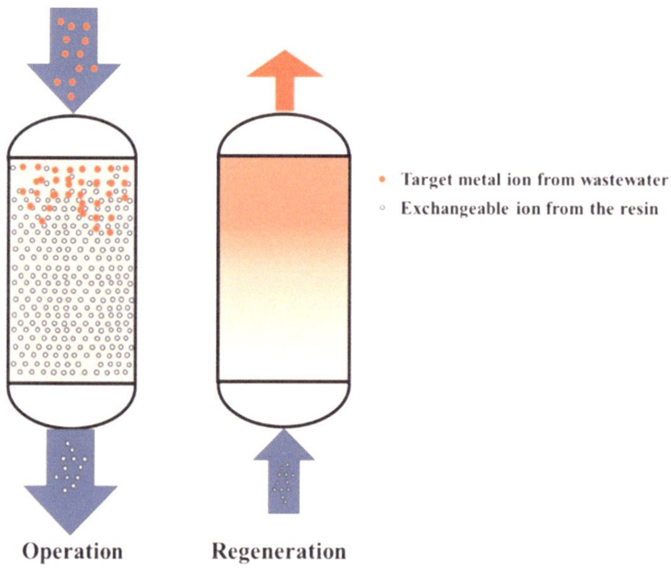

Figure 4.
Scheme of the ion exchange process.

Ion exchanger	Water type	Methodology	Obtained results	Ref.
Dowex M4195, and Lewatit® MonoPlus TP 220	Acidic streams with Cu^{2+} content	Batch tests	The highest removal capacity of Cu (86.44 mg/g) was obtained using Lewatit® MonoPlus TP 220 in chloride form, at pH around 2, resin dose of 10 g/L, and treatment time of 60 min.	[48]
C107E, MTS9301 and TP214 resins	Synthetic acetic acid leachate containing Cu^{2+}, Fe^{2+}, Pb^{2+} and Zn^{2+}	Batch tests	C107E developed the highest sorption capacity for the studied metal ions (q_{mCu} = 5 mmol/g; q_{mFe} = 2.1 mmol/g; q_{mPb} = 9 mmol/g; q_{mZn} = 20 mmol/g)	[50]
Clinoptilolite	Synthetic waters containing Pb(II), Cd(II), and Cr(VI)	Batch experiments and fixed-bed reactors	Pb and Cd were efficiently removed (>95%) from the single and binary system but were significantly affected by the presence of Cr ions or organic contaminants.	[52]
Clinoptilolite	Synthetic waters containing Cu^{2+}, Fe^{3+}, and Cr^{3+}	Batch tests, C_i = 0.01 N, T = 25°C, treatment time = 4 days, pH = 2–2.5.	The presence of SO_4^{2-} and HPO_4^{2-} anions affect the ion exchange of Cu^{2+} and in a lower measure of Fe^{3+} and Cr^{3+}.	[53]
Thomsonite	Synthetic waters containing Pb^{2+}, Fe^{3+}, and Ni^{2+}	Batch tests, C_{iMe} = 100 ppm	Extraction efficiency of ~100%, 99.9%, and 98.6%, for Pb^{2+}, Fe^{3+}, and Ni^{2+}, respectively.	[54]
Zeolite	Synthetic waters containing Cu^{2+}, Cd^{2+}, Cr^{3+}, Ni^{2+}, and Pb^{2+}	Batch tests S:L ratio = 5 g: 45 mL C_{iMe} = 5, 10, 30 mg/L, T = 23°C, treatment time = 5–60 min.	Removal efficiency 99%.	[56]
Strong base anion exchange resin	Synthetic water containing heavy metal (Cu^{2+}, Cd^{2+}, Zn^{2+}) and cyanide ions.	Semi-fluidized and fluidized beds using a transparent acrylic resin tube, d = 20 mm h = 600 mm, m_{resin} = 20 g	The ion exchange efficiency of the studied heavy metalcyanide systems decreases as the concentration ratio of cyanide and heavy metal increases.	[58]
Dowex HCR-S synthetic resin	Synthetic water containing Cd^{2+}, Ni^{2+}, and Zn^{2+}	Batch experiments at different initial solution pH, stirring speeds, temperatures, initial concentrations, and resin dosages	The exchange level of 99.76%, 93.66%, and 83.1% for Cd^{2+}, Ni^{2+} and Zn^{2+}, respectively, at T = 293 K, C_i = 250 mg/L, t = 60 min, pH = 2, resin dose = 2 g.	[60]
Amberlite IRC-50/IRC-86 and Amberlite IRA-67	Synthetic water containing Pb^{2+}, Cu^{2+} and Cd^{2+}	Column experiments, d = 17.2 mm h = 490 mm.	The most efficient resin was IRC-86. The order of adsorption capacity for the metal ions tested was $Pb^{2+} > Cd^{2+} > Cu^{2+}$. A 99% desorption of the studied metal ions was achieved using CO_2.	[61]

Table 3.
Efficiency of various ion exchangers applied on different wastewater treatment.

regeneration, smaller volumes of solutions containing the target metal ions in much higher concentrations are obtained so that they can be recovered, capitalized, or further processed using other methods [46–49].

$$R\text{–}A + Me \quad\quad R\text{–}Me + A \tag{9}$$

where:

R: ion exchanger matrix.

A: the mobile ion of the ion exchanger.

Me: the target metal ion from the residual water of the same sign as A.

Depending on their nature, IX can be: (i) inorganic-natural (zeolites, clinoptilolite) or synthetic (layered double hydroxides) [50–56]; (ii) organic-natural (cellulose, alginic acid, chitin), or synthetic (polycondensation ion exchangers; polymerization ion exchangers) [48, 57–63]. Depending on the type of mobile ion they can exchange, IX are classified as cationic, anionic, or amphoteric. The most important property of IX is the exchange capacity. The exchange capacity represents the number of ions exchanged during the process, equal to the number of functional groups capable of exchange. It is expressed in equivalents per dry product unit or wet product volume unit. In practice, two notions are used: total capacity and useful exchange capacity. The exchange capacity is influenced by constructive factors (the ratio between the column height and diameter), functional factors (percolation speed, specific load, and regeneration level), and chemical factors (the chemical composition of the water to be treated). The disadvantage of ion exchange treatment consists in the fact that the method is sensitive to the pH variations of the influent, and a pre-treatment must be carried out both to remove the particles in suspension and to avoid the precipitation of metal ions on the used resin to avoid the clogging of the exchange column. **Table 3** summarize the experimental conditions and the results of various ion exchanger applied on different wastewater treatments and the obtained metal ions removal efficiencies.

Not all ion exchangers could be used for the removal of metal ions, therefore further studies must be carried on regarding the stability and reusability of various ion exchangers [64, 65]. The challenge in the heavy metals removal from aqueous solution is the development of nontoxic, easily available, and low-cost ion exchangers. For a good design and modeling of the ion exchange process is necessary for each case to study the effect of the main parameters, such as pH, time, flow rate, dosage, initial concentration, bed height, and types of resin.

6. Adsorption

Another advanced water treatment that is efficient use in the removal of trace amounts of metal ions from water and wastewater is adsorption. Adsorption is a separation process consisting of the adhesion of metal ions dissolved in aqueous solutions to the surface of a solid, called adsorbent (**Figure 5**). Separation mechanisms of metal ions from water through adsorption are influenced both by their characteristics and by the characteristics of the adsorbent, being determined by the interactions between these two. These interactions can be of a physical nature, exerted by forces with low energies (e.g., van der Waals-type forces). In this case, the metal ions are adsorbed in the pores of the adsorbent without the involvement of electron transfer. In this case, the process is reversible, the metal ion molecules retained on the

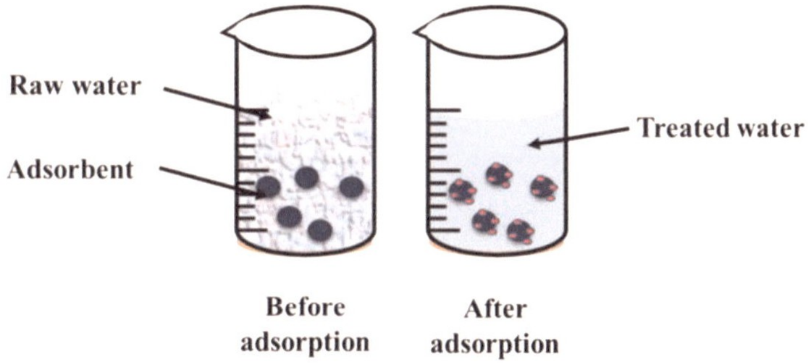

Figure 5.
Scheme of the adsorption process.

adsorbent surface can be removed by desorption, regenerating the adsorbent. If the interaction between metal ions and adsorbents involves the transfer of electrons and the formation of chemical bonds, then the process is called chemical adsorption, or chemosorption. In this case, the metal ions are not attracted to the entire surface of the adsorbent materials, but only to the active zones, which contain functional groups that react with the metal, involving higher energies than those of physical adsorption, a fact that explains the greater selectivity of the chemical adsorption [11, 66, 67].

In adsorption processes, the used adsorbent material represents the essential element for obtaining effective separations. Thus, a multitude of adsorbent materials was studied. The most frequently used is activated carbon both in its powder form and in its granular form, due to its high specific surface area [68–71]. The powder-activated carbon (PAC) could be used together with other treatment processes, for example, in the coagulation-flocculation processes, for obtaining a better separation of the heavy metals. This presents an advantage that it is not necessarily an alternative equipment, those the separation costs are reduced. Unfortunately, in this case, it is not possible to recover the activated carbon from the resulted sludge. Another disadvantage is represented by the fact that is required a great amount of PAC for a specific volume of water to be treated. The use of granular activated carbon (GAC) requires a smaller amount for the treatment of a specific volume with heavy metals content, but in this case, it is necessary to use separate devices for water treatment and for the regeneration of the exhausted GAC. Sometimes the adsorbed metal ions could be instantaneously removed from the surface of GAC appearing in the effluent at higher concentrations than they were initially in the influent [11, 68–71]. With the recent development of nanotechnologies, researchers have turned their attention to the uses of carbon nanotubes as adsorbent materials [72, 73]. Numerous studies have been carried out regarding the use of activated carbon obtained from vegetable waste as an adsorbent material, or various natural adsorbent materials [74–82], or on different wastes [83, 84]. They have the advantage of low costs, but also the disadvantage of developing low adsorption capacities and low selectivity. To obtain high selectivity and increased efficiency for certain specific adsorption processes, various materials were synthesized and studied such as ferrites or oxides [85–87]. Adsorbent materials with magnetic properties have also been developed to improve the separation stage [88–92]. Since the structure of the adsorbent greatly influences the adsorption process, compounds with well-ordered structures, such as layered double hydroxides and metal-organic frameworks, have also been intensively studied [93–97]. To obtain materials with high selectivity and improved absorbent properties, functionalized materials have been developed by impregnating numerous solid supports with compounds containing various functional groups [51, 98–105]. **Table 4** summarizes the experimental conditions and the efficiency of various adsorbent materials used for the removal of various metal ions from different wastewater.

Adsorbent materials	Water type	Methodology	Obtained results	Ref.
Microporous (CS-1501) mesoporous activated carbon cloths	Synthetic wastewater containing Cu^{2+} and Pb^{2+}	Batch mode adsorption studies using mono or bicomponent systems (V = 250 ml, m = 500 mg, C_i = 0.1–1.4 mmol/L, t = 12 h, pH = 3.5/5).	The adsorption capacities of the studied metal ions are between 0.080 and 0.175 mmol/g. In binary system, a decrease of adsorption capacities is observed.	[68]
Activated carbon produced from the biomass recovered from the medicinal plant residue	Synthetic wastewater containing Zn^{2+}, Cd^{2+}, and Pb^{2+} and leachate waters.	Batch adsorption studies C_i = 0.1–50 mg/L; pH = 4, T = 25°C, t = 15 min-5 h.	Optimum condition in synthetic waters for adsorption of Pb^{2+}, Zn^{2+}, and Cd^{2+} were pH = 5, T = 25°C, adsorbent dose = 5 g/L, t = 1.5 h. The rate of heavy metal adsorption from landfill leachate was lower than that from artificial wastewater.	[69]
Carbon nanosheet (CNS) functionalized with amine derivatives	Seawater containing Zn^{2+}, Cd^{2+}, and Hg^{2+}	Quantum mechanics studies.	The studied adsorbent present a negative adsorption energy (range: −75 to −93 kJ/mol). The metal ions adsorption was confirmed by the natural bond orbital and quantum theory of atoms in molecules.	[72]
Multi-walled carbon nanotubes (MWCNTs)	Synthetic waters containing Cu^{2+}, Pb^{2+}, Cd^{2+}, and Zn^{2+}	Kinetic experiments, C_i = 0.5 mg/L, pH = 7, T = 278–333 K, adsorbent dose = 125 mg/100 mL.	At 298 K, the binding of the metal ions by MWCNTs follows the order: Cu(II) > Pb(II) > Zn(II) > Cd(II).	[73]
Activated carbon obtained from seeds of Albizia lebbeck and Melia azedarach trees	Synthetic waters containing Pb^{2+} and Cd^{2+}	Batch adsorption studies C_i = 10–50 mg/L, adsorbent dose = 0.1–0.5 g/50 ml, pH = 2–7), T = 20–70°C	Removal efficiency of 62% form Pb and 66% for Cd using t = 120 min, pH = 5, T = 20°C adsorbent dose = 0.2 g/100 mL, C_{iMe} = 40 mg/L.	[74]
Keratin biomasses (human hair—HH and sheep fur—SF)	Waste waters conyaining Cu^{2+}, Cr^{6+}, Cd^{2+}, and Pb^{2+}	Batch adsorption varying the operation parameters (pH, contact time, and temperature)	Adsorption kinetic rates between 0.054 and 0.261 g/mmol·min have been obtained.	[75]
Treated ajwa date pits (TADP) Produced ajwa date pits (ADP)	Synthetic waters containing Cu^{2+}	Batch mode varying the operation parameters	q_{mTADP} = 1428.57 mg Cu/g q_{mADP} = 1111.1 mg Cu/g at pH = 6.5, T = 328 K	[77]

Adsorbent materials	Water type	Methodology	Obtained results	Ref.
Surface-modified sawdust-based cellulose nanocrystals	Aqueous solutions containing Cu^{2+}, Fe^{2+}, and Pb^{2+}	Combination of adsorption and coagulation techniques	q_m = 111.1 mg Cu/g q_m = 2.82 mg Pb/g q_m = 81.96 mg Fe/g	[84]
Cobalt ferrite nanoparticles	Aqueous solutions containing Pb^{2+}, Zn^{2+}, and dyes.	Fixed bed column adsorption	q_m = 275 mg Pb/g q_m = 390 mg Zn/g	[85]
Nickel ferrite nanoparticles (NFN's)	Aqueous solutions containing Cr^{6+}, Pb^{2+},, and Cd^{2+}	Batch mode varying the operation parameters (pH 3–7, dose: 10, 20, 30, 40, and 50 mg and contact time: 30, 60, 90, and 120 min)	Removal efficiencies of Cr(VI), Pb(II), and Cd (II) were obtained at 89%, 79%, and 87%, respectively, under optimal conditions.	[86]
Sol–gel derived nano metal ferrites, MFe2O4 (M = Ni, Zn, Cu)	Aqueous solutions containing Cr^{6+}	Batch mode adsorption varying the operation parameters.	$q_{mZnFe2O4}$ = 1.46 mg Cr/g $q_{mNiFe2O4}$ = 1.34 mg Cr/g $q_{mCuFe2O4}$ = 1.14 mg Cr/g The efficiency is kept for three adsorption-desorption cycles.	[87]
Ultrafine mesoporous magnetite (Fe_3O_4) nanoparticles	River water containing Pb^{2+}, Cd^{2+}, Cu^{2+} and Ni^{2+}	Batch mode adsorption varying the operation parameters	Pb^{2+}, Cd^{2+}, Cu^{2+}, and Ni^{2+}, have been successfully removed simultaneously from river water, obtaining a removal efficiency>50% for Ni and > 80% for the other metals.	[88]
Sultone-modified magnetic activated carbon	Aqueous solutions containing Pb^{2+}, As^{3+}, and Cd^{2+}	Batch mode adsorption varying the operation parameters	q_m = 147.05 mg Pb/g q_m = 151.51 mg As/g q_m = 119.04 mg Cd/g	[92]
Ni-based phosphonate metal–organic frameworks using various phosphonic acid (phosphonoacetic (CP), vinyl phosphonic acid (VP) and N,N-bis(phosphonomethyl) glycine (Gly)	Aqueous solutions containing Cr^{6+}	Batch mode adsorption varying the operation parameters: pH (range: 1–7), contact times (15–180 min), Cr^{6+} initial concentration (range: 2.5–250 mg/L)	$q_{mNi\text{-}CP}$ = 45.9 mg Cr/g $q_{mNi\text{-}Gly}$ = 76.9 mg Cr/g $q_{mNi\text{-}Ni\text{-}VP}$ = 77.0 mg Cr/g	[93]
MgFe layered double hydroxide from iron-containing acidic residual solution.	Aqueous solutions containing Mo^{6+}	Batch mode adsorption varying the operation parameters	q_{mMg4Fe} = 42.1 mg Mo/g $q_{mMg4Fe\text{-}450}$ = 55.2 mg Mo/g	[94]
Cu- and Co-based metal organic frameworks.	Aqueous solutions containing Cs^+ and Tl^+	Batch mode adsorption varying the operation parameters	$q_{mCo\text{-}SP}$ = 4.25 mg Cs/g $q_{mCo\text{-}CP}$ = 3.71 mg Cs/g $q_{mCu\text{-}CP}$ = 3.64 mg Cs/g $q_{mCo\text{-}SP}$ = 3.27 mg Tl/g $q_{mCo\text{-}CP}$ = 2.96 mg Tl/g $q_{mCu\text{-}CP}$ = 2.76 mg Tl/g	[95]

Adsorbent materials	Water type	Methodology	Obtained results	Ref.
Ionic liquidfunctionalized Zn-Al layered double hydroxides (Zn_3Al-IL)	Aqueous solutions containing Pd ions	Batch mode adsorption varying the operation parameters	$q_{mZn3Al\text{-}IL}$ = 100 mg Pd/g	[97]
Cross-linked polyethylene-graft-polystyrene sulfonic acid (PE-g-PSSA)	Aqueous solutions containing Co^{2+}, Cu^{2+}, Ni^{2+}, and Ag^{+}	Batch mode adsorption varies the operation parameters (contact time, initial metal ion concentration, pH, and temperature)	q_m = 73 mg Ni/g q_m = 59 mg Co/g q_m = 67 mg Cu/g q_m = 61 mg Pb/g q_m = 53 mg Ag/g	[99]
Ionic liquid impregnated Florisil	Aqueous solutions containing Tl^{+}	Batch mode adsorption varying the operation parameters.	The phosphonium-based ionic liquid impregnated with Florisil developed a higher efficiency in the removal process of thallium ions from an aqueous solution then the materials obtained through impregnation of Florisil with imidazolium-based ionic liquids.	[102]
Polyethyleneimine-functionalized alumina nanopowder	Aqueous solutions containing Pd ions.	Batch mode adsorption varying the operation parameters.	q_m = 97.7 mg Pd/g	[104]
Mg_3Al-LDH functionalized with ionic liquid (IL)	Aqueous solutions containing Pd ions	Batch mode adsorption varying the operation parameters.	q_m = 277.8 mg Pd/g	[105]

Table 4.
Research results of various adsorbent materials used for the removal of various metal ions from different wastewater.

Adsorption is a low-cost and easy-to-operate method for the treatment of water with metal ion content. It presents economic feasibility to be scaled up to the industrial application. The existence of various adsorbent materials and low-cost adsorbent materials increases the versatility of this treatment method. Unfortunately, this method is not suitable for automation. Due to the multitude of adsorbent materials with different adsorptive properties (see **Table 4**), intensive studies are continuously carried out, and the selection of one must be based on a preliminary study performed for the specific effluent that is to be treated.

7. Membrane separation processes

Membrane separation processes are also an advanced water treatment that could be used for the removal of trace concentrations of metal ions from water and wastewater. Membrane separation processes (**Figure 6**) consist of passing the water flow through semipermeable membranes with certain properties (pore size or electrical

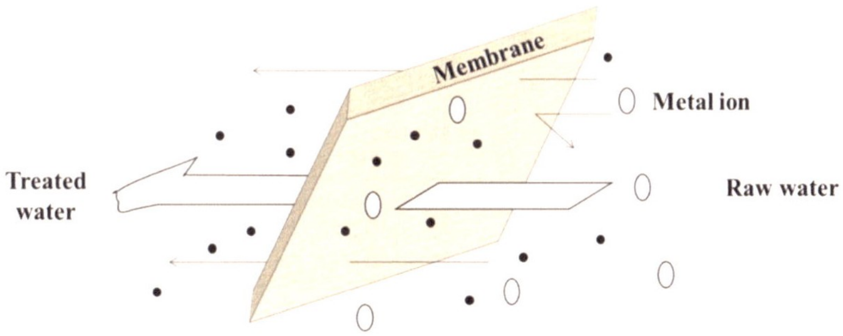

Figure 6.
Scheme of the membrane separation processes.

charge) that allow the passage of water molecules but retain the dissolved metal ions [11, 106–109].

Depending on the forces-actions used for separation and the size of the separated particles, the membranes can be classified into.

- Membranes that work under pressure (MF—microfiltration, UF—ultrafiltration, NF—nanofiltration, RO—reverse osmosis).

- Membranes actuated by electrical voltage (ED—electrodialysis, RED—reverse electrodialysis).

Since membrane treatment technologies require operation at high pressures, the membranes must present compressive strength. The properties that characterize membranes are permeability, porosity, hydrophilicity, surface charge, and thermal/mechanical stability.

Most of the studied membranes are composite materials based on polymer supports [110–112]. The use of these processes presents the following advantages: they constantly ensure good water quality, no chemical reagents are used, they lend themselves to automation, and they are compact. At the same time raise technical issues: insufficient selectivity, relatively weak transmembrane fluxes, exploitation problems, possibility of membrane fouling, requiring pretreatment step and periodic membrane cleaning, and also economic problems: many and different manufacturers with different technologies and prices.

8. Electrochemical separation processes

Electrochemical treatment processes were first used for ores electrorefining in the mineralogical industry. Researchers were reluctant to use them due to the need for energy consumption and initial investment in special equipment, which leads to increased processing costs. After studies that demonstrated their efficiency and the need for low maintenance of the equipment, the use of these methods in the water treatment processes with metal ions content became more and more promising. Electrochemical processes for treating water with metal ions content can be divided into electrocoagulation, electro flocculation, and electrodeposition.

The electrocoagulation process takes place in a simple electrolysis cell, which contains sacrificial electrodes, which could be from the same or different materials (**Figure 7**). In most of cases, the electrodes are formed from Fe^{3+} or Al^{3+}. The

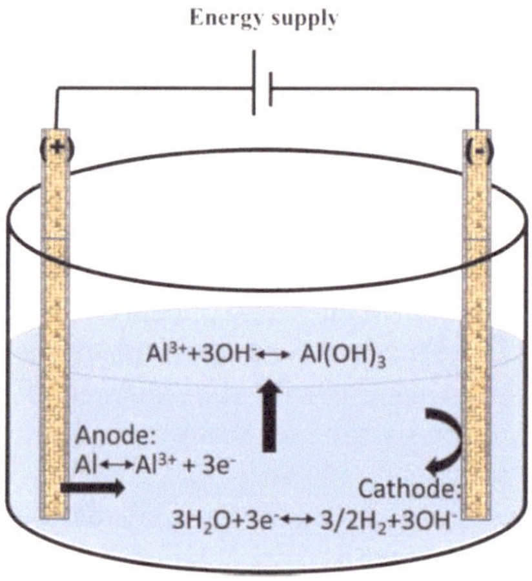

Figure 7.
Scheme of electrocoagulation processes.

electrocoagulation process consists of the dissolution of the anode, H_2 and HO^- generation at the cathode, coagulant ($Al(OH)_3$) formation, destabilization and neutralization of metal ions by introducing electric current, aggregation of the destabilized metal ions and flocks/sludge formation [113–118]. The main advantage of this process consists in the fact that no additional reagents are required, and the coagulant is formed *in situ* due to the electrical dissolution of the sacrificial electrodes. The obtained flock/sludge is stable and easy to be removed. The hydrogen formation contributes to the removal of tiny particles. The main disadvantages are represented by the possibility of cathode passivation and the need of the periodical replacement of the sacrificial anode.

In the case of electroflotation, the destabilized heavy metals adhere to the oxygen and hydrogen molecules released by the reactions from the electrodes and float to the surface of the liquid from where they are removed. The formed sludge presents better stability in this case and the process requires a shorter time. Most of the time, a combination of electrocoagulation and electroflotation is used [119–121].

Electrodeposition is an effective method of selective recovery of dissolved metals in order to recycle/reuse them. It is advantageous because it does not require additional reagents and no sludge is formed. Dissolved metals from wastewater are deposited at the cathode according to the reaction (10). In this case, it is preferable for the anode to be insoluble in order not to contaminate the recovered metals. At the anode, the reaction takes place according to the reaction (11). Side reactions such as the formation of hydrogen gas may occur during the process, reaction (12). This method is selective but is sensitive to the composition of water to be treated, and the efficiency is negatively influenced by the side reaction of hydrogen formation [11, 122, 123].

$$\text{Catod } (-): \quad M^{n+} + ne^- = M \tag{10}$$

$$\text{Anod } (+): \quad 4OH^- = O_2 + 2H_2O + 4e^- \tag{11}$$

$$H^+ + e^- = 1/2H_2 \tag{12}$$

9. Conclusions, remarks, and future perspectives

It can be seen from the literature study that each method of removing metal ions from water has both advantages and disadvantages. The choice of one or another is made following several conditions, such as knowledge in the field, experience in a certain method, the composition of the influent, the desired removal efficiency to be achieved, and the operating conditions. Sometimes two or more methods are combined and used to obtain the desired results. Chemical precipitation is frequently used in the treatment of waters with a high content of heavy metals. It presents low capital costs, simple operation condition, and a high treatment efficiency, it can be easily automated, but in some cases, even if high removal degree is obtained, residual concentrations below the maximum allowed concentrations are still not reached, so it is necessary to be followed by an advanced treatment method. The main disadvantages of chemical precipitation are underlined by the need for chemical reagents addition for pH adjustment and by the fact that great quantities of sludge are obtained, which need further treatment or special disposal. The same disadvantages are encountered in the case of using coagulation-flocculation processes. The efficiency of chemical precipitation or coagulation-flocculation processes also depends on the efficiency of the method used to remove the resulting sludge. Ion exchange processes and adsorption processes are used to treat large volumes of wastewater with a low content of metal ions. In the case of ion exchange processes, the initial investment in ion exchange columns is necessary, the problems raised in this case are given by the costs of ion exchange resins and their selectivity, stability, and reusability. Adsorption processes are the most promising treatment process of heavy metals removal from aqueous solutions due to low costs, easy operation, and the multitude of existing adsorbent materials, especially low-cost adsorbent. Also, in this case, the problem of regeneration and reuse of adsorbent materials arises. Membrane separation processes develop very high removal degrees of metal ions, but in this case, the costs of the technology and the possibility of membrane fouling limit their use at a large scale. Electrochemical processes have the advantage of selective removal of metal ions and the possibility of their recovery and reuse without the need for additional consumption of chemical reagents. At the same time, it presents the disadvantage of high energy consumption. Accordingly, for each case, studies must be carried out starting from the laboratory scale and then updated to the pilot scale to determine the optimal treatment method. Most of the studies were carried out on synthetic water. To clearly specify the efficiency of a certain treatment process or another, for different situations encountered in practice, it is necessary to carry out studies on real waters. Besides the metal of interest, the sample matrix also presents a major influence on the performance of the process treatment. Future studies must be focused on the development of cost-effective materials and methods that involve low treatment costs, high efficiency, and minimal impact on the environment.

Acknowledgements

This work was supported by a grant from the Romanian Ministry of Education and Research, CNCS - UEFISCDI, project number PN-III-P1-1.1-TE-2019-1555, within PNCDI III.

Author details

Lavinia Lupa* and Laura Cocheci
Faculty of Industrial Chemistry and Environmental Engineering, Politehnica University Timișoara, Timisoara, Romania

*Address all correspondence to: lavinia.lupa@upt.ro

References

[1] Agarwal M, Chaudhry K, Kumari P. Heavy Metal Sources Impacts & Removal Technologies. International Journal of Engineering Research & Technology. 2015;**3**:1-7

[2] Mishra J, Saini R, Singh D. Review paper on removal of heavy metal ions from industrial waste water effluent, IOP Conf. Series. Materials Science and Engineering. 2021;**1168**:012027. DOI: 10.1088/1757-899X/1168/1/012027

[3] Qasem NAA, Mohammed RH, Lawal DU. Removal of heavy metal ions from wastewater: a comprehensive and critical review. Clean Water. 2021;**4**:36. DOI: 10.1038/s41545-021-00127-0

[4] Gunatilake SK. Methods of Removing Heavy Metals from Industrial Wastewater. Journal of Multidisciplinary Engineering Science Studies (JMESS). 2015;**1**:12-18

[5] Vidu R, Matei E, Predescu AM, Alhalaili B, Pantilimon C, Tarcea C, et al. Removal of Heavy Metals from Wastewaters: A Challenge from Current Treatment Methods to Nanotechnology Applications. Toxics. 2020;**8**:101. DOI: 10.3390/toxics8040101

[6] Fu F, Wang Q. Removal of heavy metal ions from wastewaters: A review. Journal of Environmental Management. 2011;**92**:407-418. DOI: 10.1016/j.jenvman.2010.11.011

[7] Noor SFM, Ahmad N, Khattak MA, Mukhtar A, Badshah S, Khan RU. Removal of Heavy Metal from Wastewater: A Review of Current Treatment Processes. Journal of Advanced Research in Materials Science. 2019;**58**:1-9

[8] World Health Organization. Guidelines for Drinking Water Quality. 4th ed. Geneva: Switzerland; 2017

[9] Serrano LZ, Lara NO, Vera RR, Cholico-González D. Removal of Fe(III), Cd(II), and Zn(II) as Hydroxides by Precipitation–Flotation System. Sustainability. 2021;13:11913. DOI: 10.3390/ su132111913

[10] Brbooti MM, Abid BA, Al-Shuwaiki NM. Removal of Heavy Metals Using Chemicals Precipitation. Engineering Technology Journal. 2011;**29**:595-612

[11] Howe KJ, Hand DW, Crittenden JC, Trussell RR. Principles of water treatment. Hoboken, New Jersey: John Wiley & Sons Inc.; 2012

[12] Zhang Y, Duan X. Chemical precipitation of heavy metals from wastewater by using the synthetical magnesium hydroxy carbonate. Water Science and Technology. 2020;**81**: 1130-1136. DOI: 10.2166/wst.2020.208

[13] Wu R. Removal of Heavy Metal Ions from Industrial Wastewater Based on Chemical Precipitation Method. Ekoloji. 2019;**28**:2443-2452

[14] Patterson JW, Allen HE, Scala JJ. Carbonate Precipitation for Heavy Metals Pollutants. Journal (Water Pollution Control Federation). 1977;**49**:2397-2410

[15] Sadeghi M, Karimi H, Hadi M. Removal of lead ions from industrial wastewater using precipitation process. Environmental Engineering and Management Journal. 2017;**16**:1563-1568. DOI: 10.30638/eemj.2017.169

[16] Hu H, Zhang Q, Yuan W, Li Z, Zhao Y, Gu W. Efficient Pb removal through the formations of (basic) carbonate precipitates from different sources during wet stirred ball milling with $CaCO_3$. Science of The Total

Environment. 2019;**664**:53-59. DOI: 10.1016/j.scitotenv.2019.01.424

[17] Pohl A. Removal of Heavy Metal Ions from Water and Wastewaters by Sulfur-Containing Precipitation Agents. Water, Air, and Soil Pollution. 2020;**231**:503. DOI: 10.1007/s11270-020-04863-w

[18] Prokkola H, Nurmesniemi ET, Lassi U. Removal of Metals by Sulphide Precipitation Using Na_2S and HS^-Solution. ChemEngineering. 2020;**4**:51. DOI: 10.3390/chemengineering4030051

[19] Zou Q, Wang W, Zhang T, Liu Y. Simultaneous removal of Cr(VI), Cd, and Pb from aqueous solution by iron sulfide nanoparticles: Influencing factors and interactions of metals. Chinese Journal of Chemical Engineering. 2021; **40**:245-255. DOI: 10.1016/j.cjche.2020.10.021

[20] Mohd Yatim SR, Kasmuri SNH, Syahjidan HN, Mokhtar NS, Zainuddin NA. Removing copper, chromium and nickel in industrial effluent using hydroxide precipitation versus sulphide precipitation. Health Scope. 2020;**3**:54-60

[21] Verma B, Balomajumder C. Hexavalent chromium reduction from real electroplating wastewater by chemical precipitation. Bulletin of the Chemical Society of Ethiopia. 2020;**34**: 67-74. DOI: 10.4314/bcse.v34i1.6

[22] Kucher M, Babic D, Kind M. Precipitation of Barium Sulfate: Experimental Investigation about the Influence of Supersaturation and Free Lattice Ion Ratio on Particle Formation. Chemical Engineering and Processing: Process Intensification. 2006;**45**: 900-907. DOI: 10.1016/j.cep.2005.12.006

[23] Ronquim FM, Cotrim MEB, Guilhen SN, Bernardo A, Seckler MM. Improved barium removal and supersaturation depletion in wastewater by precipitation with excess sulfate. Journal of Water Process Engineering. 2018;**23**:265-276. DOI: 10.1016/j.jwpe.2018.04.007

[24] Gusa AV, Tomani A, Zhang Z, Vidic RD. Sulfate precipitation in produced water from Marcellus Shale for the control of naturally occurring radioactive material. Water Research. 2020;**177**:115765. DOI: 10.1016/j.watres.2020.115765

[25] Gu J, Liang J, Chen C, Li K, Zhou W, Jia J, et al. Treatment of real deplating wastewater through an environmental friendly precipitation-electrodeposition-oxidation process: Recovery of silver and copper and reuse of wastewater. Separation and Purification Technology. 2020;**248**:117082. DOI: 10.1016/j.seppur.2020.117082

[26] Lu J, Dreisinger DB, Cooper WC. Cobalt precipitation by reduction with sodium borohydride. Hydrometallurgy. 1997;**45**:305-322. DOI: 10.1016/S0304-386X(96)00086-2

[27] Sithole NT, Ntuli F, Mashifana T. The Removal of Cu (II) from Aqueous Solution using Sodium Borohydride as a Reducing Agent. IOP Conference Series: Earth and Environmental Science. 2018; **120**:012022. DOI: 10.1088/1755-1315/120/1/012022

[28] Jaradat AQ, Telfah DB, Ismail R. Heavy metals removal from landfill leachate by coagulation/flocculation process combined with continuous adsorption using eggshell waste materials. Water Science & Technology. 2021;**84**:3817. DOI: 10.2166/wst.2021.493

[29] Amuda OS, Amoo IA, Ipinmoroti KO, Ajayi OO. Coagulation/

flocculation process in the removal of trace metals present in industrial wastewater. Journal of Applied Sciences and Environmental Management. 2006; **10**:159-162. DOI: 10.4314/jasem.v10i3.17339

[30] Teh CY, Budiman PM, Shak KPY, Wu TY. Recent advancement of coagulation–flocculation and its application in wastewater treatment. Industrial & Engineering Chemistry Research. 2016;**55**:4363-4389. DOI: 10.1021/acs.iecr.5b04703

[31] Zouboulis AI, Tzoupanos ND. Polyaluminium Silicate Chloride - A Systematic Study for the Preparation and Application of an Efficient Coagulant for Water or Wastewater Treatment. Journal of Hazardous Materials. 2009; **162**:1379-1389. DOI: 10.1016/j.jhazmat.2008.06.019

[32] Plattes M, Bertrand A, Schmitt B, Sinner J, Verstraeten F, Welfring J. Removal of tungsten oxyanions from industrial wastewater by precipitation, coagulation and flocculation processes. Journal of Hazardous Materials. 2007; **148**:613-615. DOI: 10.1016/j.jhazmat.2007.03.016

[33] Bojic AL, Bojic D, Andjelkovic T. Removal of Cu^{2+} and Zn^{2+} from model wastewaters by spontaneous reduction-coagulation process in flow conditions. Journal of Hazardous Materials. 2009; **168**:813-819. DOI: 10.1016/j.jhazmat.2009.02.096

[34] Haydar S, Aziz JA. Coagulation-flocculation studies of tannery wastewater using cationic polymers as a replacement of metal salts. Water Science & Technology. 2009;**59**:381-390. DOI: 10.2166/wst.2009.864

[35] Tokuyama H, Hisaeda J, Nii S, Sakohara S. Removal of Heavy Metal Ions and Humic Acid from Aqueous Solutions by Co-Adsorption onto Thermosensitive Polymers. Separation and Purification Technology. 2010;**71**: 83-88. DOI: 10.1016/j.seppur.2009.11.005

[36] Nharingo T, Zivurawa MT, Guyo U. Exploring the Use of Cactus Opuntia Ficus Indica in the Biocoagulation–flocculation of Pb(II) Ions from Wastewaters. International journal of Environmental Science and Technology. 2015;**12**:3791-3802. DOI: 10.1007/s13762-015-0815-0

[37] Zhang C, Zhang M, Chang Q. Preparation of mercaptoacetyl chitosan and its removal performance of copper ion and turbidity. Desalination and Water Treatment. 2013;**53**:1909-1916. DOI: 10.1080/19443994.2013.870743

[38] Heredia BJ, Martín SJ. Removing heavy metals from polluted surface water with a tannin-based flocculant agent. Journal of Hazardous Materials. 2009;**165**:1215-1218. DOI: 10.1016/j.jhazmat.2008.09.104

[39] Assaad E, Azzou A, Nistor D, Ursu AV, Sajin T, Miron DN, et al. Metal removal through synergic coagulation–flocculation using an optimized chitosan–montmorillonite system. Applied Clay Science. 2007;**37**:258-274. DOI: 10.1016/j.clay.2007.02.007

[40] Mohd-Salleh SNA, Mohd-Zin NS, Othman N. A Review of Wastewater Treatment using Natural Material and Its Potential as Aid and Composite Coagulant. Sains Malaysiana. 2019;**48**: 155-164. DOI: 10.17576/jsm-2019-4801-18

[41] Chang L, Cao Y, Fan G, Li C, Peng W. A review of the applications of ion floatation: Wastewater treatment, mineral beneficiation and

hydrometallurgy. RSC Advances. 2019;**9**: 20226-20239. DOI: 10.1039/c9ra02905b

[42] Polat H, Erdogan D. Heavy metal removal from waste waters by ion flotation. Journal of Hazardous Materials. 2007;**148**:267-273. DOI: 10.1016/j.jhazmat.2007.02.013

[43] Shen L, Hankins NP. Removal of metallic ions at the parts per billion level from aqueous solutions using the polymer-surfactant aggregate process. Journal of Water Process Engineering. 2019;**30**:100486. DOI: 10.1016/j.jwpe.2017.09.001

[44] Jafari M, Abdollahzadeh AA, Aghababaei F. Copper Ion Recovery from Mine Water by Ion Flotation. Mine Water and the Environment. 2016;**36**:323-327. DOI: 10.1007/s10230-016-0408-2

[45] Kozlowski CA, Ulewicz M, Walkowiak W, Girek T, Jablonska J. The effect of tautomeric rearrangement on the separation of Zn(II) and Cd(II) in ion flotation process with 4-thiazolidinone derivatives. Minerals Engineering. 2002;**15**:677-682. DOI: 10.1016/S0892-6875(02)00166-8

[46] Budak TB. Removal of Heavy Metals from Wastewater Using Synthetic Ion Exchange Resin. Asian Journal of Chemistry. 2013;**25**:4207-4210. DOI: 10.14233/ajchem.2013.13902

[47] Hubicki Z, Kołodyńska D. Selective Removal of Heavy Metal Ions from Waters and Waste Waters Using Ion Exchange Methods. In: Kilislioğlu A, editor. Ion Exchange Technologies. London, UK, London: IntechOpen; 2012. DOI: 10.5772/51040

[48] Kołodyńska D, Sofińska-Chmiel W, Mendyk E, Hubicki Z. DOWEX M 4195 and LEWATIT®MonoPlus TP 220 in Heavy Metal Ions Removal from Acidic Streams. Separation Science and Technology. 2014;**49**:2003-2015. DOI: 10.1080/01496395.2014.908920

[49] Gaikwad RW, Sapkal VS, Sapkal RS. Ion exchange system design for removal of heavy metals from acid mine drainage wastewater. Acta Montanistica Slovaca Ročník. 2010;**15**:298-304

[50] Bezzina JP, Robshaw T, Dawson R, Ogden MD. Single metal isotherm study of the ion exchange removal of Cu(II), Fe(II), Pb(II) and Zn(II) from synthetic acetic acid leachate. Chemical Engineering Journal. 2020;**394**:124862. DOI: 10.1016/j.cej.2020.124862

[51] Berber-Mendoza MS, Leyva-Ramos R, Alonso-Davila P, Mendoza-Barron J, Diaz-Flores PE. Effect of pH and temperature on the ion-exchange isotherm of Cd(II) and Pb(II) on clinoptilolite. Journal of Chemical Technology & Biotechnology. 2006;**81**: 966-973. DOI: 10.1002/jctb.1483

[52] Mier MV, Callejas RL, Gehr R, Jiménez Cisneros BE, Alvarez PJJ. Heavy metal removal with mexican clinoptilolite: multi-component ionic exchange. Water Research. 2001;**35**:373-378. DOI: 10.1016/S0043-1354(00)00270-0

[53] Inglezakis VJ, Zorpas AA, Loizidou MD, Grigoropoulou HP. Simultaneous removal of metals Cu^{2+}, Fe^{3+} and Cr^{3+} with anions ${SO_4}^{2-}$ and ${HPO_4}^{2-}$ using clinoptilolite. Microporous and Mesoporous Materials. 2003;**61**:167-171. DOI: 10.1016/s1387-1811(03)00364-0

[54] Almalih MDA, Talballa MD. Sudanese Thomsonite as Ion Exchanger to Removal Pb^{2+}, Fe^{3+} and Ni^{2+} from Artificial Wastewater. Journal of Environmental Analytical Chemistry. 2016;**3**:1000181. DOI: 10.4172/2380-2391.1000181

[55] Zamzow MJ, Eichbaum BR, Sandgren KR, Shanks DE. Removal of Heavy Metals and Other Cations from Wastewater Using Zeolites. Separation Science and Technology. 1990;**25**: 1555-1569. DOI: 10.1080/ 01496399008050409

[56] Senila M, Neag E, Cadar O, Kovacs ED, Aschilean I, Kovacs MH. Simultaneous Removal of Heavy Metals (Cu, Cd, Cr, Ni, Zn and Pb) from Aqueous Solutions Using Thermally Treated Romanian Zeolitic Volcanic Tuff. Molecules. 2022; **27**:3938. DOI: 10.3390/ molecules27123938

[57] Al-Enezi G, Hamoda MF, Fawzi N. Ion Exchange Extraction of Heavy Metals from Wastewater Sludges. Journal of Environmental Science and Health, Part A. 2004;**39**:455-464. DOI: 10.1081/ese-120027536

[58] Kim SJ, Lim KH, Joo KH, Lee MJ, Kil SG, Cho SY. Removal of Heavy Metal-Cyanide Complexes by Ion Exchange. Korean Journal of Chemical Engineering. 2002;**19**:1078-1084. DOI: 10.1007/BF02707236

[59] Wang Z, Feng Y, Hao X, Huang W, Feng X. A novel potential-responsive ion exchange film system for heavy metal removal. Journal of Materials Chemistry A. 2014;**2**:10263-10272. DOI: 10.1039/c4ta00782d

[60] Fil BA, Yilmaz AE, Boncukcuoğlu R, Bayar S. Removal of divalent heavy metal ions from aqueous solutions by Dowex HCR-S synthetic resin. Bulgarian Chemical Communications. 2012;**44**: 201-207

[61] Silva DL, Brunner G. Desorption of heavy metals from ion exchange resin with water and carbon dioxide. Brazilian Journal of Chemical Engineering. 2006; **23**:213-218. DOI: 10.1590/ S0104-66322006000200008

[62] Verbych S, Hilal N, Sorokin G, Leaper M. Ion Exchange Extraction of Heavy Metal Ions from Wastewater. Separation Science and Technology. 2005;**39**:2031-2040. DOI: 10.1081/ss-120039317

[63] Zewail TM, Yousef NS. Kinetic study of heavy metal ions removal by ion exchange in batch conical air spouted bed. Alexandria Engineering Journal. 2015;**54**:83-90. DOI: 10.1016/j. aej.2014.11.008

[64] Bashir A, Malik LA, Ahad S, Manzoor T, Bhat MA, Dar GN, et al. Removal of heavy metal ions from aqueous system by ion-exchange and biosorption methods. Environmental Chemistry Letters. 2019;**17**:729-754. DOI: 10.1007/s10311-018-00828-y

[65] Dabrowski A, Hubicki Z, Podkoscielny P, Robens E. Selective removal of the heavy metal ions from waters and industrial wastewaters by ion-exchange method. Chemosphere. 2004;**56**:91-106. DOI: 10.1016/j. chemosphere.2004.03.006

[66] Mahdu Agarwal R, Singh K. Heavy metal removal from wastewater using various adsorbents: a review. Journal of Water Reuse and Desalination. 2016;**7**: 387-419. DOI: 10.2166/wrd.2016.104

[67] Mishra J, Saini R, Singh D. Review Paper on Removal of Heavy Metal Ions from Waste Water Using Adsorption Process. International Research Journal of Engineering and Technology (IRJET). 2021;**8**:1553-1556

[68] Faur-Brasquet C, Kadirvelu K, Le Cloirec P. Removal of metal ions from aqueous solution by adsorption onto activated carbon cloths: adsorption

competition with organic matter. Carbon. 2002;**40**:2387-2392. DOI: 10.1016/S0008-6223(02)00117-3

[69] Yan CZ, Kim MG, Hwang HU, Nzioka AM, Sim YJ, Kim YJ. Adsorption of Heavy Metals Using Activated Carbon Synthesized from the Residues of Medicinal Herbs. Theoretical Founda tions of Chemical Engineering. 2020;**54**: 973-982. DOI: 10.1134/ S0040579520050474

[70] Deliyanni EA, Kyzas GZ, Triantafyllidis KS, Matis KA. Activated carbons for the removal of heavy metal ions: A systematic review of recent literature focused on lead and arsenic ions. Open Chemistry. 2015;**13**:699-708. DOI: 10.1515/chem-2015-0087

[71] Zaimee MZA, Sarjadi MS, Rahman ML. Heavy Metals Removal from Water by Efficient Adsorbents. Water. 2021;**13**:2659. DOI: 10.3390/ w13192659

[72] Javarani L, Malakootian M, Hassani AH, Javid AH. Removal of Heavy Metal Ions from Water Using Functionalized Carbon Nanosheet: A Density-Functional Theory Study. Jundishapur Journal of Health Sciences. 2022;**14**:e122840. DOI: 10.5812/jjhs.122840

[73] Abdel SM. Removal of heavy metal ions from aqueous solutions with multi-walled carbon nanotubes: Kinetic and thermodynamic studies. International Journal of Environmental Science and Technology. 2012;**10**:677-688. DOI: 10.1007/s13762-012-0127-6

[74] Ullah M, Nazir R, Khan M, Khan W, Shah M, Afridi SG, et al. The effective removal of heavy metals from water by activated carbon adsorbents of Albizia lebbeck and Melia azedarach seed shells. Soil and Water Research. 2019;**15**:30-37. DOI: 10.17221/212/2018-SWR

[75] Olawale SA, Bonilla-Petriciolet A, Mendoza-Castillo DI, Okafor CC, Sellaoui L, Badawi M. Thermodynamics and Mechanism of the Adsorption of Hcavy Metal Ions on Keratin Biomasses for Wastewater Detoxification. Adsorption Science & Technology; 2022. p. 7384924. DOI: 10.1155/2022/7384924

[76] Chakraborty R, Asthana A, Singh AK, Jain B, Susan ABH. Adsorption of heavy metal ions by various low-cost adsorbents: a review. International Journal of Environmental Analytical Chemistry. 2020;**102**:342-379. DOI: 10.1080/03067319.2020.1722811

[77] Azam M, Wabaidur SM, Khan MR, Al-Resayes SI, Islam MS. Heavy Metal Ions Removal from Aqueous Solutions by Treated Ajwa Date Pits: Kinetic, Isotherm, and Thermodynamic Approach. Polymers. 2022;**14**:914. DOI: 10.3390/polym14050914

[78] Manzoor F, Karbassi A, Golzary A. Heavy Metal Contaminants Removal from wastewater by using Chlorella vulgaris: A Review. Current Environmental Engineering. 2019;**6**: 174-187. DOI: 10.2174/ 2212717806666190716160536

[79] Oboh OI, Aluyor EO. The removal of heavy metal ions from aqueous solutions using sour sop seeds as biosorbent. African Journal of Biotechnology. 2008; **7**:4508-4511

[80] Ugwu EI, Tursunov O, Kodirov D, Shaker LM, Al-Amiery AA, Yangibaeva I, et al. Adsorption mechanisms for heavy metal removal using low cost adsorbents: A review. IOP Conference Series: Earth and Environmental Science. 2020;**614**: 012166. DOI: 10.1088/1755-1315/614/1/ 012166

[81] Malik DS, Jain CK, Yadav AK. Removal of heavy metals from emerging

cellulosic low-cost adsorbents: a review. Applied Water Science. 2016;7: 2113-2136. DOI: 10.1007/s13201-016-0401-8

[82] Ghaedi M, Mosallanejad N. Removal of Heavy Metal Ions from Polluted Waters by Using of Low Cos Adsorbents: Review. Journal of Chemical Health Risks. 2013;**3**:07-22. DOI: 10.22034/jchr.2018.544016

[83] Wołowiec M, Komorowska-Kaufman M, Pruss A, Rzepa G, Bajda T. Removal of Heavy Metals and Metalloids from Water Using Drinking Water Treatment Residuals as Adsorbents: A Review. Minerals. 2019;**9**:487. DOI: 10.3390/min9080487

[84] Oyewo OA, Ramaila S, Mavuru L, Leswifi T, Onyango MS. Adsorptive and Coagulative Removal of Trace Metals from Water Using Surface Modified Sawdust-Based Cellulose Nanocrystals. J. 2021;**4**:193-205. DOI: 10.3390/j4020016

[85] Jayalakshmi R, Jeyanthi J, Aswin Sidhaarth KR. Versatile application of cobalt ferrite nanoparticles for the removal of heavy metals and dyes from aqueous solution. Environmental Nanotechnology, Monitoring & Management. 2022;**17**:100659. DOI: 10.1016/j.enmm.2022.100659

[86] Khoso WA, Haleem N, Baig MA, Jamal Y. Synthesis, characterization and heavy metal removal efficiency of nickel ferrite nanoparticles (NFN's). Scientific Reports. 2021;**11**:3790. DOI: 10.1038/s41598-021-83363-1

[87] Nizam T, Thomas M, George M, Joseph A. Adsorption efficiency of sol–gel derived nano metal ferrites, MFe2O4 (M = Ni, Zn, Cu) on the removal of Cr (VI) ions from aqueous solution. Journal of Sol-Gel Science and Technology. 2022; **101**:618-629. DOI: 10.1007/s10971-022-05736-w

[88] Fato FP, Li DW, Zhao LJ, Qiu K, Long YT. Simultaneous Removal of Multiple Heavy Metal Ions from River Water Using Ultrafine Mesoporous Magnetite Nanoparticles. ACS Omega. 2019;**4**:7543-7549. DOI: 10.1021/acsomega.9b00731

[89] González Vázquez OF, Moreno Virgen M d R, Hernández Montoya V, Tovar Gómez R, Alcántara Flores JL, Pérez Cruz MA, et al. Adsorption of Heavy Metals in the Presence of a Magnetic Field on Adsorbents with Different Magnetic Properties. Industrial & Engineering Chemistry Research. 2016;**55**:9323-9331. DOI: 10.1021/acs.iecr.6b01990

[90] Tamjidi S, Esmaeili H, Kamyab MB. Application of magnetic adsorbents for removal of heavy metals from wastewater: A review study. Materials Research Express. 2019;**6**:102004. DOI: 10.1088/2053-1591/ab3ffb

[91] Rajczykowski K, Loska K. Stimulation of Heavy Metal Adsorption Process by Using a Strong Magnetic Field. Water Air & Soil Pollution. 2018; **229**:20. DOI: 10.1007/s11270-017-3672-2

[92] Nejadshafiee V, Islami MR. Adsorption capacity of heavy metal ions using sultone-modified magnetic activated carbon as a bio-adsorbent. Materials Science and Engineering: C. 2019;**101**: 42-52. DOI: 10.1016/j.msec.2019.03.081

[93] Maranescu B, Lupa L, Visa A. Heavy metal removal from waste waters by phosphonate metal organic frameworks. Pure and Applied Chemistry. 2018;**90**: 35-47. DOI: 10.1515/pac-2017-0307

[94] Golban A, Lupa L, Cocheci L, Pode R. Synthesis of MgFe layered

double hydroxide from iron-containing acidic residual solution and its adsorption performance. Crystals. 2019; **9**:514. DOI: 10.3390/cryst9100514

[95] Maranescu B, Lupa L, Visa A. Synthesis, characterization and rare earth elements adsorption properties of phosphonate metal organic frameworks. Applied Surface Science. 2019;**481**:83-91. DOI: 10.1016/j.apsusc.2019.03.067

[96] Lupa L, Maranescu B, Visa A. Equilibrium and kinetic studies of chromium ions adsorption on Co (II)-based phosphonate metal organic frameworks. Separation Science and Technology. 2018;**53**:1017-1026. DOI: 10.1080/01496395.2017.1340953

[97] Cocheci L, Lupa L, Ţolea NS, Muntean C, Negrea P. Sequential use of ionic liquid functionalized Zn-Al layered double hydroxide as adsorbent and photocatalyst. Separation and Purification Technology. 2020;**250**: 117104. DOI: 10.1016/j.seppur.2020.117104

[98] Prabhu PP, Prabhu B. A Review on Removal of Heavy Metal Ions from Waste Water using Natural/Modified Bentonite. In: MATEC Web of Conferences, International Conference on Research in Mechanical Engineering Sciences (RiMES 2017). Vol. 144. 2018. p. 02021. DOI: 10.1051/matecconf/201814402021

[99] Nasef MM, Saidi H, Ujang Z, Dahlan KZM. Removal of metal ions from aqueous solutions using crosslinked polyethylene-graftpolystyrene sulfonic acid adsorbent prepared by radiation grafting. Journal of the Chilean Chemical Society. 2010;**55**:421-427. DOI: 10.4067/S0717-97072010000400002

[100] Mahmud HNME, Huq AKO, Rb Y. The removal of heavy metal ions from wastewater/aqueous solution using polypyrrole-based adsorbents: a review. RSC Advances. 2016;**6**:14778-14791. DOI: 10.1039/C5RA24358K

[101] Trikkaliotis DG, Ainali NM, Tolkou AK, Mitropoulos AC, Lambropoulou DA, Bikiaris DN, et al. Removal of Heavy Metal Ions from Wastewaters by Using Chitosan/Poly (Vinyl Alcohol) Adsorbents: A Review. Macromolecules. 2022;**2**:403-425. DOI: 10.3390/macromol2030026

[102] Lupa L, Negrea A, Ciopec M, Negrea P, Raluca V. Ionic liquids impregnated onto inorganic support used for thallium adsorption from aqueous solutions. Separation and Purification Technology. 2015;**55**:75-82. DOI: 10.1016/j.seppur.2015.06.043

[103] Lupa L, Popa A, Drăgan ES, Ciopec M, Negrea A, Negrea P. Adsorption performance of the organic solid support impregnated with ionic liquid in the removal process of Tl(I) from aqueous solutions. Process Safety and Environmental Protection. 2017;**108**:67-73. DOI: 10.1016/j.psep.2016.08.015

[104] Nagarjuna R, Sharma S, Rajesh N, Ganesan R. Effective Adsorption of Precious Metal Palladium over Polyethyleneimine-Functionalized Alumina Nanopowder and Its Reusability as a Catalyst for Energy and Environmental Applications. ACS Omega. 2017;**2**:4494-4504. DOI: 10.1021/acsomega.7b00431

[105] Cocheci L, Lupa L, Tolea NS, Lazău R, Pode R. IL-Functionalized Mg3Al-LDH as New Efficient Adsorbent for Pd Recovery from Aqueous Solutions. International Journal of Molecular Sciences. 2022;**23**:9107. DOI: 10.3390/ijms23169107

[106] Khulbe KC, Matsuura T. Removal of heavy metals and pollutants by membrane adsorption techniques. Applied Water Science. 2018;**8**:19. DOI: 10.1007/s13201-018-0661-6

[107] Xiang H, Min X, Tang CJ, Sillanpää M, Zhao F. Recent advances in membrane filtration for heavy metal removal from wastewater: A mini review. Journal of Water Process Engineering. 2022;**49**:103023. DOI: 10.1016/j.jwpe.2022.103023

[108] Blöcher C, Dorda J, Mavrov V, Chmiel H, Lazaridis NK, Matis KA. Hybrid flotation-membrane filtration process for the removal of heavy metal ions from wastewater. Water Research. 2003;**37**:4018-4026. DOI: 10.1016/S0043-1354(03)00314-2

[109] Al-Rashdi B, Somerfield C, Hilal N. Heavy Metals Removal Using Adsorption and Nanofiltration Techniques. Separation & Purification Reviews. 2011;**40**:209-259. DOI: 10.1080/15422119.2011.558165

[110] Vo TS, Hossain MM, Jeong HM, Kim K. Heavy metal removal applications using adsorptive membranes. Nano Convergence. 2020;**7**:36. DOI: 10.1186/s40580-020-00245-4

[111] Radzyminska-Lenarcik E, Ulewicz M, Pyszka I. Application of Polymer Inclusion Membranes Doped with Alkylimidazole to Separation of Silver and Zinc Ions from Model Solutions and after Battery Leaching. Materials. 2020;**13**:3103. DOI: 10.3390/ma13143103

[112] Zahid M, Rashid A, Akram S, Rehan ZA, Razzaq W. A Comprehensive Review on Polymeric Nano-Composite Membranes for Water Treatment. Journal of Membrane Science & Technology. 2018;**08**:179. DOI: 10.4172/2155-9589.1000179

[113] Jagwani M, Dohare ED. Electro Coagulation Applications in Water and Wastewater Treatment: A Review. International Journal of Scientific & Engineering Research. 2018;**9**:1430-1436

[114] Shim HY, Lee KS, Lee DS, Jeon DS, Park MS, Shin JS, et al. Application of Electrocoagulation and Electrolysis on the Precipitation of Heavy Metals and Particulate Solids in Washwater from the Soil Washing. Journal of Agricultural Chemistry and Environment. 2014;**3**:130-138. DOI: 10.4236/jacen.2014.34015

[115] da Mota IO, da Mota PO, de Oliveira Filho JG, da Silva LM. Removing heavy metals by electrocoagulation using stainless steel mesh electrodes: a study of wastewater from soil treated with metallurgical residues. Cadernos UniFOA. 2016;**11**:47-57. DOI: 10.47385/cadunifoa11.31

[116] Bazrafshan E, Mohammadi L, Ansari-Moghaddam A, Mahvi AH. Heavy metals removal from aqueous environments by electrocoagulation process– a systematic review. Journal of Environmental Health Science and Engineering. 2015;**13**:74. DOI: 10.1186/s40201-015-0233-8

[117] Khosa MK, Jamal MA, Hussain A, Muneer M, Zia KM, Hafeez S. Efficiency of Aluminum and Iron Electrodes for the Removal of Heavy Metals [Ni (II), Pb (II), Cd (II)] by Electrocoagulation Method. Journal of the Korean Chemical Society. 2013;**57**:316-321. DOI: 10.5012/JKCS.2013.57.3.316

[118] Mao M, Yan T, Shen J, Zhang J, Zhang D. Capacitive Removal of Heavy Metal Ions from Wastewater via an Electro-Adsorption and Electro-Reaction Coupling Process. Environmental Science & Technology. 2021;**55**:3333-3334. DOI: 10.1021/acs.est.0c07849

[119] Belkacem M, Khodir M, Abdelkrim S. Treatment characteristics of textile wastewater and removal of heavy metals using the electroflotation technique. Desalination. 2008;**228**: 245-254. DOI: 10.1016/j.desal.2007.10.013

[120] Casqueira RG, Torem ML, Kohler HM. The removal of zinc from liquid streams by electroflotation. Minerals Engineering. 2006;**19**: 1388-1392. DOI: 10.1016/j.mineng.2006.02.001

[121] Khelifa A, Moulay S, Naceur AW. Treatment of metal finishing effluents by the electroflotation technique. Desalination. 2005;**181**:27-33. DOI: 10.1016/j.desal.2005.01.011

[122] Oztekin Y, Yazicigil Z. Recovery of metals from complexed solutions by electrodeposition. Desalination. 2006; **190**:79-88. DOI: 10.1016/j.desal.2005.07.017

[123] Yang CC. Recovery of heavy metals from spent Ni-Cd batteries by a potentiostatic electrodeposition technique. Journal of Power Sources. 2003;**115**:352-359. DOI: 10.1016/S0378-7753(03)00015-6

Chapter 22

Removal of Heavy Metals from Wastewater with Special Reference to Groundnut Shells: Recent Advances

Asima Shafi, Faizan Ahmad and Sadaf Zaidi

Abstract

Wastewater contains organic pollutants and heavy metals which presents a significant threat to aquatic life and impacts human health and animals. In the past few years, the incomplete remediation of wastewater has made living beings suffer from various problems, and many health diseases are being noticed at a peak rate. Different methods have been employed to remove heavy metals from wastewater to date. However, the adsorption technique is the most efficient and eco-friendly for removing heavy metals and pollutants in wastewater remediation. Many agricultural wastes have been used as adsorbents for removing toxic pollutants and heavy metals from wastewater. Groundnut shell is widely considered agro-industrial waste. Groundnut shells account for nearly 20% of the dried peanut pod by weight, and millions of tons of its quantity are wasted every year. An increase in groundnut production leads to accumulating these groundnut shells in colossal quantities, which is not utilized; thus, they are either burnt or buried. Groundnut shells undergo slow degradation in the natural environment because they are rich in lignin content. Therefore, these shells can be converted into a valuable bio-product to produce less waste. Groundnut shells and groundnut shell-derived biochar act as good biosorbents in the wastewater treatment.

Keywords: heavy metal, groundnut shell, wastewater treatment, adsorption, biochar

1. Introduction

To begin with, freshwater is a prerequisite for human beings as well as animals. In addition, availability of clean water for maintaining a healthy life is a critical factor. However, different water sources have become polluted with increased global water demand [1]. Furthermore, researchers have found that the impacts of climate change, such as rise in temperature and water cycle changes, also aggravate the water issues, which potentially result in floods, severe droughts, and contamination of water bodies with various pollutants and toxic heavy metals [2–4]. The contaminated water sources can be harmful and perilous to humans as well as wildlife, also the aquatic life is negatively affected [4]. However, in developing countries, human health is mainly affected by the direct consumption of contaminated water because they are destitute in the efficient technologies for remediation of water sources, resulting in increased water pollution.

The World Health Organization (WHO) estimates that nearly 844 million population lack primary drinking water sources out of which approximately 230 million people spend more than 30 min/d for collecting water from an improved water source, which includes piped water, protected wells and springs, boreholes, rainwater and stored water [5, 6]. The lacking of efficient water sources and its accessibility in developing countries increases waterborne ailments. According to WHO, approximately 1.6 million people succumb to curable waterborne diseases, of which 90% of deaths are of children under 5 years of age [7].

Landfilling and mining sites have become the main reason for the introduction of heavy metals into freshwater bodies, eventually leading to water pollution. Due to their toxic nature, these heavy metals can bioaccumulate in living cells [8]. There are many conventional methods and technologies for removing heavy metals from an aqueous environment to protect human beings, wildlife, and aquatic life. However, their utilization for remediation is confronted with high costs, waste disposal issues, and effectiveness in removal. However, adsorption is a unique and promising technique and has been found as one of the most suitable methods for the removal and recovery of heavy metals from water economically and efficiently [9].

Groundnut shell is a cheaper agricultural waste and occupies an important position in agricultural waste and biomass because it is produced in larger quantities worldwide. Groundnut shell and groundnut shell-derived biochar play a crucial role in eliminating heavy metals from water sources [10, 11]. This chapter highlights the results of several scientific studies illustrating the efficacy of groundnut shells for the removal of heavy metals from wastewater. Furthermore, the chapter also discusses the characteristics of groundnut shell-derived biochar.

2. Chemical characteristics of heavy metals

Heavy metals contaminate water sources through various industrial activities, agricultural practices, and improper waste disposal [12]. Heavy metals are non-biodegradable in nature and accumulate in humans and animals when they consume contaminated food and water. Many researchers have investigated the contamination of water sources with heavy metals [13]. **Table 1** provides the characteristics of some common heavy metals found in a water sources.

Among the heavy metals, arsenic, cadmium, and lead were extensively studied for their toxicity in water sources. Studies have estimated various effects of heavy metals in drinking water [14, 15]. Arsenic and cadmium have been considered cancer-causing agents in humans [16]. Arsenic also leads to skin damage. It has also been investigated that 1 L of drinking water contaminated with 50 μg of arsenic can cause liver, kidney, bladder, and lung cancer. Lead has been examined to affect the central nervous system, cardiovascular system, gastrointestinal system, reproductive system, neurological system, renal system, endocrine system, and immune system [14]. Exposure to the low concentration levels of lead can also reduce neurodevelopment. The presence of lead in bloodstream alters the functioning of neurons and interferes

Heavy metal	Human health effects	Common sources
Arsenic (As)	Skin damage and circulatory system problems	Electronics production
Cadmium (Cd)	Kidney damage, causing cancer	Chemical industries
Chromium (Cr)	Skin allergies, nausea, vomiting, diarrhea	Steel manufacturing industries
Copper (Cu)	Gastrointestinal damage, liver damage	House plumbing systems
Lead (Pb)	Kidney damage, neurodevelopment problems	Lead-based products
Mercury (Hg)	Nervous system damage	Fossil fuel combustion

Table 1.
Characteristics of common heavy metals [13].

Heavy metal	Molecular weight (g mol-1)	Oxidation state	Van der Waals radius (10–12 m)	Electronegativity (Pauling Scale)
Arsenic (As)	74.9	3, +3, +5	119	2.18
Cadmium (Cd)	112.4	+2	158	1.69
Chromium (Cr)	52	0, +2, +3, +6	200	1.66
Copper (Cu)	63.5	+1, +2	140	1.9
Lead (Pb)	207.2	+2, +4	202	2.33
Mercury (Hg)	200.6	+1, +2	155	2

Table 2.
Chemical properties of heavy metals [13].

with the trimming and pruning of synapses during brain development, which may result in permanently altered functions [14].

Cadmium is the most often reported heavy metal in the water sources [14, 15, 17]. Cadmium-contaminated drinking water leads to chronic renal failure, anemia, cardiovascular diseases, osteoporosis, and also hypertension [14]. There are also other heavy metals that can also put a living being in jeopardy. Researchers have estimated that 1 L of drinking water contaminated with nearly 8.29–51 μg chromium can cause liver, lung, and kidney cancer and also affect genitourinary organs among women [12]. Studies have shown that the stagnation of water in hot water tanks and polypropylene water pipes can significantly increase the concentration of various heavy metals. **Table 2** provides the chemical properties of common heavy metals.

3. Effect of the parameters of water on heavy metal removal

There are various water quality parameters that play an important role in the removal of heavy metals. The most important among them are; pH, temperature, natural organic matter (NOM), and ionic strength.

3.1 Effect of pH

The pH of the water source significantly affects the presence of heavy metals and their properties. The formation of heavy metals in water sources is primarily affected by pH. Heavy metals are cationic in nature at neutral to low pH and possess more solubility and mobility in an aqueous solution. pH also affects the surface charge, ionization state, and the concentration of ions on the functional groups of the adsorbent [18]. Several studies have reported the effect of pH on the formation of heavy metals and their removal. The stability and mobility of copper in an aqueous solution have been reported to increase with a decrease in pH value [19].

With the increase in pH, heavy metals form complexes with hydroxyl ions which in turn affect the oxidation state of the heavy metals. Heavy metals tend to form solids when the pH of the aqueous solution exceeds its neutral value and gets precipitated from the solution. The oxidation state for chromium (Cr) in its stable form is Cr(III), and it changes to Cr(IV) with the increase in pH in this oxidation state, chromium has been found to be more toxic [20]. Lower pH (<4) increases the concentration level of H^+ ions in the aqueous solution, which interferes with soluble metal ions and adsorbent surface interaction, thus reducing overall heavy metal removal [21, 22]. However, with the increasing pH (between 5 and 7), adsorption increases, and the adsorbent surface becomes more negatively charged and interacts readily with the positively-charged heavy metals [23].

The probability of removing maximum heavy metal ions by adsorption is minimal at lower pH values (<3) [24]. Furthermore, with the increase in pH, the concentration of H^+ ions is decreased, and a greater number of adsorption sites become available for heavy metal ions for adsorption, which thereby increases the heavy metal removal from water sources [18]. In the case of chromium removal, it becomes anionic in nature as the pH increases (e.g., $HCrO_4^-$, CrO_4^{2-}). Hence, the adsorption of chromium has been shown to decrease with the increasing pH of the solution. This is mainly due to the electrostatic repulsion resulting from negative surface charges on the adsorbent, inhibiting chromium adsorption [25, 26]. Overall, pH is a significant parameter that affects the behavior and removal of heavy metals from water sources.

3.2 Effect of temperature

Temperature is another important parameter affecting heavy metals' behavior and subsequent removal from water sources. Mechanisms employed for the removal of heavy metals are enhanced at higher temperatures in which surface complexation reactions and various forms of ion exchange are also included [27]. It has been reported that an increase in the removal of Cr(VI) using hull wastes is achieved with an increase in the temperature from 5 to 40°C, which is attributed to the formation of additional adsorption sites on the surface of the adsorbent [28].

The adsorption process has been observed to increase with the increase in temperature due to the increased driving force of diffusion across the boundary layer and an increased rate of diffusion within the adsorbent [29]. However, in various cases, increased temperatures have also been observed to result in a decrease in the heavy metal removal from the water sources. For example, in one experimental study, the removal of total chromium by red algae has been reported to reduce from 90 to 78% with the increase in temperature, which has been possibly observed due to the tendency for ions to remain in the aqueous phase [30]. Furthermore, researchers have also reported a reduction in the heavy metal removal, such as Pb(II) and Ni(II) with increasing temperature, which was attributed to decreased surface activity [18, 31].

An experimental study had reported an increase in the adsorption of Cr(III) and Cu(II) onto peanut shells as the temperature increased to 50°C, and a decrease in the adsorption when the temperature increased to 60°C, which has been observed due to potential damage to the adsorption sites on the peanut shells [32]. Another study evaluated a 32% decrease in the adsorption capacity of the olive cake with the increased temperature from 28 to 45°C in the removal of Cd(II) from an aqueous solution [33]. Therefore, it concluded with a fact that when estimating the effects of temperature on the removal of heavy metals from water sources, each adsorbent, and the corresponding metal ion must be explicitly evaluated in order to determine the overall impact of temperature changes on the adsorption process [13].

3.3 Effect of ionic strength

The ionic strength of the water source has also been reported to significantly affect the removal of heavy metals. It has been observed that the presence of chloride in water sources can form neutral or negatively-charged chloride complexes that have a low affinity for adsorption. These complexes are soluble and difficult to remove from water. With the increase in the formation of chloride complexes, ionic strength tends to increase due to a decrease in the removal efficiency of Cu(II) and Ni(II) [34]. Researchers have also observed that increased salinity and increased concentration of dissolved metals, such as copper, cadmium, and zinc, possess a strong correlation [35].

Interactions between heavy metals and other surfaces are strongly affected by electrostatic forces, and increased ionic strength in a solution has a significant effect on the behavior and removal of heavy metals [13]. Zhang (2011) investigated the impact of ionic strength on heavy metal removal, including Cu(II), Pb(II), and Zn(II), by using dairy manure compost and reported a decrease in the removal of heavy metals with the increase in ionic strength [36]. However, various researchers have estimated that heavy metal removal increases with the increase in ionic strength. For example, Yang et al. (2016) reported a 25% increase in the removal of As(III) and Ni(II) with the increase in ionic strength from 0.01 to 1 M Cl^- ions of the solution due to the formation of an inner-sphere surface complex [37].

3.4 Effect of natural organic matter

Natural organic matter consists of humic and fulvic acids that are formed by the decomposition of plant and animal matter [38]. Natural organic matter is a complex mixture of organic compounds and is highly reactive with heavy metals, which alter the reactivity of the heavy metals and affect their mobility, bioavailability, and toxicity [38]. The particular impact of natural organic matter on heavy metals can be difficult to discover, primarily due to the wide mixture of additional factors, such as pH, humification of a particular natural organic matter, and oxidation state of heavy metals that contribute to the manner in which natural organic matter affects heavy metal removal [19]. Arsenic has been found to form complexes with both humic and fulvic acids, which leads to an increase in the immobilization of arsenic [39].

Metals, such as copper and zinc, have been observed to form complexes with natural organic matter [35]. An experimental study has reported that the removal of Cd(II), Pb(II), and Zn(II) by mollusk shells is increased by the presence of organic matter [40]. Moreover, the presence of natural organic matter can reduce the toxic form of chromium, Cr(VI), to its less harmful and more stable form, Cr(III) [19]. However, natural organic matter can reduce arsenic from its less toxic form, As(V), to its more toxic and mobile form, As(III) [19]. These research studies estimate that natural organic matter can reasonably affect heavy metal removal from water sources.

4. Groundnut shells in heavy metal adsorption

Groundnut shell is an abundant agro-industrial waste product that is mainly obtained after the removal of groundnut seed from its pod. Groundnut shell is rich in lignin content, due to which it undergoes very slow degradation [41]. Groundnut shell plays a vital role in the adsorption of heavy metals from water sources. Groundnut shell contains cellulose, organic acids, lignin, pentosan, and tannins which aid in binding heavy metal ions [42]. Efficiency of groundnut shells in heavy metal removal from water sources has been estimated by various researchers. For example, in an experimental study, the efficacy of groundnut shells for removing heavy metals from wastewater was tested by using several parameters at a standard temperature, such as the effect of pH, contact time, and dosage of groundnut shell. It has been found that the removal efficiency of groundnut shells for copper and lead is 68.19 and 77.81%, respectively [43].

Groundnut shell has been proven to be an efficient and effective adsorbent for the removal of toxic metals from zinc and chromium-plated water in which the initial concentration of zinc and chromium has been observed to reduce to the values 0.75 and 0.85, respectively. It has been concluded with the manifestation that the ability to remove toxic metals from zinc and chromium-plated water can be effectively achieved by using a groundnut shell as the adsorbent [44].

Groundnut shell has been utilized for removing copper from a water source, using batch experiments and considering different parameters, such as pH, contact time, initial concentration of metal ion, and dosage. The efficiency of groundnut shells at pH 6 for removing copper metal from water sources has been found to be approximately 94% [45]. Groundnut shell-activated carbon can also be used for the adsorption of metal ions such as Cu (II), Zn (II), Ni (II), and Cr (IV) from the aqueous solution at different parameters, such as pH and metal ion concentration of the aqueous solution. It has been estimated that the groundnut shell activated carbon has the highest adsorption capacity for removing chromium ions than its adsorption capacity for other metal ions from water sources [46].

Groundnut shells can also be positively utilized for the removal of lead ions from the aqueous solution. The adsorption capacity of groundnut shells for lead ions has been estimated at nearly 39 mg/g. Significant removal of lead ions by groundnut shells from water sources has been observed at different temperature conditions and different pH values [18]. The maximum adsorption capacity of groundnut shells at low pH values for chromium (VI) has been estimated at around 4.3 mg/g [25].

The effective removal of chromium(III) and copper(II) using groundnut shells has been evaluated at the maximum adsorption capacities of 27.89 and 25.39 mg g-1, respectively [32]. Heavy metal removal by using groundnut husk has also been significantly observed with maximum adsorption capacities of 7.69, 10.21, and 29.11 mg g-1 for Cr(III), Cu(II), and Pb(II), respectively [47]. The maximum adsorption capacity of groundnut shells for copper, nickel, and zinc has been calculated as 5, 8.1, and11 mg/g, respectively [48]. These adsorption capacities have been increased by 40–70% by using reactive dye in chemically modifying the groundnut shells using [48].

Groundnut shells can be effectively used as a raw material for preparing activated carbon to remove chromium and nickel from dye effluents based on the contact time. It has been observed that groundnut shell-activated carbon leads to an increase in chromium and nickel adsorption with an increase in contact time [49].

The adsorption of heavy metals, such as copper, zinc, lead, manganese, and cadmium, from wastewater by groundnut shells has been observed at a concentration

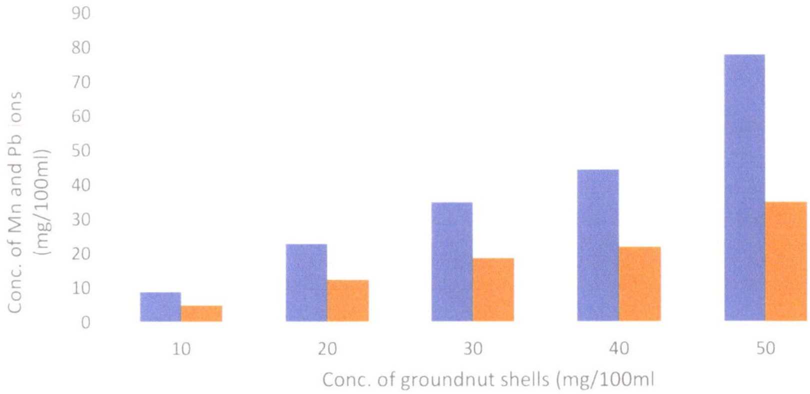

Figure 1.
Mn(II) and Pb(II) adsorption by using groundnut shells at different concentrations [50].

range of 10–50 mg/100 ml [50]. It has been evaluated that at a contact time of 100 min, maximum adsorption of groundnut shells for lead and manganese can be found. Manganese(II) and lead(II) adsorption from wastewater by using groundnut shells at different concentrations has also been observed at different concentrations at 33°C for 90 mins (**Figure 1**).

From the **Figure 1**, it is clearly observed that the adsorption of Mn(II) and Pb(II) increases with the increase in the concentration of groundnut shells [50].

5. Groundnut shells derived biochar in heavy metal adsorption

Groundnut shell-derived biochar has been used as an alternative cost-effective adsorbent to remove various heavy metals from water sources. The surface of groundnut shell biochar possesses functional groups, such as COOH and OH [51, 52]. Biochar with high efficiency can be derived from groundnut shell since no pre-treatment of biochar surfaces for adsorption is needed [53]. The production of biochar from groundnut shells by pyrolysis has been considered an eco-friendly and efficient method for the disposal of agricultural solid waste [54]. Groundnut shell biochar produced under slow and fast pyrolysis temperatures has been observed to possess a strong affinity for the removal of heavy metals from water sources. Various research experiments have been carried out to determine the adsorption efficiency of groundnut shell-derived char for different heavy metals at different parameters, such as pH of the aqueous solution, constant biochar dosage, particle size, concentration levels of heavy metal ions, constant contact time, and temperature [53].

Groundnut shell-derived biochar has been utilized for the removal of heavy metal ions from the aqueous solution at 350 ± 5°C and 700 ± 5°C. The adsorption efficiency of groundnut shell biochar for the removal of cadmium ions from the aqueous solution with concentration levels of 0.04, 0.08, and 0.20 mg/l has been estimated at 100, 99.99 and 100% at pyrolysis temperature of 350 ± 5°C, respectively. The adsorption efficiency of groundnut shell biochar for cadmium with the same concentration levels at a pyrolysis temperature of 700 ± 5°C has been estimated 100, 99.67, and 99.93%, respectively (**Table 3**). Groundnut shell biochar adsorption efficiency has been found approximately greater than 99.60% for cadmium [53]. A similar research study has reported 99.2% cadmium removal efficiency from wastewater at room temperature by using 40 g/l of adsorbent dosage, 200 mg/l of initial concentration of cadmium and pH 5 [55].

The adsorption efficiency of lead ions has been calculated 99.12% at the concentration of 0.10 mg/l, 100% each at 0.20 and 0.50 mg/l at pyrolysis temperature 350 ± 5°C, respectively, while the adsorption efficiency of lead ions in aqueous solution with same concentrations has been calculated 100% at pyrolysis temperature 700 ± 5°C was 100% (**Table 3**). The removal efficiency of lead ions in a mono-component system by groundnut biochar has been found to be higher than 99.1% [53].

The removal efficiency of mercury ions in the aqueous solution with a concentration of 0.10, 0.20, and 0.50 mg/l at 350 ± 5°C has been calculated 99.99, 100 and 100%, respectively (**Table 3**), while removal efficiency of mercury by groundnut shell biochar with same concentration levels of mercury ions has been estimated 100% at high pyrolysis temperature (**Table 3**). The removal efficiency of Hg^{2+} by low and high pyrolysis temperatures in mono-component systems with different maximum contamination limits has been reported to be almost 100% [53].

It has been found that groundnut shell-derived biochar at 400°C exhibits the highest capability for the removal of heavy metals from water sources because of its high specific surface area and rich functional groups [54]. The physicochemical properties of groundnut shell biochar prepared at varying pyrolysis temperature is shown in **Table 4**. With the temperature elevation from 350 to 400°C, the surface area and total pore volume has been observed to increase from 3.77 to 6.45 m^2/g and from 0.0097 to 0.0161 cm^3/g, respectively [54]. However, as the temperature increases more than 400°C, surface area and pore volume have been found to decrease. This phenomenon has been attributed to too-high temperatures that accelerated the deformation and collapse of micropores in biochar. Therefore, groundnut shell biochar at 400°C has the highest surface area and pore volume [56]. With the increase in temperature from 350 to 600°C, the pH value of the produced biochar has been observed to increase from 9.11 to 10.35, and the biochar yield has declined from 47.9 to 33.6%. The rise in pH has been observed by the decomposition of organic acid and carbonate into alkali

Metal ions	Initial concentration (mg/l)	350 ± 5°C	700 ± 5°C
		Adsorption efficiency (%)	
Cd^{2+}	0.04	100	100
	0.08	99.99	99.67
	0.20	100	99.93
Pb^{2+}	0.10	99.12	100
	0.20	100	100
	0.50	100	100
Hg^{2+}	0.10	99.99	100
	0.20	100	100
	0.50	100	100

Table 3.
Adsorption efficiency of groundnut shell biochar for Cd^{2+}, Pb^{2+}, Hg^{2+} ions [53].

Pyrolysis temperature (°C)	Surface area (m^2/g)	Total pore volume (cm^3/g)	pH	Yield (%)
350	3.77	0.0097	9.11	47.9
400	6.45	0.0161	9.34	41.4
500	4.78	0.0156	9.58	36.3
600	4.36	0.0153	10.35	33.6

Table 4.
The physicochemical properties of groundnut shell biochar prepared at varying pyrolysis temperatures [54].

salts. Therefore, it is beneficial for the heavy metal ions to precipitate in an alkaline environment [54].

The initial concentration of the metal ions in an aqueous solution has been considered to play an important role in the removal of metal ions in aqueous solution by the groundnut shell biochar. For example., the removal of Pb(II) ions in an aqueous solution has been reported to decrease with the increasing initial concentration of Pb(II) ions from 100 to 400 mg/l [57]. The experimental results are shown in **Figure 2**.

Similarly, pH of the solution is considered as one of the most important parameters in the removal of heavy metals from aqueous solutions. For example, the effect of pH on the adsorption of Pb(II) ions on the biochar derived from the groundnut shell has been carried out by changing the initial pH of the solution from 2 to 6. The removal of Pb(II) ions has been evaluated to increase with the increasing pH. The removal of Pb(II) ions by the groundnut shell biochar has been found to be very low at a pH value of 2.0. The increase in pH has been reported to promote the removal of heavy metals [54]. The effect of pH on the removal of Pb(II) ions from the aqueous solution by groundnut shell biochar is shown in **Figure 3**.

From **Figure 3**, it can be concluded that the pH plays a vital role in the removal of Pb(II) ions from the aqueous solution by groundnut shell-derived biochar. With an

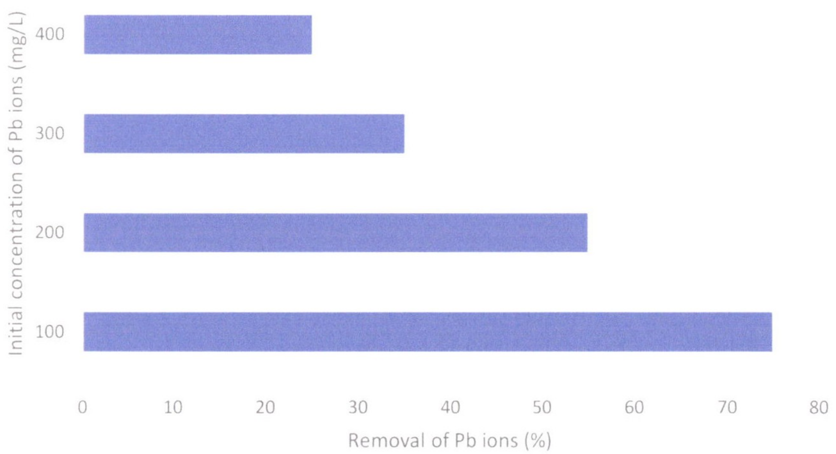

Figure 2.
Effect of initial concentration of the Pb(II) ions in aqueous solution on the removal of lead ions by the groundnut shell biochar.

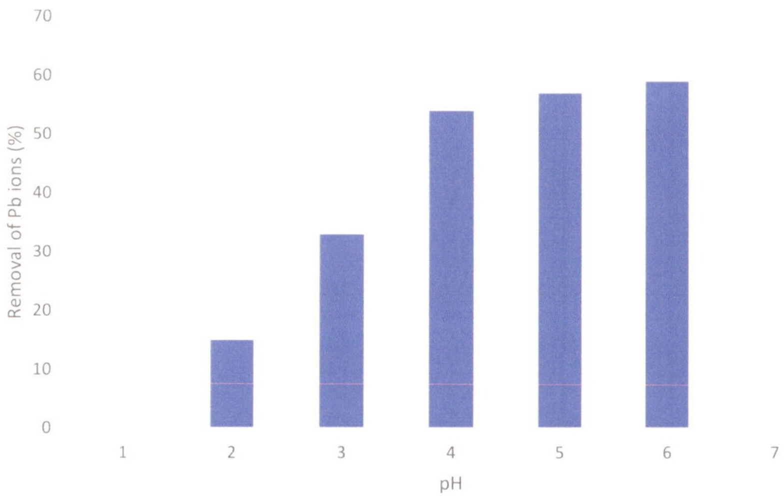

Figure 3.
Effect of pH on the removal of Pb(II) ions from the aqueous solution by groundnut shell biochar [54].

increase in the pH value, the covered H_3O^+ leaves the biochar surface and makes the sites available to Pb(II) ions [57]. Some studies have also reported that pH-dependent surface charges play a crucial role in controlling the surface adsorption of heavy metal ions through electrostatic interactions [58].

6. Conclusions

In developing countries, the increase in water scarcity and pollution significantly leads to less accessibility to clean drinking water. Heavy metal contamination in drinking water sources is a growing concern in the present era. Moreover, there is a lack of wastewater treatment methods in developing countries that would remove heavy metals from wastewater. As a result, various research studies have been conducted to investigate the use of low-cost adsorbents to remove heavy metals from water sources.

Groundnut shell is generally considered an agro-industrial waste; millions of tons of its quantity are produced every year as a leftover. Groundnut shell is rich in lignin, due to which these shells undergo slow degradation in a natural environment. Groundnut shell biomass has a wide range of applications. Groundnut shells can be converted into a valuable bio-product that can be efficiently and cost-effectively utilized in heavy metal removal from water sources. The chapter highlights the results of several scientific studies illustrating the adsorption efficiency of groundnut shells and biochar derived from the shell for the removal of various heavy metals from wastewater.

Various research studies have reported that groundnut shells can effectively remove heavy metals from water sources. The effectiveness of groundnut shells and biochar produced from groundnut shells at removing heavy metals depends on various parameters, such as pH, ionic strength, temperature, natural organic matter, initial concentration of heavy metal ions, etc. Groundnut shell-derived biochar exhibits a strong affinity for heavy metals in water sources at slow and fast pyrolysis temperatures.

Acknowledgements

The authors acknowledge the support of the Department of Post Harvest Engineering and Technology, Aligarh Muslim University, Aligarh.

Author details

Asima Shafi, Faizan Ahmad* and Sadaf Zaidi
Faculty of Agricultural Sciences, Department of Post Harvest Engineering and Technology, Aligarh Muslim University, Aligarh, India

*Address all correspondence to: f4faizahmad1989@gmail.com

References

[1] UN-Water. The United Nations World Water Development Report. Nature Based Solutions for Water. Paris, France: UNESCO; 2018

[2] Milly PC, Wetherald RT, Dunne KA, Delworth TL. Increasing risk of great floods in a changing climate. Nature. 2002;**415**(6871):514-517

[3] Noyes PD, McElwee MK, Miller HD, Clark BW, Van Tiem LA, Walcott KC, et al. The toxicology of climate change: Environmental contaminants in a warming world. Environment International. 2009;**35**(6):971-986

[4] Xi J, Duan Q, Luo Y, Xie Z, Liu Z, Mo X. Climate change and water resources: Case study of eastern monsoon region of China. Advanced Climate Change Research. 2017;**8**(2):63-67

[5] WHO. Guidelines for Drinking-Water Quality: Fourth Edition Incorporating the First Addendum. Geneva: World Health Organization; 2017

[6] WHO. Progress on Drinking Water, Sanitation and Hygiene: Update and SDG Baselines. Geneva: (UNICEF), United Nations Children's Fund; 2017

[7] Pandit AB, Kumar JK. Clean water for developing countries. Annual Review of Chemical and Biomolecular Engineering. 2015;**6**:217-246

[8] Ghorai S, Sarkar AK, Pal S. Rapid adsorptive removal of toxic Pb ion from aqueous solution using recyclable, biodegradable nanocomposite derived from template partially hydrolyzed xanthan gum and nanosilica. Bioresource Technology. 2014;**170**:578-582

[9] Awual MR. A novel facial composite adsorbent for enhanced copper (II) detection and removal from wastewater. Chemical Engineering Journal. 2015;**266**:368-375

[10] Xue Y, Gao B, Yao Y, Inyang M, Zhang M, Zimmerman AR, et al. Hydrogen peroxide modification enhances the ability of biochar (hydrochar) produced from hydrothermal hydrothermal carbonization of peanut hull to remove aqueous heavy metals: Batch and column tests. Chemical Engineering Journal. 2012;**200-202**:673-680

[11] An Q, Jiang YQ, Nan HY, Yu Y, Jiang JN. Unraveling sorption of nickel from aqueous solution by KMnO4 and KOH-modified peanut shell biochar: Implicit mechanism. Chemosphere. 2019;**214**:846-854

[12] Shakhawat Chowdhury MA, Mazumder J, Al-Attas O, Husain T. Heavy metals in drinking water: Occurrences, implications, and future needs in developing countries. Science of the Total Environment. 2016;**569-570**:476-488. DOI: 10.1016/j.scitotenv.2016.06.166

[13] Joseph L, Jun B-M, Joseph RV, Flora CM, Park YY. Removal of heavy metals from water sources in the developing world using low-cost materials: A review. Chemosphere. 2019;**229**:142-159. DOI: 10.1016/j.chemosphere.2019.04.198

[14] ATSDR (Agency for Toxic Substances and Disease Registry). Toxicologial Profiles, Toxic Substances Portal. 2015. Available from: http://www.atsdr.cdc.gov/toxprofiles/index

[15] USEPA (U.S. Environmental Protection Agency). USEPA Integrated Risk Information System (IRIS) online

database. 2015 . Available from: http://www.epa.gov/iris

[16] IARC (International Agency for Research on Cancer). IARC monographs on the evaluation of carcinogenic risks to humans. 2016, Volumes 1-115 . Available from: http://monographs.iarc.fr/ENG/Classification/latest_classif.php

[17] NTP (National Toxicology Program). Report on Carcinogens. 11th ed. Research Triangle Park, NC: U.S. Department of Health and Human Services, Public Health Service; 2015. Available from: http://ntpserver.niehs.nih.gov/ntp/roc/toc11.html

[18] Tasar S, Kaya F, Özer, A. Biosorption of lead (II) ions from aqueous solution by peanut shells: Equilibrium, thermodynamic and kinetic studies. Journal of Environmental Chemical Engineering. 2014;**2**(2):1018-1026

[19] Kumpiene J, Lagerkvist A, Maurice C. Stabilization of As, Cr, Cu, Pb and Zn in soil using amendments e a review. Waste Management. 2008;**28**:215-225

[20] Pantsar-Kallio M, Reinikainen S, Oksanen M. Interactions of soil components and their effects on speciation of chromium in soils. Analytica Chimica Acta. 2001;**439**:9-17

[21] Li X, Zhang D, Sheng F, Qing H. Adsorption characteristics of copper (II), zinc (II) and mercury (II) by four kinds of immobilized fungi residues. Ecotoxicology and Environmental Safety. 2018b;**147**:357-366

[22] Thirumavalavan M, Lai Y, Lin L, Lee J. Cellulose-based native and surface modified fruit peels for the adsorption of heavy metal ions from aqueous solution. Journal of Chemical & Engineering Data. 2010;**55**:1186-1192

[23] Tan G, Yuan H, Liu Y, Xiao D. Removal of lead from aqueous solution with native and chemically modified corncobs. Journal of Hazardous Materials. 2010;**174**:740-745

[24] Bozbas SK, Boz Y. Low-cost biosorbent: Anadara inaequivalvis shells for removal of Pb (II) and Cu (II) from aqueous solution. Process of Safety Environmental Protection. 2016;**103**:144-152

[25] Ahmad A, Ghazi ZA, Saeed M, Ilyas M, Ahmad R, Khattaka AM, et al. A comparative study of the removal of Cr (VI) from synthetic solution using natural biosorbents. New Journal of Chemistry. 2017;**41**:10799-10807

[26] Georgieva VG, Tavlieva MP, Genieva SD, Vlaev LT. Adsorption kinetics of Cr (VI) ions from aqueous solutions onto black rice husk ash. Journal of Molecular Liquids. 2015;**208**:219-226

[27] Chen H, Zhao J, Dai G, Wu J, Yan H. Adsorption characteristics of Pb (II) from aqueous solution onto a natural biosorbent, fallen Cinnamomum camphora leaves. Desalination. 2010;**262**:174-182

[28] Moussavi G, Barikbin B. Biosorption of chromium (VI) from industrial wastewater onto pistachio hull waste biomass. Chemical Engineering Journal. 2010;**162**:893-900

[29] Weng C, Lin Y, Hong D, Sharma YC, Chen S, Tripathi K. Effective removal of copper ions from aqueous solution using base treated black tea waste. Ecological Engineering. 2014;**67**:127-133

[30] Sari A, Mendil D, Tuzen M, Soylak M. Biosorption of Cd (II) and Cr (III) from aqueous solution by moss (Hylocomium splendens)

biomass—Equilibrium, kinetic and thermodynamic studies. Chemical Engineering Journal. 2008;**144**:1-9

[31] Senthil Kumar PS, Ramalingam S, Dinesh Kirupha SD, Murugesan A, Vidhyadevi T, Sivanesan S. Adsorption behavior of nickel (II) onto cashew nut shell: Equilibrium, thermodynamics, kinetics, mechanism and process design. Chemical Engineering Journal. 2011;**167**:122-131

[32] Witek-Krowiak A, Szafran RG, Modelski S. Biosorption of heavy metals from aqueous solutions onto peanut shell as a low-cost biosorbent. Desalination. 2011;**265**:126-134

[33] Al-Anber ZA, Matouq MA. Batch adsorption of cadmium ions from aqueous solution by means of olive cake. Journal of Hazardous Materials. 2008;**151**(1):194-201

[34] Villaescusa I, Fiol N, Martinez M, Miralles N, Poch J, Serarols J. Removal of copper and nickel ions from aqueous solutions by grape stalks wastes. Water Research. 2004;**38**(4):992-1002

[35] Wang W, Chen M, Guo L, Wang WX. Size partitioning and mixing behavior of trace metals and dissolved organic matter in a South China estuary. Science Total Environment. 2017;**603-604**:434-444

[36] Zhang M. Adsorption study of Pb (II), Cu (II) and Zn (II) from simulated acid mine drainage using dairy manure compost. Chemical Engineering Journal. 2011;**172**:361-368

[37] Yang S, Wu Y, Aierken A, Zhang M, Fang P, Fan Y, et al. Mono/competitive adsorption of arsenic (III) and nickel (II) using modified green tea waste. Journal of the Taiwan Institute of Chemical Engineers. 2016;**60**:213-221

[38] Merdy P, Huclier S, Koopal LK. Modeling metal-particle interactions with an emphasis on natural organic matter. Environmental Science & Technology. 2006;**40**(24):7459-7466

[39] Wang S, Mulligan CN. Natural attenuation processes for remediation of arsenic contaminated soils and groundwater. Journal of Hazardous Materials. 2006;**138**(3):459-470

[40] Du Y, Lian F, Zhu L. Biosorption of divalent Pb, Cd and Zn on aragonite and calcite mollusk shells. Environmental Pollution. 2011;**159**(7):1763-1768

[41] Zheng W, Phoungthong K, Lü F, Shao LM, He PJ. Evaluation of a classification method for biodegradable solid wastes using anaerobic degradation parameters. Waste Management. 2013;**33**(12):2632-2640

[42] Ding Y, Jing D, Gong H, Zhou L, Yang Y. Biosorption of aquatic cadmium (II) by unmodified rice straw. Bioresource Technology. 2012;**114**:20-25

[43] Shruthi KM, Pavithra MP. A study on utilization of groundnut shell as biosorbent for heavy metals removal. International Journal of Engineering Technology. 2018;**4**(3):411-415

[44] Aarti Sowmya T, Gayavajitha E, Kanimozhi R, Subalakshmi R. Removal of toxic metals from industrial wastewater using groundnut shell. International Journal of Pure Applied Mathematics. 2018;**119**(15):629-634

[45] Prabha RT, Udayashankara TH. Adsorption of copper metal ions from aqueous solution using rice husk and groundnut shell. International Journal of Scientific Research. 2014;**3**(8):705-709

[46] Romero LC, Antonio B, Elio EG. Peanut shell activated carbon:

Adsorption capacities for copper (II), zinc (II), nickel (II) and chromium (VI) ions from aqueous solutions. Adsorption Science and Technology. 2004;**22**(3):237-243

[47] Li Q, Zhai J, Zhang W, Wang M, Zhou J. Kinetic studies of adsorption of Pb (II), Cr (III) and Cu (II) from aqueous solution by sawdust and modified peanut husk. Journal of Hazardous Materials. 2007;**141**(1):163-167

[48] Shukla SR, Pai RS. Adsorption of Cu (II), Ni (II) and Zn (II) on dye loaded groundnut shells and sawdust. Separation and Purification Technology. 2005;**43**(1):1-8

[49] Idris S, Iyaka YA, Dauda BEN, Ndamitso MM, Umar MT. Kinetic study of utilizing groundnut Shell as an adsorbent in removing chromium and nickel from dye effluent. American Chemical Science Journal. 2012;**2**(1):12-24

[50] Maduabuchi MN. Agricultural waste materials as a potential adsorbent for removal of heavy metals in waste water. Journal of Waste Manage Xenobio. 2018;**1**(1):000104

[51] Ahmad M, Rajapaksha AU, Lim JE, Zhang M, Bolan N, Mohan D, et al. Biochar as a sorbent for contaminant management in soil and water: A review. Chemosphere. 2014;**99**:19-33

[52] Khan S, Waqas M, Ding F, Shamshad I, Arp HPH, Li G. The influence of various biochars on the bioaccessibility and bioaccumulation of PAHs and potentially toxic elements to turnips (Brassica rapa L.). Journal of Hazardous Materials. 2015;**300**:243-253

[53] Cobbina SJ, Duwiejuah AB, Quainoo AK. Single and simultaneous adsorption of heavy metals onto groundnut shell biochar produced under fast and slow pyrolysis. International Journal of Environmental Science and Technology. 2019;**16**:3081-3090. DOI: 10.1007/s13762-018-1910-9

[54] Shan R, Shi Y, Jing G, Wang Y, Yuan H. Single and competitive adsorption affinity of heavy metals toward peanut shell-derived biochar and its mechanisms in aqueous systems. Chinese Journal of Chemical Engineering. 2020;**28**:1375-1383. DOI: 10.1016/j.cjche.2020.02.012

[55] Cheng Q, Huang Q, Khan S, Liu Y, Li G, Liao Z, et al. Adsorption of Cd by peanut husks and peanut husk biochar from aqueous solutions. Ecological Engineering. 2016;**87**:240-245

[56] Abdelhadi SO, Dosoretz CG, Rytwo G, Gerchman Y, Azaizeh H. Production of biochar from olive mill solid waste for heavy metal removal. Bioresource Technology. 2017;**244**:759-767

[57] Chunxia H, Qiu M. Characterization of the biochar Derived from Peanut Shell and adsorption of Pb (II) from aqueous solutions. An International Quarterly Scientific Journal. 2019;**18**:225-230

[58] Wang SS, Gao B, Zimmerman AR, Li YC, Ma L, Harris WG, et al. Removal of arsenic by magnetic biochar prepared from pinewood and natural hematite. Bioresource Technology. 2015b;**175**:391-395

Section 3

Toxic Heavy Metals

Chapter 23

Perspective Chapter: Rapid Measurement of Potentially Toxic Elements (PTEs) in Petroleum Hydrocarbons Polluted Soils by X-Ray Fluorescence (XRF) Spectroscopy

Reward Kokah Douglas

Abstract

Potentially toxic elements (PTEs) contamination in soils threats human wellbeing and ecological health because of their toxicity and bioaccumulation. This research presents a portable Olympus Delta Premium 6000 Series XRF Analyser (Olympus, USA) as a rapid measurement tool (RMT) for PTEs: Cr, Cu, Fe, Pb, Mn, and Zn in contaminated soils in the Niger Delta, Nigeria. A total of 45 crude oil-contaminated soils were collected from three genuinely oil spill sites. The range of measured PTEs concentrations (mg/kg) in the study sites are as follows: Site 1: chromium (Cr) 54–75, copper (Cu) 5.4–16.6, iron (Fe) 14,841–23,404, lead (Pb) 13.5–21.4, manganese (Mn) 158–555, and zinc (Zn) 32.6–47.2; Site 2: (35–66), (5–16.1), (10166–20,967), (12–17.8), (209–440), (17.6–33.6); and Site 3: (32–115), (6.5–20.8), (7538–22,800), (12–135), (98–338), (19.9–177). The trend of PTEs across the three sites follows the same order: Fe > Mn > Cr > Zn > Pb > Cu. The average concentration values of PTEs in all the 3 sites were higher than background concentration values. Thus, crude oil spill spiked the PTEs concentrations. XRF spectroscopy is recommended as a cost-effective and RMT for PTEs in soils.

Keywords: soil, crude oil, pollution, elements, X-ray fluorescence

1. Introduction

Soil is a great reservoir for contaminants as well as a natural buffer for transportation of chemical materials and elements in the environment. There has been an increasing concern in many countries of the world about the levels of potentially toxic elements (PTEs) in the soil environment [1–3]. There are over 40 chemical elements in the soil [4]; out of which, 21 elements are commonly considered as PTEs, which are zinc, Zn; vanadium, V; uranium, U; tungsten, W; tin, Sn; thallium, TI; silver, Ag; selenium, Se;

molybdenum, Mo; mercury, Hg; manganese, Mn; lead, Pb; gold, Au; copper, Cu; cobalt, Co; chromium, Cr; cadmium, Cd; barium, Ba; arsenic, As, and antimony, Sb [5]. Among these, Pb, Cr, As, Zn, Cd, Cu, Hg, and Ni are most commonly found at contaminated sites [6].

Petroleum hydrocarbons contain PTEs such as cobalt (Co), copper (Cu), lead (Pb), iron (Fe), magnesium (Mg), manganese (Mn), zinc (Zn), cadmium (Cd), chromium (Cr), nickel (Ni), arsenic (As), titanium (Ti), silver (Ag) among others. PTEs contamination in soil has attracted significant ecological concerns because of their toxic, bioaccumulative, and persistence nature in the existing environment. Unlike most pollutants (e.g. petroleum hydrocarbons), PTEs cannot be degraded and have long-lasting effects in soil as a result of strong adsorption of many metal ions on humic and clay colloids in soils [7]. PTEs have been reported to have physiological effects on living organisms as they are not degradable [8]. Vehicle emissions, metal plating/finishing operations, disposal of industrial waste, fertilizer applications, and fly ash from incineration/combustion processes, among others are also sources of PTEs in soils [9]. It is pertinent to also mention that mining, smelting, chemical production, and factory emissions release large quantities of Cd and Pb into soils, causing significant soil pollution [10, 11].

In Nigeria, the Niger Delta region is the heart of the oil and gas Industry (OGI) and has contributed enormously to the growth and development of the country. However, since the beginning of the establishment of the OGI in the region, several oil spill incidents have been reported; and, to date, it has been estimated that 13 million tons of hydrocarbons have been spilled in the region due to pipeline fatigue, well blowout, pipeline vandalism, and sabotage [12, 13]. Additionally, Ite et al. [14] reported that the number of contaminated sites in the Niger Delta region is in excess of 2000. Furthermore, the United Nation Environment Programme (UNEP) reported in 2011 that in Ogoniland alone (a small part of the Niger Delta), over 69 sites were heavily contaminated with crude oil (concentration exceeding 139,000 mg/kg) affecting soil, air, and water quality criteria and posing a serious human health threat. This, in turn, impacts the quality of water resources, directly affecting the health of local communities, which are drinking contaminated water [15–17]. Therefore, there is urgent research need to assess and quantify PTEs in polluted soils using simple, rapid, inexpensive, and accurate analytical methods to help appraise the environmental risk of PTEs to ensure food security, environment safety, and public health safety in the Niger Delta region of Nigeria, and anywhere in the globe faced with such challenges.

Numerous analytical techniques are in use for PTEs detection and quantitative measurement in soils, including atomic absorption spectrophotometry (AAS), inductively coupled plasma-mass spectroscopy (ICP-MS), inductively coupled plasma atomic emission spectroscopy (ICP-AES), atomic fluorescence spectrophotometry (AFS), X-ray fluorescence (XRF) spectroscopy, and laser-induced breakdown (LIB) spectroscopy. Also included are optical techniques [18], electrochemical [19, 20], and voltammetry [21].

It is pertinent to mention that generally laboratory methods (in this case, laboratory methods are referred to the "chemical methods") are known for quantitative measurement with good detection limits and have been commonly used for the detection of various PTEs in soils, with high sensitivity, selectivity, and accuracy [18, 21]. However, chemical methods require comparatively costly instrumentation, relatively lengthy measurement protocols, and specialized operators needed to achieve the correct measurements [21]. Given the relative disadvantage and cost of the laboratory methods mentioned above, there is a need to assess and use analytical devices that can offer rapid, inexpensive measurements, and requires little or no sample preparation. It is worth mentioning that to date, there is no published work yet comparing the measurement accuracy between the candidate chemical methods including AAS, ICP-MS, ICP-AES, etc., and analytical techniques that do not involve wet chemistry methods on measurement accuracy of PTEs in soil and in sediment samples. In the absence of such a crucial study that is key for method selection for environmental analyses, and knowing that the first step toward decision making on the selection of the best analytical techniques for the measurement of contaminants in environmental sample(s) is driven by time, cost, and the measurement accuracy [22]; methods that can offer timely and cost-effective analysis of environmental contaminants can be applied. As a result, this study aims at using a portable X-ray fluorescence (PXRF) spectroscopy as a cost-effective and rapid measurement tool (RMT) for PTEs in petroleum hydrocarbon-contaminated soils collected from crude oil spill sites in the Niger Delta region of Nigeria.

2. Materials and methods

2.1 Study area and soil sampling

The study area located in Bayelsa and Rivers State, Niger Delta, Southern Nigeria has a tropical rain forest climate characterized by two seasons: The rainy season lasts for about 7 months between April and October with an overriding dry period in August (known as August break); and the dry season lasts for about 5 months, between November and March. The temperature varies between 25 and 35°C. The regional geology of the Niger Delta is relatively simple, consisting of Benin, Agbada (the kitchen of kerogen), and Akata formations, overlain by various types of quaternary deposits [23, 24]. Soils of the area studied were classified according to the United State Department of Agriculture (USDA) [25] soil taxonomy into two orders, that is, inceptisols and entisols, which include four subgroups of typic dystrudepts, aeric endoaquepts, typic udipsammerts, and typic psammaqnents [26]. A total of 45 representative spot sample points were collected from three oil-contaminated sites (Site 1 = Ikarama: 15 samples; Site 2 = Kalabar: 15 samples; and Site 3 = Joinkrama: 15 samples) in August 2015. The three sites were selected for sampling due to their similar exploration activities and oil spill history. The soil samples (approx. 5 kg) were collected from visible "hot-spots" in the top 15 cm soil layer using a shovel. **Figure 1** shows the sampling location map. Soil samples were kept in airtight centrifuge tubes and stored at 4°C using ice block to avoid hydrocarbon volatilization and preserve field-moist status until transported to Cranfield University in the United Kingdom. The samples were then stored in a freezer at –20°C prior to PTEs analysis by XRF spectroscopy.

2.2 Sample preparation and XRF analysis of PTEs in soils

The concentrations of PTEs: Cr, Cu, Fe, Pb, Mn, and Zn in petroleum hydrocarbon-contaminated soils were determined using a portable Olympus Delta Premium 6000 Series XRF Analyser (Olympus, USA). In diffuse reflectance mode, the Delta XRF analyzer is three beams configured, where each beam was programmed to scan soil samples for 30 seconds. Prior to soil scanning, the instrument's setting and operational conditions were done in accordance with the manufacturer's specifications, and the analyzer was calibrated with alloy 316 stainless steel coupon. Fresh soil samples were thoroughly mixed and scanned using single open-ended and snap-post venting (30.7 mm O.D x 23.1 mm High) sample cups, sealed by Prolene Thin-Film (Diam. 63.5 mm) (Chemplex, USA). Each soil sample was analyzed for Cr, Cu, Fe, Pb, Mn, and Zn concentrations.

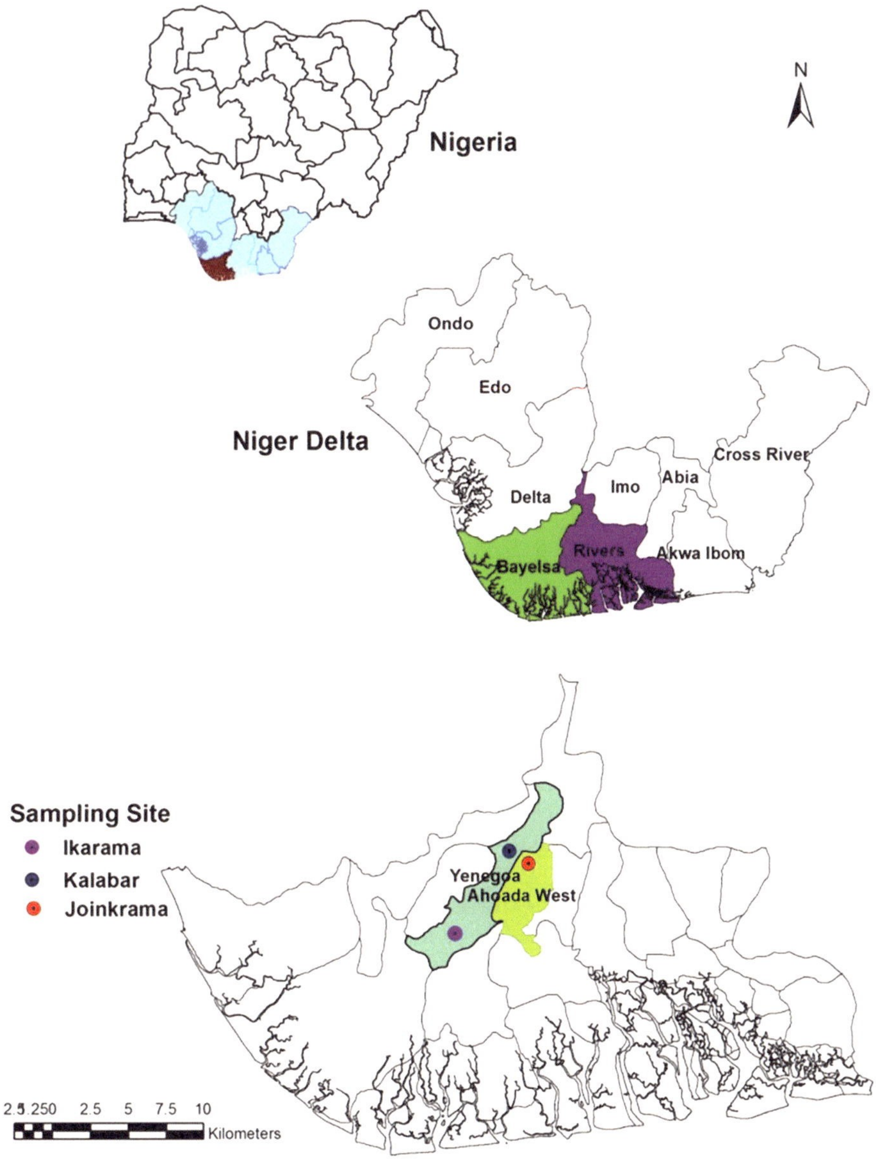

Figure 1.
Soil sampling locations for the three petroleum hydrocarbon-released sites in the Niger Delta region of Nigeria. Source: [27].

3. Results and discussion

3.1 PTEs concentrations in soils

Table 1 reports on the 6 PTEs (Cr, Cu, Fe, Pb, Mn, and Zn) concentrations in 45 genuinely crude oil-contaminated soil samples collected from three petroleum hydrocarbon-contaminated sites in the Niger Delta, Nigeria. Summary statistics of the 6 PTEs concentrations for all the three study sites range as follows: Cr ranged from 54 to 75, Cu (5.4–16.6), Fe (14841–23,404), Pb (13.5–21.4), Mn (158–555), Zn (32.6–47.2)

Site 1 (number of samples = 15)															
Fe	22,368	23,040	20,393	21,392	23,344	23,073	23,404	21,912	23,251	17,951	18,201	18,555	19,146	19,495	14,841
Mn	312	401	445	463	555	506	517	488	511	210	193	179	158	222	171
Cr	62	66	55	61	66	63	65	59	54	66	72	67	59	59	75
Zn	44.2	32.6	33.9	37.3	45.8	34.4	38.4	32.6	34.1	46.9	47.2	43.3	39.6	44.6	46.6
Pb	18.5	14.9	17.5	21.4	20.9	16.2	16.6	17.3	13.5	18.6	18.5	17.9	15.8	14.3	16.8
Cu	16.6	7.6	11.7	12.4	15.6	11.5	13.3	11.6	12.8	8.6	11.7	8.9	10.4	5.4	8.5
Site 2 (number of samples = 15)															
Fe	11,721	15,595	10,924	13,899	10,166	14,278	20,967	14,017	12,097	13,812	10,313	16,361	13,577	19,285	13,699
Mn	216	358	259	292	209	372	372	343	210	286	278	440	325	415	300
Cr	40	66	36	43	41	48	53	40	39	43	35	49	40	48	38
Zn	21	27.8	19.8	33.6	17.6	22.7	31.3	24.4	30.6	30.6	22.3	30.4	19.4	30.9	28
Pb	15.7	15.9	16.7	15	14.9	17.8	16	12.9	12	15.3	15.8	16.6	16.1	17	14.1
Cu	8.3	11.1	7.3	10.6	6.2	7.8	16.1	7.1	6.8	5	7.8	10.5	nd	9.5	6.6
Site 3 (number of samples = 15)															
Fe	13,538	17,296	15,194	19,087	22,213	21,148	13,042	22,800	13,848	7538	19,418	19,667	15,182	20,222	13,730
Mn	234	270	146	255	251	284	191	338	209	98	236	263	212	276	205
Cr	37	73	32	65	115	49	34	62	52	47	49	58	61	61	39
Zn	26.9	29.3	41.7	78.4	177	41.1	28.9	60.6	43.3	19.9	61.7	47.7	63.7	68.9	64.9
Pb	16.9	12.7	35.8	80.4	135	14.5	12	29.6	47.2	21.4	24	28.6	34.9	56.5	69.9
Cu	7.6	11.5	13.9	20.1	20.8	12.2	7.6	12.1	6.6	6.5	16.2	14.9	13.4	13.3	9

Fe = iron, Mn = manganese, Cr = chromium, Zn = zinc, Pb = lead, Cu = copper, and nd = not detected. Source: [27].

Table 1.
Potentially toxic elements concentrations (mg/kg) in soils collected from three oil spill sites in the Niger Delta, Nigeria.

Metal	Site 1			Site 2			Site 3			A	B	C
	Mean	Min	Max	Mean	Min	Max	Mean	Min	Max			
Fe	20691.07	14,841	23,404	14047.40	10,166	20,967	16928.20	7538	22,800	NR	61,625	NR
Mn	355.4	158	555	311.7	209	440	231.2	98	338	26.1	199.6	201.8
Cr	63.3	54	75	43.9	35	66	55.6	32	115	2	17.8	28.8
Zn	40.1	32.6	47.2	26	17.6	33.6	56.9	19.9	177	23	38.1	29.3
Pb	17.3	13.5	21.4	15.5	12	17.8	41.3	12	135	11	7.4	25
Cu	11.1	5.4	16.6	8.1	5	16.1	12.4	6.5	20.8	4	7.8	11.2

Results were compared with those of previous studies (A, B, C) in the same region. A = background level of PTEs in soils [29]; B = mean concentration of PTEs in oil field soils [28]; C = mean concentration of PTEs in crude oil impacted soils [7] and NR = not reported.

Table 2.
Descriptive statistics of PTEs concentrations (mg/kg) in 45 crude oil-contaminated soils collected from three spill sites in the Niger Delta, Nigeria.

in Site 1; Cr (35–66), Cu (5–16.1), Fe (10166–20,967), Pb (12–17.8), Mn (209–440), Zn (17.6–33.6) in Site 2; in Site 3, Cr (32–115), Cu (6.5–20.8), Fe (7538–22,800), Pb (12–135), Mn (98–338), Zn (19.9–177). As it can be seen, the trend of heavy metal contamination in the three sites is in the order: Fe > Mn > Cr > Zn > Pb > Cu. Fe contamination level was found to be the highest in site 1. Results (mean values) were compared with other studies [7, 28] conducted in the Niger Delta, Nigeria, the case study area of the current research. Except Fe that had no records at the moment to be compared with, most of the metal values were higher than those reported by [7, 28]. Also, the mean concentrations of Mn, Cr, Cu, Zn, and Pb were observed to be higher than the background concentrations reported by [29]. These results are presented in **Table 2**. Results show that crude oil spill is the source of PTEs pollution in the study sites.

4. Conclusions

This chapter presents a portable XRF Analyser (Olympus, USA) as a rapid measurement tool (RMT) for PTEs (Cr, Cu, Fe, Pb, Mn, and Zn) in crude oil-contaminated soils. A total of 45 field samples collected from three genuinely PHC released sites in the Niger Delta, Nigeria were analyzed. The trend of PTEs contamination across the three sites follows the same order: Fe > Mn > Cr > Zn > Pb > Cu. While the mean values of some PTEs obtained in this study (sites 1 and 2) ranked higher than those reported in previous studies; the mean values of all the PTEs in Site 3 were higher than all the previously published results in the region. Consequently, results conclude that crude oil spills on land sites contributed to the higher concentrations of the PTEs relative to the natural background values. Since PTEs are bioaccumulants, they may pose a threat to environment and human well-being. Thus, there is research need to assess the site-specific risks of PTEs contamination in both the areas where soils are potentially polluted, and sites that have recorded series of oil spill incidents in the Niger Delta, Nigeria. Furthermore, the following recommendations are made:

1. Research on the comparison between well-implemented techniques involving chemical methods and analytical techniques that do not involve chemical methods in the measurement accuracy of PTEs in soil and in sediment samples should be carried out.

2. In the interim, if time is not a crucial factor and accuracy is more appealing, techniques for soil PTEs involving chemical methods are the most appropriate option, as accuracy is higher than analytical techniques that do not use chemical methods.

Acknowledgements

The authors gratefully acknowledge the Petroleum Technology Development Fund (PTDF) of Nigeria (PTDF/OSS/PHD/DRK/711/14).

Nomenclature

PTEs	potentially toxic elements
PHC	petroleum hydrocarbon

RMT	rapid measurement tool
XRF	X-ray fluorescence
AAS	atomic absorption spectrophotometry
AFS	atomic fluorescence spectrophotometry
ICP-MS	inductively coupled plasma-mass spectroscopy
ICP-AES	inductively coupled plasma atomic emission spectroscopy
LIB	laser-induced breakdown
UNEP	United Nation Environment Programme
USDA	United State Department of Agriculture

Author details

Reward Kokah Douglas
Department of Chemical Engineering, Niger Delta University, Wilberforce Island, Bayelsa, Nigeria

*Address all correspondence to: douglasreward@ndu.edu.ng

References

[1] Khan S, Cao Q, Zheng YM, Huang YZ, Zhu YG. Health risks of heavy metals in contaminated soils and food crops irrigated with wastewater in Beijing China. Environment Pollution. 2008;**152**(3):686-692

[2] Zhang XY, Lin FF, Wong MTF, Feng XL, Wang K. Identification of soil heavy metal sources from anthropogenic activities and pollution assessment of Fuyang County China. Environment Monitoring Assessment. 2009;**154**(1-4):439-449

[3] Shi T, Chen Y, Liu Y, Wu G. Visible and near-infrared reflectance spectroscopy- An alternative for monitoring soil contamination by heavy metals: A review. Journal of Hazardous Materials. 2014;**265**:166-176

[4] Vodyanitskii YN. Standards for the contents of heavy metals in soils of some states. Annals of Agrarian Science. 2016;**14**(3):257-263

[5] Alloway BJ. Sources of heavy metals and metalloids in soils. In: Alloway BJ, editor. Heavy Metals in Soils. Berlin, Heidelberg: Springer; 2013. pp. 11-50

[6] GWRTAC. Remediation of Metals-Contaminated Soils and Groundwater: Technology Evaluation Report. Pittsburgh, PA: Technology Evaluation Report Series TE-97-01615; 1997. Available from: http://citeseerx.ist.psu.edu/viewdoc/download?doi1/410.1.1.401.3768&rep1/4rep1&type1/4pdf. [Accessed 26 September 2022]

[7] Iwegbue CMA, Williams ES, Isirimah NO. Study of heavy metal distribution in soils impacted with crude oil in Southern Nigeria. Soil and Sediment Contamination. 2009;**18**:136-143

[8] Esmaeilzadeh M, Jaafari J, Ali AM, Panahandeh M, Alahbakhsh J, Safoura J. Investigation of the extent of contamination of heavy metals in agricultural soil using statistical analyses and contamination indices. Human and Ecological Risk Assessment: An International Journal. 2018;**25**(5):1125-1136. DOI: 10.1080/10807039.2018.1460789

[9] Alhassan M. Potentials of rice husk ash for soil stabilisation assumption university. Journal of Technology. 2008;**14**(4):245-250

[10] Yang Q, Li Z. Lu: A review of soil heavy metal pollution from industrial land agricultural regions in China: Pollution and risk assessment. Science of the Total Environment. 2018;**624**:690-700

[11] Cai LM, Wang QS. Luo: Heavy metal contamination and health risk assessment for child near a Cu-smelter in Central China. Science of the Total Environment. 2019;**650**:725-733

[12] Ambituuni A, Amezaga J, Emeseh E. Analysis of safety and environmental regulations for downstream petroleum industry operations in Nigeria: Problems and prospects. Environment and Development. 2014;**9**:43-60. DOI: 10.1016/j.envdev.2013.12.002

[13] Nwilo PC, Badejo OT. Impacts and management of oil spill pollution along the Nigerian coastal areas. Administering Marine Spaces: International Issues. 2006:119

[14] Ite AE, Ibok UJ, Ite MU, Petters SW. Petroleum exploration and production: Past and present environmental issues in the Nigeria's Niger Delta. American Journal of Environmental Protection. 2013;**1**:78-90

[15] Davies O, Abolude D. Polycyclic aromatic hydrocarbons of surface water from Oburun Lake, Niger Delta, Nigeria. Applied Scientific Research. 2016;**13**:20-24

[16] Nganje TN, Hursthouse AS, Edet A, Stirlin D, Adamu CT. Hydrochemistry of surface and groundwater in the shale bedrock, Cross River Basin and Niger Delta region Nigeria. Applied Water Science. 2015;7:961-985

[17] United Nations Environment Programme. Environmental Assessment of Ogoniland. Switzerland: UNEP; 2011

[18] Soodan RK, Pakade YB, Nagpal A, Katnoria JK. Analytical techniques for estimation of heavy metals in soil ecosystem: A tabulated review. Talanta. 2014;**125**:405-410

[19] Cui L, Wu J, Ju H. Electrochemical sensing of heavy metal ions with inorganic, organic and bio-materials. Biosensors & Bioelectronics. 2015;**63**:276-286

[20] Ransod B, Kumar T, Thakur R, Rana S, Singh I. A review on various electrochemical techniques for heavy metal ions detection with different sensing platforms. Biosensors & Bioelectronics. 2017;**94**:443-455

[21] Yuanyuan L, Xinqiang L, Niyungeko C, Junjie Z, Guangming T. A review of the identification and detection of heavy metal ions in the environment by voltammetry. Talanta. 2017;**178**:324-338

[22] Douglas RK, Nawar S, Alamar MC, Coulon F, Mouazen AM. Almost 25 years of chromatographic and spectroscopic analytical method development for petroleum hydrocarbons analysis in soil and sediment: state-of-the-art, progress and trends. Critical Reviews in Environmental Science and Technology. 2017;**47**(16):1497-1527. DOI: 10.1080/10643389.2017.1385368

[23] Kogbe CA. The cretaceous and paleogene sediments of Southern Nigeria. In: Kogbe CA, editor. Geology of Nigeria. Lagos: Elizabethan Press; 1989. pp. 311-334

[24] Wright JB, Hasting DA, Jones WB, Williams HK. Geology and Mineral Resources of West Africa. UK: Allen and Unwin Limited; 1985. p. 107

[25] Soil Survey Staff. Soil Taxonomy - A basic system of soil classification for making and interpreting soil surveys. In: Agricultural Handbook 436; Natural Resources Conservation Service. 2nd ed. Washington DC, USA: USDA; 1999

[26] Udoh BT, Esu IE, Ibia TO, Onweremadu EU, Unyienyin SE. Agricultural potential of the beach ridge soils of the Niger Delta Nigeria. Malaysian Journal of Soil Science. 2013;**17**:17-37

[27] Douglas RK, Obhuo M, Opukumo AW. Rapid estimation of heavy metals in crude oil contaminated soils by X-ray fluorescence (XRF) spectroscopy. Global Journal of Earth and Environmental Science. 2020;**5**(2):51-57

[28] Iwegbue CMA, Egobueze FE, Opuene K. Preliminary assessment of heavy metals levels of soils of an oil field in the Niger Delta, Nigeria. International Journal of Environmental Science Technology. 2006;**3**(2):167-172

[29] Agip. Environmental Impact Assessment for Drilling and Development of Keenokpo "A" Location. Port Harcout, Nigeria: Nigeria Agip Oil Company; 2000

Chapter 24

Heavy Metals' Poisoning in Farm Animals

Selina Acheampong

Abstract

Heavy metals are metallic elements with a high density compared to water that are found in trace amounts in a variety of matrices. Mercury poisoning can cause brain damage, skin damage, and increase risk of cancer development. Mercury-poisoned animals cannot produce meat, liver, or kidneys fit for human consumption. Heavy metals can cause cell dysfunction and toxicity by attaching to protein sites and displacing the original metals from their native binding sites. Reducing input/output ratio of heavy metals in animals should be the main objective of effective solutions. Electro-remediation involves running an electric current through liquid manure to cause metal ions to precipitate on an electrode.

Keywords: heavy metals, toxicity, farm animals, poisoning, bioremediation, animal health, animal feed

1. Introduction

Metals are naturally occurring elements of the earth's crust, and through wind and water-induced erosion, they are dispersed as powders or dissolved into rivers [1]. However, compared to human activity, these natural processes release fewer metals into the environment. These substances spread widely in the environment, which causes them to move up the food chain. Heavy metals are metallic elements with a high density compared to water that are found in trace amounts in a variety of matrices [2]. Examples include Fe, Co, Cu, Mn, Mo, Se, Zn, Cr, and As. Since heavy metals can cause toxicity at low levels, their weight and toxicity are connected [3]. In order to maintain certain biochemical and physiological processes in humans, animals, plants, and other organisms, certain metals are necessary [3]. These trace elements, sometimes known as microelements, include cobalt (Co), copper (Cu), chromium (Cr), iron (Fe), manganese (Mn), molybdenum (Mo), selenium (Se), and zinc (Zn). Their nutritional requirements are typically minimal. Although they have varying bioavailability, they are present in a variety of matrices in trace amounts (ppb or ppm). The term "heavy metals" refers to metallic elements with an atomic number greater than 20 and possessing metallic characteristics [4]. There are trace elements, or elements necessary for the proper growth, development, and operation of living creatures, among them (such as copper, zinc, chromium, and iron) (including cadmium, lead, and mercury) [5]. However, they all share the trait of being poisonous and extremely deadly for humans, animals, and plants beyond a specific

Group of heavy metals	Examples
Macronutrient elements	Cobalt, iron
Micronutrient elements	Copper, nickel, chromium, iron, manganese, molybdenum
Highly toxic elements	Mercury, cadmium, lead, silver, gold, palladium, bismuth, arsenic, platinum, selenium, tin, zinc
Precious elements	Platinum, silver, gold, palladium, ruthenium
Radio nuclides	Uranium, thorium, radium, cerium, praseodymium

Table 1.
Classification of heavy metals with examples.

level [3]. Toxic heavy metals are generally determined by the level of pollution, but they can also be toxic depending on the species and age of the organism, the route taken to enter the body, the chemical's structure, how it interacts with other metals, or the body's physiological state [6]. Metals enter the body through the respiratory or gastrointestinal systems, where they are concentrated and stored before being carried by the blood to the tissues and organs [7]. According to European Union (EU) Regulation 1881/2006, essential trace elements are typically added to animal feed as nutritional supplements to enhance health and maximize output. However, prolonged exposure to these substances at higher concentrations has been associated with cellular or systemic problems and may be a source of pollution [8]. Other metals, such as As, Cd, Pb, and Hg, are regarded as pollutants and undesirable chemicals in animal feed (Regulation 2002/32/EC) since they lack biologically recognized roles [5]. Additionally, the preceding public health hazards—As, Cd, Cr, Pb, and Hg—show a high toxicity since they can cause organ damage even at low exposure levels [8]. According to their biological significance, heavy metals can be divided into four main groups: necessary, non-essential, less toxic, and highly toxic heavy metals (**Table 1**) [9]. Heavy metals can be both beneficial and harmful to the organism.

2. Sources of heavy metal pollution

According to the FAO, the following are the sources of heavy metals: mining effluents, industrial effluents, domestic effluents, urban stormwater, leaching of metals from solid waste dumps and garbage, metal inputs from rural areas, batteries, pigments, paints, glass, fertilizers, textiles, dental and cosmetic products, atmospheric sources, and petroleum industrial activities [9]. More causes of heavy metal contamination include the preparation of nuclear fuels, the smelting of copper, and the electroplating of chromium and cadmium [10, 11]. Tiny particles of cadmium, lead, and zinc that have been liberated allow these dangerous metals to float on the wind and land on top of soil or edible plants [12]. PVC goods, color pigments, various alloys, and today's most frequently rechargeable nickel-cadmium batteries all use cadmium compounds as stabilizers [13]. Cadmium metal is mostly utilized as an anti-corrosion substance and also be found in phosphate fertilizers as a contaminant [14]. Agricultural sewage sludge and fertilizer application are two anthropogenic sources of cadmium that can contaminate soil and increase crop cadmium uptake [15]. All living creatures' primary source of heavy metals is food [16]. In general, methylmercury exposure in fish is the main way that living things are exposed to it, along with dental amalgam [11, 16]. The electrochemical method used to produce chlorine uses mercury as an electrode, and the chlor-alkali business is a significant source of mercury use in this process [17]. Methylmercury, an extremely stable form of organic mercury, is present and builds up in the food chain [18, 19].

3. General effects of heavy metals

Heavy metals have been used for a variety of purposes since before recorded history, and they have been vital to the advancement and prosperity of civilizations. Animal populations may be subjected to dangerous amounts of heavy metals at levels very close to those experienced by humans [20, 21]. A substantial number of animals found in heavily metal-polluted areas ingest metal-affected grasses, feed materials, vegetables, and rice plants in addition to contaminated drinking water, which is another potential source of exposure to heavy metals [1, 17]. Heavy metals are widely used and naturally found in the environment, which exposes both humans and animals to them to varying degrees [1, 5]. In addition, because metals are ubiquitous and have been there throughout life's evolution, organisms must contend with them because they may be harmful [22]. Heavy metal traces, including copper (Cu), cobalt (Co), manganese (Mn), iron (Fe), and zinc (Zn), are necessary for a variety of vital physiological processes, including the regulation and operation of several enzyme systems, oxygen and electron transport, hormone synthesis, antioxidant defense, immunity, and fertility [1, 5]. In addition to negatively affecting growth and physiological processes, a lack of essential metals can also make non-essential heavy metals more hazardous [1, 5]. However, even necessary metals will turn hazardous with prolonged contact [14]. Lead (Pb), cadmium (Cd), and mercury (Hg) are examples of toxic heavy metals that are dangerous even at very low doses and have no known biological benefits [23]. Non-essential toxicant metals frequently imitate essential metals to enter the body and potentially disturb important cellular processes [24]. The bioaccumulation of hazardous metals can also be explained by this. Furthermore, because detoxification systems cannot break down an atomic species into a subcomponent with lower toxicity, the elemental nature of metals influences their biotransformation and toxicity [25]. Since metals are elements, their indestructibility and bioaccumulation together lead to a significant concern for metal as a toxicant [14, 22]. The level of exposure, type of heavy metal and its form, age, sex, physiological and nutritional health of the exposed animal, and method of poisoning all affect how toxic heavy metals are to animals [26].

The majority of metals are concentrated in the liver and kidney, along with other essential organs, where they can have toxic or non-toxic effects such as oxidative stress, immunotoxicity, cardiotoxicity, teratogenicity, enzyme inhibition, birth abnormalities, and endocrine disruption [27]. Due to the wide range of chemical characteristics and toxic endpoints, the precise chemical basis of metal toxicity is poorly understood, and there is no universal mechanism for all dangerous metals [22]. It is true that heavy metals cannot degrade into other substances since they are elements, their forms, however, can be converted to free metal ions, which will modify their biological availability, activity, and toxicity [28]. Metals in their ionic form can interact with biological systems and toxicological targets in a variety of ways, making them chemically very reactive [14]. This can lead to a variety of toxic effects and damage to different organs, such as the kidney, nervous system, liver, respiratory system,

Heavy metals	Sources	Side effects	References
Arsenic	Dipping and spraying of animal to control ecto-parasites	Skin damage, circulatory problem, and increase risk of cancer development	[14, 25]
Lead	Industrial polluted waste, Highway crops	Major organ like Liver, Kidney, Brain Damage.	[11, 25, 27]
Cadmium	Mining, smelting, and manufacturing of batteries, pigments, stabilizers, and alloys	Lungs, Kidney, Bone Damage	[18, 27]
Chromium	Tannery facilities, chromate, ferrochrome and chrome pigment production, stainless steel welding.	Kidney damage	[18]

Table 2.
Sources and toxicological side-effects of some heavy metals.

endocrine and reproductive systems, and gastrointestinal tract (**Table 2**) [11, 25, 27]. Targets for heavy metals typically include biological molecules, macromolecules, membranes, or organelles, and interactions between free metal ions and these targets are what cause toxicity [29]. Hazardous metals commonly act by inhibiting enzymes, subcellular organelles, interactions with DNA that cause mutagenesis and cancer, covalent alteration of proteins, displacement of other essential metals-dependent proteins, and the production of free radicals [28, 29]. The toxic metals are grouped into four based on their toxic effects: metals (copper and iron) acting as Fenton reaction catalysts and contributing to the production of free radicals and oxidative stress; metals (nickel, cadmium, and chromium) linked to cancer; metals (aluminum, lead, and tin) linked to neurotoxicity; and generally toxic metals like mercury [30]. However, a growing body of research indicates that most heavy metals can cause oxidative stress in a variety of animal species, including buffalo [31]. This can have an impact on the oxidative stress quotient [25]. Free radical overproduction and oxidative stress damage biomolecules, subcellular structures, and even entire cells, such as neurons, which not only compromise immunity but can result in a variety of illnesses [30]. Oxidative stress has a significant impact on farm animals' ability to produce and reproduce, and it may result in significant financial losses [30, 31]. The endocrine systems of animals may be disrupted by toxic heavy metals, which can also affect animal reproduction and productivity [27]. Heavy metal-induced co-selection of antibiotic resistance genes (ARGs) has emerged as a new environmental concern as there is mounting evidence that heavy metals might affect antibiotic resistance [32]. It has been discovered that the presence of heavy metals in the environment, such as arsenic, copper, and zinc, even at low levels increases bacterial resistance to tetracycline [33–35]. Due to their usage in feed and exposure to heavy metal contamination in the environment, livestock and the systems used to produce them are seen as a major source of heavy metals [9]. As a result, the environment around cattle may be contaminated with heavy metals and antibiotics, which could increase the fast-expanding worry over antibiotic resistance [35]. The vulnerabilities of both humans and animals as a result of compound resistance are highlighted by the confirmed positive link between heavy metal resistance and coexisting methicillin-resistant *Staphylococcus aureus* (MRSA) [35, 36]. Mercury-poisoned animals cannot produce meat, liver, or kidneys fit for human consumption [18]. Depending on the type of mercury poisoning, milk might also be dangerous [37].

3.1 Ruminants

Cattle, particularly young calves, are more susceptible to heavy metals [25, 37]. Owing to their natural curiosity, licking habits, and indiscriminate eating habits, cattle can ingest lead-bearing objects present in their environment as domestic, industrial, or agricultural waste and suffer from accidental acute lead poisoning [37]. Contamination of vegetation and pastures nearby secondary lead smelters (battery recycling units) and lead-zinc smelters [17] was the source of acute lead toxicity in cattle and buffaloes and subclinical toxicity in goats affecting the essential trace mineral profile [27]. The liver and kidneys of the fetus of a lead-poisoned pregnant heifer contained 0.425 and 4.84 ppm of lead, respectively, which were 72% and 84% of the lead concentrations in the respective organs of the dam [25, 37]. Most findings indicate a comparatively higher tolerance in sheep and goats to toxic metals like lead and cadmium than cattle [25, 38–40]. Sheep excrete higher concentrations of lead, chromium, and nickel in their excrement than cows [41]. This may be a reason for the comparatively higher lead concentrations in the milk of ewes than cows [42]. Sheep mostly show subacute toxicity that simulates signs of lead toxicity in adult cattle [37, 41, 42]. Goats, though comparatively more tolerant to lead (chronic toxicity dose of 400 mg per kg body weight) than cattle and sheep, can also exhibit cumulative lethal toxicity with predominant signs of CNS involvement following long-term exposure to lead [25, 27]. Cattle might accidentally absorb lead-containing things from their environment, such as household, industrial, or agricultural trash, due to their natural curiosity, licking tendencies, and indiscriminate feeding habits. This can result in acute lead poisoning [37, 43]. Acute lead toxicity in cattle and buffaloes, as well as subclinical toxicity in goats affecting essential trace mineral profiles, was caused by contamination of vegetation and pastures near secondary lead smelters (battery recycling units) and lead-zinc smelters [44, 45]. The majority of research shows that sheep and goats are more tolerant than cattle to hazardous metals like lead and cadmium. Compared to cows, sheep produce more lead, chromium, and nickel in their excretions [27, 39, 46]. This could explain why sheep milk has somewhat higher lead amounts than cow milk [45]. Most sheep exhibit subacute toxicity, which mimics adult cattle's lead toxicity symptoms [37, 45]. Even though they are more tolerant of lead than cattle and sheep (chronic toxicity dose: 400 mg per kg body weight), goats can nevertheless develop cumulative fatal toxicity, with CNS involvement being the main symptom after prolonged lead exposure [27, 47]. Despite the fact that unintentional acute or chronic poisoning from organic mercury or inorganic compounds can occur in domestic animals, mercury poisoning is uncommon [48–50]. It's possible that phenyl-mercury, which is present in the treated grains as organic mercury, is a more frequent cause of chronic cumulative poisoning [50, 51]. Only when massive amounts of grains are fed to cattle over long periods of time does clinical disease develop. Accidental administration of mercury-containing medications and licking or cutaneous absorption of mercuric oxide-containing skin dressings can also result in sporadic occurrences of poisoning in horses [37, 52]. Mercury poisoning is rare in domestic animals, but accidental acute or chronic poisoning can occur following ingestion of organic mercury or inorganic compounds [53]. Oral ingestion of organic mercury present in the form of phenyl-mercury in the treated grains may be a more common source of cumulative chronic poisoning [47]. However, a ration containing up to 10% of treated grains was not harmful; even feeding a single large amount of grain was incapable of causing toxicity in ruminants [49]. The clinical illness may occur in cattle only when large amounts of grains are fed for long periods [51]. Sporadic cases of poisoning in horses can occur due to accidental administration of drugs containing mercury and licking or cutaneous absorption of skin dressings containing mercuric oxide [27, 37]. In a study, mercury levels were found to be significantly higher in the blood (7.410 g/kg), milk (4.750 g/kg), and urine (2.80 g/kg) of nursing cows raised within a 5-kilometer radius of a thermal power plant [37]. The exposed cows' hemoglobin levels were significantly lower, and their blood urea nitrogen, serum creatinine, albumin, and serum glutamate oxaloacetate transaminase values were all higher, showing effects of mercury on animal health [54, 55]. The study came to the conclusion that long-term exposure of the cows to fly ash mercury may have an effect on the human population, either directly or indirectly through the food chain [55]. While chronic administration of inorganic mercuric chloride (0.8 gm/kg body weight for 14 weeks) in horses caused mercury toxicity, toxic effects in sheep can be evident with an intake of 17.4 mg/kg body weight [37]. Due to the frequent discharge of cadmium and lead together from many industrial sources, the clinical symptoms of spontaneous poisoning commonly combine the two metals [48]. Cadmium levels in feed greater than 50 mg/kg dry matter are linked to toxicity in cattle and sheep [56]. Large amounts of cadmium accidentally consumed can harm the liver and induce acute nephrotoxicity [57]. In animals, chronic intake is linked to metal accumulation, particularly in the kidneys, liver, lungs, and testes [25, 58]. Inappetence, weakness, weight loss, poor hoof keratinization, dry, brittle horns, matting of the hair, and keratosis are a few examples of clinical symptoms in cattle [59]. Significant necropsy abnormalities included degenerative alterations in most organs and hyperkeratosis of the stomach epithelium [37, 59]. Cattle and buffaloes from an industrial area have been observed to have vascular degeneration and necrotic alterations in the liver, kidneys, and lungs and frequently have tissue cadmium levels above 2 ppm [25, 58, 60]. Anemia, nephropathy, and bone demineralization were the results of experimentally poisoning sheep with 2.5 mg of cadmium per kg of body weight [27, 59, 61]. Congenital flaws, stillbirths, and abortion are further potential impacts [61]. The ruminant species that is most vulnerable to chronic copper poisoning is the sheep, and poisoning cases in sheep have been reported all over the world [62, 63]. Contrarily, cattle were thought to be significantly more tolerant of copper accumulation in the past, and up until recently, copper poisoning in cattle was very rare [64]. On the other hand, copper poisoning and the mortality it causes are on the rise everywhere in the world, especially in dairy cattle [64, 65]. Since ruminants have poor homeostatic control over copper absorption, which makes them more sensitive, they have evolved mechanisms for storing excess copper in the liver by decreasing copper in the liver [64, 66]. However, when exposed to copper levels greater than those required for health, they are unable to manage their excretion skills and could become copper toxic [66]. Acute poisoning may happen from consuming 20–100 mg of copper per kilogram of body weight in sheep and young calves and 200–800 mg of copper per kilogram of body weight in mature cattle [67]. Chronic copper poisoning in sheep may occur with daily consumption of 3.5 mg of copper per kg when their grazing pasture contains 15–20 ppm copper (DM) and low levels of molybdenum [25, 68, 69]. Goats have substantially higher copper requirements (15–25 mg per kg, DM) than sheep (640 mg per kg, DM) and can tolerate a far higher dietary copper intake than sheep [69]. Goats' great tolerance to copper may be due to low hepatic absorption [67]. Goats may be able to tolerate a higher concentration of the copper antagonist molybdenum compared to sheep and cattle [66]. The clinical symptoms of acute copper poisoning that are most frequently seen are anorexia, stomach pain, diarrhea, dehydration, unsteadiness, salivation, and collapse before death, which typically occurs within 24 hours. Animals that survive acquire icterus and dysentery [70]. Calves that survive for three or more days are known to have ascites, hydrothorax, hemoglobinuria, head pushing, opisthotonus, aimless roaming, bruxism, circling, and ataxia [71]. A haemolytic disease is chronic poisoning [72]. Sheep that are affected exhibit anorexia, thirst, sadness, jaundice, haemolytic anemia, and hemoglobinuria, as well as accelerated breathing and heart rate [73]. Sheep also exhibit nervous indications such as sadness, blindness, and tetraparesis [25, 74].

3.2 Poultry

Among animal products, eggs are also a possible source of heavy metal contamination. The transferability of heavy metals between hens raised in improved cages and those raised outdoors and found that the extensive soil contamination with these pollutants resulted in free-range eggs having greater heavy metal concentrations than conventionally produced eggs [1, 75, 76]. A study revealed that Cd levels were 0.018 vs. 0.023 ppm in the free-range group, essential Cu levels were 2.591 vs. 2.734 ppm, and essential Zn levels were 5.386 vs. 5.522 ppm in improved cages [1, 76]. By attaching to protein sites and displacing the original metals from their native binding sites, heavy metals can lead to cell dysfunction and toxicity [77, 78]. Additionally, the binding of heavy metals to macromolecules like DNA and nuclear proteins causes oxidative distress [7, 79]. Albumin, the most prevalent protein in plasma, binds to their ions [77]. According to a study, [79] they attach to the free sulfhydryl group of terminal cysteine residues and to histidine residues, which disrupt mitosis, cell respiration, and cell enzymes, especially when arsenic is present [80]. In chicken meat, higher quantities of cadmium have been found [78, 81]. Exposure to cadmium can cause oxidative stress and change the antioxidant enzyme activity in the erythrocytes of adult poultry birds [82]. Cadmium primarily builds up in the proximal tubular cells, where it damages bones or interferes with kidney function to promote bone mineralization [83]. As metallothionein, cadmium binds to proteins that are high in cysteine [84]. As they have similar oxidation states, it can take the place of the zinc in metallothionein and prevent it from acting as a cell scavenger of free radicals, producing hepatotoxicity in the liver and circulating to the kidney, accumulating in the renal tissue, and causing nephrotoxicity [85–87]. Poisoning by lead can occur in poultry [88]. A report from a study revealed that, lead in the feed can significantly stunt chicken growth and decrease blood Delta-aminolevulinic acid dehydratase levels [89]. In hens, the bone has the highest concentrations of deposited lead, followed by the kidney, liver, and skeletal muscle [90]. Majority of investigations revealed that the most often found heavy metals in chicken liver were arsenic, cadmium, mercury, and lead residues. Due to its ability to bind to and deactivate vital enzymes, lead is the most hazardous element [1, 7, 25, 37]. The liver and kidneys bio-transform arsenic, and the methylated metabolites are distributed throughout the body [91]. The amount of accumulated element in the organs varies on the duration of exposure, the amount of ingested element, the animal's age, and breed [92, 93]. Since these effects are dose-dependent, prolonged exposure to heavy metals through contaminated feed is likely to result in more harmful changes to tissues [93].

3.3 Fish

Fish may be a source of heavy metal exposure for people because it contains important proteins and n-3 polyunsaturated fatty acids [94]. Fish living in

Heavy metal	Concentration, mg/kg
Co	0.015
Cd	0.030
Pb	0.050
Fe	0.300
Cu	1.000

Table 3.
Levels of some heavy metals in fishes as reported in WHO.

contaminated waters have a tendency to accumulate heavy metals from the petroleum industry, air sources, and cosmetics in their tissues [11]. In general, accumulation activities depend on the amount of metal present, the length of exposure, the method used to prepare the metal for absorption, the environment (water, nuclear fuels, chromium and cadmium electroplating, temperature, pH, hardness, salinity), and intrinsic factors such as fish age and feeding habits [92, 93]. Rubber tire dust on the road surface contains cadmium, lead, and zinc [95]. Different fish tissues have a predilection for these tiny sizes [96]. Majority of these particles allow the harmful metals to rise on the wind, where they will primarily concentrate in the liver, kidney, and gills [97]. They may also be ingested or transferred to the soil or edible plants. Heavy metal build-up in fish tissues (**Table 3**) is mostly influenced by their concentrations in the water, in food, or in commercial feed [98]. The visceral tissues of fish, such as the liver, kidney, and intestines, which are typically discarded throughout the production procedures, tend to accumulate more metals than the muscles when compared to the other tissues in fish [1, 97, 99]. Significant correlations between lipid levels and concentrations of critical Cu and Zn [100] in two species of farmed fish (pompano and snapper). According to the research, lipid content may play a significant role in controlling the bioaccumulation of certain metals. Although copper is a necessary metal for both types of fish, its toxicity causes damage to the fish's gills, liver, and kidneys, which can result in death [101]. Lead has physiological and biochemical effects by acting as a mimicking agent to replace necessary metabolic components like calcium, iron, and zinc [102]. For example, it directly inhibits the action of sulfhydryl group-rich protein enzymes as well as zinc and iron in the production of heme [103]. Lead binds to a variety of transport proteins, including calcium-ATPase, calmodulin, transferrin, metallothionein, and metallothionein [104].

4. Analytical methods of detection in animal production

4.1 Determination of heavy metals in fish species of the Mediterranean Sea (Libyan coastline) using atomic absorption spectrometry

The toxicity of heavy metals when their concentration exceeds the allowable limit has made their presence in our environment a major source of worry [105]. The World Health Organization (WHO) has established the approved values of the Co, Cd, Pb, Fe, and Cu concentrations in fish, which are shown in **Table 3**. The document presents the Co concentrations in various tissues. The concentration of this metal was found to range from 0.570 mg/kg to 44.693 mg/kg, meaning that the levels of Co in all tissues that were investigated were higher than those that had been previously reported.

4.2 A titrimetric method for the quantitative estimation of lead in biological materials

As a result of Fischer's 1929 announcement of the exceptional affinity of dithizone (diphenylthiocarbazone) solutions for lead [106], other laboratories have been inspired to look for useful micro-methods that may be used to detect lead in biological materials.

4.3 Ion chromatographic and voltammetric determination of heavy metals in soils. Comparison with atomic emission spectroscopy

The aim is to compare different analytical techniques, aspectroscopic, an electrochemical and a chromatographicone to determine total heavy metals concentrations in soils and to establish a correlation between these three techniques even though soil solutions are complex matrices providing a lot of interference problems [107]. Atomic emission spectroscopy is the method of reference for heavy metals concentrations analysis.

4.4 Speciation of heavy metal binding non-protein thiols in Agropyronelongatum by size-exclusion HPLC: ICP-MS

In order to quantify the major of heavy metals (Pb, Cu, Cd, Co, Zn and Ni), three ionic separation column systems were evaluated [108]: (1) a cationic column (HPIC-CS2, Dionex) tested with two eluents (10 mMoxalic acid–7.5 mM citric acid; and 40 mM D-tartaric acid–12 mM citric acid); (2) an anionic column (HPIC-AS4, Dionex) evaluated with 25 mM oxalic acid as eluent; and (3) a bifunctional ion-exchange column (Ionpac CS5, Dionex) which was also tested with two eluents (6 mM pyridine, 2,6-dicarboxylic acid; and 50 mM oxalic acid/95 mM lithium hydroxide

5. Strategies to control heavy metal pollution

In order to minimize soil pollution issues and heavy metal contamination, efforts must be made to close nutrient cycles on farms by recycling nutrients in livestock manure [109]. As a result, numerous studies have been conducted to discuss methods for reducing heavy metal levels in soil and livestock waste [110–112]. Reducing the input/output ratio of heavy metals in animals should be the main objective of effective solutions [113]. For this reason, several multidisciplinary strategies should be taken into consideration to reduce animal intake, excretion in feces, and concentration in manure [1, 114, 115]. Since nutrients found in manure or in compounds come from the portion of feed that is not absorbed by the animals, adjusting the diet may be an effective strategy to influence the amount of manure generated as well as its composition [1, 116]. A formulated diet is required to increase the efficiency of nutrient retention by animals and decrease their excretion in feces [117]. For instance, using commercial amino acids to minimize nitrogen excretion in pigs and poultry is a very effective method. As a result, imports of feedstuffs high in protein, including soybean meal, are significantly reduced at the farm level. In several species, it has been proven effective to add enzymes to the meal to increase the biological availability of some particular nutrients [118]. Animal nutrition is a key factor in controlling nutrient flows on livestock farms [119]. Utilizing trace mineral supplements may help stop the "waste." Alternative innovative compounds to antibiotics as well as to Zn and Cu should be used to control enteric diseases, and the maximum permitted level should not be thought of as the ideal level for animal requirements [120]. There is a need for various technologies to remove the content of heavy metals from contaminated soil and water in agricultural land because an excessive amount of heavy metal output from manures can still seep into the soil and water [121]. Different manure treatment methods have been researched and can be used in the field to lower the heavy metal output from animals [112].

Metal concentrations can be lowered using electro-remediation, which involves running an electric current through liquid manure to cause metal ions to precipitate on an electrode [122]. The technique has not been tested on farms yet, so it's unlikely to be economical at this time. In order to reduce the environmental impact and guarantee high nutrient efficiency, the recycling loop of manure back into food production should, from a whole-farm viewpoint, be as brief as feasible [123]. A group of researchers claim that phytoremediation, a straightforward clean-up technique, has the potential to get rid of metals from agricultural land by using plants that accumulate significant amounts of heavy metal contamination [124] . The discovery that plants might metabolize harmful pesticides led to the development of this technique a few decades ago [125]. It is seen by the communities as an acceptable, efficient, and novel technology that is also cost-effective. The term "phytoremediation" refers to a group of methods that use plants and related bacteria to remove contaminants from matrices by transfer, confinement, accumulation, or dissipation [126]. Its cost-effectiveness and potential to limit the exposure of the polluted substrate to people, animals, and the environment are both facilitated by the fact that phytoremediation is typically carried out in situ [124]. The four types of phytoremediation are phytoextraction, phytofiltration, phytostabilization, and phytovolatilization, depending on the circumstances, the extent of clean-up necessary, the plants used, and the contaminants [127]. Since nutrients found in manure or in compounds come from the portion of feed that is not absorbed by the animals, adjusting the diet may be an effective strategy to influence the amount of manure generated as well as its composition [124]. Phytoremediation is a straightforward clean-up technique that makes use of plants that accumulate significant levels of heavy metal pollutants and offers hope for the removal of metals from agricultural land. The discovery that plants might metabolize harmful pesticides led to the development of this technique a few decades ago [125]. It is seen by the communities as an acceptable, efficient, and novel technology that is also cost-effective. The ideal plants for heavy metal removal ought to possess the following traits:

i. rapid growth;

ii. a deeply branched and widely dispersed root system;

iii. good climatic and environmental adaptation;

iv. ease of cultivation and harvest;

v. production of more above-ground biomass;

vi. resistance to pathogens and pests;

vii. increased accumulation of the desired heavy metals from soil; and

viii. translocation of the accumulated heavy metals from the soil.

A formulated diet is required that decreases the efficiency of nutrient retention by animals, increases their excretion in for instance, using commercial amino acids to minimize nitrogen excretion in pigs and poultry is a very effective method [117]. Alternative innovative compounds to antibiotics as well as to Zn and Cu should be used to control enteric diseases, and the maximum permitted level should not be thought of as the ideal level for animal requirements [128].

Heavy metals are absorbed, precipitated, and concentrated by plant or seed roots that have been raised in aerated water [129]. In order to stop contaminants from migrating to groundwater or entering the food chain, phytostabilization is employed to minimize their mobility and bioavailability in the environment [130]. Phytovolatilization is the process by which pollutants are taken up by plants from the soil, transformed into a volatile state, and then released into the atmosphere [131]. The primary and most effective method for removing heavy metals and metalloids from contaminated soils or water is phytoextraction [132]. The bioavailability of metals in soil, which is affected by a variety of factors such as chemical composition, pH, geochemical properties of metals, environmental variables, and agricultural soil management, has a significant impact on phytoremediation effectiveness [133]. Bioavailability can be increased by lowering the pH of the soil, using fertilizers, soil microorganisms, and root exudates, and adding chelating agents [134]. In order to accomplish the homeostasis of agriculture with natural habitats and to maintain balanced production systems, it is crucial to control environmental losses and the spread of toxins from livestock manure [1]. The unintentional discharge of farm waste into water has led to outbreaks of dangerous infections, even though it is unlawful to spread manure close to surface water and on frozen land, as it is in the majority of European nations [135]. Technologies and approaches are consequently required to manage these environmental issues.

6. Conclusions

Heavy metals are shown in the commercial agricultural sector as both mineral fertilizers and contaminants/unwanted chemicals. Although there are set extensive regulations to prevent their pollution, their distribution at various levels makes it impossible to completely avoid the presence of heavy metals in the environment and the food chain. Controlling the animal input may be a useful tactic for lowering the dangers to human health from consuming items with animal origins and from manure's environmental contamination. It is possible to change the diets of animals in order to lower the potential amount of minerals and nutrients that are not absorbed and end up in the manure. Effective solutions against heavy metals must take into account the intricate linkages between rural activities, the vast range of farming practices, the soil, and climatic variables. It should be advised to use additives more precisely in order to prevent contaminating the environment.

Author details

Selina Acheampong
University College of Agriculture and Environmental Studies, Bunso and Kwame Nkrumah University of Science and Technology, Kumasi, Ghana

*Address all correspondence to: paaby12@yahoo.com

References

[1] Hejna M, Gottardo D, Baldi A, Dell'Orto V, Cheli F, Zaninelli M, et al. Nutritional ecology of heavy metals. Animal. 2018;**12**(10):2156-2170

[2] Belay K, Abisa Z. Developing a method for trace metal analysis in spices using spectroscopic techniques: A review. International Journal of Chemistry Nature Science. 2015;**3**:195-199

[3] Singh R, Gautam N, Mishra A, Gupta R. Heavy metals and living systems: An overview. Indian Journal of Pharmacology. 2011;**43**(3):246

[4] Ali H, Khan E. What are heavy metals? Long-standing controversy over the scientific use of the term 'heavy metals'–proposal of a comprehensive definition. Toxicological & Environmental Chemistry. 2018;**100**(1):6-19

[5] Aliasgharpour M, Rahnamaye FM. Trace elements in human nutrition: A review. International journal of medical investigation. 2013;**2**(3)

[6] Vardhan KH, Kumar PS, Panda RC. A review on heavy metal pollution, toxicity and remedial measures: Current trends and future perspectives. Journal of Molecular Liquids. 2019;**290**:111197

[7] Briffa J, Sinagra E, Blundell R. Heavy metal pollution in the environment and their toxicological effects on humans. Heliyon. 2020;**6**(9):e04691

[8] Pandey G, Madhuri S. Heavy metals causing toxicity in animals and fishes. Research Journal of Animal, Veterinary and Fishery Sciences. 2014;**2**(2):17-23

[9] Ali H, Khan E, Ilahi I. Environmental chemistry and ecotoxicology of hazardous heavy metals: Environmental persistence, toxicity, and bioaccumulation. Journal of Chemistry. 2019;**2019**

[10] Baby J, Raj JS, Biby ET, Sankarganesh P, Jeevitha MV, Ajisha SU, et al. Toxic effect of heavy metals on aquatic environment. International Journal of Biological and Chemical Sciences. 2010;**4**(4)

[11] Zeitoun MM, Mehana EE. Impact of water pollution with heavy metals on fish health: Overview and updates. Global Veterinaria. 2014;**12**(2):219-231

[12] Lazor P, Tomáš J, Tóth T, Tóth J, Čéryová S. Monitoring of air pollution and atmospheric deposition of heavy metals by analysis of honey. Journal of Microbiology, Biotechnology and Food Sciences. 2021;**2021**:522-533

[13] Hayat MT, Nauman M, Nazir N, Ali S, Bangash N. Environmental hazards of cadmium: Past, present, and future. In Cadmium Toxicity and Tolerance in Plants. Academic Press; 2019. pp. 163-183

[14] Brusseau ML, Artiola JF. Chemical contaminants. In: Environmental and Pollution Science. Academic Press; 2019. pp. 175-190

[15] Alloway BJ, Steinnes E. Anthropogenic additions of cadmium to soils. Cadmium in soils and plants. 1999:97-123

[16] El-Kady AA, Abdel-Wahhab MA. Occurrence of trace metals in foodstuffs and their health impact. Trends in food science & technology. 2018;**75**:36-45

[17] Crook J, Mousavi A. The chlor-alkali process: A review of history and

pollution. Environmental Forensics. 2016;**17**(3):211-217

[18] EFSA Panel on Contaminants in the Food Chain (CONTAM). Scientific opinion on the risk for public health related to the presence of mercury and methylmercury in food. EFSA Journal. 2012;**10**(12):2985

[19] Gojkovic Z, Skrobonja A, Funk C, Garbayo I, Vílchez C. The role of microalgae in the biogeochemical cycling of methylmercury (MeHg) in aquatic environments. Phycology. 16 Sep 2022;**2**(3):344-362

[20] Nriagu JO. A silent epidemic of environmental metal poisoning? Environmental Pollution. 1988;**50**(1-2):139-161

[21] Huntingford FA, Adams C, Braithwaite VA, Kadri S, Pottinger TG, Sandøe P, et al. Current issues in fish welfare. Journal of Fish Biology. 2006;**68**(2):332-372

[22] Liu Y, Hu Z, Gao S, Günther D, Xu J, Gao C, et al. In situ analysis of major and trace elements of anhydrous minerals by LA-ICP-MS without applying an internal standard. Chemical Geology. 2008;**257**(1-2):34-43

[23] Anjulo TK, Mersso BT. Assessment of dairy feeds for heavy metals. American Academic Scientific Research Journal for Engineering, Technology, and Sciences. 2015;**11**(1):20-31

[24] Martinez-Finley EJ, Chakraborty S, Fretham SJ, Aschner M. Cellular transport and homeostasis of essential and nonessential metals. Metallomics. 2012;**4**(7):593-605

[25] Gupta AR, Bandyopadhyay S, Sultana F, Swarup D. Heavy metal poisoning and its impact on livestock health and production system. Indian Journal of Animal Health. 2021;**60**(2):1-23

[26] Gall JE, Boyd RS, Rajakaruna N. Transfer of heavy metals through terrestrial food webs: A review. Environmental Monitoring and Assessment. 2015;**187**(4):1-21

[27] Raghuvanshi R. Amelioration of heavy metal induced toxicity using probiotic Escherichia coli strain in rats (Doctoral dissertation). India: Maharaja Sayajirao University of Baroda

[28] Ochiai EI. Toxicity of heavy metals and biological defense: Principles and applications in bioinorganic chemistry-VII. Journal of Chemical Education. 1995;**72**(6):479

[29] Kumar A, Singh N, Pandey R, Gupta VK, Sharma B. Biochemical and molecular targets of heavy metals and their actions. In: Biomedical Applications of Metals. Cham: Springer; 2018. pp. 297-319

[30] Kozlowski H, Kolkowska P, Watly J, Krzywoszynska K, Potocki S. General aspects of metal toxicity. Current Medicinal Chemistry. 2014;**21**(33):3721-3740

[31] Yeotikar PV, Nayyar S, Singh C, Mukhopadhyay CS, Kakkar SS, Jindal R. Seasonal variation in oxidative stress markers of Murrah buffaloes in heavy metal exposed areas of Ludhiana. Indian Journal of Animal Research. 2019;**53**(10):1310-1315

[32] Ding J, An XL, Lassen SB, Wang HT, Zhu D, Ke X. Heavy metal-induced co-selection of antibiotic resistance genes in the gut microbiota of collembolans. Science of the Total Environment. 2019;**683**:210-215

[33] Chen S, Li X, Sun G, Zhang Y, Su J, Ye J. Heavy metal induced antibiotic resistance in bacterium LSJC7. International Journal of Molecular Sciences. 2015;**16**(10):23390-23404

[34] Yu Z, Gunn L, Wall P, Fanning S. Antimicrobial resistance and its association with tolerance to heavy metals in agriculture production. Food Microbiology. 2017;**64**:23-32

[35] Mazhar SH, Li X, Rashid A, Su J, Xu J, Brejnrod AD, et al. Co-selection of antibiotic resistance genes, and mobile genetic elements in the presence of heavy metals in poultry farm environments. Science of The Total Environment. 2021;**755**:142702

[36] Imran M, Das KR, Naik MM. Co-selection of multi-antibiotic resistance in bacterial pathogens in metal and microplastic contaminated environments: An emerging health threat. Chemosphere. 2019;**215**:846-857

[37] Constable PD, Hinchcliff KW, Done SH, Grunberg W. Veterinary Medicine: A Textbook of Diseases of Cattle, Horses, Sheep, Pigs and Goats. 11th ed. Missouri: Elsevier Ltd; 2017. pp. 421-425

[38] Abdel-Salam NM, Ahmed S, Basir A, Rais AK, Bibi A, Ullah R, et al. Distribution of heavy metals in the liver, kidney, heart, pancreas and meat of cow, buffalo, goat, sheep and chicken from Kohat market Pakistan. Life Science Journal. 2013;**10**(7s):937-940

[39] Ogbomida ET, Nakayama SM, Bortey-Sam N, Oroszlany B, Tongo I, Enuneku AA, et al. Accumulation patterns and risk assessment of metals and metalloid in muscle and offal of free-range chickens, cattle and goat in Benin City, Nigeria. Ecotoxicology and Environmental Safety. 2018;**151**:98-108

[40] Bhardwaj H, Singh C, Kumar BS, Singh D. Role of metallothionein to moderate heavy metals toxicity in animals: A review. Cancer. 2021;**8**(9)

[41] Zeinali T, Salmani F, Naseri K. Dietary intake of cadmium, chromium, copper, nickel, and lead through the consumption of meat, liver, and kidney and assessment of human health risk in Birjand, southeast of Iran. Biological trace element research. 2019;**191**(2):338-347

[42] Hussein AA, Omer SS, Ali IS, Suleiman BF. The assessment of hepatic mineral composition in sheep, cattle, chicken, and fish in Erbil City, Kurdistan region-Iraq. Kurdistan Journal of Applied Research. 2021:46-55

[43] Alkmim Filho JF, Germano A, Dibai WL, Vargas EA, Melo MM. Heavy metals investigation in bovine tissues in Brazil. Food Science and Technology. 2014;**34**:110-115

[44] Rumbeiha WK, Braselton WE, Donch D. A retrospective study on the disappearance of blood lead in cattle with accidental lead toxicosis. Journal of Veterinary Diagnostic Investigation. 2001;**13**(5):373-378

[45] Swarup D, Patra RC, Naresh R, Kumar P, Shekhar P. Blood lead levels in lactating cows reared around polluted localities; transfer of lead into milk. Science of the Total Environment. 2005;**347**(1-3):106-110

[46] Reis LS, Pardo PE, Camargos AS, Oba E. Mineral element and heavy metal poisoning in animals. Journal of medicine and medical Sciences. 2010:560-579

[47] Bilandžić N, Sedak M, Đokić M, Šimić B. Wild boar tissue levels of

cadmium, lead and mercury in seven regions of continental Croatia. Bulletin of Environmental Contamination and Toxicology. 2010;**84**(6):738-743

[48] Goyer RA, Clarkson TW. Toxic effects of metals. Casarett and Doull's Toxicology: the Basic Science of Poisons. 1996;**5**:691-736

[49] Oruc HH. Fungicides and their effects on animals. In: Carisse O, editor. Fungicides. London, UK: In-Tech Publishers; 2010. pp. 349-362

[50] Clifton JC II. Mercury exposure and public health. Pediatric Clinics of North America. 2007;**54**(2):237-2e1

[51] Syversen T, Kaur P. The toxicology of mercury and its compounds. Journal of Trace Elements in Medicine and Biology. 2012;**26**(4):215-226

[52] Clarkson TW. The toxicology of mercury. Critical Reviews in Clinical Laboratory Sciences. 1997;**34**(4):369-403

[53] Ozuah PO. Mercury poisoning. Current Problems in Pediatrics. 2000;**30**(3):91-99

[54] Russell KE, Roussel AJ. Evaluation of the ruminant serum chemistry profile. Veterinary Clinics of North America: Food Animal Practice. 2007;**23**(3):403-426

[55] Mahajan VE, Yadav RR, Dakshinkar NP, Dhoot VM, Bhojane GR, Naik MK, et al. Influence of mercury from fly ash on cattle reared nearby thermal power plant. Environmental Monitoring and Assessment. 2012;**184**(12):7365-7372

[56] Naja GM, Volesky B. Toxicity and sources of Pb, Cd, Hg, Cr, As, and radionuclides in the environment. In: Handbook of Advanced Industrial and Hazardous Wastes Management. Crc Press; 2017. pp. 855-903

[57] Almeer RS, AlBasher GI, Alarifi S, Alkahtani S, Ali D, Abdel Moneim AE. Royal jelly attenuates cadmium-induced nephrotoxicity in male mice. Scientific Reports. 2019;**9**(1):1-2

[58] Sarkar A, Ravindran G, Krishnamurthy V. A brief review on the effect of cadmium toxicity: From cellular to organ level. International Journal of Biotechnology Research. 2013;**3**(1):17-36

[59] Gumasta P, Dubey A, Swamy M, Verma Y. Study on cadmium associated microscopic lesions in cattle and buffaloes

[60] Gumasta P, Dubey A, Swamy M, Verma Y, Malapure CD. A demographic study on cadmium level in blood of bovine

[61] Randa AH, Dawlat MA, Nariman AR, Hatem ME, Dessouky MI. Clinicopathological, histopathological and immunological studies on animals exposed to lead and cadmium under experimental conditions. New York Science Journal. 2012;**5**:12

[62] Minervino AH, López-Alonso M, Barrêto Júnior RA, Rodrigues FA, Araújo CA, Sousa RS, et al. Dietary zinc supplementation to prevent chronic copper poisoning in sheep. Animals. 2018;**8**(12):227

[63] Banerjee S. Acute copper toxicity in garole sheep-a case study. World Applied Sciences Journal. 2009;**7**:1547-1551

[64] López-Alonso M, Miranda M. Copper supplementation, a challenge in cattle. Animals. 2020;**10**(10):1890

[65] Thrusfield M. Veterinary Epidemiology. John Wiley & Sons; 2018

[66] Suttle NF. Copper imbalances in ruminants and humans: Unexpected common ground. Advances in nutrition. 2012;**3**(5):666-674

[67] Araya M, Koletzko B, Uauy R. Copper deficiency and excess in infancy: Developing a research agenda. Journal of pediatric gastroenterology and nutrition. 2003;**37**(4):422-429

[68] Lukhele MS. The chemical composition and nutritive value of leaves of indigenous fodder trees (Doctoral dissertation). University of Pretoria

[69] EFSA Panel on Additives and Products or Substances used in Animal Feed (FEEDAP). Revision of the currently authorised maximum copper content in complete feed. EFSA Journal. 2016;**14**(8):e04563

[70] Glennon HM. Effect of copper oxide needles on gastrointestinal parasites in grazing meat goats

[71] Picco S, Ponzzinibio M, Mattioli G, Rosa D, Minatel L, Fazzio L, et al. Physiological and genotoxic effects of molybdenum-induced copper deficiency in cattle. Agrociencia. 2012;**46**(2):107-117

[72] Lee JJ, Kim YK, Cho SH, Park KS, Chung IJ, Cho D, et al. Hemolytic anemia as a sequela of arsenic intoxication following long-term ingestion of traditional Chinese medicine. Journal of Korean Medical Science. 2004;**19**(1):127-129

[73] Newcomer BW, Cebra C, Chamorro MF, Reppert E, Cebra M, Edmondson MA. Diseases of the hematologic, immunologic, and lymphatic systems (multisystem diseases). Sheep, Goat, and Cervid Medicine. 2021:405

[74] Giadinis ND, Papaioannoy N, Kritsepi-Konstantinou M, Roubies N, Raikos N, Karatzias H. Acute encephalopathy and clinical pathology findings in a sheep with chronic copper poisoning. Turkish Journal of Veterinary & Animal Sciences. 2009;**33**(4):363-366

[75] Piskorska-Pliszczynska J, Struciński P, Kan CA. Contaminants in eggs: Dioxins/PCBs and other toxic substances and their possible health implications. Chemical Hazards in Foods of Animal Origin. 2019;**7**:415

[76] Radu-Rusu CG, Pop IM, Albu A, Bologa M, Radu-Rusu RM. Transferability of certain heavy metals from hens feed to table eggs laid within different rearing systems. Lucrări Ştiinţifice-Seria Zootehnie. 2013;**59**(18):218-222

[77] Jaishankar M, Tseten T, Anbalagan N, Mathew BB, Beeregowda KN. Toxicity, mechanism and health effects of some heavy metals. Interdisciplinary toxicology. 2014;**7**(2):60

[78] Penido ES, Melo LC, Guilherme LR, Bianchi ML. Cadmium binding mechanisms and adsorption capacity by novel phosphorus/magnesium-engineered biochars. Science of the Total Environment. 2019;**671**:1134-1143

[79] Flora SJ, Mittal M, Mehta A. Heavy metal induced oxidative stress & its possible reversal by chelation therapy. Indian Journal of Medical Research. 2008;**128**(4):501

[80] Valko M, Jomova K, Rhodes CJ, Kuča K, Musílek K. Redox-and non-redox-metal-induced formation of free radicals and their role in human disease. Archives of Toxicology. 2016;**90**(1):1-37

[81] Orisakwe OE, Oladipo OO, Ajaezi GC, Udowelle NA. Horizontal and

vertical distribution of heavy metals in farm produce and livestock around lead-contaminated goldmine in Dareta and Abare, Zamfara state, northern Nigeria. Journal of Environmental and Public health. 2017;**2017**

[82] Kar I, Patra AK. Tissue bioaccumulation and toxicopathological effects of cadmium and its dietary amelioration in poultry—A review. Biological Trace Element Research. 2021;**199**(10):3846-3868

[83] Johri N, Jacquillet G, Unwin R. Heavy metal poisoning: The effects of cadmium on the kidney. Biometals. 2010;**23**(5):783-792

[84] Klaassen CD, Liu J, Choudhuri S. Metallothionein: An intracellular protein to protect against cadmium toxicity. Annual Review of Pharmacology and Toxicology. 1999;**39**(1):267-294

[85] McClain CJ, Marsano L, Burk RF, Bacon B. Trace metals in liver disease. In: Seminars in Liver Disease. Vol. 11, No. 04. 1991 by Thieme Medical Publishers, Inc.; 1991. pp. 321-339

[86] Nisanian M, Holladay SD, Karpuzoglu E, Kerr RP, Williams SM, Stabler L, et al. Exposure of juvenile Leghorn chickens to lead acetate enhances antibiotic resistance in enteric bacterial flora. Poultry Science. 1 Apr 2014;**93**(4):891-897

[87] Castagnetto JM, Hennessy SW, Roberts VA, Getzoff ED, Tainer JA, Pique ME. MDB: The metalloprotein database and browser at the Scripps research institute. Nucleic Acids Research. 2002;**30**(1):379-382

[88] Baloš MŽ, Jakšić S, Pelić DL. The role, importance and toxicity of arsenic in poultry nutrition. World's Poultry Science Journal. 2019;**75**(3):375-386

[89] Bersényi A. Study of toxic metals (Cd, Pb, Hg and Ni) in rabbits and broiler chickens

[90] Trampel DW, Imerman PM, Carson TL, Kinker JA, Ensley SM. Lead contamination of chicken eggs and tissues from a small farm flock. Journal of Veterinary Diagnostic Investigation. 2003;**15**(5):418-422

[91] Vahter M. Genetic polymorphism in the biotransformation of inorganic arsenic and its role in toxicity. Toxicology letters. 2000;**112**:209-217

[92] Gasparik J, Massanyi P, Slamecka J, Fabis M, Jurcik R. Concentration of selected metals in liver, kidney, and muscle of the red deer (Cervus elaphus). Journal of Environmental Science and Health, Part A. 2004;**39**(8):2105-2111

[93] Leggett RW. An age-specific kinetic model of lead metabolism in humans. Environmental Health Perspectives. 1993;**101**(7):598-616

[94] Muinde VM, Nguu EK, Ogoyi DO, Shiundu PM. Effects of heavy metal pollution on omega-3 polyunsaturated fatty acids levels in tilapia fish from Winam gulf of Lake Victoria. The Open Environmental Engineering Journal. 2013;**6**(1)

[95] Radziemska M, Fronczyk J. Level and contamination assessment of soil along an expressway in an ecologically valuable area in Central Poland. International Journal of Environmental Research and Public Health. 2015;**12**(10):13372-13387

[96] Parker BW, Beckingham BA, Ingram BC, Ballenger JC, Weinstein JE, Sancho G. Microplastic and tire wear particle occurrence in fishes from an urban estuary: Influence of feeding characteristics on exposure risk. Marine Pollution Bulletin. 2020;**160**:111539

[97] Vajargah MF. A review on the effects of heavy metals on aquatic animals. Journal of ISSN. 2021;**2766**:2276

[98] Feldlite M, Juanicó M, Karplus I, Milstein A. Towards a safe standard for heavy metals in reclaimed water used for fish aquaculture. Aquaculture. 2008;**284**(1-4):115-126

[99] Kato LS, Ferrari RG, Leite JV, Conte-Junior CA. Arsenic in shellfish: A systematic review of its dynamics and potential health risks. Marine Pollution Bulletin. 2020;**161**:111693

[100] Hemmadi V. A critical review on integrating multiple fish biomarkers as indicator of heavy metals contamination in aquatic ecosystem. International Journal of Bioassays. 2017;**6**(9):5494-5506

[101] Fonseca VF, França S, Serafim A, Company R, Lopes B, Bebianno MJ, et al. Multi-biomarker responses to estuarine habitat contamination in three fish species: Dicentrarchus labrax, Solea senegalensis and Pomatoschistus microps. Aquatic Toxicology. 2011;**102**(3-4):216-227

[102] Shadab GH, Afzal M. Lead and zinc interactions–An influence of zinc over lead related toxic manifestations. Journal of Trace Elements in Medicine and Biology. 2021;**64**:126702

[103] Zhang S, Sun L, Zhang J, Liu S, Han J, Liu Y. Adverse impact of heavy metals on bone cells and bone metabolism dependently and independently through anemia. Advanced Science. 2020;**7**(19): 2000383

[104] Aldoghachi MA, Azirun MS, Yusoff I, Ashraf MA. Ultrastructural effects on gill tissues induced in red tilapia Oreochromis sp. by a waterborne lead exposure. Saudi Journal of Biological Sciences. 2016;**23**(5):634-641

[105] Khalifa KM, Hamil AM, Al-Houni AQA, Ackacha MA. Determination of heavy metals in fish species of the Mediterranean Sea (Libyan coastline) using atomic absorption spectrometry. International Journal of PharmTech Research. 2010;**2**:1350-1354

[106] Horwit MK and Cowgill George RA. Titrimetric Method for the Quantitative Estimation of Lead in Biological Materials. www.jbc.org. 2011. pp. 553-564

[107] Gunkel P, Fabre B, Prado G, Baliteau JY. Ion chromatographic and voltammetric determination of heavy metals in soils. Comparison with atomic emission spectroscopy. EDP Wiley-VCH Sciences. 1999;**27**:823-828

[108] Wei MS. Zhenggui, Wong Jonathan Woonchung, Chen Dengyun. Speciation of Heavy Metal Binding Non-Protein thiols in Agropyronelongatum by Size-Exclusion HPLC–ICP Microchemical Journal. 2003;**74**:207-213

[109] Sharma B, Vaish B, Singh UK, Singh P, Singh RP. Recycling of organic wastes in agriculture: An environmental perspective. International journal of environmental research. 2019;**13**(2):409-429

[110] Su G, Ong HC, Zulkifli NW, Ibrahim S, Chen WH, Chong CT, et al. Valorization of animal manure via pyrolysis for bioenergy: A review. Journal of Cleaner Production. 2022:130965

[111] Hou D, O'Connor D, Nathanail P, Tian L, Ma Y. Integrated GIS and multivariate statistical analysis for regional scale assessment of heavy metal soil contamination: A critical review. Environmental Pollution. 2017;**231**:1188-1200

[112] Guo G, Zhou Q, Ma LQ. Availability and assessment of fixing additives for the in situ remediation of heavy metal contaminated soils: A review. Environmental Monitoring and Assessment. 2006;**116**(1):513-528

[113] Halberg N, Verschuur G, Goodlass G. Farm level environmental indicators; are they useful?: An overview of green accounting systems for European farms. Agriculture, Ecosystems & Environment. 2005;**105**(1-2):195-212

[114] Acutis M, Alfieri L, Giussani A, Provolo G, Di Guardo A, Colombini S, et al. ValorE: An integrated and GIS-based decision support system for livestock manure management in the Lombardy region (northern Italy). Land Use Policy. 2014;**41**:149-162

[115] Hejna M, Moscatelli A, Onelli E, Baldi A, Pilu S, Rossi L. Evaluation of concentration of heavy metals in animal rearing system. Italian Journal of Animal Science. 2019;**18**(1):1372-1384

[116] Ferket PR, Van Heugten E, Van Kempen TA, Angel R. Nutritional strategies to reduce environmental emissions from nonruminants. Journal of Animal Science. 2002;**80**(E-suppl_2):E168-E182

[117] Nahm KH. Efficient feed nutrient utilization to reduce pollutants in poultry and swine manure. Critical Reviews in Environmental Science and Technology. 2002;**32**(1):1-6

[118] Glencross BD, Booth M, Allan GL. A feed is only as good as its ingredients–a review of ingredient evaluation strategies for aquaculture feeds. Aquaculture Nutrition. 2007;**13**(1):17-34

[119] Petersen SO, Sommer SG, Béline F, Burton C, Dach J, Dourmad JY, et al. Recycling of livestock manure in a whole-farm perspective. Livestock science. 2007;**112**(3):180-191

[120] Hejna M, Onelli E, Moscatelli A, Bellotto M, Cristiani C, Stroppa N, et al. Heavy-metal phytoremediation from livestock wastewater and exploitation of exhausted biomass. International Journal of Environmental Research and Public Health. 2021;**18**(5):2239

[121] Kanwar VS, Sharma A, Srivastav AL, Rani L. Phytoremediation of toxic metals present in soil and water environment: A critical review. Environmental Science and Pollution Research. 2020;**27**(36):44835-44860

[122] Modin O, Aulenta F. Three promising applications of microbial electrochemistry for the water sector. Environmental Science: Water Research & Technology. 2017;**3**(3):391-402

[123] Oberč BP, Arroyo SA. Approaches to sustainable agriculture. Exploring the pathways. 2020

[124] Marques AP, Rangel AO, Castro PM. Remediation of heavy metal contaminated soils: Phytoremediation as a potentially promising clean-up technology. Critical Reviews in Environmental Science and Technology. 2009;**39**(8):622-654

[125] Van Aken B. Transgenic plants for enhanced phytoremediation of toxic explosives. Current Opinion in Biotechnology. 2009;**20**(2):231-236

[126] Cunningham SD, Berti WR. Remediation of contaminated soils with green plants: An overview. In Vitro Cellular & Developmental Biology-Plant. 1993;**29**(4):207-212

[127] Ashfaq H, Abubakar M, Ghulzar H, Farid M, Yaqoob S, Komal N, et al.

Phytoremediation potential of oilseed crops for lead-and nickel-contaminated soil. In: Plant Ecophysiology and Adaptation under Climate Change: Mechanisms and Perspectives II. Singapore: Springer; 2020. pp. 801-820

[128] Bonetti A, Tugnoli B, Piva A, Grilli E. Towards zero zinc oxide: Feeding strategies to manage post-weaning diarrhea in piglets. Animals. 2021;**11**(3):642

[129] Dushenkov V, Kumar PN, Motto H, Raskin I. Rhizofiltration: The use of plants to remove heavy metals from aqueous streams. Environmental Science & Technology. 1995;**29**(5):1239-1245

[130] Bolan NS, Park JH, Robinson B, Naidu R, Huh KY. Phytostabilization: A green approach to contaminant containment. Advances in Agronomy. 2011;**112**:145-204

[131] Zayed A, Pilon-Smits E, de Souza M, Lin ZQ, Terry N. Remediation of selenium-polluted soils and waters by phytovolatilization. In: Phytoremediation of Contaminated Soil and Water. CRC press; 2020. pp. 61-83

[132] McGrath SP, Zhao FJ. Phytoextraction of metals and metalloids from contaminated soils. Current opinion in Biotechnology. 2003;**14**(3):277-282

[133] Emenike CU, Jayanthi B, Agamuthu P, Fauziah SH. Biotransformation and removal of heavy metals: A review of phytoremediation and microbial remediation assessment on contaminated soil. Environmental Reviews. 2018;**26**(2):156-168

[134] Violante A, Cozzolino V, Perelomov L, Caporale AG, Pigna M. Mobility and bioavailability of heavy metals and metalloids in soil environments. Journal of Soil Science and Plant Nutrition. 2010;**10**(3):268-292

[135] Menzi H, Oenema O, Burton C, Shipin O, Gerber P, Robinson T, et al. Impacts of intensive livestock production and manure management on the environment. Livestock in a Changing Landscape. 2010;**1**:139-163

Chapter 25

Toxic Heavy Metals in Soil and Plants from a Gold Mining Area, South Africa

Iyioluwa Busuyi Raji and Lobina Gertrude Palamuleni

Abstract

The mining of mineral deposits abundant in South Africa has led to the release of toxic heavy metals into the environment. The aim of this study was to investigate heavy metal pollution from a gold mining area. The concentrations of arsenic, cadmium, lead, and zinc were analyzed in soil and plants found within 500 m and 1000 m radius of a gold mine established in North-West. The concentrations of these heavy metals were determined using Inductively Coupled Plasma Mass Spectrometry. The results showed that the concentrations of the studied heavy metals were all below the national and international threshold but higher than the concentrations found several kilometers away from the mining area, the natural background concentration. The results from the pollution load index shows that the topsoil is contaminated for the selected heavy metals. There was accumulation of the studied heavy metals in the studied *Eragrostis hypnoides* plant's leaves and roots. In this study, it was revealed that zinc and cadmium bioaccumulated in the plant via the soil. These findings suggests that the consumption of agricultural products from farms within the 1 km radius of the mining site could be detrimental to the wellbeing of direct and indirect consumers.

Keywords: arsenic, cadmium, lead, zinc, heavy metals, gold mining, South Africa, pollution load index, concentration factor, pollution

1. Introduction

South Africa as a country is regarded as one of the richest countries with abundance of mineral resources such as gold, platinum, coal, cobalt and many more [1, 2]. The exploration of gold started in the country in the 19th century [3], when the world largest gold deposit was discovered in the Witwatersrand Basin, South Africa [4]. The discovery of gold has contributed to the development of South Africa as one of the most developed and largest economy in the African continent [5]. However, the mining of gold and other minerals has led to the increase in the concentrations of toxic heavy metals in the biosphere, atmosphere, and hydrosphere [6–10]. Heavy metals are found naturally in all spheres of life at a minimal concentration. Natural sources include bedrock weathering, volcanic activities, and atmospheric fallout.

However, these concentrations have been increased due to anthropogenic activities for example, exploration of mineral deposits [11, 12].

Heavy metals are classified as metals with specific density more than 5 g/cm^3 which negatively impact the environment and living organisms [13]. Heavy metals can be classified into two types; essential heavy metals and non-essential heavy metals [14]. Essential heavy metals such as iron and zinc are important for human metabolism at low concentrations; however at higher concentrations, they are toxic [15]. For example, iron is a type of needed protein for the red blood cells that carries oxygen from the lungs to all parts of the human body [16, 17]. However, at high concentrations, iron becomes toxic to humans leading to hemochromatosis which can cause serious damage to the human heart, liver and pancreas [18, 19]. Zinc is a major player in the development of DNA, growth of the body cells, building proteins and healing damaged body cells at low concentrations [20, 21]. At high concentrations, it causes nausea, vomiting, loss of appetite, stomach cramps, diarrhea, and headaches [22–24].

The non-essential heavy metals, for example, cadmium (Cd), arsenic (As), lead (Pb), uranium (U), are toxic to humans even at low concentrations [14, 25, 26]. Cadmium is known to cause lung damage, kidney damage and fragile bones which can result in death [27, 28]. Exposure to arsenic from food and water leads to cancer, skin lesions, cardiovascular disease, and diabetes [29–32]. Lead reduces brain development, causes anemia, body weakness, kidney, and brain damage [33–35]. Uranium can cause lung cancer, liver and kidney damage [36, 37]. Most of these non-essential heavy metals are carcinogenic; they could lead to the development of cancer of vital human organs which can eventually lead to death [14, 38]. This is even worse when these toxic heavy metals are ingested indirectly by immuno-deficient humans [39].

There are several pathways for heavy metals such as through air, water, and sediments [40, 41]. While some of these heavy metals are highly mobile, such as uranium (U), arsenic (As), some are of less mobility like lead (Pb), and cadmium (Cd). Depending on the mobility of the heavy metals, high concentration of the metal can be found several miles away from the mining environment. High concentrations of heavy metals have been recorded in sediments within and outside the perimeter of a mining area in South Africa. Fashola et al. [42] reported concentrations of As, Cd, Pb and Zn above the recommended levels by South African guidelines for soils and sediment qualities guidelines from abandoned gold mining sites. In the study of the spatial assessment of heavy metals contamination in household soils in rural Limpopo Province, South Africa, Kapwata et al. [43] reported high concentration of Pb, Cd, As and mercury (Hg), exceeding the Canadian reference levels of these heavy metals in soils characterized by abandoned mines and artisan mining activities in Limpopo Province, South Africa.

Concentrations of heavy metals have also been reported in plants around a mining environment. Flefel et al. [44] reported high concentrations of Cd, 0.85–30.30 mg/kg and Pb, 21.50–68.00 mg/kg in aquatic plants higher than the sampled water. The researchers concluded that the concentrations of heavy metals in plants were above the acceptable limits of Cd, 0.02 mg/kg, Pb, 2 mg/kg set by World Health organization (WHO). In the study by Kausar et al. [45] high concentrations of Cd, Pb, and Zn were recorded in crops irrigated by heavy metal polluted water. The accumulation of Cd, Pb and Zn was recorded in the crops' leaves and root of carrot and spinach. While some of these plants are consumed directly by humans, some are used to feed animals such as cattle which exposes humans to direct and indirect carcinogenic health risks.

In mines nearby streams, high concentration of heavy metals has been recorded by several researchers [10, 46, 47]. Shapi et al. [46] reported heavy metal concentrations from water samples that have accumulated in wetlands due to the past gold mining operations in Krugersdorp, South Africa. The maximum concentrations recorded for As, Pb and Zn are 32.20 mg/L, 6.30 mg/L and 783 mg/L respectively [46]. In the study by Chetty et al. [47] on the transportation and accumulation of heavy metals in the Klip River's catchment, elevated concentrations of different heavy metals were recorded which included As, Cd, Pb. The authors concluded that the extensive accumulation of these heavy metals in the water body is the legacy of past and extensive gold mining in the Witwatersrand Basin [47]. Raji et al. [10] reported high concentration of uranium and other heavy metals in Rietspruit system because of gold mining activities that was operation at the headwater of the Rietspruit, Far West Rand goldfield. Residents of informal settlements that depend on nearby surface water from streams and dams polluted by heavy metals for their basic domestic needs eventually ingest dissolved heavy metals. The availability of these heavy metals will continuously pose an enormous health risk to downstream water users.

Many researchers have also reported high concentrations of heavy metals in air around the mining environment. Mining wastes such as unrehabilitated tailing-dams are rich in heavy metals which are fine-grained. The top-layers of these tailing-dams can be transported by wind. Local residents within close proximity inhale this air which can lead to different respiratory diseases [48, 49]. Residents of Soweto, South Africa, residing near tailing-dams have reported several respiratory illnesses which has been linked to the presence of tailing-dams close to their community [50]. Heavy metals in air in due course settle down on land and plants. Plants uptake these heavy metals and bio-accumulate the toxic elements in their leaves, stems, and roots [51–53]. These plants are later consumed by humans either directly or indirectly through cattle that fed on those plants. In all the findings, the general conclusion was that there is an obvious decrease in the concentration of the heavy metals as the distance from the source pollutant (mining site) increases.

In this study, the concentrations of As, Cd, Pb, and Zn within a 1 km radius of a gold mining environment in North-West Province, South Africa was studied. These heavy metals are some of the most toxic heavy metals highlighted by the World Health Organization, WHO, [54–56] and they have been associated with gold mining operations [42, 57–59]. This study is important because there is currently limited published literature about the concentration of these heavy metals from the gold mine located in the North-West Province of South Africa. Considering the several agricultural activities within the proximity of the mines such as maize plantation and cattle husbandry, it is very important to determine the concentration of these selected heavy metals within a 1 km radius of the gold mine if crop safety measures need to be implemented.

The main objective of this study is to determine the concentrations of As, Cd, Zn and Pb in soils and plants within the 1 km radius of a gold mine. The specific objectives are highlighted below.

1. Determine the concentrations of selected heavy metals in the soil.

2. Determine the pollution load index of the selected heavy metals in the soil.

3. Examine the bioaccumulation of heavy metals in *Eragrostis hypnoides* within the 1 km radius of the gold mine.

4. Determine the uptake of the selected heavy metals in *E. hypnoides*.

1.1 Study area

The study area is in the Ratlou Local Municipality, North-West Province, South Africa surrounding an open pit mine (**Figure 1**). The gold mine was established in 1996 and gold is mined from the gold bearing ore in a banded ironstone formation of the Kraaipan Greenstone Belt. North-West Province is a water scarce Province because of the high-water demand and low precipitation. The rate of evaporation is more than double the rate of precipitation [58]. An extreme drought is experienced in the province.

The North-West Province is dominated by the Savannah Biome and the rest falls in the Grassland Biome. The climate of the Province is categorized by hot temperature reaching about 38°C in the summer and cold sunny winter.

Besides mining, agriculture is another mainstay of the economic activities in this region, hence, the province is regarded as the food basket of the nation. Large maize plantation and sunflowers are located within the 2 km radius of the gold mine. Grazing cattle and ranches were witnessed during fieldwork. The studied plant (*Eragrostis hypnoides*) is among the observed plants that the cattle fed on.

This study site was selected because it characterizes distinct land use landscapes. With the gold mine within proximity of established commercial farmlands, it is crucial to examine the concentrations of As, Cd, Pb and Zn in the soil and plant found within 1 km of an active gold mine.

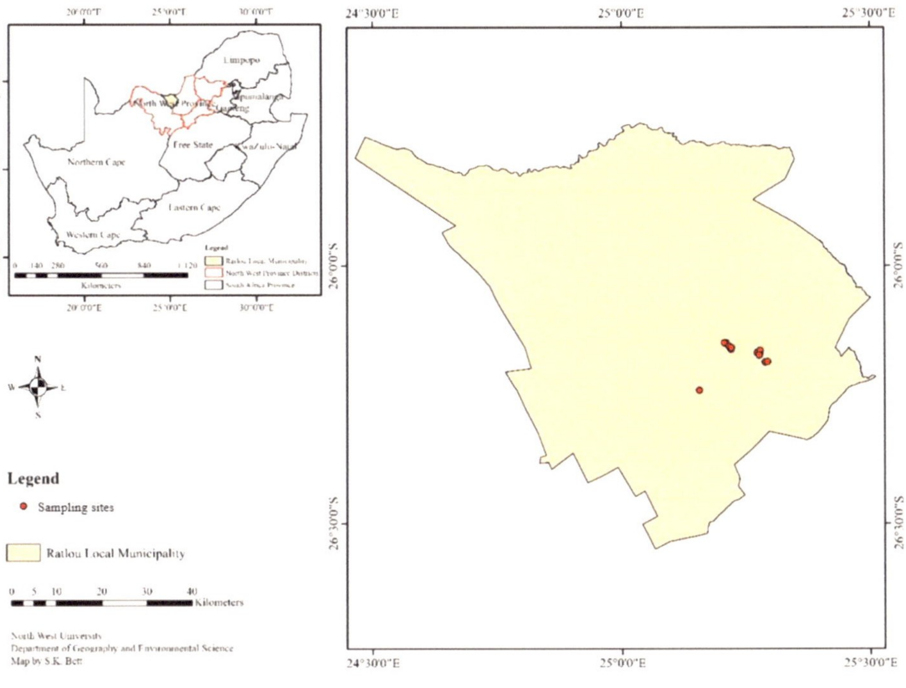

Figure 1.
Map of the study area showing the sampled sites.

2. Methodology

This report involves an experimental qualitative design which included heavy metals laboratory testing and statistical analysis.

2.1 Identification of soil and plant sampling sites

To identify sampling sites, several climatic conditions were investigated. This included the physical conditions of the study area such as the geology of the environment, type of soil as well as rainfall. Sampled sites were evenly distributed to allow the determination of possible pollution sites and to know the degree of heavy metal pollution.

Other considered factors are:

1. Sampled sites must provide a standard geostatistical fit.

2. Sampled sites must represent the different activities of land use predominate in the area.

3. Samples must be collected from flat areas to avoid sampling eroded soil materials.

4. Sample were collected several meters away from the road and residential areas to reduce human prints.

Considering the relationship between the parent rocks and non-point source pollution, it is believed that 32 samples for four metals will be sufficient to represent the study area and perform statistical and geostatistical evaluation.

2.2 Soil sampling

According to Branquiho et al. [60], spatial dust effects are localized and it ranges less than 1 km from the source pollutant. To allow even distribution of soil samples, soil samples were taken at 500 m and 1000 m away from the gold mining site in various directions – north, northeast, northwest, south, southwest, southeast, east, and west. At each distance, two samples were taken using an auger at a depth of 0–10 cm (topsoil) and 10–2- cm (subsurface). These depths were used in previous studies [61, 62] and they suggest the disparity in land contamination at varied distances [61].

A site devoid of any physical human activity was selected to collect a soil sample referred to as the control or background sample. This Background sample was taken 10 km away from the study area ($26^0\ 15'\ 08''$ S $25^0\ 11'\ 32''$ E) at the same depth used for other collected samples. This was essential to have an uncompromised Background sample. The climatic and physical conditions of the study area and the background site were identical.

Three samples above 1 kg were collected at each of the sampled site for precision purposes. A total of 28 soil samples were collected, stored in a plastic bag, and taken to the laboratory for further analysis. Each of the soil samples coordinate were recorded using a handheld Garmin GPSMAP 65 s Global Positioning System, manufactured by Garmin, South Africa.

2.3 Plant sampling

Plant samples were collected at the same spots where soil samples were taken. A total of four plant samples were collected at the northern and eastern location of the mine at same spot where soil samples were taken. At other sampled location, the plant of interest was not found at the specified distance, 500 m, and 1000 m. Plants rely on the physical and chemical conditions of the topsoil, hence, they are disturbed by these circumstances.

In this study, a specific plant species, *E. hypnoides*, was collected. Collected plant samples were thoroughly cleaned with tap water, followed by 0.1 mol/L of hydrochloric acid (HCl) and deionized water. The washing of the plant samples was important to remove every soil and dust particles from the sample. If the plant samples were left unwashed, the soil and dust particle will influence the determined concentration of the studied heavy metals in the sample.

In line with the method of Ma et al. [63], plant samples were placed inside an oven at 70°C for three days. This was done to remove moisture content of the plant sample. After drying, all the plant samples were then pulverized in a mill and packed in a well labeled sealable plastic bag before proceeding to aqua regia digestion of the samples.

2.4 Determination of heavy metals

The analysis of cadmium (Cd), lead (Pb), zinc (Zn), and mercury (Hg) were done using the inductively coupled plasma mass spectrometry (ICP-MS) for both the collected soil and plant samples. These selected heavy metals are toxic to humans, animals, and plants. They have been reportedly found at high concentrations within the environment of a gold mining and processing.

Before using the ICP-MS to determine the concentrations of the studied heavy metals, the collected and labeled plants and soil samples were digested using 0.6 mL of concentrated sulfuric acid (H_2SO_4), 0.6 mL of concentrated nitric acid (HNO_3) and 1.8 mL of concentrated HCl for two hours at 95°C.

After digestion and the cooling of the machine, each sample volume was brought up to 10 mL by adding deionized water. Each sample was then arranged in the ICP-MS machine accordingly to determine the concentration of the studied heavy metal in each sample. This procedure has been used in the study of Kamunda et al. [64]. The detection limit of Zn, Cd, Pb and As using the ICP-MS are 1.173 mg/kg, 0.006 mg/kg, 0.045 mg/kg and 0.026 mg/kg respectively.

2.5 Quality control

Every equipment used were firstly calibrated with reference standard. Glassware used for heavy metal analyses were rinsed in dilute HNO3 before usage. All reagents and heavy metal standards used were of analytical grades. Analyses were done in duplicate to ensure precision and accuracy of the obtained data.

2.6 Determination of pollution levels in the study area

To assess the pollution levels in the study area, the soil contamination factor (CF) was used. Therefore, the standard background value which represents the value of the elements, measured relative to the amount of the Upper Continental Crust (UCC) was used as the reference material [65].

$$CF = \frac{C_a}{C_{ref}} \tag{1}$$

Where C_a is the metal concentration in the soil (total), and C_{ref} is the background value of the pristine environment.

The contamination levels were classified based on the following classes: low contamination ($CF < 1$), moderate contamination ($1 \leq CF < 3$), high contamination ($3 \leq CF < 6$) and very high contamination ($CF \geq 6$).

Pollution load index (PLI) was calculated using Eq. 2 to assess the overall contamination at each site and to distinguish natural origin from anthropogenic sources [66].

$$PLI = \sqrt[n]{CF_1 \times CF_2 \times .....CF_n} \tag{2}$$

Where CF1, CF2 are CF of elements 1, 2, 3, ..., n; When the PLI > 1, denotes significant deterioration in the system, 0 < PLI < 1, indicates baseline level of contamination [66].

2.7 Determination of heavy metals uptake by plant

Concentration factor (CF) was introduced to calculate the relationship between the uptake of As, Cd, Pb and Zn from soil by plants. This is a measure of soil-plant transfer that supports the understanding of plant uptake signature [62]. The determined concentration of each metal in the plant (M_{plant}) was divided by the concentration of each metal determined in the soil (M_{soil}). A quotient greater than 1 means that the plant has been influenced by the metal (accumulator). However, if the quotient is less than 1, it means that the plant has not been influenced by the metal (excluder).

$$CF = \frac{M_{plant}}{M_{soil}} \tag{3}$$

Where CF is the concentration factor, M_{plant} is the metal concentration in the plant and M_{soil} is the metal concentration in the soil.

2.8 Data analysis

Data obtained from laboratory analysis were subjected to basic descriptive statistics (i.e., mean, and standard deviations) tabulated using SPSS software. In addition, the concentrations of the selected heavy metals were compared with South African agricultural soil standards. Also, SPSS software was used for all the statistical analysis which include Chi-square and one-way ANOVA. The Chi-square evaluated the deviation between the determined concentration of the studied heavy metals from the sample site and the control site. One-way ANOVA was used to determine the significant difference in the determined heavy metal concentration while considering all the sampled sites.

3. Results and discussion

3.1 Heavy metal concentration in soil samples

The results of the described field sampling are summarized in **Tables 1–4**. The results depicted the varying concentrations of each heavy metals at each of the

Zn concentration (mg/kg)											
Location	Distance	500 m		1000 m		Background sample 1		Background sample 2		South African limit
	Depth (cm)	0–10	10–20	0–10	10–20	0–10	10–20	0–10	10–20	
North		8.9	5.7	12.8	14.7	7.3	17.1	10.1	9.2	200
Northeast		7.9	32.8	8.0	10.5					
Northwest		6.9	10.2	11.9	6.6					
East		8.8	6.5	8.2	6.7					
Southeast		8.7	12.9	6.9	8.7					
West		9.2	6.7	9.1	6.3					
Southwest		8.8	5.7	N/A	N/A					
South		4.6	8.1	N/A	N/A					
Mean		7.9	11.1	9.5	8.9					
SD		1.6	9.1	2.4	3.3					
Range		4.6–9.2	5.7–32.8	6.9–12.8	6.3–14.7					

N/A = not accessible.

Table 1.
Concentration of zinc.

respective sampling points and depths. Results were compared with the upper limit threshold for agricultural soils in South Africa [67].

3.1.1 Zinc concentration

The ICP-MS result of Zn concentration indicated that the concentration of Zn was from 4.6 mg/kg to 9.2 mg/kg at a depth of 0–10 cm (topsoil). At a depth of 10–20 cm (subsurface), it ranged from 6.9 mg/kg to 12,8 mg/kg. These were from 500 m from the gold mine.

At 1 km radius from the mining site, the maximum recorded concentration for Zn was 12.8 mg/kg and the minimum was 6.9 mg/kg at the topsoil. At 10–20 cm, the maximum recorded concentration was 14.7 mg/kg and the lowest was 6.3 mg/kg (**Table 1**). The concentration of Zn at all the sampled site were below the permissible limit of Zn in South African agricultural soil, 200 mg/kg [67].

When the concentrations of Zn from the study area were compared with the Zn concentration from the control site, the mean of Zn concentration at the sampled sites were more than the concentration of Zn from the Background sample at 10–20 cm within the 0.5 km radius of the mine (**Table 1**).

Based on the result from 500 range of the mine, the concentrations of Zn are lower at the topsoil than at the subsurface. Akin to the findings of Ekweu et al. [68] where higher concentration of Zn were reported at a depth of 15–20 cm than at a depth of 0–15 cm. Leaching effect was reported to be responsible. However, in the study of Raulinaitis et al. [69], the concentration of Zn at the topsoil, 36.8 mg/kg was higher than at the subsurface, 18.3 mg/kg.

Cd concentration (mg/kg)										
Location	**Distance**	**500 m**		**1000 m**		**Background sample 1**		**Background sample 2**		**South African limit**
	Depth (cm)	**0–10**	**10–20**	**0–10**	**10–20**	**0–10**	**10–20**	**0–10**	**10–20**	
North		0.014	0.016	0.028	0.039	0.014	0.039	0.025	0.024	3.00
Northeast		0.019	0.396	0.020	0.027					
Northwest		0.023	0.026	0.031	0.017					
East		0.013	0.011	0.019	0.014					
Southeast		0.020	0.034	0.011	0.017					
West		0.012	0.023	0.015	0.019					
Southwest		0.008	0.010	N/A	N/A					
South		0.011	0.024	N/A	N/A					
Mean		0.015	0.068	0.021	0.022					
SD		0.005	0.133	0.008	0.01					
Range		0.008–0.023	0.010–0.395	0.011–0.031	0.014–0.039					

N/A = not accessible.

Table 2.
Concentrations of cadmium.

Pb concentration (mg/kg)										
Location	Distance	500 m		1000 m		Background sample 1		Background sample2		South African limit
	Depth (cm)	0–10	10–20	0–10	10–20	0–10	10–20	0–10	10–20	
North		3.5	3.8	4.5	4.9	0.058	0.7	0.1	1.08	100
Northeast		3.3	0.3	3.2	4.4					
Northwest		3.8	4.9	5.8	3.2					
East		3.1	2.6	4.2	3.4					
Southeast		3.8	5.1	4.5	3.12					
West		3.9	2.6	3.4	4.6					
Southwest		3.3	4.9	N/A	N/A					
South		2.9	4.7	N/A	N/A					
Mean		3.4	3.6	4.2	3.9					
SD		0.3	1.7	0.9	0.8					
Range		2.97–3.855	0.334–5.077	3.191–5.787	3.155–4.987					

N/A = not accessible.

Table 3.
Concentration of lead.

As concentration (mg/kg)										
Location	**Distance from mine**	**500 m**		**1000 m**		**Background sample 1**		**Background sample 2**		**South Africa limit**
	Depth (cm)	**0–10**	**10–20**	**0–10**	**10–20**	**0–10**	**10–20**	**0–10**	**10–20**	
North		0.8	1.1	1.3	3.5	0.7	2.0	1.0	1.4	5.8
Northeast		0.8	0.2	1.0	1.4					
Northwest		0.7	1.4	1.2	0.6					
East		0.9	1.1	1.1	0.7					
Southeast		0.8	3.7	0.7	0.7					
West		0.8	1.0	0.6	1.0					
Southwest		0.7	0.7	N/A	N/A					
South		0.8	1.1	N/A	N/A					
Mean		0.8	1.3	0.9	1.3					
SD		0.1	1.0	0.3	1.1					
Range		0.7–0.9	0.2–3.7	0.6–1.3	0.6–3.5					

N/A = not accessible.

Table 4.
Concentration of arsenic.

In comparison with the permissible limit of Zn in other countries, the concentration of Zn reported in this study is lower in many folds than the following countries - Austria (111 mg/kg), China (74.2 mg/kg), Germany (225 mg/kg) and USA (60 mg/kg) [70]. As result, we can conclude that the soil within the study area is not polluted.

3.1.2 Cadmium concentrations

The maximum concentration of Cd recorded at the topsoil, 500 m away from the mining site was 0.023 mg/kg and the minimum was 0.008 mg/kg. At the subsurface, the maximum concentration recorded was 0.395 mg/kg and the lowest was 0.010 mg/kg. Similar concentrations of Cd were also recorded at 1 km radius of the mine at the specific sampling depths (**Table 2**).

The average concentration of Cd is lower at the topsoil than at the subsurface at both distances from the mine. The mean concentration of Cd at both distances showed that the study area is not polluted with Cd because they are lower than the permissible limit of Cd in South African soil used for agriculture, 3.00 mg/kg [67] and the mean concentration of Cd reported in China, 0.1 mg/kg, Japan, 0.41 mg/kg and in the United Kingdom, 0.62 mg/kg [70].

When the concentration of Cd from the study area is compared with the concentration from the Background samples, the mean concentration of Cd is more at the subsurface than the concentration of Cd at the background sample site 500 m away from the mining site (**Table 2**). Based on depths (0 10 cm and 10–20 cm), the mean concentration of Cd at the topsoil is lower is lower than the mean concentration of Cd at the subsurface. This is similar to the findings of Ekwue et al. [68] and Raulinaitis et al. [69] where they both reported higher concentration of Cd at the subsurface than the concentration of Cd at the topsoil.

3.1.3 Lead concentrations

The highest concentration of Pb recorded was 3.9 mg/kg and the lowest was 2.97 mg/kg at 0.5 km distance away from the mine at the topsoil. At the subsurface, the maximum concentration recorded was 5.1 mg/kg while the lowest recorded concentration of Pb was 0.3 mg/kg.

1 km away from the mine, the concentration of Pb was between 3.2 mg/kg and 5.1 mg/kg at the topsoil and 3.2 mg/kg and 4.9 mg/kg at the subsurface. Lead's average concentration in the topsoil 0.5 km away from the mine is lower than at the subsurface and vice-versa when compared with the mean concentration of Pb at 1 km away from the mine (**Table 3**). This is a result of atmospheric deposition from vehicular activity [71] because most of the sampled sites about 1 km from the mine are closer to roads leading to the farms.

Overall, the study area is not polluted because the mean concentration of Pb is lower than the permissible limit of Pb in South African soil used for agriculture, 100 mg/kg [67]. The mean concentration of Pb is also lower than the mean reported in China, 26 mg/kg, Japan, 20.4 mg/kg and the UK, 29.2 mg/kg at every sampled location [70].

In addition, the average concentration of Pb at all the sampled sites and depths are lower than the concentration of Pb from the control sites. This means human activities are responsible for the elevation of Pb at the study area such as gold mining and farming [72].

3.1.4 Arsenic concentrations

The maximum concentration of As recorded at the topsoil was 0.99 mg/kg and 3.66 mg/kg at the subsurface. The minimum concentration of As recorded was 0.66 mg/kg at the topsoil and 0.15 mg/kg at the subsurface (**Table 4**) at 500 m away from the mining area.

At 1 km away from the mine, the maximum concentration recorded at the topsoil and subsurface are 1.3 mg/kg and 3.5 mg/kg respectively. The lowest recorded concentration of As were approximately 0.6 mg/kg at both the topsoil and subsurface (**Table 4**).

The mean concentration of As were higher at the subsurface than at the topsoil at both 500 m and 1000 m away from the gold mine. This is comparable to the findings of Wahl [73] which reported the same trend in As concentration in soils around a gold mine in Kwa-Zulu-Natal Province, South Africa. The mean concentration of As at both locations and depths are below the permissible limit of As in South African soils used for agriculture, 5.8 mg/kg [67]. This means that the soil is not contaminated.

At the control sites, the higher concentrations of As recorded at a depth of 10–20 cm than at a depth of 0–10 cm. the recorded concentration of As at the control sites were all higher than the mean concentration of As at both depths and distance except the mean concentration of the topsoil at a distance of 500 m from the mine (**Table 4**). Similar to the study of Ekwue et al. [68], the concentration of As increase from the topsoil to the subsurface.

Higher As concentrations have been reported in other countries - Germany (50 mg/kg) [74], Australia (20 mg/kg), [75], China (30 mg/kg), [76]; and Canada (12 mg/kg), [77] at every sampled location.

Generally, the total soil concentrations of As, Cd, Pb and Zn are below the upper limit threshold in comparison to the background samples and to other country's concentration. It suggests that the mining activity has not yet impacted the concentration of heavy metals in the soil because the concentration of the background sample has similar soil concentration as that obtained around the mine.

3.2 Pollution load index

The eq. 1 and 2 were used to calculate the contamination factor, CF, and pollution load index, PLI, of each studied heavy metal, respectively. The results are shown in **Tables 5–8**. The CF of Zn ranged from 0 to 2 within the 500 m radius of the mine. It means the area contamination ranged from a low contamination, $CF < 1$, to moderate contamination, $1 \leq CF < 6$, at all the sampled sites (**Table 5**). The PLI of Zn indicated that all the sampled sites are polluted (**Table 5**).

For Pb, the CF shows that the area is very highly contaminated, $CF > 6$, at the subsurface (0–10 cm) at both distances (**Table 6**). At 10–20 cm depth, the CF ranged from low contamination to high contamination, $3 \leq CF < 6$ (**Table 6**). The PLI for Pb shows that the study area is polluted.

The CF for As ranged from low contamination to moderate contamination at all the sites (**Table 7**). The PLI result indicated that the studied area is polluted except at the depth of 10–20 cm with $PLI < 1$.

The CF of Cd results show 82% of the sampled sites are moderately contaminated while 14% are lowly contaminated. However, at a depth of 10–20, the Northeastern site has a very high contamination, $CF > 6$ (**Table 8**). The PLI result indicated that all the sampled sites are polluted except at a depth of 10–20 cm with a PLI below 1 (**Table 8**).

3.3 Heavy metal concentrations in plant samples

E. hypnoides, was investigated to establish the metal concentrations and the results are presented in **Tables 9** and **10**. The accumulation of heavy metals in plants show the site's heavy metal pollution status and also the potential of the plant species to uptake heavy metal from the soil [62].

Sampled sites	CF (0–10 cm) at 500 m	CF (10–20 cm) at 500 m	CF (0–10 cm) at 1000 m	CF (10–20 cm) at 1000 m
North	1	0	1	2
Northeast	1	2	1	1
Northwest	1	1	1	1
East	1	0	1	1
Southeast	1	1	1	1
West	1	0	1	1
Southwest	1	0	N/A	N/A
South	1	0	N/A	N/A
PLI	**1**	**1**	**1**	**1**

N/A = means no access to take samples.

Table 5.
The contamination factor and pollution load index of zinc at various sites and depths.

Sampled site	CF (0–10 cm) at 500 m	CF (10–20 cm) at 500 m	CF (0–10 cm) at 1000 m	CF (10–20 cm) at 1000 m
North	60	5	32	5
Northeast	57	0	23	4
Northwest	66	7	42	3
East	54	4	31	3
Southeast	65	7	33	3
West	66	4	24	4
Southwest	57	7	N/A	N/A
South	51	6	N/A	N/A
PLI	**59**	**4**	**30**	**4**

N/A = means no access to take samples.

Table 6.
The contamination factor and pollution load index of lead at various sites and depths.

Sampled sites	CF (0–10 cm) at 500 m	CF (10–20 cm) at 500 m	CF (0–10 cm) at 1000 m	CF (10–20 cm) at 1000 m
North	1	1	1	2
Northeast	1	0	1	1
Northwest	1	1	1	0
East	1	1	1	1
Southeast	1	2	1	0
West	1	0	1	1
Southwest	1	0	N/A	N/A
South	1	1	N/A	N/A
PLI	**1**	**0**	**1**	**1**

N/A = means no access to take samples.

Table 7.
The contamination factor and pollution load index of arsenic at various sampled sites and depths.

3.3.1 Zinc concentration

The maximum concentration of Zn in the root of *E. hypnoides* was 77.3 mg/kg and the minimum was 21 mg/kg. In the root, the maximum concentration of Zn was 76 mg/kg and 20 mg/kg were the lowest (**Table 9**). Higher concentration of Zn was recorded in the root of the plant than the leaf with the mean concentration of Zn in the root being 44.98 mg/kg while the average concentration of Zn in the leaf was 44.1 mg/kg. The concentration of Zn in the parts of plants were all more than the concentration of Zn in the soil where the plant sample was taken. This shows that plant bioaccumulate heavy metals in many folds than the soil under normal growing condition [78].

The recommended permissible limit of Zn in plant is 50 mg/kg according to the World Health Organization (WHO) [79]. This is higher than all the concentration of Zn in the roots and leaves of *E. hypnoides* except the eastern plant sample at 500 m (**Table 9**).

Sampled sites	CF (0–10 cm) at 500 m	CF (10–20 cm) at 500 m	CF (0–10 cm) at 1000 m	CF (10–20 cm) at 1000 m
North	1	0	1	2
Northeast	1	10	1	1
Northwest	2	1	1	1
East	1	0	1	1
Southeast	1	1	0	1
West	1	1	1	1
Southwest	1	0	N/A	N/A
South	1	1	N/A	N/A
PLI	**1**	**1**	**1**	**1**

N/A = means no access to take samples.

Table 8.
The concentration factor and pollution load index of cadmium at various sampled sites and depths.

Location	Distance (m)	Zn Concentration (mg/kg)	
		Root	Leaf
North	500	37.7	35.2
	1000	21.0	20.0
East	500	**77.3**	**76.0**
	1000	47.9	45.1
Mean		44.9	44.1
SD		23.7	23.7
Range		21.0–77.3	20.0–76.0
WHO limit		50	

Bold to indicate concentrations above the permissible limit.

Table 9.
Concentration of zinc in plant sample.

The proximity of the eastern side to mine waste and the dispersion of Zn by wind is responsible for the high concentration of Zn recorded at 500 m away from the mine [80].

3.3.2 Cadmium concentration

The concentration of Cd recorded in the roots and leaf of the plant were all higher than the permissible limit of Cd in plant, 0.025 mg/kg [81]. This means that the plant is polluted and therefore not suitable for consumption. The grazing of cattle in this environment must be avoided at all costs.

Overall, the average concentration of Cd in the root is more than the average concentration of Cd in the leaf. This is similar to Zn concentration in the plant which means the plant cannot translocate heavy metals from the root to the leaf [82].

The concentration of Cd in the plant's root and leaf are all more than the concentration of Cd recorded in the soil where the plant samples was taken. Lower concentrations of Cd were also recorded 1000 m away from the mine than at 500 m

Location	Distance (m)	Cd Concentration (mg/kg)	
		Root	Leaf
North	500	0.06	0.056
	1000	0.03	0.02
East	500	**0.18**	**0.16**
	1000	**0.15**	**0.14**
Mean		0.11	0.09
SD		0.07	0.07
Range		0.02–0.18	0.02–0.16
WHO limit		0.02	

Bold to indicate concentrations above the permissible limit.

Table 10.
Concentration of cadmium in plant sample.

(**Table 10**). The deposition of cadmium oxide in the air due to mining activities in the rea is responsible for the wide spread of Cd in the area. It has been reported that Cd is dispersed widely by aid from melting and smelting activities plus additional anthropogenic pathways [83].

3.3.3 Lead concentration

The concentration of Pb in the plant's leaf and root were all higher than the permissible limit of Pb in plant, 2 mg/kg according to WHO [84] except the concentration of Pb in plant grown 1 km away from the mine in the northern part of the mine (**Table 11**).

Punshon et al. [85], reported a direct proportion between the concentration of heavy metal in soil and plant. Similar to the trend noticed in Zn and Cd, the concentration of Pb in the leaf is lower than the recorded concentration of Pb in the root.

3.3.4 As concentration

In the root of E. hypnoides, the maximum concentration of As recorded was 2.8 mg/kg and the lowest was 0.2 mg/kg. The minimum concentration recorded in the

Location	Distance (m)	Pb Concentration (mg/kg)	
		Root	Leaf
North	500	0.7	0.5
	1000	**15.0**	**14.8**
East	500	0.9	0.8
	1000	0.7	0.6
Mean		4.3	4.1
SD		7.1	7.1
Range		0.7–15.0	0.5–14.8
WHO limit		2	

Bold to indicate concentrations above the permissible limit.

Table 11.
Concentration of lead in plant sample.

Location	Distance (m)	As Concentration (mg/kg)	
		Root	Leaf
North	500	**0.2**	0.2
	1000	**2.8**	**1.5**
East	500	**0.3**	**0.3**
	1000	**0.3**	**0.3**
Mean		0.9	0.6
SD		1.2	0.6
Range		0.2–2.8	0.2–1.5
WHO limit		0.2	

Bold to indicate concentrations above the permissible limit.

Table 12.
Concentration of arsenic in plant samples.

plant's leaf was 0.2 mg/kg and 1.5 mg/kg as the maximum concentration (**Table 12**). Overall, the mean concentration of As in the soil is higher than the concentration recorded in the plant. This shows that the plant has a low potential of taking As from the soil.

The permissible limit of As in plant is 0.2 mg/kg [85]. With the recorded concentration of As in the studied plant all above 0.2 mg/kg, except the leaf concentration of the north within 500 m which is approximately 0.2 mg/kg. It means that the plant is contaminated therefore not suitable to feed cattle.

In order to determine the dependency of As, Cd, Pb and Zn concentration in the root and leaves of *E. hypnoides*, Chi-square test was used. Using SPSS to evaluate the result of heavy metal concentration in plant samples, it was revealed that the P-value, 0.05, is lower than the significance level of 0.213. This means that the null hypothesis can be rejected, and we can conclude that the concentration of As, Cd, Pb, and Zn in the roots and leaf of *E. hypnoides* are independent.

In conclusion, the results of all the evaluation are the same for all the metals. This means the studied heavy metals bioaccumulated in the plant by absorption through the roots and are translocated to the leaf. Hence, the reason for the consistency observed in all the studied heavy metals.

3.4 Heavy metal uptake by plants

Importantly, it must be noted that plants have the potential and mechanism to uptake heavy metals by absorption from the soil through the roots [86]. To calculate the uptake of As, Cd, Pb and Zn by *E. hypnoides* eq. 3 was used and the result is in table **Figures 2–5** below.

3.4.1 Zinc uptake by plant

The CF of Zn based on the recorded concentration of Zn in the plant and soil indicated that the plant is a good accumulator of Zn. The CF value of Zn is greater than 1 (**Figure 2**). This means that *E. hypnoides* can be useful to phyto-remediate a Zn polluted soil. Unfortunately, the consumption of plants with a high concentration of Zn is not healthy for both human and animals [87].

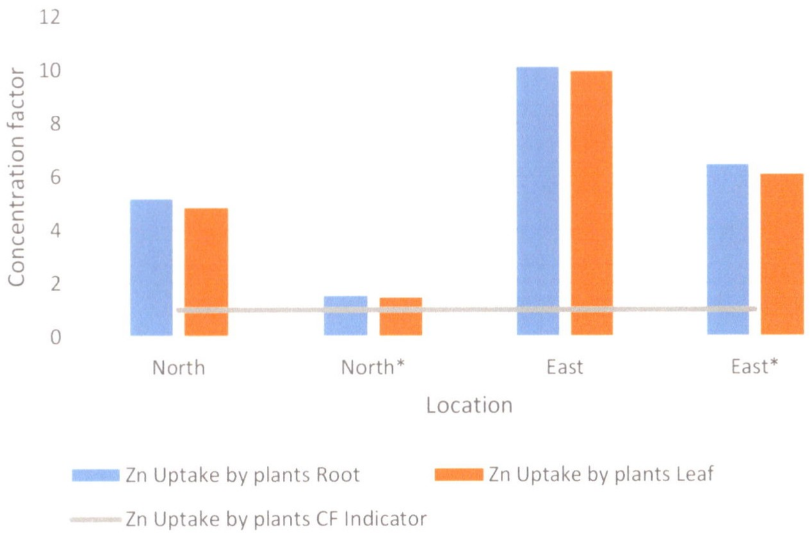

Figure 2.
Zinc concentration factor in Eragrostis hypnoides. (north and east* were taken at 1000 m while north and east were taken at 500 m).*

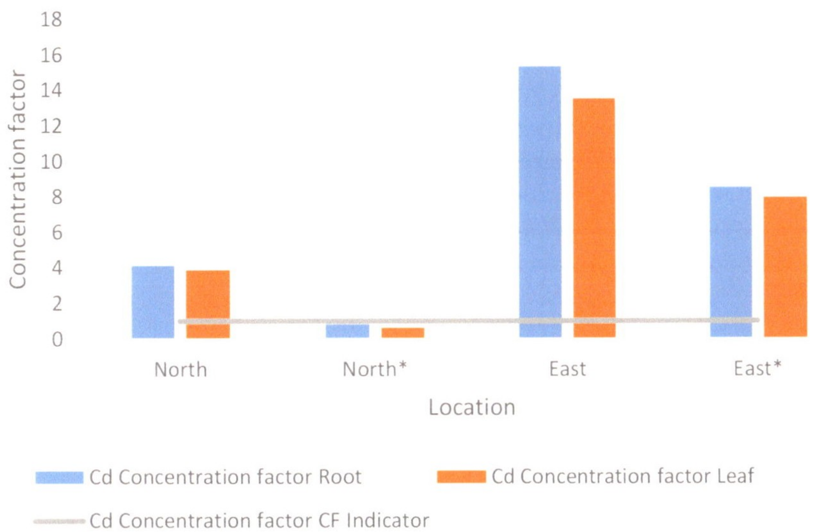

Figure 3.
Cadmium concentration factor in Eragrostis hypnoides. (north and east* were taken at 1000 m while north and east were taken at 500 m).*

3.4.2 Cadmium uptake by plant

The CF result indicated that 75% of the sampled sites have Cf greater than 1 in the root and leaf. The CF of the plant from the northern location of the sampled area is the only exception (**Figure 3**).

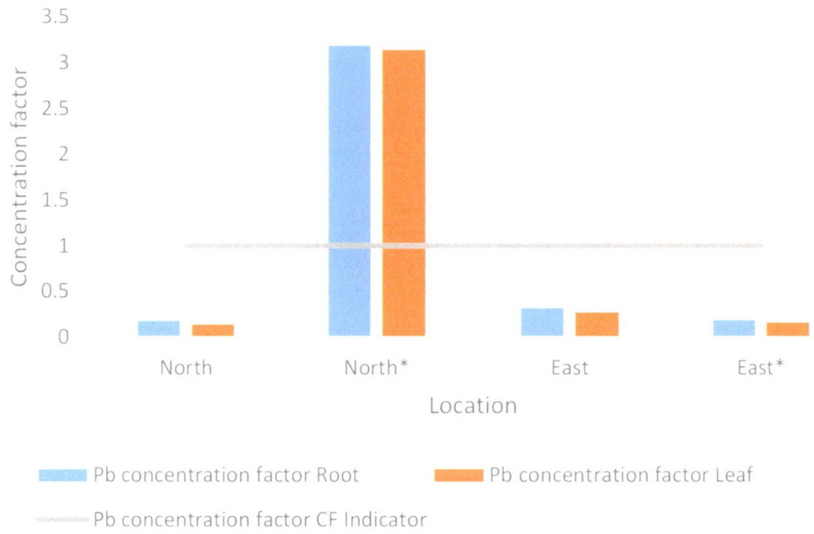

Figure 4.
Lead concentration factor in Eragrostis hypnoides. (north and east* were taken at 1000 m while north and east were taken at 500 m).*

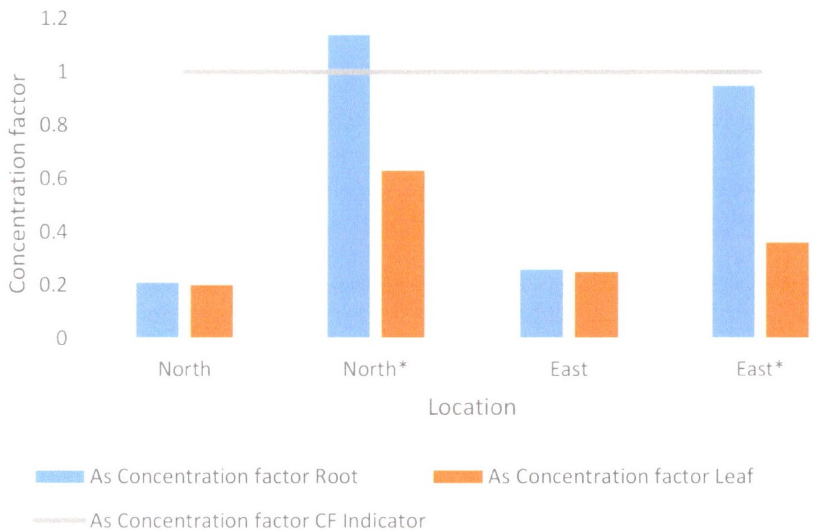

Figure 5.
Arsenic concentration factor in Eragrostis hypnoides. (north and east* were taken at 1000 m while north and east were taken at 500 m).*

Similar to Zn, this confirms that *E. hypnoides* can be used for the phytoremediation of Cd polluted soil. One of the most important factor considered in order to evaluate the ability of a plant to be used for phytoremediation is that the plant must have a high tolerance to high heavy metal concentration and must be able to bioaccumulate these heavy metals in their stem, leaves and fruits [88].

3.4.3 Lead uptake by plant

Due to the ability of Pb to bind to organic matters in soil, the ability of plants to uptake Pb is limited [89]. This is confirmed by the result of Pb CF which wsere all below 1 except for the northern sample at 1 km from the mining site (**Figure 4**). This exception is warranted due to the high concentration of Pb recorded in the plant (**Table 11**) and soil (**Table 3**) at this site.

It can be concluded that *E. hypnoides* cannot be used for the phyto-extraction of Pb from the soil. According to Islam et al. [90], they reported that a plant can be used for the phyto-extraction if the plant species can bioaccumulate toxic metals in the soil.

3.4.4 Arsenic uptake by plant

Similar to the uptake of Pb by the studied plant, the CF of As is below 1 in all the sampled sites besides in the root of the northern site at 1 km away from the gold mine (**Figure 5**). This means that the studied plant cannot bioaccumulate As and as a result cannot be used for the phytoremediation of As polluted soil [91].

4. Conclusions

From the results of the concentrations of As, Cd, Pb and Zn in both the soil and plants samples, the following conclusions have been reached to meet the objectives of this study:

1. There is a direct relationship between the concentrations of the studied heavy metals in soil and plants. This means *E. hypnoides* bioaccumulate Zn and Cd within the mining area. This constitutes a health risk to both humans and animals (cattle that grade in this environment)

2. The concentration of the studied heavy metals in soil within the gold mine are all below the permissible limit of soil used for agriculture as approved by the South African government agency. Therefore, the soil is not polluted.

3. The concentration of the studied heavy metals in plants are more than the permissible limit s proposed by the WHO. This can negatively compromise the health of animals and human that feed directly and indirectly on the plant. This means that the feeding of cattle with the *E. hypnoides* from the area should be discouraged and the grazing of animals within the 1 km radius of the gold mine must be discontinued.

4. *E. hypnoides* has the potential to bioaccumulate Zn and Cd. Due to this, the plant can be used for the phytoremediation of Zn and Cd polluted soil.

It is highly recommended that thorough research should be done within the proximity of gold mines to ascertain the concentration of various toxic heavy metals in the environment. This is essential because commercial farms and animal husbandry are commonly established within mining areas. This study is important to safeguard the health of the citizens and food security.

Author details

Iyioluwa Busuyi Raji[1*] and Lobina Gertrude Palamuleni[2]

1 Faculty of Natural and Agricultural Sciences, Unit of Environmental Sciences and Management, North-West University, Potchefstroom, South Africa

2 Faculty of Natural and Agricultural Sciences, Unit of Environmental Sciences and Management, North-West University, Mmabatho, South Africa

*Address all correspondence to: rajiiyioluwa@gmail.com

References

[1] Darkwa E, Acquah B. Small scale mining and rural livelihoods: How is small scale mining affecting livelihoods in Ghana? Asian Journal Social Science Management Technology. 2022;**4**(2):13-22

[2] Marais L, Cloete J, Lenka M. The plight of mining cities in South Africa: Planning for growth and closure. Cities. 2022;**130**:103965

[3] Mate G. Colonial Mining: A Global Historical Context. Mining the Landscape: Springer; 2022. pp. 51-82

[4] Okanigbe DO, Popoola AP, Malatji N, Lesufi T, Sekgobela G. Bionanomining: A Revised Insight into Processing of South Africa's Complex Gold Ores. Rare Metal Technology 2022. Cham: Springer; 2022. pp. 189-200

[5] Larsen TH, Hansen UE. Sustainable industrialization in Africa: The localization of wind-turbine component production in South Africa. Innovation and Development. 2022;**12**(2):189-208

[6] Shams M, Tavakkoli Nezhad N, Dehghan A, Alidadi H, Paydar M, Mohammadi AA, et al. Heavy metals exposure, carcinogenic and non-carcinogenic human health risks assessment of groundwater around mines in Joghatai, Iran. International Journal of Environmental Analytical Chemistry. 2022;**102**(8):1884-1899

[7] Muthusamy L, Rajendran M, Ramamoorthy K, Narayanan M, Kandasamy S. Phytostabilization of metal mine tailings—A green remediation technology. In: Phytoremediation Technology for the Removal of Heavy Metals and Other Contaminants from Soil and Water. Elsevier; 2022. pp. 243-253

[8] Sithole S, Agboola O, Mugivhisa L, Amoo S, Olowoyo J. Elemental concentration of heavy metals in oyster mushrooms grown on mine polluted soils in Pretoria, South Africa. Journal of King Saud University-Science. 2022;**34**(2):101763

[9] Tran TS, Dinh VC, Nguyen TAH, Kim K-W. Soil contamination and health risk assessment from heavy metals exposure near mining area in bac Kan province, Vietnam. Environmental Geochemistry and Health. 2022;**44**(4):1189-1202

[10] Raji IB, Hoffmann E, Ngie A, Winde F. Assessing uranium pollution levels in the Rietspruit River, far west Rand goldfield, South Africa. International Journal of Environmental Research and Public Health. 2021;**18**(16):8466

[11] Shezi B, Street RA, Webster C, Kunene Z, Mathee A. Heavy metal contamination of soil in preschool facilities around industrial operations, Kuils River, Cape Town (South Africa). International Journal of Environmental Research and Public Health. 2022;**19**(7):4380

[12] Bakare BF, Adeyinka GC. Evaluating the potential health risks of selected heavy metals across four wastewater treatment water works in Durban, South Africa. Toxics. 2022;**10**(6):340

[13] Järup L. Hazards of heavy metal contamination. British Medical Bulletin. 2003;**68**(1):167-182

[14] Fasae KD, Abolaji AO. Interactions and toxicity of non-essential heavy metals (Cd, Pb and Hg): Lessons from Drosophila melanogaster. Current Opinion in Insect Science. 2022;**51**:100900

[15] Zaynab M, Al-Yahyai R, Ameen A, Sharif Y, Ali L, Fatima M, et al. Health

and environmental effects of heavy metals. Journal of King Saud University-Science. 2022;**34**(1):101653

[16] Mehboob R. Importance of iron metabolism: Importance of iron metabolism. Pakistan BioMedical Journal. 2022;**5**(3):01

[17] Kumar SB, Arnipalli SR, Mehta P, Carrau S, Ziouzenkova O. Iron deficiency anemia: Efficacy and limitations of nutritional and comprehensive mitigation strategies. Nutrients. 2022;**14**(14):2976

[18] Sheth S. Transfusional iron overload. Rossi's Principles of Transfusion Medicine. 2022:587-597

[19] Allameh A, Amini-Harandi A, Osati-Ashtiani F, O'Brien PJ. Iron overload induced apoptotic cell death in isolated rat hepatocytes mediated by reactive oxygen species. Iranian Journal of Pharmaceutical Research. 2022;**7**(2):115-121

[20] Dumrongwongsiri O, Winichagoon P, Chongviriyaphan N, Suthutvoravut U, Grote V, Koletzko B. Zinc and iron adequacy and relative importance of zinc/iron storage and intakes among breastfed infants. Maternal & Child Nutrition. 2022;**18**(1):e13268

[21] Ho E, Wong CP, King JC. Impact of zinc on DNA integrity and age-related inflammation. Free Radical Biology and Medicine. 2022;**178**:391-397

[22] Wahiduzzaman M, Islam MM, Sikder AHF, Parveen Z. Bioaccumulation and heavy metal contamination in fish species of the Dhaleswari River of Bangladesh and related human health implications. Biological Trace Element Research. 2022;**200**(8):3854-3866

[23] Mitra S, Chakraborty AJ, Tareq AM, Emran TB, Nainu F, Khusro A, et al. Impact of heavy metals on the environment and human health: Novel therapeutic insights to counter the toxicity. Journal of King Saud University-Science. 2022;**34**(3):101865

[24] Sperdouli I, Adamakis I-DS, Dobrikova A, Apostolova E, Hanć A, Moustakas M. Excess zinc supply reduces cadmium uptake and mitigates cadmium toxicity effects on chloroplast structure, oxidative stress, and photosystem II photochemical efficiency in Salvia sclarea plants. Toxics. 2022;**10**(1):36

[25] Pederiva S, Crescio MI, Ingravalle F, Abete MC, Marchis D, Squadrone S. Processed animal proteins (PAPs) in animal nutrition: Assessment of the chemical risk of essential and non-essential elements. Journal of Trace Elements in Medicine and Biology. 2022;**71**:126959

[26] Koch W, Czop M, Iłowiecka K, Nawrocka A, Wiącek D. Dietary intake of toxic heavy metals with major groups of food products—Results of analytical determinations. Nutrients. 2022;**14**(8):1626

[27] Sikdar A, Jeyasundar PGSA, Debnath B, Hossain M, Islam M, Ahammed GJ. Cadmium contamination in the soil environment: Impact on plant growth and human health. In: Agrochemicals in Soil and Environment. Singapore: Springer; 2022. pp. 367-408

[28] Zhao D, Wang P, Zhao F-J. Dietary cadmium exposure, risks to human health and mitigation strategies. Critical Reviews in Environmental Science and Technology. 2022:1-25

[29] Argos M, Ahsan H, Graziano JH. Arsenic and human health: Epidemiologic progress and public health implications. Reviews on Environmental Health. 2012;**27**(4):191-195

[30] Saha J, Dikshit A, Bandyopadhyay M, Saha K. A review of arsenic poisoning and its effects on human health. Critical Reviews in Environmental Science and Technology. 1999;**29**(3):281-313

[31] Prakash S, Verma AK. Arsenic: It's toxicity and impact on human health. International Journal of Biological Innovations, IJBI. 2021;**3**(1):38-47

[32] Jaydhar AK, Pal SC, Saha A, Islam ARMT, Ruidas D. Hydrogeochemical evaluation and corresponding health risk from elevated arsenic and fluoride contamination in recurrent coastal multi-aquifers of eastern India. Journal of Cleaner Production. 2022;**369**:133150

[33] Mohanraj N, Joshi NS, Poulose R, Patil RR, Santhoshkumar R, Kumar A, et al. A proteomic study to unveil lead toxicity-induced memory impairments invoked by synaptic dysregulation. Toxicology Reports. 2022;**9**:1501-1513

[34] Mishra KP, Singh VK, Rani R, Yadav VS, Chandran V, Srivastava SP, et al. Effect of lead exposure on the immune response of some occupationally exposed individuals. Toxicology. 2003;**188**(2):251-259

[35] Mishra S, Bharagava RN, More N, Yadav A, Zainith S, Mani S, et al. Heavy metal contamination: An alarming threat to environment and human health. In: Environmental Biotechnology: For Sustainable Future. Singapore: Springer; 2019. pp. 103-125

[36] Balaram V, Rani A, Rathore D. Uranium in groundwater in parts of India and world: A comprehensive review of sources, impact to the environment and human health, analytical techniques, and mitigation technologies. Geosystems and Geoenvironment. 2022;**1**(2):100043

[37] Prasad M, Semwal P, Panwar P, Gusain G, Ramola R. Uranium contamination in drinking water as a health concern in Uttarakhand, India. Journal of Radioanalytical and Nuclear Chemistry. 2022;**331**(4):1933-1940

[38] Taylor S, Terkildsen M, McQuilty R, Lee D, Wing-Simpson A, Gray R. Non-essential heavy metals and protective effects of selenium against mercury toxicity in endangered Australian sea lion (Neophoca cinerea) pups with hookworm disease. Environment International. 2022;**169**:107521

[39] Lehmann I, Sack U, Lehmann J. 8 metal ions affecting the immune system. Metal Ions in Toxicology: Effects, Interactions, Interdependencies. 2015;**8**:157-186

[40] Kicińska A, Pomykała R, Izquierdo-Diaz M. Changes in soil pH and mobility of heavy metals in contaminated soils. European Journal of Soil Science. 2022;**73**(1):e13203

[41] Tripathi P, Singhal A, Jha PK. Metal Transport and its Impact on Coastal Ecosystem. Coastal Ecosystems: Springer; 2022. pp. 239-264

[42] Fashola MO, Ngole-Jeme VM, Babalola OO. Physicochemical properties, heavy metals, and metal-tolerant bacteria profiles of abandoned gold mine tailings in Krugersdorp, South Africa. Canadian Journal of Soil Science. 2020;**100**(3):217-233

[43] Kapwata T, Mathee A, Sweijd N, Minakawa N, Mogotsi M, Kunene Z, et al. Spatial assessment of heavy metals contamination in household garden soils in rural Limpopo Province, South Africa. Environmental Geochemistry and Health. 2020;**42**(12):4181-4191

[44] Flefel H, Nokhrin D, Donnik I. Determine heavy metals in water, aquatic plants, and sediment in water systems. E3S Web of Conferences. 2020;**222**:02028

[45] Kausar S, Faizan S, Haneef I. Effect of wastewater irrigation on heavy metal

accumulation, growth and yield of vegetables. International Journal of Plant and Environment. 2017;**3**(1):65-66

[46] Shapi M, Jordaan MA, Mbambo AT, Davies TC, Chirenje E, Dube M. Determination of potentially harmful element (PHE) distribution in water bodies in Krugersdorp, a Mining City in the west Rand, Gauteng Province, South Africa. Minerals. 2021;**11**(10):1133

[47] Chetty S, Pillay L, Humphries MS. Gold mining's toxic legacy: Pollutant transport and accumulation in the Klip River catchment, Johannesburg. South African Journal of Science. 2021;**117**(7-8):1-11

[48] Nkosi V, Wichmann J, Voyi K. Chronic respiratory disease among the elderly in South Africa: Any association with proximity to mine dumps? Environmental Health. 2015;**14**(1):1-8

[49] Mpanza M, Adam E, Moolla R. Perceptions of external costs of dust fallout from gold mine tailings: West Wits Basin. Clean Air Journal. 2020;**30**(1):1-12

[50] Benchmarks-Foundation. Soweto Report: Waiting to Inhale: A Survey of Household Health in Four Mines Soweto Report: Waiting to Inhale: A Survey of Household Health in Four Mine-affected Communities'. Policy Gap 12.-affected Communities 2017 [Policy Gap 12. Available from: http://www.benchmarks.org.za/ [accessed: 3]

[51] Eissa MA, Negim OE. Heavy metals uptake and translocation by lettuce and spinach grown on a metal-contaminated soil. Journal of Soil Science and Plant Nutrition. 2018;**18**(4):1097-1107

[52] Kumar R, Ivy N, Bhattacharya S, Dey A, Sharma P. Coupled effects of microplastics and heavy metals on plants: Uptake, bioaccumulation, and environmental health perspectives. Science of The Total Environment. 2022;**836**:155619

[53] Ngugi MM, Gitari HI, Muui CW, Gweyi-Onyango JP. Growth tolerance, concentration, and uptake of heavy metals as ameliorated by silicon application in vegetables. International Journal of Phytoremediation. 2022;**24**(14):1-14

[54] Tchounwou PB, Yedjou CG, Patlolla AK, Sutton DJ. Heavy metal toxicity and the environment. Molecular, Clinical and Environmental Toxicology. 2012;**101**:133-164

[55] Jaishankar M, Tseten T, Anbalagan N, Mathew BB, Beeregowda KN. Toxicity, mechanism and health effects of some heavy metals. Interdisciplinary Toxicology. 2014;**7**(2):60

[56] Balali-Mood M, Naseri K, Tahergorabi Z, Khazdair MR, Sadeghi M. Toxic mechanisms of five heavy metals: Mercury, lead, chromium, cadmium, and arsenic. Frontiers in Pharmacology. 2021;**12**:643972

[57] Durand JF. The impact of gold mining on the Witwatersrand on the rivers and karst system of Gauteng and north West Province, South Africa. Journal of African Earth Sciences. 2012;**68**:24-43

[58] Bakatula E, Cukrowska E, Chimuka L, Tutu H. Characterization of cyanide in a natural stream impacted by gold mining activities in the Witwatersrand Basin, South Africa. Toxicological & Environmental Chemistry. 2012;**94**(1):7-19

[59] Hansen RN. Contaminant leaching from gold mining tailings dams in the Witwatersrand Basin, South Africa: A new geochemical modelling approach. Applied Geochemistry. 2015;**61**:217-223

[60] Branquinho C, Gaio-Oliveira G, Augusto S, Pinho P, Máguas C, Correia O.

Biomonitoring spatial and temporal impact of atmospheric dust from a cement industry. Environmental Pollution. 2008;**151**(2):292-299

[61] Al-Khashman OA, Shawabkeh RA. Metals distribution in soils around the cement factory in southern Jordan. Environmental Pollution. 2006;**140**(3):387-394

[62] Olowoyo JO, Odiwe AI, Mkolo NM, Macheka L. Investigating the concentrations of different elements in soil and plant composition from a mining area. Polish Journal of Environmental Studies. 2013;**22**(4):1135-1141

[63] Ma JF, Ryan PR, Delhaize E. Aluminium tolerance in plants and the complexing role of organic acids. Trends in Plant Science. 2001;**6**(6):273-278

[64] Kamunda C, Mathuthu M, Madhuku M. Health risk assessment of heavy metals in soils from Witwatersrand gold Mining Basin, South Africa. International Journal of Environmental Research and Public Health. 2016;**13**(7):663

[65] Wedepohl KH. The composition of the continental crust. Geochimica et Cosmochimica Acta. 1995;**59**(7):1217-1232

[66] Chetty S, Pillay L. Assessing the influence of human activities on river health: A case for two south African rivers with differing pollutant sources. Environmental Monitoring and Assessment. 2019;**191**(3):1-11

[67] Herselman JE. The Concentration of Selected Trace Metals in South African Soils. Stellenbosch: University of Stellenbosch; 2007

[68] Ekwue Y, Gbadebo A Arowolo T, Adesodun J. Assessment of metal contamination in soil and plants from abandoned secondary and primary goldmines in Osun state, Nigeria. Journal of Soil Science and Environmental Management. 2012;**3**(11):262-274

[69] Raulinaitis M, Ignatavičius G, Sinkevičius S, Oškinis V. Assessment of heavy metal contamination and spatial distribution in surface and subsurface sediment layers in the northern part of Lake Babrukas. Ekologija. 2012;**58**(1):33-43

[70] Wu J, Teng Y, Lu S, Wang Y, Jiao X. Evaluation of soil contamination indices in a mining area of Jiangxi, China. PLoS One. 2014;**9**(11):e112917

[71] Zhang W, Liu X, Cheng H, Zeng EY, Hu Y. Heavy metal pollution in sediments of a typical mariculture zone in South China. Marine Pollution Bulletin. 2012;**64**(4):712-720

[72] Accounts S-CE, Division S. Human Activity and the Environment. Statistics Canada; 2010

[73] Wahl JJ. Soil Ecological Risk Assessments of Selected South African Soils. Doctoral Dissertation. South Africa: North-West University; 2014

[74] Lee D, Lee C. Regulatory Standards of Heavy Metal Pollutants in Soil and Groundwater in Taiwan. Taipei, Taiwan: National Taiwan University; 2011

[75] Chen S-b, Meng W, Li S-s, Zhao Z-q, Wen-di E. Overview on current criteria for heavy metals and its hint for the revision of soil environmental quality standards in China. Journal of Integrative Agriculture. 2018;**17**(4):765-774

[76] He Z, Shentu J, Yang X, Baligar VC, Zhang T, Stoffella PJ. Heavy Metal Contamination of Soils: Sources, Indicators and Assessment. Journal of Environmental Indicators. 2015;**9**:17-18

[77] Provoost J, Cornelis C, Swartjes F. Comparison of soil clean-up standards for trace elements between countries:

Why do they differ? Journal of Soils and Sediments. 2006;**6**(3):173-181

[78] Kim IS, Kang KH, Johnson-Green P, Lee EJ. Investigation of heavy metal accumulation in Polygonum thunbergii for phytoextraction. Environmental Pollution. 2003;**126**(2):235-243

[79] Nazir R, Khan M, Masab M, Rehman HU, Rauf NU, Shahab S, et al. Accumulation of heavy metals (Ni, Cu, Cd, Cr, Pb, Zn, Fe) in the soil, water and plants and analysis of physico-chemical parameters of soil and water collected from Tanda dam Kohat. Journal of pharmaceutical sciences and research. 2015;7(3):89

[80] Keith LS, Faroon OM, Fowler BA. Chapter 59 - uranium∗. In: Nordberg GF, Fowler BA, Nordberg M, editors. Handbook on the Toxicology of Metals. Fourth ed. San Diego: Academic Press; 2015. pp. 1307-1345

[81] Zulfiqar U, Ayub A, Hussain S, Waraich EA, El-Esawi MA, Ishfaq M, et al. Cadmium toxicity in plants: Recent progress on morpho-physiological effects and remediation strategies. Journal of Soil Science and Plant Nutrition. 2021;**22**:1-58

[82] Bu-Olayan AH, Thomas BV. Translocation and bioaccumulation of trace metals in desert plants of Kuwait governorates. Research Journal of Environmental Sciences. 2009;**3**(5):581-587

[83] ATSDR. Case Studies in Environmental Medicine (CSEM) Cadmium Toxicity. 2010. Available from: https://www.atsdr.cdc.gov/csem/cadmium/docs/cadmium.pdf [accessed 18 December 2022]

[84] Ogundele D, Adio A, Oludele O. Heavy metal concentrations in plants and soil along heavy traffic roads in north Central Nigeria. Journal of Environmental & Analytical Toxicology. 2015;**5**(6):1

[85] Punshon T, Jackson BP, Meharg AA, Warczack T, Scheckel K, Guerinot ML. Understanding arsenic dynamics in agronomic systems to predict and prevent uptake by crop plants. Science of the Total Environment. 2017;**581**:209-220

[86] Pandey AK, Zorić L, Sun T, Karanović D, Fang P, Borišev M, et al. The anatomical basis of heavy metal responses in legumes and their impact on plant–rhizosphere interactions. Plants. 2022;**11**(19):2554

[87] Calabró MR, Roqueiro G, Tapia R, Crespo DC, Bargiela MF, Young BJ. Chronic toxicity, bioavailability and bioaccumulation of Zn, Cu and Pb in Lactuca sativa exposed to waste from an abandoned gold mine. Chemosphere. 2022;**307**:135855

[88] Tournay RJ, Doty SL. Microbial endophytes for clean-up of pollution. Good Microbes in Medicine, Food Production, Biotechnology, Bioremediation, and Agriculture. 2022:358-371

[89] Collin S, Baskar A, Geevarghese DM, Ali MNVS, Bahubali P, Choudhary R, et al. Bioaccumulation of lead (Pb) and its effects in plants: A review. Journal of Hazardous Materials Letters. 2022;**3**:100064

[90] Islam M, Kormoker T, Khan R, Proshad R, Kabir M, Idris AM. Strategies for heavy metals remediation from contaminated soils and future perspectives. In: Soil Health and Environmental Sustainability. Cham: Springer; 2022. pp. 615-644

[91] Kumar P, Koul B, Sharma M. Phytoremediation of heavy metals. In: Heavy Metals in Plants Physiological to Molecular Approach. United States: CRC Press; 2022. pp. 369-387

Chapter 26

Ameliorative Effects of Zinc and Vitamin E on Physiological Changes after Exposure to Heavy Metal

Samuel A. Seriki and Charles C. Mfem

Abstract

Heavy metals have been known to have great deteriorative impacts on the physiology of the body, altering the normal functioning of the body. These impacts cut across the various systems of the body including cardiopulmonary, endocrine, neurological, gastrointestinal, hematological, etc. However, not every exposure will leave such effects in the aftermath. The level of exposure to one heavy metal that is considered harmful may not be with another metal. This chapter examines the various levels of exposure that may be considered unhealthy to the human body, and the mechanisms by which the metals exert their impacts, with the aim of educating readers on how to keep exposure below such threshold level. This chapter also explains that not all heavy metals are considered unhealthy as there are essential heavy metals that may have some beneficial effects to the physiology of the human system.

Keywords: heavy metals, physiological changes, anxiety disorder, toxicity, essential and non-essential heavy metals

1. Introduction

Heavy metals are defined as metallic elements that have a relatively high density compared to water. These metals may be toxic or poisonous even at low concentrations. They are described as those elements having atomic number greater than 20 and atomic density above 5 g cm^{-3} and must exhibit the properties of metals [1, 2].

Examples include cadmium (Cd), mercury (Hg), zinc (Zn), copper (Cu), chromium (Cr), lead (Pb), arsenic (As), and nickel (Ni).

Natural phenomena like weathering and volcanic eruptions have also been implicated in heavy metal pollution [3, 4]. Some of them are exploited for various industrial and economic purposes. They are grouped into essential and non-essential heavy metals.

While essential heavy metals such as iron (Fe), magnesium (Mg), copper (Cu), and the like, are essential nutrients required for various biochemical and physiological functions such as growth, metabolism, and development of different organs, non-essential heavy metals such as cadmium (Cd), antimony (Sb), lead (Pb), vanadium (V) have no established biological functions, yet they still find their way into the body system, and have been reported to affect cellular organelles and components in biological systems [5].

There are numerous essential heavy metals required by plants as they form cofactors that are structurally and functionally vital for enzymes and other proteins. Essential elements are often required in trace amounts in the level of 10–15 ppm and are known as micronutrients.

Due to their high levels of toxicity, arsenic, cadmium, chromium, lead, and mercury are among the priority metals that are important for public health. Even at the modest exposure levels, these metallic elements are known to cause numerous organ damage and are regarded as systemic toxicants.

Despite the fact that some of these metals only affect human physiology at high amounts, others, including cadmium, mercury, lead, chromium, silver, and arsenic, have significant effects on the body even in minute quantities, leading to acute and chronic toxicities in humans [6].

Exposure to these heavy metals has been associated with certain physiological changes ranging from mental, hematological, and hormonal.

This chapter discusses major physiological changes that exposure to these metals can cause to the human body, as well as the risk factors that can lead to changes in human physiology. The impact of cadmium on the central nervous system (CNS) is used as a case study. It also discusses how these changes could be ameliorated.

2. Effects of cadmium on some physiological parameters in human

Cadmium chloride is a colorless heavy metal that can dissolve in ethanol, methanol, and water. It is considered a major environmental pollutant as a result of its widespread industrial use. It is present not only in soil and food, but also in water and air. So, it could be contaminated through food intake and could be released into water as a by-product. Combustion of coal and oil could also expose individuals to it [6]. It has a long half-life of between 15 and 30 years in humans due to its low rate of excretion from the body [7, 8].

International Agency for Research on Cancer (IARC) has identified cadmium as a known or probable human carcinogen. It has also been listed in the International Register of Potentially Toxic Chemicals (IRPTC) of the United Nations Environment Program (IRPTC), even as the World Health Organization (WHO) estimated 500 micrograms per week cadmium as the safe level for human ingestion [9].

Fish, liver, grains, and vegetables remain major sources of dietary cadmium [10].

Cadmium chloride has various lines of applications and is mostly used industrially. The major industrial applications of cadmium include the production of alloys, pigments, and batteries [11].

Invariably, people are exposed to cadmium on a daily basis, with common exposure in industrial work places, plants, soils, and from smoking. Due to its low permissible exposure to humans, over exposure may occur even in situations where trace quantities of cadmium are found [12]. Shortness of breath, pneumonitis, and pulmonary edema can all be signs of more serious respiratory system injuries [13, 14].

Long-term accumulation of cadmium in a number of tissues, including the kidneys, liver, CNS, and peripheral neuronal systems, may have hazardous effects at the peripheral level. It could cross the blood-brain barrier at the CNS and enter the CNS through the nasal mucosa or olfactory pathways. Exposure to cadmium is implicated in hyperactivity, increased aggression, impaired social memory processes, and altered drinking behavior [15, 16].

2.1 Mechanism of action

Cadmium acts as catalysts for biochemical reactions, regulators of gene expression, second messengers in signaling pathways, and co-factors for vital enzymes notorious for regulating physiological, pathological, and behavioral functions [7, 17].

In comparison with other brain regions, the hippocampus collects the divalent metals to a larger amount. The hippocampus impairment that results from heavy metal exposure has been linked to behavioral changes. Animal studies using Cd exposure also show behavioral changes in this approach. Reduced memory and altered anxiety and fear responses have been seen in rats exposed to Cd [18].

2.2 Zinc and Vitamin E

2.2.1 Zinc

Zinc (Zn), a trace element necessary for live cells and an important heavy metal for many enzymes, is involved in DNA replication, transcription, and protein synthesis, which all have an impact on cell division and differentiation [19]. It performs the task of attaching particular genes to tetrahedral bonds, causing transcription, and is thus directly implicated in the translation stage of DNA element gene expression.

Zinc deficiency may prevent the production of new proteins, which would reduce the amount of protein and cause a buildup of amino acids. This is due to zinc, a ribosome structural element that maintains the structural integrity of the ribosomes. In the absence of it, ribosomes break down [19].

By the antioxidant system's action, it stops cell damage. It performs many different roles and is a crucial part of the antioxidant defense system [20, 21].

2.2.2 Vitamin E

The collection of eight fat-soluble compounds includes vitamin E. It can be discovered in many foods and oils. Alpha-tocopherol is mostly found in nuts, seeds, vegetable oils, fortified cereal, and green vegetables. Significant levels are also present in green leafy vegetables and fortified cereals. Food-based vitamin E is not known to be harmful. However, there is proof that extremely high doses of vitamin E supplementation might cause pro-oxidant damage [22, 23].

Vitamin E plays a role in the prevention of diseases like cancer, Alzheimer's disease, HIV/AIDS, and others by preventing oxidative stress, protecting cell membranes, controlling platelet aggregation, and activating protein kinase C. According to other theories, vitamin E regulates gene expression and cell signal transmission [24–26].

2.2.3 Anxiety

Many conditions that produce trepidation, fear, concern, and worrying are together referred to as anxiety. It is described as a feeling that is accompanied by tense sensations, anxious thoughts, and physical changes like raised blood pressure. Fear is a reaction to an immediate threat, actual or perceived; anxiety is the anticipation of an impending threat [27, 28].

Muscle tension, agitation, exhaustion, and attention issues are frequently present in conjunction with it. Although experiencing anxiety occasionally is normal, a person may develop an anxiety disorder. Drug addiction, drug withdrawal, and genetic factors are all possible causes of anxiety disorders [27].

Therapy, medication, and lifestyle modifications are all potential treatment options. Worry, which is considered to be a result of metacognitive beliefs, is something that metacognitive treatment aims to eliminate [29].

This chapter addresses how cadmium chloride impacts the CNS to create anxiety as well as the role of zinc and vitamin E in reducing its effects on anxiety levels. Exposure to cadmium chloride may change the physiology of biological organs and systems.

3. Materials and methods

3.1 Animal preparation

For the investigation, 25 healthy CD1 mice, 8–10 weeks old, and weighing 18–30 g were employed. The animals were given unlimited access to food and water. For the course of the trial, the food and water troughs were replaced daily, and the beddings were also changed every three to five days. They were kept in a room with standard temperature and humidity levels (between 18 and 23°C and 40 and 60%, respectively), as well as a 12/12-hour light/dark cycle.

3.2 Experimental design

The 25 mice were randomly assigned into four (5) groups of five (5) animals;
Group A: Control
Group B: $CdCl_2$ (14 days)
Group C: $CdCl_2$ (28 days)
Group D: $CdCl_2$ + Zinc
Group E: $CdCl_2$ + Vitamin E

3.3 Drug administration

The test drugs were reconstituted into appropriate concentrations as follows: 500 mg of $CdCl_2$ was dissolved in 10,000 ml of distilled water; (50 ppm). 400 mg of vitamin E was dissolved in 5 ml of castor oil. 100 mg of zinc was dissolved in 69 ml of normal saline.

The $CdCl_2$ was administered orally with the aid of an orogastric cannula to the mice in group B for a period of 14 days (short-term exposure), while to the mice in group C, it was administered for 28 days (long-term exposure). Mice in groups D and E received same dose of $CdCl_2$ for 14 days but in addition they were given zinc and vitamin E, respectively. Following the Light and Dark Transition Box (LDTB) and Elevated Plus Maze (EPM) paradigms, anxiety-like behaviors were assessed at the conclusion of the treatment session.

3.4 Determination of anxiety

The test for anxiety was done using:

1. Elevated Plus Maze (EPM)

2. Light and Dark Transition Box (LDTB)

3.4.1 Elevated plus maze (EPM)

Rodents are used in the Elevated Plus Maze (EPM), a test for detecting anxiety in lab animals, as a general research tool in the study of neurobiological anxiety as well as a screening test for potential anxiolytic or anxiogenic substances. The animal in the EPM displays this concern by spending greater time in the enclosed arms [30, 31].

A raised, plus-shaped (+) apparatus with two open and two enclosed arms is used for the test. The behavioral model is based on rodents' typical dislike of open areas. Due to this aversion, a behavior known as thigmotaxis develops, which is the desire for staying in enclosed areas or close to the boundaries of a confined region. This results in the animals restricting their movement to the confined arms of the EPM. An increase in the proportion of time spent in the open arms (time in open arms/ total time in open or closed arms) and an increase in the proportion of entries into the open arms (entries into open arms/total entries into open or closed arms) are indicators of reduced anxiety in the plus maze. Occasionally, the total number of closed-arm entries is used as a gauge of overall activity [31, 32].

The EPM was created in accordance with Lister's specifications (1987). From a center square (5 x 5 cm), the maze contains two open arms (45 5 cm^2) with 0.25 cm high borders and two closed arms (40 5 cm^2) with 15 cm high walls. There is a small ledge in the open arms (4 mm high) to stop the mice from losing their footing and going off the edge. Because they are contained, like most anxiety tests, the closed arms give a feeling of security. This job takes advantage of mice's natural desire to investigate novel surroundings and their aversion to wide open spaces. Anxiety is also quantified by the open arm avoidance score [33].

To remove olfactory cues as well as feces and urine, the surfaces and closed sides of the plus maze arms were washed with methylated spirit prior to the test. The mouse was positioned in the plus maze's middle square so that it initially faced an open arm away from the experimenter. Mouse was given five minutes to examine the device after placement before a silent stopwatch was started. The testing procedure was documented. Open arm movements and head dipping were deemed exploratory behaviors, and a higher frequency of these actions indicates a higher level of investigation [34].

Following are the behaviors scored:

1. The animal's total distance traveled while participating in the test.

2. The animal's open arm entries: the frequency with which it did so. For a mouse to be eligible for entrance, all four of its paws have to be inside the arm.

3. Closed arm entries: the animal's frequency of entry into the closed arms. For a mouse to be eligible for entrance, all four of its paws have to be inside the arm.

4. The animal's time in the open arms was measured in open arm duration.

5. Time spent in closed arms: how long the animal was held there.

6. Center square entries: how frequently the animal used all four paws to enter the center square.

7. Time spent in the core square: the amount of time the animal was there.

8. Head dipping: the animal frequently dropped its head over the sides of the open arm and down toward the ground.

9. Stretch attend postures: the animal frequently extends its head and shoulders forward before retracting back to its starting position.

10. Rearing: the number of times an animal stands on its hind legs or leans its front paws against a maze wall.

11. Grooming: the amount of time an animal spends stationarily licking or scratching itself.

12. Urination: the quantity of urine pools or streaks.

13. Number of fecal boli formed during defecation.

3.4.2 Light/Dark transition box (LDTB)

Two compartments make up the LDTB device. The light compartment occupies two-thirds of the box and is both open and well-lit. A covered and dark compartment makes up one-third of the entire box. The two chambers are connected by a 7-cm door. Rodents choose shadowy environments than bright ones. Rodents, on the other hand, show a propensity to explore when placed in an unfamiliar habitat. There are visible indications of anxiety as a result of these two opposing feelings. The dark compartment is often where rodents spend more time than the bright one. The percentage of time spent in the light compartment will rise in animals given anxiolytic injections. Rearing, or when a rodent raises up on its hind legs, is an indication of motion and nighttime exploration increase in compartment as well. The amount of time spent in the dark compartment increases after receiving anxiogenic injections. There is no prerequisite training for the LDTB. No food or water is restricted and only natural stressors like as light are utilized [35, 36].

The wooden light/dark box (45 x 27 x 27 cm) has two compartments that are of different sizes. Two-fifths of the box is painted white for the bigger compartment (27 x 27 cm), while two-fifths of the box is painted black for the smaller compartment (18 x 27 cm). A door (7.5 x 7.5 cm) that is situated in the middle of the wall between the two compartments at floor level connects them. The Plexiglas-covered floor is separated into 9 x 9 cm squares. The covers of both sections are made of transparent Plexiglas.

The apparatus's light box is filled with a mouse, which is given free rein to move about. The mouse will typically explore the compartment's edges before discovering the door. The mouse is given five minutes to investigate the device, and the rodent's actions inside the box recorded. To be deemed an entry, all four paws must be inserted into the opposing chamber. The mouse is then taken out, and the box is cleaned with cotton wool and 70% ethyl alcohol. The box is then allowed to dry in between experiments [37].

3.4.3 Behaviors score

1. Number of transitions: how often the animal enters the opposing compartment (The mouse's four paws must enter the new compartment for it to be scored and to be regarded as having been entered.) [38].
2. Number of times the animal stepped over a line marked on the box's floor.
3. Rearing: the regularity with which the animal stands up straight or leans its front paws against the box wall.
4. Stretch attend postures: the animal frequently exhibits forward extension of the head and shoulders, followed by retraction to the starting position.
5. The length of time the animal spent stationarily licking the body.
6. The animal's time in the box's dark side is measured in terms of how long it was there.
7. The animal's time in the light side of the cage, measured in minutes.
8. Defecation: the quantity of fecal boli that are formed (light vs. dark).
9. The quantity of pee pools or streaks (light vs. dark).

4. Statistical analysis

Data from the tests were analyzed, and the outcomes were displayed as graphs of means and standard error of means (SEM). To determine whether there was any significant variation between the test and control groups, analysis of variance (ANOVA) and a post-hoc Student's t-test were utilized. $P < 0.05$ was adopted as the threshold for significance.

5. Results

5.1 Comparison of frequency of rearing among the different experimental groups in the light/dark transition box

The mean ± SEM rearing frequency for the control, $CdCl_2$ (14 days), $CdCl_2$ (28 days), $CdCl_2$ + Zinc, and $CdCl_2$ + vitamin E groups were 59.60 ± 5.12, 35.40 ± 3.41, 26.50 ± 5.11, 55.40 ± 3.04, and 53.60 ± 3.66, respectively.

The results revealed that as compared to the control group, the rearing frequency in the $CdCl_2$ group was significantly lower ($p < 0.05$). However, compared to the $CdCl_2$ group, the rearing frequency of the $CdCl_2$ + Zinc and $CdCl_2$ + vitamin E groups was significantly greater ($p < 0.05$) (see **Table 1**).

5.2 Comparison of stretch attend posture frequency (SAPLDT) in the light/dark transition box among the experimental groups

The mean ± SEM SAP frequency for the control, $CdCl_2$ (14 days), $CdCl_2$ (28 days), $CdCl_2$ + Zinc, and $CdCl_2$ + vitamin E groups were 2.00 ± 0.71, 3.60 ± 0.51, 4.70 ± 0.41, 1.60 ± 0.51, and 1.60 ± 0.51 respectively.

Groups	Rearing	Stretch Attend Posture	Transition between Light and Dark	Dark Duration in Light and Dark	Light Duration in Light and Dark
Control	59.60 ± 5.12	2.00 ± 0.71	21.60 ± 2.01	85.00 ± 12.98	215.00 ± 12.98
$CdCl_2$ (short-term exposure)	35.40 ± 3.41	3.60 ± 0.51	7.40 ± 1.03	213.50 ± 17.06	86.55 ± 17.09
$CdCl_2$ (long-term exposure)	26.50 ± 5.11	4.70 ± 0.41	5.30 + 1.13	227.66 ± 15.87	72.48 ± 13.87
$CdCl_2$ + Zn	55.40 ± 3.04	1.60 ± 0.51	17.40 ± 1.63	81.63 ± 15.87	218.37 ± 15.87
$CdCl_2$ + Vitamin E	53.60 ± 3.66	1.60 ± 0.51	21.40 ± 2.01	63.76 ± 18.35	236.24 ± 18.35

Table 1.
Comparing the occurrence of Rearing, Stretch Attend Posture (SAP), Transition between Light and Dark (TLD), Dark Duration in Light and Dark (DDLD) Transition Box Test, and Light Duration in Light and Dark (LDLD) Transition Box Test among various groups.

The results showed that there was no discernible change in the SAP frequency between the $CdCl_2$ group and the control group. Yet when compared to the $CdCl_2$ group, the SAP frequency in the $CdCl_2$ + Zinc and $CdCl_2$ + vitamin E groups was considerably lower ($p < 0.05$) (see **Table 1**).

5.3 Comparison of frequency of transition between the light/dark (TLD) compartments among the different experimental groups in the light/dark transition box

The mean ± SEM frequency of transition for the control, $CdCl_2$ (14 days), $CdCl_2$ (28 days), $CdCl_2$ + Zinc, and $CdCl_2$ + vitamin E groups were 21.60 ± 2.01, 7.40 ± 1.03, 5.30 + 1.13, 17.40 ± 1.63, and 21.40 ± 2.01 respectively.

The findings demonstrated that the frequency of transition was substantially lower ($p< 0.05$) in the $CdCl_2$ group than in the control group. When compared to the $CdCl_2$ group, the frequency of transition was significantly higher ($p < 0.05$) in the $CdCl_2$ + Zinc and $CdCl_2$ + vitamin E groups (see **Table 1**).

5.4 Comparison of dark duration in light/dark (DDLD) transition box test among the different experimental groups

The mean ± SEM dark duration for the control, $CdCl_2$ (14 days), $CdCl_2$ (28 days), $CdCl_2$ + Zinc, and $CdCl_2$ + vitamin E groups were 85.00 ± 12.98, 213.50 ± 17.06, 227.66 ± 15.87, 81.63 ± 15.87, and 63.76 ± 18.35 respectively.

The results showed that the $CdCl_2$ group's time spent in the dark chamber was substantially longer ($p < 0.05$) than that of the control group. When compared to the $CdCl_2$ group, the dark chamber duration was considerably shorter ($p < 0.05$) in the $CdCl_2$ + Zinc and $CdCl_2$ + vitamin E groups (see **Table 1**).

5.5 Comparison of light duration during light/dark (LDLD) transition box test among the different experimental groups

The mean ± SEM light duration for the control, $CdCl_2$ (14 days), $CdCl_2$ (28 days), $CdCl_2$ + Zinc, and $CdCl_2$ + vitamin E groups were 215.00 ± 12.98, 86.55 ± 17.09, 72.48 ± 13.87, 218.37 ± 15.87, and 236.24 ± 18.35, respectively.

Results showed that the $CdCl_2$ group's time spent in the light chamber was substantially shorter ($p < 0.05$) than that of the control group. The light chamber duration was substantially longer ($p < 0.05$) for the $CdCl_2$ + Zinc and $CdCl_2$ + vitamin E groups than for the $CdCl_2$ group (see **Table 1**).

5.6 Comparison of grooming frequency during light/dark transition box test among the different experimental groups

The mean ± SEM grooming frequency for the control, $CdCl_2$ (14 days), $CdCl_2$ (28 days), $CdCl_2$ + Zinc, and $CdCl_2$ + Vitamin E groups were 1.80 ± 0.37, 7.20 ± 0.86, 9.33 ± 0.38, 2.60 ± 1.08, and 4.20 ± 0.58 respectively.

The findings revealed that the $CdCl_2$ group's grooming frequency was substantially higher ($p < 0.05$) than that of the control group. However, when compared to the $CdCl_2$ group, the grooming frequency in the $CdCl_2$ + Zinc group was considerably lower ($p < 0.05$) (see **Table 2**).

5.7 Comparison of the elevated plus maze (REPM) test's frequency of rearing among the several experimental groups

The mean ± SEM rearing frequency for the control, $CdCl_2$ (14 days), $CdCl_2$ (28 days), $CdCl_2$ + Zinc, and $CdCl_2$ + Vitamin E groups were 40.60 ± 3.89, 23.20 ± 3.73, 19.42 ± 3.86, 35.00 ± 3.54, and 40.40 ± 1.86, respectively.

Groups	Grooming (Light/Day Transition)	Regaining consciousness during Elevated Plus Maze	Stretch Attend Posture during the Elevated Plus Maze	Head Dips during the Elevated Plus Maze	Grooming during the Elevated Plus Maze Freq.
Control	1.80 ± 0.37	40.60 ± 3.89	7.60 ± 1.54	11.00 ± 2.70	2.60 ± 0.40
$CdCl_2$ (short-term exposure)	7.20 ± 0.86	23.20 ± 3.73	11.00 ± 1.26	14.40 ± 1.72	7.20 ± 0.58
CdC_2 (long-term exposure)	9.33 ± 0.38	19.42 ± 3.86	13.00 ± 1.65	17.00 ± 1.12	10.40 ± 0.28
$CdCl_2$ + Zn	2.60 ± 1.08	35.00 ± 3.54	7.00 ± 1.64	8.00 ± 1.22	2.20 ± 0.58
$CdCl_2$ + Vitamin E	4.20 ± 0.58	40.40 ± 1.86	4.00 ± 0.71	9.80 ± 1.16	1.80 ± 0.37

Table 2.
Comparing the frequency of grooming during the Light/Dark Transition Box Test, the occurrence of regaining consciousness during the Elevated Plus Maze, the occurrence of the Stretch Attend Posture, the occurrence of Head Dips, and the occurrence of grooming during the Elevated Plus Maze among the various experimental groups.

The results revealed that as compared to the control group, the rearing frequency in the $CdCl_2$ group was considerably lower ($p < 0.05$). However, when compared to the $CdCl_2$ group, the rearing frequency in the $CdCl_2$ + Vitamin E group was considerably greater ($p < 0.05$) (see **Table 2**).

5.8 Comparison of the elevated plus maze test's stretch attend posture (SAPEPM) frequency in the various experimental groups

The mean ± SEM SAP frequency for the control, $CdCl_2$ (14 days), $CdCl_2$ (28 days), $CdCl_2$ + Zinc, and $CdCl_2$ + Vitamin E groups were 7.60 ± 1.54, 11.00 ± 1.26, 13.00 ± 1.65, 7.00 ± 1.64, and 4.00 ± 0.71, respectively.

The findings showed that there was no discernible change in the SAP frequency between the $CdCl_2$ group and the control group. In contrast to the $CdCl_2$ group, the SAP frequency in the $CdCl_2$ + Vitamin E group was considerably lower ($p < 0.05$) (see **Table 2**).

5.9 Comparison of the variable experimental groups' head dip frequency in the elevated plus maze test

The mean ± SEM head dip frequency for the control, $CdCl_2$ (14 days), $CdCl_2$ (28 days), $CdCl_2$ + Zinc, and $CdCl_2$ + Vitamin E were 11.00 ± 2.70, 14.40 ± 1.72, 17.00 ± 1.12, 8.00 ± 1.22, and 9.80 ± 1.16, respectively.

The results showed that there was no discernible difference in the frequency of head dips between groups (see **Table 2**).

5.10 Comparison of the various experimental groups' grooming frequency throughout the elevated plus maze test

The mean ± SEM grooming frequency for the control, $CdCl_2$ (14 days), $CdCl_2$ (28 days), $CdCl_2$ + Zn, and $CdCl_2$ + Vitamin E groups were 2.60 ± 0.40, 7.20 ± 0.58, 10.40 ± 0.28, 2.20 ± 0.58, and 1.80 ± 0.37, respectively.

Results showed that the $CdCl_2$ group's grooming frequency was substantially higher ($p < 0.05$) than that of the control group. When compared to the $CdCl_2$ group, grooming frequency was considerably reduced ($p < 0.05$) in the $CdCl_2$ + Zinc and $CdCl_2$ + Vitamin E groups (see **Table 2**).

5.11 Comparison of grooming duration in the elevated plus maze test among the different experimental groups

The mean ± SEM grooming duration for the control, $CdCl_2$ (14 days), $CdCl_2$ (28 days), $CdCl_2$ + Zinc, and $CdCl_2$ + Vitamin E groups were 10.71 ± 2.79, 38.08 ± 5.61, 45.02 ± 2.68, 11.49 ± 3.33, and 7.39 ± 1.99, respectively.

The results revealed that the $CdCl_2$ group's grooming time was substantially longer ($p < 0.05$) than that of the control group. In contrast to the $CdCl_2$ group, the grooming time was considerably shorter in the $CdCl_2$ + Zinc and $CdCl_2$ + Vitamin E groups ($p < 0.05$) (see **Table 3**).

Groups	$Grooming_{(EPM)}$ duration	Freq. CAE_{EPM}	CAE_{EPM} Duration	Freq. OAE_{EPM}	OAE_{EPM} Dur.
Control	10.71 ± 2.79	3.80 ± 0.80	112.27 ± 10.92	9.40 ± 0.75	185.73 ± 10.66
$CdCl_2$ (short-term exposure)	38.08 ± 5.61	5.80 ± 0.66	238.29 ± 18.21	2.80 ± 0.37	61.71 ± 18.21
$CdCl_2$ (long-term exposure)	45.02 ± 2.68	7.01 ± 0.58	310.25 ± 12.81	1.48 ± 0.51	39.46 ± 18.66
$CdCl_2$ + Zinc	11.49 ± 3.33	3.00 ± 0.45	144.29 ± 7.96	8.20 ± 0.86	155.71 ± 7.96
$CdCl_2$ + Vitamin E	7.39 ± 1.99	3.20 ± 0.58	106.34 ± 26.04	8.60 ± 0.51	193.66 ± 26.04

Table 3.
Compares the length of grooming in the Elevated Plus Maze Test, frequency of closed arm entry during the test, duration of closed arm entry during the test (CAE Dur.), frequency of open arm entry during the test (Freq. OAE), and duration of open arm entry during the test among the various experimental groups.

5.12 Comparison of the frequency of closed arm entry during elevated plus maze test among the different experimental groups

The mean ± SEM frequency of closed arm entry for control, $CdCl_2$ (14 days), $CdCl_2$ (28 days), $CdCl_2$ + Zinc, and $CdCl_2$ + Vitamin E groups were 3.80 ± 0.80, 5.80 ± 0.66, 7.01 ± 0.58, 3.00 ± 0.45, and 3.20 ± 0.58, respectively.

The results showed that the frequency of closed arm entry in $CdCl_2$ group had no significant difference when compared with the control group. However, the $CdCl_2$ + Zinc and $CdCl_2$ + Vitamin E groups had significantly lower ($p < 0.05$) frequency of closed arm entry when compared with the $CdCl_2$ group (see **Table 3**).

5.13 Comparison of the duration in the closed arm entry during (CAE Dur.) elevated plus maze test among the different experimental groups

The mean ± SEM closed arm duration for the control, $CdCl_2$ (14 days), $CdCl_2$ (28 days), $CdCl_2$ + Zinc, and $CdCl_2$ + Vitamin E were 112.27 ± 10.92, 238.29 ± 18.21, 310.25 ± 12.81, 144.29 ± 7.96, and 106.34 ± 26.04, respectively.

The results showed that the $CdCl_2$ group's duration in the closed arm was substantially longer ($p < 0.05$) than that of the control group. When compared to the $CdCl_2$ group, the duration in the closed arm was considerably shorter for the $CdCl_2$ + Zinc and $CdCl_2$ + Vitamin E groups ($p < 0.05$) (see **Table 3**).

5.14 Comparison of the frequency of open arm entry (Freq. OAE) during elevated plus maze test among the different experimental groups

The mean ± SEM frequency of open arm entry for the control, $CdCl_2$ (14 days), $CdCl_2$ (28 days), $CdCl_2$ + Zinc, and $CdCl_2$ + Vitamin E groups were 9.40 ± 0.75, 2.80 ± 0.37, 1.48 ± 0.51, 8.20 ± 0.86, and 8.60 ± 0.51, respectively.

In comparison with the control group, the $CdCl_2$ group's open arm entry frequency was significantly lower, according to the data ($p < 0.05$). The open arm entry frequency was significantly higher ($p < 0.05$) in the $CdCl_2$ + Zinc and $CdCl_2$ + Vitamin E groups as compared to the $CdCl_2$ group (see **Table 3**).

5.15 Comparison of the elevated plus maze test time in the open arm among the several experimental groups

The mean ± SEM open arm duration for the control, $CdCl_2$ (14 days), $CdCl_2$ (28 days), $CdCl_2$ + Zinc, and $CdCl_2$ + Vitamin E were 185.73 ± 10.66, 61.71 ± 18.21, 39.46 ± 18.66, 155.71 ± 7.96, and 193.66 ± 26.04, respectively.

Results showed that the $CdCl_2$ group's open arm duration was considerably shorter ($p < 0.05$) than that of the control group. However, compared to the $CdCl_2$ group, the open arm time was considerably longer ($p < 0.05$) in the $CdCl_2$ + Zinc and $CdCl_2$ + Vitamin E groups (see **Table 3**).

6. Discussion

The elevated plus maze (EPM) and light and dark transition box (LDTB) tests were employed to measure anxiety. In order to assess anxiety, signs such as grooming frequency and duration, rearing, and stretch attend posture were also used [27].

Longer duration in the light box compartment shows decreased anxiety and greater length in the dark box compartment reveals increased anxiety [36].

The results of this experiment's light/dark transition box test revealed that the control group's dark chamber duration was shorter and its light chamber duration was longer than that of the experimental group. This suggests that the mice who were not exposed to $CdCl_2$ experienced less anxiety.

However, there was a longer time in the dark room and a shorter time in the light chamber in the group exposed to $CdCl_2$, indicating that exposure to $CdCl_2$ enhanced anxiety. This is demonstrated in **Table 1** as a longer time of exposure had a more significant effect.

Also, the zinc and vitamin E treated groups that had previously been exposed to $CdCl_2$ had a shorter time in the dark and a longer time in the light. This means that even after exposure to situations that had previously elevated anxiety, zinc and vitamin E may be able to reduce it.

Increased grooming frequency indicates higher levels of anxiety [39].

According to the experiment's findings, the $CdCl_2$ exposed group groomed more frequently than the control group. Even more frequent grooming was observed in the group exposed for a longer period of time, indicating a proportionate rise in anxiety with $CdCl_2$ exposure, compared to less frequent grooming in the zinc and vitamin E treated group. This supports **Table 2**'s finding that zinc and vitamin E lessen anxiety.

The animal spends more time in the enclosed arm when it is anxious, indicating that an increase in duration in the closed arm represents an increase in anxiety level and an increase in duration in the open arm suggests a drop in anxiety levels [31].

When compared to the control group, the $CdCl_2$ group had a longer duration in the closed arm of the labyrinth and a shorter length in the open arm, according to the results of this experiment's elevated plus maze test. While the zinc and vitamin E treated group had a reduced duration in the closed arm and a higher duration in the open arm of the labyrinth, these effects are much more significant in the group

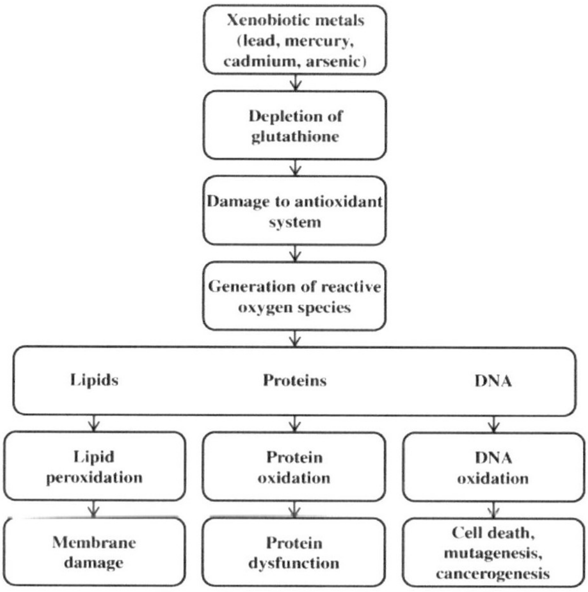

Figure 1.
Showing mechanisms of heavy metal toxicity.

with longer period of exposure. Also, it revealed that the $CdCl_2$ group groomed more frequently and for longer than the control group, whereas the zinc and vitamin E treated groups groomed less frequently and for shorter periods of time.

This supports the finding that anxiety is caused by cadmium chloride ($CdCl_2$) exposure in mice, and that the longer the exposure, the more anxiety is caused. As shown in **Table 3** [20, 24], it also suggests that zinc and vitamin E, two important antioxidants, aid in reducing anxiety and stress.

The primary mechanisms of heavy metal toxicity include free radical production, which leads to oxidative stress, damage to biological molecules such as enzymes, proteins, lipids, and nucleic acids, as well as damage to DNA, which is crucial for both neurotoxicity and carcinogenesis. See **Figure 1**.

7. Conclusion

The results of this study suggest that exposure to cadmium chloride ($CdCl_2$) causes anxiety in CD1 mice. This anxiety is even more pronounced with longer period of exposure. Zinc and vitamin E, through their antioxidant property, show ameliorative effect by lowering anxiety levels in CD1 mice after exposure to cadmium chloride ($CdCl_2$).

Funding

This work did not receive any form of funding or financial support whatsoever from anyone or any group. It is solely the personal efforts of the authors. It also did not receive any form of non-financial support from anyone.

Conflict of interest

Authors declare that there is no conflict of interest over the manuscript.

Ethical clearance

Ethical approval was obtained: UCFMSECAE1047.

Author details

Samuel A. Seriki[1]* and Charles C. Mfem[2]

1 Department of Human Physiology, College of Medical Sciences, Edo State University, Uzairue, Edo State, Nigeria

2 Department of Human Physiology, College of Medical Sciences, University of Calabar, Calabar, Cross Rivers State, Nigeria

*Address all correspondence to: seriki.adinoyi@gmail.com; seriki.samuel@edouniversity.edu.ng

References

[1] Fergusson JE. The Heavy Elements: Chemistry, Environmental Impact and Health Effects. Oxford: Pergamon Press; 1990

[2] He ZL, Yang XE, Stoffella PJ. Trace elements in agroecosystems and impacts on the environment. Journal of Trace Elements in Medicine and Biology. 2005;**19**(2-3):125-140

[3] Nriagu JO. A global assessment of natural sources of atmospheric trace metals. Nature. 1989;**338**:47-49

[4] Bradl H. Heavy Metals in the Environment: Origin, Interaction and Remediation. Vol. 6. London: Academic Press; 2002

[5] Wang S, Shi X. Molecular mechanisms of metal toxicity and carcinogenesis. Molecular Cell Biochemistry. 2001;**222**:3-9

[6] Burger J. Assessment and management of risk to wildlife from Cadmium. The Science of the Total Environment. 2008;**389**:37-45

[7] Zadorozhnaja TD, Little RE, Miller RK, Mendel NA, Taylor RJ, Presley BJ, et al. Concentrations of arsenic, cadmium, copper, lead, Mercury and zinc in human placentas from two cities in Ukraine. Journal of Toxicology and Environmental Health. 2000;**61**:255-263

[8] Benoff S, Jacob A, Hurley IR. Male infertility and environmental exposure to lead and cadmium. Human Reproduction Update. 2000;**6**:107-121

[9] WHO. Cadmium. Geneva: WHO; 1992

[10] Jarup L, Akesson A. Current status of cadmium as an environmental health problem. Toxicology and Applied Pharmacology. 2009;**238**:201-208

[11] Wilson DN. Association cadmium. Cadmium market trends and influences. In: Cadmium 87 Proceedings of the 6th International Cadmium Conference. London: International programme on chemical safety (World Health Organization); 1988. pp. 9-16

[12] Takaki A, Jimi S, Segewa M, Hisano S, et al. Long-term cadmium exposure accelerates age related mitochondrial changes in renal epithelial cells. Toxicology. 2004;**203**:145-154

[13] Seidal K, Jorgensen N, Elinder CG, Sjogren B, et al. Fatal cadmium - induced pneumonitis. Scandinavian Journal of Work, Environment, Health. 1993;**19**:429-431

[14] Godt J, Scheidig F, Grosse-Siestrup C, Esche V, et al. The toxicity of cadmium and resulting hazards for human health. Journal of Occupational Medical Toxicology. 2006;**1**:22

[15] Okuda B, Iwamoto Y, Tachibana H, Sugita M. Parkinsonism after acute cadmium poisoning. Clinical Neurology and Neurosurgery. 1997;**99**:263-265

[16] Cao Y, Chen A, Radcliffe J, Dietrich KN, Jones RL, Caldwell K, et al. Postnatal cadmium exposure, neurodevelopment and blood pressure in children at 2, 5 and 7 years of age. Environmental Health Perspectives. 2009;**117**:1580-1586

[17] Shukla A, Shukla GS, Srimal R. Cadmium-induced alterations in Blood-Brain Barrier permeability and its possible correlation with decreased micro vessel antioxidant potential in rat.

Human and Experimental Toxicology. 1996;**15**:400-405

[18] Brenneman K, Wong B, Buccellato M, Costa ER, Gross E, Dorman DC. Direct olfactory transport of inhaled manganese[(54) MnCl (2)] to the rat brain: Toxicokinetic investigation in a unilateral Nasal occlusion model. Toxicology and Applied Pharmacology. 2000;**169**:238-248

[19] Tsui C. The role of zinc in auxin synthesis in the tomato plant. American Journal of Botany. 1948;**35**:172-179

[20] Powell SR. Antioxidant properties of zinc. Journal of Nutrition. 2000;**130**:1447s-1454s

[21] Ozdemir G, Inanc F. Zinc may protect remote ocular injury caused by intestinal ischaemia re perfusion in rats. The Tohoku Journal of Experimental Medicine. 2005;**206**:247-251

[22] Niki E, Traber MG. A history of vitamin E. Nutritional Metabolism. 2012;**61**:207-212

[23] Di Mascio P, Murphy ME, Sies H. Antioxidant defense systems: the role of carotenoids, tocopherols and thiols. American Journal of Clinical Nutrition. 1991;**53**:194s-200s

[24] Galli F, Azzi A, Birringer M, Cook-Mills JM, Eggersdorfer M, Frank J, et al. Vitamin E: Emerging aspects and new directions. Free Radical Biologie et Médecine. 2017;**102**:16-36

[25] Rimbach G, Moehring J, Huebbe P, Lodge JK. Gene-regulating activity of alpha tocopherol. Molecules. 2010;**15**(3):1746-1761

[26] Azzi A. Many tocopherols, one vitamin E. Molecular Aspects of Medicine. 2018;**61**:92-103

[27] American Psychiatric Association. Diagnostic and Statistical Manual of Mental Disorders. 5th ed. Arlington, VA: American Psychiatric Publishing; 2013. p. 189

[28] Bouras N, Holt G. Psychiatric and Behavioural Disorders in Intellectual and Developmental Disabilities. 2nd ed. Cambridge: Cambridge University Press; 2007

[29] Wells A. Metacognitive Therapy for Anxiety and Depression. New York: Guilford Press; 2011

[30] Ojo J, Mouzon B. Chronic Repetitive Mild traumatic brain injury results in reduced cerebral blood flow, axonal injury, Gliosis and increased T-Tau and Tau oligomers. Journal of Neuropathology and Experimental Neurology. 2016;**75**:636-655

[31] Carobrez AP, Bertoglio LJ. Ethological and temporal analyses of anxiety-like behaviour: the elevated plus-maze model 20 years on. Neuroscience and Biobehavioral Reviews. 2005;**29**:1193-1205

[32] Hogg SA. Review of the validity and variability of the elevated plus-maze as an animal model of anxiety. Pharmacological Biochemical Behaviour. 1996;**54**:21-30

[33] Tapiero H, Tew KD. Trace elements in human physiology and pathology, zinc and Metallothioneins. Biomedicine & Pharmacotherapy. 2003;**57**:399-411

[34] Brown RE, Corey SC, Moore AK. Differences in measures of exploration and fear in MHC - Congenic C57BL/6J and B6-H-2k mice. Behavioural Genetics. 1999;**29**:263-271

[35] Ennaceur A. Tests of unconditioned anxiety — Pitfalls and disappointments. Physiology & Behavior. 2013;**135**:55-71

[36] Bourin M, Hascoet M. The mouse light/dark box test. European Journal of Pharmacology. 2003;**463**:55-65

[37] Hascoet M, Bourin M, Dhonnchadha B. The mouse light-dark paradigm: A review. European Journal of Pharmacology. 2001;**25**:141-166

[38] Berglund M, Åkesson A, Nermell B, Vahter M. Intestinal absorption of dietary cadmium in women depends on body iron stores and fiber intake. Environmental Health Perspectives. 1994;**102**:1058-1066

[39] Hall CS. Emotional behaviour in the rat; defecation and urination as measures of individual differences in emotionality. Journal of Comparative Psychology. 1934;**18**:382-403

Section 4

Impact of Heavy Metals

Chapter 27

Environmental Impact of Heavy Metals

Muhammad Mudassir Usman, Shamsu Shuaibu Bala and K. Hamza

Abstract

Heavy metals cause toxicity in biological systems by bonding to Sulfhydryl groups and producing reactive oxygen species (ROS). Many international organizations established a standard regarding the presence of heavy metals in the environment, food, and drinking water as a result of numerous harmful effects on humans and animals observed. This study aimed to determine the level of heavy metal concentrations and water quality in the Romi River, where some heavy metals concentrations (Iron 0.89 mg/L and Nickel 0.36 mg/L) exceeded the WHO maximum standard limits (Nickel 0.02 mg/L and 0.030 mg/L). The mean concentrations of the metal's chromium, iron, nickel, and zinc with standard deviation were found to be: 0.100.1 mg/L, 0.890.1 mg/L, 0.060.1 mg/L, and 0.200.1 mg/L, respectively. This study revealed that the contamination occurs as a results of effluents release into the river thereby causing many harmful effect to the community around them.

Keywords: contamination, concentration, environment, heavy metals & toxicity, environmental pollution

1. Introduction

Water pollution occurs in both rural and urban areas in Nigeria. Many factories in Nigeria are located on river banks and use the rivers as open sewers for their effluents. It is important to note that some of these heavy metals are required for proper biochemical function. Metals such as lead, chromium, and arsenic, on the other hand, can be toxic when consumed in small or large amounts. In general, the ionic form of a metal is more toxic because it can form toxic compounds with other ions. Electron transfer reactions with oxygen can produce toxic oxyradicals [1, 2].

Water contamination endangers other resources, such as fisheries, and land resources, for example, have already suffered significantly. The majority of environmental pollution is caused by anthropogenic sources, specifically domestic and industrial activities [1, 2]. Failure to halt further deterioration of environmental quality may jeopardize the health of a large proportion of the population, with serious political and socioeconomic consequences [3, 4].

Heavy metal exposure has increased as a result of anthropogenic, industrial, and agricultural activity as well as modern industrialization, all of which have negative impacts on human health. The environmental concern of hazardous metal contamination of water and air affects hundreds of millions of people worldwide. Heavy metal pollution in food is a problem for both human and animal health. In this context, the concentration of heavy metals in air, food, and water sources is evaluated [3]. Among the numerous toxins in the environment, metals can exist naturally and stay in the ecosystem. As a result, human exposure to metals is unavoidable, and some researchers have found that the toxicity of metals varies depending on gender [4]. Biological systems may typically react to them by losing one or more electrons, releasing metal cations that are affine to the nucleophilic sites of essential macromolecules. Several acute and long-term harmful effects of heavy metals have an impact on several human organs. Examples of the adverse effects of heavy metal toxicity include cancer, gastrointestinal and kidney dysfunction, nervous system diseases, skin lesions, vascular damage, immune system malfunction, and birth defects. The cumulative effects of simultaneous exposure to two or more metals have been reported [5–7].

Heavy consequences such abdominal cramping, bloody diarrhea, and kidney failure can occur after exposure to high doses of heavy metals, especially lead and mercury [5, 6]. Contrarily, low-dose exposure poses a subtle and unnoticed risk unless it is consistently experienced, at which point its side effects, such as neuropsychiatric illnesses characterized by exhaustion, anxiety, and negative effects, may be identified.

2. Research review

2.1 Toxic effect of some heavy metals

Out of the 92 naturally occurring elements, 30 are recognized to be potentially harmful to humans. These are created by anthropogenic or natural processes, but the industrial discharge that is of concern in this case is of particular note [7]. Additionally, it is well recognized that the heavy metal pollution chain moves in a circular pattern from industry to atmosphere to soil to water to food to people. Heavy metals can be harmful even at relatively low levels, despite the fact that toxicity is a function of concentrations. The importance of human exposure, consumption, and absorption was emphasized, particularly in industrialized nations.

2.2 Chromium (Cr)

One of the heavy metals whose concentration continually rises as a result of industrial expansion, particularly the growth of the chemical and tanning industries, is chromium. Electroplating, leather tanning, wood preservation, pulp processing, steel manufacture, and many other operations release chromium into the environment, and the concentrations of chromium and nickel in the environment vary greatly. The greater use of these two metals in emerging nations and their non-degradability raise serious concerns [1]. The human body is carcinogenic and highly soluble in hexavalent chromium. It is also well known that the metallurgies, refractory, chemical, and tannery sectors employ this same hexavalent chromium extensively.

2.3 Iron (Fe)

A heavy metal in the first row of transition metals, iron is one of them. Although Fe^{2+} is also detected, Fe^{3+} is the main form that is seen. Iron serves as an oxidizing and reducing agent in the porphyrin enzyme of respiration (Vines and Rees).

Seawater contains roughly 3.5 ppm of iron, the fourth most prevalent element in the earth's crust [1]. It reacts fairly quickly. The transition metal (heavy metal) iron is by far the most common and significant one that has a purpose in living systems. Proteins that contain iron take part in the transport of oxygen and the transfer of electrons, respectively. There are other molecules, [4] whose job it is to transmit and store iron. Ferritin and albumin serve as the storage proteins in humans and many other higher animals.

2.4 Nickel (Ni)

Nickel is the most useful element in soil and plant research. Nickel appears to be required for the growth of marine micro algae. The effect of food containing very low concentrations of nickel (e.g., 40.00 ug/g) includes impaired liver metabolism, decreased iron absorption, and decreased activity of many enzymes. The average concentration of nickel in the world's soil is 40.00 mg/g [7]. In the absence of the emission effect, dietary nickel intake was estimated to be 16,511 ug/day on average [8].

The toxicity of nickel is determined by the route of exposure and the solubility of the nickel compound. Epidemiology studies have shown that occupational inhalation exposure to nickel (Ni) dust can result in an increase in pulmonary and nasal cancer [9].

2.5 Zinc (Zn)

Zinc is a heavy metal in the periodic table's first row of transition metals. Zinc is found everywhere and has been shown to be a growth factor in plants and some rodents. Its absence causes mottled leaf disease in fruit trees. Zinc is found in mammalian enzymes such as carbonic anhydrase. It is required for protein metabolism and appears to be involved in the production of chlorophyll in some way. Zinc is essential for plant growth due to its role in auxin formation and as a component of certain enzymes [9]. Zinc is required for the synthesis of the molecule tryptophan, from which auxin is produced.

Several crop disorders were reported in the early 1900s that have now been identified as zinc deficiencies [9]. A thorough investigation of zinc deficiency in all plants reveals, among other symptoms, some form of leaf chlorosis, mostly on veins and ranging in color from white to light green.

Zinc deficiency is common in soils with abnormally high levels of soluble or total phosphates. An early study on Tung's tree zinc deficiency in fluoride concluded that high phosphate in soils was an important factor reducing available zinc [10]. A relatively low concentration of the element in the body can cause heavy metal toxicity, most commonly intestinal distress.

3. Environmental pollution

Through their bonds with sulfhydryl groups and the production of ROS, heavy metals cause toxicity in biological systems. In addition to oxidative stress and glutathione depletion, this results in the inactivation of important macromolecules. There are a number of events that take place once hazardous metals enter the body and are exposed to them, including interactions with or inhibitions of certain metabolic pathways [10, 11]. Multiple negative consequences on both people and animals are consequently seen. Congenital disorders, immune system problems, hormone changes, particular organ dysfunctions, metabolic abnormalities, cancer, and congenital disorders are a few of these [11]. The presence of metals in the environment, food supply, and drinking water is therefore regulated by a number of international organizations. Studies on risk assessment examine if heavy metals are present in food and water. Nearly 21% of them had amounts that could be detected, it was discovered.

4. Solubility of metals and metal compounds

Chemical speciation affects the environment's metal and metal compound solubility, bioavailability, and persistence; for some metals, speciation may affect the pattern of toxicity (e.g., inorganic arsenic versus organic compounds, inorganic and organic mercury compounds). The papers on exposure concerns and bioavailability and bioaccumulation explore the function of speciation in bioavailability and bioaccumulation within the environment as well as bio-accessibility to human receptors. It is typically believed that the potential toxicity of inorganic species is connected to the cat-existence ions in bodily tissues (in most cases, bound to a tissue ligand). The potential or availability of the metal for interacting at a particular biological target, such as may depend on the intracellular environment and kind of ligand or protein binding [2].

Solubility is one of the most important factors influencing metal and metal compound bioavailability and absorption. A metal compound's solubility is determined by its chemical species, the pH of its medium (H+ ions), and the presence of other chemical species in the medium (see the environmental chemistry paper) [12]. Except for silver, mercury, and lead, nitrates, acetates, and all chlorides of most metals are soluble. Except for barium and lead, most metal sulphates are also soluble. Most hydroxides, carbonates, oxalates, phosphates, and sulphones, on the other hand, are poorly soluble. Particle size is another factor that influences the absorption of poorly soluble compounds: fine particles are usually more soluble. Metallic lead in body tissues (as may occur after gunshot wounds) is most likely absorbed [2].

5. Measures of exposure to metals

In terms of health assessment, the extent of a metal's exposure is best determined by measuring its internal concentration, and even better, the biologically effective dose at the target organ (as opposed to environmental concentration). However, for a variety of reasons, determining the internal or biologically effective dose of the metal at the target tissue is not always feasible. For example, the activity of the heme-synthesizing enzyme aminolaevulinic acid dehydrate (ALAD) in red blood cells is directly related to blood lead concentration and thus may be used as a surrogate for blood lead measurement. The use of biological indicators or markers of exposure, also known as "biomarkers of exposure, " is a method of linking a person's external exposure [13].

5.1 Material

Polyethylene (plastic) bottles, 14 volumetric flasks (100.00 ml), glass funnel, filter paper, 14 beakers (500.00 ml), hot plate pipette, measuring cylinder, hydrochloric acid (HCl), nitric acid (HNO_3), Atomic Absorption Spectroscopy (AAS), Conductivity Meter, Turbidity Meter, and pH Meter were used for the analysis.

5.2 Methods

To avoid the risk, the sample was prepared and digested using standard analytical methods with nitric acid (HNO3) and hydrochloric acid (HCl) at relatively low temperatures, as reported by [1].

The powerful solvent used was aqua regia, a mixture of hydrochloric acid and nitric acid (10:1 V/V). 100.00 ml of each sample was measured and transferred to a 500.00 ml beaker, followed by 10.00 ml of hydrochloric acid (HCl) and 1.00 ml of nitric acid (HNO3). The mixture was then heated on a hot plate for about 3 hours at a relatively low temperature (200C-500C) (NB: Do not allow it to boil) until it was reduced to about 20.00 ml.

The mixtures were then cooled and filtered in a 100.00 ml volumetric flask using a glass funnel and filter paper, and then diluted to volume (i.e., distilled water was added to the mark of the volumetric flask).At the quality control laboratory, the mixtures were tested using Atomic Absorption Spectroscopy (AAS). Based on the above procedure, we reanalyzed the five different samples for the following heavy metals: chromium (Cr), iron (Fe), nickel (Ni), and zinc (Zn).

6. Results and discussion

6.1 Discussion

Table 1 shows the conductivity values for January samples analyzed at five different points: 593.00, 618.00, 828.00, 507.00, and 690.00 µohms/cm. The conductivity values were generally greater than the WHO maximum permissible limit of 500.00 µohms/cm. This is evident from the measured mean of 647.20 ohms/cm. Point 3 had the highest conductivity of 828.00 µohms/cm, while point 1 had the lowest conductivity of 593.00 µohms/cm. Furthermore, when compared to the other points, the conductivity at point 4 (507.00 µohms/cm) was slightly higher than the WHO standard limit.

Table 2 shows the conductivity (µohms/cm) values for February samples analyzed at five different points: 511.00, 499.00, 497.00, 500.00, and 513.00 µohms/cm. In comparison to the mean of 504.00 µohms/cm recorded, the conductivity values were generally greater than the maximum value of 500.00 µohms/cm of WHO standards limits. Points 5 and 1 had the highest conductivities of 513.00 µohms/cm and 511.00 µohms/cm, respectively, while points 3, 2, and 4 were all within the WHO standard limits.

The pH values for January, February, and March samples at five different points are as follows: 7.60, 6.53, 7.52, 6.89, 6.50, 6.66, 6.86, 6.56, 6.76, 6.84, 6.93, 7.10, 7.09, 7.25, and 7.06, as shown in **Tables 1** and **2**. The pH values were found to be generally within the WHO acceptable limits of 6.50–9.20. This is evident from the mean pH value of 7.01, which falls within the WHO threshold range of 6.50–9.20. This implies that the effluent discharged into the river by Kaduna Refining and Petrochemical Company has no effect on the river's pH concentration.

The temperature values recorded for the five different samples over three consecutive months (January, February, and March) at five different points are as follows: 21.30, 20.80, 20.20, 20.90, 20.30, 20.80, 20.50, 20.40, and 20.70°C, which are all less than the room temperature of 25.00°C. While those of March were 27.60, 29.30, 29.80, 30.10, and 30.00°C higher than the room temperature, as shown in **Tables 1–3**.

The mean temperature values of 20.80°C, 20.54°C, and 29.36°C demonstrate this. The increase in temperature values observed in March could be attributed to seasonal changes rather than Kaduna Petrochemical and Refining Company's activities.

As shown in **Table 1**, the turbidity values for January samples at five different points are: 28.70, 14.45, 43.69, 103.00, and 14.01 NTU, which are higher than the maximum limits of WHO and Federal Environmental Protection Agency (FEPA) acceptable standards of 10.00 NTU. Considering the recorded mean turbidity value of 40.77 NTU, which is significantly higher than the acceptable value.

As shown in **Table 2**, the turbidity values for February samples at five different points are: 26.35, 18.36, 23.85, 25.93, and 31.93 NTU, which are higher than the maximum limits of WHO and Federal Environmental Protection Agency (FEPA) standard acceptable standards of 10.00 NTU. Taking into account the recorded mean turbidity value of 25.28 NTU, which is higher than the acceptable value.

The turbidity values for March samples at five different points, as shown in **Table 3**, are greater than the maximum limits of WHO and Federal Environmental Protection Agency (FEPA) standard acceptable standards of 10.00 NTU. Considering the recorded mean turbidity value of 34.16NTU, which is significantly higher than the acceptable value. As a result, the increase in mean turbidity values of 40.77, 25.28, and 34.16 NTU above the acceptable standard in January, February, and March could be attributed to effluents released by the Kaduna Petrochemical and Refining Company.

6.2 Chromium (Cr)

The amount of chromium (Cr) in the five (5) different samples at different points in January, February, and March samples was less than 0.0100 mg/kg (<0.0100 mg/L), as shown above. This is evidence from the mean (<0.0100), standard deviation (0.0112), and variance (0.0001), which show how far apart each metal is from its arithmetic mean. The coefficient of correlation between the three (3) months was discovered to be zero (0), indicating that there was no correlation between them. The WHO and FEPA drinking water standards were 0.05–2.00 mg/L (**Table 4**).

6.3 Iron (Fe)

The iron (Fe) content of these five samples at various points is shown in **Table 5** for January samples in mg/L. At point 1, the amount is high, but at point 2, the amount decreases, while at point 3, the amount increased dramatically, and at point 4, the amount was very low, while at point 5, the amount decreased dramatically.

For February samples, the amount decreases at points 1, 2, and 3, but increases slightly higher at point 4, and dramatically increases at point 5.

For March samples, the amounts at points 1, 2, 3, and 4 decrease, but the amount at point 5 increases to be greater than the amounts at points 3 and 4 but less than the amounts at points 1 and 2. This is evidenced by the three-month mean of 0.2587, 0.8663, and 1.5467, the standard deviation of 0.1996, 0.2585, and 0.0605, and the variance of 0.0398, 0.668, and 0.668. The coefficient of correlation for January and February was found to be +1, indicating a perfect positive correlation between the two months, whereas the correlation for January and March and February and March was found to be zero (0), indicating no correlation between the months.

The amount of Iron (Fe) allowed by the American Public Health (APH) and World Health Organization [14] for drinking water was 0.30 mg/L.

6.4 Nickel (Ni)

The amount of Nickel (Ni) in the five (5) different samples at five different points, as shown in **Table 6**, decreases at points 1, 2, 3, and 4, while the amount decreases dramatically at point 5.

The amounts at all points (i.e., points 1–5) were less than 0.0100 mg/L in February and March samples. The mean of the three months is 1.0502, 0.0100, and 0.0100, the standard deviation is 0.5094, 0.0112, and the variance is 0.2595, 0.0001, and 0.0001. The coefficient of correlation between the three (3) months was discovered to be zero (0), indicating that there was no correlation between them. WHO has recommends a nickel concentration of 0.02 mg/L for drinking water.

6.5 Zinc (Zn)

The amount of Zinc (Zn) in the five (5) different samples at five different points in mg/kg, as presented in **Table 7**, for January samples, decreased at points 1 and 2, but increased at points 3 and 4, and decreased at point 5, while increasing at points 1, 2, 3, and 4.

For February samples, the amount decreases from point 1 to point 2, while the amount decreases from point 3 to point 5, and there is a sudden increase in the amount at point 5 that is greater than the amount at points 1, 2, 3, and 4.

For March samples, the amounts at points 1 and 2 increase, but the amount at point 3 decreases to less than that at points 1 and 2, while at point 4, there was a drastic decrease in the amount compared to that at points 1, 2, and 3, but at point 5, there was

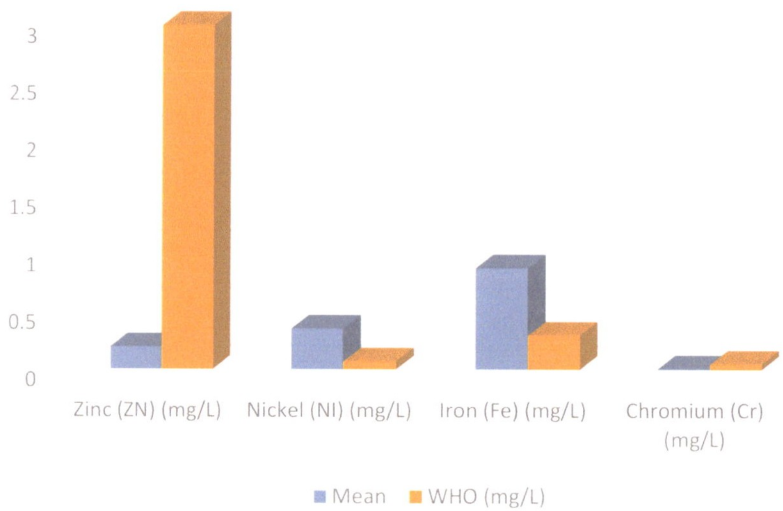

Figure 1.
Comparison between mean samples and WHO maximum concentration limit.

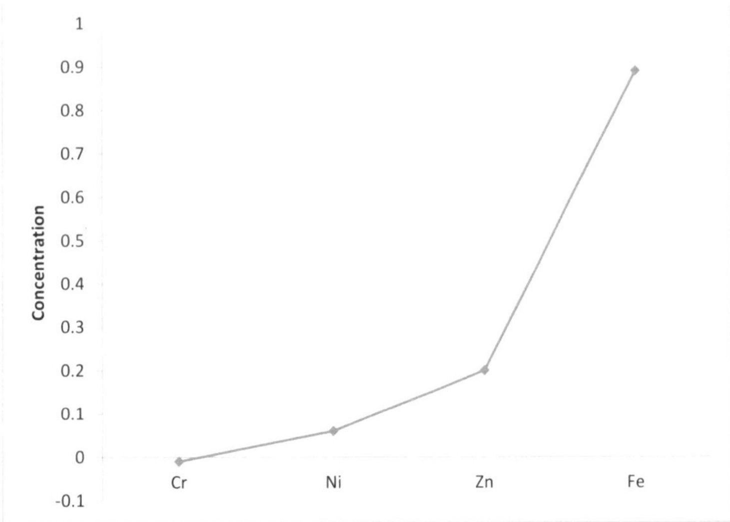

Figure 2.
A graph of concentration (mg/L) of heavy metals.

a drastic increase in the amount that was more than that at points 1, 3, and 4, but less than that at point 2. The mean of the three months is 0.1194, 0.2137, and 0.2596, the standard deviation is 0.0182, 0.0219, and the variance is 0.0003, 0.0005, and 0.0021. The coefficient of correlation between the three (3) months was discovered to be zero (0), indicating that there was no correlation between them.

The heavy metals' mean concentrations (mg/L) (chromium, iron, nickel, and zinc), and a calibration curve (**Figure 1**) was plotted using the table to indicate the heavy metals' concentrations (mg/L) against the metals.

The curve demonstrated an increase in heavy metal concentration (mg/L) in the samples as follows. Iron (Fe) > Zinc (Zn) > Nickel (Ni) > Chromium (Cr). This indicates that iron (Fe) has the highest concentration value in the samples (**Figure 2** and **Table 8**).

7. Results

Samples/Parameters	pH	Temperature (°C)	Conductivity (μohms/cm)	Turbidity (NTU)
Point 1	7.60	21.30	593.00	28.70
Point 2	6.53	20.80	618.00	14.45
Point 3	7.52	20.80	828.00	43.69
Point 4	6.89	20.20	507.00	103.00
Point 5	6.50	20.90	690.00	14.01

Table 1.
Parameters analyzed for January samples.

Samples/Parameters	pH	Temperature(°C)	Conductivity (μohms/cm)	Turbidity (NTU)
Point 1	6.66	20.30	511.00	26.35
Point 2	6.86	20.80	499.00	18.36
Point 3	6.56	20.50	497.00	23.85
Point 4	6.76	20.40	500.00	25.93
Point 5	6.84	20.70	513.00	31.93

Table 2.
Parameters analyzed for February samples.

Samples/Parameters	pH	Temperature (°C)	Conductivity (μohms/cm)	Turbidity (NTU)
Point 1	6.93	27.60	416.00	49.47
Point 2	7.10	29.30	418.00	42.99
Point 3	7.09	29.80	418.00	36.74
Point 4	7.25	30.10	413.00	41.61
Point 5	7.06	30.00	415.00	39.31

Table 3.
Parameters analyzed for the of March samples.

Samples	January	February	March
Point 1	< 0.0100	< 0.0100	< 0.0100
Point 2	< 0.0100	< 0.0100	< 0.0100
Point 3	< 0.0100	< 0.0100	< 0.0100
Point 4	< 0.0100	< 0.0100	< 0.0100
Point 5	< 0.0100	< 0.0100	< 0.0100

Table 4.
The concentration of chromium (mg/L) in the samples.

Samples	January	February	March
Point 1	0.2354	0.7622	1.6467
Point 2	0.0837	0.5056	1.6454
Point 3	0.9546	0.6152	1.4924
Point 4	0.0357	0.6843	1.3636
Point 5	<0.0100	1.7753	1.5765

Table 5.
The concentration of Iron (mg/L) in the samples.

Samples	January	February	March
Point 1	0.9136	< 0.0100	< 0.0100
Point 2	0.0690	< 0.0100	< 0.0100
Point 3	0.0634	< 0.0100	< 0.0100
Point 4	0.0142	< 0.0100	< 0.0100
Point 5	< 0.0100	< 0.0100	< 0.0100

Table 6.
The concentration of nickel (mg/L) in the samples.

Samples	January	February	March
Point 1	0.0870	0.1933	0.2255
Point 2	0.0900	0.1595	0.3876
Point 3	0.1370	0.2290	0.2116
Point 4	0.1091	0.2087	0.1527
Point 5	0.1741	0.2782	0.3071

Table 7.
The concentration of zinc (mg/L) in the samples.

Metals/Months	January	February	March	Mean	WHO (mg/L)
Zinc (ZN) (mg/L)	0.1194	0.2137	0.2569	0.1967	3
Nickel (NI) (mg/L)	1.0502	<0.0100	<0.0100	0.3567	0.07
Iron (Fe) (mg/L)	0.2587	0.8663	1.5467	0.8906	0.3
Chromium (Cr) (mg/L)	<0.0100	<0.0100	<0.0100	0.0100	0.05

Table 8.
Mean concentration of metals in the samples in comparison with WHO standard.

8. Conclusions

Heavy metals enter the human body through a variety of routes, including drinking water, air, food, and, on rare occasions, dermal exposure. Following absorption, heavy metals are retained and accumulate in the human body. When the research findings are compared to WHO standards, it indicates that the level of contamination in the Romi river with some heavy metal concentrations such as (Iron 0.89 mg/L and Nickel 0.36 mg/L) has exceeded the WHO maximum standard limits (Nickel 0.02 mg/L and 0.030 mg/L). The mean concentrations of the metals: Chromium, Iron, Nickel, and Zinc with the standard deviation was found to be: $< 0.01 \pm 0.1\, mg/L$, 0.89 ± 0.1 mg/L, 0.06 ± 0.1 mg/L, and0.20 ± 0.1 mg/L. This study found that the mean concentrations of chromium and zinc were lower than the World Health Organization's acceptable limits, while the higher concentration values of nickel were higher(0.06 ± 0.01 mg/kg) and iron (0.89 ± 0.01 mg/kg) was discovered to be greater than the WHO standard acceptable limits for heavy metals. This contamination is attributed to the Kaduna Petrochemical and Refining Company's effluent discharge into the river.

Author details

Muhammad Mudassir Usman[1,2*], Shamsu Shuaibu Bala[1,2] and K. Hamza[1,2]

1 Federal University of Kashere, Gombe State, Nigeria

2 Rabiu Musa Kwankwaso College of Advance and Remedial Studies, Nigerian Nuclear Regulatory Authority, Tudun Wada Kano, Nigeria

*Address all correspondence to: mudather1197@gmail.com

References

[1] Bokare AD. Advanced oxidation process based on the Cr (III)/Cr (VI) redox cycle. Environmental Science Technology;**45**(21):9332-9338

[2] Usman MM, Abdullahi M, Abdulkareem MN, Hamza K. Environmental impact assessment of heavy metals effluents (Chromium & Nickel). FUDMA Journal of Science (FJS). 2020;**4**(3):704-707

[3] Borowska S, Brzóska M, Gałazyn-Sidorczuk M, Rogalska J. Effect of ˙ an extract from Aroniamelanocarpa L. berries on the body status of zinc and copper under chronic exposure to cadmium: An in vivo experimental study. Nutrients. 2017;**9**(12):1374

[4] Cefalu WT, Hu FB. Role of chromium in human health and in diabetes. Diabetes Care. 2004;**27**(11):2741-2751. DOI: 10.2337/diacare.27.11.2741

[5] Cao ZR, Cui SM, Lu XX, Chen XM, Yang X, Cui JP, et al. Effects of occupational cadmium exposure on workers' cardiovascular system. Zhonghua Lao Dong Wei Sheng Zhi Ye Bing Za Zhi. 2018;**36**(6):474-477. DOI: 10.3760/ cma.j. Issn.1001-9391.2018.06.025

[6] Bodaghi-Namileh V, Sepand MR, Omidi A, Aghsami M, Seyednejad SA, Kasirzadeh S, et al. Acetyl- l -carnitine attenuates arsenic-induced liver injury by abrogation of mitochondrial dysfunction, inflammation, and apoptosis in rats. Environmental Toxicology and Pharmacology. 2018;**58**: 11-20. DOI: 10.1016/j.etap. 2017.12.005

[7] Ferreira LMR, Cunha-Oliveira T, Sobral MC, Abreu PL, Alpoim MC, Urbano AM. Impact of carcinogenic chromium on the cellular response to proteotoxic stress. International Journal of Molecular Sciences. 2019;**20**(19): 4901. DOI: 10.3390/ ijms20194901

[8] Gorini F, Muratori F, Morales MA. The role of heavy metal pollution in neurobehavioral disorders: A focus on autism. Review Journal of Autism and Developmental Disorders. 2014;**1**(4):354-372. DOI: 10.1007/s40489-014-0028-3

[9] Gupta DK, Tiwari S, Razafindrabe B, Chatterjee S. Arsenic contamination from historical aspects to the present. In: Arsenic Contamination in the Environment. Berlin, Germany: Springer; 2017. pp. 1-12

[10] Salama A, Hegazy R, Hassan A. Intranasal chromium induces acute brain and lung injuries in rats: Assessment of different potential hazardous effects of environmental and occupational exposure to chromium and introduction of a novel pharmacological and toxicological animal model. PLoS One. 2016;**11**(12):e0168688. DOI: 10.1371/ journal.pone.0168688

[11] Zhitkovich A. Chromium in drinking water: Sources, metabolism, and cancer risks. Chemical Research in Toxicology. 2011;**24**(10):1617-1629. DOI: 10.1021/ tx200251t

[12] Deng Y, Wang M, Tian T, Lin S, Xu P, Zhou L, et al. The effect of hexavalent chromium on the incidence and mortality of human cancers: A meta-analysis based on published epidemiological cohort studies. Frontiers in Oncology. 2019;**9**:24. DOI: 10.3389/fonc.2019.00024

[13] Effluent Analysis. Area of Kano. Nigeria Research Journal of Science;**2**(2): 121

[14] WHO. International Standard for Water. Geneva: WHO; 2004

Chapter 28

Occurrence and Impact of Heavy Metals on Groundwater Sources: A Case Study of Two Communities in Nasarawa State, Nigeria

Opaluwa Obaje Daniel

Abstract

Heavy metals are major contaminants in groundwater, and they have potentials for toxicity even at low concentrations with health hazards. This led to the determination of heavy metal concentrations and evaluation of HMPI and MI of groundwater sources in Keffi and Karu to ascertain the suitability for domestic usage. Heavy metals, Cd, Cr, Cu, Fe, Mn, Ni, Pb and Zn, were determined using AAS. The results obtained showed that Cd, Cr and Pb had mean concentrations higher than the SON recommended limit. Statistical analyses at $p < 0.05$ showed that there were no significant differences between heavy metal contents in water from boreholes and wells in Keffi and Karu. HMPI were 77.33 and 92.08 for borehole and hand dug well water respectively in Keffi and 105.27 and 127.41 for borehole and hand dug well water respectively in Karu. The values from Keffi are lower than the critical value of 100, while those of Karu are slightly higher. MI for borehole and hand dug well waters was 0.564 and 0.606 respectively in Keffi and for Karu, the values were 0.634 and 0.723 respectively and are all less than 1. These suggest that the water sources are not affected with heavy metal pollution when compared to the water quality classification scheme adopted.

Keywords: heavy metals, water, contamination, pollution, metal index

1. Introduction

One of the most important substances in the life of living organisms is water [1]. It is of very great value because of the roles it play in living organism: aid digestion, flushing of wastes from the body, transportation of nutrients, regulation of body temperature and maintenance of other bodily functions [2]. The establishment of stable human settlements, rural, semi-urban and urban areas, is dependent largely on the readily available portable water sources [3]. Over the decades, there has been an increased demand for water that is of good quality, and this could be attributed to ever increasing human population, urbanisation and industrialisation [3].

In developing countries, these increased demands for water lead to the use of any available water sources to meet the daily water need of inhabitants. The known sources of water are either surface water (rivers, dams, lakes, pond and a few others) or groundwater (boreholes, hand-dug wells and spring) [4]. The surface water sources are readily accessible, because to have access to the groundwater sources, energy in some forms would have to be expended and in some cases requires funds needed for digging. However, there are many settlements where the surface water sources are not available as well as the pipe-borne supply; therefore, the inhabitants are left with no option but to resort to groundwater sources for their daily water need. Groundwater forms a vital domestic and agricultural water sources in rural and urban communities of most developing countries, Nigeria inclusive [5]. Groundwater storage is better with the sedimentary aquifer compared to crystalline basement. Boreholes fail at a high rate in the basement and in areas with basement complex; hand-dug well is always the main source of water [6].

The source of groundwater is the water cycle, and the water is always stored in the aquifer beneath the earth surface. Water that falls as precipitation (rain or snow) flows along the surface of the ground, and it infiltrates into the ground of the surface and is stored in the spaces of soil pores and the fractures of rock formations as groundwater. Due to this reason, groundwater is always believed to be comparatively cleaner than surface water systems that receive contaminants directly [6].

In the recent past, the contamination of groundwater sources has become an issue of very serious environmental concern [7]. There is a vast range of contaminants that affect groundwater sources, and heavy metals are of a particular interest since they have potentials for toxicity even at low concentrations. Although some are essential mineral elements and very important to life (Fe, Cu, Zn and a host of others), at elevated levels, they could become deleterious with associated health risks [8]. Heavy metals are chemical elements, metals or metalloid with a specific gravity five times that of water and have potentials for toxicity. The major source of heavy metals in potable water is the contamination of groundwater and surface water sources [9]. Many heavy metals are natural constituents of the environment. However, elevated levels of these heavy metals in the environment could be attributed to the anthropogenic activities of man. Groundwater contamination could come from industrial sewage, proliferation of dumpsites for domestic wastes without due consideration to government rules, mining, agricultural run-off and a few other factors [9].

It is of utmost importance to assess the metal contents in water from either surface or groundwater sources. This is because even though the trace element could be very essential to animals and humans for biochemical activities, when present at levels higher than recommended limits, it could lead to some morphological disorders in humans, such as mutagenic effect, reduced growth, increased mortality and a host of others [10, 11]. The use of dumpsites as farmland is a common practice in urban and sub-urban centres in Nigeria because decayed and composted wastes enhance soil fertility [12]. These wastes often contain heavy metals in various forms and at different contamination levels. Some heavy metals like As, Cd, Hg and Pb, which are particularly hazardous to plants, animals and humans [13], could find their ways into these water bodies through leaching as well as runoff waters. Also, from farmlands that had agrochemicals and fertilisers applied on them, these agro-based products contain metals like Cu, Mg, Mn, Pb or Zn, which could eventually be found in groundwater through run-off and infiltration [14].

Heavy metals are given serious concern because they are not biodegradable, largely immobile in soil and tend to concentrate and persist for a long time in the environment [15]. Due to the non-biodegradable nature of heavy metals at elevated levels, they could easily undergo bioaccumulation. There is a bioaccumulation of chemical specie in a biological system when there is an increase in the concentration of specie in the biological system compared to its concentration naturally in the environment [16]. One of the easiest means by which metals get to living organism is through water and, if present at elevated levels, is associated with serious health risks. For instance, some are carcinogenic and could cause neurological disorder, liver and kidney dysfunctions and a lot of other serious health challenges; hence, there is the need to assess the metal contents of groundwater sources [17].

Several research works have been done within and outside the shores of Nigeria to assess the metal contents of groundwater sources. These include the evaluation of heavy metals in groundwater, South of Najaf, Iraq [4], the evaluation of heavy metals in groundwater around Keshere and its environs, upper Benue Trough, North-eastern Nigeria [18], the determination of heavy metals in borehole and hand-dug well in selected areas of Mubi, Adamawa state, Nigeria [9] and a host of others.

The government is solely responsible for the provision of potable water in Nasarawa State just like in almost all states in Nigeria, and in most cases, it is characterised by low productivity and few areas covered, and there is always inefficiency and ineffectiveness in the delivery. Most citizenry that do not benefit from the provisions by the government, therefore, depend on the available sources, groundwater (wells and boreholes) whose quality cannot be certified satisfactorily. And, as such, cases of water-associated health problems abound due to drinking and other domestic usage of water from these sources. This study is geared towards investigating the heavy metal contents in water from these sources in accordance with the requirements of the Standards Organisation of Nigeria (SON) [19] to ascertain the suitability of water from boreholes and hand-dug well in Keffi, Karu and their environs for domestic purpose.

2. Materials and methods

2.1 Study area

Keffi and Karu Local Government Areas (LGAs) are part of the 13 local government areas in Nasarawa State, Nigeria. The location of Nasarawa State on the World Map is latitude 8° 00′ to 8° 30′N and longitude 8° 30′ to 9° 00′E. The temperature of the study area could be described to be generally very warm with high humidity (**Figure 1**). There are two seasons in the area: dry season (October to March) and the rainy season (April to September). A high percentage of the population from the communities in these areas depends largely on groundwater sources (boreholes and hand-dug wells) for their daily water need. Then, others that reside in the headquarters of the local government areas rely on boreholes and deep hand-dug wells sank by individuals and the government for their daily water supply. Pipe-borne water supply is available in some places with epileptic operations. The communities used for this study are AgwanLambu/High Court, DadinKowa/Angwan Kwara, Yelwa and City centre all in Keffi and Uke, Auta-Balefi, Masaka, Ado/New Nyanya/Karu and Mararaba all in Karu Local Government Area.

2.2 Sample collection, treatment and preservation

Samples were taken from boreholes and hand-dug wells in both Keffi and Karu Local Government Areas. Samples of water were collected two times in a day for a period of two weeks. Variation in the day and time samples were collected was

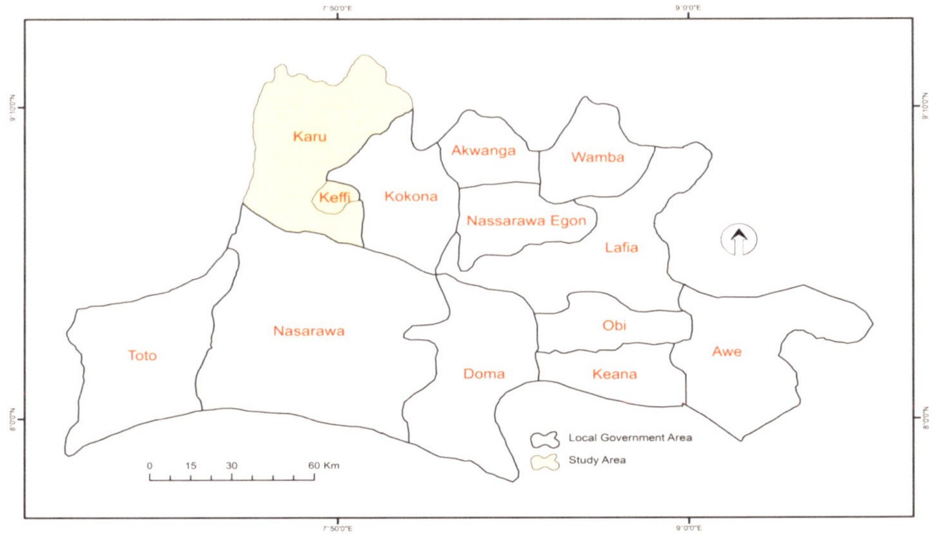

Figure 1.
Map of Nasarawa State showing Keffi, Karu (study Areas) and other LGAs.

observed in order to take care of changes that might occur at irregular intervals at the point of collection. Samples of water from borehole were collected from the taps bringing water from the borehole, and those of the hand-dug wells were collected with the use of a fetcher. All the samples collected were stored in clean 1500-cm^3 white plastic containers that had been properly washed by soaking overnight in 0.1-mol/dm^3 nitric acid solution and washed with soap solution. They were then rinsed with deionised water and concentrated nitric acid before being filled with deionised water to the locations designated for sampling. The containers were then emptied and rinsed with the samples to be collected severally before the sample collection proper. The sample containers were covered (air tight), properly labelled and immediately transported to the laboratory awaiting digestion and metal analysis [20].

2.3 Digestion of water samples for metal analysis

Water sample, 250 cm^3, was measured into an evaporating dish and concentrated HNO_3 measuring 5.0 cm^3 was added. The mixture was digested for about 1 hour on a heating mantle in a fume cupboard at temperatures of 90–95°C, and the quantity was reduced to 25 cm^3 with a colour that is characteristics of a complete digestion. The clear digest was brought down and allowed to cool. It was filtered using Whatman filter paper no. 1 into a 50-cm^3 volumetric flask that had been washed with an acid and properly rinsed with deionised water. The filtrate was made up to the mark with deionised water and kept awaiting metal analysis with atomic absorption spectrophotometer (AAS) [21].

2.4 Metal analysis

Heavy metals such as Cd, Cr, Cu, Fe, Mn, Ni, Pb and Zn in the digested water samples were determined using atomic absorption spectrophotometer (AAS) (ICE 3000AA0213410.VI.30 System).

2.5 Statistical analysis

Data obtained from the study were subjected to statistical tools such as mean, standard deviation, and correlation using SPSS.

2.6 Heavy metal pollution index (HMPI)

Countries, institutions and organisations have always provided recommended standard limits for different heavy metals in water for the purpose of being able to ascertain the quality of water with respect to individual metal contents. However, this does not give any information on the pollution level of metals in water with regard to all the metals that are detectable [18]. Heavy metal pollution index (HMPI) is a technique that determines the quality of water by providing information on the influence of each detectable metal on the overall quality of water. The values for ranking are 0–1, and the importance of each quality considerations is inversely proportional to the standard permissible limits [18].

The stages involved in the calculation of unit weight of the *i*th parameter, the calculation of the quality rating for each parameter and the addition of these sub-indices in the overall index are as follows:

$$Wi = \frac{k}{Si} \tag{1}$$

where Wi is the unit weight for the ith parameter, Si is the recommended standard limit for the ith parameter and k is the proportionality constant. The quality rating for individual parameter Qi can be evaluated as follows:

$$Qi = \frac{100 \times Mi}{Si} \tag{2}$$

where Qi is the sub-index of the ith parameter, Mi is the concentration of the ith heavy metal, Si is the recommended standard limit for the ith parameter and 100 is the critical pollution index value

$$\text{HMPI} = \frac{\sum(Qi \times Wi)}{\sum Wi} \tag{3}$$

where HMPI is the heavy metal pollution index, Qi is the sub-index of the ith parameter and Wi is the unit weight for the ith parameter [18].

2.7 Metal index (MI)

Metal index when evaluated for drinking water gives information of the likely additive effects of all the detectable heavy metals in water on the health of humans and that greatly aid the determination of the overall quality of water. MI can be computed as follows:

$$\text{MI} = \frac{Mi}{Si} \tag{4}$$

where MI is the metal index, Mi is the observed metal level in water and Si is the highest recommended permissible limit. MI is a tool used to determine the quality and suitability of water that is meant for drinking. The classification of water quality using metal index are; <3.0 very pure, 0.3–1.0 pure, 1.0–2.0 slightly affected, 2.0–4.0 moderately affected, 4.0–6.0 strongly affected and >6.0 seriously affected [18].

3. Results and discussion

Table 1 shows the sample locations in the study area, **Tables 2** and **3** show the concentrations of heavy metals in samples of water from boreholes and hand-dug wells, respectively, in Keffi and **Tables 4** and **5** show the concentrations of heavy metals in samples of water from boreholes and hand-dug wells, respectively, in Karu.

The mean values of cadmium in water samples from boreholes and hand-dug wells in Keffi were 0.039 and 0.039 mg/L, respectively, while for Karu the mean cadmium values were 0.037 and 0.04 mg/L, respectively. All these values are higher than the range of 0.00–0.011 mg/L reported for cadmium in Azare groundwater system [22] but are lower than the mean value of 0.08 mg/L of cadmium in water reported for stream water in Bauchi metropolis [23]. But the values are higher than the value of 0.003 mg/L recommended as the acceptable permissible limits by the SON [19] for cadmium in water that can be used domestically and industrially. Cadmium in water may be due to the mineralisation of cassiterite and also as a result of the use of insecticides, fertilisers and pesticides for farming and could be leached into groundwater systems. It could also come from wastes and effluents generated domestically like used batteries and other waste materials and eventually find its way into the water systems. High intake of cadmium via water is associated with toxicity to the kidney [19].

The mean values of chromium in water samples from boreholes and hand-dug wells in Keffi were 0.05 and 0.055 mg/L, respectively, while for Karu the mean cadmium values were 0.051 and 0.058 mg/L, respectively. All these values are within the range of 0.00–0.29 mg/L for chromium in boreholes and hand-dug well in both dry and wet seasons reported for selected rivers in Nasarawa State [24]. These

	Location	Borehole water sample	Hand-dug well water sample
Keffi	Angwan Lambu/High Court	S_1	S_6
	Dadin Kowa/ Angwan Kwara	S_2	S_7
	Tudun Amama	S_3	S_8
	Yelwa	S_4	S_9
	City Centre	S_5	S_{10}
Karu	Uke	S_{11}	S_{16}
	AutaBalefi	S_{12}	S_{17}
	Masaka	S_{13}	S_{18}
	Ado/New Nyanya/Karu	S_{14}	S_{19}
	Mararaba	S_{15}	S_{20}

Table 1.
Sampling locations in Keffi and Karu, Nasarawa State, Nigeria indicating sample numbers.

Location							
Parameters	S_1	S_2	S_3	S_4	S_5	**Mean**	**SON Standard [19]**
Cd	0.061	0.032	0.037	0.035	0.032	0.039	0.003
Cr	0.050	0.045	0.052	0.048	0.056	0.050	0.050
Cu	ND	ND	ND	ND	ND	-	1.00
Fe	0.097	0.176	0.096	0.117	0.084	0.114	0.30
Mn	0.056	0.054	0.049	0.058	0.055	0.054	0.200
Ni	ND	ND	ND	ND	ND	-	0.020
Pb	0.015	0.009	0.012	0.005	0.003	0.009	0.01
Zn	0.090	0.184	0.089	0.078	0.084	0.105	3.00

Table 2.
Mean values of heavy metal contents in borehole water samples in Keffi LGA.

Location							
Parameters	S_6	S_7	S_8	S_9	S_{10}	**Mean**	**SON Standard [19]**
Cd	0.035	0.046	0.037	0.037	0.044	0.039	0.003
Cr	0.050	0.056	0.064	0.049	0.055	0.055	0.050
Cu	ND	ND	ND	ND	ND	-	1.00
Fe	0.067	0.075	0.099	0.086	0.247	0.115	0.30
Mn	0.056	0.062	0.057	0.049	0.065	0.058	0.200
Ni	ND	ND	ND	ND	ND	-	0.020
Pb	0.015	0.012	0.014	0.011	0.009	0.012	0.01
Zn	0.105	0.067	0.099	0.086	0.247	0.121	3.00

Table 3.
Mean values of heavy metal contents in hand-dug well water samples in Keffi LGA.

Location							
Parameters	S_{11}	S_{12}	S_{13}	S_{14}	S_{15}	**Mean**	**SON Standard [19]**
Cd	0.056	0.033	0.041	0.035	0.022	0.037	0.030
Cr	0.05	0.045	0.052	0.048	0.056	0.051	0.050
Cu	0.055	0.059	0.054	0.061	0.065	0.059	1.00
Fe	0.097	0.186	0.096	0.119	0.084	0.116	0.30
Mn	0.056	0.054	0.049	0.058	0.058	0.055	0.200
Ni	0.018	0.019	0.017	0.016	0.019	0.018	0.020
Pb	0.018	0.009	0.016	0.005	0.008	0.011	0.01
Zn	0.090	0.184	0.089	0.078	0.094	0.107	3.00

Table 4.
Mean value of heavy metal contents in borehole water samples in Karu LGA.

Location							
Parameters	S_{16}	S_{17}	S_{18}	S_{19}	S_{20}	Mean	SON Standard [19]
Cd	0.039	0.046	0.036	0.039	0.041	0.040	0.03
Cr	0.057	0.065	0.064	0.049	0.055	0.058	0.050
Cu	0.055	0.065	0.054	0.06	0.075	0.062	1.00
Fe	0.067	0.075	0.012	0.086	0.257	0.099	0.30
Mn	0.046	0.062	0.075	0.051	0.068	0.060	0.200
Ni	0.019	0.021	0.017	0.018	0.019	0.019	0.020
Pb	0.015	0.012	0.014	0.017	0.019	0.015	0.01
Zn	0.109	0.065	0.0959	0.106	0.247	0.125	3.00

Table 5.
Mean value of heavy metal contents in hand-dug well water samples in Karu LGA.

values are lower than majority of the mean values for water samples from three different streams that had not detected, 0.93, 1.58 and 0.65, 0.49, 1.12 mg/L for dry and wet seasons, respectively, reported for groundwater sources in Okene Local Government Area, Kogi State [25]. Apart from the mean value of chromium in water from boreholes in Keffi, all the other values are slightly higher than the value of 0.05 mg/L recommended as the acceptable permissible limits by the SON [19] for chromium in water that can be used domestically and industrially. Chromium comes from wastes and effluents generated domestically. The deficiency of chromium causes impaired insulin function, hence increased insulin secretion and the risk of diabetes mellitus. High intake of chromium that is more than the recommended standards causes cancer [19].

Copper was not detected in water samples from both boreholes and hand-dug wells in Keffi but was, however, detected in boreholes and hand-dug wells in Karu and had mean concentrations of 0.59 and 0.62 mg/L, respectively. These values are higher than the mean value of 0.51 mg/L for copper in water reported for water sources in Bauchi metropolis [23] as well as the mean value of 0.1669 ± 0.1414 for copper in the groundwater systems of Azare [22]. The mean values from this study are lower than the value of 1.00 mg/L recommended as the permissible tolerable limits, and a concentration higher than this limit is always associated with gastrointestinal disorder [19].

The mean values of iron in water samples from boreholes and hand-dug wells in Keffi were 0.114 and 0.115 mg/L, respectively, while for Karu the mean iron values were 0.116 and 0.099 mg/L, respectively. The mean values from this study are lower than the value of 0.30 mg/L recommended as the permissible tolerable limits [19]. Iron is one of the components of haemoglobin responsible for the transport of oxygen in the body. It also aids in the oxidation of carbohydrates, proteins and fats as well as helping to prevent anaemia [26]. The concentration of iron, which is more than the acceptable limits, supports iron-dependent bacteria to cause deterioration in water quality [27]. Iron sources are domestically generated wastes, runoffs and probably the geological formations of the location under investigation. The level of iron can be reduced or removed completely from water through aeration [28].

The mean values of manganese in water samples from boreholes and hand-dug wells in Keffi were 0.054 and 0.058 mg/L, respectively, while for Karu the mean iron values were 0.055 and 0.060 mg/L, respectively. These values are lower than the mean values recorded for boreholes and hand-dug wells in dry and wet seasons, which were 0.42 mg/L for boreholes and 0.34 mg/L for hand-dug wells in dry season and 0.36 mg/L fore boreholes and 0.44 mg/L for hand-dug wells in wet season [25]. The values are also lower than 0.34 mg/L for manganese in water reported for water sources in Bauchi metropolis [23]. The mean values from this study are lower than the value of 0.20 mg/L recommended as the permissible tolerable limits [19]. Manganese is found in the environment due to the activities and from domestic wastes. High level of manganese in water above the recommended tolerable limits causes neurological disorder [19].

Nickel was not detected in water samples from both boreholes and hand-dug wells in Keffi but was, however, detected in boreholes and hand-dug wells in Karu and had mean concentrations of 0.018 and 0.019 mg/L, respectively. These mean values are slightly lower than mean value of 0.02 mg/L for nickel in water reported for water sources in the metropolis of Bauchi, Nigeria [23]. The mean values from this study are lower than the value of 0.020 mg/L recommended as the permissible tolerable limits, and a level higher than this is associated with possible carcinogenic effects [19]. Nickel comes from activities that originate from mechanic workshops, dumpsites and fertiliser-rich sewage sludge [23].

The mean values of lead in water samples from boreholes and hand-dug wells in Keffi were 0.009 and 0.012 mg/L, respectively, while for Karu the mean iron values were 0.011 and 0.015 mg/L, respectively. All these values are lower than the mean concentration of lead, 0.1100 $\pm$ 0.1097 mg/L, in groundwater systems of Azare [22] as well as the mean concentration of 0.048 mg/L for lead in water sources [23]. The concentrations of lead from this study are all higher than the recommended tolerable limits of 0.01 mg/L except for lead level in boreholes in Keffi, and anything above this limit can cause cancer, interfere with vitamin D metabolism, affect mental development in infants and is toxic to the central and peripheral nervous system. Sources of lead in the environment are mechanic, battery charger workshops as well as car wash spots and wastes generated domestically.

The mean concentrations of zinc in water samples from boreholes and hand-dug wells in Keffi were 0.105 and 0.121 mg/L, respectively, while for Karu the mean iron values were 0.107 and 0.125 mg/L, respectively. These mean concentrations are within the same range with the mean concentrations for zinc in boreholes from three different locations, which are 0.10, 0.11 and 0.12 mg/L reported for groundwater sources in Benin City, Edo State and Agbor, Delta State, all in Nigeria [20]. The mean values from this current work are lower than the recommended permissible limits of 3.00 mg/L [19].

Tables 6 and 7 shows the results of statistical analysis for test of significance between heavy metal contents of borehole water samples and hand-dug well water samples from Keffi Local Government Area and Karu Local Government Area, respectively. From the results shown, there were no significant differences between all the heavy metal contents in borehole and hand-dug well in both Keffi and Karu. The p-values for all heavy metal contents did not conform to $p < 0.05$, with the implication that for all the parameters, there were no significant differences between the borehole water samples and the hand-dug well water samples from Keffi Local Government Area and Karu Local Government Area. This could be attributed to the fact that in each location or study area, the geologic formation of the soil is the same for places where boreholes or hand-dug well had been drilled.

Table 8 shows the mean HMPI and MI of water from Keffi and Karu Local Government Areas. HMPI was used in the characterisation of water from boreholes

Metals	Borehole water (Mean of mean ± SD) mg/L	Well water (Mean of mean ± SD) mg/L	t-value	p-value
Cd	0.039 ± 0.122	0.039 ± 0.005	–0.068	0.948
Cr	0.050 ± 0.004	0.055 ± 0.006	–1.414	0.195
Cu	0.000 ± 0.000	0.000 ± 0.000	–	–
Fe	0.114 ± 0.037	0.115 ± 0.075	–0.021	0.983
Mn	0.054 ± 0.003	0.058 ± 0.006	–1.086	0.309
Ni	0.000 ± 0.000	0.000 ± 0.000	–	–
Pb	0.009 ± 0.005	0.012 ± 0.002	–1.390	0.202
Zn	0.105 ± 0.044	0.121 ± 0.072	–0.417	0.687

Table 6.
Heavy metals in borehole and well water samples in Keffi LGA ($p < 0.05$).

and hand-dug wells from Keffi and Karu Local Government Areas. The characterisation gave values that were compared with the critical values to assess the extent of heavy metal pollution [18]. The mean HMPI values calculated were 77.33 and 92.08 for borehole water and hand-dug well water, respectively, in Keffi and 105.27 and 127.41 for borehole water and hand-dug well water, respectively, in Karu. The values from Keffi are lower than the critical value of 100, while those of Karu are slightly higher. Higher values are indication of pollution of water from the Karu. The discrepancy could be attributed to the fact that some parameters, such as Cu and Ni, were not detected in water samples from Keffi but were detected in the ones from Karu. The HMPI for both water sources is high, and it signifies that the source of the contaminants could be infiltration of runoffs from dumpsites of domestic wastes.

MI values for borehole and hand-dug well waters were 0.564 and 0.606, respectively, in Keffi, and for Karu the values were 0.634 and 0.723, respectively. These values, however, suggest that the water sources from the study area are pure and not affected with heavy metal pollution when compared to the water quality classification scheme adopted [18].

Metals	Borehole Water (Mean of mean ± SD) mg/L	Well Water (Mean of mean ± SD) mg/L	t-value	p-value
Cd	0.037±0.012	0.040 ± 0.004	–0.482	0.643
Cr	0.051 ±0.004	0.058 ± 0.007	–2.229	0.056
Cu	0.059 ± 0.005	0.062 ± 0.009	–0.692	0.508
Fe	0.116 ± 0.378	0.099 ± 0.093	1.093	0.306
Mn	0.055 ± 0.004	0.060 ± 0.012	–0.966	0.362
Ni	0.018 ± 0.001	0.019 ± 0.002	–1.132	0.290
Pb	0.011± 0.006	0.015 ± 0.003	–1.524	0.166
Zn	0.107 ± 0.043	0.125 ± 0.071	–0.475	0.648

Table 7.
Heavy metals in borehole and well water samples in Karu LGA ($p < 0.05$).

Parameters	Keffi		Karu	
	BHW	HDWW	BHW	HDWW
HMPI	77.33	92.08	105.27	127.41
MI	0.564	0.606	0.634	0.723

BHW: Borehole water; HDWW: Hand-dug well water; HMPI: Heavy metal pollution index; MI: Metal index.

Table 8.
HMPI and MI of water samples from Keffi and Karu LGAs.

4. Conclusion

Water is very important in the lives of humans and other living organisms, and accessibility and availability of this water that is of good quality is always a serious problem. This informed the research to ascertain the suitability of water from boreholes and hand-dug wells in the study area for domestic and industrial purposes.

Results of metal analysis for all the water samples from the study areas showed that cadmium, chromium and lead had mean concentrations higher than the recommended tolerable limits prescribed by regulatory bodies, while all the other metals studied had mean values lower than the recommended concentrations. The statistical analyses done at 95% confidence limit ($p < 0.05$) showed that there were no significant differences between heavy metal contents in water samples from boreholes and wells from Keffi and Karu. HMPI evaluated for the water samples revealed that Keffi water sources were not polluted, but the Karu ones were slightly polluted. However, the MI evaluated for the samples revealed that groundwater sources in Keffi and Karu are pure and not affected by heavy metals.

Author details

Opaluwa Obaje Daniel
Faculty of Natural and Applied Sciences, Department of Chemistry, Nasarawa State University, Keffi, Nigeria

*Address all correspondence to: obajeo@nsuk.edu.ng

References

[1] Mahipal SS, Rajeev K. Contamination of heavy metals in groundwater and its toxic effects on human health and environment. International Journal of Environmental Sciences and Natural Resources. 2019;**18**(5):555996

[2] Thiyagarajan V: The importance water in our daily life. 2019. Available from: https://www.indiahomehealthcare.com

[3] Opaluwa OD, Muhammad AI, Madaki KS, Usman A, Mohammad IA. Water quality assessment of River Adema, Nasarawa, Nasarawa State, Nigeria. Journal of Applied Sciences and Environmental Management. 2022; **26**(8):1433-1442

[4] Ahmed JM. Evaluation of heavy metals with some physicochemical and biological parameters for groundwater in South of Najaf City, Iraq. World Journal of Pharmaceutical. 2016;**5**(3):129-138

[5] Shalom NC, Obinna CN, Adetayo YO, Vivienne NE. Assessment of water quality in Canaanland, Ota, Southwest Nigeria. Agriculture and Biology Journal of North America. 2011;**2**(4):577-583

[6] Dada TE, Awokunmi EE, Falegan CR, Olaninipeku EO. Status of ground water quality in selected areas of Ado-Ekiti, Nigeria. WIT Transactions on Ecology and the Environment. 2011;**145**:339-348

[7] Vodela JK, Lenz SD, Renden JA, McElhenney WH, Kemppainen BW. Drinking water contaminants (arsenic, cadmium, lead, benzene and trichloroethylene). 2. Effects on reproductive performance, egg quality, and embryo toxicity in broiler breeders. Poultry Science. 1997;**76**(11):1493-1500

[8] Marcovecchio JE, Botte SE, Frejie RH. Heavy metals, major metals, trace elements. In: Nollet LM, editor. Handbook of Water Analysis. 2nd ed. London: CRC Press; 2007. pp. 275-311

[9] Dusa AA, Timothy N, Magili ST, Tukur S. Determination of heavy metals in boreholes, hand dug wells and surface water in some selected areas of Mubi North Local Government Area, Adamawa State, Nigeria. International journal of Chemistry and Chemical sciences. 2017;**4**(1):75-81

[10] Asaaolu SS. Determination of Some Heavy Metals in Orechromis nicoticus, Claria genicpirues and Synontic Spp from the Coastal Water of Ondo State, Nigeria. Pakistan Journal of Science and Industrial Research. 2002;**45**:17-19

[11] Adeyeye ET, Abulude ED. Analytical assessments of some surface and groundwater resources in Ile-Ife, Nigeria. Journal of Chemical Society of Nigeria. 2004;**2004**:98-103

[12] Oluyemi EA, Feuyit G, Oyekunle JAO, Ogunfowokan AO. Seasonal variations in heavy metal concentrations in soil and some selected crops at a landfill in Nigeria. African Journal of Environmental Science and Technology. 2008;**2**:89-96

[13] Alloway BJ, Ayres DC. Chemical Principles of Environmental Pollution. 2nd ed. London: Black Academic and Professional Chapman and Hall; 1997. pp. 190-242

[14] Pate KP, Panda RM, George L. Heavy metal content of different effluents water around major industrial cities of Gujurat. Journal of India Society of soil Science. 2001;**59**(1):89-94

[15] Adeseye MA. Heavy metal composition of some solid minerals in

Nigeria and their health implications to the environment. Pakistan Journal of Biological Sciences. 2007;**10**(1): 4438-4443

[16] Oyatayo KT, Songu GA, Amos GA, Ndabula C. Assessment of heavy metal concentration in hand dug well water from selected land uses in Wukari Town, Wukari, Taraba State, Nigeria. Journal of Geoscience and Environmental Protection. 2015;**3**:1-10

[17] Lenntech WT. Chemical Properties, Health and Environmental Effects of Copper. The Netherlands: Lenntech Water Treatment and Purification Holding BV; 2009. Available from: https://www.lenntech.com/periodic/elements/cu.htm

[18] Yusuf A, Olasehinde A, Mboringong MN, Tabale RP, Daniel EP. Evaluation of heavy metals concentration in groundwater around Kashere and its environs, upper Benue Trough, North eastern Nigeria. Global Journal Geological Science. 2018;**16**: 25-36

[19] SON: Standards Organisation of Nigeria (Nigerian Industrial Standards for Drinking Water Quality- NIS): 554. Lekki, Lagos. 2007

[20] Akpoveta OV, Okoh BE, Osakwe SA. Quality assessment of borehole water used in the vicinity of Benin, Edo State and Agbor, Delta State of Nigeria. Current Research in Chemistry. 2011;**3**:62-69

[21] Atolaiye BO, Aremu MO, Shagye D, Pennap GRI. Determination and concentartion of some mineral elements in soil in soil sediment, ambient water and the body parts of Clarias Gariepinus and Tilapia geneensis fishes in River Tammah, Nasarawa State, Nigeria. Current World Environment. 2006; **1**(20):95-100

[22] Akinola LK, Gimba CE, Salihu L, Egwaikhide PA, Lawal U. Total heavy metal concentrations in Azare groundwater system in Nigeria. International Journal of Chemical Sciences. 2012;**5**(2):277-283

[23] Tanko IY, Dagauda U, Moumouni A, Shettima B, Ibrahim U. Heavy metal distribution in soil, stream sediment and water in Bauchi metropolis and its environs. International Journal of Chemical Sciences. 2008;**1**(2):222-229

[24] Aremu MO, Gav BL, Opaluwa OD, Atolaiye BO, Madu PC, Sangari DU. Assessment of physicochemical contaminants in waters and fishes from selected Rivers in Nasarawa State. Nigeria. Research Journal of Chemical Sciences. 2011a;**1**(4):6-17

[25] Aremu MO, Majabi GO, Nghargbu KA, Abiola KA, Ogah AT, Magaji JI. Baseline concentration of metals in water samples from streams, wells and boreholes within Okene Local Government Area, Kogi State, Nigeria. International Journal Chemical Science. 2011b;**4**(2):234-243

[26] Bender A. Meat and meat products in human nutrition in developing countries. In: FAO, Foods and Nutrition Paper 53. Rome, Italy: FAO; 1992. pp. 46-47

[27] Aremu MO, Sangari DU, Musa BZ, Chaanda MS. Assessment of ground water and stream water quality for trace metals and physicochemical contaminations in toto local government area of Nasarawa State, Nigeria. International Journal of Chemical Sciences. 2008;**1**(1):8-19

[28] Okedi C, Oni OO. Basic Water Treatment Operations. Kaduna: Published by National Water Resources Institute; 1997. pp. 1-90

Chapter 29

Effect of Heavy Metal Pollution on Invertebrates

Samir Ghannem, Sonia Dhaouadi and Samir Touaylia

Abstract

Metal pollutants are widespread in air, soil and water causing a decline in invertebrates worldwide. The increase of environmental pollution by heavy metals has a negative impact to organisms and influence their diversity, distribution, physiology and behavior. Contrary to other pollutants, metals are non-degradable and can potentially bio-accumulate and be biomagnified in the trophic chain. Because soil invertebrates tend to be strongly affected by environmental disturbances, high concentrations of these metals can become hazardous to invertebrates. Noxious effects can affect all biological levels, and toxins affect all ecological interactions. In this brief chapter, we have tried to develop a comprehensive understanding of the influence of metal contamination on ecosystem disturbance. We give examples of studies on the effects of pollutants on invertebrates.

Keywords: impact, heavy metals, invertebrate decline, environmental pollution, bioindicators

1. Introduction

Pollution is a mixture of contaminants in the environment that has serious consequences on the environment. It can originate from chemicals in any environmental medium, with deleterious consequences on living species. The damage expressed to living organisms can lead to damage to their health or interference with the ecological systems of which they are a part [1]. However, the ecological risk of heavy metals has become a major concern in developing countries. The multiple industrial, domestic, agricultural, medical and technological applications of heavy metals have led to their wide distribution in the environment, raising increasing concerns about their potential effects on human health and the balance of ecosystems. Heavy metals are also part of the contaminants that can be found in residential areas. Heavy metals present in the atmosphere, soil and water today come from various sources such as landfills, domestic and industrial waste, mineral and oil extraction sites as well as atmospheric pollution. These products are accumulated by the fauna and flora and magnified along the food chain, which multiplies the impact of this pollution. Each year the levels of pollutants, in particular metals, increase and cause environmental threats that lead to the imbalance of the natural system [2]. Toxic chemicals in the environment can enter ecosystems and end up throughout the biosphere. Ecosystems can be affected by chemical contamination, disturbing the activities of living organisms or changing the physical properties of ecosystems [3].

Disturbed environments and excessively loaded with metals, can lead to a decrease in the biological activity of invertebrates. Currently the regression of the diversity of invertebrates in the world is very remarkable. According to [4], the effects of toxic metal pollution on terrestrial invertebrate species appear to be extremely significant and widespread. Such a revision of regulatory thresholds to better protect terrestrial invertebrates, which seem more sensitive to metal pollution than vertebrates, is necessary and needs urgent attention from scientists and stakeholders [5]. The pollution of ecosystems exposes an astonishing danger to invertebrate's species through direct and/or indirect contact with contaminated soil [6, 7]. The contamination of ecosystems is the complication that strongly affects biodiversity, the environment and human health all over the world via soil and water pollution [8]. Bioindicators have been considered an essential tool for monitoring and discovering changes in the environment [9]. The choice of the most adequate indicator is linked to the objective of the survey and the characteristics of the indicator. In the invertebrate group, several species or groups of species have been used as indicators. Bioindicators species are sensitive to environmental changes, such as high levels of metals [10, 11], this sensitivity is expressed by the absence or presence of species, or by altered physiological and/or morphological changes [12]. The relationships between toxic chemical elements, the biotope and living organisms can cause the degradation of these compounds during their modification, leading to transformations of the environment and leading to harmful damage to living organisms [3]. The response of invertebrates to metal pollution is closely related to species as well as pH conditions, exposure time, and metal form and concentration, which are deeply tied to metal bioavailability [13, 14]. It should be noted that the use of certain species of invertebrates as better indicators than others is linked to their different ecological requirements. As part of biomonitoring programs, measurements of metal residues have been established on invertebrate species [15]. The utility of studies performed on invertebrates can be essential for such an assessment of the impact of toxic metal pollution on environmental behavior and the strategies adopted to anticipate such an effect [10]. They show an important role in terrestrial environments linked to their vast abundance, biomass, and diversity [16]. Invertebrates living in polluted environments have been regularly revealed to accumulate heavy metals [17–19]. Using sensitive species invertebrates lets for an improved valuation of the threats environmental and for more informative and efficient test systems [10].

2. Materials and methods

This study focuses to evaluate the potential heavy metals impacts on invertebrates. The results cited come from the literature of 26 years from 1997 until 2022.

3. Bioindicators concept

In recent decades, the use of alternative biological activities for monitoring the state of the environment has been increased due to the high cost associated with traditional instrumental measurement methods. The use of bio-indicators is considerably more economical since it generally makes it possible to avoid using expensive technological equipment and to save time [20, 21]. Bio-indication therefore refers to a process of analysis of various biological indicators that is part of the biomonitoring of the quality of ecosystems [22]. The indicators that can provide ideas on the qualitative state of the environment are defined as bioindicators. These bioindicators are frequently used in scientific investigations to assess the progressive impacts of pollutants due to anthropogenic activities. Bioindicators are sensitive to changes and disturbances in biotopes. However, the quality of an environment can be assessed by living species, which have an essential role in monitoring its disturbances [23]. Moreover, certain bio-indicators for their capacity of bioaccumulation, allow an early detection of pollutants or disturbances [24]. Furthermore, bio-indicators provide information "on the bioavailability of pollutants rather than on their total concentration in the environment" [20]. This distinction is significant when looking at the effects of pollutants on organisms [25]. Finally, unlike instrumental measurements which take parameter values instantaneously and locally, bio-indicators provide an integrated indication of the spatiotemporal effects of pollutants on biota since they "reflect the total time of exposure to the pollutant" [20]. According to [26], a bioindicator (or bio-indicator or biological indicator) is an organism or set of organisms which, by reference to biochemical, cytological, physiological, ethological or ecological variables, makes it possible, in a practical and safe way, to characterize the state of an ecosystem or an ecocomplex and to bring to light as early as possible their modifications, natural or provoked". The choice of indicator species is based on their easy monitoring, and that they can predict the conditions of the environment to which they belong [27]. In addition, indicator species must be easily collected, very abundant, widely distributed, available all year round, easily recognized by non-specialists, possess a high sensitivity to environmental constraints and should be profitable [3].

4. Effect of heavy metal pollution on invertebrates

Metallic elements are considered systemic toxins known to induce neural and multi-organ damage, and contribute to cancer in animals and humans, even at low levels of exposure. Their toxicity depends on several factors, including dose, route of exposure and chemical species, as well as the age, sex, genetics and nutritional status of those exposed. However, in some invertebrates, physiological and ecological changes following pollution by heavy metal can be detected by neurological dysfunctions and sensory, endocrine or metabolic disturbances [28]. Following exposure to a series of soils contaminated with different concentrations of cadmium and zinc on the talitrid amphipod species *Orchestia gammarellus*, [29] detect significant changes in the morphology and in the ultrastructural organization of hepatopancreatic cells as well as the rhythm of locomotor activity. However, the results show that the damage observed at depends on the concentration at which the individuals were expressed. Concerning the behavioral response, a great intervariability of the locomotor rhythm was observed. Patterns were in majority bimodal for the uncontaminated individuals and became unimodal and multimodal when exposed under Cd and the Zn respectively. In addition, the circadian period lengthened after Zn exposure. Referring to [30] pollution intensities could have negative effects on the restriction of the distribution area or even the disappearance of certain species. In a polluted environment morphological, behavioral, tissue or physiological alterations, as well as the abundance, reproductive success and mortality of animal species are observable and measurable variables reflecting the state of their habitat [12, 24, 25]. In line with that, toxic metals interfere with cytoplasmic membranes and lead to pathological results [31, 32]. These pathological ramifications contain the nuclear and cytoplasmic corruptions. The follicular epithelial cells showed signs of damage which is a remarkable appearance seen in the ovarian cells from the polluted site. This can delay oocyte maturation and result in imperfect yolk deposition [33, 34]. The distorted brush borders of the microvilli may block the passage of the materials toward the oocyte. Also, the deformity of yolk granules could obstruct vitellogenesis and result in a lower fecundity and egg viability [34].

Referring to [12], following pollution by Cd, Ni and Zn, the response of invertebrate species of *Pterostichus oblongopunctatus* (ground beetles) results in severe intestinal degeneration. In the case of Zn pollution, a decrease in the number of eggs laid by *Poecilus cupreus* has been observed [35]. Furthermore, the results obtained by [36] indicate an increase in egg production and a reduction in egg quality (hatching) in zones polluted by metals compared to unpolluted areas. In [37], it was shown that in polluted areas the body weight of *P. oblongopunctatus* is higher compared to unpolluted areas. It was explained by advanced metal tolerance [38]. Several publications have appeared in recent years documenting that exposure of invertebrates to toxic elements causes sublethal consequences, sometimes difficult to evaluate, such as impaired fertility observed in (grasshoppers: [39]; springtail: [40]; earthworm: [41], resistance to pathogens; ant: [42]; honey bee: [43]), developmental abnormalities fly [44]; moth: [45]; ant: [46]) and similarly altered feeding behavior (aphid: [47]; honey bee: [48]). In addition, metal pollution can cause cell damage or death in the brains of invertebrates, as their nervous systems are small in size and contain relatively few neurons [49, 50]. According to [51], toxic elements, particularly zinc, can disturb molting cycles, alter the digestive glands in isopods and can affect food consumption, reproduction and the composition of the community. Previous studies indicate that metal pollution has severe consequences on pollinator behavior by affecting foraging activity, food perception and memory abilities necessary for energetic foraging [5, 48, 52, 53]. The results obtained by [54] on tenebrionid beetles (*Blaps polycresta* and *Trachyderma hispida*) inhabiting polluted soil shows several abnormalities in oocytes and trophocytes. These baseline abnormalities included exfoliation and vacuolation of follicular epithelium, vacuolated trophocyst, nuclear abnormalities, and morphological changes in cytoplasmic organelles. Co-exposure of honey bees (*A. mellifera*) to cadmium and copper resulted in increased developmental time, high mortality, and decreased food intake and response to sucrose [55]. Referring to [56], the study of the impact of Cadmium on species of Tenebrionidae, reveals a decrease in population density, a reduction in body weight, an increase in the mortality rate and an increase in the sex ratio insects. Moreover, the results also revealed a striking decrease in the body length of polluted insects with a marked increase in the percentage of deformed gonads and digestive tract.

5. Results and discussion

One of the main characteristics of the different bioindicator species is, among other things, the ability to respond in an observable and measurable way to disturbances or to accumulate the pollutant. The distribution of invertebrates in all types of terrestrial and aquatic makes them excellent for ecotoxicological analysis and are efficient biomarkers in detecting environmental pollution. However, the internal study of the compartment of metals in different subcellular fractions, allows us to better interpret the mechanisms of accumulation of toxic elements in organisms and their transfer through trophic chains [12]. In general, toxic metals, even at low concentrations, have a polluting nature with harmful effects for living organisms. Heavy metals have a cumulative effect. Their selective attachment to sensitive organs and tissues can be dangerous when their concentrations are high [57]. It is also likely that heavy metal ions act on Ca^{2+} receptors [58] and metal ions enter the interior of the cell, from where they initiate free radical oxidation [59]. An oxidative stress reaction then develops which in turn influences the lysosomes. Furthermore, variation in the concentration of heavy metals affects the susceptibility of insects to entomopathogenic fungi [60, 61]. According to [62], the humoral immunity of invertebrates could be affected by the stress of toxic elements, and the response of the humoral immunity of insects was found to affect the susceptibility of insects to entomopathogenic microorganisms [62, 63]. In addition, insects reduce their feeding or decrease their digestion and absorption as a result of heavy metals accumulated in food [64]. Referring to [65], a reduction in intermediate carbohydrate metabolism, was observed in chironomid (*Chironomus tepperi*), including glucose 6-phosphate, fructose 6-phosphate and disaccharides following exposure to Zn stress. Thus, to better understand the effect of heavy metal stress on insect disease resistance and their corresponding regulatory mechanisms, studies on several insect species and different types of toxic elements should be done [66].

6. Conclusions

The use of invertebrates as bioindicators to assess environmental risks occupies a high priority for researchers and environmental observers due to their efficient and rapid responses to various external causes. However, toxic elements are capable of penetrating the body and accumulating in certain tissues, causing cellular alterations as well as damage to energy metabolism, growth and reproduction. Further study of the issue would be of interest to improve international guidelines for metal pollutants and to develop maintenance plans to protect invertebrates and ecosystem services.

Conflict of interest

The authors declare no conflict of interest.

Author details

Samir Ghannem*, Sonia Dhaouadi and Samir Touaylia
Faculty of Sciences of Bizerte, LR01ES14 Laboratory of Environment Bio-monitoring, University of Carthage, Bizerte, Tunisia

*Address all correspondence to: ghan_samir@yahoo.fr

References

[1] Appannagari RR. Environmental pollution causes and consequences: A study. North Asian International Research Journal of Social Science and Humanities. 2017;**3**(8):151-161

[2] Al-Khazraji HI, Thakir BM, El-Hadeeti SAK. Bioindicators of pesticides pollution in the aquatic environment: A review. Plant Archives. 2020;**20**:1607-1618

[3] Abdel-Aziz N, Shalaby S, El-Bakry A. Using of Bioindicators for Monitoring Pesticides Pollution. 2nd ed. LAP LAMBERT Academic Publishing; 2021. p. 72

[4] Thimmegowda GG, Mullen S, Sottilare K, Sharma A, Mohanta SS, Brockmann A, et al. A field-based quantitative analysis of sublethal effects of air pollution on pollinators. Proceedings of the National Academy of Sciences. 2020;**117**(34):20653-20661

[5] Monchanin C, Devaud JM, Barron AB, Lihoreau M. Current permissible levels of metal pollutants harm terrestrial invertebrates. Science of the Total Environment. 2021;**779**:146398

[6] Ciadamidaro L, Puschenreiter M, Santner J, Wenzel WW, Madejón P, Madejón E. Assessment of trace element phytoavailability in compost amended soils using different methodologies. Journal of Soils and Sediments. 2017;**17**(5):1251-1261

[7] Touceda-González M, Prieto-Fernández A, Renella G, Giognoni L, Sessitsch S, Brader G, et al. Microbial community structure and activity in trace element contaminated soils phytomanaged by Gentle Remediation Options (GRO). Environmental Pollution. 2017;**231**:237-251

[8] Jan S, Rashid B, Azooz MM, Hossain MA, Ahmad P. Genetic strategies for advancing phytoremediation potential in plants: A recent update. In: Plant Metal Interaction. 2nd ed. Elsevier; 2016. pp. 431-454

[9] Rainio J, Niemelä J. Ground beetles (Coleoptera: Carabidae) as bioindicators. Biodiversity and Conservation. 2003;**12**(3):487-506

[10] Ghannem S, Khazri A, Sellami B, Boumaiza M. Assessment of heavy metal contamination in soil and *Chlaenius* (*Chlaeniellus*) *olivieri* (Coleoptera, Carabidae) in the vicinity of a textile factory near Ras Jbel (Bizerte, Tunisia). Environmental Earth Sciences. 2016;**75**(5):1-10

[11] Parmar TK, Rawtani D, Agrawal YK. Bioindicators: The natural indicator of environmental pollution. Frontiers in Life Science. 2016;**9**(2):110-118

[12] Bednarska AJ, Świątek Z. Subcellular partitioning of cadmium and zinc in mealworm beetle (Tenebrio molitor) larvae exposed to metal-contaminated flour. Ecotoxicology and Environmental Safety. 2016;**133**:82-89

[13] Ashraf MA, Maah MJ, Yusoff I. Bioaccumulation of heavy metals in fish species collected from former tin mining catchment. International Journal of Environmental Research. 2012;**6**(1):209-218

[14] Rengel Z. Availability of Mn, Zn and Fe in the rhizosphere. Journal of Soil Science and Plant Nutrition. 2015;**15**(2):397-409

[15] Nico M, Van Straalen NM, Ruslan O, Butovsky RO, Andrei D,

Pokarzhevskii AD, et al. Metal concentrations in soil and invertebrates in the vicinity of a metallurgical factory near Tula (Russia). Pedobiologia. 2001;**45**:451-466

[16] Zödl B, Wittmann KJ. Effects of sampling, preparation and defecation on metal concentrations in selected invertebrates at urban sites. Chemosphere. 2003;**52**(7):1095-1103

[17] Braun M, Simon E, Fábián I, Tóthmérész B. Elemental analysis of pitfall-trapped insect samples: Effects of ethylene glycol grades. Entomologia Experimentalis et Applicata. 2012;**143**:89-94

[18] Simon E, Harangi S, Baranyai E, Braun M, Fábián I, Mizser S, et al. Distribution of toxic elements between biotic and abiotic components of terrestrial ecosystem along an urbanization gradient: Soil, leaf litter and ground beetles. Ecological Indicators. 2016;**60**:258-264

[19] Ghannem S, Touaylia S, Boumaiza M. Beetles (Insecta: Coleoptera) as bioindicators of the assessment of environmental pollution. Human and Ecological Risk Assessment: An International Journal. 2018;**24**(2):456-464

[20] Markert BA, Breure AM, Zechmeister HG. Bioindicators and Biomonitors. Principle, Concepts and Applications. 2nd ed. Vienne: Elsevier; 2003. p. 997

[21] Carignan V, Villard MA. Selecting indicator species to monitor ecological integrity: A review. Environmental Monitoring and Assessment. 2002;**78**(1):45-61

[22] Chabot B. Les facteurs de sélection des bio-indicateurs de la qualité des écosystèmes aquatiques: Elaboration d'un outil d'aide à la décision. Québec: Université de Sherbrooke; 2014

[23] Asif N, Malik M, Chaudhry FN. A review of on environmental pollution bioindicators. Pollution. 2018;**4**(1):111-118

[24] Kaiser J. Bioindicators and Biomarkers of Environmental Pollution and Risk Assessment. 2nd ed. Hartford: Science Publishers; 2001. p. 304

[25] Kerckhove O. Espèces ou association d'espèces de poissons en tant que bioindicateur de l'état de santé des récifs coralliens. Sherbrooke, Québec: Université de Sherbrooke; 2012

[26] Banaru D, Perez T. Bioindicateurs, Biomarqueurs, Notes de cours. Paris: Université de Marseille; 2010. p. 24

[27] Siddig AA, Ellison AM, Ochs A, Villar-Leeman C, Lau MK. How do ecologists select and use indicator species to monitor ecological change? Insights from 14 years of publication in ecological indicators. Ecological Indicators. 2016;**60**:223-230

[28] Amiard-Triquet C, Amiard JC. L'écotoxicologie du comportement. In: Amiard JC, Amiard-Triquet C, editors. Les biomarqueurs dans l'évaluation de l'état écologique des milieux aquatiques. 2008. pp. 211-240

[29] Jelassi R, Khemaissia H, Ayari A, Bohli-Abderrazek D, Ghemari C, Raimond M, et al. Physiological and behavioral responses of Orchestia gammarellus (Amphipoda, Talitridae) towards trace elements contamination soil. Microscopy Research and Technique. 2022;**85**(5):1713-1722

[30] Acevedo-Whitehouse K, Duffus AL. Effects of environmental change

on wildlife health. Philosophical Transactions of the Royal Society B: Biological Sciences. 2009;**364**(1534): 3429-3438

[31] Kheirallah DAM, El-Moaty ZA, El-Gendy DA. Impact of cement dust on the testis of *Tachyderma hispida* (Forskal, 1775) (Coleoptera: Tenebrionidae), inhabiting Mariout region (Alexandria, Egypt). Journal of Entomology. 2016;**13**(3):55-71

[32] Shonouda M, Osman W. Ultrastructural alterations in sperm formation of the beetle, *Blaps polycresta* (Coleoptera: Tenebrionidae) as a biomonitor of heavy metal soil pollution. Environmental Science and Pollution Research. 2018;**25**(8):7896-7906

[33] Sreelatha KB, Geetha PR. Disruption of oocyte development and vitellogenesis in Oryctes rhinoceros treated with methanolic extract of Eupatorium odoratum leaves. Journal of Biopesticides. 2010;**3**(Special Issue):253

[34] Osman W, Shonouda M. X-ray metal assessment and ovarian ultrastructure alterations of the beetle, Blaps polycresta (Coleoptera, Tenebrionidae), inhabiting polluted soil. Environmental Science and Pollution Research. 2017;**24**(17):14867-14876

[35] Kramarz P, Laskowski R. Effect of zinc contamination on life history parameters of a ground beetle, *Poecilus cupreus*. Bulletin of Environmental Contamination and Toxicology. 1997;**59**(4):525-530

[36] Lagisz M, Laskowski R. Evidence for between-generation effects in carabids exposed to heavy metals pollution. Ecotoxicology. 2008;**17**(1):59-66

[37] Zygmunt PM, Maryański M, Laskowski R. Body mass and caloric value of the ground beetle (Pterostichus oblongopunctatus) (Coleoptera, Carabidae) along a gradient of heavy metal pollution. Environmental Toxicology and Chemistry: An International Journal. 2006;**25**(10):2709-2714

[38] Tőzsér D, Magura T, Simon E, Mizser S, Papp D, Tóthmérész B. Pollution intensity-dependent metal accumulation in ground beetles: A meta-analysis. Environmental Science and Pollution Research. 2019;**26**(31):32092-32102

[39] Schmidt GH, Ibrahim NM, Abdallah MD. Toxicological studies on the long-term effects of heavy metals (Hg, Cd, Pb) in soil on the development of *Aiolopus thalassinus* (Fabr.) (Saltatoria: Acrididae). Science of the Total Environment. 1991;**107**:109-133

[40] Crouau Y, Pinelli E. Comparative ecotoxicity of three polluted industrial soils for the Collembola Folsomia candida. Ecotoxicology and Environmental Safety. 2008;**71**(3):643-649

[41] Konečný L, Ettler V, Kristiansen SM, Amorim MJB, Kříbek B, Mihaljevič M, et al. Response of Enchytraeus crypticus worms to high metal levels in tropical soils polluted by copper smelting. Journal of Geochemical Exploration. 2014;**144**:427-432

[42] Sorvari J, Rantala LM, Rantala MJ, Hakkarainen H, Eeva T. Heavy metal pollution disturbs immune response in wild ant populations. Environmental Pollution. 2007;**145**(1):324-328

[43] Polykretis P, Delfino G, Petrocelli I, Cervo R, Tanteri G, Montori G, et al. Evidence of immunocompetence reduction induced by cadmium exposure in honey bees (*Apis mellifera*). Environmental Pollution. 2016;**218**:826-834

[44] Nascarella MA, Stoffolano JG Jr, Stanek EJ III, Kostecki PT, Calabrese EJ. Hormesis and stage specific toxicity induced by cadmium in an insect model, the queen blowfly, Phormia regina Meig. Environmental Pollution. 2003;**124**(2):257-262

[45] Van Ooik T, Rantala MJ, Saloniemi I. Diet-mediated effects of heavy metal pollution on growth and immune response in the geometrid moth Epirrita autumnata. Environmental Pollution. 2007;**145**(1):348-354

[46] Skaldina O, Peräniemi S, Sorvari J. Ants and their nests as indicators for industrial heavy metal contamination. Environmental Pollution. 2018;**240**:574-581

[47] Stolpe C, Krämer U, Müller C. Heavy metal (hyper) accumulation in leaves of Arabidopsis halleri is accompanied by a reduced performance of herbivores and shifts in leaf glucosinolate and element concentrations. Environmental and Experimental Botany. 2017;**133**:78-86

[48] Burden CM, Morgan MO, Hladun KR, Amdam GV, Trumble JJ, Smith BH. Acute sublethal exposure to toxic heavy metals alters honey bee (Apis mellifera) feeding behavior. Scientific Reports. 2019;**9**(1):1-10

[49] Niven JE, Farris SM. Miniaturization of nervous systems and neurons. Current Biology. 2012;**22**(9):323-329

[50] Klein S, Cabirol A, Devaud JM, Barron AB, Lihoreau M. Why bees are so vulnerable to environmental stressors. Trends in Ecology & Evolution. 2017;**32**(4):268-278

[51] Drobne D. Terrestrial isopods—A good choice for toxicity testing of pollutants in the terrestrial environment. Environmental Toxicology and Chemistry: An International Journal. 1997;**16**(6):1159-1164

[52] Sivakoff FS, Gardiner MM. Soil lead contamination decreases bee visit duration at sunflowers. Urban Ecosystem. 2017;**20**(6):1221-1228

[53] Xun E, Zhang Y, Zhao J, Guo J. Heavy metals in nectar modify behaviors of pollinators and nectar robbers: Consequences for plant fitness. Environmental Pollution. 2018;**242**:1166-1175

[54] Kheirallah DA, El-Samad LM. Oogenesis anomalies induced by heavy metal contamination in two tenebrionid beetles (*Blaps polycresta* and *Trachyderma hispida*). Folia Biologica (Kraków). 2019;**67**(1):9-23

[55] Di N, Zhang K, Hladun KR, Rust M, Chen YF, Zhu ZY, et al. Joint effects of cadmium and copper on Apis mellifera forgers and larvae. Comparative Biochemistry and Physiology Part C: Toxicology & Pharmacology. 2020;**237**:108839

[56] OsmanW, El-SamadM, MokhamerEH, El-Touhamy A, Shonouda M. Ecological, morphological, and histological studies on *Blaps polycresta* (Coleoptera: Tenebrionidae) as biomonitors of cadmium soil pollution. Environmental Science and Pollution Research. 2015;**22**(18):14104-14115

[57] Lallogo H. Concentration de certains métaux d'importance médicale dans les poisons: cas des poissons de la lagune de Lomé. In: Mémoire de Technicien Supérieur en Génie Sanitaire (EAM). 1992

[58] Nemeth EF. Ca2+– Receptor–dependent regulation of cellular functions. News in Physiological Sciences. 1995;**10**(2):1-5

[59] Berridge MJ, Heslop JP, Irvine RF, Brown KD. Inositol trisphosphate formation and calcium mobilization in Swiss 3T3 cells in response to platelet-derived growth factor. Biochemical Journal. 1984;**222**(1):195-201

[60] Dubovskiy IM, Grizanova EV, Ershova NS, Rantala MJ, Glupov VV. The effects of dietary nickel on the detoxification enzymes, innate immunity and resistance to the fungus Beauveria bassiana in the larvae of the greater wax moth Galleria mellonella. Chemosphere. 2011;**85**(1):92-96

[61] Nanda KP, Kumari C, Dubey M, Firdaus H. Chronic lead (Pb) exposure results in diminished hemocyte count and increased susceptibility to bacterial infection in Drosophila melanogaster. Chemosphere. 2019;**236**:124349

[62] Pragya P, Shukla AK, Murthy RC, Abdin MZ, Kar Chowdhuri D. Characterization of the effect of Cr (VI) on humoral innate immunity using Drosophila melanogaster. Environmental Toxicology. 2015;**30**(11):1285-1296

[63] Afify A, Ghezawy NA, Ali MM, El-Tayeb TA. Effects of copper chlorophylline and magnesium chlorophylline on the humeral immune response, ultrastructural changes of midgut and DNA damage to Culex pipiens larvae. African Entomology. 2019;**27**(1):146-158

[64] Li K, Chen J, Jin P, Li J, Wang J, Shu Y. Effects of Cd accumulation on cutworm Spodoptera litura larvae via Cd treated Chinese flowering cabbage *Brassica campestris* and artificial diets. Chemosphere. 2018;**200**:151-163

[65] Long SM, Tull DL, Jeppe KJ, De Souza DP, Dayalan S, Pettigrove VJ, et al. A multi-platform metabolomics approach demonstrates changes in energy metabolism and the transsulfuration pathway in *Chironomus tepperi* following exposure to zinc. Aquatic Toxicology. 2015;**162**:54-65

[66] Jiang D, Tan M, Guo Q, Yan S. Transfer of heavy metal along food chain: A mini-review on insect susceptibility to entomopathogenic microorganisms under heavy metal stress. Pest Management Science. 2021;**77**(3): 1115-1120

Section 5

Detection of Heavy Metals

Chapter 30

Electrochemical Techniques for the Detection of Heavy Metals

Afrasiab Ur Rehman, Abdul Hakim Shah, Atta Ur Rahman, Fida Ur Rahman, Sher Ali, Atta Ur Rehman, Raza Ullah, Ikram Ullah, Muhammad Fayaz and Keying Shi

Abstract

The present chapter emphases on the approach of electrochemical sensor of metallic oxide nanocomposites to sense/detect heavy metal ions. Several methods have been incorporated with modified electrode for the sensing/detection of heavy metal ions. Among these methods square wave anodic stripping voltammetry method and differential normal pulse voltammetry method have been tested and being recommended for the individual analysis as well as simultaneous analysis of heavy metal ions by various researchers. We also endorse the said methods as the best choice for sensing of heavy metal ions however the material phase (plane) is also of specific importance in this regard. We suggest that these methods may be practiced by chemical industries the main sources of heavy metal ions waste. Furthermore, the statistical approach for the detection limit (3σ method) has been illustrated in the last paragraph of this unit.

Keywords: electrochemical sensing, rGO/MOx, heavy metal ions, material phase (plane), statistical approach for the detection limit

1. Introduction

1.1 Research background

As we know that the heavy metals in solution in the form of heavy metal ions (HMIs) contaminating water in a little quantity not only be risky to public health, but also may disturb aquatic life, i.e. the lowest level of heavy metal ions (HMIs) and its compounds are very toxic, dangerous for liver, brain, heart muscle, kidneys, human nervous system, blood circulation system and can damage skeletal system resulting in skeletal disease. In response to the needs of modern society and rapid industrial development, it is necessary to design high-efficiency, environmentally friendly, and low-cost electrochemical sensor. Nevertheless, major issues are associated with these sources, such as rapidly increasing prices, environmental consequences, and global climate change. These serious problems have necessitated the development of alternative energy sources.

In the past several decades, heavy metal ions (HMIs) have been of great attention, as they are enormously injurious in the biosphere and even their minute aggregate possesses an unfavorable threat to human health [1–3].

1.2 Methods used for the detection of heavy metal ions

The sensing and quantification of heavy metal ions (HMIs) are important in many applications, including waste management, environmental monitoring, developmental biology, and clinical toxicology. Several techniques/methods have been incorporated over the years for heavy metal ion (HMIs) such as atomic absorption spectroscopy (AAS), [4] inductively coupled plasma-mass spectrometry (ICP-MS), [5] inductively coupled plasma atomic emission spectrometry (ICP-AES) [6] X-ray fluorescence (XRF) spectrometry, [7] and so on. As these spectroscopic practices are time-consuming, their instruments are expensive and complicated in operation. Furthermore, the individual as well as simultaneous sensing/detection of heavy metal ions (HMIs) of great sensitivity and selectivity is the need of today. In this favor, electrochemical techniques/methods especially anodic stripping voltammetric (ASV) has been reflected to be a powerful, most sensitive, extremely rapid, and cost-effective method [8–12].

1.3 Electrochemical methods in sensing/detection of heavy metal ions

So many electrochemical techniques have been practiced for sensing chemical biomolecules and contaminants [13–15]. Normally, voltammetric methods for instance cyclic voltammetry (CV), linear sweep voltammetry (LSV), square wave voltammetry (SWV), differential pulse voltammetry (DPV) etc., potentiometric techniques and electrochemical impedance (EIS) techniques are employed in the sensing/detecting of analytes. In this regard, electrochemical behavior for the detecting of analytes are extremely applicable in micro fluidics and wider field of separation/partition science for the purposes of detection, valving and pumping. Few of the microscopic techniques are joined with electrochemical principles, such as scanning electrochemical microscopy (SECM) and chemically selective scanning tunneling microscopy (CSSTM), which are actually spatially resolved electrochemical sensors, even though they are classified as microscopic techniques [16]. An electrochemical sensor comprises of two constituents: (i) a recognition/perception element also called target receptor may be chemical or biological; and (ii) a material transducer/sensor usually a modified electrode that convert the sensing signals to an electronic signal. Collaborations between the detecting/sensing element of the substrate (HMIs) and the analytes are calculated through the quick reply, sensitivity, discrimination, and flexibility of the modified sensors (electrode) [17]. Durable contacts/interactions are usually linked with greater sensitivity and selectivity; however exemplary adjustability needs weak interactions. The serious factors like sensitivity, selectivity, response time, detection limit, signal-to-noise ratio, linearity, and stability are responsible for the performance of electrochemical sensors [18].

Among these methods square wave anodic stripping voltammetry method and differential normal pulse voltammetry method have been tested and are recommended for the individual analysis as well as simultaneous analysis of heavy metal ions (HMIs) by various researchers. We also endorse the said methods as the best choice for sensing of heavy metal ions however the material phase (plane) is also of specific importance in this regard.

The facet-dependent electrochemical behavior of Co_3O_4 nanoplates and nanocubes based on their adsorption/sensing behaviors toward heavy metal ions (HMIs) has been practiced. The Co_3O_4 nanoplates with plane (111) were better in electrochemical sensing than Co_3O_4 nanocubes with plane (001). Both adsorption quantities and density-functional theory (DFT) calculations were in accordance with the concept that the variance in electrochemical properties was due to the sensing of heavy metal ions (HMIs) [19]. It is prominent from the study that sensing interface modified electrodes play a key role in the sensing/detection of heavy metal ions (HMIs).

The graphene-analogue carbon nitride (GA-C_3N_4) has been recommended as suitable sensor for Cu^{2+} purpose. So far graphene base nanocomposites are considered promising candidate for heavy metal ions (HMIs) determination in water environment [20].

The electrochemical system for the sensing/detection of heavy metals in soil has also been reported. The electrochemical sensor with screen-sprinted electrode (SPE) adapted by ionic liquid (IL) n-octylpyridinum hexafluorophosphate (OPFP) and graphene (GR) was tested for sensing/detection of Cd (II) in soil [21]. The system was further tested for sensitive detection of trace cadmium ions by square wave anodic stripping voltammetry (SWASV).

We are trying to describe the electrochemical stand by combing the reduced graphene oxides/metallic oxides (rGO/MOx) nanocomposites for the analysis of heavy metal ions (HMIs) in solution by electrochemical methods. The detection simultaneous limit (3σ method) used for HMIs of the rGO/MOx nanocomposite modified electrode can be calculated for electrochemical methods on individual analysis as well as simultaneous analysis.

For electrochemical sensing Electrochemical Workstation (Potentiostat/Galvanostat) with three electrode system or multiple channel system may be used as shown in **Figure 1(a)**. The pH of the solution may be maintained from (5–12) for different metal ions. The electrode composition also plays important role in the sensing/detection of HMIs as illustrated in **Figure 1(b)** larger surface area greater will be adsorption etc. **Figure 1(c, d)** describe the active planes playing important roles in sensing/detection. The Co_3O_4 nanoplates with plane (111) were better in electrochemical sensing than Co_3O_4 nanocubes with plane (001) as described in introduction. So far we experienced the reduced graphene oxide/Metallic oxides (rGO/MOx) nanocomposites are the best choice for the sensing/detection of HMIs of boilers (high temperature). Where reduced graphene oxide (rGO) act as a base and prevent metallic oxides (MOx) from aggregation. The reduced Graphene oxide/Metallic oxides (rGO/MOx) nanocomposites facets (planes) also have influence on the sensing mechanism **Figure 1(d)**. **Figure 1(e, f)** illustrate the sensing/detection through square wave anodic stripping voltammetry (SWV) method and differential normal pulse voltammetry (DNPV) voltammetric peaks of reduced graphene oxides/metallic oxides (rGO/MOx) nanocomposites modified electrode for the analysis of heavy metal ions (HMIs) or if someone may practice will get similar peaks at different potential specified for each ion. On the other hand, reduced graphene oxide/Metallic oxides with conducting polymers may be practiced with cold water solution system as conducting polymers are not stable at high temperature. Simply in **Figure 1(a–f)** we are try to explain that the sensing/detection of HMIs by nanocomposites are influenced by composition, active planes of the nanocomposites, the Co_3O_4 nanoribbons keep chemical sensitivity than Co_3O_4 nanoparticles with active (110), (220) planes and play key role to adsorb heavy metal ions than the latter [22].

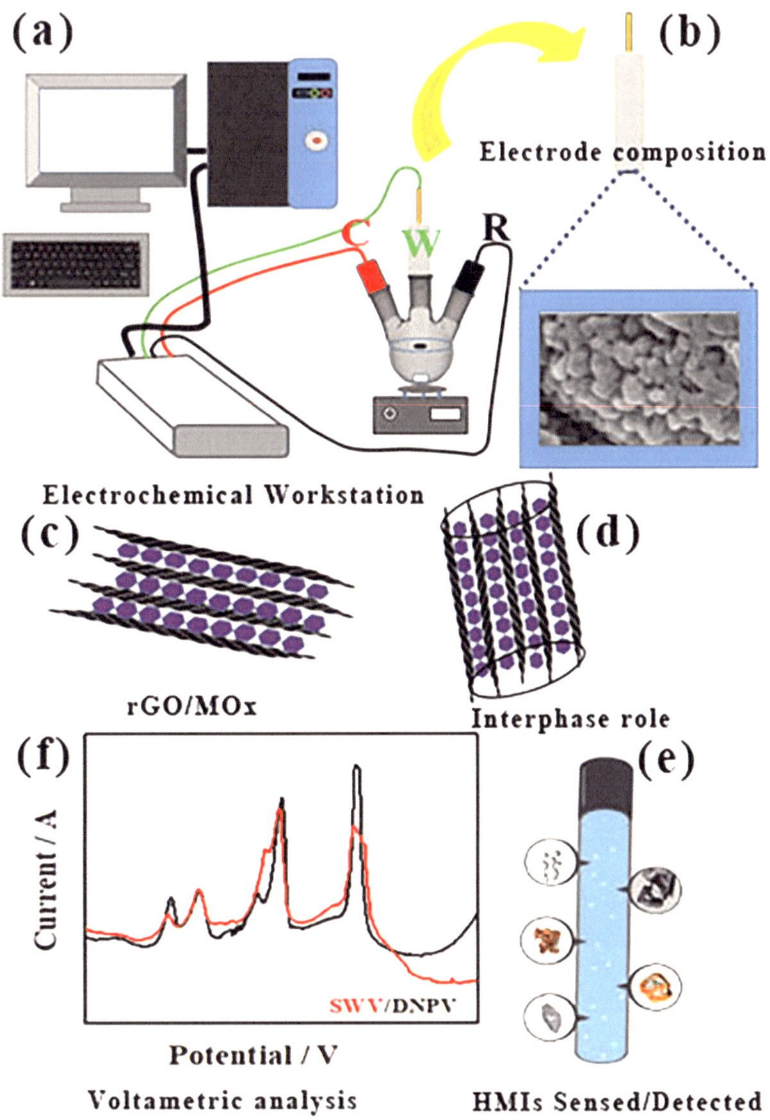

Figure 1.
Electrochemical workstation, (a) nanocomposite modified electrode, (b) reduced graphene oxide/metallic oxide phase (plane), (c, d) sensing mechanism (e) Voltammetric signals of square wave anodic stripping voltammetry (SWASV) and differential normal pulse voltammetry response of rGO/MOx modified electrode for the simultaneous analysis of heavy metal ions (HMIs) over a specified concentration range (1-10 μM) in acetate buffer (pH 5–12) (f).

2. Synthesis, modification and electrochemical use of rGO/MOx modified electrode

The reduced graphene oxide/metallic oxide (rGO/MOx) nanocomposites may be synthesized by the hydrothermal method or by chemical vapor deposition (CVD) method.

Electrochemical experiments may be performed at Electrochemical Analyzer/ Workstation (Potentiostat/Galvanostat) with a conventional three or multi-electrode system. The working electrode may be glassy carbon (coated with reduced graphene oxide/Metallic oxides nanocomposites electrode. The silver/silver chloride (Ag/AgCl) or Calomel (Hg/Hg_2Cl_2) electrode and a platinum electrode may be tested as the reference and the auxiliary electrode, respectively.

2.1 Modified electrode preparation

The carbon electrode (CE) as a working electrode may be practiced after sonication in alcohol and soaking in deionsed water. The suspension of rGO/MOx nanocomposites after dispersing with ultrasonic agitation into alcohol may be coated on the electrode. So the nanocomposites film modified electrodes will sense/detect heavy metal ions.

2.2 Heavy metal ions detection

Among the voltammetry methods Square wave anodic stripping voltammetry (SWASV) and differential normal pulse voltammetry (DNPV) method if practiced for the individual analysis as well as simultaneous analysis of heavy metal ions will give better results. The heavy metals may be sensed at the definite potential range (V), frequency (Hz), amplitude (V), increment potential (mV) for the simultaneous and selective detection of heavy metal ions.

2.3 Square wave anodic stripping voltammetry (SWASV)

Among electroanalytical techniques stripping voltammetry (SV) is considered to be the most sensitive and extensively used method for the sensing/detection of heavy metal ions (HMIs). This type of analytical investigation comprises a two-step practice: pre-concentration buildup and voltammetry sensing/detection. In the initial stage, the working electrode is plunged in metal ions solution of desired concentration. The metal ions after assortment on the surface of working electrode are reduced with voltage optimization. In the final stage, the potential is drifted to positive value to re-oxidize the metal into metal ions and voltammogram means peak current and peak potential (i-E) are recorded. Each peak of voltammogram indicate the metal and the height of peak indicate the concentration metal ion. Various factors such as synthesis method, working of electrode, practical method etc. may affect the detection limit and the sensitivity of heavy metal sensors.

Wither square wave voltammetry (SWV) or differential normal pulse voltammetry (DNPV) electrochemical analysis are practiced to detect heavy metal ions (HMIs) of various concentration (μM) solution with increased potential of millivolt (mV) for SWV and DNPV at (pH 5–12) various peaks will appear. For the stripping analysis of HMIs using both SWV and DNPV the obtained results in the form of voltammetric peaks are shown in **Figure 2** with rGO/MOx nanocomposites when experienced for the simultaneous analysis of HMIs.

The different peaks at different value (−1.0 to +0.8 V) are assigned for of potential (V) and may be attributed to HMIs on SWV and DNPV respectively. **Figure 2(a)** shows SWV with red line and DNPV with black line. Where X and Y represent the potential of various metal ions, similarly, some other peaks may arise when concentration is increased due to metallic interaction [23]. **Figure 2(b)** Linearization equations, Adj. R^2 response may be developed for rGO/MOx nanocomposites and then at the last voltammetry analysis while changing concentration (A, B and C)

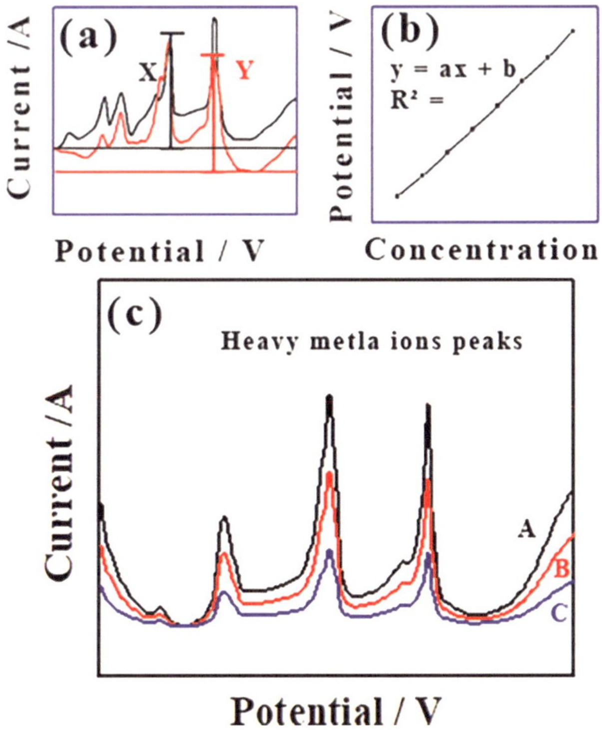

Figure 2.
The respective calibration curves of simultaneous detection of heavy metal ions (HMIs) with current intensity (X, Y) (a) linearization equations, Adj. R^2 response of rGO/MOx nanocomposites modified glassy carbon electrode. (b) Stripping voltammetry intensity peaks of various concentration (A, B & C) (c).

(i.e. 1-10 μM) in acetate buffer (pH 5–12) of the selected metal ions and are recorded as **Figure 2(c)**. The elongated peaks in **Figure 2(c)** show the different metal ions sensed/detected.

2.4 Individual as well as stripping behavior toward HMIs using SWV and DNPV

The square wave voltammetry (SWV) or differential normal pulse voltammetry (DNPV) response of rGO/MOx modified electrode or any modified electrode for the individual voltammetry investigation or simultaneous detection of heavy metal ions (HMIs) can be measured in (μM), their linearization equations and calibration curves and the limit of detection (LOD) can be easily calculated (**Tables 1** and **2**).

2.5 Sensitivity, limit of detection (LOD) and limit of quantitation (LOQ) calculation for the rGO/MOx nanocomposite

As shown in **Figure 2(a)** the corresponding calibration curves (−1.0 to +0.8 V) potential range for heavy metal ions (HMIs) simultaneous analysis are recorded for various concentration (1-10 μM). Similarly, the inset of **Figure 2(b)** and **Tables 1** and 2 represent the linearization equations and the corresponding correlation coefficients for rGO/MOx for both individual as well as simultaneous analysis on square wave voltammetry (SWV) or differential normal pulse voltammetry (DNPV).

S. No.	Concentration	Potential (V)
1	A	X
2	B	Y
3	C	Z
SE of Intercept		
SD of Intercept = SE of Intercept * $\sqrt{N}$		
LOD = 3.3*(SD of Intercept/Slope)		
LOQ = 10*(SD of Intercept/Slope)		
Slope		
$\sqrt{N}$		

Table 1.
Demonstrate the statistical calculation table of limit of detection (LOD) and limit of quantitation (LOQ) for simultaneous sensing/detection of heavy metal ions (HMIs) using SWV and DNPV voltammetry technique.

Metal ions	y = ax+b	Correlation coefficient (R^2)	Linearization eq. (I/μA)	LOD (M)	LOQ (M)
M^+	y =?	R^2 =?	I = (μM/mM) etc.	3.3*(SD of Intercept/ Slope)	10*(SD of Intercept/Slope)
M^+	y =?	R^2 =?	I = (μM/mM) etc.	3.3*(SD of Intercept/ Slope)	10*(SD of Intercept/Slope)
M^+	y =?	R^2 =?	I = (μM/mM) etc.	3.3*(SD of Intercept/ Slope)	10*(SD of Intercept/Slope)

Table 2.
Statistical calculation of nanocomposites (rGO/MOx) for individual as well as simultaneous analysis of heavy metal ions (HMIs).

The limit of detection (LOD) and limit of quantitation (LOQ) for individual as well as simultaneous detection of heavy metal ions (HMIs) via rGO/MOx modified electrodes may be calculated/measured respectively.

The Sensitivity (μA/μM) for individual as well as simultaneous analysis may be calculated. The results gained by the use of rGO/MOx for limit of detection (LOD) and limit of quantitation (LOQ) both on DNPV and SWV voltammetry analysis may be compared with the World Health Organization (WHO) data for different heavy metal ions (HMIs).

The limit of detection (LOD) and limit of quantitation (LOQ) of the nanocomposites can be calculated using calibration standards. The limit of detection (LOD) and limit of quantitation (LOQ) may be determined by 3.3σ/S and 10σ/S respectively, where S is the slope of the calibration curve and σ is the standard deviation of reaction.

The slope can be assessed from the calibration curve of the selection. The estimate of σ is typically the root mean squared error (RMSE) or standard deviation of the residuals taken from the regression line. The slope is used to convert the variation in the response back to the scale of the theoretical concentration.

3. Electrochemical methods industrial application

As we know that Electrochemical methods have a wide spread application so can be practiced for the treatment of community, industrial waters and wastewaters. The requirement for the practice of electrochemical methods is to eliminate large particles from water. The electrochemical methods can take out pollutants (organic and inorganic), ions and microorganisms acquire for clean water of distilled water quality. The electrochemical reduction intended for metals salvage and conversion of determined organic compounds to a reduced amount of toxic forms.

Recently, special attention has been given to the treatment of industrial wastewater by using advanced treatment technologies, among all methodologies, electrochemical methods appears to be one of the most promising methods for treatment organic pollutant-containing wastewater. The profits of electrochemical technology include durability, low cost, easy operation, energy efficiency, automation, fast response, high sensitivity and environmental compatibility [24–26].

Furthermore, from the above discussion it can be illustrated that rGO/MOx nanocomposites modified electrode may be helpful to practice on industrial scale

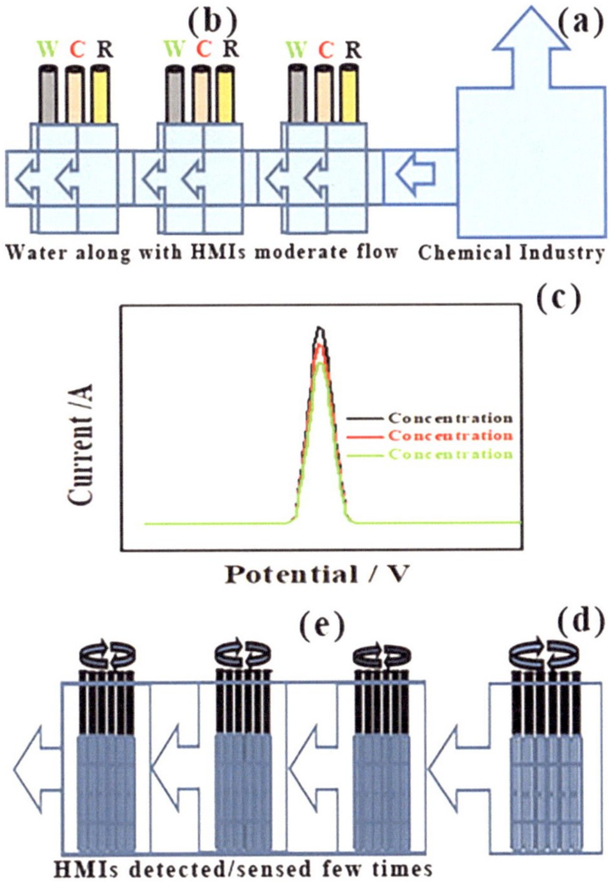

Figure 3.
Chemical industry contaminating water and detection of heavy metal ions (a, b), stripping voltammetry approach of rGO/MOx modified electrodes for the individual analysis (c) HMIs sensed few time in a flow from a chemical industry.

as well as in laboratory. Here in the last paragraph we have schemed to illustrate a chemical industry **Figure 3(a)** continuously contaminating the water (environmental and underground) along with HMIs. As has been reported by various researchers that nanocomposites modified electrode have the capacity to sense/absorb HMIs. If system of three or multiple electrode with modified nanocomposite is incorporated, then some amount of heavy metal ions can be sensed/detected **Figure 3(b)**. Therefore, we suggest that if from various concentration (1-10 μM) few amount of HMIs is sensed/ detected by rGO/MOx modified electrode **Figure 3(c)**, in this way by using multiple nanocomposites modified electrodes when practiced may sense/detect few amount of heavy metal ions **Figure 3(d)**.

Therefore, if a plant having multiple modified electrodes **Figure 3(e)** connected together then HMIs sensed few times may decrease the concentration of HMIs from the waste water of the chemical industry. We suggest that in this regard rGO/MOx nanocomposites will be a best choice for the sensing/detection of heavy metal ions (HMIs) coming out from chemical industries contaminating drinking water.

4. Brief summary

In this unit, we have emphases on the reduced graphene Oxide/metallic oxide (rGO/MOx) nanocomposites for both individual and simultaneous detection of heavy metal ions (HMIs) in solution with the help of analysis on square wave voltammetry (SWV) or differential normal pulse voltammetry (DNPV). The nanocomposite may be synthesized by hydrothermal or chemical vapor deposition (CVD) method. The expandable graphite (EG) was reduced to graphene oxide GO during hydrothermal treatment and reduced graphene oxide during chemical vapor deposition (CVD) method to enhance the flow of electron on modified electrode. Furthermore, the MOx nanoparticles dispersed on graphene sheets will tend to accumulate the HMIs on the electrode surface. The Sensitivity (μA/μM) for individual as well as simultaneous analysis, limit of detection (LOD) and limit of quantitation (LOQ) both on DNPV and SWV voltammetry analysis and can recorded and calculated (**Tables 1** and **2**). That why Electrochemical techniques have the advantages of low cost, easy operation, fast response, high sensitivity and specificity, which are suitable for ion sensing.

5. Conclusion

In this Unit, we are trying to describe the electrochemical stand by combing the reduced graphene oxide/metallic oxides (rGO/MOx) nanocomposites for the analysis of heavy metal ions (HMIs) in solution by electrochemical methods. The detection simultaneous limit (3σ method) used for HMIs of the rGO/MOx nanocomposite modified electrode can be calculated for electrochemical methods on individual analysis as well as simultaneous analysis.

The enhanced electrochemical performance can be ascribed to three factors (1) rGO could be used to prevent the aggregation of MOx nanocomposites, resulting in fast migration of electrons; the MOx nanocomposite with active planes recollect chemical sensitivity and adsorb heavy metal ions.

Acknowledgements

This work was supported by the Program for Interim Placement of Fresh PhDs Program (Phase-II) Batch-I; the National Academy of Higher Education, Higher Education Commission of Pakistan (HEC), (Ref. No. 484/IPFP-II (Batch-I)/NAHE/HEC/2020/320, 336/IPFP-II (Batch-I)/SRGPP/NAHE/HEC/2020/318, 11080/IPFP-II (Batch-I)/SRGPP/NAHE/HEC/2020/319).

Notes/thanks/other declarations

This work is dedicated to my honorable Ph.D. supervisor, Prof. Dr. Keying Shi, my colleague, my Lab fellows for their support and their unconditional help and valuable discussions, my students, my family members, loving father, great mother, sweet uncles, my brothers who were all a source of strength and inspiration for me. May Almighty Allah bless those long lives, happiness and prosperity!

And above all, I am eternally grateful to Almighty Allah who blessed me with the strength to write this unit.

Author details

Afrasiab Ur Rehman[1*], Abdul Hakim Shah[2], Atta Ur Rahman[2], Fida Ur Rahman[2], Sher Ali[1], Atta Ur Rehman[1], Raza Ullah[1], Ikram Ullah[1], Muhammad Fayaz[1] and Keying Shi[3]

1 Department of Chemistry, Khushal Khan Khattak University, Karak, Khyber Pakhtunkhawa, Pakistan

2 Department of Physics, Khushal Khan Khattak University, Karak, Khyber Pakhtunkhawa, Pakistan

3 Key Laboratory of Functional Inorganic Material Chemistry, Ministry of Education, School of Chemistry and Material Science, Heilongjiang University, Harbin, China

*Address all correspondence to: afrasiabchem83@gmail.com; dr.afrasiab@kkkuk.edu.pk

References

[1] Battistel D, Baldi F, Marchetto D, Gallo M, Daniele S. A rapid electrochemical procedure for the detection of Hg (0) produced by mercuric-reductase: Application for monitoring Hg-resistant bacteria activity. Environmental Science & Technology. 2012;**46**:10675-10681. DOI: 10.1021/es301444a

[2] Gao C, Yu XY, Xu RX, Liu JH, Huang XJ. AlOOH-reduced graphene oxide nanocomposites: One-pot hydrothermal synthesis and their enhanced electrochemical activity for heavy metal ions. ACS Applied Materials & Interfaces. 2012;**4**:4672-4682. DOI: 10.1021/am3010434

[3] Yao XZ, Guo Z, Yuan QH, Liu ZG, Liu JH, Huang X. J: Exploiting differential electrochemical stripping behaviors of octahedral Fe3O4 nanocrystals toward heavy metal ions by crystal cutting. ACS Applied Materials & Interfaces. 2014;**6**(15):12203-12213. DOI: 10.1021/am501617a

[4] Bagheri H, Afkhami A, Saber-TehraniM,KhoshsafarH.Preparation and characterization of magnetic nanocomposite of Schiff Base/silica/ Mmagnetite as a pre concentration phase for the trace determination of heavy metal ions in water, food and biological samples using atomic absorption spectrometry. Talanta. 2012;**97**:87-95. DOI: 10.1016/j.talanta.2012.03.066

[5] Ataro A, RI MC, Botha BM, CME MC, Ndibewu PP. Quantification of trace elements in raw Cow's Milk by inductively coupled plasma mass spectrometry (ICP-MS). Food Chemistry. 2008;**111**:243-248. DOI: 10.1016/j.foodchem.2008.03.056

[6] Zougagh M, de Torres AG, Alonso EV, JMC P. Automatic on line preconcentration and determination of lead in water by ICP-AES using a TS-microcolumn. Talanta. 2004;**62**(3):503-510. DOI: 10.1016/j.talanta.2003.08.033

[7] Sogut O, Bali T, Baltas H, Apaydin G. Determination of trace elements in ashes of Milk samples by using XRF technique. Asian Journal of Chemistry. 2013;**25**(8):4385-4388. DOI: 10.14233/ajchem.2013.13985

[8] Madhu R, Sankar KV, Chen SM, Selvan RK. Eco-friendly synthesis of activated carbon from dead mango leaves for the ultrahigh sensitive detection of toxic heavy metal ions and energy storage applications. RSC Advances. 2014;**4**:1225-1233. DOI: 10.1039/C3RA45089A

[9] Cui L, Wu J, Ju H. Nitrogen-doped porous carbon derived from metal-organic gel for electrochemical analysis of heavy-metal ion. ACS Applied Materials & Interfaces. 2014;**6**:16210-16216. DOI: 10.1021/am504367t

[10] Wei Y, Gao C, Meng FL, Li HH, Wang L, Liu JH, et al. SnO2/reduced graphene oxide nanocomposite for the simultaneous electrochemical detection of cadmium(II), Lead(II), copper(II), and mercury(II): An interesting favorable mutual interference. Journal of Physical Chemistry C. 2012;**116**:1034-1041. DOI: 10.1021/jp209805c

[11] Shang Q, Zhou Z, Shen Y, Zhang Y, Li Y, Liu S, et al. Potential-modulated Electrochemiluminescence of carbon nitride Nanosheets for dual-signal sensing of metal ions. ACS Applied Materials & Interfaces. 2015;**7**(42):23672-23678. DOI: 10.1021/acsami.5b07405

[12] Veerakumar P, Veeramani V, Chen SM, Madhu R, Liu SB. Palladium nanoparticle incorporated porous activated carbon: Electrochemical detection of toxic metal ions. ACS Applied Materials & Interfaces. 2016;**8**(2):1319-1326. DOI: 10.1021/acsami.5b10050

[13] Kirsch J, Siltanen C, Zhou Q, Revzin A, Simonian A. Biosensor technology: Recent advances in threat agent detection and medicine. Chemical Society Reviews. 2013;**42**:8733-8786. DOI: 10.1039/C3CS60141B

[14] Ren-Xia X, Xin-Yao Y, Gao C, Liu J-H, Compton RG, Huang X-J, et al. Enhancing selectivity in stripping voltammetry by different adsorption behaviors: The use of nanostructured Mg-Al-layered double hydroxides to detect Cd(ii). The Analyst. 2013;**138**:1812-1818. DOI: 10.1039/C3AN36271J

[15] Tang D, Tang J, Su B, Chen G. Ultrasensitive electrochemical immunoassay of staphylococcal enterotoxin B in food using enzyme-Nanosilica-doped carbon nanotubes for signal amplification. Journal of Agricultural and Food Chemistry. 2010;**58**:10824-10830. DOI: 10.1021/jf102326m

[16] Patrick Sophie C, Hein R, Paul D, Beer J, Davis J. Continuous and polarization-tuned redox capacitive anion sensing at electroactive interfaces. Journal of the American Chemical Society. 2021;**143**(45):19199-19206. DOI: 10.1021/jacs.1c09743

[17] Hierlemann A, Gutierrez-Osuna R. Higher-order chemical sensing. Chemical reviews. 2008;**108**(2):563-613. DOI: 10.1021/cr068116m

[18] Bonizzoni M, Anslyn EV. Combinatorial methods for chemical and biological sensors. Journal of the American Chemical Society. 2009;**131**(40):14597-14598. DOI: 10.1021/ja907230d

[19] Xin-Yao Y, Qiang-Qiang M, Tao L, Yong J, Bai S, Qun-Xiang L, et al. Facet-dependent electrochemical properties of Co3O4 nanocrystals toward heavy metal ions. Scientific Reports. 2013;**3**:2886. DOI: 10.1038/srep02886

[20] Hui X, Jia Y, She Xiaojie X, Li XJ, Yuanguo X, Yanhua S, et al. Graphene-analogue carbon nitride: Novel exfoliation synthesis and its application in photocatalysis and photoelectrochemical selective detection of trace amount of Cu2+. Nanoscale. 2014;**6**:1406-1415. DOI: 10.1039/C3NR04759H

[21] Liu X, Yao Y, Ying Y, Ping J. Recent advances in nanomaterial-enabled screen-printed electrochemical sensors for heavy metal detection. TrAC Trends in Analytical Chemistry. 2019;**115**:187-202. DOI: 10.1016/j.trac.2019.03.021

[22] Rehman AU, Fayaz M, Lv H, Liu Y, Zhang J, Wang Y, et al. Controllable synthesis of a porous PEI-functionalized Co3O4/rGO nanocomposite as an electrochemical sensor for simultaneous as well as individual detection of heavy metal ions. ACS Omega. 2022;**7**(7):5870-5882. DOI: 10.1021/acsomega.1c05989

[23] Cui Lin W, Huangxian JJ. Electrochemical sensing of heavy metal ions with inorganic, organic and bio-materials. Biosensors and Bioelectronics. 2015;**63**:276-286. DOI: 10.1016/j.bios.2014.07.052

[24] Harshika S, Sangal Vikas K, Manish V. Treatment of tannery industry effluent by electrochemical methods: A review. Materials Today: Proceedings. 2021;**47**(7):1438-1444. DOI: 10.1016/j.matpr.2021.03.300

[25] Martínez-Huitle Carlos A, Enric B. Decontamination of wastewaters containing synthetic organic dyes by electrochemical methods: A general review. Applied Catalysis B: Environmental. 2009;**87**(3-4):105-145. DOI: 10.1016/j.apcatb.2008.09.017

[26] Ding Q, Li C, Wang H, Xu C, Kuang H. Electrochemical detection of heavy metal ions in water. Chemical Communications. 2021;**57**(59):7215-7231. DOI: 10.1039/D1CC00983D

Chapter 31

A Study on the Methyl and Ethylmercury Artifacts in Biological Samples Using Sodium Tetra(n-Propyl)Borate as a Derivatizing Agent

Abdelkarem A.S. Elgazali, Youssef F. Lawgali and Hatem Fawzi Gharour

Abstract

Sodium tetra (n-propyl) borate was used as derivatizing agent to measure methyl and ethylmercury compounds. This study investigated the artifact formation of methyl and ethylmercury compounds during derivatization using $NaBPr_4$, simultaneously with the influence of this artifact on methylmercury analysis in biological samples (chlor alkali hair samples). The artifact methylmercury and ethylmercury compounds during derivatization using $NaBPr_4$ were evident and depended strongly on the amount of inorganic mercury (Hg^{2+}) present in the sample solution for derivatization and depended on the purity of sodium tetra (n-propyl) borate reagent. The high formation rate of artifact Et-Hg (0.76–0.81% of high-level Hg^{2+} present) interferes strongly with the ethylmercury analysis. The rate of artifact formation of Me-Hg is small and constant at the different concentration ranges of In-Hg (0.012% of In-Hg present) and does not affect on Me-Hg analysis and it can be subtracted from this Me-Hg artifact ratio from the measured value of Me-Hg in the biological samples. However, the mathematical correction for Me-Hg measurement can be done only when the Et-Hg peak is already appearing in the chromatogram samples.

Keywords: inorganic mercury, monomethyl ethylmercury, sodium tetra (n-propyl) borate, artifacts derivatization, mercury compounds

1. Introduction

Mercury has been well known as an environmental toxin and pollutant for several decades. The environmental cycling of mercury is a very complex distribution, involving a large variety of physical and chemical processes that affect its toxicity and mobility [1–4]. The lengthy mercury transport cycle in the atmosphere, it is deposition, bioaccumulation, and the concentration of extremely hazardous methylmerury methylmercury (Me-Hg) molecules in the aquatic food chain represent a severe environmental concern, even in distant places, poisoning people [3–7].

Analytical techniques for the separation of methylmercury (Me-Hg) are well documented [8]. After extraction from solid matrices and derivatization, the methylmercury (Me-Hg) is frequently measured using hyphenated techniques [8]. Mercury speciation analysis is usually performed by resorting to hyphenated techniques, based on the coupling of an effective separation technique to a sensitive element-specific detector. Capillary gas chromatography (CGC), liquid chromatography (LC), or more recently capillary electrophoresis (EC) can be interfaced with specific atomic detection including atomic absorption spectrometry (AAS), atomic fluorescence spectrometry (AFS), electron capture, or inductively coupled plasma mass spectrometry (ICPMS) [9–17]. Recently research has shown that the coupling of GC to ICP-MS appears to be a more suitable hyphenated technique to carry out the mercury speciation analysis because of its high sensitivity, multi-isotopic, and multi-elemental capabilities [8, 9, 15]. However, the GC-ICP-MS is only suitable for volatile species like mercury species [11, 15]. The isotope dilution ICP-MS is a new powerful approach and offers great potential for very small uncertainties since quantitative recoveries are not required and rearrangement reactions are easily detected [10, 17–21]. The main advantage of this technique (IDMS) is that chemical separation if required for accurate ratio determination need not be quantitative. Moreover, concentrations of chemical species can be measured very precisely because ratios can be measured very reproducibly [8, 11, 16].

Quality results are sometimes associated with sample pre-treatment; the analysis of solids such as biological and environmental samples requires leaching (alkaline or acid)/digestion step to liberate mercury species from the sample matrix before detection with GC-ICP-MS. However, for ionic mercury species, derivatization reactions are required to achieve good results [11, 22].

In earlier studies, monomethyl mercury (Me-Hg) was the most investigated organomercury compound, and measurement of monomethylmercury (Me-Hg) in environmental samples using sodium tetraethylborate ($NaBEt_4$) was one of the most used methods for methylmerury analysis [23, 24]. However, in some cases during the ethylation (Eth) with sodium tetraethyl borate ($NaBEt_4$), the Hg^{2+} is transformed to $HgEt_2$, while MeHg forms MeHgEt Eqs. (1) and (2).

$$Hg^{2+} + 2NaB\ (C_2H_5)_4 \rightarrow Hg\ (C_2H_5)_2 + 2Na^+ + 2B(C_2H_5)_3 \tag{1}$$

$$CH_3Hg^+ + NaB\ (C_2H_5)_4 \rightarrow CH_2HgC_2H_5 + Na^+ + B(C_2H_5)_3 \tag{2}$$

As mentioned above, isotope dilution ICP-MS is a new powerful approach to solving the problems with the matrix and non-quantitative derivatization. A drawback of the ethylation (Eth) procedure is the impossibility to distinguish between Hg^{2+} and $EtHg^+$, both species that often coexist in the environment [25]. It was observed that derivatization using ethylation reagent ($NaBEt_4$) induced the formation of MeHg from inorganic mercury (InHg) if inorganic mercury was present at high concentrations and also the presence of dissolved organic matrix in the sample strongly interferes with ethylation process [18, 26, 27]. Therefore, ignoring this effect of artifact formation may lead to systematic errors in methylmercury analysis. Recently, an alternative is the use of the propylation as a derivatization technique with sodium tetra-propyl borate as the derivatizing agent which is more tolerant to interferences from chlorides [11, 18, 26, 28]. However, it was found that the artifact of methylmercury (Me-Hg) and ethylmercury (Et-Hg) compounds during $NaBPr_4$ derivatization was evident and depended strongly on the concentration of inorganic mercury (Hg^{+2}) presence in the solution for derivatization. For example, Jen-How Huang [26] observed a transformation of In-Hg into ethylmercury (Et-Hg) and methylmerury (Me-Hg) during derivatization using $NaBPr_4$, and he reported that the artifact formation rates of EtHg and MeHg are 0.99–2.9% and 0.03–0.28%, respectively. This conclusion may ignore the artifact formation of monomethylmercury (Me-Hg) and monoethylmercury (Et-Hg) during derivatization by $NaBPr_4$ similar to $NaBEt_4$. Therefore, without taking this effect of artifact formation into account, the artifact may lead to an overestimation of organomercury species concentrations and a false impression of organomercury speciation.

This study aims to investigate the formation of Me-Hg and Et-Hg artifacts in hair samples with the high level of In-Hg in hair workers of ICL factory in Pakistan by comparing the Hg artifacts in un-spiked and spiked blank samples, different concentrations of normal abundance In-Hg solution, enriched ^{199}In-Hg solution and hair sample (normal hair) with low level (0.98 mg/kg) of In-Hg during the derivatization step.

The objectives are [1] to examine the artifact formation of methyl and ethylmercury from inorganic mercury (Hg^{2+}) during propylation using $NaBPr_4$, [2] to identify the factors which govern the artifact formation of MeHg and EtHg, and [3] to evaluate the influence of MeHg and EtHg artifact information on the determination of actual monomethylmercury (Me-Hg) concentrations in chlor alkali hair samples with high inorganic mercury concentrations (Up to 0.9%).

2. Materials and methods

2.1 Devices and instrument

Microwave digestion oven model MARS-5 from CEM Instrument, UK, was used for digestion and decomposition of hair samples. The microwave operating conditions are listed in **Table 1**. A gas chromatograph (GC) model HP 6850 outfitted with a capillary column was connected to an Agilent model HP-7500 ICP mass spectrometer through a heated steel transfer capillary for speciated isotope dilution analysis (SIDMS). The heated steel transfer capillary was inserted into the ICP torch injector, and connection to the torch was realized through a glass T-piece. A conventional Meinhard concentric nebulizer and low volume water-cooled cyclonic spray chamber

Microwave instrument	MARS-5 from CEM instrument, UK
Power	800 W
%	100
Ramp	3.0 minutes
Temperature control	55°C for 20 min and 60°C for 20 min
Pressure (psi)	0.00
Temperature programme	65°C for 40 minutes
Stage	2

Table 1.
Microwave operating conditions for hair samples digestion.

were connected to the heated steel transfer capillary line connected ICP torch, and this enabled continuous aspiration of a standard thallium solution (25μgl^{-1}). This configuration allowed optimization of instrument performance and simultaneous measurement of ^{203}Tl and ^{205}Tl for mass bias correction during the chromatographic run [9]. Operating conditions for the GC-ICP-MS coupling system are listed in **Table 2**.

2.2 Reagents and standards

All chemicals used were of analytical reagent grade unless stated otherwise. Tetramethylammonium hydroxide (TMAH, 25% w/w in water) and ethylmercury chloride were purchased from Alfa Aesar (UK). Methanol, sodium acetate, and acetic acid glacial (super grade) were purchased from VWR (BDH, UK). Sodium tera-n-propylborate ($NaBpR_4$, ≥98% purity) was purchased from Chemos GmbH (Germany). 2, 2, 4 trimethylpentane (isooctane, spectrophotometric grade, ≥99% purity) and methylmercury (II) chloride were purchased from Sigma Aldrich (UK). Inorganic mercury (In-Hg) standard solution for ICP (934 ± 3.0 mg/kg) was purchased from Fluka (UK).

The derivatization solution was prepared by dissolving 1 g of sodium tetrapropylborate ($NaBpR_4$) in 100 ml of deionized water. The solution was stored at −20°C in a refrigerator and protected from light. Buffer acetate (0.1 M) in deionized water was prepared by mixing 0.1 M sodium acetate solution (90 ml) with 0.1 M acetic acid

ICP-MS Instrument	Agilent 7500 series
Hg isotope acquired	199, 200, 201, 202
Acquired mode	Time-resolved
Dwell time	0.035 sec/point
RF power	1380 W
RF matching	1.53 V
Sample depth	6.3 mm
Torch-H	1.1 mm
Torch-V	0.4 mm
Carrier gas	Argon/0.79 l min^{-1}
Makeup gas	Argon/0.17 l min^{-1}
Extract 1	−2 V
Internal standard	Tl (25 ppb)
Nebulizer pump flow rate	0.20 rps
Spray chamber temperature	2 °C
GC Instrument	Agilent HP 6850
Injection	Split/splitless—1 μl
Oven program	50 °C (1 min), 50 °C/min 220°C (5 min
Carrier gas	Helium
Transfer line temp	200°C
GC injector temp	220°C

Table 2.
GC and ICP-MS operating parameters.

solution (410 ml) and adjusted the final volume to 1000 ml (1 L) with deionized water and adjusting to pH 3.9. Inorganic mercury working standard solutions (1.0 and 10 mg/kg as Hg) were prepared from appropriate dilution of inorganic mercury standard stock solution (934 ± 3.0 mg/kg). Enriched inorganic mercury (1.0 mg/kg ^{199}In-Hg199 as Hg) and enriched methylmercury (1.0 mg/kg ^{201}MeHg as Hg) working solutions were prepared from appropriate dilution of their standard stock solutions. Milli-Q quality water (Millipore) was used throughout.

2.3 Derivatization by sodium tetrapropylborate and analytical procedures

Blank (TMAH) and hair samples with a low level (0.98 mg/kg) of In-Hg and one hair sample from ICL with a high level of In-Hg (1000 mg/kg) were spiked with the same amount of ^{201}MeHg and ^{199}InHg (double spike, 70 μl from 1.0 ppm of each enriched Hg standard). The spiked and un-spiked hair samples were digested using a microwave device. The spiked sample solution (blank and digested spiked hairs) and un-spiked hair solution samples in cleaned and dried glass vials (1 ml of each) were then adjusted to pH 3.9 with acetate buffer, and then, 1 ml of 1% $NaBPr_4$ was added in the glass vials for derivatization to form the corresponding peralkylated mercury (Hg) species such as.

$$Hg^{2+} + 2NaB\,(C_3H_5)_4 \rightarrow Hg\,(C_3H_5)_2 + 2Na^{+} + 2B\,(C_3H_5)_3 \quad (3)$$

$$CH_3Hg^{+} + NaB\,(C_3H_5)_4 \rightarrow CH_2HgC_3H_5 + Na^{+} + B\,(C_3H_5)_3 \quad (4)$$

Extraction of derivatized Hg species (peralkylated Hg) was done by vigorous shaking for 5 min with 1 ml isooctane, the isooctane extract was afterwards centrifuged for 10 min at 3000 rpm, and then, the extracted Hg species into isooctane layer were transferred to GC vials and analyzed with a coupling of GC-ICP-MS. In addition, the conc. Normal abundance In-Hg, conc. Enriched In-Hg^{199}, and hair sample from chlor alkali plant with similar Hg conc. to both In-Hg conc. (about 1000 mg/kg) and different concentrations of normal abundance In-Hg standard solutions (20, 40, 60, 80, and 100 mg/kg) were derivatized and extracted as described above. The samples were diluted after derivatization and extraction steps (D.F 1:10 for the concentrated Hg standards and chlor alkali hair sample).

3. Results and discussion

3.1 Hg isotope ratio calculation for spiked and un-spiked samples

The mercury isotope ratios (IR) were calculated for four measured Hg isotopes (199, 200, 201, 202) in spiked, un-spiked blank (TMAH), hair samples, and Hg standard solution (normal abundance In-Hg and enriched ^{199}In-Hg) to compare the formation of mercury artifact in spiked and un-spiked samples during propylation with $NaBPr_4$. The artifact formation of monomethy lmercury (Me-Hg) and ethylmercury (EtHg) from inorganic mercury (In-Hg) was observed during propylation with $NaBPr_4$ for spiked blank with enriched ^{199}InHg and enriched ^{199}In-Hg standard solution comparing with un-spiked blank (**Figures 1** and **2**). The extent of artifact formation for the organomercury compounds was in the order: Et-Hg > -Me-Hg > Hg(0). Moreover, the artifact formation of monomethyl–ethylmercury

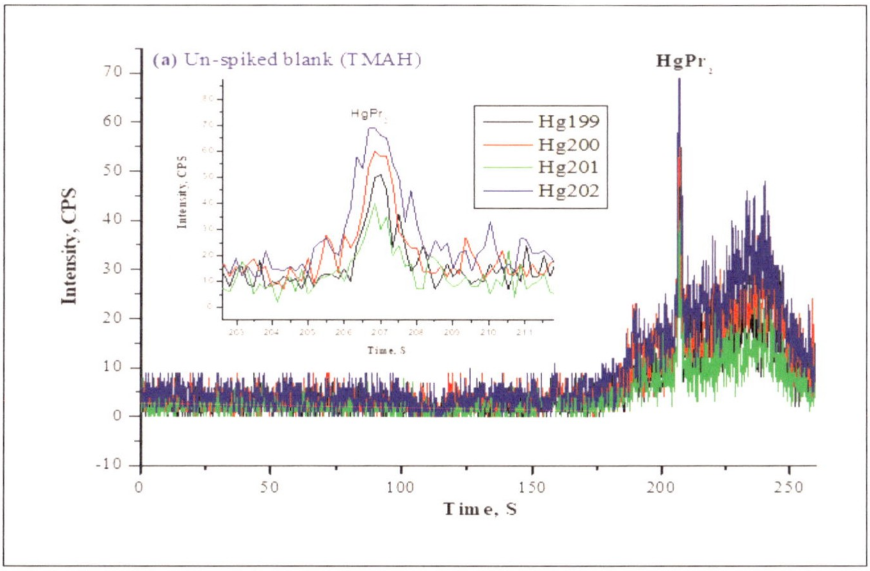

Figure 1.
Typical chromatogram of un-spiked blank (1.0 ml TMAH) obtained during derivatization using $NaBPr_4$.

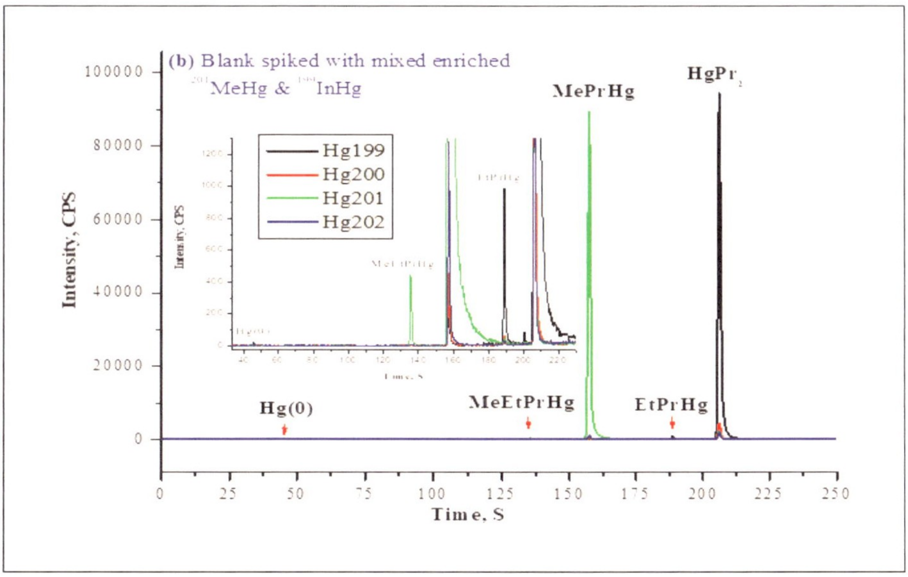

Figure 2.
Typical chromatogram of spiked blank (TMAH) with mixed enriched ^{201}MeHg & ^{199}In-Hg standard solutions obtained during derivatization using $NaBPr_4$.

(MeEt-Hg) from methylmercury was observed also during propylation with $NaBPr_4$ for **a** spiked blank (TMAH) with mixed enriched mercury standards (^{201}Me-Hg and ^{199}In-Hg) as shown in **Figure 2**.

For mercury isotope ratio (IR) calculation results for un-spiked blank (TMAH) compared with spiked blank with the same amount of mixed enriched mercury standards (^{201}MeHg and ^{199}InHg, 70 µl from 1.0 ppm of each into 1 ml of TMAH) after

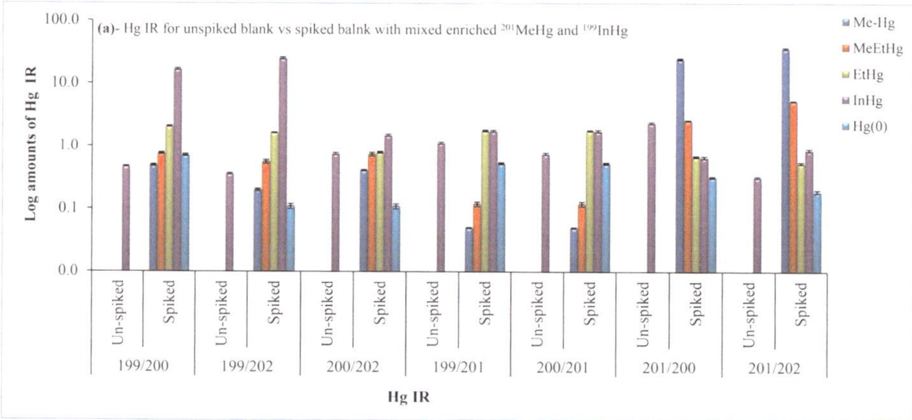

Figure 3.
Compression of calculated mercury isotope ratios (Hg-IR) in un- spiked blank with spiked blank with mixed enriched ^{201}MeHg & ^{199}In-Hg standard solutions during the derivatization using $NaBPr_4$.

propylation with $NaBPr_4$ as shown in **Figure 3**. It can be seen that the signal in each of the mercury isotopic ratios (Hg199/200, Hg199/201, Hg199/202, Hg200/201, Hg200/202, and Hg201/200) for spiked blank is increased in the order: In-Hg > Et-Hg > -MeEt-Hg > Me-Hg, but for un-spiked blank was observed only the similar ratio for all Hg isotope ratios for In-Hg only. This means that the artifact formation of an organomercury compound is increased with increasing amounts of inorganic mercury (InHg) when spiked the blank with enriched mercury standards (^{201}Me-Hg and ^{199}In-Hg) and propylated with $NaBPr_4$. However, this is indicating that the artifact formation of MeHg and EtHg from a high concentration of inorganic mercury is caused by $NaBPr_4$.

In addition, it can be seen from **Figure 4** that the results of IR calculation for spiked blank (TMAH) with enriched ^{199}InHg are similar to those calculated in

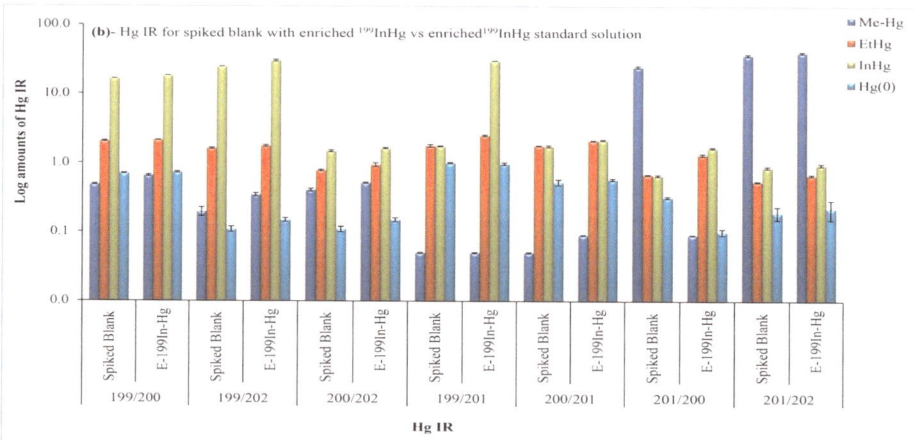

Figure 4.
Compression of calculated mercury isotope ratios (Hg-IR) in spiked blank with enriched ^{199}InHg spiked blank versus enriched ^{199}In-Hg standard solution during derivatization using $NaBPr_4$.

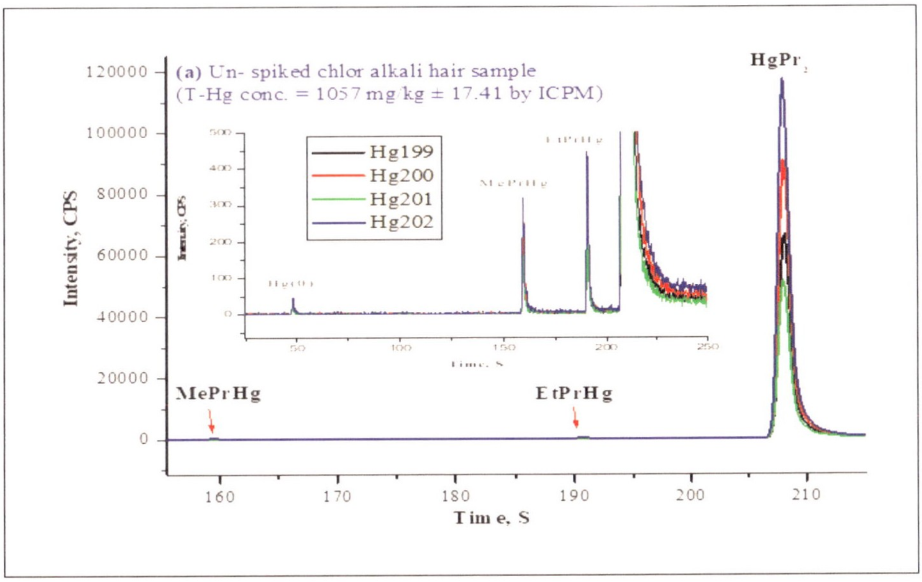

Figure 5.
Typical chromatogram of un- spiked chlor alkali hair sample (CA hair sample with T-Hg conc. = 1000 mg/kg) obtained during derivatization using $NaBPr_4$.

enriched ^{199}InHg standard solution after propylation by $NaBPr_4$. Moreover, it can be deduced from IR calculations that MeHg and EtHg artifact creation originate exclusively from InHg, but MeEtHg artefact formation is not detected. This indicates that MeEtHg artefacts only originate from enriched ^{201}MeHg when the blank or sample is spiked with high levels of enriched ^{201}MeHg (greater than 50 l of 1.0 ppm to 1.0 ml of blank or 0.01.

For un-spiked hair samples from one chlor alkali plant with a high concentration of inorganic mercury (In-Hg) and normal abundance inorganic mercury (In-Hg) with similar Hg concentration to those found in selected chlor alkali plants hair sample, the mercury isotope ratios (IR) were calculated for four measured Hg isotopes (199, 200, 201, and 202) same as in spiked, un-spiked blank (TMAH), and enriched ^{199}InHg. The observation of artifact formation of methylmercury (MeHg) and ethylmercury (EtHg) from inorganic mercury (InHg) was similar to those found in the spiked blank with enriched ^{199}InHg and enriched ^{199}InHg standard solution during propylation with $NaBPr_4$ (**Figures 5** and **6**). However, as shown in **Figures 5** and **6**, the transformation of MeHg and EtHg from In-Hg in both chlor alkali hair and normal abundance InHg is similar and the extent of artifact formation for the organomercury compounds was in the order: EtHg>MeHg>Hg(0).

Mercury isotope ratio (IR) calculation results for un-spiked chlor alkali hair compared with spiked same hair sample with mixed enriched mercury standards (^{201}MeHg and ^{199}InHg, 70 ul from 1.0 ppm of each into 1 ml of TMAH) and normal abundance InHg after propylation with $NaBPr_4$ as shown in **Figures 7** and **8** below showed similar amounts of all mercury isotope ratio (IR) in both spiked and un-spiked hair samples. This indicates that the methylmercury (MeHg) and ethylmercury (EtHg) compounds are artifacts of high amounts of inorganic mercury in hair samples owing to the propylation with NaBPr4 and are not reliant on the spiking quantity of

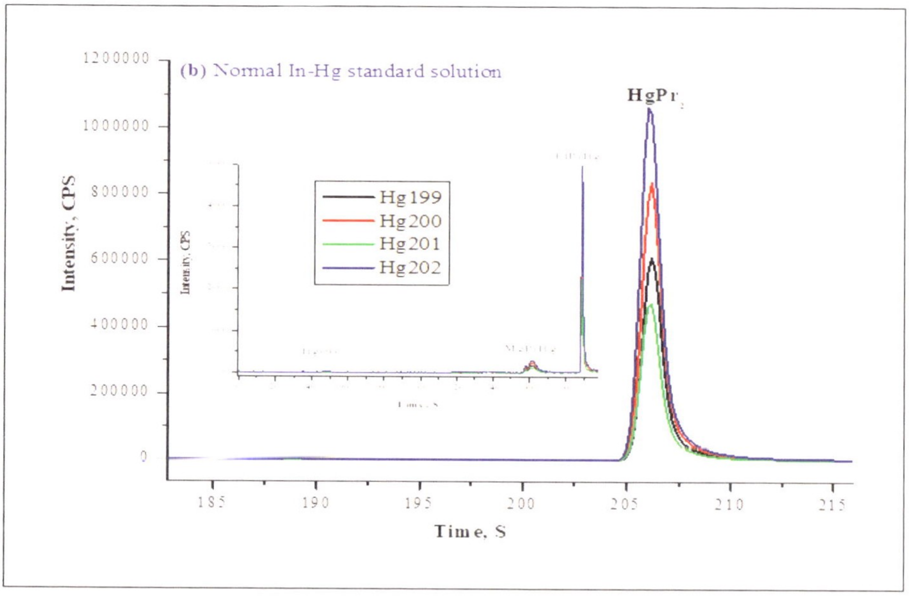

Figure 6.
Typical chromatogram of normal abundance In-Hg (1000 mg/kg as Hg^{2+}) obtained during derivatization using $NaBPr_4$.

enhanced mercury standards, as well as depending on the purity of the propylation reagent.

Moreover, when Hg isotope ratios (IR) calculating results in spiked normal hair the sample (control hair sample) is compared to spiked chlor alkali hair (both hair samples were spiked with the same quantity of mixed enriched ^{201}MeHg and ^{199}InHg standard solutions), as shown in **Figure 9** below, it can be noted that the Hg isotope ratio calculation findings are identical in both spiked hair samples except for In-Hg isotope ratios (IR).

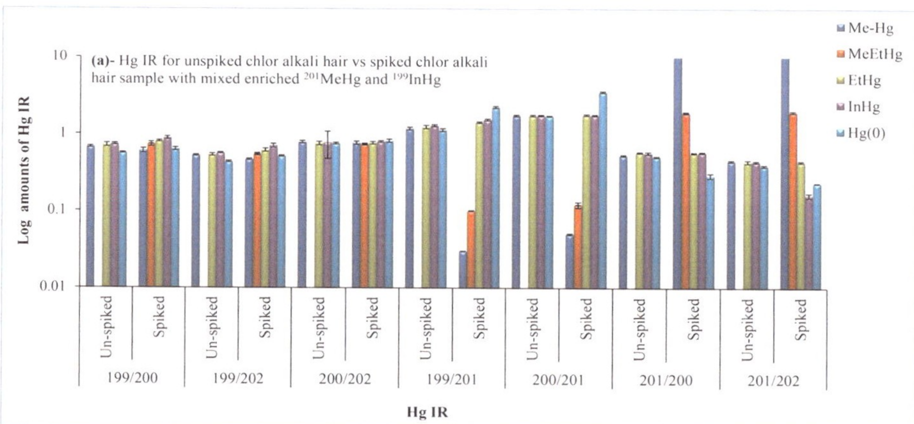

Figure 7.
Compression of calculated mercury isotope ratios (Hg-IR) in un- spiked CA hair spiked CA hair mixed enriched ^{201}MeHg & ^{199}In-Hg standards during derivatization using $NaBPr_4$.

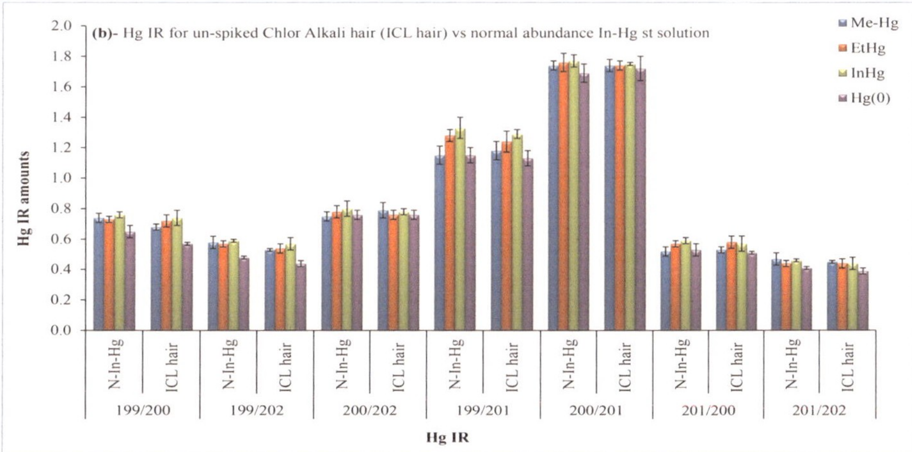

Figure 8.
Compression of calculated mercury isotope ratios (Hg-IR) in un- spiked CA hair verses similar In-Hg concentration of normal abundance In-Hg standard solution during derivatization using $NaBPr_4$.

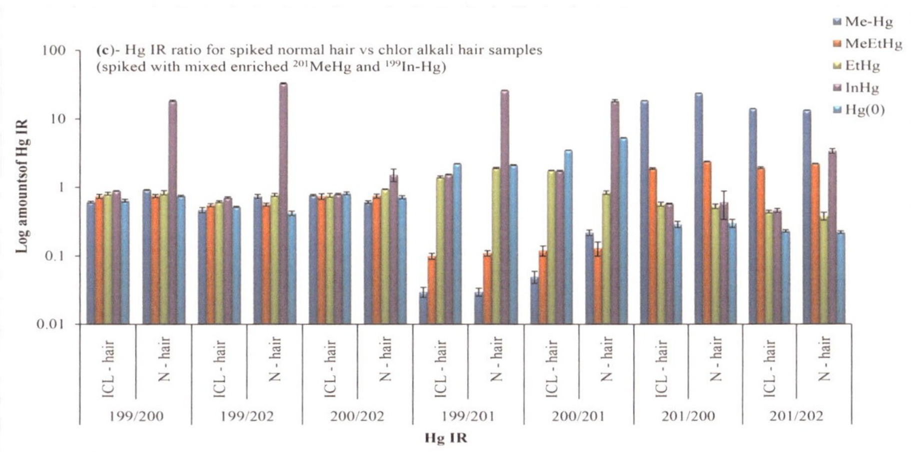

Figure 9.
Comparing the compression of (Hg-IR) in spiked normal hair (N-hair) with mixed enriched 199In-Hg and 201MeHg standards to spiked CA-hair mixed standard solutions utilizing NaBPr4.

3.2 Methyl (MeHg) and Ethylmercury (EtHg) percentage (%) artifact formation in the blank (TMAH) and the hair samples during derivatization by $NaBPr_4$

To calculate the percentage amounts of artifacts formation of methyl and ethylmercury in spiked blank and hair samples, the normal abundance inorganic mercury standard solution, and enriched inorganic mercury (^{199}In-Hg) during derivatization by NaBPr4, the artifact percentage (%) calculation was done as follows:

3.2.1 Comparison of MeHg and EtHg % artifact formation in un-spiked blank (TMAH) with spiked blank and enriched abundance ^{199}InHg

It can be observed from **Table 3** and **Figure 10** that the average artifact formation of MeHg was 0.012% for both spiked blank (TMAH) with the low amount of enriched ^{199}InHg (70 ul of 1 mg/kg of standard) and high amount of ^{199}InHg standard solution (20 mg/kg), while there is no artifact present in un-spiked blank (TMAH). Also, the average artifact formation of EtHg (0.82%) was higher about 68 times than the average artifact formation of MeHg in both spiked blank and enriched InHg standard solution.

3.2.2 Comparison of MeHg and EtHg % artifact formation in un-spiked chlor alkali hair sample with spiked CA hair and normal abundance InHg standard solution

To compare the rate of MeHg and EtHg percentage (%) artifact formation in chlor alkali hair sample (un-spiked and spiked CA hair sample) and normal

Sample ID	MeHg and EtHg % artifact formation				
	Hg (0) (%)	MeEtHg (%)	MeHg (%)	EtHg (%)	InHg (%)
Un-spiked blank (TMAH)	0.00	0.00	0.00	0.00	100
Spiked blank with enriched ^{199}InHg	0.00	0.00	0.012	0.82	99.10
Enriched ^{199}InHg standard solution	0.00	0.00	0.012	0.82	99.10

Table 3.
Comparison of MeHg and EtHg % artifact formation in the un-spiked blank (TMAH) with spiked blank and enriched abundance InHg standard during derivatization using NaBPr$_4$.

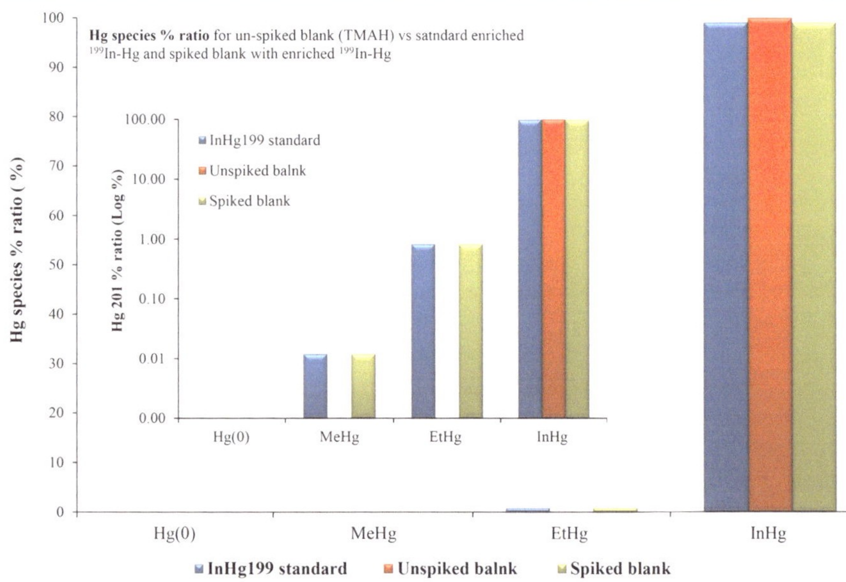

Figure 10.
Compression of calculated mercury isotope ratios (Hg-IR) in spiked blank with enriched ^{199}InHg the spiked blank versus enriched ^{199}In-Hg standard solution during derivatization using NaBPr$_4$.

Sample ID	MeHg and EtHg % artifact formation				
	Hg (0) (%)	MeEtHg (%)	MeHg (%)	EtHg (%)	InHg (%)
Un-spiked CA hair	0.012	0.000	0.220	0.76	99.04
Spiked CA hair with mixed enriched ^{201}MeHg &^{199}InHg	0.013	0.24	0.232	0.76	98.78
Normal abundance InHg standard solution	0.010	0.000	0.012	0.76	99.20

Table 4.
Comparison of MeHg and EtHg % artifact formation in un-spiked CA hair with spiked CA hair and normal abundance InHg during derivatization using $NaBPr_4$.

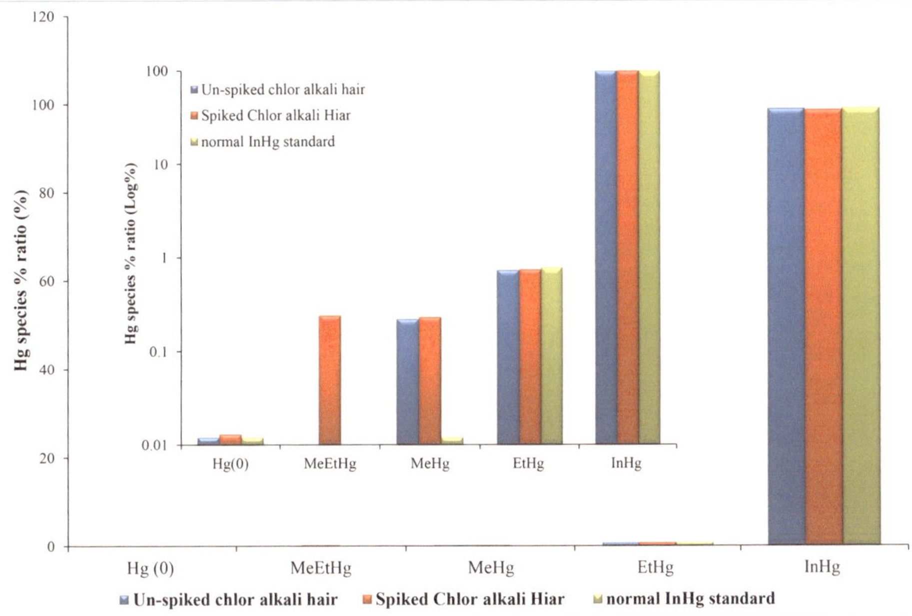

Figure 11.
Comparison of MeHg and EtHg % artifact formation in un-spiked CA hair versus spiked CA hair with mixed enriched ^{201}MeHg & ^{199} InHg standards and the same amount of normal abundance InHg during derivatization using $NaBPr_4$.

abundance In-Hg standard solution, further calculations were done as shown in **Table 4** and **Figure 11**, and the percentage results were similar to those found in spiked and spiked blank. However, it can be observed that the rate of MeHg % artifact formation is increased by the factor of 0.012% (from 0.220% in un-spiked CA hair to 0.232% in spiked CA hair sample with mixed enriched ^{201}MeHg and ^{199}InHg), while the percentage artifact formation rates of EtHg were constant (0.76% for both of each). In addition, for normal abundance standard solution, the percentage (%) artifact formation of MeHg and EtHg was recorded similar to those found in spiked chlor alkali hair sample (CA hair). Moreover, the % arifact formation rate of MeEtHg (0.24%) was recored only in spiked CA hair sample with mixed enriched ^{201}MeHg and

^{199}InHg standard solutions (70 ul of 1.0 ppm from each standard). This is meaning that the artifact formation of MeEtHg comes from enriched ^{201}MeHg in comparison with un-spiked CA hair and normal abundance In-Hg standard solution.

3.3 Influence artifact formation on determining of methylmercury (Me-Hg) in biological (hair) and environmental samples

To estimate the influence of the artifact on the methyl mercury (MeHg) and ethylmercury (Et-Hg) analysis, the artifact during extraction and derivation was investigated by comparing of un-spiked chlor alkali hair sample (CA hair, with a high level of In-Hg) with different concentration of normal abundance In-Hg (20, 40, 60, 100 mg/kg In-Hg as Hg^{2+}). The un-spiked chlor alkali hair sample (0.02 g) was digested in 5 ml TMAH using microwave device under temperature programme of 55°C for 20 min and 60°C for 20 min. The extract for CA hair and five normal abundance In-Hg standard solutions were then derivatized with $NaBPr_4$, extracted, and analyzed with the same procedure as described in Section 2.3. From the results as shown in **Table 5** and **Figure 12**, it can be seen that there is no substantial enhancement of artifact of Me-Hg and Et-Hg observed as compared to the amounts of artifact MeHg and EtHg shown in **Tables 3** and **4** and **Figures 10** and **11**. This result indicates that the percentage of artifact formation of methylmercury (MeHg) and ethylmercury (EtHg) from all different concentrations of normal abundance In-Hg standard solution was constant at the rates of 0.012 and 0.80%, respectively, during derivatization using $NaBPr_4$. However, as shown in **Table 5** and **Figure 12**, the most of MeHg found in the un-spiked CA hair extract more likely originated from the CA hair sample. Taking 0.012% as average constant formation rate for artifact MeHg, the CA hair sample showed that artifact MeHg might result in less than 6% of the measured MeHg value. Moreover, **Table 5** and **Figure 12** indicate that all EtHg found in un-spiked CA and five different concentrations of normal abundance In-Hg standard solutions are artifacts at the same artifacts formation percentage (0.80%) from the high level of In-Hg in the samples during derivatization using $NaBPr_4$. Despite of this, this method is useful to measure the actual amount of MeHg in hair samples by subtracting of percentage MeHg artifact formation constant ratio (0.012%) from that found in chlor alkali hair samples or any samples contains high levels of inorganic mercury (In-Hg).

Sample ID	% Artifact formation from InHg			
	Hg (0) (%)	MeHg (%)	EtHg (%)	InHg (%)
In-Hg (20 mg/l)	0.012	0.012	0.80	99.18
In-Hg (40 mg/l)	0.011	0.012	0.81	99.12
In-Hg (60 mg/l)	0.012	0.012	0.80	99.14
In-Hg (80 mg/l)	0.011	0.012	0.80	99.13
In-Hg (100 mg/l)	0.012	0.013	0.81	99.12
Un-spiked chlor alkali hair	0.011	0.220	0.77	99.09

Table 5.
Comparison of MeHg and EtHg % artifact formation in un-spiked CA hair with different amounts of normal abundance InHg during derivatization using $NaBPr_4$.

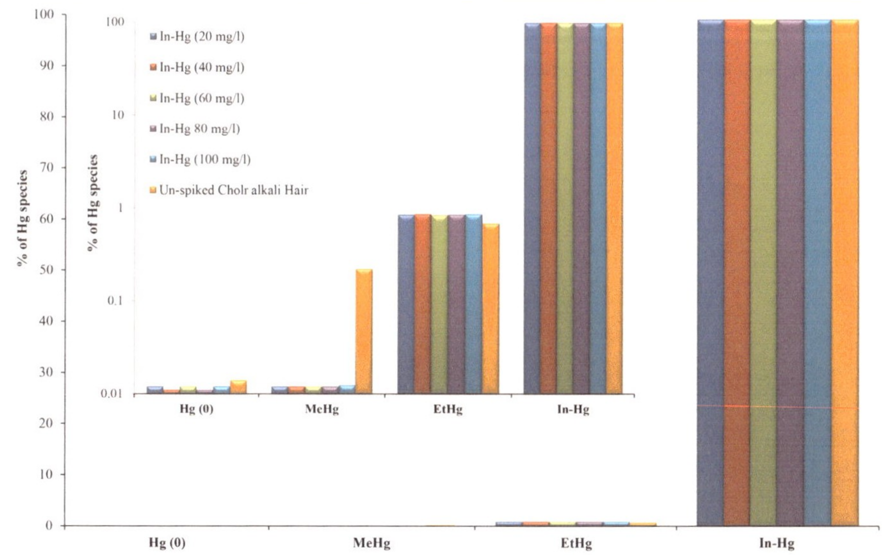

Figure 12.
Comparison of MeHg and EtHg % artifact formation in un-spiked CA hair versus different concentrations of normal abundance InHg during derivatization using $NaBPr_4$.

4. Conclusions

Artificial mercury speciation and quantification errors of organomercury compounds are caused by the artifact generation of organomercury compounds like Me-Hg and Et-Hg during the analytical methods. There were obvious artifact formation of methylmercury (Me-Hg) and ethylmercury (Et-Hg) compounds from a high level of inorganic mercury (more than 20 mg/kg) during NaBPr4 derivatization, and so this highly depends on the amount of inorganic mercury (Hg2+) present in the derivatization solution and the purity of sodium tetra (n-propyl) borate (NaBP). The high rate of artifact Et-Hg formation (0.76 to 0.81% of high-level Hg2+ present) seriously impairs Et-Hg analysis. This demonstrates that the sodium tetra (n-propyl) borate (NaBPr4) reagent is not suitable for the analysis of EtHg when inorganic mercury (In-Hg) concentrations in samples are higher than 20 mg/kg. The rate of methylmercury (MeHg) artifact creation is low and steady (0.012% of InHg present), and it has no impact on the analysis of methylmercury (MeHg) since the MeHg artifact ratio can be removed from the observed value of MeHg in the samples. However, the EtHg peak must be visible in the samples' chromatograms to do the mathematical correction for MeHg measurement. Additionally, the majority of the inorganic mercury (In-Hg) from the solid samples can be removed using acid leashing procedures before the derivatization step to prevent the formation of organomercury compounds (Me-Hg and Et-Hg) as an artifact during the derivatization process using NaBPr4 [29, 30].

Author details

Abdelkarem A.S. Elgazali[1,2], Youssef F. Lawgali[2,3]* and Hatem Fawzi Gharour[4]

1 Faculty of Arts and Sciences, Department of Chemistry, Benghazi University, Tocra, Libya

2 Environmental and Biological Chemistry Research Centre (EBCRC), University of Benghazi, Tocra, Libya

3 Faculty of Sciences, Department of Microbiology University of Benghazi, Benghazi, Libya

4 Faculty of Medicine, Department of Biochemistry, University of Benghazi, Benghazi, Libya

*Address all correspondence to: yogaly6@yahoo.com

References

[1] Li P, Feng X, Qiu G, Shang L, Li G. Human hair mercury levels in the Wanshan mercury mining area, Guizhou Province, China. Environmental Geochemistry and Health. 2009;**31**(6): 683-691

[2] Zhang Z, Wang Q, Zheng D, Zheng N, Lu X. Mercury distribution and bioaccumulation up the soil-plant-grasshopper-spider food chain in Huludao City, China. Journal of Environmental Sciences. 2010;**22**(8): 1179-1183

[3] Rahman GMM, Fahrenholz T, Kingston HM. Application of speciated isotope dilution mass spectrometry to evaluate methods for efficiencies, recoveries, and quantification of mercury species transformations in human hair. Journal of Analytical Atomic Spectrometry. 2009;**24**(1):83-92

[4] Delgado A, Prieto A, Zuloaga O, de Diego A, Madariaga JM. Production of artifact methylmercury during the analysis of certified reference sediments: Use of ionic exchange in the sample treatment step to minimise the problem. Analytica Chimica Acta. 2007;**582**(1):109-115

[5] Kim SA, Jeon CK, Paek DM. Hair mercury concentrations of children and mothers in Korea: Implication for exposure and evaluation. Sci Total Environ. 2008;**402**(1):36-42

[6] Agusa T, Kunito T, Iwata H, Monirith I, Tana TS, Subramanian A, et al. Mercury contamination in human hair and fish from Cambodia: Levels, specific accumulation and risk assessment. Environmental Pollution. 2005;**134**(1):79-86

[7] Chen S, Chou S, Hwang D. Determination of methylmercury in fish using focused microwave digestion following by Cu2+ addition, sodium tetrapropylborate derivatization, n-heptane extraction, and gas chromatography–mass spectrometry. Journal of Chromatography A. 2004; **1024**(1–2):209-215

[8] Bancon-Montigny C, Yang L, Sturgeon RE, Colombini V, Mester Z. High-yield synthesis of milligram amounts of isotopically enriched methylmercury (CH3198HgCl). Applied Organometallic Chemistry. 2004;**18**(2):57-64

[9] Krupp EM, Donard OFX. Isotope ratios on transient signals with GC-MC-ICP-MS. International Journal of Mass Spectrometry. 2005;**242**(2-3): 233-242

[10] Mizanur Rahman GM, Skip Kingston HM, Bhandari S. Synthesis and characterization of isotopically enriched methylmercury (CH3201Hg+). Applied Organometallic Chemistry. 2003;**17**(12): 913-920

[11] Martín-Doimeadios RCR, Krupp E, Amouroux D, Donard OFX. Application of isotopically labeled methylmercury for isotope dilution analysis of biological samples using gas chromatography/ ICPMS. Analytical Chemistry. 2002; **74**(11):2505-2512

[12] Tseng CM, Amouroux D, Brindle ID, Donard OFX. Field cryofocussing hydride generation applied to the simultaneous multi-elemental determination of alkyl-metal(loid) species in natural waters using ICP-MS detection. Journal of Environmental Monitoring. 2000;**2**(6):603-612

[13] Gao Y, Shi Z, Long Z, Wu P, Zheng C, Hou X. Determination and speciation of mercury in environmental and biological samples by analytical

atomic spectrometry. Microchemical Journal. 2012;**103**(0):1-14

[14] Reyes LH, Mar JLG, Hernández-Ramírez A, Peralta-Hernández JM, Barbosa JMA, Kingston HMS. Microwave assisted extraction for mercury speciation analysis. Microchimica Acta. 2011;**172**(1):3-14

[15] Rodrigues JL, Alvarez CR, Fariñas NR, Berzas Nevado JJ, Barbosa F Jr, Rodriguez Martin-Doimeadios RC. Mercury speciation in whole blood by gas chromatography coupled to ICP-MS with a fast microwave-assisted sample preparation procedure. Journal of Analytical Atomic Spectrometry. 2011; **26**(2):436-442

[16] Clough R, Truscatt J, Belt ST, Hywel Evans E, Fairman B, Catterick T. Isotope dilution ICP-MS for speciation studies. Applied Spectroscopy Reviews. 2003; **38**(1):101-132

[17] Hill SJ, Pitts LJ, Fisher AS. High-performance liquid chromatography–isotope dilution inductively coupled plasma mass spectrometry for speciation studies: An overview. TrAC Trends in Analytical Chemistry. 2000;**19**(2–3): 120-126

[18] Rodríguez-González P, Marchante-Gayón JM, García Alonso JI, Sanz-Medel A. Isotope dilution analysis for elemental speciation: A tutorial review. Spectrochimica Acta Part B: Atomic Spectroscopy. 2005;**60**(2):151-207

[19] Qvarnström J, Frech W. Mercury species transformations during sample pre-treatment of biological tissues studied by HPLC-ICP-MS. Journal of Analytical Atomic Spectrometry. 2002; **17**(11):1486-1491

[20] Lambertsson L, Lundberg E, Nilsson M, Frech W. Applications of enriched stable isotope tracers in combination with isotope dilution GC-ICP-MS to study mercury species transformation in sea sediments during in situ ethylation and determination. Journal of Analytical Atomic Spectrometry. 2001;**16**(11):1296-1301

[21] Hintelmann H, Evans RD. Application of stable isotopes in environmental tracer studies—Measurement of monomethylmercury (CH3Hg+) by isotope dilution ICP-MS and detection of species transformation. Fresenius' Journal of Analytical Chemistry. 1997;**358**(3):378-385

[22] Yang L, Colombini V, Maxwell P, Mester Z, Sturgeon RE. Application of isotope dilution to the determination of methylmercury in fish tissue by solid-phase microextraction gas chromatography–mass spectrometry. Journal of Chromatography A. 2003; **1011**(1–2):135-142

[23] Horvat M, Bloom NS, Liang L. Comparison of distillation with other current isolation methods for the determination of methyl mercury compounds in low level environmental samples: Part 1 Sediments. Analytica Chimica Acta. 1993;**281**(1):135-152

[24] Horvat M, Liang L, Bloom NS. Comparison of distillation with other current isolation methods for the determination of methyl mercury compounds in low level environmental samples: Part II water. Analytica Chimica Acta. 1993;**282**(1):153-168

[25] Cai Y, Jaffé R, Jones R. Ethylmercury in the soils and sediments of the Florida everglades. Environmental Science & Technology. 1997;**31**(1):302-305

[26] Huang J. Artifact formation of methyl- and ethyl-mercury compounds from inorganic mercury during

derivatization using sodium tetra(n-propyl)borate. Analytica Chimica Acta. 2005;**532**(2):113-120

[27] Mao Y, Liu G, Meichel G, Cai Y, Jiang G. Simultaneous speciation of monomethylmercury and monoethylmercury by aqueous phenylation and purge-and-trap preconcentration followed by atomic spectrometry detection. Analytical Chemistry. 2008;**80**(18):7163-7168

[28] Demuth N, Heumann KG. Validation of methylmercury determinations in aquatic systems by alkyl derivatization methods for GC analysis using ICP-IDMS. Analytical Chemistry. 2001;**73**(16):4020-4027

[29] Bermejo-Barrera P, Verdura-Constenla EM, Moreda-Piñeiro A, Bermejo-Barrera A. Rapid acid leaching and slurry sampling procedures for the determination of methyl-mercury and total mercury in human hair by electrothermal atomic absorption spectrometry. Analytica Chimica Acta. 1999;**398**(2–3):263-272

[30] Gao Y, De Galan S, De Brauwere A, Baeyens W, Leermakers M. Mercury speciation in hair by headspace injection–gas chromatography–atomic fluorescence spectrometry (methylmercury) and combustion-atomic absorption spectrometry (total Hg). Talanta. 2010;**82**(5):1919-1923

Chapter 32

Recent Advances in Evaluating Insects as Bioindicators of Heavy Metal Pollution

Iram Liaqat, Noor Virk and Nazish Mazhar Ali

Abstract

Natural ecosystems are adversely affected by man-made interventions. Among living organisms, insects are regarded as susceptible to environment disruption as delicate body confirms the presence or absence of polluted environment thus found as suitable indicators of the aquatic and terrestrial ecosystem. Insects are being considered indicators of environmental pollution because different taxa of different localities provide robust information, provide a comparison of various communities, and quantitative data associated with indicators etc. Most of them present the quick reliable influence to heavy metal accumulation as pronounced disruptions were observed at molecular and biochemical level hence considered as best opted indicators of environmental pollution.

Keywords: heavy metals, bioindicators, insects, environmental pollution, bioaccumulation factor

1. Introduction

Environmental pollution by heavy metal is a wide ranging problem due to the urbanization and industrial development. For recent years, scientific world had shown the great interest in heavy metals as they had detrimental effect on atmosphere, aquatic and terrestrial environment [1]. Heavy metals are important class of pollutant with mass density greater than 4 g/cm^3 or more such as, lead (Pb), chromium (Cr), cadmium (Cd), mercury (Hg), copper (Cu), zinc (Zn), arsenic (As) nickel (Ni) and manganese (Mn) [2]. These are discharged in the surroundings as an outcome of industrial activities, agricultural practices and atmospheric deposition.

Metals have important physiological and biochemical role in organism, symbolized as carcinogenic and toxic when concentration increased the required level as enlisted in **Table 1**. For instance, iron, copper and zinc have essential role in maintaining the structure of organism but show acute toxicity when exceed the concentration as in certain genetic disorders. Some heavy metals are not essential even in small concentration such as arsenic, mercury and lead become carcinogenic. For example, ingestion of arsenic, generally occur as arsenate, may cause cancer of various part of human body [10]. Lead and mercury are responsible for the autoimmunity thus

Metal	Symbol	Effect on insects	Reference
Lead	Pb	Metamorphosis. Reduce environmental fitness	[3] [4]
Cadmium	Cd	Inhibit developmental period of insect	[5]
Chromium	Cr	Ovipositional response and developmental effect. Larval duration and pupation rate.	[6] [7]
Arsenic	As	Reduce Population. Reduce environmental fitness	[4]
Mercury	Hg	Oxidative stress in insects.	[8]
Zinc	Zn	Metamorphosis	[3]
Manganese	Mn	Foraging activity of bees. Contractibility and excitability of visceral muscles	[9]

Table 1.
Effect of heavy metal on insects.

leading to disease state of nervous system, renal, circulatory system and arthritis [11]. Cadmium exposure might cause kidney issues [12]. Exposure to mercury also causes various psychological and neurological damage such as panic, irregular sleep wake cycle and restlessness [12].

According to the report of WHO (World Health Organization), a million of people died every year in developing countries due to pollution induced diseases [13, 14]. The agricultural soil is contaminated by metal pollutant as a result of industrial product used in modern farming including insecticides, fertilizers, herbicide and sewage [15, 16]. Meanwhile, the pollutants are uptake to plant tissue from roots in soil therefore contaminated crop consumption becoming hazardous to living organism [17]. Pollutants toxicity has impacted the biological processes

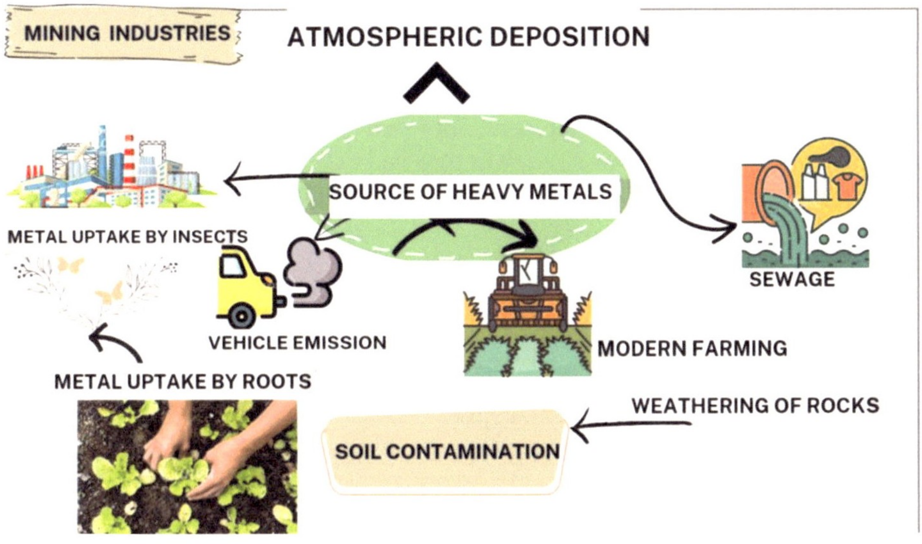

Figure 1.
Various means of heavy metal pollution and their transfer to insects.

as well as biotic interaction among living organism such as ecological structure, parasitism and predator prey relationship [18] as shown in **Figure 1**.

2. Bioindicator

Bioindicators are described as living entities such as planktons, microbes, animals and plants that are operated for screening the environmental health of ecosystem. They have immense potential of qualitative analysis of health of environment and biogeographical variation in their surrounding [19]. The balance of intracellular and extracellular caused the membrane permeability to alter as ions transverse the cell membrane [20]. Thus, insects are considered as potent heavy metal bioindicator of environmental pollution.

2.1 Classification of bioindicator

Mac Farlane *et al.* [21] described the classification of bioindicators on the basis of mode of action and origin of organism (**Figure 2**) [21].

2.1.1 On the basis of mode of action

Accumulative bioindicators: A type of bioindicators of environment that accumulate single or more element and chemical compound are called as accumulative bioindicators.

Sensitive bioindicators: A type of bioindicators that represent certain modification due to chemicals or elements exposure are called as sensitive bioindicators. The modifications may be related to tissue, morphology, cytology and organism or population behavior.

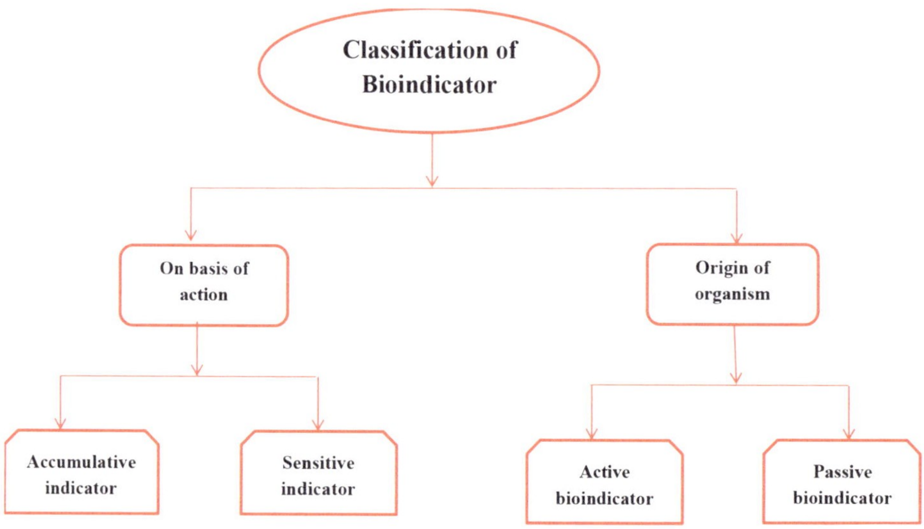

Figure 2.
Schematic illustration of classification of bioindicators.

2.1.2 On the basis of origin of organism

It is classified as active and passive bioindicators.

Active bioindicators: A type of bioindicators that are exposed at particular place for specific time interval thus used to examine the concentration of compounds and elements are called as active bioindicators.

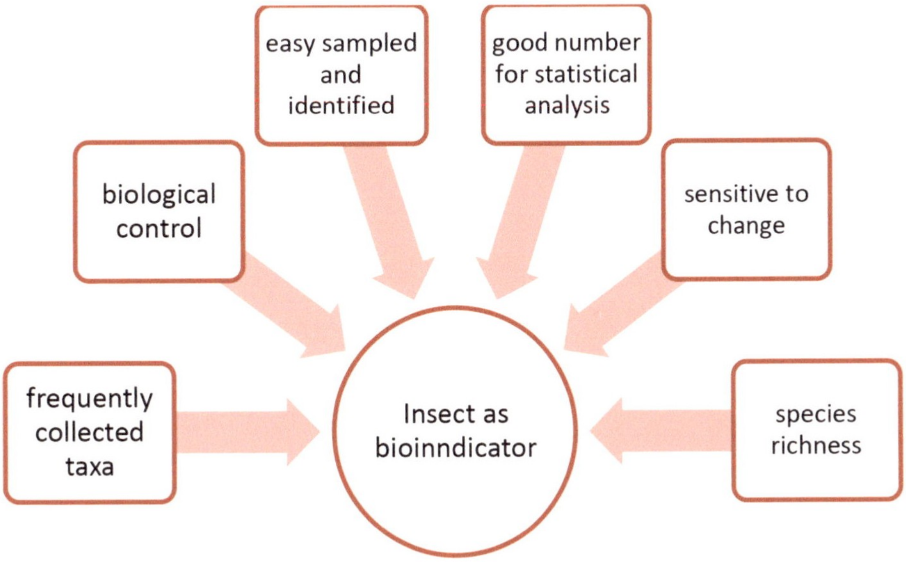

Figure 3.
Characteristics of insect as bioindicator.

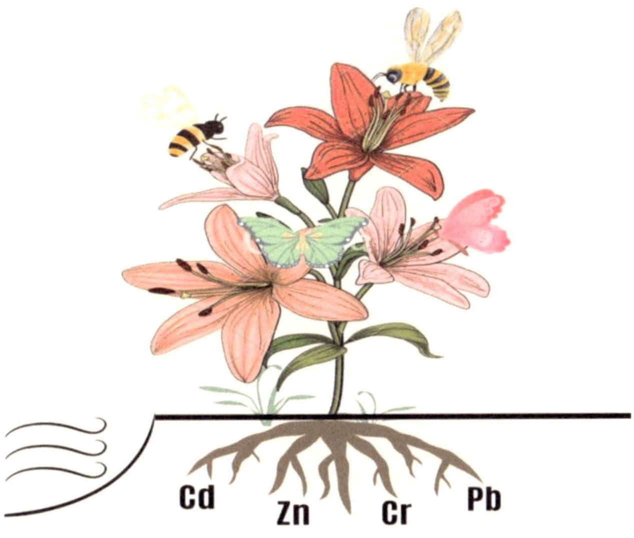

Figure 4.
Transfer of heavy metals to insects through contaminated soil.

Passive bioindicators: A type of bioindicators collected from their natural ecological community used to analyze the concentration of compounds and elements and their direct and indirect effect are called as passive bioindicators.

2.2 Insect as bioindicator

Insect is utilized as an efficient bioindicator of heavy metal pollution because of their diverse richness of species, easy handling and traps are good enough for effective statistical analysis. They are generally collected for their role as predator and significant for biological control such as spider and beetles (**Figure 3**). Heavy metals have negative influence on insect impacting their fecundity, weight, mortality and developmental stages [22, 23]. Furthermore, insects enable the selection of behavioral and demographic factor with strong association to the predefined abiotic factor that can be observed and quantified in particular environment [24]. Insects can be impacted directly by various means as associated with polluted soil and air deposition (**Figure 4**). Parasites and predator are also affected if they consume the insect that have greater amount of the heavy metals [25].

3. Bioaccumulation in insects

Metals are non-degradable as compared to other contaminant elements thus show bioaccumulation in the trophic chain [26] as **Figure 4** displaying bioaccumulation of heavy metals in insects. Insects are potent source of bioindicators to determine the toxicity of heavy metals and human activities in terrestrial ecosystem as they are closely associated with the sediments [27].

Bioaccumulation of the pollutant is also influenced with the feeding behavior of insects as herbivorous and omnivorous usually have lower heavy metal concentration than predatory and carnivorous insects [28]. Corbi *et al.* [29] had emphasized the

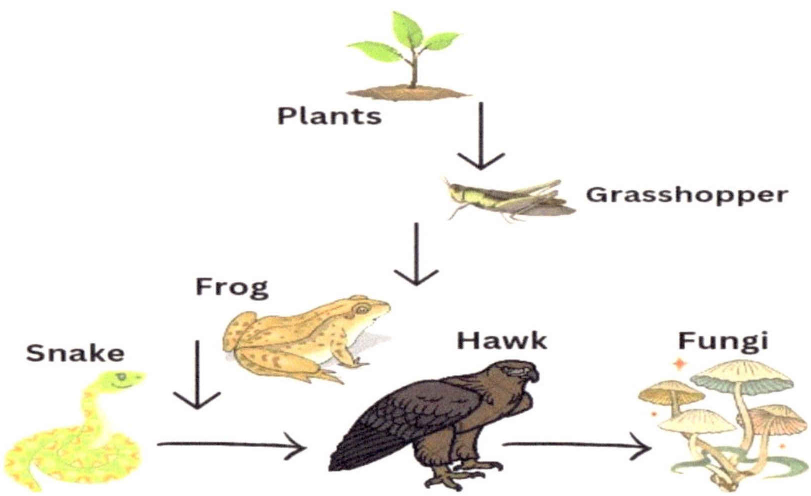

Figure 5.
Schematic presentation of bioaccumulation of heavy metals via food chain.

contamination of aquatic ecosystem [29] as a consequence of utilization of different heavy metal accumulation including lead, cadmium, chromium, zinc, nickel [30] as fertilizers for agricultural practices that had major impact on the water supply of surrounding cultivated areas. Meanwhile, the lack of riparian vegetation is another major cause of bioaccumulation of toxic metals from the surrounding cultivation [30, 31]. Owing to the significance and placement in food chain, heavy metal accumulations are detrimental to global health. Metal accumulation of aquatic invertebrates and sediment deposition provide a potential link to the upper trophic level of the food chain [32] as illustrated in **Figure 5**.

3.1 Bioaccumulation factor

The bioaccumulation factor for aquatic species was tested [33]. The ratio between metal concentrations in organism to that of environment is referred as bioaccumulation factor. When metal concentrations in organism exceed those in aquatic environment this is referred as bioaccumulation. Thus, the bioaccumulation is recognized for value lower than 1.

$$\text{Bioaccumulation factor} = \frac{\text{metal concentration in organism}}{\text{metal concentration in sediment}}$$

Major insects used as bioindicator of heavy metal pollution are

- Honey bee
- Beetles
- Grasshopper
- Termite
- Butterfly
- Dragonfly
- Ant
- Housefly
- Parasitic wasp

4. Honey bee as bioindicator

Honey bee has been studied extensively as bioindicator of metal pollution [34] due to their diverse foraging activity. Di Fiore *et al*. (2022) had suggested the utilization of honey bees as bioindicator of metal pollution to assess the air qualities of different region [35]. Various characteristics of honey bee including sensitivity to the pollutant elements, extensive flying capacity, elevated rate of reproduction and utilization of product (honey) make them as bioindicator species. They are the pollinated insects that presented efficient monitoring protocols at low cost [36].

5. Beetles as bioindicator

Ground beetles as efficient bioindicator are commonly used by researchers because they show response to ecological variation as a consequence of anthropogenic activities including overgrazing, soil and land pollution [37]. In relation to their cosmopolitan distribution in land, Carabid beetles are commonly utilized to assess the heavy metal pollution in soil [38]. Previously, it was discovered the significant bioaccumulation factor range of mercury and arsenic in a research of *Carabus lefebvri* indicating that beetles were favorable for assessing mercury and arsenic in the environment [39].

6. Ant as bioindicator

Owing to the species richness and vast diversity, ants are employed as bioindicators as these characteristics correspond to efficient monitoring of biological processes and landscape disturbance in various environments [40, 41]. Ants are effective for monitoring the foliage inhabiting regions, open habitat and toxicity of heavy metal pollution in terrestrial environment as their substantial role at ground level [42]. In mining region, ants have been utilized as bioindicators of restoration effectiveness [43]. For the research of heavy metal environmental impact, combine monitoring of ants and forager bees is a reliable method [44].

7. Grasshopper as bioindicator

Grasshoppers are herbivorous insects, therefore, have significant part in bioaccumulation as well as transport heavy metal to upper trophic levels via food chain [45] as predators such as mantid accumulates grasshopper thus transmitted heavy metals to further species in higher trophic level of food chain. Furthermore, Soliman and El-Shazly (2017) studied the average concentration of cadmium, lead, zinc, chromium, nickel and iron in grasshopper species [46]. Several researches have been conducted on heavy metal analysis in grasshopper, plants and soils having metal concentration higher in region closest to polluted areas [38, 47]. Guria *et al.* [4] described the reduction of grasshopper fitness to environment due to toxicity of arsenic and lead thus vulnerable to predation and infection [4].

8. Termite as bioindicator

Termites are detritivorus insects that feed on debris of plant and about 75% feed on soil [48]. These social insects are used as bioindicatiors of land fertility and provide basis of nutrient recycling, nitrogen fixation and transport of soil material. Likewise, expansion of carbon content, nutrient and clay they are also considered as ecological engineer. Aljama *et al.* (2019) assessed accumulation as regard to various metals (mercury, lead, chromium, cadmium, zinc and copper) in termites along with associated soil [49].

9. Butterfly as bioindicator

Butterflies are prominent bioindicators of ecological and metal pollution due to their easy recognition, conspicuous nature and capacity of accumulating heavy metal from surroundings. Kobiela and Snell- Rood [50] indicated the presence of transgenerational effect of nickel contamination in butterfly [50]. Azam *et al.* [51] had assessed the various heavy metals (Ni, Cu, Zn, Cr) accumulation in butterfly *Danaus chrysippus* near industrial areas of Gujrat and concluded as good indicator of metal pollution [51].

10. Dragonfly as bioindicator

Ecologically dragonfly is considered as useful bioindicator of aquatic and terrestrial environment. Ninety percent of invertebrate fauna comprises of aquatic insects representing the lotic and lentic food web, controlling nutrient cycle and energy flow [52]. Dragonflies are considered as the most susceptible to habitat disruption among aquatic insects. Noor *et al.* [53] demonstrated iron as suitable indicator for the lead, iron and zinc where iron showed positive association of species richness with sediments whereas negative relation when iron detection in water [53].

11. Housefly as bioindicator

Housefly (*Musca domestica*) is a philanthropic species and thus shows intimate association with humans and their surroundings. The housefly is a cosmopolitan insect, and a significant human and animal sanitary pest. It is the mechanical carrier of over 100 pathogens that include the antibiotic resistant ones [54]. Heavy metals like zinc, cadmium, copper, lead usually accumulate in the abdominal tissues of housefly [55]. The digestive tract injury had been reported due to accumulation of cadmium in larvae of housefly [56] however minimum concentration of cadmium has less effect on development and growth of housefly. Furthermore, the effect of cadmium on metamorphosis had been stated that showed dramatic variation as cadmium level gradually increased during larval phase but reduced dramatically following pupariation.

12. Parasitic wasp as bioindicator

Parasitic wasp had been devoted as bioindicator of woodland environment [57] due to certain biological characteristics and feeding habits [58]. Their habitat conditions are specified and intricated as they appeared higher in tropic level, restricted host dimensions and complex behavior [59]. Aguiar *et al.* [60] reported 103,000 species of hymenopteran including 70% as parasitic wasp operating as pest of agriculture [60]. Lead concentration was found in fecal mass of wasp larvae as stated [61].

13. Analytical procedure of heavy metal assessment in insects

These are the following analytical procedure of heavy metal assessment in insects reported by different authors.

13.1 Atomic absorption spectrophotometer

13.1.1 Graphite furnace and acetylene flame

Azam *et al.* [51] had demonstrated the chemical procedure of determining the heavy metal in different group of insects [51]. Insects were weighed after manually dried in oven and digested in solution of four ratios of percholric acid to one ratio of supra pure nitric acid. A PU 93090X graphite furnace and acetylene flame of atomic absorption spectrophotometry was utilized to assess various metal concentrations of insects.

13.1.2 Electrothermal atomic absorption spectroscopy

Using a microwave, 1.5 mL of nitric acid (2.5 percent) and 430 micro liter of hydrogen peroxide (30 percent) and 570 micro liter ultrapure water were applied to honey samples of honey bee atleast 0.6 g for determination of toxicity analysis as stated [44]. Following the digestion, the solution was mixed with one percent of Triton X 100 to reduce the viscosity before transferring it to the 25 mL flask. For the quantification of cadmium and lead, samples diluted with ultrapure water to 25 mL for chromium or 1 percent nitric acid and 3.30 micro liter of magnesium nitrate. The heavy metal analysis was regulated by the electrothermal atomic absorption spectroscopy using triplicate graphite tubes for various matrixes.

13.2 Plasma atomic emission spectrometry.

13.2.1 Nitric acid and hydrogen peroxide

Corbi *et al.* [62] mentioned the analytical method to determine the insect as heavy metal bioindicator by utilizing the plasma atomic emission spectrometry [62]. Deionized double distilled was used in this method. Aquatic insects that had been frozen were defrosted at room temperature and concerted to get 0.20 g dry weight. Insects placed in the 100 mL beaker having 5 mL nitric acid that were processed at ninety degree centigrade on hotplate until completely dried, for complete digestion I mL of hydrogen peroxide was also added. Processed samples were normalized at room temp. While filtered through filter paper in 50 mL flask. Standards run along with samples were analyzed for heavy metal detection using plasma atomic emission spectrometry.

13.2.2 Nitric acid and hydrochloric acid

Alajmi *et al.* [63] stated another method of heavy metal detection in insects by using plasma atomic emission spectrometer [63]. Aphid, leaves and soil samples were oven dried for one hour at 105 degrees, digested for six hours at 550 degrees in a furnace with use of 3 mL of hydrochloric acid then diluted with the deionized water. The targeted heavy metals dried weight was estimated as mg/g and quantified utilizing plasma atomic emission spectrometer.

13.3 Plasma optical emission spectroscopy

Alajmi *et al.* [49] illustrated another method for the quantification of heavy metal concentration in termites and associated soil sample using plasma Optical Emission

Spectroscopy [49]. Samples were oven dried for six to twelve hours at 105 degrees centigrade and measured utilizing microbalance followed by digestion of 0.5 g in 5 mL of nitric acid for six to twelve hours at 105 degree centigrade then added 2 mL of perchloric acid and volume adjusted by deionized water to 10 mL.

14. Molecular analysis of heavy metals on insects

El- Samad *et al.* [64] analyzed the molecular approach of heavy metal effect in beetles [64]. According to the dispersive energy X-ray microanalysis, testicular tissues and soil samples gathered from the metal contaminated area contained higher level of heavy metal as compared to the beetles of reference area. The transcription level of heat shock protein as well as seminal fluid proteins of accessary gland (AcPC01) of testicular were measured to investigate the genotoxicity in beetle sample taken from polluted area. Heat shock proteins 90, 70 and 60 gene expressions had been significantly less 1.5 fold in samples taken from polluted area but Hsp 60, Hsp 70 and HSP 90 expressed more as greater than 2 fold. Researcher also noted the occurrence of micronuclei development in testicular cells. The prevalence of was noticeably higher in samples gathered from contaminated area and also extracellular matrix aberration including nuclear and cytoplasmic disruption was also found.

15. Enzymatic biomarkers in aquatic insects

Aquatic invertebrates that accumulate the heavy metals were subjected to oxidative stress that leads to activation of the different antioxidant enzymes including acetylcholinesterase and glutathione S transferase [65]. Acetylcholinesterase, a neurotransmitter, had been observed as biomarker for detection of pollutant exposure in invertebrates [66]. Many antioxidant factors such as glutathione, catalase, lipid peroxidase and glutathione transferase were utilized as biomarkers in various living organism [67] as enzymes work with antioxidants for removal of free radical. Moreover some so-called biomarkers such as heat shock proteins and metallothioneins increase in concentration when exposed to heavy metals [68].

16. Comet assay

DNA damage of insects living in contaminated environment has been evaluated by the comet assay [69]. Using the comet assay, earlier investigation regarding DNA damage due to environmental contaminant can be established as DNA damage shows the molecular abnormalities that can cause pathogenic adversity. The environmental potential for genotoxicity in terrestrial insects can be measured by the comet assay, therefore, it aids in providing the early management of xenobiotic hazards which is evaluated by this assay [70, 71]. Kheirallah *et al.* [72] had reported the beetles as biomonitoring model to evaluate the DNA damage in a contaminated terrestrial environment [72]. Over recent years, chromosomal research had been of great importance due to increased demand to assess the genotoxicity of aquatic of environmental toxicity [73]. Various organisms at the end of trophic level of food chain are at severe risk of toxicology because of genotoxin presence in aquatic environment [74] as exposure causes DNA damage.

17. Heavy metals and immunological response

Various studies had reported effect of heavy metal pollution on immunocompetent cells. Dynamics of environment, administrative method and concentration of heavy metals are indicators with respect to heavy metals toxicity on these cells [75]. Borowska and Pyza [55] examined the activity of heavy metals like cadmium, lead, copper zinc on immune response of *M. domestica* by change in number and morphology of hemocytes such as granulocytes lessened whereas stem cells increased showing the effect on immune cell concentration [55].

18. Conclusion

Environmental pollution by heavy metals is the global hazard due to increased industrialization, urbanization as well as anthropogenetic activities that show detrimental effect on ecosystem. The analysis has shown that heavy metals accumulation along the food chain increased the bioaccumulation factor with increase in trophic level. However, it is not directly transferred from plants but need insects for the transfer to higher trophic level. The biochemical, physiological and behavioral alteration in insects confirms the accumulation of heavy metals activity thus they act as good bioindicators of metal pollution to demonstrate the ecological impact of metals. Recent advances at biochemical and molecular levels are being done to determine the toxicity of heavy metals in insects. Steps should be taken to lessen the metal pollution and enhance the survival of insects with accumulation at the cost of additional energy.

Conflict of interest

There is no conflict of interests as declared by the authors.

Author details

Iram Liaqat*, Noor Virk and Nazish Mazhar Ali
Department of Zoology, Government College University, Lahore, Pakistan

*Address all correspondence to: dr.iramliaqat@gcu.edu.pk

References

[1] Cunningham MM, Tran L, McKee CG, Polo RO, Newman T, Lansing L, et al. Honey bees as biomonitors of environmental contaminants, pathogens, and climate change. Ecological Indicators. 2022;**134**:108457. DOI: 10.1016/j.ecolind.2021.108457

[2] Hawkes SJ. What is a "heavy metal"? Journal of Chemical Education. 1997;**74**(11):1374. DOI: 10.1021/ed074p1374

[3] Lidman J, Jonsson M, Berglund ÅM. The effect of lead (Pb) and zinc (Zn) contamination on aquatic insect community composition and metamorphosis. Science of the Total Environment. 2020;**734**:139406. DOI: 10.1016/j.scitotenv.2020.139406

[4] Guria S, Parveen S, Goswami DS. Alteration of morphology, phagocytic behaviour and aggregation of insect haemocytes exposed to contaminated food with arsenic and lead. International Journal of PharmTech Research. 2016;**9**(5):177-186

[5] Luo M, Cao HM, Fan YY, Zhou XC, Chen JX, Chung H, et al. Bioaccumulation of cadmium affects development, mating behavior, and fecundity in the Asian Corn Borer. Ostrinia furnacalis. Insects. 2019;**11**(1):7. DOI: 10.3390/insects11010007

[6] Trumble JT, Jensen PD. Ovipositional response, developmental effects and toxicity of hexavalent chromium to *Megaselia scalaris*, a terrestrial detritivore. Archives of Environmental Contamination and Toxicology. 2004;**46**(3):372-376. DOI: 10.1007/s00244-003-3007-8

[7] Gao Q, Wang X, Wang W, Lei C, Zhu F. Influences of chromium and cadmium on the development of black soldier fly larvae. Environmental Science and Pollution Research. 2017;**24**(9):8637-8644. DOI: 10.1007/s11356-017-8550-3

[8] Zaman K, MacGill RS, Johnson JE, Ahmad S, Pardini RS. An insect model for assessing mercury toxicity: Effect of mercury on antioxidant enzyme activities of the housefly (*Musca domestica*) and the cabbage looper moth (*Trichoplusia ni*). Archives of Environmental Contamination and Toxicology. 1994;**26**(1):114-118. DOI: 10.1007/BF00212802

[9] Ben-Shahar Y. The impact of environmental Mn exposure on insect biology. Frontiers in Genetics. 2018;**9**:70. DOI: 10.3389/fgene.2018.00070

[10] Ng JC, Wang J, Shraim A. A global health problem caused by arsenic from natural sources. Chemosphere. 2003;**52**(9):1353-1359. DOI: 10.1016/S0045-6535(03)00470-3

[11] Barakat MA. New trends in removing heavy metals from industrial wastewater. Arabian Journal of Chemistry. 2011;**4**(4):361-377. DOI: 10.1016/j.arabjc.2010.07.019

[12] Järup L. Hazards of heavy metal contamination. British Medical Bulletin. 2003;**68**(1):167-182. DOI: 10.1093/bmb/ldg032

[13] World Health Organization. 7 million Premature Deaths Annually Linked to Air Pollution. Geneva, Switzerland: World Health Organization; 2014

[14] World Health Organization. WHO guidelines for indoor air quality: Household fuel combustion. Geneva, Switzerland: World Health Organization; 2014

[15] Zarić NM, Brodschneider R, Goessler W. Honey bees as biomonitors–Variability in the elemental composition of individual bees. Environmental Research. 2022;**204**:112237. DOI: 10.1016/j.envres.2021.112237

[16] Gall JE, Boyd RS, Rajakaruna N. Transfer of heavy metals through terrestrial food webs: A review. Environmental Monitoring and Assessment. 2015;**187**(4):1-21. DOI: 10.1007/s10661-015-4436-3

[17] Zhuang P, McBride MB, Xia H, Li N, Li Z. Health risk from heavy metals via consumption of food crops in the vicinity of *Dabaoshan mine*, South China. The Science of the Total Environment. 2009;**407**(5):1551-1561. DOI: 10.1016/j.scitotenv.2008.10.061

[18] Walker CH, Sibly RM, Peakall DB. Principles of Ecotoxicology. Boca Raton: CRC Press; 2005. DOI: 10.1201/9781439894224

[19] Parmar TK, Rawtani D, Agrawal YK. Bioindicators: The natural indicator of environmental pollution. Frontiers in Life Science. 2016;**9**(2):110-118. DOI: 10.1080/21553769.2016.1162753

[20] Armstrong N, Ramamoorthy M, Lyon D, Jones K, Duttaroy A. Mechanism of silver nanoparticles action on insect pigmentation reveals intervention of copper homeostasis. PLoS One. 2013;**8**(1):e53186. DOI: 10.1371/journal.pone.0053186

[21] MacFarlane GR, Pulkownik A, Burchett MD. Accumulation and distribution of heavy metals in the grey mangrove, *Avicennia marina* (Forsk.) Vierh.: Biological indication potential. Environmental Pollution. 2003;**123**(1):139-151. DOI: 10.1016/S0269-7491(02)00342-1

[22] Wu F, Zhang G, Yu J. Interaction of cadmium and four microelements for uptake and translocation in different barley genotypes. Communications in Soil Science and Plant Analysis. 2003;**34**(13-14):2003-2020. DOI: 10.1081/CSS-120023233

[23] Hayford BL, Ferrington LC. Biological assessment of Cannon Creek, Missouri by use of emerging chironomidae (Insecta: Diptera). Journal of the Kansas Entomological Society. 2005;**78**(2):89-99. DOI: 10.2317/0405.20.1

[24] da Rocha JR, de Almeida JR, Lins GA, Durval A. Insects as indicators of environmental changing and pollution: A review of appropriate species and their monitoring. Holos Environment. 2010;**10**(2):250-262. DOI: 10.14295/holos.v10i2.2996

[25] Vickerman DB, Trumble JT. Biotransfer of selenium: Effects on an insect predator, *Podisus maculiventris*. Ecotoxicology. 2003;**12**(6):497-504. DOI: 10.1023/B:ECTX.0000003036.81351.31

[26] Kontas A. Trace metals (cu, Mn, Ni, Zn, Fe) contamination in marine sediment and zooplankton samples from Izmir Bay.(Aegean Sea, Turkey). Water, Air, and Soil Pollution. 2008;**188**(1):323-333. DOI: 10.1007/s11270-007-9547-1

[27] Ghannem S, Touaylia S, Boumaiza M. Beetles (Insecta: Coleoptera) as bioindicators of the assessment of environmental pollution. Human and Ecological Risk Assessment: An International Journal. 2018;**24**(2):456-464. DOI: 10.1080/10807039.2017.1385387

[28] Ali H, Khan E. Trophic transfer, bioaccumulation, and biomagnification of non-essential hazardous heavy metals and metalloids in food chains/webs—Concepts and implications for wildlife and human health. Human and Ecological

Risk Assessment: An International Journal. 2019;**25**(6):1353-1376. DOI: 10.1080/10807039.2018.1469398

[29] Corbi JJ, Costa CG, Gorni GR, Colombo V, Rios L. Environmental diagnosis of metals in streams near sugarcane cultivation areas: Current and historical analysis in the central region of the State of São Paulo. An. Acad. Brasil. Ciênc. 2018;**90**:2711-2719. DOI: 10.1590/0001-3765201820170808

[30] Angelotti-Netto A, Crestana S, De Oliveira SC, Barbosa RV. Metais pesados provenientes de atividade agrícola: formas, prevenção e controle. Bacia Hidrográfica. 2004:1-4

[31] Primavesi O, Freitas AR, Primavesi AC, Oliveira HT. Water quality of the Canchim's Creek watershed in São Carlos, SP, Brazil, occupied by beef and dairy cattle activities. Brazilian Archives of Biology and Technology. 2002;**45**:209-217. DOI: 10.1590/S1516-89132002000200013

[32] Duran M, Kara Y, Akyildiz GK, Ozdemir A. Antimony and heavy metals accumulation in some macroinvertebrates in the Yesilirmak River (N Turkey) near the Sb-mining area. Bulletin of Environmental Contamination and Toxicology. 2007;**78**(5):395-399. DOI: 10.1007/s00128-007-9183-x

[33] Souto RD, Corbi JJ, Jacobucci GB. Aquatic insects as bioindicators of heavy metals in sediments in Cerrado streams. Limnetica. 2019;**38**(2):575-586. DOI: 10.23818/limn.38.33

[34] Bogdanov S. Contaminants of bee products. Apidologie. 2006;**37**(1):1-8. DOI: 10.1051/apido:2005043

[35] Di Fiore C, Nuzzo A, Torino V, De Cristofaro A, Notardonato I, Passarella S, et al. Honeybees as bioindicators of heavy metal pollution in urban and rural areas in the South of Italy. Atmosphere. 2022;**13**(4):624. DOI: 10.3390/atmos13040624

[36] Quigley TP, Amdam GV, Harwood GH. Honey bees as bioindicators of changing global agricultural landscapes. Current Opinion in Insect Science. 2019;**35**:132-137. DOI: 10.1016/j.cois.2019.08.012

[37] Avgın SS, Luff ML. Ground beetles (Coleoptera: Carabidae) as bioindicators of human impact. Munis Entomology and Zoology. 2010;**5**(1):209-215

[38] Conti E. Ecotoxicological evaluation of *Parallelomorphus laevigatus* (Coleoptera, Carabidae) as a useful bioindicator of soil metal pollution. Advanced Technology in Biology Medicine. 2017;**5**:1-3. DOI: 10.4172/2379-1764.1000228

[39] Talarico F, Brandmayr P, Giulianini PG, Ietto F, Naccarato A, Perrotta E, et al. Effects of metal pollution on survival and physiological responses in *Carabus* (*Chaetocarabus*) *lefebvrei* (Coleoptera, Carabidae). European Journal of Soil Biology. 2014;**61**:80-89. DOI: 10.1016/j.ejsobi.2014.02.003

[40] Löffler F, Fartmann T. Effects of landscape and habitat quality on Orthoptera assemblages of pre-alpine calcareous grasslands. Agriculture, Ecosystems and Environment. 2017;**248**:71-81. DOI: 10.1016/j.agee.2017.07.029

[41] Dziock F. Life-history data in bioindication procedures, using the example of hoverflies (Diptera, Syrphidae) in the Elbe floodplain. International Review of Hydrobiology. 2006;**91**(4):341-363. DOI: 10.1002/iroh.200510889

[42] Gerlach J, Samways M, Pryke J. Terrestrial invertebrates as bioindicators: An overview of available taxonomic groups. Journal of Insect Conservation. 2013;**17**(4):831-850. DOI: 10.1007/s10841-013-9565-9

[43] Andersen AN, Majer JD. Ants show the way down under: Invertebrates as bioindicators in land management. Frontiers in Ecology and the Environment. 2004;**2**(6):291-298. DOI: 10.1890/1540-9295(2004)002[0292:ASTWDU]2.0.CO;2

[44] Satta A, Verdinelli M, Ruiu L, Buffa F, Salis S, Sassu A, et al. Combination of beehive matrices analysis and ant biodiversity to study heavy metal pollution impact in a post-mining area (Sardinia, Italy). Environmental Science and Pollution Research. 2012;**19**(9):3977-3988. DOI: 10.1007/s11356-012-0921-1

[45] Zhang ZS, Lu XG, Wang QC, Zheng DM. Mercury, cadmium and lead biogeochemistry in the soil–plant–insect system in Huludao City. Bulletin of Environmental Contamination and Toxicology. 2009;**83**(2):255-259. DOI: 10.1007/s00128-009-9688-6

[46] Soliman MM, El-Shazly MM. Bioaccumulation of heavy metals by grasshoppers and a mantid along a pollution gradient. Ecologia balkanica. 2017;**9**(2):7-21

[47] Zhuang P, Huiling ZO, Wensheng SH. Biotransfer of heavy metals along a soil-plant-insect-chicken food chain: Field study. Journal of Environmental Sciences. 2009;**21**(6):849-853. DOI: 10.1016/S1001-0742(08)62351-7

[48] Nithyatharani R, Kavitha US. Termite soil as bio-indicator of soil fertility. International Journal for Research in Applied Science and Engineering Technology. 2018;**6**(1):659-661. DOI: 10.22214/ijraset.2018.1099

[49] Alajmi R, Abdel-Gaber R, AlOtaibi N. Characterization of the 12S rRNA gene sequences of the harvester termite *Anacanthotermes ochraceus* (Blattodea: Hodotermitidae) and its role as A bioindicator of heavy metal accumulation risks in Saudi Arabia. Insects. 2019;**10**(2):51. DOI: 10.3390/insects10020051

[50] Kobiela ME, Snell-Rood EC. Nickel exposure has complex transgenerational effects in a butterfly. Integrative and Comparative Biology. 2018;**58**(5):1008-1017. DOI: 10.1093/icb/icy096

[51] Azam I, Afsheen S, Zia A, Javed M, Saeed R, Sarwar MK, et al. Evaluating insects as bioindicators of heavy metal contamination and accumulation near industrial area of Gujrat, Pakistan. BioMed Research International. 2015;**2015**. DOI: 10.1155/2015/942751

[52] Nicacio G, Juen L. Chironomids as indicators in freshwater ecosystems: An assessment of the literature. Insect Conservation and Diversity. 2015;**8**(5):393-403. DOI: 10.1111/icad.12123

[53] Noor-Ul-Islam H, Khan K, Zia SA, Naeem M, Shams WA. Heavy metals accumulation in dragonflies (Odonata) and their habitats in district Swabi, Khyber Pakhtunkhwa, Pakistan: Assessing dragonfly bionomics in the region. Bulletin of Environmental Contamination and Toxicology. 2021;**107**(5):838-847. DOI: 10.1007/s00128-021-03338-w

[54] Macovei L, Miles B, Zurek L. Potential of houseflies to contaminate ready-to-eat food with antibiotic-resistant enterococci. Journal of Food Protection. 2008;**71**(2):435-439. DOI: 10.4315/0362-028X-71.2.435

[55] Borowska J, Pyza E. Effects of heavy metals on insect immunocompetent cells. Journal of Insect Physiology. 2011;**57**(6):760-770. DOI: 10.1016/j.jinsphys.2011.02.012

[56] Chang-ying NI, Yond JI, Chao-liang LE, Cui HU. Effects of cadmium on housefly: Influence on growth and development and metabolism during metamorphosis of housefly. Insect Science. 2002;**9**(1):27-33. DOI: 10.1111/j.1744-7917.2002.tb00139.x

[57] Hilszczański J, Gibb H, Hjältén J, Atlegrim O, Johansson T, Pettersson RB, et al. Parasitoids (Hymenoptera, Ichneumonoidea) of saproxylic beetles are affected by forest successional stage and dead wood characteristics in boreal spruce forest. Biological Conservation. 2005;**126**(4):456-464. DOI: 10.1016/j.biocon.2005.06.026

[58] Skaldina O, Sorvari J. Biomarkers of ecotoxicological effects in social insects. In: Perspectives in Environmental Toxicology. Champions: Springer; 2017. pp. 203-214. DOI: 10.1007/978-3-319-46248-6_10

[59] Abdul M, Maeto K, Ishii T. Arthropods as bioindicators of sustainable forest management, with a focus on plantation forests. Applied Entomology and Zoology. 2009;**44**(1):1-11. DOI: 10.1303/aez.2009.1

[60] Aguiar AP, Deans AR, Engel MS, Forshage M, Huber JT, Jennings JT, et al. Order Hymenoptera. In: Zhang, Z.-Q. (Ed.) Animal biodiversity: An outline of higher-level classification and survey of taxonomic richness (Addenda 2013). Zootaxa. 2013;**3703**(1):51-62

[61] Urbini A, Sparvoli E, Turillazzi S. Social paper wasps as bioindicators: A preliminary research with *Polistes dominulus* (Hymenoptera Vespidae) as a trace metal accumulator. Chemosphere. 2006;**64**(5):697-703. DOI: 10.1016/j.chemosphere.2005.11.009

[62] Corbi JJ, Froehlich CG, Strixino ST, Santos AD. Bioaccumulation of metals in aquatic insects of streams located in areas with sugar cane cultivation. Química Nova. 2010;**33**:644-648. DOI: 10.1590/S0100-40422010000300030

[63] Alajmi R, Al-Shammari M, Abdel-Gaber R, Metwally D, El-Khadragy MF, Alrajeh A. Effectiveness of naturally occurring *Aphis gossypii* on tomato plants as a bio-indicator for heavy metals in Riyadh and Hafar Al-Batin, Saudi Arabia. Saudi Journal of Biological Sciences. 2021;**28**(7):4096-4101. DOI: 10.1016/j.sjbs.2021.04.027

[64] El-Samad LM, El-Ashram S, Kheirallah DA, Abdul-Aziz KK, Toto NA, Mokhamer EH. Relative gene expression, micronuclei formation, and ultrastructure alterations induced by heavy metal contamination in *Pimelia latreillei* (Coleoptera: Tenebrionidae) in an urban-industrial area of Alexandria. Egypt. Plos One. 2021;**16**(6):e0253238. DOI: 10.1371/journal.pone.0253238

[65] Gunning RV, Moores GD, Devonshire AL. Biochemical resistance detection in *Helicoverpa armigera* in Australia. Recent Resistance Development in Entomology. 1997;**1**:203-213

[66] Lavado R, Ureña R, Martin-Skilton R, Torreblanca A, Del Ramo J, Raldua D, et al. The combined use of chemical and biochemical markers to assess water quality along the Ebro River. Environmental Pollution. 2006;**139**(2):330-339. DOI: 10.1016/j.envpol.2005.05.003

[67] Li W, Kong AN. Molecular mechanisms of Nrf2-mediated

antioxidant response. Molecular Carcinogenesis: Published in cooperation with the University of Texas MD Anderson Cancer Center. 2009;**48**(2): 91-104. DOI: 10.1002/mc.20465

[68] Said Ali K, Ferencz Á, Nemcsók J, Hermesz E. Expressions of heat shock and metallothionein genes in the heart of common carp (*Cyprinus carpio*): Effects of temperature shock and heavy metal exposure. Acta Biologica Hungarica. 2010;**61**(1):10-23. DOI: 10.1556/abiol.61.2010.1.2

[69] de Lapuente J, Lourenço J, Mendo SA, Borràs M, Martins MG, Costa PM, et al. The Comet Assay and its applications in the field of ecotoxicology: A mature tool that continues to expand its perspectives. Frontiers in Genetics. 2015;**6**:180. DOI: 10.3389/fgene.2015.00180

[70] Abdelfattah EA, Augustyniak M, Yousef HA. Biomonitoring of genotoxicity of industrial fertilizer pollutants in *Aiolopus thalassinus* (Orthoptera: Acrididae) using alkaline comet assay. Chemosphere. 2017;**182**:762-770. DOI: 10.1016/j.chemosphere.2017.05.082

[71] Karpeta-Kaczmarek J, Kubok M, Dziewięcka M, Sawczyn T, Augustyniak M. The level of DNA damage in adult grasshoppers *Chorthippus biguttulus* (Orthoptera, Acrididae) following dimethoate exposure is dependent on the insects' habitat. Environmental Pollution. 2016;**215**:266-272. DOI: 10.1016/j.envpol.2016.05.032

[72] Kheirallah DA, El-Samad LM, Mokhamer EH, Abdul-Aziz KK, Toto NA. DNA damage and oogenesis anomalies in *Pimelia latreillei* (Coleoptera: Tenebrionidae) induced by heavy metals soil pollution. Toxicology and Industrial Health. 2019;**35**(11-12):688-702. DOI: 10.1177/0748233719893200

[73] Kushwaha B, Srivastava S, Singh B, Nagpure N, Ponniah AG. Evaluation of comet assay and micronuclei test as genotoxic assays in *Channa punctatus*. National Academy Science Letters. 2000;**23**(11-12):177-179

[74] Depledge MH, Fossi MC. The role of biomarkers in environmental assessment (2). Invertebrates. Ecotoxicology. 1994;**3**(3):161-172. DOI: 10.1007/BF00117081

[75] Olabarrieta I, L'azou B, Yuric S, Cambar J, Cajaraville MP. In vitro effects of cadmium on two different animal cell models. Toxicology in Vitro. 2001;**15**(4-5):511-517. DOI: 10.1016/S0887-2333(01)00056-X

Chapter 33

Electrochemical Detection of Heavy Metals Based on Nanostructured, or Film-Modified Paper Electrodes

Alberto Sánchez Calvo and María del Carmen Blanco Lopez

Abstract

Heavy metals have a huge impact on the environment due to their toxicity and bioaccumulation capacity. A great variety of methods can be used to the determination. The electrochemical methods are one of the best options due to their sensitivity, selectivity, and rapidness. They are based on the use of electrochemical cells made of different materials depending on the analyte to determine. An eco-friendly cheap option that has earned big importance is paper-based electrodes, which are formed by a cellulose matrix modified with conductive inks. Carbon ink is the most used, and it can be modified with nanoparticles to increase sensitivity. Alternatively, metallic surfaces or "films" such as mercury or bismuth can improve the determination of heavy metals because of their interactions with the film. This chapter focuses on the methods to determination of heavy metals based on their affinity with different nanomaterials or films on low-cost electrode substrates.

Keywords: paper-based, heavy metals, nanomaterials, films, bismuth

1. Introduction

Heavy metals group are a group of metals, semimetals, or even non-metals, which can be related because of their density or molecular weight [1, 2]. They all have in common high toxicity and nonbiodegradable nature. Each one of them has a wide range of uses (pesticides, paints, fertilizers, batteries, or industrial manufacturing). However, their toxicity and bioaccumulation capacity are too dangerous not only for the environment but also for human health. The heavy metals usually being analyzed to test new monitoring methods are mercury, lead, cadmium, and arsenic.

Mercury is possibly one of the most studied heavy metal. Its toxicity in elemental or inorganic form can affect dangerously human health, damaging brain, kidney, and lungs, and its bioaccumulation capacity in methylmercury form makes it mandatory to be controlled in the environment (soil, water). Europe legislation limits the use of mercury in products and the exportation of it to other countries for mercury and mercury compounds [3]. In the case of mercury in drinking waters, the

Environmental protection agency (EPA) in 1991 has set a maximum concentration of Hg^{2+} of 2 ppb, while the Commission Regulation (EU) 2022/617 has set limits of 0.50 mg/kg in fish [4].

Lead has an impact on human health, interfering with enzymes of the cells in the brain, kidneys, heart, acting as a carcinogen, and causing cardiovascular and brain damage. The European Commission Regulation (EU) 2021/1317 has set limits of 0.10 mg/kg in cases of bovine and pig meat. Moreover, lead can also develop neurotoxicity in young children, and consequently, a limit of 0.02 mg/kg has been set for raw and 0.01 mg/kg for infant milk [5].

Cadmium toxicity is related to lungs, liver, and kidneys, generating chronic poisoning because of its slow release from the body. Cadmium can be present in water, soil, and food with limits of 0.005 mg/L in water and values between 0.02 and 0.2 mg/kg for different types of foods [5, 6]. There are also regulations of cadmium in the case of fertilizers [7].

Arsenic in inorganic form is highly toxic affecting skin, lungs, and livers causing seizures, and in high levels coma and death. Limits for arsenic (inorganic) in rice have been settled to 0.20 mg/kg and 0.10 in the case of food for young children [8].

There are a huge variety of analytical techniques for the determination of heavy metals, such as atomic fluorescence spectrometry, inductively coupled plasma with mass spectrometry (ICP-MS), or electrochemical analysis. All of them have high sensitivity and selectivity. However, electrochemical methods have also advantages, such as portability to make *in situ* analysis, and miniaturization, needing less volume of reagents to work, becoming therefore a cheaper and sustainable alternative.

Electrochemical methods are based on the use of a working electrode, which is a conductive transducer in which the heavy metal can be oxidized or reduced. Mercury was the first material used in a dropping mercury electrode. It has a renewable electrode surface with a broad cathodic potential window and a good affinity for metals, such as lead or cadmium, which increases their sensitivity, but its high toxicity has triggered the search for safer and more eco-friendly alternatives. Nowadays, carbon is usually the most used material for electrochemical sensors [9] but there are other alternatives, such as gold, silver, or platinum. One of the main advantages of carbon is the possibility of easy modification with different types of materials or polymers, depending on the objectives [10, 11]. It can be modified with nanomaterials to increase their conductivity and sensitivity, or in other cases, with films of materials such as bismuth to increase the selectivity and sensitivity of the analyte based on a strong interaction with it.

Different techniques can be applied depending on the range of concentrations needed or the selectivity required. Potentiometry is considered highly selective because it relies on the use of ion-selective electrodes (ISEs), which are composed of a membrane with a high affinity for the analyte. The interaction of the membrane with the sample and an inner solution of the ion generates a potential. This can be measured and related to the concentration by Nikolsky equation (Eq. (1)).

$$E = L + \beta\left(0.05916/z_i\right)\log\left[a_i + \sum_j\left(k_{i.j}^{pot}.a_j^{zi/zj}\right)\right] \tag{1}$$

where E is the potential, L and β are constants, a_i is the activity of the ion to determinate and a_j is the activity of an interference; z_i is the charge of the ion, and $k^{pot}{}_{i \cdot j}$ is the potentiometric selective coefficient for the ion analyte and the interference.

Anodic stripping voltammetry is also used in the determination of heavy metals. With this technique, a preconcentration step is done previously to an oxidation step, in order to increase the sensitivity [12]. The stripping voltammetry can be done in linear sweep voltammetry (LSV), cyclic voltammetry (CV), differential pulse voltammetry (DPV), and square wave voltammetry (SWV). Different equations are used depending on the technique. For example, in lineal and cyclic voltammetry, a lineal potential sweep is applied to the electrode, and Randles-Ševčík equation (Eq. (2)) is used, where A is the electroactive area, C is the concentration of the analyte, v is the scan rate, n is the number of electrons, and D is the diffusion coefficient.

$$i_p = 2.69\,x\,10^5\,n^{\frac{3}{2}}ADd^{\frac{1}{2}}Cv^{\frac{1}{2}} \tag{2}$$

At DPV and SWV, different modulations are used in order to apply the potential to the electrode (see **Figure 1**): small pulses of fractions of a second at DPV, and a square function combining pulses of oxidation a reduction at SWV. The latter is especially favorable for reversible processes. In all cases, the peak intensity is also proportional to the area of the electrode and the concentration. By preconcentrating the analyte on the area of the electrode, an increase in the concentration is observed, which makes an increase in the intensity studied.

The new trends in the electrochemical determination of heavy metals are aligned with the Sustainable Development Goals, especially with the 6th (clean water and sanitation) and 15th (life on land), to develop new fast analysis methods to control the waste in the environment. Thus, one of the objectives for the development of new electrochemical transducers is their miniaturization, in order to save reagents. Initially, electrochemical cells were composed of three macroscopic electrodes with volumes of mL. However, in the last years, electrochemical cells with microliters volumes have become widely popular.

Apart from miniaturization, there is an interest in green and cheaper alternatives as it is being described in Sustainable Development Goal 12 (Responsible consumption and production). This would be attractive to extend the use of electrochemical sensors in under developed countries, but also in developed countries. Electrodes made of cellulose or paper-based electrodes have earned a significative relevance [13]. Paper is a cheap and abundant material, composed mainly of cellulose fibers or nitrocellulose, whose porosity allows hydrophilic samples to diffuse along the matrix by capillarity. Besides, paper can be modified with materials such as wax to create hydrophobically barriers and circuits, where samples diffuse and react with reagents previously immobilized within the cellulose matrix can be designed [14]. As a result, cheap and easy-to-use platforms with a rapid response for different applications, such as point-of-care-testing, can be obtained. Optical paper-based sensors are usually the most used, especially those with colorimetric detection. However, electrochemical paper-based sensors are becoming an improved alternative. Paper can be modified not only with biomarkers or wax but also with conductivity inks to form an electrochemical transducer. The resulting platform will have the capacity to determine analytes electrochemically active, such as heavy metals. Moreover, conductive markers can be used in voltammetric or amperometric biosensors, increasing the sensitivity in comparison with colorimetric biosensors [15].

Paper electrodes can be modified in the same way as commercial screen-printed electrodes. Nanomaterials are one of the most common options. Their optical properties make them be used as an indicator for colorimetric analysis, while their conductive properties help to increase the sensitivity in electrochemical paper-based

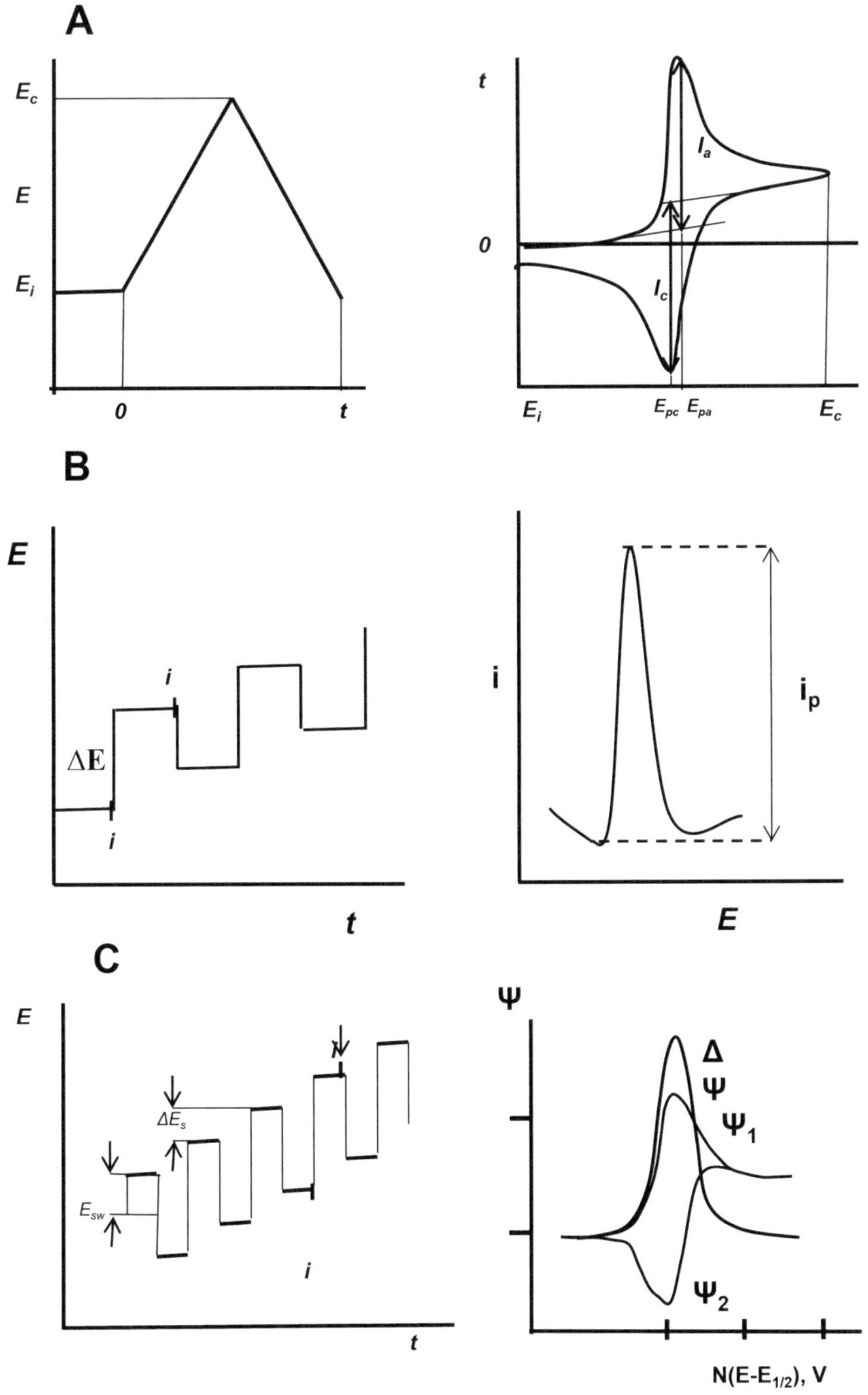

Figure 1.
Modulations applied and voltamperograms obtained respectively in (A) cyclic voltammetry, (B) differential pulse voltammetry, (C) square wave voltammetry.

electrodes [16–18]. This is due to their capacity to increase the signal-to-area ratio. Finally, their compatibility with bioreagents allows them to be functionalized with antigens, antibodies, or proteins. Nanomaterials can be directly deposited into the matrix cellulose or electrogenerated on the working paper electrode. On the other hand, paper electrodes can be modified by metallic surfaces or "films" by electrodeposition. Films can be made from different materials, such as mercury or bismuth, which improve the sensitivity and selectivity of heavy metals based on their affinity with the metal film.

In this chapter, we focus on paper-based electrodes used to measure heavy metals and the different modifications were done to improve the sensitivity or selectivity, such as the modification with nanomaterials (graphene and gold nanoparticles) or the electrodeposition of metallic films.

2. Heavy metals determination by paper-based electrochemical cells

Electrochemical cells based on paper can be entirely designed manually, or just a part and they can be coupled to commercial platform. Different configurations can be found in the literature for the final device, depending on the analytes of interest, the detection procedure, the portability capacity, or the easy-to-use requirements.

Two different approaches can be followed: partial or total paper-based platforms.

2.1 Partially electrochemical paper-based platforms

The first part of this chapter focuses on paper devices with a function and an attachment to a commercial platform. There are sustainable interests to choose a paper substrate as a transducer: in this case, the supporting cell can be reused without any contamination, applying cleaning steps after the measurement, and saving reagents and time. In addition, paper working electrodes could be used to test how viable is the protocol before designing a full paper-based platform.

Reports with different protocols and analytes can be found in the literature. For instance, Kingkan Pungjunun designed a paper-based analytical device by an origami method coupled to a boron-doped diamond electrode (BDD) [19]. The BDD was used as a working electrode, whereas the paper-based device was placed on top to form a disposable electrochemical cell. Their main objective was the determination of total arsenic (As (III) + As (V)). Using a wax printer, the paper matrix was modified with wax, which was melted to create hydrophobic zones. After that, conductive inks (silver and carbon) were added to form reference and counter-electrodes. In the end, a paper device with three different chambers was created, each one with a purpose. First, there is the working electrode zone coupled to the BDD, the modification zone, and the detection zone. These last two zones consisted of an auxiliary and reference electrode, so they can form an electrochemical cell with the BBD. With these designs, the working electrode could be modified with nanoparticles and the sample could be analyzed just by unfolding and refolding the different zones. This way, *ex situ* steps that can complicate the measure were excluded. The electrodeposition of gold nanoparticles on the BDD with the modification chamber generated a more conductive transducer with higher selectivity for arsenic, which was then measured in the determination zone by square wave voltammetry with limits of detections of 0.02 μg/mL. Commercial rice samples were measured, and their results compared by an ICP-OES analysis showed no significative difference between both methods, confirming the accuracy. An interferences study was also done, and cupper was identified as a possible interference. However, it could be corrected by adding ferricyanide, obtaining a low-cost paper-based device with high sensitivity for arsenic determination.

Paper electrodes can also be modified with metal layers instead of nanomaterials. Rochelle Silva et al. designed a potentiometric sensor on a glassy carbon electrode coupled to a paper substrate, with the aim of using its filtering properties for the complex matrix of environmental samples. This way they protected the ion-selective membrane of the electrode [20]. However, the negative charges of cellulose can interact with the metallic ions obtaining a super-Nernstian response in the potentiometric analysis. In order to correct the response, paper substrates were modified with gold layers as a spacer between heavy metals and cellulose fibers. Palladium and platinum were also tested showing similar improvements in comparison with unmodified paper-based electrodes.

These last reports were based on a paper electrochemical cell coupled to a commercial working electrode or a selective membrane electrode. However, there are research works based on the opposite: paper working electrodes coupled to a commercial cell. E. Nunez-Bajo et al. designed a paper-based working electrode to be used on a commercial screen-printed electrode. Wax printing and carbon inks were also used, but in this case, a disposable working paper electrode was created [21]. The paper electrode was coupled to the screen-printed transducer by means of an adhesive, so the electrochemical cell was constructed by using its auxiliary and pseudo-reference electrodes. After the measurements, the paper electrode can be disposed of without altering the screen-printed electrode, and therefore, it can be reused again. The conductivity of the carbon paper electrode was improved by the electrodeposition of gold nanoparticles by chronoamperometry and cyclic voltammetry. After the electrodeposition, the paper electrodes were washed and peeled off from the screen-printed platform. Subsequently, they were placed on another clean screen-printed electrode to confirm the presence of nanoparticles in cellulose fibers instead of the working electrode of the previously used screen-printed working electrode.

In this case, total arsenic and arsenic (III) were measured by chronoamperometry by preconcentrating first arsenic (III) in a reduction step to arsenic (O), and then reoxidating it. The limit of detection obtained for total arsenic was (LOD) 2.4 μg/L and 2.2 μg/L for As (III). Real wine samples were measured and the results were validated by comparison with those of an ICP-MS method. No significant differences between them were found, and the viability of this quicker, cheaper, and easy-to-use method that consumes less reagents than the reference ICP-MS method. The paper working electrode coupling avoids the contamination of the screen-printed platform and can be easily disposed of away without any further treatment.

These two mentioned reports for arsenic determination with gold nanoparticles showed that different devices can be adapted to the same principle with good results, demonstrating the versatility of paper electrodes [19, 21]. Paper electrodes are not only limited to a single analysis but they can also be modified in different ways for multiplexed determinations.

The design described above was also used in other reports. A. Sánchez-Calvo et al. investigated the modification properties with carbon nanomaterials (graphene oxide and nanofibers) and compared the results with those of the nanostructuration of paper electrodes with gold nanoparticles (AuNPs), in order to develop a method for mercury determination in river water samples [22]. Different combinations of nanomaterials were tested. The best option was the use of carbon nanomaterials as a conductive matrix combined with gold nanoparticles, as a selective transducer for mercury. Anodic stripping voltammetry was used. Mercury was preconcentrated on the surface on the basis of its high affinity for gold nanoparticles, and then it was redissolved. Three different oxidation peaks were observed, which could be assigned to the interaction of mercury with gold nanoparticle, gold nanoparticle nearly covered, or mercury without any interaction with gold nanoparticle, respectively [23]. The limits of detection obtained were 30 nM and spiked river samples could be analyzed with recoveries of 91%.

In this study, not only gold nanoparticles and carbon nanomaterials were tested. Cellulose fibers were also modified with metals like mercury or bismuth to form a conductive "film" with applications for the determination of heavy metals [24]. Solutions of bismuth (III) and mercury (II) were used to form a film by electrodeposition and compare their affinity for Cd (II), Pb (II), In (III), and Cu (II). The results show that mercury has higher sensitivity than bismuth films. Moreover, a bismuth oxidation peak overlapped the copper signal, impeding its determination. Water samples were spiked with Cd (II), Pb (II), and Zn (II) salts to avoid the formation of intermetallic compounds. Recoveries obtained on mercury and bismuth films were around 99 and 88%, respectively, confirming the better sensitivity of mercury films. Nevertheless, the use of bismuth films could be more attractive since they have not been found toxic to the environment. With this option, the lineal range of concentrations for heavy metal determination is around mg/L or "ppm." This limitation on the sensitivity would mean that those sensors would not be useful for trace-level analysis without a preconcentration treatment.

2.2 Full made paper-based platforms

There are different techniques to create entire paper-based platforms. Screen printing is the most frequently used for electrochemical cells. M. Medina-Sánchez *et al.* developed a lateral flow paper-based sensor with a flow channel in which the sample diffuses and gets in contact with a screen-printed electrode. This was formed by a graphite area acting as a working and counter electrode, with a silver/silver chloride as a reference electrode [25]. This design benefits from the capillarity of the cellulose acting as a filter of solid particles in water samples and increases the sensitivity because of the continuous flow of the sample along the electrodes during the deposition step.

In this case, the sensor must be disposed of after the measurement, because a successful cleaning step could not be developed. However, this work is still interesting as a source of inspiration for low-cost paper-based sensor technology. Cadmium (II) and lead (II) were determined with limits of detection of 11 and 7 ppb, respectively. Solutions with both metals showed changes in potential probably due to the formation of intermetallic species or alloys, but sensitivity was not affected. Samples of complex matrixes like mud or seawater were analyzed in order to test the filtering capacity, confirming that the protocol was efficient and there was no need for a pretreatment step. Therefore, this disposable quick sensor was valid as a single-use device. In another report, Yue-Hong Pang et al. designed a portable sample cell made of 1 mm thick double-sided adhesive tape, which was 8 mm punched and aligned with a platinum sheet with an 8 mm hole. On the bottom layer of the portable cell, an auxiliary electrode (carbon) and a reference electrode (Ag/AgCl) were integrated [26]. Apart from the cell another layer formed by a carbon paper acting as a working electrode was modified by gold nanoparticles (AuNPs) by electrodeposition followed by addition of a metal-organic framework (Co-MOF-NH_2) material. After the modification procedure, the two parts were combined in a sensor and optimized to quantify Cd (II) and Pb (II). The limits of detection (LOD) found were 0.07 ng/mL and 0.011 ng/mL, respectively, and these metals could be successfully determined in real food samples (grain, juice, and vegetables). The reusability was found acceptable, with no high deviation for repeated measurements.

Full paper platforms were also used by Huang et al. [27]. These authors designed a paper-based electrochemiluminescence biosensor adapted to the determination of Ni (II) and Hg (II) [27]. The biosensor had two different channels, each one adapted to the determination of an analyte. In this case, silver nanoparticles (AgNPs) and cubic Cu_2O were added to increase the conductivity and as a catalyst for hydrogen peroxide in each case. The origami design has the possibility of an auto-cleaning procedure, lowering the number of steps needed and reducing the measurement time. The limits of detection found for Ni (II) and Hg (II) were 3.1 nM and 3.9 pM, respectively. Real river samples were spiked and analyzed with recovery values from 96 to 104% confirming the reliability of the platform. The different channels prepared confirmed the good features of the cellulose matrix and the suitability of the capillarity as a driving force for testing different analytSes in different procedures with the same sensor at the same time.

Qiu-Mei Feng et al. designed a disposable electrode to determine heavy metals by anodic stripping voltammetry [28]. The electrode was composed of conductive carbon tape on an indium tin oxide (ITO) glass as a way to enhance the conductivity. The carbon working electrode was modified with bismuth by electrodeposition, in order to increase the stripping analysis with successful measurements of Zn, Cd, and Pb ions. Apart from the good sensitivity and capacity to distinguish among heavy ions, the low-cost and low volumes are also important highlights to consider. Another full paper-based electrode modified with bismuth film was developed by D. Soulis. In this report, the channels for fluid flow were marked by a pen instead of melted wax and combined with a graphite screen-printed electrochemical cell for the determination of Pb (II) and Cd (II) [29]. Here, instead of forming the bismuth film in a procedure apart from measuring the sample or what it can be called in a "*ex situ*" way, bismuth and heavy metals were stored in the same solution. By applying a reduction potential, the three elements were simultaneously reduced together, forming metal alloys that were subsequently oxidized. The limits of detection obtained were 2.4 μg/L for Cd (II) and 4.2 for Pb (II), which are lower than the limits for drinking water in the EU and USA. This method has the advantage of saving time and reagents because all the requirements for preconcentration are included in a single step.

Paper-based sensors are usually coupled to a single detection system. However, their versatility and flexibility are suitable for the design of multiple detection systems. Habdias A. Silva-Neto et al. designed a paper-based plug-and-play platform with both colorimetric and electrochemical sensors for the detection of metals [30]. The sensor consists of two different paper platforms, each one with one detection method and a polymer substrate. The polymer was prepared by die cutting, so it could be connected with the colorimetric and the electrochemical sensor on each extreme of the template. The colorimetric paper platform was designed in a tree shape containing three channels with three different zones (pretreatment, colorimetric, and waste zone), whereas the electrochemical zone is formed by a self-made screen-printed paper-electrode composed of graphite ink containing multi-walled carbon nanotubes (MWCNT). Those platforms can be reversibly assembled using a plug-and-play mechanism and using the same drop of sample. In conclusion, both detection measurements can be done at the same time. This way not only heavy metals, such as Cd (II), Pb (II), Zn (II), and Ni (II) can be determined, but also other metals, such as Fe (II) and Cu (II). The analysis of river samples was validated by comparison with results obtained by atomic absorption spectrometry. This method combined the flexibility of using paper platforms with their modification properties to detect multiple analytes by different principles. Miniaturization is also an important advantage, allowing to use lower volumes than the reference method.

Platform	Type	Analytes	Linear range	Limit of detection (LoD)
AuNPs paper filter/ boron-doped diamond electrode	Partially electrochemical paper based platform	Total As	0.1–1.5 μg/mL	20 ng/mL
AuNPs + Working paper electrode /SPCE commercial		As (III)	2–50 μg/L	2.2 μg/L
		Total As	2–50 μg/L	2.4 μg/L
AuNPs paper filter/ion membrane electrode		Pb (II)	10^{-5}–$10^{-2.2}$ M	10^{-5} M
Carbon nanomaterial + AuNPs working paper electrode /SPCE commercial		Hg (II)	0.1–1.2 μM	30 nM
Mercury or bismuth films working paper/SPCE commercial		Cd (II)	0.5–10 /2.5–10 μg/mL	0.4/1 μg/mL
		Pb (II)	0.5–10/1–10 μg/mL	0.1/0.7 μg/mL
		In (III)	0.1–5/1–4 μg/mL	0.04/0.6 μg/mL
		Cu (II)	0.25–6.35/X μg/mL	0.2/X μg/mL
Full paper electrode	Full made paper-based platform	Pb (II)	10–100 μg/L	7 μg/L
		Cd (II)	10–100 μg/L	11 μg/L
Co-MOF-NH_2/AuNPs full paper electrode		Pb (II)	0.5–20 ng/mL	0.011 ng/mL
		Cd (II)	0.75–35 ng/mL	0.07 ng/mL
AgNPs + Cu_2O-Au Full paper electrode		Ni (II)	10 nm–0.2 mM	3.1 nM
		Hg (II)	1 pm–1 μM	3.8 pM
Bismuth film Full paper electrode		Pb (II)	0–500 μg/L	2 μg/L
Bismuth film Full paper electrode		Pb (II)	10–1000 μg/L	2.4 μg/L
		Cd (II)	5–800 μg/L	4.2 μg/L
Tin working electrode Full paper electrode		Cd (II)	5–40 μg/L	0.9 μg/L
		Zn (II)	5–40 μg/L	1.1 μg/L
MWCNT Full paper electrode		Zn (II)	100–1400 μg/L	10.5 μg/L
		Cd (II)	10–1400 μg/L	1.3 μg/L
		Pb (II)	10–1400 μg/L	0.9 μg/L

Table 1.
Comparison of results of electrochemical analysis on paper-based platforms.

Most of paper electrodes are made of carbon, but there are alternatives. C. Kokkinos et al. designed a microfluidic paper electrode with a working electrode composed of a sputtered film of tin (Sn) to determine Cd (II) and Zn (II) at trace levels with limits of detection of 0.9 and 1.1 μg/L, respectively [31]. The tin working electrode and the auxiliary and reference electrode were made by sputtering over the paper substrate. A microfluidic channel was printed on reverse paper. This helps to diffuse the samples, taking advantage of the capillarity. Volumes as low as 10 μL could be used.

All mentioned reports show a variety of options for designing paper electrochemical cells for heavy metal determination. **Table 1** resumes all the results previously commented. All of them contribute to the goal of low-cost, sustainable, easy-to-use, and green analytical platforms. This is especially attractive for use in underdevelopment countries or in places in which portability and rapidity are the main requirements.

3. Conclusions

Heavy metal determination by electrochemical methods has evolved over time, following the analytical trends of miniaturization and, more recently, sustainable and eco-friendly processes. Paper-based platforms are the next alternative to low-cost and easy-to-use electrodes. In this chapter, the main designs reported in the literature have been critically analyzed. The modification with nanomaterials or metal films in order to increase their conductivity properties was found successful strategy. This confirms that cellulose-based platforms are the foundations of a new generation of versatile and low-cost electrochemical sensors.

Acknowledgements

This work was founded by the Consejería de Educación y Ciencia del Principado de Asturias, Spain (Ref. SV-PA-21-AYUD/2021/52132) and the Agencia Estatal de Investigación, Spain (Ref MCI-21-PDC2021-121444-100).

Conflict of interest

The authors declare no conflict of interest.

Author details

Alberto Sánchez Calvo* and María del Carmen Blanco Lopez
Department of Physical and Analytical Chemistry, University of Oviedo, Oviedo, Spain

*Address all correspondence to: sanchezcalberto@uniovi.es

References

[1] Pourret O, Bollinger J, Hursthouse A, Pourret O, Claude BJ, Hursthouse A, et al. Heavy metal: A misused term? Acta Geochimica. 2021;**2021**:466-471

[2] Duffus JH. "Heavy metals"—a meaningless term? (IUPAC technical report). Pure and Applied Chemistry. 2002;**74**(5):793-807

[3] European Union. Regulation (EU) 2017/852 of the European Parliament and of the Council of 17 May 2017 on mercury, and repealing Regulation (EC) No 1102/2008. Official Journal of the European Union. 2017;**2017**:1386

[4] European Commission. Commission Regulation (EU). Amending Regulation (EC) No 1881/2996 as regards maximum levels of mercury and fish in salt. Official Journal of the European Union. 2022;**2022**:1-4

[5] Commision Regulation (EU) 1317. Commission Regulation (EU) 2021/1317 of 9 August 2021 amending Regulation (EC) No 1881/2006 as regards maximum levels of lead in certain foodstuffs. Official Journal of the European Union. 2021;**8**(1881):13-16

[6] Council Directive. On the quality of water intended for human consumption. Official Journal of the European Communities. 1998;**1998**:1-23

[7] Regulation EU. Regulation of the european parliament and of the council laying down rules on the making available on the market of EU fertilising products and amending regulations (EC) No 1069/2009 and (EC) No 1107/2009 and repealing Regulation (EC) No 2003/2003. Official Journal of the European Union. 2003;**2019**:114

[8] European Commission. Commission Regulation (EU) 2015/1006 amending Regulation (EC) No 1881/2006 as regards maximum levels of inorganic arsenic in foodstuffs. Official Journal of the European Union. 2015;**14**:14-16

[9] Yosypchuk B, Barek J, Fojta M. Carbon powder based films on traditional solid electrodes as an alternative to disposable electrodes. Electroanalysis. 2006;**18**(11):1126-1130

[10] Serrano N, Díaz-Cruz JM, Ariño C, Esteban M. Ex situ deposited bismuth film on screen-printed carbon electrode: A disposable device for stripping voltammetry of heavy metal ions. Electroanalysis. 2010;**22**(13):1460-1467

[11] Sochr J, Machková M, Machyňák Ľ, Čacho F, Švorc Ľ. Heavy metals determination using various in situ bismuth film modified carbon-based electrodes. Acta Chimica Slovaca. 2016;**9**(1):28-35

[12] Pitsou M, Kokkinos C, Economou A, Fielden PR, Baldock SJ, Goddard NJ. "Green" three-electrode sensors fabricated by injection-moulding for on-site stripping voltammetric determination of trace In(III) and Tl(I). Chemosensors. 2021;**9**(11)

[13] Arduini F. Electrochemical paper-based devices: When the simple replacement of the support to print ecodesigned electrodes radically improves the features of the electrochemical devices. Current Opinion in Electrochemistry. 2022;**35**:101090

[14] Jin Y, Aziz A, Ur R, Wu B, Lv Y, Zhang H, et al. The road to unconventional detections: Paper-based microfluidic chips. Micromachines. 2022;**13**(11):1835

[15] Kuswandi B, Hidayat MA, Noviana E. Paper-based sensors for rapid important biomarkers detection. Biosensors and Bioelectronics. 2022;**12**:1-22

[16] Ding R, Cheong YH, Ahamed A, Lisak G. Heavy metals detection with paper-based electrochemical sensors. Analytical Chemistry. 2021;**93**(4): 1880-1888

[17] Meng C, Liu C, Fan S. Flexible carbon nanotube/polyaniline paper-like films and their enhanced electrochemical properties. Electrochemistry Communications. 2009;**11**(1):186-189

[18] Pang R, Zhu Q, Wei J, Meng X, Wang Z. Enhancement of the detection performance of paper-based analytical devices by nanomaterials. Molecules. 2022;**27**(2):1-31

[19] Pungjunun K, Chaiyo S, Jantrahong I, Nantaphol S, Siangproh W, Chailapakul O. Anodicstrippingvoltammetricdetermination of total arsenic using a gold nanoparticle-modified boron-doped diamond electrode on a paper-based device. Microchimica Acta. 2018;**185**(7)

[20] Silva R, Ahamed A, Cheong YH, Zhao K, Ding R, Lisak G. Non-equilibrium potentiometric sensors integrated with metal modified paper-based microfluidic solution sampling substrates for determination of heavy metals in complex environmental samples. Analytica Chimica Acta. 2022;**1197**:339495

[21] Nunez-Bajo E, Blanco-López MC, Costa-García A, Fernández-Abedul MT. Electrogeneration of gold nanoparticles on porous-carbon paper-based electrodes and application to inorganic arsenic analysis in white wines by Chronoamperometric stripping. Analytical Chemistry. 2017;**89**(12):6415-6423

[22] Sánchez-Calvo A, Fernández-Abedul MT, Blanco-López MC, Costa-García A. Paper-based electrochemical transducer modified with nanomaterials for mercury determination in environmental waters. Sensors & Actuators, B: Chemical. 2018;**2019**(290):87-92

[23] Martín-Yerga D, González-García MB, Costa-García A. Use of nanohybrid materials as electrochemical transducers for mercury sensors. Sensors Actuators B: Chemical. 2012;**165**(1):143-150

[24] Sánchez-Calvo A, Blanco-López MC, Costa-García A. Paper-based working electrodes coated with mercury or bismuth films for heavy metals determination. Biosensors. 2020;**10**(5):17-19

[25] Medina-Sánchez M, Cadevall M, Ros J, Merkoçi A. Eco-friendly electrochemical lab-on-paper for heavy metal detection. Analytical and Bioanalytical Chemistry. 2015;**407**(28):8445-8449

[26] Pang YH, Yang QY, Jiang R, Wang YY, Shen XF. A stack-up electrochemical device based on metal-organic framework modified carbon paper for ultra-trace lead and cadmium ions detection. Food Chemistry. 2022;**2022**(398):133822

[27] Huang Y, Li L, Zhang Y, Zhang L, Ge S, Yu J. Auto-cleaning paper-based electrochemiluminescence biosensor coupled with binary catalysis of cubic Cu2O-Au and polyethyleneimine for quantification of Ni^{2+} and Hg^{2+}. Biosensors & Bioelectronics. 2018;**126**:339-345

[28] Feng QM, Zhang Q, Shi CG, Xu JJ, Bao N, Gu HY. Using nanostructured conductive carbon tape modified with bismuth as the disposable working

electrode for stripping analysis in paper-based analytical devices. Talanta. 2013;**115**:235-240

[29] Soulis D, Kokkinos C, Economou A, Prodromidis MI. Single-use fluidic electrochemical paper-based analytical devices fabricated by pen plotting and screen-printing for on-site rapid voltammetric monitoring of Pb (II) and Cd (II). Sen. 2021;**21**:1-11

[30] Silva-Neto HA, Cardoso TMG, McMahon CJ, Sgobbi LF, Henry CS, Coltro WKT. Plug-and-play assembly of paper-based colorimetric and electrochemical devices for multiplexed detection of metals. The Analyst. 2021;**146**(11):3463-3473

[31] Kokkinos C, Economou A, Giokas D. Paper-based device with a sputtered tin-film electrode for the voltammetric determination of Cd(II) and Zn(II). Sensors Actuators B: Chemical. 2018;**260**:223-226

Chapter 34

Heavy Metal Speciation, and the Evaluation and Remediation of Polluted Mine Wastes and Soils

Arturo Aguirre Gómez and Margarita Eugenia Gutiérrez Ruiz

Abstract

The chapter exposes how a sound methodology can be instrumented to both, biogeochemically speciate heavy metal (HM) polluted mine wastes and soils, and to develop solid strategies to agriculturally stabilize and remediate HM-polluted terrestrial environments. Using single- and sequential extraction procedures, polluted environments can be chemically speciated to successfully remediate impacted sites. Once metal(loid) toxic levels are determined, common amendments (compost, P-fertilizers, lime, gypsum) can be added to abate HM levels, and to re-sustain vegetation, based on bioassay results of HM-sensitive plants. The approach addresses first: a) a discussion of concepts and relevant chemistry that apply to study mine tailing materials and soils, via single or multiple HM-fractionation schemes; b) characterizing chemically mine tailings and soils, in terms of the metal(loid)-sorption-complexing affinities, and c) creating a "fertile environment" by agriculturally reconditioning the HM-polluted acidic mine waste to allow the vegetation regrowth, based on bioassay test performance. Results of two successful cases of study are included; one showing the use of single extraction procedures to evaluate phytoavailable/toxic HM levels to agriculturally remediate polluted sites, and another showing the role of sequential extraction procedures to discriminate heavy metal (loid)s of a spill from other metal deposits of the same ore.

Keywords: soil pollution, metal mine tailings, chemical speciation, heavy metalloid bioremediation, heavy metal bioassays

1. Introduction

Mining extraction of heavy metals from sulfidic materials produces considerable levels of potential acidity which eventually, if not prevented and neutralized generate the so-called metalliferous acidic mine drainage, resulting in a potential mobilization of soluble heavy metals [1, 2]. The quantity of acid-forming minerals found in many mines of Central Mexico around the neo-volcanic axis [3, 4] and in Northern Mexico contains dominantly pyrite (FeS_2), galena (PbS), sphalerite (ZnS), pyrrhotites ($Fe_{1-x}S$), chalcopyrite ($CuFeS_2$), arsenopyrite (AsFeS), bornite (Cu_5FeS_4), and many other metallic sulfosalts [3]. After oxidation, these minerals generate the H^+-producing redox and hydrolytic processes of the components (e.g., S, Fe, Mn, Zn, Cu, Pb, etc.) left behind in the mine tailings. Mine wastes, polluted- and pristine-soils must then be studied and chemically speciated [5], usually by applying simple or multiple-sequential extraction procedures [5–8], this to fractionate the HM species according to their expected chemical interaction with the various solid phase compartments present in soils or mining wastes. Based on that, HMs can be grouped into one or several of the following categories: a) water-soluble (free metal ions, M^{2+}, inorganic and organic metal complexes, ML, whether labile or not, etc.); b) exchangeable (non-specifically adsorbed); c) ligand extractable (bioavailable); d) acid-extractable (carbonate-precipitated); e) organically and sulfide bound (oxidizing fraction); f) chemisorbed on Fe-, and Mn oxides (reducible fraction) and g) lattice-retained (occluded or residual). Metal speciation and fractionation to assess any remediation strategy must then rely on finding the right methodology to evaluate which of the extraction procedures may serve as a metal available/toxic fraction indicator [5–8]. This will eventually assure that a metal-sensitive plant can grow once any other undesirable physicochemical characteristics are resolved, such as low pH values, low nutrient status, and high EC). The proposal must then include the addition of ameliorating materials (lime, phosphates, compost, gypsum) that both, help to abate the HM-phytoavailable levels, while serving to neutralize the acidified metal-polluted site, and to mend the growing media, so that plant regrowth will not be impeded. Once these two aspects are guaranteed, results must, all in all, meet the requirements of national or international standards and norms and pass the chosen specific bioassay so that the site can be considered agrostabilized [9–11]. Other necessary physical and chemical characterization procedures shall include the total metal content, $[M_T]$, original minerals identification, electrical conductivity, pH, HM-sorption-complexing affinities, etc. [12, 13]. The chapter addresses in Section 2, the relevant chemistry to be considered when researching HMs in mine tailings and polluted soils to chemically characterize wastes and pristine soils. The most commonly chemical speciation and fractionation terms and protocols used are discussed in terms of validating which one best adapts to the specific purpose. In Sections 3 and 4, two cases of study are presented: A) One including an example of how a single extraction procedure was used to study-remediate a Cd-, Cu-, Pb- and Zn- polluted mine waste from "Mina La Negra", at Zimapan, Hidalgo State, in Central Mexico, following a successful application of a methodology to: a) sufficiently abate the levels of phytoavailable-toxic HM-contents of the mine waste by adding lime, gypsum, phosphates or compost; b) ameliorate the waste material to create a "fertile" environment and; c) test a) and b) by applying and passing a well-established bioassay using an HM-sensitive plant, this to assure a successful regrow of vegetation; and B) Another, including a successful application of a sequential extraction protocol to evaluate the effect a spill of a Cu-mine into the Sonora River basin in Northern Sonora State, Mexico, on the bio-accessible levels of metal(loid)s such as Fe, Al, Cu, Mn, and minor amounts of As, Cd, Zn, Pb, and Cr.

2. Chemical speciation and extraction procedures

It is well known that the chemical speciation of heavy metal(loid)s depends not on their total concentrations, but more on the form in which they are found in the environment [5]. Chemical forms depend not only on the reactions that control HM solubility [14–16], and thus on their availability [17–19] and toxicity to plants [20], but, on their mobility and distribution in the environment [21], which in turn are dominated by their physicochemical interactions (complexation, redox-chemistry, and sorption) [14] with the different solid phase compartments (minerals lattices, oxides, organics, carbonates). Based on this, it is important first to discuss the relevant chemistry of these processes, and secondly, to which extent these will give a good indication of the availability/toxic levels to plants or other living organisms. Because many international standards [22] and national norms [23–28] base their threshold-legal values on certain metal-extractabilities, and in few cases, on the performances of certain bioassays and biological tests [29–34], a good selection of extracting procedures and pertinent bioassays must be correlated to assure both, the successful remediation strategy followed, and the health of the terrain to sustain vegetation regrowth. Although several research papers and reviews have been published in the past decades regarding the importance of chemical speciation studies and sequential extraction protocols [5–8] to evaluate sediments, soils, biosolids, etc. [5] there is still a strong need to establish a sound methodology, in terms of their applicability, to examine the chemistry of HM-polluted mine wastes. To do so, it is important first then, to present a focused review of the relevant chemical aspects of the interactions implicated among the diverse metal(loid)s present, and the various solid phases found on these wastes.

2.1 Chemical speciation; definitions and concepts

Although the term "heavy metal" could be imprecise and sometimes misleading, as stated by IUPAC [35], over the past four decades, the term "heavy metal" continues to be widely used and applied to a group name of metal(loid)s and that are associated with pollution and/or potential (eco)toxicity. Besides, legal regulations often refer to the term "heavy metal" not only not specifying which metals are included, but under which chemical basis it is assumed that HM and their compounds have (eco)toxic effects, or pose analogous physicochemical, biological, and toxicological properties. Thus, any new name, definition, or classification of metals would be better based on the chemical properties of metals, and such a categorization should reflect our understanding of the chemical basis of reactivity and toxicity, so that their toxic effects can be predicted. More appropriate terms seem to be sound if based on the relevant chemical behavior of the metal, such as those considered by the hard and soft acid and base Pearson's-Lewis theory [36–41], or the "s", "p", "d", and "f" character of metals, according to the orbitals used in the HM-bonding, or on their relevant periodic properties involved in the interaction (electronegativity, ionic potential, ionic radius, metal hydrolysis) [14] that governs their ultimate fate. Other names have been suggested [5–8, 42] presumably being more appropriate, such as potentially toxic elements, PTEs [42]. Yet, although the name heavy metal persists and continues to be used in literature, policies, and regulations, and we will utilize it, in this chapter, in correspondence with the title of the book, and following the general use and acceptance of many researchers in the literature, reference will be made to the chemical aspects of metal reactivity.

2.1.1 Chemical, functional, and operational speciation

Several attempts have been made to clarify unambiguously the term chemical speciation (of heavy metals) in environmental and agricultural sciences [5, 7, 35, 42], especially when used to characterize the relevant chemistry of a specific element that can be toxic or nutrimental to living organisms. However, when a metal is distributed in the environment, several ways are used to describe its behavior in the environment, or its chemical form and activity, among other properties. In the following sections, we will focus not only on the metal chemical forms by itself, in terms of the phases where the metal is distributed (aqueous or solid), but unambiguosly on the nature of the specific bonding involved (van der Waals forces, ionic or covalent bonds, inner or outer sphere complexes, etc.), and on the type of compartment that the specific HM occupies, according to its interaction with the solid phase, such as water-soluble, exchangeable (non-specifically adsorbed), chemisorbed (specifically adsorbed), chelate extractable (available), acid- extractable (carbonate-precipitated), organically +sulfide bound (oxidizable), Fe/Mn oxides (reducible), occluded-residual (lattice-retained), or even more, on the ability to be taken up, ingested, bioaccumulated, etc., by living organisms. In a document on chemical speciation terminology published by IUPAC [43], it is recommended the term 'chemical species' for describing the form of an element defined as isotopic composition, electronic or oxidation state, and/or complex or molecular structure. This definition has been considered as inappropriate for most studies on solid materials such as soils, sediments, and other materials [5, 42] and transfers the selective extraction procedures relevant to such solid phases to the category of fractionation methodologies. Broader definitions to include soils and mine wastes, for instance, are defined in terms of the so-called functional or operational speciation. In this respect, the term may be better defined to identify, describe, and quantify the amounts of the species forms and phases present in each material [7, 42, 43]. Thus, a general consensus points toward adopting a definition that includes both, forms or phases, so that speciation seems to be better defined as: a) referring to a specific chemical compound or oxidation state in environmental samples, even though this type of speciation in which the precise chemical form of an element is measured is the most difficult to achieve, since very sensitive and selective analytical techniques are required; and; b) functional speciation, for specific usages as for terms like phytoavailable (plant-available species) or bio-accessible (for animals or humans), etc., or c) operational speciation, defined in terms of the extraction procedure utilized to refer to the physical or chemical fraction characterized (water-soluble, acid-extractable, etc.) in soils, or mine wastes. Physical procedures to divide samples by particle size, or fractions separated by filtration, centrifugation, or dialysis, etc., are also considered an operational speciation [7, 42, 44], as the distinction between soluble and insoluble species is based on the ability to pass a sample through a 0.45 μm filter. In fact, many operational procedures are often used to fraction metals based on particle size: dissolved (<1 kDa), colloidal (1 kDa-0.45 μm), and particulate (> 0.45 μm).

2.1.2 The heavy metal-solid phase interaction: Extraction protocols and fractions

It has been long recognized that knowledge of both, the elemental composition of the solid materials (sediments, soils, biosolids, mine tailings, etc.), and the total concentration of HMs present in the environment, are of little use in assessing the availability and toxicity to biota. For these reasons, chemical tests have relied more on measurement of extractable or "labile" fractions of these potentially toxic elements. Such tests, however, have provided little basis to relate HM-extractability in mine wastes, to the chemical forms that can be toxic to organisms and deleterious to the environment. To evaluate and plan a remediation strategy for an HM-polluted site, a fundamental understanding of the processes that control heavy metal solubility and availability to biota is needed. Thus, the relevant physicochemical aspects of the interactions between HM and the solid phases present in the mine waste deposits must be reviewed to successfully correlate their extractability with plant tolerance, so that both, the requirements imposed by standards and norms are fulfilled, and the site can be agriculturally stabilized to allow revegetation. Chemical interactions among HMs and solid phases in unaltered minerals of soils, and in altered mine wastes left behind after ore exploitation, pose and/or create very diverse physical and chemical conditions that influence speciation, such as 1) ionic strength-electrical conductivity of media, 2) presence of dissolved organic matter and complexing ligands, 3) pH and potential acid-forming equilibria, 4) redox potential, E_h-*pe*-values, 5) hard/soft-acid/base character of the metals (charge, ionic radii, ionic potential, metal hydrolysis, etc.), and 6) reaction kinetics. The combined action of these factors, plus those of the edaphogenetic characteristics of the original soil, the biogeochemical processes occurring, and climate will favor the formation of different metal species, resulting in an innocuous/toxic HM species with a higher or lower bioavailability [45]. Thus, analyzing metal concentrations of aqueous or solid phases is not sufficient to determine its toxicity as biological effects [46], so that, chemical knowledge might provide a more effective diagnosis of environmental conditions [47]. Several biological factors that may influence the bioavailability of metals, include the route of exposure, the mechanism of sequestration and transport of metals by organic ligands, and the exposed organism [48]. According to Rainbow and Luoma [49], in metal ecotoxicology, the term bioavailability, corresponds, both, to the metal that is available for capture by a living organism and is integrated into the metabolic processes, and the fraction of the concentration of the metal that is absorbed and/or adsorbed by the organism. The assimilated fraction may then interact with receptors and physiological sites, causing toxic effects [49]. In the following sections a brief review of the relevant chemistry that will otherwise tilt the balance toward a given single or sequential extraction procedures to speciate HM in mine wastes and soils, will be discussed. Based on Tessier et al. [6], and other studies [5, 42, 50] and protocols [7, 8], the most common soil- and mine waste species and phases to be single or sequentially extracted may include: a) the soil- and mine waste-water-soluble fraction; b) the exchangeable/non-specifically adsorbed species; c) the acid-extractable (carbonate+specifically chemisorbed species) phase; d) the (oxidizable) organically complexed+sulfide metal species; e) the (reducible) hydrous Fe/Mn oxides fraction; and f) the residual-occluded and strong acid-extractable species.

2.2 The water-soluble fraction and solution speciation

In the soil, sediment, and mine waste solutions, the chemical speciation of heavy metals (HM) must consider the solvation process and complexes formed with organic and inorganic ligands [51]. In solution, ligands can form inner- or outer sphere complexes with metal cations [52]. Information on solution speciation is required for predicting bioavailability because the free metal ion, M^{2+}, is the most toxic species for biota and the most reactive one that interacts with the solid phases [53]. Depending on the metal, the free ion may be the dominant species ($Cd^{2+}_{(aq)}$, $Zn^{2+}_{(aq)}$, etc.) or may account for only a minor fraction of the total metal in solution [54], due to the formation of stable metal complexes forms (CuL, PbL, etc.). Metals in soil solution may be present as $M^{2+}_{(aq)}$, or as kinetically labile/non-labile metal- complexes, with inorganic (Cl^-, SO_4^{2-}, CO_3^{2-}, etc.), or organic ligands, or associated with mineral colloids [52, 53]. Thus, the analytical evaluation of the free metal activity, (M^{2+})-value, is an important step in the process of chemical speciation of pollutant metals in aqueous solutions [38, 55–58]. However, in the past, the experimental determination of (M^{2+}) was restricted due to several limitations regarding sensitivity and selectivity of the applied methodologies. Traditionally, the strict evaluation of (M^{2+}) was only possible using solid-state metal ion-selective electrodes (M-ISE), but some other analytical techniques have also been used to estimate 'free ion' concentrations of metals in solution (Donnan dialysis, resin exchange methods, and chromatographic techniques [54, 59], voltammetry (e.g., Anodic Stripping Voltammetry, ASV [56], or Cathodic-Stripping Voltammetry)). However, HM ion fraction determinations, for instance, might become unreliable if concentrations in solution are below the detection limits (e.g., Donnan dialysis, resin exchange methods) necessary for the use of very sensitive techniques such as ASV [55, 56]. Measurements with ion-selective electrodes (ISEs) in natural samples may be affected by the fouling of the electrode by organic matter [60]. The use of ISEs under large Cl^{-} concentrations are also not advisable, and other interferences may occur, resulting in an overestimation of the free metal ion concentrations [61]. Although during the last decades, much progress has been made in reducing the detection limits of ISEs [62]. Most studies to measure trace metal speciation in soil solutions have been on Cu^{2+}, for which detection limits of 10^{-13} M or even smaller have been reported [63]. Voltametric techniques, specifically ASV, have been successfully used to determine not only the concentrations of labile metal species from a current measured in solution as a metal is dissolved or released from a Hg-electrode, but the (M^{2+})-value too. Although analytical problems have been claimed regarding overlapping of stripping peaks, adsorption of surface-active organic compounds on the Hg-surface inhibiting the metal deposition, or formation of insoluble intermetallic compounds that affect peak size and position, other studies showed that these inconveniences can be overcome, so that free Cd^{2+}, Cu^{2+}, Pb^{2+}, and Zn^{2+} activities [55, 56] can be measured in natural polluted soil solution samples.

2.2.1 The free metal activity measurement

Traditionally, the strict evaluation of (M^{2+})-values was only possible using solid-state metal ion-selective electrodes (M-ISE), but except perhaps for Cu^{2+}, no M-ISE has adequate sensitivity and specificity for evaluating trace (free) metals in solutions where many metals coexist. It is well known that the M-ISE for Cd^{2+} and Pb^{2+}, for instance, respond similarly to both metals; hence, Cd and Pb interfere with each other [64, 65]. Aguirre et al. [55, 56] developed a robust method to determine (M^{2+})-values of Cd, Cu, Pb, and Zn by ASV. The method was tested using metal-buffer solutions to control (M^{2+}), by complexing metals with weak, medium, and strong ligands, and varying pH, total metal aqueous concentration (10^{-6}–10^{-7} M), metal–ligand ratios of 1:20 M, and 0.010 M acetate medium. For the studied metals, Cd, Cu, Zn, and Pb, the agreement was found among theory, experimental ASV measurements, and (M^{2+}), predicted by using a speciation chemical equilibria program and stability constants reported in the literature. Good agreement was found between the theory and calculated (M^{2+}), and between experimental ASV results and calculated (M^{2+}). Free metal activities in the order of $pCd \leq 12$, $pCu \leq 18$, $pPb \leq 10$, and $pZn \leq 9$ were measurable under the established experimental conditions. Results (not shown) also revealed good agreement between Cu-ISE and Cu-ASV when measuring the free Cu^{2+} activity in aqueous extracts of four soils. Values of soil-(Cu^{2+}) measured were in the order of 10^{-5} to 10^{-9}. The calibration curves for each metal were prepared according to speciation calculations of the metal–ligand–pH equilibrium systems (M–L–pH) in 0.1 M acetate as an indicator ligand, since fulvic acids in soils contain appreciable amounts of carboxylic compounds of low molecular weights, such as acetates, oxalates, and citrates, among others. Synthetic solutions of final concentrations of total metal $[M_T] = 1 \times 10^{-5}$ M, total ligand $[L_T] = 2–4 \times 10^{-4}$ M, and pH values in the 4.0–7.5 range were tested. The calibration curves $(\Delta E_p)_c$ versus $\log(M^{2+})$ were generated, with $(\Delta E_p)_c$ being the displacement of the peak potential due to metal complexation, ML_p, and $\log(M^{2+})$ is the logarithm of theoretical chemical activity. Calculations were made with the MINEQL+ program [66, 67], using stability constants reported in the literature [55, 56, 68], corresponding for each of the metals Cd, Cu, Pb, and Zn, in their aqueous free form (M^{2+}). The parameter $(\Delta E_p)_c$ was calculated with eq. (1)

$$(\Delta E_p)_C = \Delta E_p + \frac{RT}{nF} ln \frac{i_{D,ML}}{\phi_{ML_p}} = \frac{RT}{nF} ln \left(M^{2+}\right)_b \quad (1)$$

where ΔE_p is the observed experimental value of the displacement in half-wave potential due to complexation, i.e. the difference in half-wave potentials of the complex, ML_p minus that of the free metal (in acetate), $(E_p)_c - (E_p)_{M2+}$, ϕ is the sensitivity of the determination, obtained from the linear calibration curve (not shown) of $i_{D,M2+}$ vs. $(M^{2+})_{std}$, (μA per units of chemical activity, (A u_{aq}^{-1}): Substitution of the common R, T, and F values, and converting *ln* to *log*, gives eq. (2):

$$(\Delta E_p)_C = 29.7 \times log \left(M^{2+}\right)_b \quad (2)$$

Figure 1 shows: a) at upper left, the calibration working curves to estimate the free metal activities of Zn, Cd, Pb and Cu in solution; b) at upper right, the information of the selected ligands and pH used to generate specific levels of (M^{2+})-values, calculated by Mineql+, for each metal–ligand system, besides the relevant parameters used for calculations of $(\Delta E_p)_C$; c) at lower left, current-potential curves containing five voltammograms generated for each metal, under the predicted (M^{2+})-values (from left to right, respectively) for Zn, Cd, Pb, and Cu, of 9.7×10^{-6}, 7.1×10^{-6}, 8.2×10^{-7}, 5.9×10^{-7}, and 1.5×10^{-7} in 0.01 M acetic acid, pH = 4.5; and d) at lower right, voltammograms of the real samples, $[M^{2+}]_{ac}$-OM: 20% Soil:80% mine waste solution. From left to right, voltammograms correspond to Zn, Cd, Pb, and Cu, respectively. Free metal activities measured under low (5%) and high (20%) doses of compost added to mixtures (w/w) are presented in red and green, respectively. Levels of metal measured were Zn = 3×10^{-6} (red); Cd = 2.4×10^{-10} (red); Pb = 2.4×10^{-8} (red); and Cu = 3.5×10^{-15} (green) and 4.4×10^{-7} (red).

2.2.2 The total HM soluble fraction

As mentioned before, an important step in the process of chemical speciation of pollutant metals in aqueous solutions is the analytical evaluation of the (M^{2+})-value [57, 58]. Although this parameter helps in the assessment and remediation of polluted sites, it is well known that a fraction of the sorbed metals may also contribute to the bioavailable fraction by replenishing into the solution, part of the exhausted ions that plants take up from the solution. Extractable fractions, i.e., the exchangeable and the readily acid-soluble precipitates (e.g., sulfates, carbonates, etc.), can also substantially contribute to the nutrition of plants, as well as a small fraction of the metals non-specifically adsorbed by organic matter and the Si, Mn, Fe and Al-oxides. Thus, although the free metal ion is the most toxic of metal species, its determination is not the only important one when evaluating the phytoavailable-toxic HM levels. Thus, the

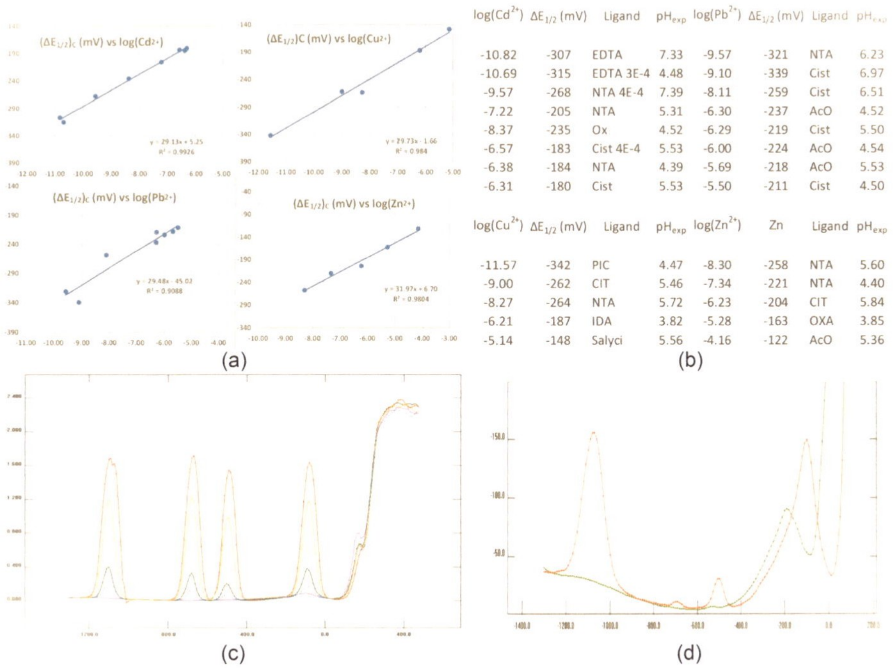

$log(Cd^{2+})$	$\Delta E_{1/2}$ (mV)	Ligand	pH_{exp}	$log(Pb^{2+})$	$\Delta E_{1/2}$ (mV)	Ligand	pH_{exp}
-10.82	-307	EDTA	7.33	-9.57	-321	NTA	6.23
-10.69	-315	EDTA 3E-4	4.48	-9.10	-339	Cist	6.97
-9.57	-268	NTA 4E-4	7.39	-8.11	-259	Cist	6.51
-7.22	-205	NTA	5.31	-6.30	-237	AcO	4.52
-8.37	-235	Ox	4.52	-6.29	-219	Cist	5.50
-6.57	-183	Cist 4E-4	5.53	-6.00	-224	AcO	4.54
-6.38	-184	NTA	4.39	-5.69	-218	AcO	5.53
-6.31	-180	Cist	5.53	-5.50	-211	Cist	4.50

$log(Cu^{2+})$	$\Delta E_{1/2}$ (mV)	Ligand	pH_{exp}	$log(Zn^{2+})$	Zn	Ligand	pH_{exp}
-11.57	-342	PIC	4.47	-8.30	-258	NTA	5.60
-9.00	-262	CIT	5.46	-7.34	-221	NTA	4.40
-8.27	-264	NTA	5.72	-6.23	-204	CIT	5.84
-6.21	-187	IDA	3.82	-5.28	-163	OXA	3.85
-5.14	-148	Salyci	5.56	-4.16	-122	AcO	5.36

Figure 1.
a) Upper left linear graphs show $(\Delta E_p)_c$-log(M^{2+}) calibration curves for Cd, Cu, Pb, and Zn as free metals, $(M^{2+})_{aq}$, obtained with eq. 2. (M^{2+}) was calculated using Mineql+; b) upper right columns show calculated log(M^{2+})-values, measured ASV-$\Delta E_{1/2}$-values (as conditioned by ligands and pH_{exp}) and log-$(\Delta E_p)_c$ graphs; c) Voltammograms at lower left, show experimental peak-current curves for Zn, Cd, Pb, and Cu, respectively, for five free Zn^{2+}, Cd^{2+}, Pb^{2+}, Cu^2 activities, corresponding to 9.7×10^{-6}, 7.1×10^{-6}, 8.2×10^{-7}, 5.9×10^{-7}, and 1.5×10^{-7}, for each metal, in 0.01 M acetic acid pH 4.5; d) lower right voltammograms show current-potential curves for $[M^{2+}]_{ac}$-OM-treated mixture 20%-soil:80%-mine waste solutions. Peak currents in red and green represent (M^{2+})-values measured under low (5% w/w) and high (20%) compost doses added to mixtures. Metal activities were Zn = 3×10^{-6}; Cd = 2.4×10^{-10}; Pb = 2.4×10^{-8}; and Cu = 4.4×10^{-7} (in red) and Cu = 3.5×10^{-15} (green). For ASV-conditions, see [55, 56].

exchangeable-, acid-soluble-, and chelate-extractions, must be considered too. Thus, in the water-soluble fraction both, the (M^{2+})-value, and the total soluble metal must be evaluated when studying HM-polluted mine wastes.

2.3 Extraction procedures for solid phase-bound-heavy metals

One of the most widely used protocols to extract HMs sequentially was proposed by Tessier et al. [6]. Elements were separated into five "operationally" defined fractions: exchangeable; acid-soluble (carbonates); reducible (Fe/Mn oxides); oxidizable (organic matter); and residual. Other authors have referred differently to similar fractions, and even suggesting different order of sequence (chelate extractable, sulfide-associated, etc.), and even modifying concentrations, reaction times, separation procedures, etc. (BCR [7], modified BCR [8]; Geological Society of Canada (GCS)-procedure [69–71]). Based on these, diverse fractions can be visualized to include most HM-containing phases. Although many attempts to unify terms and criteria have been published, the most popular protocols and concepts will be reviewed in terms of chemical relevance to be applied to HM-polluted sites.

2.3.1 The exchangeable and chemisorbed fractions

Heavy metals extracted in the exchangeable fractions comprise both, inner and outer sphere adsorbed species. Whereas the outer sphere weakly adsorbed metal species include those retained on the solid surfaces by relatively weak electrostatic interactions (e.g., van der Waals forces) that can be released by ion-exchange processes, those metals strongly sorbed (chemisorbed and precipitated), are retained covalently by (inner-sphere-) complex interactions. Reagents used for these purposes include mostly the rather strong Mg^{2+} ion-exchange capacity. The most popular reagents used for these extraction procedure are $MgCl_2$, $Mg(NO_3)_2$, $CaCl_2$, $Ca(NO_3)_2$, KNO_3, KCl, NH_4Cl, CH_3COONH_4, CH_3COOH (see Section 2.3.2), and $Ba(NO_3)_2$, among others. These reagents do not attack organic matter, silicates, or metal sulfides [6, 72], although some dissolution of carbonates has been reported [6]. Slight decrease in pH has also been reported during the extraction [73], most probably because heavy metals may displace chemisorbed-H^+ ions (salt effect), or polymeric Al-ions [14] which might hydrolyze leading to a partial dissolution of carbonates and manganese oxide fractions [14, 16]. Extraction with acetate salts, particularly NH_4OOCCH_3, has also been used frequently in soil studies. Divalent cations, in general, are more effective than monovalent cations in ion-exchange processes, but K^+ and NH_4^+ promote the replacement of chemisorbed metal ions in the interlayer exchange sites of some clay minerals (illite and vermiculite). Acetate ions are slightly more stable than chloro-metal-complexes reducing the readsorption and precipitation of the extracted metals and limiting pH variations because of the buffering capacity of the solution [72]. Other reagents showing similar properties have also been used, such as nitrate salts (to avoid complexation) or calcium salts (Ca^{2+} being sometimes more effective than Mg^{2+} or NH_4^+ in removing exchangeable ions, but showing precipitation risks with, e.g., sulfates or phosphates). Results obtained with most of these reagents have shown a good correlation with plant uptake [74]. Permanent charge sites of layer silicate clays also retain metal cations by nonspecific electrostatic forces and, in the absence of conditions that would favor metal hydrolysis (e.g., high pH), divalent (M^{2+}) and trivalent (M^{3+}) transition- and HM-cations show typical ion-exchange behavior on layer silicates [14]. Several studies have confirmed this for ions such as Cu^{2+}, Co^{2+}, Ni^{2+}, Mn^{2+}, etc. which retain their inner hydration sphere offering direct support for the involvement of electrostatic forces only [51]. The strength of metal bonding then, should only depend on charge, ionic radii, and hydration properties of the cation. Thus, the ionic radii series for M^{2+}-ions seem to apply:

$$Ba^{2+} > Sr^{2+} > Ca^{2+} > Mg^{2+} > Hg^{2+} > Cd^{2+} > Zn^{2+} > Mn^{2+} > Fe^{2+} > Co^{2+} > Ni^{2+} > Cu^{2+}.$$

Based on this sequence, it results clear the usefulness of using Mg^{2+}-ions [6] to exchange HM-divalent ions from the nonspecific adsorption sites on clay minerals and other solid phases present on soils (Fe, Mn, Al, and Si oxides and organic matter) [50]. Other ions such as K^+ [75–77], NH_4^+ [78–81], Ca^{2+} [78, 82], Ba^{2+} [76], and even H^+ (from CH_3COOH, [8]) have also been used with this purpose. However, chances of precipitation of Ca^{2+} and Ba^{2+} with specific anions (e.g., CO_3^{2-}, PO_4^{3-}, SO_4^{2-}) must be considered. Increasing the concentration of K^+ and NH_4^+ (to 1 M–2 M), lowering the concentration to 0.01 M for $CaCl_2$, buffering pH, and adding complexing ions for Ba^{2+} may avoid overestimating this fraction. Transition and HMs in soils, when present at trace levels, are retained largely in non-exchangeable forms [14–16]. Schemes for complete metal extraction require extreme treatments, including the oxidative degradation of organic matter and dissolution of Fe and Mn oxides [83, 84]. Even the preferential adsorption of polymeric hydroxy-metal cations by layer silicates would not seem to account for the stability of these sorbed form of metals. Hydr-(oxides) of Si, Al, Fe, and Mn, as well as amorphous aluminosilicates offer surface sites for HM-chemisorption. According to McBride [14], evidence for the formation of surface-metal bonds includes; a) a stoichiometry of $2H^+$ ions released for each M^{2+} ion adsorbed [85]; b) a high degree of specificity shown by Al- and Fe-oxides [86], humic substances for some metals; c) changes in the surface charge properties of the oxide as a result of adsorption [38, 58], this last effect attributed to the increased surface positive charge developed by chemisorption. A generally accepted affinity series for the specific adsorption of HMs by solid phases present in soils and sediments relates directly to their increasing ability to form hydroxy complexes (metal hydrolysis). The expected order of adsorption would then be Hg > Pb > Cu> > Zn > Co > Ni > Cd [87]. Whereas this series correlates well, but not identical, with goethite and hematite, however, several authors have reported different affinity sequences [14, 50, 86, 88]. These sequences indicate that oxides and organic fractions adsorb preferentially Pb, Cu, and Zn, as compared with Cd, Ni, and Co [89, 90]. **Tables 1** and **2** show the relative adsorption selectivity of solid phases for metal ions, and metal ions preferences for adsorption, respectively, if based on their chemical properties. Predicted metal affinity sequences based on their chemical properties are shown in **Table** 2.

Solid phase	Affinity sequence	Reference
Amorphous Al-oxides	Cu > Pb > Zn > Ni > Co > Cd	Kinniburgh et al. [86]
Amorphous Fe-oxides	Pb > Cu > Zn > Ni > Cd > Co	Kinniburgh et al. [86]
Goethite (FeOOH)	Cu > Pb > Zn > Co > Ni > Mn	McKenzie [88]
Hematite	Pb > Cu > Zn > Co > Ni > Mn	McKenzie [88]
Mn-oxide (birnesite)	Pb > Cu > Mn = Co > Zn > Ni	McKenzie [88]
Fulvic acid pH = 5	Cu > Pb > Zn	Schnitzer and Skinner [89]

Table 1.
Heavy metal affinities for some soil fractions (adapted from Ross [50]).

Chemical property	Predicted order of affinity
Electronegativity	Cu > Ni > Co > Pb > Hg > Ag > Fe > Cd > Zn > Mn > Mg > Ca > Sr.
Ionic potential (charge/ radius)	Ni > Mg > Cu > Co > Zn > Cd > Sr. > Pb
Ionic radii	$Cs^+ > Rb^+ > K^+ > Na^+ > Li^+ > Ba^{2+} > Sr^{2+} > Ca^{2+} > Mg^{2+} > Hg^{2+} > Cd^{2+} > Zn^{2+}$
Pearson's Hardness	Sn > Pb > Co > Ni > Fe > Mn > Zn > Cu > Cd > Hg
Softness	Pb > Cd > Co > Cu > Ni > Zn > Sr. > Mg
Hydrolysis	$Cu^{2+} > Pb^{2+} > Ni^{2+} > Co^{2+} = Zn^{2+} > Mg^{2+} > Cd^{2+} > Sr^{2+}$
Irving-Williams series	$Ba^{2+} < Sr^{2+} < Ca^{2+} < Mg^{2+} < Mn^{2+} < Fe^{2+} < Co^{2+} < Zn^{2+} < Ni^{2+} < Cu^{2+}$

Table 2.
Chemical properties determining metal adsorption selectivity on soils, sediments, and mine wastes solid phases [16, 37].

Table 2 shows the expected preferences of adsorption on solid phases of soils, sediments, and mine wastes, of free metal ions, based on the relevant chemical properties that could determine at first instance, the selectivity of solid phases for the metal ions. These are charge, electronegativity, ionic radii [16] which together with charge potential (z/r) or ionic potential (Z^2/r), and the Pearson's hardness parameter, Hp-value [36, 39, 91], o Softness Y-value [16, 37] (polarizability and hardness-softness), directly influence the covalent-ionic character of the adsorbed-adsorbate interaction and the relative affinity of adsorption for each metal [92].

2.3.2 Acid-soluble fraction

The acid-soluble fraction attacks mainly acid-active solid phases, releasing HMs such as Mn and Cd, which usually co-precipitate with carbonates. This procedure attacks solid phases that become soluble at pH ≈ 5. A buffered acetic acid/acetate solution is used (0.1–1 M, pH 2–5). The HM fraction recovered under these conditions not only may come from coprecipitates with carbonate minerals but from parts of specifically adsorbed metals on clay surfaces and edges, organic matter, Fe/Mn oxyhydroxides [72], and some sulfosalts of lead, $PbSO_4$ [93], amorphous Fe-sulfides and Fe associated with pyrrhotite [94]. This reagent releases some trace metals remaining on the specifically adsorbed sites that would other way escape the extraction in previous steps [10]. Although large proportions of total Mn are frequently found in these extracts [95], Tessier et al. [6] concluded that Fe^{2+} and Mn^{2+} were not coming from a partial attack of Fe^{III}/Mn^{IV} oxides but from Mg/Ca carbonate coprecipitates [96, 97], and/or from Mn chemisorbed at calcite surfaces. To get a complete carbonate dissolution, a 0.5 M (pH 4.74) acetate solution can be used [98]. Complexing agents such as EDTA, are used to extract HM ions bound to organic matter too. This acid-soluble extraction procedure if used under sequential extraction protocols should be applied before the oxidation of organic matter [99].

2.3.3 The Fe and Mn hydrous oxides: The reducible fraction

Iron and Mn oxides are excellent HMs-adsorbents. By controlling the reaction E_h and pH, dissolution of metal-oxide phases can be achieved [72]. The most successful reagents to extract the total amount of metal ions associated with these oxides use both, a reducing reagent, and a ligand to retain released ions in a soluble form, the efficacy of the reagent is determined by its reduction potential and the ability to attack Fe and Mn crystalline oxyhydroxides [72]. This dissolution can take place in one to three steps, to separate amorphous and crystalline Mn and Fe-oxides. Hydroxylamine (NH_2OH), oxalic acid ($H_2C_2O_4$), and dithionite ($Na_2S_2O_4$) are the most used reagents.

a. Hydroxylamine (E_h° = 1.87 V) can dissolve different metal oxides, depending on pH, concentration, extracting time, and temperature. To differentiate the various Fe-oxides, warm NH_2OH solutions can be used at pH 2. Acetic acid or HCl is preferred over HNO_3 to avoid readsorption problems [100], taking advantage of complexing properties of ions such as Cl^- or CH_3COO^-. A complete dissolution of amorphous Fe-oxides has been reported [101], skipping the attack of the crystalline phases. Other authors preferred NH_2OH/CH_3COOH solutions for better extraction yields than $NH_2OH{\cdot}HCl$ in HNO_3 [102]. Simultaneous extraction of Mn-Fe-oxides can be achieved with 0.02–0.04 M NH_2OH solutions in 25% CH_3COOH, at high temperatures (96–100°C). Tessier et al. [6] found total dissolution of Fe-reducible fractions within 6 h. However, the protocol seemed insufficient for Fe extraction [97, 103, 104] when Fe content is high [103, 105], for which an additional Fe-specific step is advised [97, 104, 105]. Total Mn- and amorphous Fe-oxides, and partial dissolution of crystalline Fe-oxides, can be reached at low pH (1.7) and high $NH_2OH{\cdot}HCl$ concentrations (0.25 M).

b. Oxalic acid/oxalate ($H_2C_2O_4/C_2O_4^{2-}$) extractions are used due to the relatively high number of stable Fe-oxalate complexes (log K = 4.35–18.49 for Fe^{3+} and 3.20–5.15 for Fe^{2+}) that form and the low reducing properties of the solution ($E° = -0.38$ V). This protocol was originally proposed to specifically eliminate amorphous Fe-oxides from other Fe solid phases in soils, following the previous Mn oxides destruction [105]. However, not only the amorphous Si- and Al-oxides are extracted (due to the very stable Al-oxalate complexes, with log K = 15) [104], but part of the crystalline Fe-oxides (maghemite [101] and lepidocrocite and goethite [106]). HMs associated with Al-oxides cannot be distinguished from those coming from the HMs-interactions with amorphous Fe (hydr-)oxides. Furthermore, as the Fe^{II}-oxalate complex auto-catalytically promotes the Fe^{III}-reduction [107], Fe from organic complexes [101] is sometimes included too. This makes it difficult to use this extraction protocol if detailed specific speciation studies are needed. Thus, to specifically solubilize Fe (hydr-)oxides from the Fe-phases of mine wastes, a 0.2 M $H_2C_2O_4/C_2O_4^{2-}$-solution can be used. Because the reductive dissolution of crystallized Fe^{III}-(hydr-)oxides by oxalate is slow in the dark, the reaction can be UV-catalyzed [32], or a reducing agent, as ascorbate ions [30, 33] or dithionite [6] can be added. Ascorbic acid (HA), with variable $E_h°$-values ($E_h° = 0.19$ V for dehydro-HA, DHA/HA at pH 3.5), increases Fe^{III}-dissolution as pH decreases, the mechanism goes through an inner-sphere complex formation of adsorbed ascorbate on the hydrous Fe^{III}-oxide surfaces, as the electron exchange, on the $Fe^{III/II}$-O bond surfaces, favors the detachment of the more labile Fe^{2+} from the $Fe^{III/II}$-O bond surfaces [107]. Since $C_2O_4^{2-}$ ions form sparingly soluble Ca and Pb salts, the extraction causes low soluble Pb-recoveries [72].

c. Sodium dithionite ($Na_2S_2O_4$, $E_h° = -1.12$ V), a strong reducing agent, is used to dissolve even well-crystallized Fe-oxides at pH 7–8 [6]. To avoid FeS precipitation, a strong ligand is added, and the solution is buffered to stabilize pH and redox potentials. The method [108] uses a sodium citrate-sodium dithionite solution, with additions of $NaHCO_3$ to adjust pH to 7.3 (at 80°C). This method approximates to the combined extraction content of amorphous and crystalline Fe- oxides, but a substantial precipitation of trace metals may cause an underestimation of HMs associated with oxides due to sulfide-sulfate precipitates formation [109]. This reagent is still widely used to evaluate total reactive Fe fractions in pedology [106].

2.3.4 The organic matter and sulfides: The oxidizing fraction

Heavy metals interact in many forms, not only with organic matter, humified materials, and living organisms, in soils and sediments, but with organic detritus or sulfides of some old mine wastes deposits that may have sustained HM-hyper-accumulating plants [50, 72]. In freshly deposited mine wastes, although the content of humic substances can be limited, part of the original sulfidic material may remain, so that the levels of oxidizing fraction can be high. Under oxidizing conditions, organic materials and sulfides tend to be degraded, leading to the release of sorbed metals. So, oxidizing reagents such as H_2O_2 (E° = 1.77 V) or NaClO (E° = 0.90 V), and pyrophosphate ions, are frequently used in fractionation studies to extract HMs associated with organic matter and sulfidic materials. Thus, since some oxidizing agents simultaneously oxidize organic matter and sulfides, this step is more commonly named as the "oxidizing fraction".

a. Hydrogen peroxide, (H_2O_2) in dilute HNO_3 is generally used to prevent metal sequestrations by the Fe hydroxide formation at high pH values. However, under these conditions, even though the oxidation is promoted by time-heating operations [72], organic matter is not completely destroyed, and sulfides are partially dissolved [6]. Maximum efficacy of OM attack was found at 3 h and 2 M H_2O_2 [110]). To avoid readsorption of released metals, the extraction is followed by a weak complexing with NH_4OOCCH_3 in HNO_3 [6]. There have been controversies about the oxalate formation, as a major by-product of the organic matter destruction [72], which may cause a Fe-oxide dissolution and the precipitation of sparingly soluble oxalates. The amount of HM species extracted depends on pH, where the high levels of H^+ ions exchange the non-selectively adsorbed metal cations from OM and other soil fractions. An optimal pH, however, is required to provide the best estimation of trace metals bound to soil organic matter. Because H_2O_2 also has reducing abilities and reduces MnO_2 at pH < 5 [84], this fractionation scheme is placed after the metal-oxide extraction. Some authors prefer to apply it after Mn-oxide dissolution and before Fe-oxides dissolution because the latter step is somewhat able to extract organically complex metals. If the organic matter is oxidized after the exchangeable step to destroy the organic coating of Fe-oxide particles, facilitate the next steps.

b. Hypochlorite, pyrophosphate, and sodium hydroxide. Because the use of sodium hypochlorite, NaClO, in alkaline conditions leads to better destruction of organic matter, besides minimizing the attack to amorphous constituents and clay minerals, it is preferred as an oxidizing reagent, instead of the H_2O_2 procedure. However, it must be considered that some Mn oxides are converted into MnO_4^- ions [72]. The 0.7 M NaClO-extraction is carried out at pH 8.5, and at high temperatures, however, since hypochlorite ions are thermo-unstable in aqueous solutions and decompose quite rapidly, it is advisable to use short extraction times. To get efficient destruction of the organic matter requires to repeat 2–3 times the oxidation step. This procedure prevents the precipitation of metal hydroxides but induces the partial destruction of carbonates. For calcareous materials, a significant part of the high carbonate content is dissolved, so that the extraction should be placed after the acid-soluble step.

c. Sodium pyrophosphate ($Na_4P_2O_7$) in basic media promotes the dispersion of organic colloids. However, at pH 10, amorphous oxides are also extracted [72]. NaOH can dissolve organic matter but also attacks aluminosilicates and clays. This last reagent leads to hydroxide precipitation and is used more for fractionation studies in sludges, which essentially have a very high content of organic matter.

2.3.5 The residual fraction

Primary and secondary minerals containing metals in the crystalline lattice constitute the bulk of this fraction. Its destruction is achieved by digestion with strong acids, such as HF, $HClO_4$, HCl, and HNO_3.

2.4 A note on sequential extraction schemes

Sequential extraction protocols are very useful experimental tools for special cases where complete characterization and HM speciation studies are required. However, in cases where a single bioavailable-toxic fraction is required, these classical extraction procedures are of less use if applied in sequence. Nevertheless, established methodologies may become more instrumental if used as single extraction methods for evaluating HM-fractions that could correlate well with plant responses when exposed to limiting or excessive concentrations of essential trace (e.g., Cu, Mn, Zn, etc.), or toxic (Pb, Cd, etc.) metals. These HM-fractionation schemes, such as those of Tessier (five steps) [6, 111], BCR (four steps) [5, 7, 112], or modified BCR (three steps) [9] serve more to evaluate the potential mobilization of metals in polluted soils, sediments, and mine wastes, where pH fluctuations, extreme potential leaching conditions, or high-risk assessment studies that might foresee floodings and other effects of severe dispersion vectors that can affect specific environments. To reduce the complexity of the procedures but maintain similar outcomes, a three-step scheme has been proposed for HM-polluted soils, sewage sludge, and for studying sulfur in soils [111, 112] and hence sulfidic mine wastes. This three-step procedure uses: acetic acid (step 1), hydroxylamine (step 2), and hydrogen peroxide (step 3). The scheme was then applied for a certification of a sediment reference material (CRM 601), and that allowed it to be validated [113]. Sequential extraction procedures are applied not without presenting several experimental and theoretical problems, mainly due to the lack of selectivity of reagents [94, 98, 114, 115], readsorption and redistribution of metals during the extraction [94, 98, 116] sample pretreatments [73, 94, 100, 103–105], and general methodological associated methods. Regarding incomplete dissolution of some phases and changes in pH can lead to controversial results regarding readsorption and the redistribution of some metals. Many authors have reported that Cu, Zn, and Pb redistribute on Fe-oxides or on humic substances [94, 117] whereas others [118] stated that redistribution was significant only for high metal contents. Carbonated species of the various metals with different solubilities $Ks_{PbCO3} = 10^{-13.1}$; $Ks_{MnCO3} = 10^{-9.3}$; and $Ks_{ZnCO3} = 10^{-10}$, will show incomplete dissolution during this step, and an overestimation of the HMs extracted in the reducing fraction appear too, especially Pb [94, 105, 116] showed the lack of selectivity of these schemes toward sulfides and organic compounds. Extraction of OM by oxidative agents is also unsatisfactory because refractory compounds are not completely destroyed, and sulfides are also oxidized.

From this review, it appears that all the reagents used in the various schemes have advantages and disadvantages and there is no ideal reagent or an ideal protocol for general use. Therefore, the choice of procedure must be related to a definite objective, considering the nature of the sample.

2.5 Mining wastes and the functional extraction procedures selected

Mining of Pb-Zn-Cu ores commonly generates mine wastes rich in Pb, Zn, Cu, and Cd. Some of these tailings contain pyrite-rich materials which produce not only strong acidity when oxidized (pH values <2) but cause emissions of Zn, Pb, and Cd at levels which can cause adverse effects in terrestrial environments. It has been reported that strongly acidic Zn-rich mine wastes cause severe Zn phytotoxicity [20] and can prevent all plants from surviving on the soil. There is evidence of Zn phytotoxicity, potential Cd risk to humans if tobacco, or edible plants, are grown on contaminated soils, and Pb risk to children, if exposed to road and/or house dust [119]. Although there has been important progress in risk assessment strategies for soil metals, and research on methods to remediate Zn, Cd, and Pb polluted soils and sediments, by in situ treatments or by adding amendments (e.g., phosphates, compost, biochar, biosolids, lime-rich wood-ashes, etc.), which reverse phytotoxicity of Zn and Pb risk [11, 119], there is still a strong need to find sound methodologies to remediate HM-polluted mine wastes. The following sections will present examples of such methodologies to handle this type of HM-polluted terrestrial environment.

3. A single extraction study of a metal-polluted mine waste of Central Mexico

3.1 Introduction

In this study, we focused mainly on lowering the bioavailability and mobility of Cd, Cu, Pb, and Zn below official environmentally safe values and to warrant a biologically clean and sustainable ecosystem. To reach this goal, two schemes were visualized. First, we assayed the addition of widely used agronomic materials consisting of lime ($Ca(OH)_2$); gypsum ($CaSO_4{\cdot}2H_2O$); P-fertilizer (KH_2PO_4) and compost to "treat" a gradient of soil-fresh mine tailings mixtures to assess the treatment efficacy to abate the levels of the most toxic metal species available for plant growth; in solution, the free metal ion activity, (M^{2+})-value, for Cd^{2+}, Cu^{2+}, Pb^{2+}, and Zn^{2+}; and on the solid phase; the so-called DTPA-phytoavailable [17], and the acid-soluble fraction imposed by some international [22, 120] standards, and a national norm [23–28]; second, a bioassay was applied to find the conditions that allowed a sensitive indicator plant to grow in these "fertile" ameliorated media. Our studies proved to be useful in deriving soil-substrate quality criteria to establish specific strategies to verify the success of remediation processes. To evaluate the HM-toxicity abatement, both the bioavailable (acid-extractable) HM fraction and the chemical activity of the free metal ion, (M^{2+}) were measured after incubation with the agrostabilizing treatments. Acid drainage was emulated using the standardized acetic acid extraction procedure required by norms [23–28] and standards [22, 121].

3.2 Water, DTPA, and acid-extractable heavy metal levels

Extraction solutions to evaluate water-DTPA- and acid-extractable solutions consisted of 1) H_2O-CO_2 [24, 120] as saturation extract; 2) 1:2 ratio DTPA-extraction [17, 22], and 3) acetic acid (HAcO)-extraction [22], were used to obtain different species and fractions of metals from soil, mine waste, and mixtures. Extracts were analyzed for total dissolved Cd, Pb, Cu, and Zn by FAAS. Initial and in equilibrium (after incubation) extractable levels of metals, $[M]_{HAcO}$, were determined at a 1:20 solid:liquid ratio in 0.2 M CH_3COOH. Equilibrium-free HM ion activities, (M^{2+}), were determined in the aqueous extracts of the treated mixtures by ASV [55, 56]. Calculations of (M^{2+}) were carried out with MINEQL+ software [66, 67].

3.3 Experimental substrate mixtures and agrochemical treatments

Enough total mass for speciation studies and bioassays of six different substrate systems were prepared by mixing soil and mine tailings at various ratios (w/w) to emulate different degrees of soil pollution as follows: A (100:0%), B (80:20%), C (60:40%), D (40:60%), E (20:80%), and F (0:100%) soil:mine waste material. The four agrochemical treatments tested consisted in adding lime [$Ca(OH)_2$], gypsum [$CaSO_4 \cdot 2H_2O$], P-fertilizer [KH_2PO_4], and compost [OM] at three different doses. Agronomic materials were stoichiometrically formulated according to the initial sum of the Cd, Cu, Pb, and Zn extractabilities in 0.2 N HAcO (highest dose), DTPA (medium dose), and H_2O-CO_2 (lowest dose). Compost dose was added to reach 5, 10, and 20% (w/w) of OM. Blank and treated mixture systems were incubated for three weeks to reach equilibrium, adding water to keep a 1:2.5 solid:liquid ratio. For P-fertilizer, the stoichiometric addition also considered the amount of exchangeable Ca^{2+} levels. This test helped to discriminate treatments that efficiently decreased the HM-extractable contents from those shown by the untreated blank mixtures.

3.4 Toxicity bioassays

This bioassay was carried out only for the PO_4 and OM treatments following international standard instructions (ISO 1993 [29], ISO 2005 [30]). At least seven barley (*Hordeum vulgare*) plants per experimental unit were grown in 100 mL black conical plastic pots (max/min radii of 3.7/1.9 cm and 5 cm height) to contain ca. 140 g of material of each of the six soil-mine waste mixtures, including the four ameliorating treatments and the three doses to give a total 72 pot systems. Root length was measured [32] and statistically analyzed with a 95–99% Fisher test of significance against the HM extracted from the mixture systems as an indicator to evaluate the efficacy of treatments and doses to lower the HM toxic effects and to assess the cleanness of the treated polluted mixtures.

3.5 Results

The presence of high amounts of Pb and Zn is common in Zimapan and they are found in combination with As, mainly in minerals of arsenopyrite (AsFeS), scorodite ($FeAsO_4 \cdot 2H_2O$), and in association with pyrrhotite ($Fe_{1-x}S$), pyrite and marcasite (FeS_2), sphalerite (ZnS) and galena (PbS), very common minerals in the area of Zimapan [3] such that As-levels are within the reported values for this element in soils which were 19–17,384 ppm in Mexico [121–124] and within 5200–40,853 ppm in mine tailings of Zimapan [3, 122, 125]. Regarding the four metals of interest levels found were within the reported values for soils 15–7200 ppm for Cu, 31–3400 ppm for Pb, and 26–6270 ppm for Zn [68, 122, 123], whereas for mine wastes in the country, our results were within the reported ranges of 186–2787 ppm, 910–9500 ppm, and 2218–32,400 for the same metals, respectively [3, 121, 122]. Official Mexican regulations [23] established levels of Cd and Pb in soils of the order of 100–300 and 3–5 mg kg^{-1}, respectively, as hazardous to crops. These limits were not exceeded in the soil sample extracts obtained with 0.2 M HAcO. Regarding Cu and Zn, levels higher than 0.2 mg kg^{-1} and 1.0 mg kg^{-1}, respectively, are reported as adequate for these micronutrients [23]. Accordingly, levels of Cd and Pb in mine tailing must not exceed 24 mg kg^{-1} and 120 mg kg^{-1}, respectively, in the aqueous and/or HAcO extracts, so that Pb is not within the allowable levels when extracted with HAcO 0.2 M [26].

Zn and Cu are not potentially toxic elements regulated by Mexican official norms. The efficacy of the agronomic treatments was evaluated by comparing the initial and final quantities of the studied metals, based on the acid-extractable fraction for each experimental mixture. **Figure 2** shows, in contrast with reported values, which found more than 87% decrease of the HCl-extractable concentrations of Cd, Cu, Pb, and Zn in polluted soils, after a combined $CaCO_3$-$CaHPO_4$ stabilizer was added [126], in our studies when agricultural lime and gypsum were applied, the $[M]_{AcO}$-extracted did not show a significant HM level decrease, with respect to their initial concentrations, as compared with controls (see gypsum and lime graphs in **Figure 2**), moreover, redissolution process was observed for all metals except for Pb in the case of gypsum, and with some tendency to positive results for the E and F systems for Cu, Pb, and Zn. For Cd, gypsum worked well only when the soil fraction dominated (systems A, B, and C at lowest dose) but, in general, without significant differences between blank and treatments (F-test, 95%). In contrast with lime and gypsum, P-fertilizer showed excellent results (see **Figure 2** at lower left) when suppressing the acid-extractable levels of Cd, Pb and Zn at any dose getting for the latter diminutions of 92% of initial quantities. For Pb the lowest dose showed a biphasic behavior indicating there exist two distinct sites for sorption which agrees with results found elsewhere [127]. For Cu, a significant decrease of $[M]_{AcO}$ was observed only when the dose and mine

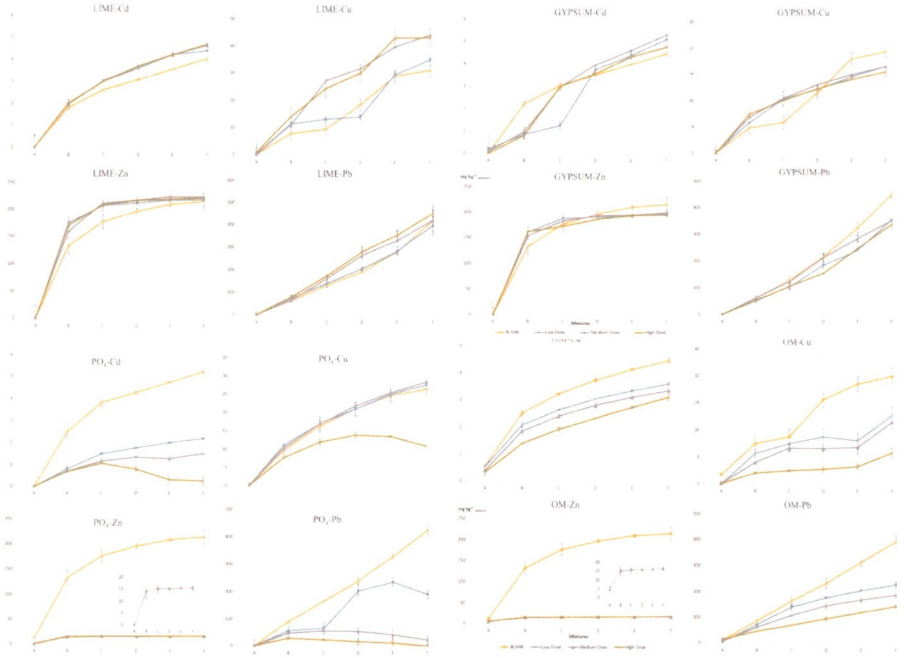

Figure 2.
Effect of adding ameliorating materials lime, $Ca(OH)_2$ (upper left), gypsum, $CaSO_4{\cdot}2H_2O$ (upper right), P-fertilizer, KH_2PO_4 (lower left) and OM-compost (lower right), at low, medium, and high doses, over the 0.2 M acid-extractable levels (mg kg^{-1}) of Cd, Cu, Zn, and Pb (Y-axis), for six experimental mixtures soil:mine waste (s:mw): A: 100% soil, B: 80:20 s:mw, C: 60:40 s:mw, D: 40:60 s:mw, E: 20:80 s:mw, F: 100% mw. Curves show in yellow-orange the blank treatment (no ameliorating material); blue the low dose (ameliorating materials added based on the sum of the four concentrations of water-soluble metals); purple the medium dose (based on the sum of the four concentrations of the DTPA-extractable metals); and red the high dose (based on the sum of the four concentrations of the acid-extractable metals). For OM-compost, low, medium, and high doses were added to reach 5% (low dose), 10% (medium), and 20% (high dose) OM levels (w/w-basis), for the water-soluble, DTPA-extractable, and acid-extractable metals, respectively.

tailing contents were highest for systems C to F (F-test, 95%). **Figure 2** also reveals that compost showed the best results of all amending materials where HM level suppression was more homogeneous. For Cd and Zn this treatment showed a significance reduction of the extractable metal levels at the three doses tested, although results for Zn were much more pronounced. For Cu and Pb the decrease of the extractable metal where mine tailing material was higher (systems D to F) the abatement was significant with respect to the blank system A. However, where the soil was pure or slightly polluted (systems A to C) the effect was not significant, especially for Cu-lowest and medium doses where even the metal extractability increased. For Pb, only the highest and medium doses showed some efficacy in suppressing these values. Note again the biphasic sorption for Cu and Pb at all doses, but more pronounced at the lowest one. These results completely correspond with those obtained by other authors [128–130]; who added composts, biosolids, manure, and peat materials effectively reducing Cd, Pb, and Zn mobility. These results were also consistent with the aqueous free metal, $[M^{2+}]_{ac}$, levels determined by ASV (not shown). Increment in the doses produced an important drop in the activity of this toxic chemical species even in the pure tailing systems, obtaining, in the best case, a diminution of three to five orders of magnitude orders, for example, Cu and Cd system-E treated with OM, respect to control.

3.6 Biotoxicity assays

Toxicity bioassay systems (A to F) and the P-fertilizer and compost treatments were tested at the medium doses, with the only intention of evaluating if there was a chance for a positive response when applying these amending materials and the indicator-sensitive plant could prove a fertile non-toxic media was created. Root length was the agronomic parameter measured [29, 30]. Fisher test was applied to root length (95% significance) to see differences among soil-mine waste mixture treatments (OM and P-fertilizer) and doses (low, medium, and high). **Figure 3** shows the results of these analyses. The effect of treatments including the null one was

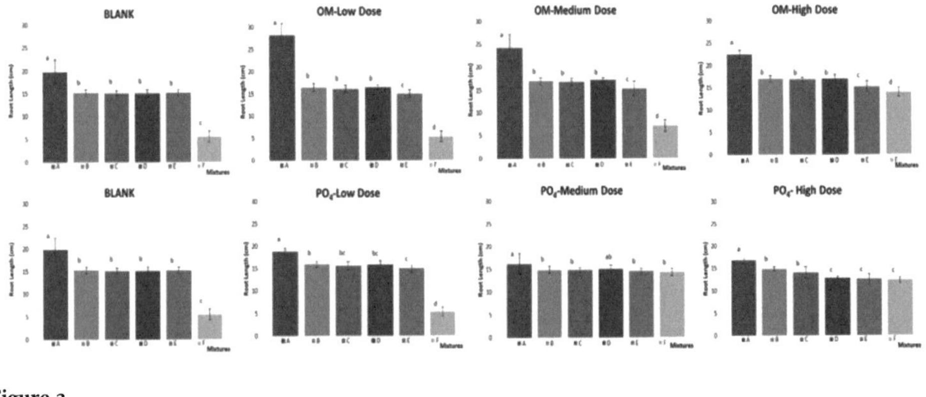

Figure 3.
*Histograms show the root length (cm) of barley (*Hordeum vulgare*) plants as affected by treatments. Upper row shows the results for the blank and compost added; lower row shows the results for blank and P-fertilizer, KH_2PO_4 added, both at low, medium, and high doses, respectively. Different lower-case letters mean significant differences among treatments, according to Fisher's test of 95% significance. Low, medium, and high PO_4-doses were added based on the sum of the four concentrations of the water-soluble, DTPA-, and acid-extractable metals, respectively). For OM-compost, low, medium, and high doses corresponded to 5%, 10% and 20% (w/w-basis), respectively.*

investigated in the six soil:mine waste systems to evaluate the effect of mine waste incorporation and predominance in these emulated scenarios of polluted soils.

Figure 3 shows that roots growing in the different mixture systems were considerably affected because of the increasing content of mine waste material added to the pristine soil, having a shortage of more than 75% of the length, when exposed to pure mine waste (blank-mixture F), respect to the pure soil (blank-mixture A). With the presence of P-fertilizer or OM treatments, a significant increment in root length with respect to control systems (PO_4- and OM-mixture F) was observed, especially for the pure mine tailing mixtures (PO_4-mixture F), where the highest doses improved the growth remarkably. The addition of P-fertilizer in medium and high doses was effective in providing a good media for the growth of the sensitive plant in pure mine tailing whereas compost as shown by the good response of root growth when the dose was highest. Based on these findings, it results clear the DTPA- and the acid-extractable levels of Cd, Cu, Pb and Zn, gave a good indication of the phytoavailability-phytotoxicity levels being suppressed by the ameliorating material added. PO_4-medium and high doses effectively corrected most of the growing problems shown on the blank (not amended) treatment and on the low doses of PO_4 added which was based on the sum of all four water-extracted metals. For the OM-compost treatments, it is important to note that low, medium, and high OM levels were chosen based on what is recommended for optimum growth of plants, according to what FAO and other OM classifications suggest which consider 5–6% as the minimum good OM-level to improve soil fertility [23].

4. Application of sequential metal(loid) extraction to evaluate the impacted area of the Sonora River acid spill

4.1 Introduction

A spill from a copper mine dam located in northern Sonora State, Mexico, occurred on August 6, 2014. Approximately 40,000 m^3 of a diluted acid solution containing iron, aluminum, copper, manganese, and minor amounts of arsenic, nickel, cadmium, zinc, lead, and chromium was released into the Sonora River basin. As part of contingency measures, the mining company built a natural gypsum dike and released 1800 tons of lime gravel along the river (190 km downstream from the source). The significant pH increase favored the precipitation of soluble ions and the sedimentation of suspended particles, mainly clays and iron oxy-hydroxides, likely with adsorbed metals [131]. These whitish sediments were removed and transported back to the mine. There was heavy rainfall the following September due to Hurricanes Norberto and Odile, which dispersed the remaining sediments along the river, possibly reaching the Rodolfo Félix Valdés Dam. However, precipitation was lower in the north, near the accident area, than in the southern portion of the basin. Additionally, pedogenic carbonates from river sediments helped raise the pH of the impacted water [132]. As a result of both natural and anthropogenic attenuation processes, in eight days the river water was already neutral in several areas [133]. The authorities consider the acid solution to have flowed 190 km downstream from the accident site to the Rodolfo Felix Valdés Dam [134]. Nevertheless, based on pH monitoring of the superficial water downstream, acidity (pH 2.6–3.7) reached no farther than 150 km downstream [135]. Total metal(loid) concentrations failed to reveal information concerning the size of the affected area, since it is naturally rich in metal(loids) and because there are waste deposits from current and historical mining in several places. Prior to the accident, significantly high metal values were found in natural compartments. The Mexican Geological Service [136] reported the geochemical composition of sediments with important metal(loid) concentrations [134]. Gonzales-Leon et al. [137] described the geological formation of the Arizpe sub-basin, reporting high natural values of several elements in the soil, including those classified as toxic to humans, such as arsenic. In 2006, the Technological Institute of Sonora reported that concentrations of arsenic, cadmium, and thallium in Sonora River basin groundwater were found to be higher than those specified in the water quality criteria for drinking water by Mexican regulations and other international institutions, such as the Environmental Protection Agency of the USA and the World Health Organization. All these data indicated high metal(loid) baselines. However, despite this fact and the important spill attenuation measures already mentioned, many residents believe that the metal(loid) concentrations found in water bodies and soils are solely due to the spill. This risk perception has caused concern and controversy among inhabitants, journalists, non-governmental organizations, and the authorities [138, 139]. Consequently, spill impact evaluation is important, but discriminating the input from metal spill from that of other sources poses a formidable challenge, since this is an area in which all the metals could have come from the same ore deposits and may share the same isotopic footprints. Total metal(loid) concentrations in sediments, their sequential extraction and bioaccessibility, and other analyses were performed to enhance information on the consequences of the acid solution spill, both for the environment and for its human inhabitants [134]. Some of these results, together with sequential extractions of selected metals in mining soil and wastes, are presented to evaluate the usefulness of chemical fractionation schemes.

4.2 Methodology

4.2.1 Sample preparation and analysis

Representative sediment samples (from impacted and non-impacted sediments in 2015), polluted soils and tailings were transported in hermetically sealed plastic containers and dried at 40°C for 48 h, ground, sieved (mesh #10, < 2 mm) and homogenized by quartering [140] Portions of 100 g were re-milled (Fritsch ball grinder), sieved (mesh #200 < 74 μm), and dried at 96°C for chemical analysis. The samples were preserved at room temperature (20°C) in hermetic containers [141]. All analyses were conducted by LABQA-UNAM (Accredited Laboratory No. R-0593-031/14 by *Entidad Mexicana de Acreditación*). Analytical reproducibility was inspected following the laboratory's QA/QC analytical procedure, using spike samples and certified international reference material, and preparing blanks. All analyses were duplicates, all reagents were analytical grade or high purity, and the water was ultrapure deionized (Nanopure).

Total concentration was measured through X-ray fluorescence (XRF) with a portable model DP-6000-CC Thermo Scientific XRF Olympus analyzer, used following the 6200-method [142]. Sequential Extraction was performed using a modified Tessier et al. [6] method. The procedure consisted of five successive extractions: Fraction I (F1): Exchangeable (1 M $MgCl_2$, pH 7, shaken 250 rpm, 1 h at room temperature (19–23°C)); Fraction II (F2): Carbonates (1 M CH_3COO^-/CH_3COOH buffer, pH 5, shaken 250 rpm, 5 h at room temperature); Fraction III (F3): Fe/Mn Oxides (0.3 M, $Na_2S_2O_4$ + 0.025 M, citric acid+0.175 M, sodium citrate, shaken at

250 rpm, 5 h at 96 ± 3°C); Fraction IV (F4) Organic matter/sulfides: (3 mL [HNO_3] + 5 mL 30% H_2O_2, pH 2, shaken 250 rpm, 5 h, at 85 ± 5°C); and Fraction V (F5): Residual phase, total concentration minus the sum of fractions I-IV. Oral gastric bio-accessibility was determined with method NOM-147 [143], analogous to SBET, RBALP, and SBRC [144]. Solid samples were mixed with glycine ($C_2H_5NO_2$) at pH 1.50 ± 0.05, reached with HCl 1 M, in a ratio (1 g:100 mL); shaken 1 h at 30 rpm, controlling the temperature with an immersion recirculation heater at 37 ± 2°C. Concentrations were measured in accordance with ICP-OES EPA 6010 [145].

4.2.2 Statistical analysis

The Mann–Whitney U test was applied to test the null hypothesis in this work, that is, there will be no statistically significant differences in metal(loid) concentrations in groups of sediments, sediments and polluted soils, or sediments and tailings. This function takes two data samples as parameters, uses the median as a measure of central tendency, and then sends the results with a p-value showing the statistical significance. All analyses use a significance level of $p = 0.05$. If the p-value ≤ 0.05, the conclusion is to reject the null hypothesis and to accept a difference between the ranks of the two groups (sediments, soil, tailings).

4.3 Results and discussion

Tables 3–5 present the results of the sequential extraction and total concentrations of three representative metal(loids) from the acid spill: As, Cu, and Fe. No anionic sequential extraction for As was made, since the results are similar to those obtained with cationic sequential extraction [134]. The recovered fractions were: Exchangeable (I); Carbonates (II); Fe and Mn-oxide/hydroxides (III); Organic Matter/Sulfides (IV); and Residual (V). **Figures 4–7** show the recovery percentages for the three representative elements in each fraction, and **Figure 4** presents the recovery fractions in wastes and polluted soils. There are no statistically significant differences in total concentrations of As, Cu, or Fe, between impacted and baseline sediments (p values = 0.19, 0.21, and 0.07 respectively). Indeed, total As and Fe concentrations means in impacted sediments were slightly lower than those of the baseline sediments (**Tables 3–5**). On the other hand, sequential extraction does provide valuable information: As is the only element for which recovery in F3-fraction was very significant, Cu was recovered in different fractions, including F1, and Fe was retrieved from impacted sediments, predominately in fraction F5 followed by F3 (**Tables 3–5**). The dominant As fractions were F3 (linked to Mn/Fe-oxides) and F5 (residual). F3 distribution was variable along the river, with higher values in the backwaters where sediments have more easily precipitated (**Figure 4**). The differences between impacted and baseline sediments were only statistically significant for Cu in F1-fraction ($p = 1.6 \times 10^{-2}$), and for Fe in F3 ($p = 3.5 \times 10^{-3}$) and F5 ($p = 1.4 \times 10^{-3}$). Therefore, Fe can also be used as a tracer of the impact of the acid solution spill, although the differences between impacted and baseline sediments are most evident in the case of As. In samples of non-impacted sediments, fraction F5 equaled 98.95%, while in impacted sediments an important percentage of As concentration belonged to F3-fraction. Significant differences between baseline and impacted sediments were statistically proved (F3, $p = 3.4 \times 10^{-5}$ and F5, $p = 3.2 \times 10^{-5}$) (**Figure 4**). Significant differences were also found when statistically comparing fraction F3 of impacted sediments with the same fraction of tailings and polluted soil not affected by the spill

As mg/kg	Samp.	S1	S2	S3	S4	S5	S6	S7	S8	S9	S10	S11	S12	S13	S14	S15	S16	S17
	Tot.	29.4	34.9	36.3	43.3	36.7	33.3	45.0	37.0	39.1	36.2	32.7	25.5	24.4	30.8	47.4	27.8	31.6
	F1	≦ LOD	≦ LOD	≦ LOD	≦ LOD	≦ LOD	≦ LOD	≦ LOD	≦ LOD	≦ LOD	≦ LOD	≦ LOD	≦ LOD	≦ LOD	≦ LOD	≦ LOD	≦ LOD	≦ LOD
	F2	≦ LOD	≦ LOD	≦ LOD	≦ LOD	≦ LOD	≦ LOD	0.79	≦ LOD	≦ LOD	≦ LOD	≦ LOD	≦ LOD	≦ LOD	≦ LOD	≦ LOD	≦ LOD	≦ LOD
	F3	22.6	25.4	23.0	29.0	5.3	5.5	9.9	6.1	9.8	9.2	6.7	4.5	4.1	5.0	25.0	18.0	16.2
	F4	≦ LOD	≦ LOD	≦ LOD	≦ LOD	≦ LOD	≦ LOD	0.79	≦ LOD	≦ LOD	≦ LOD	≦ LOD	≦ LOD	≦ LOD	≦ LOD	≦ LOD	≦ LOD	≦ LOD
	F5	6.9	9.5	13.3	14.3	31.4	27.8	35.1	30.9	29.3	27.0	26.0	21.0	20.3	25.9	22.4	9.8	15.4
	Samp.	**S18**	**S19**	**S20**	**S21**	**S22**	**S23**	**S24**	**S25**	**S26**	**BL1**	**BL2**	**BL3**	**BL4**	**BL5**	**BL6**	**BL7**	**BL8**
	Tot.	39	31	51	39	32	24	25	23	23	50	41	35	33	27	30	45	35
	F1	≦ LOD	≦ LOD	≦ LOD	≦ LOD	≦ LOD	≦ LOD	≦ LOD	≦ LOD	≦ LOD	≦ LOD	≦ LOD	≦ LOD	≦ LOD	≦ LOD	≦ LOD	≦ LOD	≦ LOD
	F2	≦ LOD	≦ LOD	≦ LOD	≦ LOD	≦ LOD	≦ LOD	≦ LOD	≦ LOD	≦ LOD	3.09	≦ LOD	≦ LOD	≦ LOD	≦ LOD	≦ LOD	≦ LOD	≦ LOD
	F3	25.9	21.4	33.3	27.4	22.4	11.6	≦ LOD	0.5	0.3	≦ LOD	≦ LOD	≦ LOD	≦ LOD	≦ LOD	≦ LOD	≦ LOD	≦ LOD
	F4	≦ LOD	≦ LOD	≦ LOD	≦ LOD	≦ LOD	≦ LOD	≦ LOD	≦ LOD	≦ LOD	≦ LOD	≦ LOD	≦ LOD	≦ LOD	≦ LOD	≦ LOD	≦ LOD	≦ LOD
	F5	13.0	9.3	17.3	11.9	9.6	12.0	24.9	22.9	22.7	46.6	40.8	34.7	32.6	26.6	29.7	45.2	35.4

LOD = Limit of detection = 0.25 mg/kg.
S = sediment, BL = base line sediment, Samp = Sample, Tot = total concentration.

Table 3.
As Sequential extraction and total concentration.

Cu mg/kg	Samp	S1	S2	S3	S4	S5	S6	S7	S8	S9	S10	S11	S12	S13	S14	S15	S16	S17
	Tot	146.7	144.3	129.7	274.7	279.7	470.3	502.0	140.3	136.0	135.0	177.7	186.0	136.0	91.7	176.7	170.7	117.0
	F1	1.8	26.6	21.2	78.0	100.0	229.4	77.6	≦ LOD	22.3	23.7	33.3	47.7	27.6	12.4	21.4	19.5	7.0
	F2	12.1	6.3	4.6	6.5	11.1	34.2	16.3	11.8	7.4	8.2	11.2	15.0	10.2	7.9	2.4	2.1	4.8
	F3	85.2	59.7	58.7	151.9	84.2	189.9	226.3	40.7	44.6	44.6	54.4	48.1	44.5	31.4	51.5	45.6	66.6
	F4	≦ LOD	≦ LOD	≦ LOD	1.4	≦ LOD	≦ LOD	≦ LOD	≦ LOD	≦ LOD	≦ LOD	≦ LOD	≦ LOD	≦ LOD	0.1	≦ LOD	1.9	3.1
	F5	47.6	51.7	45.2	36.8	84.5	16.8	181.9	87.9	61.7	58.5	78.9	75.3	53.8	39.9	101.4	101.6	35.5
	Samp	**S18**	**S19**	**S20**	**S21**	**S22**	**S23**	**S24**	**S25**	**S26**	**BL1**	**BL2**	**BL3**	**BL4**	**BL5**	**BL6**	**BL7**	**BL8**
	Tot.	177.3	163.0	98.7	280.7	792.3	298.0	76.7	199.3	20.0	334.0	277.0	269.3	84.0	155.7	125.0	279.3	355.0
	F1	13.6	23.8	5.5	7.3	≦ LOD	≦ LOD	6.2	2.3	0.6	20.6	5.4	≦ LOD	≦ LOD	1.8	1.8	21.3	28.5
	F2	17.5	8.8	3.9	8.4	33.9	25.4	4.7	4.6	0.6	22.5	7.4	≦ LOD	7.2	2.1	1.8	30.1	21.8
	F3	114.0	85.4	51.4	254.2	718.4	227.8	21.2	74.3	10.3	156.0	123.0	43.6	70.0	60.9	72.1	126.5	98.1
	F4	9.3	≦ LOD	≦ LOD	9.6	24.2	10.6	0.7	≦ LOD	0.5	7.7	4.8	≦ LOD	1.6	≦ LOD	1.4	23.0	17.6
	F5	23.0	45.0	37.9	1.3	15.9	34.2	43.9	118.1	8.1	128.0	136.5	225.7	5.3	90.9	47.9	78.4	189.0

LOD = Limit of detection = 0.13 mg/kg.
S = sediment, BL = base line sediment, Samp = Sample, Tot = total concentration.

Table 4.
Cu Sequential extraction and total concentration.

Fe mg/kg	**Samp.**	**S1**	**S2**	**S3**	**S4**	**S5**	**S6**	**S7**	**S8**	**S9**	**S10**	**S11**	**S12**	**S13**	**S14**	**S15**	**S16**	**S17**
	Tot	48,001	51,651	5221	48,977	4926	44,901	47,811	51,920	5397	4380	4809	39,901	37,562	4609	5108	41,952	4082
	F1	≦ LOD	≦ LOD	≦ LOD	≦ LOD	≦ LOD	≦ LOD	≦ LOD	5	≦ LOD	≦ LOD	≦ LOD	4	≦ LOD	≦ LOD	≦ LOD	≦ LOD	≦ LOD
	F2	5	26	27	20	159	54	316	145	360	400	245	188	180	1	56	877	3
	F3	10,747	12,391	4895	9673	2938	2600	3447	3390	4978	3957	4167	2929	3257	4183	2203	20,091	3683
	F4	5	10	15	≦ LOD	≦ LOD	≦ LOD	≦ LOD	≦ LOD	≦ LOD	4	≦ LOD	≦ LOD	4	0	1	8	≦ LOD
	F5	37,244	39,224	284	39,285	1829	42,247	44,048	48,379	59	18	397	36,780	34,121	425	2849	20,976	396
	Samp	**S18**	**S19**	**S20**	**S21**	**S22**	**S23**	**S24**	**S25**	**S26**	**BL1**	**BL2**	**BL3**	**BL4**	**BL5**	**BL6**	**BL7**	**BL8**
	Tot.	42,354	43,575	63,785	52,889	45,416	45,336	41,060	4428	47,317	57,478	49,638	49,517	49,820	44,929	45,306	17,888	10,608
	F1	≦ LOD	≦ LOD	≦ LOD	≦ LOD	≦ LOD	≦ LOD	41	≦ LOD	5	≦ LOD	5	≦ LOD	≦ LOD	13	5	7	11
	F2	30	13	26	16	5	≦ LOD	57	225	≦ LOD	6	≦ LOD	≦ LOD	≦ LOD	≦ LOD	≦ LOD	5	15
	F3	7192	8589	14,364	31,977	13,084	7236	2780	294	2820	8455	2958	2921	3537	3316	3035	1159	1779
	F4	≦ LOD	4	6	5	≦ LOD	≦ LOD	189	≦ LOD	14	12	15	≦ LOD	35	4	≦ LOD	93	49
	F5	35,133	34,969	49,388	20,891	32,327	38,100	37,993	3909	44,478	49,005	46,660	46,595	46,248	41,595	42,266	16,623	8755

LOD = Limit of detection = 0.25 mg/kg.
S = sediment, BL = base line sediment, Samp = Sample, Tot = total concentration.

Table 5.
Fe Sequential extraction and total concentration.

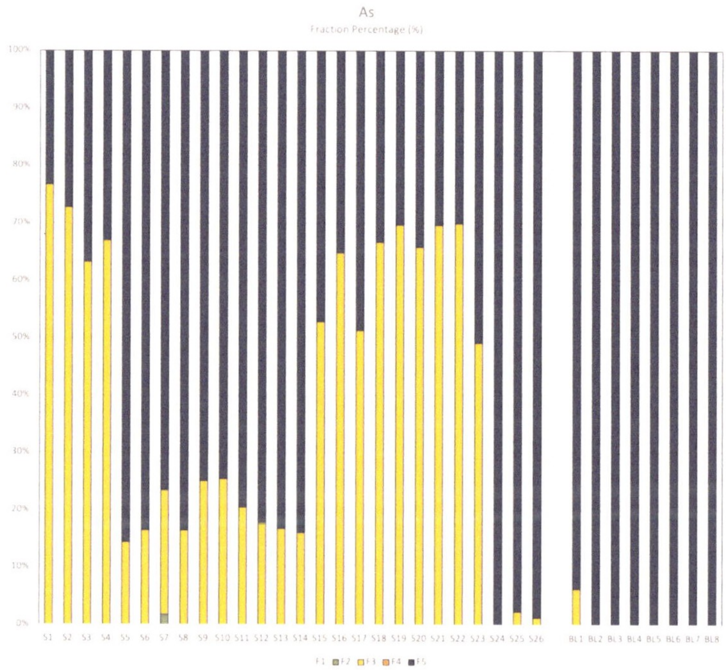

Figure 4.
Sequential extraction for As.

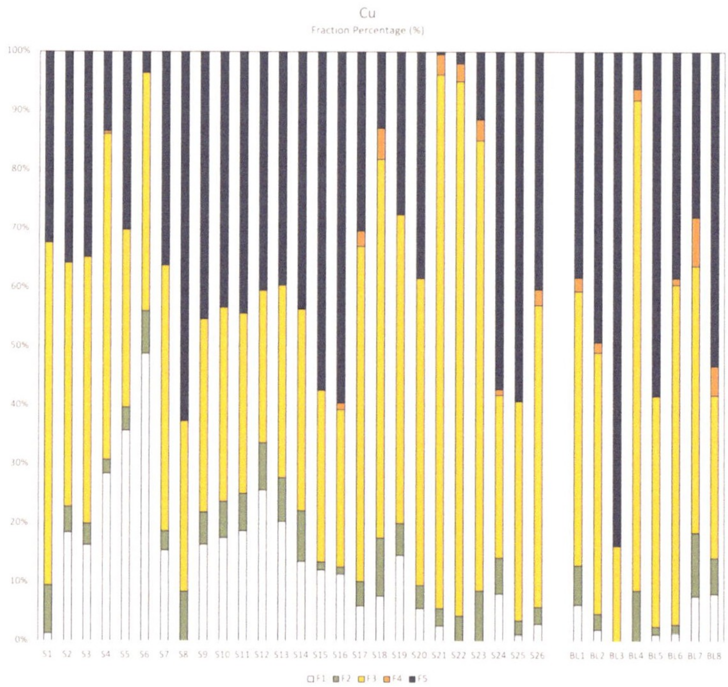

Figure 5.
Sequential extraction for Cu.

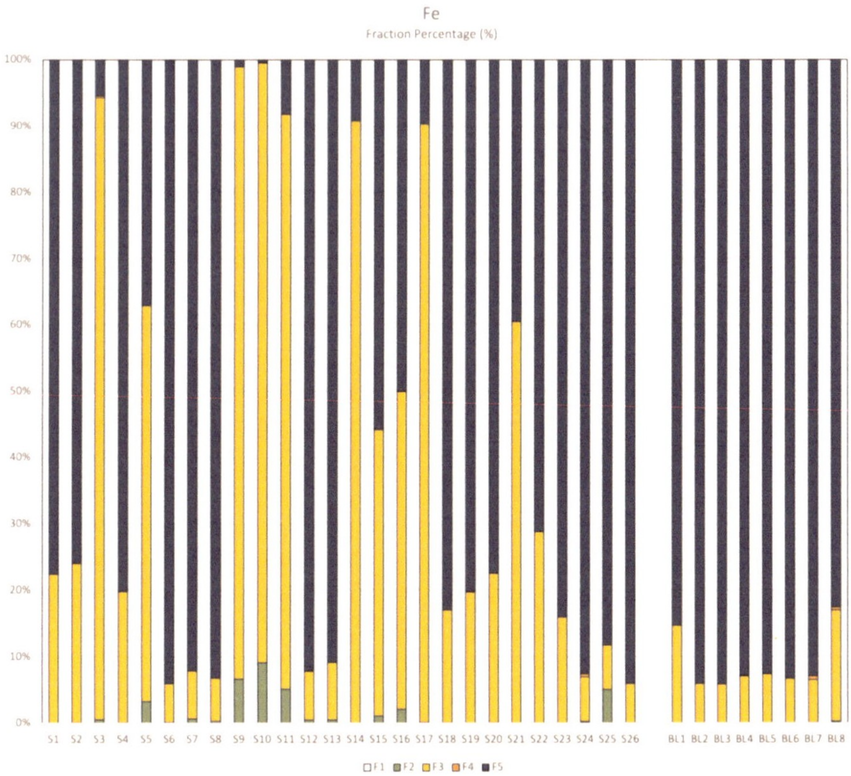

Figure 6.
Sequential extraction for Fe.

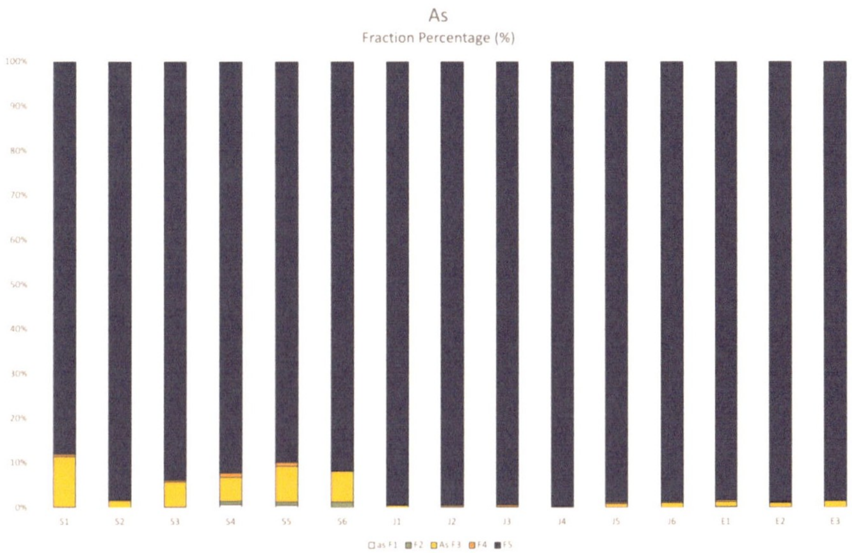

Figure 7.
Sequential extraction for As in polluted soils and tailings.

($p = 2.5 \times 10^{-4}$ and $p = 1.6 \times 10^{-3}$, respectively). This behavior indicates that they mainly contain arsenopyrite in relatively high concentrations, which is the most reported As-mineral in the area [146] and recovered in the F5-residual fraction [147]. After sediment 24, in samples 25 and 26, practically all of the As was recovered in F5-fraction. This behavior was also observed in the following samples (data not shown), indicating no acid solution impact downstream at those sites. The high As recovery in fraction F3 of impacted sediments is mainly a consequence of the chemical changes that took place between the waste rock and the dam (the leaching process), and likely in the river as well. The acid solution spilled from the dam was a lixiviate formed during the pretreatment of rock waste deposits. Those low-grade Cu minerals were doused with a weak sulfuric acid solution, to destroy the basic minerals occurring naturally in waste rock and that partially neutralize the added sulfuric acid. This process reduces future Cu leaching. In the waste rock, most of the As was in the form of arsenopyrite (FeAsS) and was possibly present in lower concentrations as scorodite [147]. Additionally, traces of As_2O_3 have been reported in the waste rock deposit. Impure sulfuric acid is added to waste rock to boost the microbiological oxidation of S-minerals. Sulfur oxidation increases Cu recuperation from Chalcocite and Chalcopyrite. Under acidic conditions, As^{III} can be partially oxidized to arsenates by the MnO_2 from waste rock [148]. Arsenates over pH 2 lose H^+-ions, forming negatively charged species which could be retained on jarosite by sulfate substitution and/or forming inner and outer sphere complexes [149]. They can also be sorbed by schwertmannite, amorphous ferrihydrite, maghemite and goethite [150]. These As retention processes could likely happen at the dam and or in the river when basic materials added to the water lead to the formation of amorphous Fe compounds with high sorption capacity. The arsenates sorbed onto Fe compounds were recovered in F3-fraction due to Fe^{III}-reduction. Non-oxidized As^{III} (arsenopyrite) was recovered on the F5-fraction, as Ankan and Schreiber [147] also observed. From sample sites 23 and up, As was recovered in F5. Although only sediments 24 and 26 are shown in **Figure 4**, the same behavior was observed for the rest of the analyzed sediments (data not presented). Thus, the segment of the river impacted by the acid solution spill was no greater than 30 km, an area significantly smaller than what was initially considered before remedial action was taken (190 km), indicating that control actions were effective. Although total concentration could not be used as a guide for the impacted area, the results show that sequential extraction allowed the distinction of the impact zone from other anthropogenically polluted materials containing natural minerals. The As in fraction F3 was the best tracer.

5. Conclusions

Based on the theoretical discussion presented, regarding the right selection of a single extraction protocol, it results clear that for functional speciation studies, it is of utmost relevance to evaluate and correlate the phytoavailable/phytotoxic HM levels determined, as affected by the ameliorating agronomic materials added, with plant growth, through the application of HM-sensitive plant bioassays.

Total concentration cannot be used as a guide to determine the area impacted by the acid spill, because no difference in total metal(loid) concentrations was noted between polluted and baseline sediments after control and remedial measures were taken. On the other hand, As fractionation is an excellent option. Two other reported strategies for the identification of areas of the river impacted by the acid solution spill are based on the detection of gypsum formed after the addition of calcite, and of jarosite formed only under the acidic conditions prevailing at the dam. However, these two methods require sophisticated equipment, as amorphous particles are practically invisible to most analytical techniques. In their stead, sequential extraction conforms to a relatively simple and inexpensive method. Monitoring metal(loid) behavior is recommended in this area to evaluate changes in sediments. The eventual attainment of an equilibrium that would form more insoluble compounds, mainly from Cu, is expected. This F1-fraction recovered metal can be desorbed, which would likely impact macro-invertebrate populations of the river.

Acknowledgements

We thank Daniel Amaro for his general support and Reyna Roldan, Esau Jimenez, Eduardo Godoy, and Edgar Bonilla from the Laboratory of Environmental Biogeochemistry of UNAM for their valuable participation in the chemical analyses. We would also like to thank Sonia Helen Ponce Wainer for proofreading and editing.

Author details

Arturo Aguirre Gómez[1*] and Margarita Eugenia Gutiérrez Ruiz[2]

1 Faculty of Superior Studies, Laboratory of Agricultural and Environmental Chemistry, National Autonomous University of Mexico, Cuautitlan, Mexico

2 Faculty of Chemistry, Laboratory of Biogeochemistry, National Autonomous University of Mexico, Cuautitlan, Mexico

*Address all correspondence to: aag@unam.mx

References

[1] Sobek AA, Schuller WA, Freeman JR, Smith RM. Field and laboratory methods applicable to overburdens and mine soils. In: US Environmental Protection Agency Environmental Protection Technology, EPA-600. Cincinnati, OH; 1978

[2] Schafer and Associates and Reclamation Research, Laboratory analytical protocol for the Streambank Tailings and Revegetation Study, Silver Bow Creek RI IFS, EPA Region VIII, Doc. No. SBC-STARS-LAPF-RI-1987;121-187

[3] Moreno TR, Barbanson L, Careño AD. Neoformación mineralógica en residuos mineros (jales) del distrito minero Zimapán, Estado de Hidalgo, México. Minería y Geología. 2009;**25**(2):1-31

[4] Armienta MA, Dótor A, Arcega-Cabrera FE, Talavera O, Aguayo A, Ceniceros N, et al. Assessing water-sediment processes for metals in rivers polluted by mining to predict environmental impacts in developing countries. IAHS Pub. 2007;**134**:155

[5] Ure AM, Davidson CM, editors. Chemical Speciation in the Environment. 2nd. ed. U.K.: Blackwell Science. London; 2002

[6] Tessier A, Campbell PGC, Bisson M. Sequential extraction procedure for the speciation of particulate trace metals. Analytical Chemistry. 1979;**51**(7): 844-851. DOI: 10.1021/ac50043a017

[7] Ure AM, Quevauviller P, Muntau H, Griepink B. Speciation of heavy metals in soils and extraction techniques undertaken under the auspices of BCR of the Commission of the European Communities. International Journal of Environmental Analytical Chemistry. 1993;**51**:135-151. DOI: 10.1080/03067319308027619

[8] Rauret G et al. Improvement of the BCR three step sequential extraction procedure prior to the certification of new sediment and soil reference materials. Journal of Environmental Monitoring. 1999;**1**(1):57-61. DOI: 10.1039/A807854H

[9] Cruz-Ruiz E. Effect of liming, manuring, and fertilizing of acid mining wastes on the chemical availability and bioaccumulation of Cu, Cd, Pb and Zn in Festuca arundinaceous. M.Sc. [Thesis] on Earth Sciences Program at National Autonomous University of México (in Spanish) 2017

[10] Guzmán RG. Strategy for the Remediation of Sites Contaminated with Cu, Cd, Pb and Zn Applying Chemical Agronomic Treatments to Mining Metallurgical Wastes and Soils. Mexico: of Environmental Engineering at National Autonomous University of Mexico; 2012

[11] Aguirre AG, Nuñez LVB, Hidalgo CM, Etchevers JDB. The Use of Biochar of High Growth Rate Plants to Agriculturally Remediate Heavy Metal Polluted Acidic Mine Wastes. London, UK, U.K.: Intechopen; 2022

[12] Jennings SR, Dollhopf DJ). Acid–base account effectiveness for determination of mine waste potential acidity. Journal of Hazardous Materials. 1995;**41**(2–3):161-175. DOI: 10.1016/0304-3894(95)00003-D

[13] Romero FM, Gutiérrez MR. Comparative study of the danger of mine wastes in two mining areas located in southern and Central Mexico. Bulletin of the Mexican Geological Society. 2010;**62**: 43-53

[14] McBride MB. Reactions controlling heavy metal solubility in soils. Advances in Soil Sciences. 1989;**10**:1-56

[15] McBride MB. Processes of Heavy and Transition Metal Sorption by Soil Minerals. Dordrecht: Kluwer Academic Publishers; 1991

[16] McBride MB. Environmental Chemistry of Soils. New York: Oxford University Press; 1994

[17] Lindsay WL, Norvell WA. Development of a DTPA soil test for zinc, iron manganese and copper. Soil Science Society of America Journal. 1978; **42**(3):421-428. DOI: 10.2136/sssaj1978.03615995004200030009x

[18] Basta NT, Ryan JA, Chaney RL. Trace element chemistry in residual-treated soil: Key concepts and metal bioavailability. Journal of Environmental Quality. 2005;**34**(1):49-63. DOI: 10.2134/jeq2005.0049dup

[19] Houba VJG, Lexmond ThM, Novozamsky I, Lee JJ van der. State of the art and future developments in soil analysis for bioavailability assessment. Science Total Environments, 1996; 178 (1–3): 21-28. DOI: 10.1016/0048-9697(95)04793-X

[20] Chaney RL. Zinc phytotoxicity. In: Robson AD, editor. Zinc in Soils and Plants. Dordretch: Kluwer Academic Publi; 1993. pp. 135-150

[21] Violante A, Cozzolino V, Perelomov L, Caporale AG, Pigna M. Mobility and bioavailability of heavy metals and metalloids in soil environments. Journal of Soil Science and Plant Nutrition. 2010;**10**(3):268-292 ISSN 0718-9516

[22] EPA. US-Environmental Protection Agency. Method 1311: Toxicity Characteristic Leaching Procedure, part of Test Methods for Evaluating Solid Waste, Physical/Chemical Methods. 1992

[23] NOM-021-RECNAT-SEMARNAT-2000, which establishes the specifications of fertility, salinity, and classification of soils. Study, sampling, and analysis, within the standard you can find definitions regarding the classification of soils and their fertility conditions and/or salinity, as well as the methods of sampling and analysis for each of these purposes and how to interpret the results of the corresponding analyzes (in Spanish)

[24] NOM-141-SEMARNAT-2003, which establishes the procedure for characterizing the mine wastes, as well as the specifications and criteria for the characterization and preparation of the site, project, construction, operation, and post-operation of tailings dams (in Spanish)

[25] NOM-147-SEMARNAT/SSA1-2004, which establishes criteria for determining the remediation concentrations of soils contaminated by arsenic, barium, beryllium, cadmium, hexavalent chromium, mercury, nickel, silver, lead, selenium, thallium and/or vanadium, this standard refers to soils known to have been contaminated due to unsustainable practices. It aims to provide methods for assessing background and reference concentrations, bioavailability and remediation concentrations of pollutants (in Spanish)

[26] NMX-AA-132-SCFI-(2006) which issues general recommendations for developing sampling plans for soils contaminated with heavy metals and metalloids, to determine the existence, location, and extent of contamination of a property (in Spanish)

[27] NOM-157-SEMARNAT-2009, which establishes the elements and procedures for implementing mining waste management plans, whose objective is to establish the elements and procedures that must be considered when formulating and applying mining waste management plans, with the purpose of promoting the prevention of waste generation and recovery (in Spanish)

[28] NOM-120-SEMARNAT-2011, which establishes the environmental protection specifications for direct mining exploration activities, in agricultural, livestock or wasteland areas and in areas with dry and temperate climates where xerophytic scrub vegetation, tropical deciduous forest, coniferous forests or oaks develop (in Spanish)

[29] ISO, 11269-1. Determination of the Effects of Pollutants on Soil Flora - Method for the Measurement Growth. Ginebra, Switzerland: International Organization for Standardization; 1993

[30] ISO. 11269–2. Soil Quality - Determination of the Effects of Pollutants on Soil Flora: Part 2: Effects of Chemicals on the Emergence and Growth of Higher Plants Edition. Ginebra, Switzerland: International Organization for Standardization; 2005

[31] OCDE. Guideline for Testing Chemicals. Terrestrial Plant Test: Seedling emergence and Seedling Growth Test 208-227. Organization for Economic Cooperation and Development. Paris: France; 2004. p. 21

[32] Ginocchio R, Rodríguez P. Validation of the Terrestrial Biotic Ligand Model: Prediction of Copper Toxicity to Barley in Soils of North-Central Chile. Chile: Centro de Investigación Minera y Metalúrgica (CIMM); 2007

[33] CEPA, California Environmental Protection Agency. Soil Toxicity and Bioassessment Test Methods for Ecological Risk Assessment. USA: Office of Environmental Health Hazard Assessment; 2008

[34] EPS. Biological Test Method: Test for Measuring Emergence and Growth of Terrestrial Plants Exposed to Contaminants in Soil EPS1/RM/45. Canada: Environmental Protection Service; 2007

[35] IUPAC. Duffus J. "Heavy metals" A meaningless term? Pure and Applied Chemistry. 2002;**74**(5):793-807. DOI: 10.1351/pac200274050793

[36] Pearson RG. Hard and soft acids and bases. Journal of American Chemical Society. 1963;**85**(22):3533-3539. DOI: 10.1021/ja00905a001

[37] Huheey JE. Inorganic Chemistry: Principles of Structure and Reactivity. NY: Harper & Row; 1972

[38] Stumm W, Morgan JJ. Aquatic Chemistry: An Introduction Emphasizing Chemical Equilibria in Natural Waters. Environmental Science and Technology Series. 2a. ed. USA: Wiley; 1981

[39] Pearson RG. The HSAB principle-more quantitative aspects. Inorganica Chimica Acta. 1995;**240**(1–2):93-98. DOI: 10.1016/0020-1693(95)04648-8

[40] Alloway BJ. Heavy Metals in Soils. 2nd ed. Reino Unido: Chapman & Hall; 1995

[41] Aguirre AG. Química de los suelos de inundación temporal y perenne. México: Universidad Nacional Autónoma de México; 2007

[42] Bacon JR, Davidson CM. Is there a future for sequential chemical

extraction? The Analyst. 2008;**133**(1): 25-46. DOI: 10.1039/B711896A

[43] IUPAC, Templeton DM, Ariese F, Cornelis R, Danielsson LG, Muntau H, et al. Guidelines for terms related to chemical speciation and fractionation of trace elements. Definitions, structural aspects, and methodological approaches. Pure and Applied Chemistry. 2000;**72** (1453)

[44] Nystrand MI, Österholm P, Nyberg ME, Gustafsson JP. Metal speciation in rivers affected by enhanced soil erosion and acidity. Applied Geochemistry. 2012;**27**(4):906-916. DOI: 10.1016/j.apgeochem.2012.01.009

[45] Bjerregaard P, Andersen O. Ecotoxicology of metals: Sources, transport, and effects in the ecosystem. In: Gunnar FN, Bruce AF, Monica N, Lars TF, editors. Handbook on the Toxicology of Metals, 3rd. Burlington: Academic Press; 2007. pp. 251-277

[46] de Paiva MD, da Costa Marques MR, Baptista DF. Metal bioavailability and toxicity in freshwaters. Environmental Chemistry Letters. 2015;**13**(1):69-87. DOI: 10.1007/s10311-015-0491-9

[47] Munne A, Tirapu L, Sola C, Olivella L, Vilanova M, Ginebreda A, et al. Comparing Chemical and Ecological Status in Catalan Rivers: Analysis of River Quality Status Following the Water Framework Directive. Emerging and Priority Pollutants in Rivers. Vol. 19. Berlin: Springer; 2012. pp. 243-265

[48] Vijver MG, van Gestel CAM, Lanno RP, Straalen van NM, Peijnenburg WJGM. Internal metal sequestration and its ecotoxicological relevance: A review. Environmental Science & Technology. 2004;**38**(18): 4705-4712. DOI: 10.1021/es040354g

[49] Rainbow PS, Luoma SN. Metal toxicity, uptake, and bioaccumulation in aquatic invertebrates-modelling zinc in crustaceans. Aquatic Toxicology. 2011; **105**(3–4):455-465. DOI: 10.1016/j.aquatox.2011.08.001

[50] Ross SM. Toxic Metals in Soil-Plant Systems. Chichester, England: John Wiley & Sons; 1994

[51] Sposito G. The Chemistry of Soils. Oxford: Oxford University Press; 1989

[52] Ritchie GSP, Sposito G. Speciation in soils. In: Ure AM, Davidson CM, editors. Chemical Speciation in the Environment. 2nd ed. London, U.K.: Blackwell Science; 2002

[53] Degryse F, Smolders E, Parker DR. Partitioning of metals (Cd, Co, Cu, Ni, Pb, Zn) in soils: concepts, methodologies, prediction, and applications – a review. European Journal of Soil Science. 2009;**60**(4): 590-612. DOI: 10.1111/j.1365-2389.2009.01142.x

[54] Nolan AL, McLaughlin MJ, Mason SD. Chemical speciation of Zn, Cd, Cu and Pb in pore waters of agricultural and contaminated soils using Donnan dialysis. Environmental Science & Technology. 2003;**37**:90-98. DOI: 10.1021/es025966k

[55] Aguirre-Gómez A, McBride MB, Norvell WA. A voltammetric method for determining free metal activities in aqueous solutions. Part 2: Anodic stripping voltammetry of Cd, Cu, Pb and Zn in synthetic and soil solutions. International Journal of. Environment and Pollution. 2006;**26**:68-89. DOI: 10.1504/IJEP.2006.009099

[56] Aguirre-Gómez A, McBride MB, Norvell WA. A voltammetric method for determining free metal activities in

aqueous solutions. Part 1: Direct current voltammetry of Cd, Cu, and Pb. International Journal of Environment and Pollution. 2006;**26**:41-67

[57] Buffle J, Tercier-Waeber ML. Voltammetric environmental trace metal analysis and speciation: From laboratory to in situ measurements. Trends in Analytical Chemistry. 2005;**24**(3): 172-191. DOI: 10.1016/j.trac.2004.11.013

[58] Stumm W, Morgan JJ. Aquatic Chemistry, Chemical Equilibria and Rates in Natural Waters. 3rd ed. New York: John Wiley & Sons, Inc; 1996

[59] Zhang H, Young SD. Characterizing the availability of metals in contaminated soils. II. The soil solution. Soil Use & Management. 2006;**21**: 459-467. DOI: 10.1079/SUM2005349

[60] Fish SJ, Brassard P. Dialysis membrane to prevent cadmium ion specific electrode fouling. Talanta. 1997; **44**(5):939-945. DOI: 10.1016/S0039-9140(96)02144-3

[61] Westall JC, Morel FMM, Hume DN. Chloride interference in cupric ion-selective electrode measurements. Analytical Chemistry. 1979;**51**(1): 1792-1798. DOI: 10.1021/ac50047a043

[62] Pretsch E. The new wave of ion-selective electrodes. Trends in Analytical Chemistry. 2007;**26**(1):46-51. DOI: 10.1016/j.trac.2006.10.006

[63] Rachou J, Gagnon C, Sauvé S. Use of an ion-selective electrode for free copper measurements in low salinity and low ionic strength matrices. Environmental Chemistry. 2007;**4**(2):90-97. DOI: doi.org/10.1071/EN06036

[64] Mascini M, Liberti A. Preparation and analytical evaluation of a new lead (II) heterogeneous membrane electrode. Analytica Chimica Acta. 1972;**60**: 405-410

[65] Mascini M, Liberti A. The preparation and analytical study of a new heterogeneous membrane electrodes. Analytica Chimica Acta. 1973;**64**:63-70

[66] Schecher WD, McAvoy DC. Mineql+: A Chemical Equilibrium Program for Personal Computers. Cincinnati: The Procter & Gamble Co; 1991

[67] Schecher WD, McAvoy DC. Mineql+: A software environment for chemical equilibrium modeling. Computer Environment and Urban Systems. 1992;**16**:65-76. DOI: 10.1016/0198-9715(92)90053-T

[68] Aguirre-Gómez A. Electrochemical Studies of Cd, Cu, Pb and Zn Complexation in Synthetic and Soil Solutions. Ithaca, NY: Cornell University; 1995

[69] Gleyzes C, Tellier S, Astruc M. Fractionation studies of trace elements in contaminated soils and sediments: A review of sequential extraction procedures. Trends in Analytical Chemistry. 2002;**21**(6–7):451-467. DOI: 10.1016/S0165-9936(02)00603-9

[70] Zimmerman AJ, Weindorf DC. Heavy metal and trace metal analysis in soil by sequential extraction: A review of procedures. International Journal of Analytical Chemistry. 2010. DOI: 10.1155/2010/387803

[71] Benitez LN, Dubois JP. Evaluation of the selectivity of sequential extraction procedures applied to the speciation of cadmium in soils. International Journal of Environmental Analytical Chemistry. 1999;**74**(1–4):289-303. DOI: 10.1080/03067319908031433

[72] Pickering WF. Metal ion speciation-soils and sediments (a review). Ore Geology Reviews. 1986;**1**(1):83-146. DOI: 10.1016/0169-1368(86)90006-5

[73] Bordas F, Bourg ACM. A critical evaluation of sample pretreatment for storage of contaminated sediments to be investigated for the potential mobility of their heavy metal load. Water, Air, and Soil Pollution. 1998;**103**:137

[74] Sheppard MI, Stephenson M. Contaminated Soils, 3rd International Conference on the Biogeochemistry of Trace Elements. Paris, France: INRA Editions; 1997

[75] Sposito G. On the use of the Langmuir equation in the interpretation of adsorption phenomena. II the "two-surface" Langmuir equation. Soil Science Society of American Journal. 1982;**46**: 1147-1152. DOI: 10.2136/sssaj1982.03615995004600060006x

[76] Emmerich WE, Lund LJ, Page AL, Chang AC. Solid phase forms of heavy metals in sewage sludge-treated soils. Journal of Environmental Quality. 1982; **11**:178-181. DOI: 10.2134/jeq1982.00472425001100020007x

[77] Sauvé S, McBride MB, Hendershot WH. Speciation of lead in contaminated soils. Environmental Pollution. 1997;**98**:149-155. DOI: 10.1016/S0269-7491(97)00139-5

[78] Houba VJG, Lexmond TM, Novozamsky I, van der Lee JJ. State of the art and future developments in soil analysis for bioavailability assessment. Science Total Environment. 1996;**178**: 21-28. DOI: 10.1016/0048-9697(95)04793-x

[79] Hickey MG, Kittrick JA. Chemical partitioning of cadmium, copper, nickel and zinc in soils, sediments containing high levels of heavy metals. Journal of Environmental Quality. 1984;**13**:372-376. DOI: 10.2134/jeq1984.00472425001300030010x

[80] Mandal L, Mandal B. Fractionation of applied zinc in rice soils at two moisture regimes and levels of organic matter. Soil Science. 1987;**144**:266-273

[81] Davies, BE, JM Lear, and NL Lewis. Plant Availability of Heavy Metals in Soils. In Coughtrey PJ

[82] Martin, and MH Unsworth, editor. Pollutant Transport and Fate in Ecosystems. Oxford: Blackwell Scientific Publishers. pp. 267-275

[83] Miller WP, Martens DC, Zelazny LW. Effect of sequence extraction of trace metals from soils. Soil Science American. 1985;**50**:598-601. DOI: 10.2136/sssaj1986.03615995005000030011x

[84] Shuman LM. Zinc, manganese, and copper in soil fractions. Soil Science. 1979;**127**:10-17

[85] Forbes EA, Posner AM, Quirk JP. The specific adsorption of divalent Cd, Co, Cu, Pb, and Zn on goethite. Journal of Soil Science. 1976;**27**:154-166. DOI: 10.1111/j.1365-2389.1976.tb01986.x

[86] Kinniburgh DG, Jackson ML, Syers JK. Adsorption of alkaline earth, transition, and heavy metal cations by hydrous oxide gels of iron and aluminum. Soil Science Society of America Journal. 1976;**40**:796-799. DOI: 10.2136/sssaj1976.03615995004000050047x

[87] Brümmer G, Tiller KG, Herms U, Clayton PM. Adsorption-desorption and/or precipitation-dissolution processes of zinc in soils. Geoderma. 1983;**31**:337-354. DOI: 10.1016/0016-7061(83)90045-9

[88] McKenzie RM. The adsorption of lead and other heavy metals on oxides of manganese and iron. Australian Journal of Soil Research. 1980;**18**:61-73. DOI: 10.1071/SR9800061

[89] Schnitzer M, Skinner SI. Organo-metallic interactions in soils. 7 stability constants of Pb^{2+}, Ni^{2+}, Mn^{2+}, Co^{2+}, Cu^{2+} and Mg^{2+} fulvic acid complexes. Soil Science. 1967;**103**:247-252

[90] Stevenson FJ. Nature of divalent transition metal complexes of humic acids as revealed by a modified potentiometric titration method. Soil Science. 1977;**123**:10-17

[91] Martell AE, Hancock RD. Metal Complexes in Aqueous Solutions: Modern Inorganic Chemistry. NY, USA: Springer Science; 1996

[92] Selim HM, Sparks DL, editors. Heavy Metals Release in Soils. 1st ed. Boca Raton, FL, USA: Lewis Publishers, CRC Press; 2001

[93] Benschoten JE, Matsumoto M, Young WH. Evaluation and analysis of soil washing for seven Lead-contaminated soils. Journal of Environmental Engineering. 1997;**123**: 217-224. DOI: 10.1061/(ASCE)0733-9372 (1997)123:3(217)

[94] Kheboian C, Bauer CF. Accuracy of selective extraction procedures for metal speciation in model aquatic sediments. Analytical Chemistry. 1987;**59**: 1417-1423. DOI: 10.1021/ac00137a010

[95] Tack FMG, Verloo MG. Chemical speciation and fractionation in soil and sediment heavy-metal analysis: A review. International Journal of Environmental Analytical Chemistry. 1995;**59**(2–4):225-238. DOI: 10.1080/03067319508041330

[96] Shan X, Chen B. Evaluation of sequential extraction for speciation of trace metals in model soil containing natural minerals and humic acid. Analytical Chemistry. 1993;**65**(6): 802-807. DOI: 10.1021/ac00054a026

[97] Arunachalam J, Emons H, Krasnodebska B, Mohl C. Sequential extraction studies on homogenized forest soil samples. Science of The Total Environment. 1996;**181**:147-159. DOI: 10.1016/0048-9697(95)05005-1

[98] Ma YB, Uren NC. Application of a new fractionation scheme for heavy metals in soils. Communications in Soil Science and Plant Analysis. 1995;**26**: 3291-3303. DOI: 10.1080/00103629509369527

[99] Ure AM. Trace element speciation in soils, soil extracts and solutions. Mikrochimica Acta. 1991;**104**:49-57

[100] Farrah H, Pickering WF. Factors influencing the potential mobility and bioavailability of metals in dried lake sediments. Chemical Speciation and Bioavailability. 1993;**5**:81-96. DOI: 10.1080/09542299.1993.11083207

[101] Chao T, Zhou L. Extraction techniques for selective dissolution of amorphous Iron oxides from soils and Sediments1. Soil Science Society of America Journal. 1983;**47**:225-232. DOI: 10.2136/sssaj1983.03615995004700020010x

[102] Bermond A, Eustache S. Hydroxylamine extraction of trace-metals in soils; kinetic aspects. Environmental Technology. 1993;**14**: 359-365. DOI: doi.org/10.1080/09593339309385300

[103] Coetzec PP. Determination of speciation of heavy metals in sediments of the Hortbees port: Dam by sequential chemical extraction. Water SA. 1993;**9**: 291-300 ISSN 0378-473

[104] La Force MJ, Fendorf S. Solid-phase ion characterization during common selective sequential extractions. Soil Science Society of America Journal. 2000;**64**:1608-1615. DOI: 10.2136/sssaj2000.6451608x

[105] Gleyzes C, Tellier S, Sabrier R, Astruc M. Arsenic characterisation in industrial soils by chemical extractions. Environmental Technology. 2001;**22**(1): 27-38. DOI: doi.org/10.1080/09593332208618313

[106] Kostka JE, Luther GW III. Partitioning and speciation of solid phase iron in saltmarsh sediments. Geochimica et Cosmochimica Acta. 1994;**58**(7): 1701-1710. DOI: 10.1016/0016-7037(94)90531-2

[107] Stumm W. Chemistry of the Solid-Water Interface in Natural Systems. New York, USA: John Wiley & Sons, Inc; 1992. p. 162

[108] Mehra OP, Jackson ML. Iron oxide removal from soil and clays by a dithionite-citrate system buffered with sodium carbonate. Clays and Clay Minerals. 2013;7:317-327. DOI: 10.1016/B978-0-08-009235-5.50026-7

[109] Shuman LM. Fractionation method for soil microelements. Soil Science. 1985;**140**:11-22

[110] Couturas S, Bourgeois S, Bermond A. A critical study of the use of hydrogen peroxide to determine trace metals bound to soil organic matter. Environmental Technology. 2000;**21**: 77-86. DOI: doi.org/10.1080/09593332108618139

[111] MerweI L, Kempster PL, van Vliet HR, van Staden JF. pH-dependent Sequential Extraction Procedure to Determine Mobilisation and Transport of Metals in Sediments, University of Pretoria, Pretoria, South Africa

[112] Quevauviller P, Rauret G, Muntau H, Rubio R, López-Sánchez JF, Fiedler H, et al. Evaluation of a sequential extraction procedure for the determination of extractable trace metal contents in sediment. Fresenius' Journal of Analytical Chemistry. 1994;**349**: 808-814

[113] Quevauviller P, Rauret G, López-Sánchez JF, Rubio R, Ure A, Muntau H. Certification of trace metal extractable contents in a sediment reference material (CRM 601) following a three-step sequential extraction procedure. Science Total Environmental. 1997;**205** (2–3):223-234. DOI: 10.1016/S0048-9697(97)00205-2

[114] Tessier A, Campbell PG, Bisson M. Particulate trace metal speciation in stream sediments and relationships with grain size: Implications for geochemical exploration. Journal of Geochemical Exploration. 1982;**16**:77-104. DOI: 10.1016/0375-6742(82)90022-x

[115] Gomez-Ariza JL, Santos MM, Morales E, Giraldez I, Sanchez-Rodas D, Vieira N. Organotin contamination in the Atlantic Ocean off the Iberian Peninsula in relation to shipping. Chemosphere. 2006;**64**(7):1100-1108. DOI: 10.1016/j.chemosphere.2005.11.068

[116] Singh SP, Tack FMG, Verloo MG. Solid-phase distribution of heavy metals as affected by single reagent extraction in dredged sediment derived surface soils. Chemical Speciation and Bioavailability. 1996;**8**:105-110. DOI: doi.org/10.1080/09542299.1996.11083265

[117] Hirner AV. Trace element speciation in soils and sediments using sequential chemical extraction methods.

International Journal of Environmental Analytical Chemistry. 1992;**46**:77. DOI: doi.org/10.1080/030673192080 26999

[118] Belzile N, Lecompte P, Tessier A. Testing readsorption of trace elements during partial chemical extractions of bottom sediments. Environmental Science & Technology. 1989;**23**:1015. DOI: 10.1021/es00066a014

[119] Chaney RL, Brown SL, Daniels WL, Henry CL, Li YM, Siebielec G, et al. Risk assessment and remediation of soils contaminated by mining and smelting of lead, zinc and cadmium. Revista Internacional de Contaminación Ambiental. 2000;**16**(4):175-192 ISSN 0188-4999

[120] ASTM. Official site of the American Society for Testing and Materials. USA: ASTM; 2011. Available from: http://www.astm.org

[121] Romero FM, Armienta MA, Gutiérrez ME, Villaseñor G. Factores geológicos y climáticos que determinan la peligrosidad y el impacto ambiental de jales mineros. Review in Contamination of Ambient. 2008;**24**:43-54 ISSN 0188-4999

[122] Monroy M, Díaz-B F, Razo I, Carrizales L. Evaluación de la contaminación por arsénico y metales pesados (Pb, Cu, Zn) y análisis de riesgo en salud en villa de La Paz-Matehuala, S. L.P. Nota informativa, Instituto de Metalurgia, U.A.S.L.P. México

[123] Puga S, Sosa M, Lebgue T, Quintana C, Campos A. Contaminación por metales pesados en suelo provocada por la industria minera. Ecología Aplicada. 2006;**5**:149-155

[124] Reyes CVE, Veloz RM, Vargas RM, Monroy FMG, Cruz GR, Briones GR, et al. Evaluación de las alternativas para la disminución del impacto ambiental por jales históricos en el distrito minero de Zimapán. Ciencias Ambientales. 2005;**21**:34-47

[125] Ortega LP, Xoconostl CB, Maldonado MIE, Carrillo GR, Hernández HJ, Díaz GM, et al. Plant and fungal biodiversity from metal mine wastes under remediation at Zimapán, Hidalgo, Mexico. Environmental Pollution. 2009;**30**:1-10. DOI: 10.1016/j.envpol.2009.10.034

[126] Wang YM, Chen TC, Yeh KJ, Shue MF. Stabilization of an elevated heavy metal contaminated site. Journal of Hazardous Materials B. 2001;**88**:63-74. DOI: 10.1016/S0304-3894(01)00289-8

[127] McGowen SL, Basta NT, Brown GO. Use of diammonium phosphate to reduce heavy metal solubility and transport in smelter-contaminated soil. Journal of Environmental Quality. 2001; **30**:493-500. DOI: 10.2134/jeq2001.302493x

[128] Basta NT, Gradwohl R, Snethen KL, Schroder JL. Chemical immobilization of lead, zinc, and cadmium, in smelter-contaminated soils using biosolids and rock phosphate. Journal of Environmental Quality. 2001;**30**:1222-1230. DOI: 10.2134/jeq2001.3041222x

[129] Brown SL, Chaney RL, Hallfrisch JA, Ryan JA, Berti WR. In situ soils treatments to reduce the phyto-and bioavailability of lead, zinc, and cadmium. Journal of Environmental Quality. 2004;**33**(522):531. DOI: 10.2134/jeq2004.5220

[130] Cao X, Wahbi A, Li LMaB,Yang Y. Immobilization of Zn, Cu and Pb in contaminated soils using phosphate rock and phosphoric acid. Journal of

Hazardous Materials 2009; 164: 555-564. DOI: 10.1016/j.jhazmat.2008.08.034

[131] Romero-Lazaro EM, Ramos-Perez D, Romero FM, Sedov S. Indirect indicators of residual contamination in soils and sediments of the Sonora River basin, Mexico. Review in International Contamination Ambient. 2019;**35**(2): 371-386. DOI: 10.20937/RICA.2019.35.02.09

[132] Rivera-Uria MY, Ziegler-Rivera FRA, Díaz-Ortega J, Prado-Pano B, Romero FM. Effect of an acid mine spill on soils in Sonora River Basin: Micromorphological indicators. Spanish Journal of Soil Science. 2018;**8**(2): 258-274. DOI: 10.3232/SJSS.2018.V8.N2.08

[133] Ceniceros-Gómez A, Gutiérrez-Ruiz ME, Romero FM, Martínez-Jardines LG, Domínguez-Martínez R. Monitored natural attenuation of heavy metals in the Sonora River: The water quality. Proceedings: 6th International Symposium on Sediment Management (Chiapas – México: June 19-23, 2018). Revista Internacional de Contaminación Ambiental. 2018;**34**:267-270. DOI: 10.20937/2018.34.M6ISSM

[134] Gutierrez-Ruiz M, Muro-Puente A, Ceniceros-Gomez AE, Amaro-Ramirez D, Perez-Manzanera L, Martinez-Jardines LG, et al. Acid spill impact on Sonora River basin. Part I. sediments: Affected area, pollutant geochemistry and health aspects. Journal of Environmental Management. 2022;**314**: 115032

[135] Laboratorios ABC Química Investigación y Análisis S.A. de C.V. Informe de resultados del muestreo y análisis de agua, sedimentos y biota de los ríos Bacanuchi y Sonora contaminados por la fuga proveniente de la mina Buenavista del Cobre en Cananea, hasta el 18 de septiembre de 2014. Fideicomiso río Sonora México. 2014

[136] SGM. Total Metal and Metalloid Concentrations on Sediment Sampled from 1995 to 2005. [Internet]. 2016. Available from: https://www.datos.gob.mx/busca/dataset/cartografia-geoquimica-escala -1-250000/resource/d3e0ffdf-56de-4444-b289-0aadcdc3a5ac [Accessed: October 14, 2022]

[137] González-León CM, Valencia V, López M, Bellon H, Valencia-Moreno MA, Calmus T. The Arizpe sub-basin: Sedimentary and magmatic evolution of the basin and range in north-Central Sonora, México. Revista Mexicana de Ciencias Geológicas. 2010;**27**:292-312

[138] Lamberti J. Análisis del Fideicomiso Río Sonora. Desarrollo, Educación e Investigación (Ciudad de México): Proyecto sobre Organización; 2018

[139] Fideicomiso río Sonora. Libro Blanco de las acciones y medidas de remediación, reparación y/o compensación de los daños al ambiente y a la salud generados por el derrame en los ríos Bacanuchi y Sonora. G.II. Reporte Limpieza del río Sonora 2014

[140] Hesse PR. A Textbook of Soil Chemical Analysis. London: John Murray; 1971

[141] U.S. EPA. Chapter Four of the SW-846 Compendium: Organic Analytes. [Internet]. 2018. Available from: https://www.epa.gov/hw-sw846/chapter-four-sw-846-compendium-organic-analytes. [Accessed October 12, 2022]

[142] U.S. EPA. Method 6200: Field Portable X-Ray Fluorescence Spectrometry for the Determination of Elemental Concentrations in Soil and Sediment. [Internet]. 2007. Available

from: https://www.epa.gov/sites/default/files/2015-12/documents/6200.pdf [Accessed October 12, 2022]

[143] SEMARNAT. NOM-147/SSA1-2004. Criterios para determinar las concentraciones de remediación de suelos contaminados por arsénico, bario, berilio, cadmio, cromo hexavalente, mercurio, níquel, plata, plomo, selenio, talio y/o vanadio. 2007. DOF, Ciudad de México

[144] Dabin C, Guigonnet-Sergent A, Algros E, Charissou AM. Bioavailability and Bioaccessibility of Pollutants in Contaminated Soils: State of Present Knowledge and Research Perspectives. France: RECORD; 2012

[145] EPA US. Method 6010D (SW-846): Inductively Coupled Plasma-Atomic Emission Spectrometry, Revision 4. Washington, D.C: U.S. Environmental Protection Agency, National Center for Environmental Assessment; 2014

[146] Kenneth RL. Preliminary deposit-type map of Northwestern Mexico. U.S. Geological Survey Open-File Report. 1989;**1989**:89-158

[147] Ankan B, Schreiber ME. Arsenic release from arsenopyrite weathering: Insights from sequential extraction and microscopic studies. Journal of Hazardous Materials. 2013;**262**:896-904. DOI: 10.1016/j.jhazmat.2012.12.027

[148] Gutierrez- Ruiz ME. Study of Dust and Effluents of the Mining-Metallurgic Complex "La Caridad" Nacozari, Sonora. México: Environmental Biogeochemistry Team, Institute of Geography, UNAM; 2013

[149] Gutiérrez-Ruiz ME, Ceniceros-Gómez AE, Villalobos M, Romero FM, Santiago P. Natural arsenic attenuation via metal arsenate precipitation in soils contaminated with metallurgical wastes: II. Cumulative evidence and identification of minor processes. Applied Geochemistry. 2012;**27**:2204-2214. DOI: 10.1016/j.apgeochem.2012.02.021

[150] Escobar-Quiroz IN, Villalobos-Peñalosa M, Pi-Puig T, Romero FM, Albornoz JAC. Identification of jarosite and other major mineral Fe phases in acidic environments affected by mining-metallurgy using X-ray absorption spectroscopy: With special emphasis on the august 2014 Cananea acid spill. Review Mexico Ciencias Geology. 2019;**36**(2):229-241. DOI: 10.22201/cgeo.20072902e.2019.2.1021

Chapter 35

Monitoring Strategies for Heavy Metals in Foods and Beverages: Limitations for Human Health Risks

Anamika Kalita Deka, Kushwaha Jashvant Kumar and Sunshri Basumatary

Abstract

Foods and beverages with heavy metal contents, their Maximum Permissible Limits (MPL), Estimated Dietary Intake (EDI), Target Hazard Quotient (THQ) to study carcinogenic effects with other human health related matters and metal remediation's are high priority issues for sustainable world-wide developments. Inductively Coupled Plasma-Optical Emission Spectroscopy (ICP-OES), Inductively Coupled Plasma-Mass Spectrometer (ICP-MS), Flame atomic absorption spectroscopy (FAAS), Total Reflection X-Ray Fluorescence (TRXF) Spectroscopy, Chemical Replacement Combined with Surface-Enhanced Laser-Induced Breakdown Spectroscopy (CR-SENLIBS), Electrochemical apt- sensors are some advanced monitoring tactics for heavy metal detection. Nanotechnology innovations, soil state-of art remediation are used now-a-days for removal of metals from foods and beverages. In addition to this, chelating ligands, plant phenolic have crucial applications in heavy metal removal from foods. Bio-absorbents like microbial cultures, fermentation wastes also play crucial role in heavy metal remediation from foods and beverages. In the present chapter various metal monitoring tactics are focused with advance metal remediation procedures associated with food and beverages. Limitations of various metals associated with human health risks are also summarized herein.

Keywords: foods, beverages, heavy metal content, remediation, monitoring tactics, bio-absorbents, nanotechnology

1. Introduction

Heavy metals are high density elements; generally non-degradable and available in earth's crust [1, 2]. Heavy metals are present in various food stuffs viz., cake, beans, fish, meat, fruits, herbal drinks, alcoholic and non-alcoholic beverages [3]. Heavy metal contamination in foods is presently a global concern for scientists. Metal contamination in food staffs are mainly associated with global industrialization and day-by-day increase in environmental pollution (air, soil, water). The raw materials used during processing of food may also be one reason for metal contamination in foods [3, 4].

The acceptability of any food and beverage depends on the toxicity load of heavy metals present in it owing to the nutritional values [5]. In other words, the quality of foods and beverages can be judged based on the content of heavy metals found in them [5–7]. Heavy metals available in our surrounding environment are: cadmium (Cd), cobalt (Co), chromium (Cr), copper (Cu), mercury (Hg), lead (Pb), selenium (Se), nickel (Ni), aluminum (Al), arsenic (As), antimony (Sb), barium (Ba), beryllium (Be), molybdenum (Mo), iron (Fe), magnesium (Mg), manganese (Mn), gallium (Ga), germanium (Ge), gold (Au), indium (In), lithium (Li), platinum (Pt), vanadium (V), etc. [8, 9]. Out of these metals, Co, Cu, Cr, Fe, Mg, Mn, Mo, Ni, Se, Zn, Na, K, Mn are essential for human life's and other are non-essential. Co, Cu, Cr, Fe, Mg, Mn, Mo, Ni, Se, Zn Na, K, Mn are considered to be essential heavy metals only if the exposure of the metals are within Maximum Permissible Limits (MPL) provided by World Health Organization (WHO), International Organization of Grapes and Wine (OIV), United States Environment Protection Agencies for risk Assessment (USEPA), Standard Organization of Nigeria (SON), FAO/WHO Expert Committee on Food Additives etc. [10–13]. The excessive intake of essential metals are also toxic for human exposure. For instance, excess exposure of Fe causes Parkinson's disease, high intake of Mn causes Mn-induced Parkinsonism, high level of Zn deals with impairment of growth and reproduction etc. [5, 13]. Instead, a small quantity of non-essential metals like Pb, As, Hg, Sb etc. are found in foods and beverages; they are very toxic. Therefore, evaluation of acceptance or ignorance range of heavy metal has an important influence before exposure to humans. EDI (Estimated Daily Intake), RDI (Recommended Daily Allowance), MPL (Maximum Permissible Limits), THQ (Target Hazard Quotient), TCR (Target Cancer Risk) are some monitoring parameters for detection of permissible limits of heavy metals by JECFA (Joint Food and Agriculture Organization Expert Committee on Food Additives), WHO (World Health Organization), USEPA (United States Environmental Protection Agencies) etc. [10, 14, 15]. These monitoring parameters tells estimated intake metal to human within maximum prescribed limit, daily limit of metal exposure, cancer causing risk etc. respectively [5, 13].

Carcinogenic effects associated with heavy metal- toxicity and various other health related issues vary from metal to metal since physico-chemical properties and features of each metal differ from each other [16]. The present chapter focuses on various evaluation parameters for metal exposure (as proposed by WHO, OIV, USEPA. SON and FAO/WHO), modern analytical method for metal detections. Various advanced remedies of heavy metals remediation from food stuffs are also summarized in the chapter.

2. Methods

2.1 Sources of heavy-metal contaminations in foods and beverages

The sustenance of life for all living organism is in need of food. The growth, development and all biological functions of all living organism including mankind is associated with quality of food [16–18]. Food can be plant-based, processed food or ready-to-eat food. Food provides essential nutrients such as vitamins, minerals, carbohydrates, proteins to human body. But human exposure to contaminated foods and beverages are toxic and longtime exposure can be life threatening. Contamination of food by both essential (Co, Cu, Cr, Fe, Mg, Mn, Mo, Ni, Se, Zn Na, K, Mn, etc.) and non-essential (Pb, As, Hg, Sb, etc.) heavy metals can causes various health problems. Cardiovascular disease, diabetes, lung-cancer, hearing disorders, visual impairment are some serious issues associated with heavy metal contamination. **Table 1** shows the potential toxicity associated with heavy contaminations based on earlier reports.

Sl. No.	Heavy metal	Potential toxicity	Possible food sources	MPL	Ref.
1	Cadmium (Cd)	Causes Cardiovascular disease, Osteoporosis and Carcinogenic	Plant based foods like wheat & rice, animal milk and fatty tissues.	0.005 mg/L by EPA 0.003 mg/L by WHO	[19–21]
2	Mercury (Hg)	Causes Damage of fetus, brain and kidney	Marine foods eg. Shark, swordfish, redfish etc.	0.002 mg/L by EPA 0.001 mg/L by WHO	[22–24]
3	Cobalt (Co)	Although Co is the metal constituent of vitamin B_{12} excessive exposure shows adverse effects like hearing & visual impairment, cardiovascular disease.	Green vegetables, sea foods, animal meat etc.	5×10^{-6} mg/m^3 by CalEPA 3×10^{-5} mg/m^3 MRL by ATSDR	[25–27]
4	Chromium (Cr)	Its exposure is associated with lung cancer, hexavalent Cr(vi) has been classified as carcinogenic element by IARC.	Green vegetables.	0.003 mg/kg bw/ day for Cr(vi)	[28, 29]
5	Lead (Pb)	Decrease growth rate of children, effects nervous system and metabolism	Plant based food.	PTWI is 0.025 mg/kg/bw by JECFA 0.01 mg/L by WHO	[21, 30, 31]
6	Arsenic (As)	It is associated with genetic toxicity, reproductive toxicity, cellular toxicity etc.	Drinking natural water, As is present in the format H_3AsO_3 (arsenous acid) & H_3AsO_5 (arsenic acid).	0.010 mg/L MCL by USEPA	[32, 33]
7	Antimony (Sb)	Cardiotoxicity (49% patients), pancreatitis, visceral leishmaniasis co-infections.	Plant-based foods near antimony mines.	5 μg/day	[34–36]
8	Magnesium (Mg)	High dose of Mg causes diarrhoea, abdominal cramp, fatal hypermagnesemia etc.	Dietary supplements in the form of magnesium oxide, citrate, vegetables etc.	350 mg/day	[37–41]
9	Manganese (Mn)	Mn-induced neurotoxicity, mitochondrial disinfection, inflammation etc. at high level of exposure.	Both veg and non-veg food.	Not ruled out by EPA & FDA 11 mg/day by NAS	[42, 43]

Sl. No.	Heavy metal	Potential toxicity	Possible food sources	MPL	Ref.
10	Iron (Fe)	Iron poising is deadly among children, failure in diagnosis can even cause multi-organ failure or even death.	Vegetable, sea-foods, legumes etc.	PTMI is 0.8 mg/kg/bw/day by WHO	[44, 45]
11	Barium (Ba)	Hypokalaemia and weakness in muscle is observed after 1-4 hours of ingestion, vomiting, abdominal pain, watery diarrhoea etc.	Drinking water, Brazil nuts, seaweed (may be toxic for long term exposure).	2.0 mg/L by EPA	[46–53]
12	Beryllium (Be)	Carcinogenic to humans.	Drinking water near garden peas, beans grown in soil rich in beryllium.	0.004 mg/L by EPA	[54, 55]

Table 1.
Heavy metals with potential toxicity to human exposure, food sources and maximum permissible limit of intake.

Therefore, heavy metal contamination in foods as a serious matter of concern now-a-days for global scientific forum.

Heavy metal contamination in foods and beverages originated from different routes. Environmental pollution, industrial waste, soil where plant-based foods are cooked, processing of foods etc. are some well-known sources of heavy metal contamination. For instance, Anderson et al. [56] reported that contamination of food chain with heavy metal is due to environmental pollution, Cabrera et al. [57] reported that some food processing techniques are responsible for heavy metal contamination also [56–58]. Furthermore, exposure of humans to contaminated soil, air and water may be another reason for contamination with heavy metals. Raw material used, water, food-processing are some leading reason associated with metal contents in foods [3]. The schematic representation of various sources of heavy metal contaminations in foods is shown in **Figure 1**.

Heavy metals may be originated in both alcoholic and non-alcoholic beverages from added plant-based material during brewing procedure (cereal, hop etc.), manufacturing protocol, storage etc. [3, 5]. Pesticides and various man-made chemical fertilizers generally used during cultivation can be the cause of metal contaminations

Figure 1.
Schematic representation of sources of heavy metal contamination to humans [12].

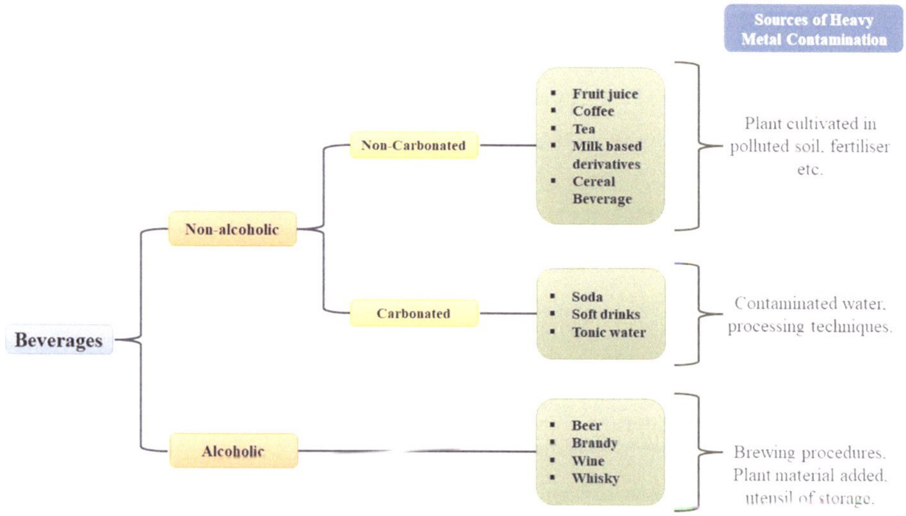

Figure 2.
Classification of beverages [25] and various sources of heavy metal contamination [3, 5].

in plants, which are transmitted to corresponding foods and beverages during processing and packaging [3, 59]. The classification of beverages and various sources responsible for heavy metal contaminations are shown in **Figure 2**. During beverage formulations although addition of any chemical and biological contaminants are strictly prohibited but sometimes unintentionally their formation occur in beverages. Prolonged exposure to low level contaminations also lead human health to potential health risks like carcinogenic, mutagenic and teratogenic effects [60, 61].

2.2 Various limitation parameters for heavy metal exposure to humans

2.2.1 Recommended daily allowances (RDA)

It is the daily exposure limits for heavy metal exposure as prescribed by WHO, FAO/WHO, EVM, USEPA [5]. For instance, RDA of Co is 100 μg/day [3, 5], RDA of Cu ranges from 15 to 500 μg/kg bw/day [11]. Heavy metals Fe and Mn can be ingested upto 10–18 mg/day/person and 2–5 mg/day/person respectively [62]. For Ni, tolerable daily intake (TDI) is 5 μg/kg bw/day. Pb has TDI 7.14 μg/kg bw/day for 60 kg adult [3, 5, 63].

2.2.2 Estimated daily intake (EDI)

Human exposure to heavy metals can be calculated by EDI as described by United States of Environmental Protection Agency (USEPA) [64].

The mathematical formula for determination of EDI is given below [65]:

$$EDI = \frac{C_{metal} \times Q_{Ig}}{b.w} \quad (1)$$

Where C is the concentration of metal in foods, vegetables and other sources (in μg, mg etc.); Q_{Ig} = quality of ingestion (kg/day); bw = body weight (generally considered 60 kg adult); EDI is generally expressed as μg/kg bw/day or mg/kg bw/day.

2.2.3 Target Hazard Quotient (THQ)

THQ as prescribed by USEPA [65] explains the carcinogenic and non-carcinogenic effects of foods, beverages etc. The mathematical equation for calculation of THQ is given below:

$$THQ = \frac{E_{fr} \times E_{Dur} \times C_{metal} \times S_{Ig}}{O_{Rf} \times bw \times ATn} \times 10^{-3} \quad (2)$$

Where E_{fr} = exposure frequency (365 days/year); E_{Dur} = duration of exposure (year); S_{Ig} = ingestion rate (g/day); C_{metal} = concentration of metal (μg or mg); O_{Rf} = oral reference dose (mg/kg bw/day, **Figure 3**, shows oral reference doses of some heavy metals as per limits suggested by USEPA, WHO etc.); bw = average body weight (60 kg); ATn = average time for non-carcinogens; 10^{-3} = unit conversion factor.

If, THQ < 1, the food or beverage is less carcinogenic or no-health carcinogenic health risk is associated with this.

If, THQ > 1, food staff is highly carcinogenic or associated with adverse health risk.

2.2.4 Hazard index (HI)

Multiple heavy metal contamination can occur in same food & beverage at same time. HI is the summation of THQ of each metal present in food staff or other. It is an estimation of more than one metal induced toxicity and calculated as [3–5, 65]:

$$HI = \sum THQ_{metals} \quad (3)$$

$$= THQ_{M1} + THQ_{M2} + THQ_{M3} + THQ_{M4} + \cdots\cdots\cdots \quad (4)$$

Where $M1$, $M2$, $M3$ are metal 1, 2, 3, respectively.

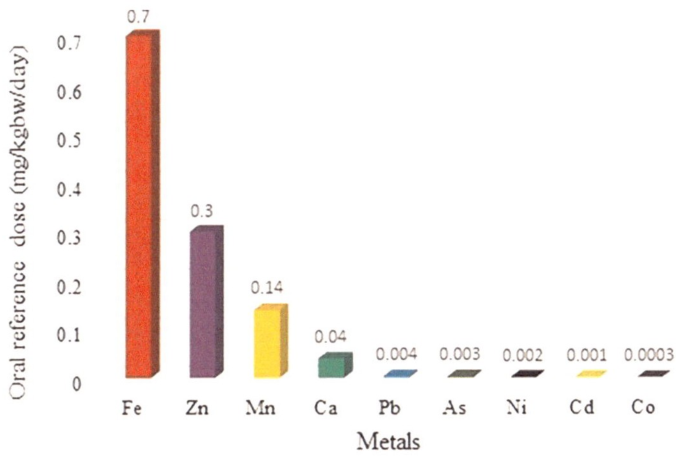

Figure 3.
Oral reference doses for some heavy metals prescribed by USEPA [5, 62, 66].

2.2.5 Total cancer risk (TCR)

This is the estimation of cancer risk associated with food staff. USEPA, 2011 prescribed the calculation for target cancer risk (TCR) as given by following mathematical equation [67]:

$$TCR = C_{SF} \times EDI \tag{5}$$

where, C_{SF} is the cancer slope factor of heavy metals. USEPA [67] prescribed CSF of various heavy metals for instance Cn for Pb is 36 mg/kg/day and for Cd it is 15 mg/kg/day.

As per estimation of New York State Department of Health (NYSDOH). If $TCR \leq 10^{-6}$, heavy metal associated health risk is low. Moderate cancer risk is associated with heavy metal exposure if is 10^{-3} to $\geq 10^{-1}$. Above this range cancer risk is too high [68].

3. Analytical methods for quantification of heavy metals in food and beverages

The quantification of heavy metals in foods, beverages and other is done by various analytical tools. The estimation of metal content in food stuff and other is dependent on the property of metal and it's concentration on to be examined sample. Pre-treatment of samples *viz.*, sample digestion by concentrated acids like nitric acid, HNO_3 and sulfuric acid, H_2SO_4 etc., dry ash digestion, digestion in acidic medium using microwave etc., prior to perform analytical experiments are needed for samples under investigation [3, 5, 69]. The accurate determination of metals is ensured by choosing an appropriate digestion technique, and it has been demonstrated that specific digestion process affects the determination of metals. Therefore, to get precise results; the adaption of right digestion technique is necessary [70]. The common analytical techniques used for quantitative determination of heavy metals are Inductively Coupled Plasma-Mass Spectrometry (ICP-MS), Atomic Absorption Spectrometry (AAS) and Inductively Coupled Plasma-Optical Emission Spectrometry (ICP-OES). Various digestion methods applicable for samples preparations for analytical quantification are tabulated in **Table 2**.

3.1 Flame atomic absorption spectrometry (AAS)

Due to its simplicity and ability to measure several metals even at trace levels such as Cd, Cr, Ni, Pb, Mn, Cu, Co, Fe, Flame Atomic Absorption Spectrometry (FAAS) is frequently employed for metal identification from food and beverages materials [81, 82]. One of the most effective methods for obtaining trace elements in various sample is chemical vapor generation in combination with atomic absorption spectroscopy, which comprises Hydride Generation Atomic Absorption Spectroscopy (HGAAS) and Cold Vapor Atomic Absorption Spectroscopy (CVAAS). In contrast to CVAAS, is the superior method for mercury analysis from various samples, HGAAS is suitable for hydride-forming metals such as As, Pb, Se, and Sn [70]. Chuachuad et al. employed an intriguing technique for the measurement of Cd in wines by flow injection Cold Vapor AAS (CVAAS) [83] and Pb by HGAAS following wine microwave digestion by combination of $HNO_3 + H_2O_2$ [84].

Sample	Country	Metals	Digestion method	Analytical technique	Ref.
Cocoa beans & products	Malaysia	As, Cd, Pb, Sb	Microwave digestion, 1 g of sample + 6 ml HNO_3 & 2 ml of H_2O_2 at 200°C for 15 min.	ICP-MS	[71]
Fruits, vegetables & cereals	Nigeria	Cr, Cd, Pb, Ni	Dry-ashing at 500°C for 1 hour than ash is treated with 25 ml 1 M HNO_3	AAS	[72]
			Wet digestion, 1 g of sample + 12 ml acid mixture HNO_3:HCl (3:1) heated at 150° C for 150 minutes.		
Green coffee bean	Ethiopia	Al, Ca, Fe, K, Mg, Na, P, S	Microwave digestion, 0.5 g sample + 7 ml HNO_3 & 1 ml H_2O_2 (30%) heated at 80°C for 5 minutes, 50°C for 5 minutes than 190°C for 20 minutes.	ICP-OES	[73]
		B, Cu, Mn, Ni, Rb, Sr, Zn		ICP-Ms	
		Hg		DMA	
Potato chips and biscuits	India	Fe, Al, Zn, Ni, Cu, Mn, Co, Cr, Pb, Cd	Wet digestion, 1 g sample + 12 ml acid mixture HNO_3:H_2SO_4 (8:4) at 130°C for 3 hours.	ICP-AES	[74]
seafood, vegetables, & stimulant drinks	Spain	Al	Acid digestion, 0.25 g sample + 5 ml HNO_3 (65%) + few mg of V_2O_5 (catalyst) heated at 120 °C for 90 minutes.	ETA-AAS	[75]
soft drink	Ghana	Fe, Co, Zn, Cd, Pb, Cu	Microwave digestion, 5 ml sample + 6 ml HNO_3 (65 %), 3 ml HCl (35 %), and 0.25 ml of H_2O_2 for 26 min	FAAS	[76]
Carbonated, flavored yogurt, juice drinks	Egypt	Fe, Mn, Cd, Pb, Ni, Cr, Cu	5 g sample dry ashing at 550°C for 8 hours than ash is treated with 25 ml diluted HCl	ICP-OES	[77]
Vegetables	Congo	l, As, Cd, Cr, Cu, Mn, Pb, Se, Zn	Microwave digestion, 0.25 g sample + 16 ml HNO_3 and 4 ml H_2O_2	ICP-OES, ICP-MS	[78]
Food	Bangladesh	Cr, Ni, Cu, As, Cd, Pb	Microwave digestion, 0.2-0.3 g sample + 1.5 mL HNO_3 (69%) and 4.5 mL HCl (35%)	ICP-MS	[79]
Soft drinks	Brazil	Ti, Cr, Sb, As, Pb	Acid digestion, 5 ml sample + HNO_3 and H_2O_2 at 120-130°C	TXRF	[80]

Table 2.
Digestion method used for heavy metals determination by various techniques.

3.2 Total reflection X-ray fluorescence (TXRF)

TXRF is a recognized analytical method for the determination of metals in a wide range of samples; particularly powdered and liquids micro samples are analyzed by this tool [85]. The advantages of TXRF are: it requires very low mass of sample with very low analysis time (100–1000 s). The primary drawbacks are caused by the potential peak overlapping, which may restrict element identification and reduce the estimation precision [80]. Drinks and beverages make really good liquid samples for TXRF analysis due to the quick and easy preparation process for qualitative analysis, which involves depositing a small amount of sample on a clean quartz-glass carrier and drying it. The noticeable impact for this analytical tools is that the internal standard added at the early state of quantification are free from the original sample at the final stage of quantification [86]. TXRF is recommended by several studies as an appropriate method for elemental analysis of wine with little to no pre-treatment [87, 88]. Direct wine drop deposition on the sample carrier, followed by internal standard deposition [89]. According to the earlier reported, the sample's digestion makes the chemical analysis more precise. In those instances, the samples were digested using a mixture of HNO_3 and H_2O_2 [90].

3.3 Inductively coupled plasma-optical emission spectroscopy (ICP-OES)

ICP-OES, which stands for inductively coupled plasma-optical emission spectrometry, is used to quickly and accurately identify trace elements in a variety of materials and is appropriate for multi-elements analysis. This method uses argon gas-created plasma for atomization and is distinguished by great sensitivity, excellent reproducibility, and minimal matrix influence. It is necessary to digest the sample before injecting it into the device since samples delivered in plasma must be liquid [91].

3.4 Inductively coupled plasma-mass spectrometer (ICP-MS)

ICP-MS, a mass spectrometer paired with inductively coupled plasma ionization, is one of the most sensitive analytical techniques for quick multi-element detection of heavy metals in trace and ultra-trace quantities in various sample matrices [92]. In the present, it is the most appropriate approach for the analysis of trace elements in bulk materials, due to its recent development as a potent technology. Few drawbacks associated with ICP-MS are: significant capital investment and a lack of recognized reference standards [93]. For the majority of elements, ICP-MS gives incredibly low detection limits, ranging from a part per billion (ppb) to a trillion (ppt). In comparison to GF-AAS and ICP-AES, it has lower detection limits and a faster multi-element scanning capabilities over a wider range of masses [92].

3.5 Chemical replacement combined with surface-enhanced laser-induced breakdown spectroscopy (CR-SENLIBS)

Due to its appealing qualities, including quick, simultaneous multi-element detection and in-situ, real-time analysis capabilities, laser-induced breakdown spectroscopy (LIBS) is one of the competitive methods for monitoring water quality [94]. Surface-enhanced LIBS (SENLIBS), a novel method for the phase transition from liquid to solid, has recently been regarded as a flexible analytical approach for liquid samples, and solid samples [95]. The liquid sample was combined with the powder sample in a viscous mixture. Chemical processing converted the solid sample into the liquid sample. The liquid sample was subsequently dried as a solid layer or applied as a gel-like layer on a surface of the non-absorbent substrate, and LIBS analysis was performed. Up till now, a variety of techniques, including liquid micro extraction [96], chemical replacement [94] have been suggested to further enhance detection sensitivity or the spectrum intensity of SENLIBS [97].

4. Results and discussion (Removal tactics for heavy metal contamination)

4.1 Nano-technological innovations

Developing urbanization, environmental pollution leads to heavy metal contamination in water. Contaminated water is hazardous to human-beings as well as to other living organisms. Nano-technological innovations like nano-polymer composites, metal-oxides nanomaterials, non-carbon nanomaterials, such as layered double hydroxides etc. have potential application in the removal of heavy metals from contaminated water [78]. Mudzielwana et al. reported that toxic metal ions from wastewater can be removed by metal oxide nano-particles [79]. **Figure 4** shows different nano-technological innovations associated to bio-remediation of heavy metal.

Titanium oxide (TiO_2) and zinc oxide (ZnO) nano-particles (NPs) are some reported semi-conductors having potential effect against heavy-metal water contamination removal [79]. It is reported earlier that, toxic heavy metal can be de-contaminated from waste water with Cu, Ag and Fe-induced NPs [98, 99]. Nano-bioremediation is also a low-cost method for pollution reduction in water and soil. The various application of nanotechnology for de-contamination of heavy metal are given below:

It is reported that NPs reduce the heavy metal stress in plants, heavy metals present in soil is absorbed by NPs which minimizes the bio-availability and mobility of metals [100]. For example, Sebastian et al. reported that application of Fe_3O_4 NPs reduce the mobility of heavy metal Cd [101]. Konate et al. and Yao et al. again reported that the antioxidant enzyme-activated NPs for e.g. CeO_2 NPs, Mn_3O_4 NPs, Fe_3O_4 NPs have ability reduce ROS (Radical Oxygen Species). Therefore, helps in the reduction of crop production loss due to the stress [100, 102].

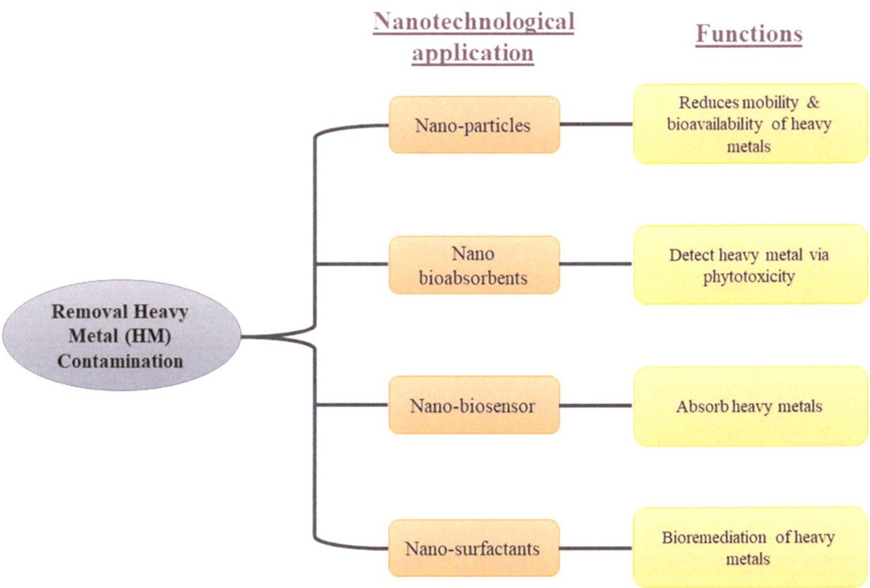

Figure 4.
Various nanotechnological application for removal of heavy metals contamination.

HMs like Hg, Cd, Pb, As etc. are accumulated in crop from polluted soil and transmitted to human after consumption [98]. Nano-biosensors can detect the heavy-metal phyto-toxicity [98]. Nano-biosensors with great specification for detection of heavy metals, can be applied to various areas like nutrient monitoring, agriculture fertilizers, pesticides etc.; boosts the crop yield [98].

'Biosorption' is a biologically derived method used for elimination of organic as well as inorganic matter [103, 104]. Nano-technological application with biosorption is known as "Nano-biosorbents"; recent technique for heavy metal removal. Carboxyl group (—COOH) and hydroxyl (—OH) groups present in biosorbents facilitate absorption of heavy metal [105]. For instance, rice husk based graphene quantum dots is an effective nano- biosorbent used for La (III) and Pb (III) removal [106].

Bio-surfactants possess both lyophilic and hydrophilic activities are molecules on living spaces or secreted by microbes [98]. Microbe-induced bio-surfactants are used for remediation of heavy metals like Zn, Cu, and Ni [98]. *Bacillus subtilis* based surfactant i.e., lipopeptide bio-surfactant play a crucial role in bio-remediation of heavy metals from soil. Nanoparticle capped bio-surfactants are called nano-bio-surfactants are also useful in the bioremediation of heavy metals. For e.g. (synthesized from *Pseudomonas aeruginosa*) [105–107] with *Rhamnolipid* capped Zn NPs removal of heavy metals can be achieved [108].

4.2 Soil state-of-art remediation of heavy metal

Heavy metal contamination in plant-based foods and beverages actually originated from contaminated soil. Some remediation technique of HMs from soil are: application of strong-chelating ligand [109], high-surface-area-absorbent [110, 111], phytoremediation [112] etc.

In bio-remediation by chelating-ligands the heavy metals present in functional groups of soil surface are liberated. But the consumption of high number of chelating ligands, nutrient loss from soil are some main draw-backs associated with the technique [109]. The mobility and bio-availability of heavy metals can be reduced by the of high-surface-area absorbent. This method requires long-term monitoring to capture immobilized HMs [113].

Phytoremediation of HMs is a high-energy efficient recent treatment. It requires long times for treatment with a probability of creating secondary pollution by accumulations metals in biomass [110].

4.3 Plant phenolic compounds for heavy metal removal

Phenolic compounds are one of the major secondary metabolite present in plant have high tendency to chelate metals, play a crucial role in growth and development of plants [112]. Hydroxyl group (—OH) and carboxyl group (—COOH) groups are present in phenolic compounds. Some examples of polyphenol compounds found in plants are: catechin, caffic acid, gallic acid, ferulic acid, syringic acid, sinapic acid, epi-catechin, epi-gallocatechin etc. The hydroxyl and carboxyl groups present in phenolic compounds can bind with heavy metals. High nucleophilic character of atomic rings of phenolic compounds may be the reason for metal-polyphenol capping [114]. Because of heavy metal exposure; the production of phenolic compounds in plant increases [112], if the exposure happens to useful metals like Cu, Fe and Zn necessary for plant growth. On other hand Cd, Pb, As are toxic for plant's life and growth.

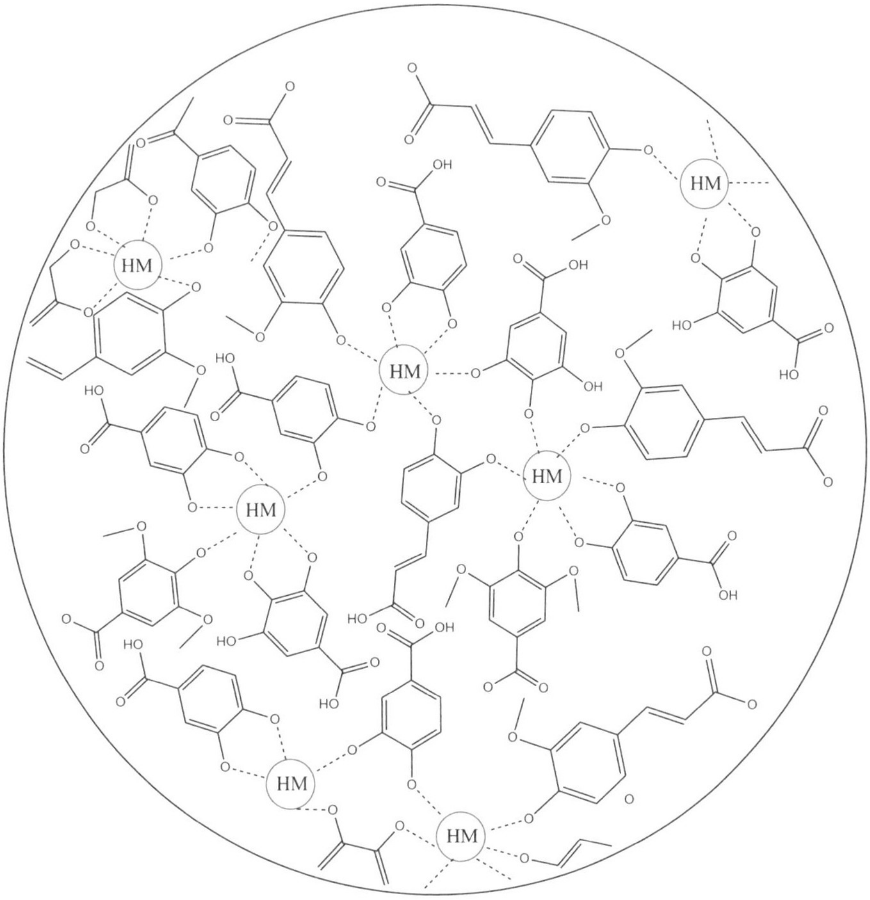

Figure 5.
Possible chelation by hydroxyl (—OH) group and carboxyl (—COOH) groups of plant phenolic compounds with heavy metal [115, 117].

Radical oxygen species (ROS) formation occurs in plant, when plants are exposed to heavy metal contamination; simultaneously responsible for physiological changes in plants [115].

Plants with lower anti-oxidant activity or with lower amount polyphenols than the amount of ROS, suffers more damage [116]. Therefore, plant phenolic compounds plays protective role depending on heavy metal stress conditions as well as on environmental conditions.

The chelation of polyphenol molecule with heavy- metal is shown **Figure 5** [115].

4.4 Role of chemical chelating ligands in heavy metal toxicity removal

Chelation therapy based on co-ordination chemistry is a most promising medical treatment for toxic heavy metal removal. Chelating ligands bind with toxic metals to form metal-complexes. The metal-complex so formed are being extracted by body further [118]. For instance, 2,3-dimercaprol has been used to remove Pb and As poisoning, meso-2,3-dimercaptosuccinic acid is used for extraction of metal etc. [118].

The chelating ligands under use for metal removal should be of low toxicity, higher solubility in water with good penetration ability through cell membrane [119]. Some example of clinical application of ligands in chelation are: EDTA (Ethylene Diamine Tetra Acetic acid), Trientive, D-pecicillamine, Deferiorone etc. [119]. It is reported that, arsenic based metal poisoning can be removed by chelating ligand BAL (British anti-Lewisite) and but fewer limit of toxicity was detected in 1960s due to presence of thiol group and further it was modified to DMSA (2,3- Dimercaptosuccinic acid) [119].

Thus, chelation therapy imposes the removal of heavy metal toxicity from human health in the form of metal-ligand complexes. Hence, patients are free from metal toxicity and at the same time free from heart attack, stroke, swelling and other health related issues associated with heavy metal toxicity [119].

4.5 Microbial culture bio-absorbents for bioremediation of heavy metal

Bio-absorption is an efficient profitable method used for water pollution removal [120]. Micro-organisms induced bio-absorbents are used effectivity for HM removal now-a-days. But the efficacy of microbial bio-absorbent is dependent on the ambient environment, absorbing material and heavy metal to be removed etc. [120]. Bacteria, yeast, fungi, algae etc., may be used as bio-absorbents on the basis of ion-exchange ability, physical absorption, complex formation capacity, precipitation, process conditions (acidity of medium. Bio-sorbent concentration), sorption center density, immobilization techniques etc., to remove toxic metals like Hg, Pb, As, Cd in addition to precious metals Au, Ag, Pt [121, 122].

Modak et al. [123] and Vijayaraghavan et al. [124] reported that dead microbial biomass shows lots of benefits over living cells like low cost, high sorption–desorption rate, absence of nutrients etc [121, 122]. The microbial cell wall with different functional groups having varying geometry like carbonyl, hydroxide, amino, sulfate etc., plays main role for removal of heavy metals from aqueous solutions [125].

Gram-positive bacteria, gram-negative bacteria, cyanobacteria, yeast can be used as bio-sorbents and these micro-biosorbents are small in size with low density and low elasticity. For example, the immobilization of Cd (III) ions by *Bacillus subtillis*, removal of Cu (II) b *Arthobacter* Sp., absorption of Pb (II) and Cu (II) by *Bacillus drentensis* MG21831T biomass was reported earlier [126].

5. Conclusions

The present chapter explains about some sources of heavy metal contaminations in foods and beverages, various parameters for heavy metal toxicity measures, analytical tools for heavy metal detection and new approaches of heavy metal remediation's from food stuffs. We observed from earlier reports that, environmental pollutions (soil and water pollution) is the main reason of heavy metal contamination in plant-based foods, which transmitted to cooked foods and other processed foods and further to humans. Again, storage of beverages in metal- based utensil or processing, brewing procedure, use of contaminated water are some other potential reasons of heavy metal contamination. The permissible limits of ingestion of heavy metals associated with food stuffs should be evaluated by different methods suggested by WHO, USEPA, SON, FAO/WHO etc. The analytical methods focused in the chapter are ICP-OES, ICP-MS, AAS etc. The various bio-remediation techniques like application of nano-technological innovations, microbial bio-absorbents, bio-surfactants are also summarizes in the chapter. We hope, the chapter will help the researchers to get some information of heavy metal remediation, sources of heavy metal contamination in foods and beverages etc.

Conflict of interest

The authors declare no conflict of interest.

Acronyms and abbreviations

EPA	Environmental Protection Agency
WHO	World Health Organization
CalEPA	California Environmental Protection Agency
ATSDR	Agency for Toxic Substances and Diseases Registry
IARC	International Agency for Research on Cancer
bw	body weight
JECFA	Joint FAO/WHO Expert Committee for Food Additives
MRL	minimal risk level
PTWI	provisional tolerable weekly intake
USEPA	United States Environmental Protection Agency
NAS	National Academy of Science
MCL	Maximum Concentration Limit
TDI	tolerable daily intake
NYSDOH	New York State Department of Health
HGAAS	Hydride Generation Atomic Absorption Spectroscopy
CVAAS	Cold Vapor Atomic Absorption Spectroscopy
LIBS	Laser-Induced Breakdown Spectroscopy
SENLIBS	Surface-Enhanced Laser-Induced Breakdown Spectroscopy
NPs	nanoparticles
ROS	radical oxygen species

Author details

Anamika Kalita Deka*, Kushwaha Jashvant Kumar and Sunshri Basumatary
Department of Chemistry, Central Institute of Technology Kokrajhar (Deemed to be University, Under MoE, Govt. of India) Kokrajhar, Assam, India

*Address all correspondence to: a.kalita@cit.ac.in

References

[1] Tchounwou PB, Yedjou CG, Patlolla AK, Sutton DJ. Heavy metals toxicity and the environment. National Institutes of Health. 2012;**101**:133-164

[2] Izah SC, Chakrabarty N, Srivastav AL. A review on heavy metal concentration in potable water sources in Nigeria: Human health effects and mitigating measures. Exposure and Health. 2016;**8**:285-304

[3] Izah SC, Inyang IR, Angaye TCN, Okowa IP. A review of heavy metal concentration and potential health implications of beverages consumed in Nigeria. Toxics. 2017;**5**(1):1-15. DOI: 10.3390/toxics5010001

[4] Nnorom IC, Osibanjo O, Ogugua K. Trace heavy metal levels of some bouillon cubes, and food condiments readily consumed in Nigeria. Pakistan Journal of Nutrition. 2007;**6**:122-127

[5] Deka AK, Handique P, Deka DC. Ethnic food beverages with heavy metal contents: Parameters for associated risk to human health, North-East India. Toxicology Reports. 2021;**8**:1220-1225. DOI: 10.1016/j.toxrep.2021.06.013

[6] Iwegbue CMA, Ojelum AL, Bassey FI. A survey of metal profiles in some traditional alcoholic beverages in Nigeria. Food Science & Nutrition. 2014; **2**:724-733

[7] Hague T, Petroczil A, Andrews PLR, Barker J, Naughton DP. Determination of metal ion content of beverages and estimation of target hazard quotients: A comparative study. Chemistry Central Journal. 2008;**2**:13

[8] Chang LW, Magos L, Suzuki T, editors. Toxicology of Metals. Boca Raton, FL, USA: CRC Press; 1996

[9] WHO/FAO/IAEA. Trace Elements in Human Nutrition and Health. Switzerland, Geneva: World Health Organization; 1996

[10] USEPA. Risk Assessment: Guidance for Superfund. in: Human Health Evaluation Manual (Part a), Interim Final, Vol 1, Office of Emergency and Remedial Response. Washington DC: U. S. Environmental Protection Agency; 1989

[11] World Health Organisation. Evaluation of Certain Food Additives and Contaminants, Technical Reports Series no. 837. Geneva: WHO; 1993

[12] Standards Organisation of Nigeria (SON). Nigeria Standard for Drinking Water Quality. Nigeria Industrial Standard NIS 554: 2007. Abuja, Nigeria: Standard Organisation of Nigeria; 2007

[13] Handique P, Kalita Deka A, Deka DC. Metal profile of traditional alcoholic beverages prepared by the ethnic communities of Assam, India. Journal of the Institute of Brewing. 2017;**123**(2): 284-288

[14] World Health Organisation. Safety Evaluation of Certain Food Additives and Contaminants: Cadmium, WHO Food Additives Series 46. Geneva: Joint FAO/WHO Expert Committee on Food Additives; 2001. Available from: https://apps.who.int/iris/handle/10665/42395

[15] World Health Organisation. Summary and conclusions of the 61st meeting of joint FAO/WHO Expert Committee on Food Additives (JECFA) JECFA/61/sc. Rome, Italy. 2003. Available from: https://cdn.who.int/media/docs/default-source/food-safety/jecfa/summary-and-conclusions/jecfa91-

1to12 march2021summary and conclusions.Pdf sfvrsn=1d79351f_5

[16] Iweala EEJ, Olugbuyiro JAO, Durodola BM, Fubara-Manuel DR, Okoli AO. Metal contamination of foods and drinks consumed in Ota, Nigeria. Research Journal ofEnvironmental Toxicology. 2014;**8**:92-97

[17] Izah SC, Aseiba ER, Orutugu LA. Microbial quality of polythene packaged sliced fruits sold in major markets of Yenagoa Metropolis, Nigeria. Point Journal of Botany and Microbiology Research. 2015;**1**:30-36

[18] Izah SC, Kigigha LT, Anene EK. Bacteriological quality assessment of Malus domestica Borkh and *Cucumis sativus* L. in Yenagoa Metropolis, Bayelsa state, Nigeria. British Journal of Applied Research. 2016;**1**:5-7

[19] Chunhabundit R. Cadmium exposure and potential health risk from foods in contaminated area, Thailand. Toxicology Research. 2016;**32**(1):65-72. DOI: 10.5487/TR.2016.32.1.065. Epub 2016 Jan 31

[20] Figueroa E. Are more restrictive food cadmium standards justifiable health safety measures or opportunistic barriers to trade; an answer from economics and public health. Science of the Total Environment. 2008;**389**:1-9

[21] Agency for Toxic Substances and Disease Registry (ATSDR). Toxicological Profile for Cobalt. Public Health Service, U.S. Atlanta, GA: Department of Health and Human Services; 1992

[22] Agency for Toxic Substance and Disease Registry (ATSDR). Toxicological Profile for Mercury U.S. Department of Health and Humans Services, Public Health Humans Services, Centers for Diseases Control. Atlanta: 2003

[23] Oehlenschläger J. Identifying heavy metals in fish. In: Bremner HA, editor. Safety and Quality Issues in Fish Processing. Cambridge: Woodhead Publishing Limited; 2002. pp. 95-113

[24] WHO. Guidelines for Drinking-water Quality. Sixty-first Meeting, Rome, 10-19 June 2003. Joint FAO/WHO Expert Committee on Food Additives. 2004. Available from: http://ftp.fao.org/es/esn/jecfa/jecfa61sc.pdf

[25] Leyssens L, Vinck B, Van Der Straeten C, Wuyts F, Maes L. Cobalt toxicity in humans—A review of the potential sources and systemic health effects. Toxicology. 2017;**387**:43-56. DOI: 10.1016/j.tox.2017.05.015

[26] California Environmental Protection Agency (CalEPA). Technical Support Document for the Determination of Noncancer Chronic Reference Exposure Levels. Draft for Public Comment. Berkeley, CA: Office of Environmental Health Hazard Assessment; 1997

[27] Arthur YAU. Vegetables and Chromium Contamination, Food Safety Focus (63rd Issue, October 2011)—Incident in Focus, Risk Assessment Section, Centre for Food Safety. Available from: https://www.cfs.gov.hk [Accessed: December 26, 2022]

[28] Sun H, Brocato J, Costa M. Oral chromium exposure and toxicity. Current Environmental Health Reports. 2015;**2**(3):295-303. DOI: 10.1007/s40572-015-0054-z

[29] Agency for Toxic Substance and Disease Registry (ATSDR). Toxicological Profile for Lead U.S. Department of Health and Humans Services, Public Health Humans Services. Atlanta: Centers for Diseases Control; 2007

[30] Castro-González MI, Méndez-Armenta M. Heavy metals: Implications

associated to fish consumption. Environmental Toxicology and Pharmacology. 2008;**26**:263-271

[31] Sawyer CN, McCarty PL, Parkin GF. Chemistry for Environmental and Engineering and Science. fifth ed. NY: McGraw Hill; 2003

[32] United States Environmental Protection Agency (USEPA). Arsenic in Drinking Water Fact Sheet. 2005. Available from: http://www.epa.gov/safewater/arsenic.html

[33] Sundar S, Chakravarty J. Antimony toxicity. International Journal of Environmental Research and Public Health. 2010;7(12):4267-4277. DOI: 10.3390/ijerph7124267. Epub 2010 Dec 20

[34] Iyengar GV, Tanner JT, Wolf WR, Zeisler R. Preparation of a mixed human diet material for the determination of nutrient elements, selected toxic elements and organic nutrients: A preliminary report. Science of the Total Environment. 1987;**61**:235-252

[35] Wu F, Fu Z, Liu B, Mo C, Chen B, Corns W, et al. Health risk associated with dietary co-exposure to high levels of antimony and arsenic in the world's largest antimony mine area. The Science of the Total Environment. 2011;**409**(18): 3344-3351. DOI: 10.1016/j.scitotenv.2011.05.033

[36] Rude RK. Magnesium. In: Coates PM, Betz JM, Blackman MR, Cragg GM, Levine M, Moss J, White JD, editors. Encyclopedia of Dietary Supplements. 2nd ed. New York, NY: Informa Healthcare; 2010. pp. 527-537

[37] Rude RK. Magnesium. In: Ross AC, Caballero B, Cousins RJ, Tucker KL, Ziegler TR, editors. Modern Nutrition in Health and Disease. 11th ed. Baltimore, Mass: Lippincott Williams & Wilkins; 2012. pp. 159-175

[38] Institute of Medicine (IOM). Food and Nutrition Board. Dietary Reference Intakes: Calcium, Phosphorus, Magnesium, Vitamin D and Fluoride. Washington, DC: National Academy Press; 1997

[39] McGuire JK, Kulkarni MS, Baden HP. Fatal hypermagnesemia in a child treated with megavitamin/megamineral therapy. Pediatrics. 2000;**105**:E18

[40] Magnesium | The Nutrition Source. Available from: https://www.hsph.harvard.edu›the nutrition source [Accessed: December 26, 2022]

[41] Pajarillo E, Nyarko-Danquah I, Adinew G, Rizor A, Aschner M, Lee E. Chapter Six—Neurotoxicity mechanisms of manganese in the central nervous system. In: Aschner M, Costa LG, editors. Advances in Neurotoxicology. Vol. 5. Academic Press; 2021. pp. 215-238. DOI: 10.1016/bs.ant.2020.11.003

[42] Finley JW, Penland JG, Pettit RE, Davis CD. Dietary manganese intake and type of lipid do not affect clinical or neuropsychological measures in healthy young women. The Journal of Nutrition. 2003;**133**:2849-2856. DOI: 10.1093/jn/133.9.2849

[43] World Health Organization Evaluations of the Joint FAO/WHO Expert Committee on Food Additives (JECFA). 1983. Available from: https://apps.who.int

[44] Yuen H-W, Becker W. Iron Toxicity, National Library of Medicine, National Centre for Biotechnology Information. 2022. Available from: https://www.ncbi.nlm.nih.gov

[45] Sigue G, Gamble L, Pelitere M, et al. From profound hypokalemia to life-

threatening hyperkalemia: A case of barium sulfide poisoning. Archives of Internal Medicine. 2000;**160**:548-551

[46] Shankle R, Keane JR. Acute paralysis from inhaled barium carbonate. Archives of Neurology. 1988;**45**:579-580

[47] Johnson CH, VanTassell VJ. Acute barium poisoning with respiratory failure and rhabdomyolysis. Annals of Emergency Medicine. 1991;**20**:1138-1142

[48] CDC. Barium toxicity after exposure to contaminated contrast solution—Goias State, Brazil, 2003. MMWR. 2003; **52**:1047-1048

[49] Dawson A. Barium. In: Nelson LS, Lewin NA, Howland MA, Hoffman RS, Goldfrank LR, Flomenbaum NE, editors. Goldfrank's Toxicologic Emergencies. 9th ed. New York, NY: McGraw-Hill; 2011. pp. 1364-1373

[50] Agency for Toxic Substances and Disease Registry. Toxicological Profile for Barium and Barium Compounds [online]. 2007. Available from: http://www.atsdr.cdc.gov/ToxProfiles/tp24.pdfpdficon

[51] Environmental Protection Agency. Barium and Compounds [online]. 2005. Available from: http://www.epa.gov/iris/subst/0010.htmexternalicon

[52] Barium and Barium Compounds. Available from: https://www.atsdr.cdc.gov

[53] Hubert A, Vaessen MG, Szteke B. Beryllium in food and drinking water—a summary of available knowledge. Food Additives and Contaminants. 2000;**17**:149-159. DOI: 10.1080/026520300283504

[54] Agency for Toxic substances and Disease Registry (ATSDR). Public Health Statement Berryllium. 2002 CAS: 7440-41-7. Available from: https://www.atsdr.cdc.gov

[55] Njoga EO, Ezenduka EV, Ogbodo CG, Ogbonna CU, Jaja IF, Ofomatah AC, et al. Detection, distribution and health risk assessment of toxic heavy metals/metalloids, arsenic, cadmium, and lead in goat carcasses processed for human consumption in South-Eastern Nigeria. Food. 2021;**10**(4):798

[56] Anderson RA, Bryden NA, Polansky MM. Dietary chromium intake. Freely chosen diets, institutional diet and individual foods. Biological Trace Element Research. 1992;**32**:117-121

[57] Cabrera C, Lloris F, Gimenez R, Olalla M, Lopez C. Mineral content in legumes and nuts: Contribution to the Spanish dietary intake. Science of the Total Environment. 2003;**308**:1-14

[58] Tchounwou PB, Yedjou CG, Patlolla AK, Sutton DJ. Heavy metal toxicity and the environment. Experientia Supplementum. 2012;**101**:133-164. DOI: 10.1007/978-3-7643-8340-4_6

[59] Magomya AM, Yebpella GG, Okpaegbe UC. An assessment of metal contaminant levels in selected soft drinks sold in Nigeria. International Journal of Innovative Science Engineering and Technology. 2015;**2**: 517-522

[60] Borah H, Dutta U. Trends in beverage packaging. In: Grumezescu AM, Holban AM, editors. Trends in Beverage Packaging. 1st ed. Vol. 16. Duxford, UK: Academic Press; 2019. pp. 1-14

[61] Thompson LA, Darwish WS. Environmental chemical contaminants in food: Review of a global problem. Journal of Toxicology. 2019;**244**:433-439

[62] United state Environmental Protection Agency (USEPA). Risk-Based Concentration Table. Available from: https://epa-prgs.ornl.gov/cgi-bin/chemicals/csl_search [Accessed: January 18, 2021]

[63] Amirah MN, Afiza AS, Faizal WI, Nurliyana MH, Laili S. Human health risk assessment of metal contamination through consumption of fish. Journal of Environment Pollution and Human Health. 2013;**1**(1):1-5

[64] USEPA Risk Assessment: Guidance for Superfund. in: Human Health Evaluation Manual (Part a), Interim Final, Vol 1 Office of Emergency and Remedial Respose. Washington DC: U.S. Environmental Protection Agency; 1989

[65] NYSDOH (New York State Department of Health). Hopewell Precision Area Contamination: Appendix C-NYSDOH. Procedure for Evaluating Potential Health Risks for Contaminants of Concern. NY, USA: Department of Health; 2007

[66] Iwegbue CM. Composition and daily intakes of some trace metals from canned beers in Nigeria. Journal of the Institute of Brewing. 2010;**116**(3): 312-315

[67] Soylak M, Tuzen M, Narin I, Sari H. Comparison of microwave, dry and wet digestion procedures for the determination of trace metal. Journal of Food and Drug Analysis. 2004;**12**(3): 254-258

[68] Scaeteanu Vasile G, Madjar RM, Mot A. An Overview of Methods Used for Quantification of Heavy Metal Contents in Vegetal Samples. 2021

[69] Sebecic B, Pavisic-Strache D, Vedrina-Dragojevic I. Trace elements in wine from Croatia. Deutsche Lebensmittel-Rundschau. 1998;**94**(10): 341-344

[70] Aceto M, Abollino O, Bruzzoniti MC, Mentasti E, Sarzanini C, Malandrino M. Determination of metals in wine with atomic spectroscopy (flame-AAS, GF-AAS and ICP-AES); a review. Food Additives & Contaminants. 2002;**19**(2):126-133

[71] Habte G, Hwang IM, Kim JS, Hong JH, Hong YS, Choi JY, et al. Elemental profiling and geographical differentiation of Ethiopian coffee samples through inductively coupled plasma-optical emission spectroscopy (ICP-OES), ICP-mass spectrometry (ICP-MS) and direct mercury analyzer (DMA). Food Chemistry. 2016;**212**: 512-520

[72] Gopalani M, Shahare M, Ramteke DS, Wate SR. Heavy metal content of potato chips and biscuits from Nagpur city, India. Bulletin of Environmental Contamination and Toxicology. 2007;**79** (4):384-387

[73] López FF, Cabrera C, Lorenzo M, López MC. Aluminum content in foods and beverages consumed in the Spanish diet. Journal of Food Science. 2000;**65** (2):206-210

[74] Ackah M, Anim AK, Zakaria N, Osei J, Saah-Nyarko E, Gyamfi ET, et al. Determination of some heavy metal levels in soft drinks on the Ghanaian market using atomic absorption spectrometry method. Environmental Monitoring and Assessment. 2014;**186** (12):8499-8507

[75] Abdel-Rahman GN, Ahmed MB, Sabry BA, Ali SS. Heavy metals content in some non-alcoholic beverages (carbonated drinks, flavored yogurt drinks, and juice drinks) of the Egyptian

markets. Toxicology Reports. 2019;**6**: 210-214

[76] Nuapia Y, Chimuka L, Cukrowska E. Assessment of heavy metals in raw food samples from open markets in two African cities. Chemosphere. 2018;**196**: 339-346

[77] Islam MS, Proshad R, Asadul Haque M, Hoque MF, Hossin MS, Islam Sarker MN. Assessment of heavy metals in foods around the industrial areas: Health hazard inference in Bangladesh. Geocarto International. 2020;**35**(3): 280-295

[78] Baby R, Hussein MZ, Abdullah AH, Zainal Z. Nanomaterials for the treatment of heavy metal contaminated water. Polymers (Basel). 2022;**14**(3):583. DOI: 10.3390/polym14030583

[79] Mudzielwana RGMW, Ndungu P. Uptake of as (V) from groundwater using Fe-Mn oxides modified kaolin clay: Physicochemical characterization and adsorption data Modeling. Water. 2019; **11**:1245. DOI: 10.3390/w11061245

[80] Anjos MJ, Lopes RT, De Jesus EF, Moreira S, Barroso RC, Castro CR. Trace elements determination in red and white wines using total-reflection X-ray fluorescence. Spectrochimica Acta Part B: Atomic Spectroscopy. 2003;**58**(12): 2227-2232

[81] Chuachuad W, Tyson JF. Determination of cadmium by flow injection atomic absorption spectrometry with cold vapor generation by a tetrahydroborate-form anion-exchanger. Journal of Analytical Atomic Spectrometry. 2005;**20**(4):273-281

[82] Chuachuad W, Tyson JF. Determination of lead by flow injection hydride generation atomic absorption spectrometry with tetrahydroborate immobilized on an anion-exchange resin. Journal of Analytical Atomic Spectrometry. 2005;**20**(4):282-288

[83] Beltrán BG, Ramos-Sanchez V, Chávez-Flores D, Rodríguez-Maese R, Palacio E. Total reflection X-ray fluorescence spectroscopy (TXRF) method validation: Determination of heavy metals in dietary supplements. Journal of Chemistry. 3 Dec 2020;**2020**:1-9

[84] Zucchi OL, Moreira S, Salvador MJ, Santos LL. Multielement analysis of soft drinks by X-ray fluorescence spectrometry. Journal of Agricultural and Food Chemistry. 2005;**53**(20):7863-7869

[85] Borgese L, Bilo F, Dalipi R, Bontempi E, Depero LE. Total reflection X-ray fluorescence as a tool for food screening. Spectrochimica Acta Part B: Atomic Spectroscopy. 2015;**113**:1-5

[86] Castineira MM, Brandt R, Von Bohlen A, Jakubowski N. Development of a procedure for the multi-element determination of trace elements in wine by ICP–MS. Fresenius' Journal of Analytical Chemistry. 2001;**370**(5): 553-558

[87] Haswell SJ, Walmsley AD. Multivariate data visualisation methods based on multi-elemental analysis of wines and coffees using total reflection X-ray fluorescence analysis. Journal of Analytical Atomic Spectrometry. 1998;**13** (2):131-134

[88] Carvalho ML, Barreiros MA, Costa MM, Ramos MT, Marques MI. Study of heavy metals in Madeira wine by total reflection X-ray fluorescence analysis. X-Ray Spectrometry. 1996;**25**(1):29-32

[89] Novaes CG, Bezerra MA, da Silva EG, dos Santos AM, da Silva Romao IL, Neto JH. A review of multivariate

designs applied to the optimization of methods based on inductively coupled plasma optical emission spectrometry (ICP OES). Microchemical Journal. 2016; **128**:331-346

[90] Rao RN, Talluri MK. An overview of recent applications of inductively coupled plasma-mass spectrometry (ICP-MS) in determination of inorganic impurities in drugs and pharmaceuticals. Journal of Pharmaceutical and Biomedical Analysis. 2007;**43**(1):1-3

[91] Moldovan M, Krupp EM, Holliday AE, Donard OF. High resolution sector field ICP-MS and multicollector ICP-MS as tools for trace metal speciation in environmental studies: A review. Journal of Analytical Atomic Spectrometry. 2004;**19**(7):815-822

[92] Yang XY, Hao ZQ, Li CM, Li JM, Yi RX, Shen M, et al. Sensitive determinations of Cu, Pb, Cd, and Cr elements in aqueous solutions using chemical replacement combined with surface-enhanced laser-induced breakdown spectroscopy. Optics Express. 2016;**24**(12):13410-13417

[93] Yang X, Li X, Cui Z, Yao G, Zhou Z, Li K. Improving the sensitivity of surface-enhanced laser-induced breakdown spectroscopy by repeating sample preparation. Frontiers in Physics. 2020;**8**:194

[94] de Jesus AM, Aguirre MÁ, Hidalgo M, Canals A, Pereira-Filho ER. The determination of V and Mo by dispersive liquid–liquid microextraction (DLLME) combined with laser-induced breakdown spectroscopy (LIBS). Journal of Analytical Atomic Spectrometry. 2014; **29**(10):1813-1818

[95] Aguirre MA, Selva EJ, Hidalgo M, Canals A. Dispersive liquid–liquid microextraction for metals enrichment: A useful strategy for improving sensitivity of laser-induced breakdown spectroscopy in liquid samples analysis. Talanta. 2015;**131**:348-353

[96] Mohamed R, Zainudin BH, Yaakob AS. Method validation and determination of heavy metals in cocoa beans and cocoa products by microwave assisted digestion technique with inductively coupled plasma mass spectrometry. Food Chemistry. 2020; **303**:125392

[97] Akinyele IO, Shokunbi OS. Comparative analysis of dry ashing and wet digestion methods for the determination of trace and heavy metals in food samples. Food Chemistry. 2015; **173**:682-684

[98] Daniel SCGK, Malathi S, Balasubramanian S, Sivakumar M, Sironmani TA. Application of nanotechnology in water research. In: Multifunctional Silver, Copper and Zero Valent Iron Metallic Nanoparticles for Wastewater Treatment. Hoboken, NJ, USA: John Wiley & Sons; 2014

[99] Mathur S, Singh D, Ranjan R. Remediation of heavy metal(loid) contaminated soil through green nanotechnology. 2022. DOI: 10.3389/fsufs.2022.932424

[100] Yao J, Cheng Y, Zhou M, Zhao S, Lin S, Wang X. ROS scavenging Mn_3O_4 nanozymes for in vivo anti-inflammation. Chemical Science. 2018;**9**: 2927-2933. DOI: 10.1039/C7SC05476A

[101] Sebastian A, Nangia A, Prasad MNV. Cadmium and sodium adsorption properties of magnetite nanoparticles synthesized from Hevea brasiliensis Muell. Arg. Bark: Relevance in amelioration of metal stress in rice. Journal of Hazardous Materials. 2019;

371:261-272. DOI: 10.1016/j.jhazmat.2019.03.021

[102] Konate A, He X, Zhang Z, Ma Y, Zhang P, Alugongo GM. Magnetic (Fe_3O_4) nanoparticles reduce heavy metals uptake and mitigate their toxicity in wheat seedling. Sustainability. 2017;**9**:790. DOI: 10.3390/su9050790

[103] Vishwakarma K, Singh VP, Prasad SM, Chauhan DK, Tripathi DK, Sharma S. Silicon and plant growth promoting rhizobacteria differentially regulate AgNP-induced toxicity in Brassica juncea: Implication of nitric oxide. Journal of Hazardous Materials. 2020;**390**:121806. DOI: 10.1016/j.jhazmat.2019.121806

[104] Rizwan K, Rasheed T, Bilal M. Alginate-based nanobiosorbents for bioremediation of environmental pollutants. In: Nano-Biosorbents for Decontamination of Water, Air, and Soil Pollution. Sahiwal: Elsevier; 2022. pp. 479-502. DOI: 10.1016/B978-0-323-90912-9.00021-6

[105] Sarubbo LA, Maria da Gloria CS, Durval IJB, Bezerra KGO, Ribeiro BG, Silva IA, et al. Biosurfactants: Production, properties, applications, trends, and general perspectives. Biochemical Engineering Journal. 2022;**181**:108377. DOI: 10.1016/j.bej.2022.108377

[106] Mahmoud ME, Fekry NA, Abdelfattah AM. A novel nanobiosorbent of functionalized graphene quantum dots from rice husk with barium hydroxide for microwave enhanced removal of lead (II) and lanthanum (III). Bioresource Technology. 2020;**298**:122514. DOI: 10.1016/j.biortech.2019.122514

[107] Kasture MB, Patel P, Prabhune AA, Ramana CV, Kulkarni AA, Prasad BLV. Synthesis of silver nanoparticles by sophorolipids: Effect of temperature and sophorolipid structure on the size of particles. Journal of Chemical Sciences. 2008;**120**:515-520. DOI: 10.1007/s12039-008-0080-6

[108] Hazra C, Kundu D, Chaudhari A, Jana T. Biogenic synthesis, characterization, toxicity and photocatalysis of zinc sulfide nanoparticles using rhamnolipids from Pseudomonas aeruginosa BS01 as capping and stabilizing agent. Journal of Chemical Technology and Biotechnology. 2013;**88**:1039-1048. DOI: 10.1002/jctb.3934

[109] Dermont G, Bergeron M, Mercier G, Richer-Laflèche M. Soil washing for metal removal: A review of physical/chemical technologies and field applications. Journal of Hazardous Materials. 2008;**152**:1-31

[110] Xu J, Liu C, Hsu PC, et al. Remediation of heavy metal contaminated soil by asymmetrical alternating current electrochemistry. Nature Communications. 2019;**10**:2440. DOI: 10.1038/s41467-019-10472-x

[111] Cervilla LM, Blasco B, Rios JJ, Rosales MA, Sanchez-Rodriguez E, Rubio-Wilhelmi MM, et al. Parameters symptomatic for boron toxicity in leaves of tomato plants. Journal of Botany. 2012. DOI: 10.1155/2012/726206

[112] Michalak A. Phenolic compounds and their antioxidant activity in plants growing under heavy metal stress. Polish Journal of Environmental Studies. 2006;**15**:523-530

[113] Guo G, Zhou Q, Ma LQ. Availability and assessment of fixing additives for the in situ remediation of heavy metal contaminated soils: A review.

Environmental Monitoring and Assessment. 2006;**116**:513-528

[114] Yadav SK. Heavy metals toxicity in plants: An overview on the role of glutathione and phytochelatins in heavy metal stress tolerance of plants. South African Journal of Botany. 2010;**76**: 167-179

[115] Sánchez-Vioque R, Polissiou M, Astraka K, De Los Mozos-Pascual M, Tarantilis P, Herraiz-Peñalver D, et al. Polyphenol composition and antioxidant and metal chelating activities of the solid residues from the essential oil industry. Industrial Crops and Products. 2013;**49**: 150-159

[116] Flora SJ, Pachauri V. Chelation in metal intoxication. International Journal of Environmental Research and Public Health. 2010;7(7):2745-2788. DOI: 10.3390/ijerph7072745. Epub 2010 Jun 28

[117] Pal S, Mondal S, Mukherjee R, Maity J. Removal of heavy metals from the body system by chelation therapy. Nanomedicine and Nanotechnology Journal. 2018;**2**(2):120

[118] Priya AK, Gnanasekaran L, Dutta K, Rajendran S, Balakrishnan D, Soto-Moscoso M. Biosorption of heavy metals by microorganisms: Evaluation of different underlying mechanisms. Chemosphere. 2022;**307**(Part 4):135957. DOI: 10.1016/j.chemosphere.2022.135957

[119] Velkova Z, Kirova G, Stoytcheva M, Kostadinova S, Todorova K, Gochev V. Immobilized microbial biosorbents for heavy metals removal. Engineering in Life Sciences. 2018;**18**(12):871-881. DOI: 10.1002/elsc.201800017

[120] Fosso-Kankeu E, Mulaba-Bafubiandi AF. Review of challenges in the escalation of metal-biosorbing processes for wastewater treatment: Applied and commercialized technologies. African Journal of Biotechnology. 2014;**13**:1756-1771

[121] Modak JM, Natarajan KA. Biosorption of metals using nonliving biomass—A review. Minerals & Metallurgical Processing. 1995;**12**: 189-196

[122] Vijayaraghavan K, Yun YS. Bacterial biosorbents and biosorption. Biotechnology Advances. 2008;**26**: 266-291

[123] Gadd GM. Biosorption: Critical review of scientific rationale, environmental importance and significance for pollution treatment. Journal of Chemical Technology and Biotechnology. 2009;**84**:13-28

[124] Ahmad MF, Haydar S, Bhatti AA, Bari AJ. Application of artificial neural network for the prediction of biosorption capacity of immobilized *Bacillus subtilis* for the removal of cadmium ions from aqueous solution. Biochemical Engineering Journal. 2014; **84**:83-90

[125] Hasan SH, Srivastava P. Batch and continuous biosorption of Cu (II) by immobilized biomass of Arthrobacter sp. Journal of Environmental Management. 2009;**90**:3313-3332

[126] Seo H, Lee M, Wang S. Equilibrium and kinetic studies of the biosorption of dissolved metals on Bacillus drentensis immobilized in biocarrier beads. Environmental Engineering Research. 2013;**18**:45-53

Chapter 36

Sequential Speciation Analysis of Metals in Geological Samples by Mass Spectrometry

Jiaquan Xu and Huanwen Chen

Abstract

Sequential analysis of metal speciation in geological samples has been demonstrated using mass spectrometry (MS). Various speciation for metal occurrences, such as water-soluble, exchangeable, oxidable, reducible, and crystalline, have been sequentially extracted from geological samples using a homemade device, followed by online ionized for MS analysis. The metal speciation in geological samples was not only useful for revealing the formation mechanism of geological samples but also beneficial for guiding the separation and exploitation of metals. Compared with the conventional methodology, the present approach takes the advantages of short analysis time (1 h), low sample consumption (1.0 mg), and high recovery (>95%), providing a useful platform for the efficient quantitative speciation analysis of metals in geological samples. This chapter introduces the mechanism and application of the MS approach for the analysis of rare earth minerals, dinosaur fossils, soil, etc.

Keywords: sequential analysis, direct analysis, metal speciation, mass spectrometry, geological samples

1. Introduction

Geological samples include a wide range of substances, such as rocks, minerals, sediments, soil, coal, oil, fossil, atmosphere, and organisms. Chemical analysis of geological samples is one of the primary techniques in earth science, which generally involved to get information on major and trace elements, stable and radioactive isotopes, biomarkers, etc. The distribution coefficients of trace elements have been applied to deduce the diagenetic process [1], while the contents and associations of trace elements are usually employed to discriminate the tectonic environment [2]. Moreover, some trace elements can be regarded as geothermometers and geobarometers to estimate the physical-chemical conditions of diagenesis and mineralization [3]. The fractionation mechanisms of stable isotopes and the decay effects of radioactive isotopes have played an irreplaceable role in solving fundamental problems of earth science [4], by providing valuable information about geological time, terrestrial heat sources, atmosphere-ocean interaction, the evolution of crust and mantle, diagenesis, mineralization, tectonism, paleoclimate, and paleoenvironment. Besides, the speciation of metals also shows great implications for guiding the exploitation of minerals [5]. Overall, chemical analysis of geological samples benefits the understanding of the earth from a microscopic perspective, with diversified application scenarios and broad application prospects.

To date, a great number of approaches have been proposed for the analysis of metal speciation in geological samples, including spectrometry, energy spectrometry, and mass spectrometry. For example, X-ray absorption spectroscopy (XAS) [6], X-ray diffraction (XRD) [7], and X-ray photoelectron spectroscopy (XPS) [8] were both applied to analyze the metal species. The advantages of these methods were that the sample pretreatment was simple and the analytical speed was fast. However, these methods always required a high concentration of metal species because of the relatively low detection sensitivity, and the quantitative performance of these methods was also deficient. On the other side, complex sample pretreatment, such as grinding, digestion, centrifugation, and filtration were used to extract the metal compounds from geological samples. Each process is time-consuming and labor-consuming, and maybe suffer from analytes losses [9]. After the sample pretreatment, the extracted compounds were then analyzed by chromatography methods (e.g., high-performance liquid chromatography, capillary electrophoresis chromatography, etc.) to identify the speciation by retention time, followed by coupling to inductively coupled plasma mass spectrometry (ICP-MS) [10] for quantitative analysis and electrospray ionization (ESI) MS for molecular structure analysis [11], which required expensive instruments and complex operation. Otherwise, a multistep offline sample pretreatment procedure was required for the sequential extraction of different heavy metal fractions according to the association form of metals in the samples [12, 13]. For example, in the Community Bureau of Reference protocol (BCR) method [12], 0.1 mol/L acetic acid (HAC), 0.1 mol/L $NH_2OH{\cdot}HCl$, 8.8 mol/L H_2O_2 and 1.0 mol/L NH_4AC were used sequentially for extraction of different metal speciation from the soil, respectively. Each extraction step took more than 16 h. In the Tessier method [13], 1 mol/L NaAc solution (pH = 8.2), 1 mol/L NaAc/HAc (pH = 5.0), 0.3 mol/L $Na_2S_2O_4$ + 0.2 mol/L Na-citrate +0.03 mol/L $NH_2OH{\cdot}HCl$ in 25% HAC, 0.02 mol/L HNO_3 + 30% H_2O_2 (pH = 2), HF-$HClO_4$ were used to extract exchangeable, bound to carbonates, bound to Fe-Mn oxides, bound to organic matter, and residual fraction or speciation, respectively. The whole procedure would take more than 40 h. The extracted fraction or speciation was then analyzed by ICP-MS [14] or by inductively coupled plasma optical emission spectrometry (ICP-OES) [15]. Thus, a method that has the advantages of without tedious sample pretreatment, high analytical speed, high detection sensitivity, and high accuracy is necessitous for the analysis of the speciation of the metal.

Herein, a new strategy for sequential analysis of metals speciation in geological samples is proposed by regulating the interaction between reagents/energy and geological samples. Assisted by multiple energy forms such as ultrasound and heat, various element species in geological samples were sequentially extracted by chemical reagents with different physicochemical properties of polarity, acidity, and redox using a small amount of samples. The extracted analytes were then ionized online by different ionization techniques (e.g., ESI, ICP, etc.) for MS analysis. This strategy takes the advantages of avoiding sample pretreatment, high analysis speed, low sample consumption, and low labor and energy consumption, which has been used in various areas, such as environment monitoring [16, 17], material science [18, 19], food science [20, 21], life science [22, 23], and geology [24, 25]. In this chapter, we will focus on the application of this sequential speciation analytical strategy in geological samples, such as mineral samples, fossils samples, and soil samples.

2. Sequential speciation analysis of metals in geological samples

2.1 Construction of the sequential ionization device

A sequential ionization device was home-built (**Figure 1**), which consists of eight systems: (1) quantitative addition and mixing system of the chemical reagents, which was used to prepare the corresponding reagents for extraction of the defined speciation from geological samples; (2) microscale liquid sample transportation system, which was used for precision transfer of the reagents and the extracted speciation; (3) microstructure morphology imaging system, which was used to monitor the morphology change in real time; (4) micro-electrolytic cell and interface system, which was used to load the sample; (5) field energy (temperature, ultrasonic, microwave) coupling and regulation system, which was used to promote the extraction of speciation from samples; (6) high-efficiency ionization system, which was used to ionize the extracted speciation; (7) timing trigger and intelligent control system, which was used to control the device; and (8) power supply and support connection system. This device can be directly coupled to the commercial mass spectrometer for direct MS analysis of geological samples.

2.2 Sequential analysis of rare earth elements (REEs) speciation in mineral samples

REEs contained 15 elements of the lanthanide group and yttrium and scandium, which are widely applied in multiple areas, including materials science, energy sources, biomedical science, chemo-catalysis, etc. Rare earth mineral is a kind of non-renewable resource and their reserves are decreasing rapidly. Thus, it is urgent to

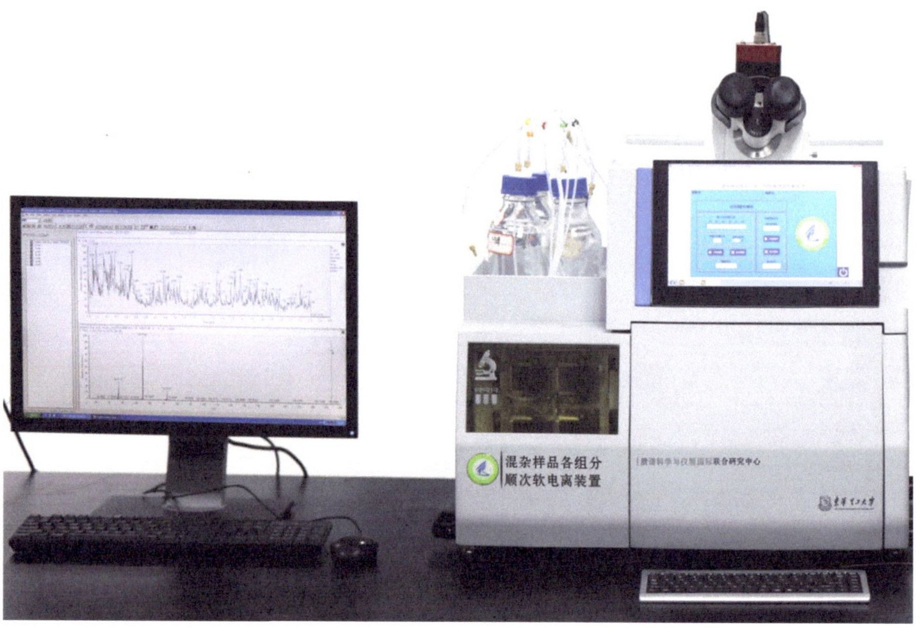

Figure 1.
The picture of a home-built sequential ionization device.

increase the exploration and exploitation efficiency of REEs which is closely related to the elemental speciation of REEs in minerals [5]. For example, ionic rare earth mines are typically exploited by the ion exchange methods [26], while the fluorocarbon cerium earth mines are typically exploited by oxidizing roast chemical methods [27]. Moreover, the information on REEs speciation in mineral samples is useful for the study of metallogenic mechanisms. Therefore, comprehensive analysis of the REEs composition and speciation in mineral samples has great significance for the rare earth industry.

Xu et al. developed a method for rapid characterization of REEs speciation in ore samples by coupling sequential speciation extraction with ICP-MS analysis [24]. As shown in **Figure 2**, six extraction solutions including ① H_2O, ② NH_4AC, ③ NH_2OH-HCl, ④ mixture of HNO_3 and H_2O_2, ⑤ HF, and ⑥ HNO_3 sequentially flowed through the sample chamber at 0.3 mL/min to extract the ① water-soluble REEs speciation, ② exchangeable REEs speciation, ③ reducible REEs speciation, ④ oxidable REEs speciation, and ⑤-⑥crystalline REEs speciation from the REEs ore sample. The extracted REEs species were mixed online with 4% HNO_3 (0.3 mL/min) at the outlet of the sample chamber for online ICP-MS analysis. The extracted ion chromatography (EIC) signals of Y (m/z = 89), La (m/z = 139), Ce (m/z = 140), Pr (m/z = 141), Nd (m/z = 146), Sm (m/z = 147), Eu (m/z = 153), Gd (m/z = 157), Tb (m/z = 159), Dy (m/z = 163), Ho (m/z = 165), Er (m/z = 166), Tm (m/z = 169), Yb (m/z = 172), and Lu (m/z = 175) were recorded in real-time. The extraction solution changed when one speciation MS signal decreased to baseline.

As shown in **Figure 3**, REEs and associated metals (e.g., Mn, Fe, etc.) with five types of speciation, were sequentially detected within 1 h. Standard curves of REEs (**Figure 4a**) and the EIC area of the corresponding REE MS signal (**Figure 3a-o**) were used to quantify the REEs content. The amount of each REE was La 391.7 μg/g, Ce 203.1 μg/g, Pr 74.1 μg/g, Nd 243.9 μg/g, Sm 41.0 μg/g, Eu 5.2 μg/g, Gd 45.9 μg/g, Tb 6.3 μg/g, Dy 28.6 μg/g, Ho 5.5 μg/g, Er 14.9 μg/g, Tm 2.1 μg/g, Yb 10.8 μg/g, Lu 2.0 μg/g, and Y 148.4 μg/g. **Figure 4b** showed the normalized ratio of each REE with the order of La (32.01%) > Nd (19.94%) > Ce (16.60%) > Y (12.13%) > Pr (6.05%) > Gd (3.76%) > Sm (3.35%) > Dy (2.33%) > Er (1.22%) > Yb (0.88%) > Tb (0.52%) > Ho (0.45%) > Eu (0.42%) > Tm (0.17%) > Lu (0.16%). The ratio of

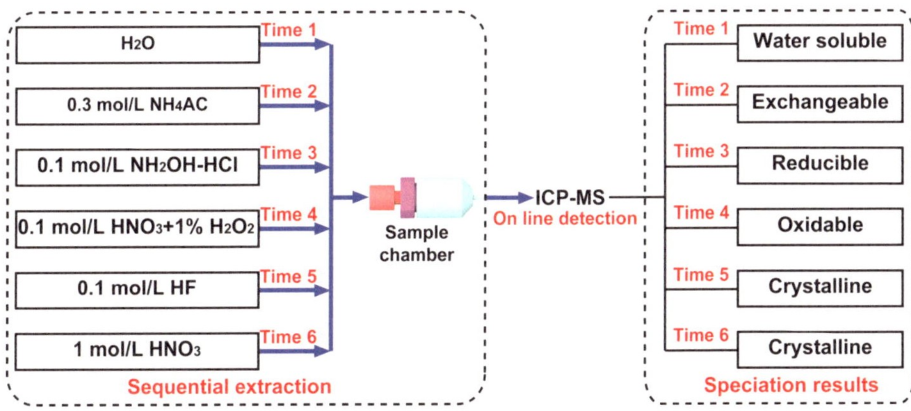

Figure 2.
The procedure for speciation analysis of REEs in ore samples.

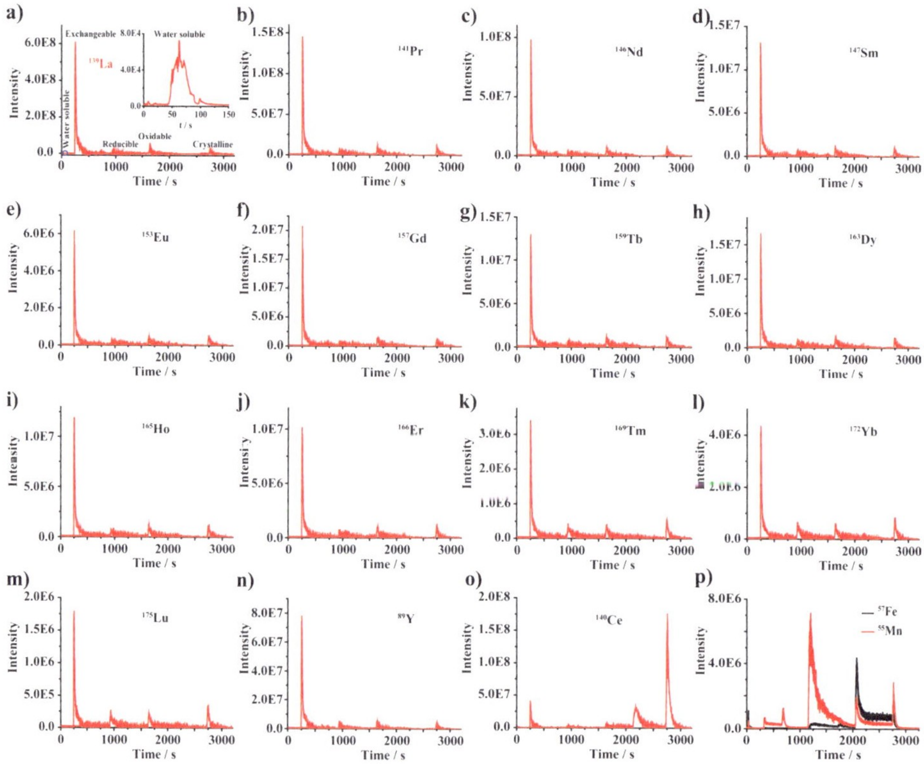

Figure 3.
Analytical results of REEs in an ore sample. a)-o) EIC of 15 kinds of REEs and p) 2 kinds of associated metal elements.

different REEs speciation was shown in **Figure 4c**. For most of the REEs, the exchangeable speciation (>50%) was dominant. However, for Ce, the dominant speciation was the crystalline (≥80%). These results provided essential information for guiding REEs exploration and exploitation.

The quantitative performance of this method was verified by standard methods (**Figure 4d**). The recoveries of most REEs (e.g., La, Ce, Pr, and Nd) were in the range of 95%–110%. However, higher recovery (>120–228%) was observed for those low abundance REEs, such as Er (14.9 μg/g), Yb (10.8 μg/g), Tb (6.3 μg/g), Eu (5.2 μg/g), Tm (2.1 μg/g), and Lu (2.01 μg/g), which could be due to the fact that this method required no sample pretreatment, thus reduced the potential analytes losses which may occur in the conventional methods.

In summary, the sequential speciation analytical method developed by Xu et al. takes the advantage of high analysis speed, high detection throughput, high efficiency, low energy consumption, and low sample consumption. For instance, the sequential speciation analytical method can obtain quantitative results of 15 REEs with five speciations (15*6 = 90 data sets) using 1.0 mg ore sample within an analytical procedure (1 h), which took about 40 s for one data set. In general, the BCR method required about 80 h to get 90 data sets using 1.0 g sample [12], and the Tessier method required about 40 h to get 90 data sets using 1.0 g sample [13]. Besides, no sample pretreatment could greatly lower labor and energy consumption.

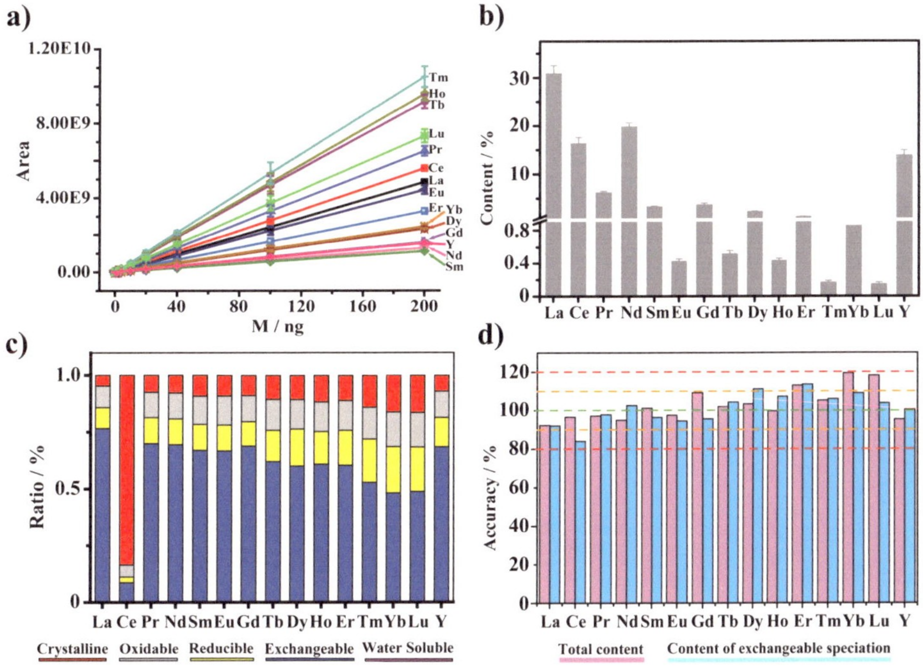

Figure 4.
Quantitative information of REEs speciation in an ore sample. (a) The standard curves of REEs (n = 3); (b) The normalized ratio of REEs; (c) The ratio of REEs speciation; (d) The accuracy of the present method varied by standard methods.

2.3 Sequential analysis of REEs speciation in fossil samples

REEs pattern in fossil samples is useful for taphonomy, provenance, and paleoenvironment [28]. It has been applied to solve some significant problems such as to identify reworked fossil samples as well as a useful indicator of bone provenance within mixed deposits [29]. Besides, the REEs pattern of fossils was also applied for the determination of stratigraphy, such as the dating formation [30]. Basically, REEs enriched in fossils were attributed to a post-mortem process [28]. Two processes were proposed during the enrichment process of REEs, which are substitution and adsorption. These two main processes corporately determined the concentration of REEs in biogenic apatite. On one side, the adsorption process can rapidly reach an equilibrium. However, because of the weak physical molecular force, such as the van der Waals force, these REEs ions can be desorbed easily. On the other side, the substitution process is up to the properties of bulk crystal-chemical, which accounts for the major contribution of concentrated REEs. Therefore, REEs speciation is vitally essential for the research of fossil bones.

Thus, Wang et al. carried out the sequential ionization MS analysis (**Figure 2**) for the quantification of REEs in dinosaur bone fossils samples [25]. Fossil particle samples were sampled in different spots on the horizontal part of the bone fossil. Drilled fossils were ground and filtered to obtain powders using a 200-mesh sieve, followed by a sequential ionization MS procedure as shown in **Figure 2**. As an example, **Figure 5** showed the results of La in a powder sample and particle sample. H_2O was first selected as an eluent to extract the water-soluble REEs. After the MS signal of water-soluble REEs (inset enlarged peaks 1) decreased to zero, 0.3 mol/L NH_4Ac, 0.1 mol/L $NH_2OH{\cdot}HCl$, and a mixture of 0.1 mol/L HNO_3 + 1% H_2O_2, were sequentially

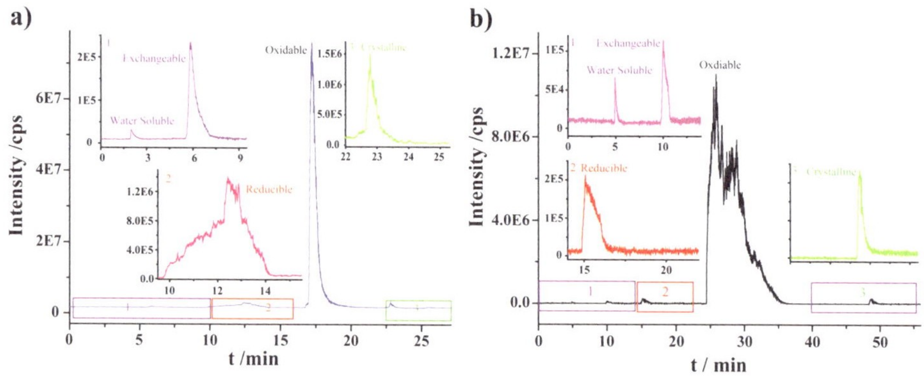

Figure 5.
Sequential analysis of REEs speciation in a dinosaur bone fossil sample. (a) EIC of La in fossil powder; (b) EIC of La in fossil particle. The insets in (a) and (b) were the enlarged REEs peaks.

flowed through the fossil sample to extract the exchangeable REEs, reducible REEs, and oxidable REEs until the MS signals of the previous speciation disappeared. Finally, the crystalline REEs was extracted by 0.1 mol/L HF and 1 mol/L HNO_3 solutions (inset enlarged peaks 3). The contents of REEs speciation were obtained by integrating the EIC area of the REEs speciation peaks and the corresponding standard curve. Compared with traditional standard digestion methods, this method gave similar results of REEs concentration in a dinosaur bone, but with the advantages of a higher analysis speed (1 h), lower sample consumption (1 mg), and more information on spatial distribution. Furthermore, the REEs in both powder samples and particle samples of dinosaur bone could be sequentially detected with high sensitivity in this work. This approach may offer us a wonderful tool for the investigation of REEs distribution in the microstructure of the fossil sample in the future.

2.4 Sequential speciation analysis of heavy metals in soil samples

Heavy metal is one of the most toxic species which exists widely in soil and is closely related to human health [31]. The heavy metals in soil can further enter drink water and plants, and then enter the human body by diet. In particular, the biotoxicity of heavy metals is closely related to their speciation [32]. For example, soluble Pb speciation are easily metastasis and assimilated by the plants, resulting in strong toxicity, while insoluble speciation Pb has weak toxicity. Thus, chemical profiling of heavy metal speciation in soil samples is of great significance for the evaluation of health risks. In addition, the information regarding heavy metal speciation is useful for tracing the pollution source.

Song et al. developed a sequential ionization MS method for analysis of the speciation of the metal in soil samples without sample pretreatment by a combination of a flow electrolytic cell to ESI-MS [17]. Four speciations of heavy metals, including water-soluble speciation, organic speciation, insoluble speciation, and elemental speciation, were extracted sequentially online by H_2O, CH_3OH, EDTA-2Na, and electrolysis for ESI-MS detection. **Figure 6a** displays the high-resolution MS results of $Pb(NO_3)_2$ which was employed as a representative of water-soluble metal compounds. EDTA-2Na was used to chelate the extracted Pb^{2+} online for ESI-MS analysis. The related signals were assigned to $[EDTA\text{-}2H]^{2-}$ (m/z = 145.0370), $[Pb^{206} + EDTA\text{-}4H]^{2-}$ (m/z = 247.0161), $[Pb^{207} + EDTA\text{–}4H]^{2-}$ (m/z = 247.5168),

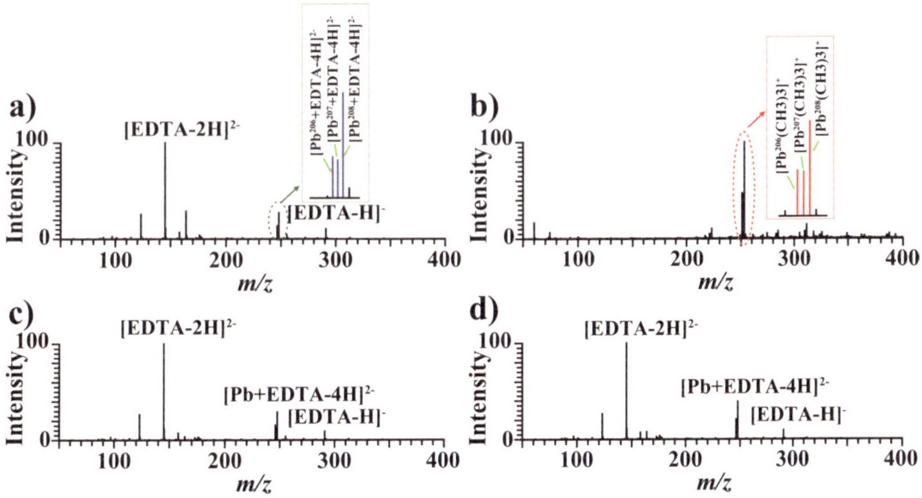

Figure 6.
Qualitative analysis of different speciation Pb by EC-MS. (a) water-soluble speciation $Pb(NO_3)_2$; (b) organic speciation trimethyl lead chloride; (c) insoluble speciation $Pb(CO_3)_2$; (d) elemental Pb.

$[Pb^{208}$ + EDTA-4H$]^{2-}$ (*m/z* = 248.0171) and [EDTA–H]$^-$ (*m/z* = 291.08230) according to the ratio of isotope and high-resolution *m/z* values. **Figure 6b** displays the MS results of trimethyl lead chloride which was employed as an organic metal compound. Obvious $[Pb^{206}(CH_3)_3]^+$ (*m/z* = 251.0443), $[Pb^{207}(CH_3)_3]^+$ (*m/z* = 252.0458), $[Pb^{208}(CH_3)_3]^+$ (*m/z* = 253.0465) were detected. Because EDTA can form a highly stable complex with most of the metal ions, EDTA-2Na was used as an extraction solution to dissolve the insoluble $PbCO_3$ for MS detection, and obvious [EDTA-2H]$^{2-}$ (*m/z* = 145.0370), $[Pb^{206}$ + EDTA-4H$]^{2-}$ (*m/z* = 247.0161), $[Pb^{207}$ + EDTA–4H$]^{2-}$ (*m/z* = 247.5168), $[Pb^{208}$ + EDTA-4H$]^{2-}$ (*m/z* = 248.0171), and [EDTA–H]$^-$ (*m/z* = 291.0823) were detected in **Figure 6c**, indicating that using EDTA-2Na to dissolve the insoluble speciation was feasible. Elemental Pb cannot be dissolved by H_2O, CH_3OH, or EDTA-2Na. Strong inorganic acids (e.g., HCl, HNO_3) are typically used for the dissolution of Pb. However, strong inorganic acids would suppress the MS signals and damage the MS instrument. Therefore, the electrolysis method was applied to mildly convert the Pb into Pb^{2+}, followed by reacting with EDTA-2Na in situ to form the EDTA-Pb complex for MS detection. As shown in **Figure 6d**, obvious signals of [Pb + EDTA–4H]$^{2-}$ were also detected.

Figure 7 showed the results of the heavy metals in the faucet sediment obtained by sequential speciation analysis. The whole analytical procedure was completed within 20 min without complex sample pretreatment. The signals of water-soluble speciation, organic speciation, insoluble speciation, and elemental speciation were sequentially detected at 0–4 min, 4–6 min, 6–11 min, and 12–18 min, respectively. The results indicated that the sediment contained water-soluble, insoluble, and elemental Pb, Ni, Zn, and Fe, insoluble and elemental Cu, water-soluble and insoluble Ca, and insoluble Mn. No organic speciation metal was detected in the sediment.

2.5 Sequential speciation analysis of arsenic in sediments and organisms

Despite great efforts have been made to study redox-sensitive elements in the environment such as arsenic (As) and antimony (Sb), there are still some challenges and difficulties

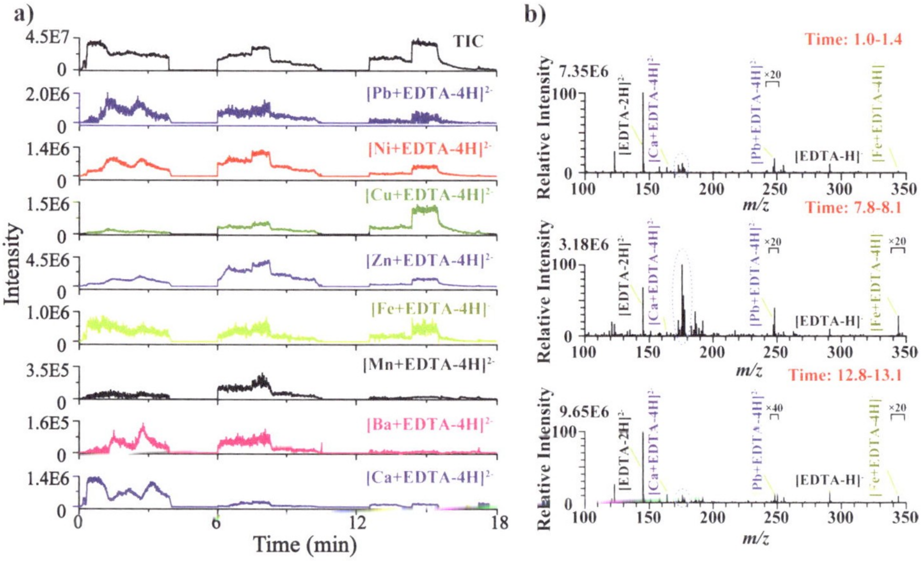

Figure 7.
Application of EC-MS in the analysis of heavy metals in the faucet sediment. (a) EIC of the different speciation of heavy metals; (b) mass spectrum of the corresponding speciation.

for the speciation analysis. As is one of the most toxic elements and exists widely in the natural environment [33]. Furthermore, the chemical behavior of As, such as adsorption, toxicity, transformation, and bioavailability, was not only related to its speciation but also related to the co-existing elements (e.g., Fe, Mn) [34]. Therefore, a robust analytical method is required that can provide information on the concentration, spatial distribution, redox states, and the inter-influence with coexisting elements. However, conventional speciation analysis methods normally required tedious sample pretreatment [12, 13].

Thus, Wu et al. proposed an online sequential fractionation analysis method for As in ferrihydrite samples [16]. As adsorption experiment was first conducted in solution with synthesized ferrihydrites as adsorbents. The collected As-adsorbed ferrihydrite was then used as a raw sample for direct MS analysis. The online fractionation analysis method consisted of eluents, an online extraction device, online acidification, and detection (**Figure 8**). Based on the association between As and iron oxides, various eluents were selected, such as H_2O, 5% NH_4NO_3, 0.2 mol/L $NH_4H_2PO_4$, 0.1 mol/L $(NH_4)_2C_2O_4$ (pH = 3), and a mixture of 0.5% HF and 5% HNO_3, respectively. Online extraction of As, which was assisted by an ultrasonic bath, was then performed and acidified by 2% HNO_3. Finally, the analytes were efficiently detected by ICP-MS.

The results showed that water-soluble fraction, non-specifically adsorbed fraction, specifically adsorbed fraction, iron oxide bonded fraction, and residual fraction were sequentially determined with different proportions (**Figure 9**). For instance, iron oxide bonded As species (83%) and specifically adsorbed As species (13%) were the dominant As(III) fractions on ferrihydrite. The method had the advantages of no sample pretreatment, low sample consumption, high detection sensitivity, short analysis time, and simple experimental manipulation, providing a useful tool for understanding As behavior in the environment. Besides, compared to conventional methods, the present method could provide more information on the As association with iron oxides on the surface of solid samples.

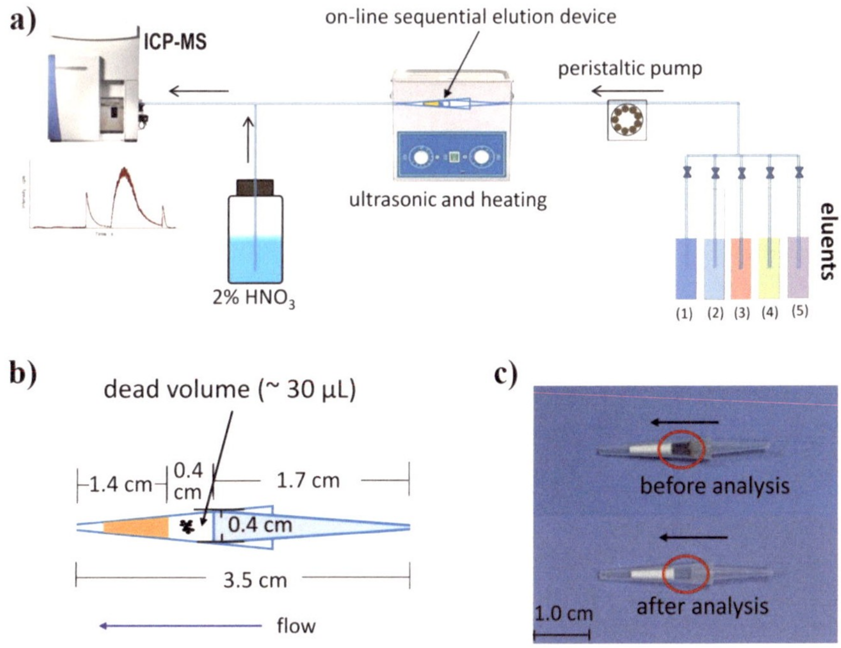

Figure 8.
Sequential analysis of As speciation in ferrihydrite sample. (a) Schematic diagram of online sequential analysis of As species on ferrihydrite. The eluents (1)–(5) are H_2O, 5% NH_4NO_3, 0.2 mol/L $NH_4H_2PO_4$, 0.1 mol/L $(NH_4)_2C_2O_4$ and a mixture of 0.5% HF and 5% HNO_3, respectively; (b) online elution component with detailed parameter; (c) the picture of the sample before and after analysis.

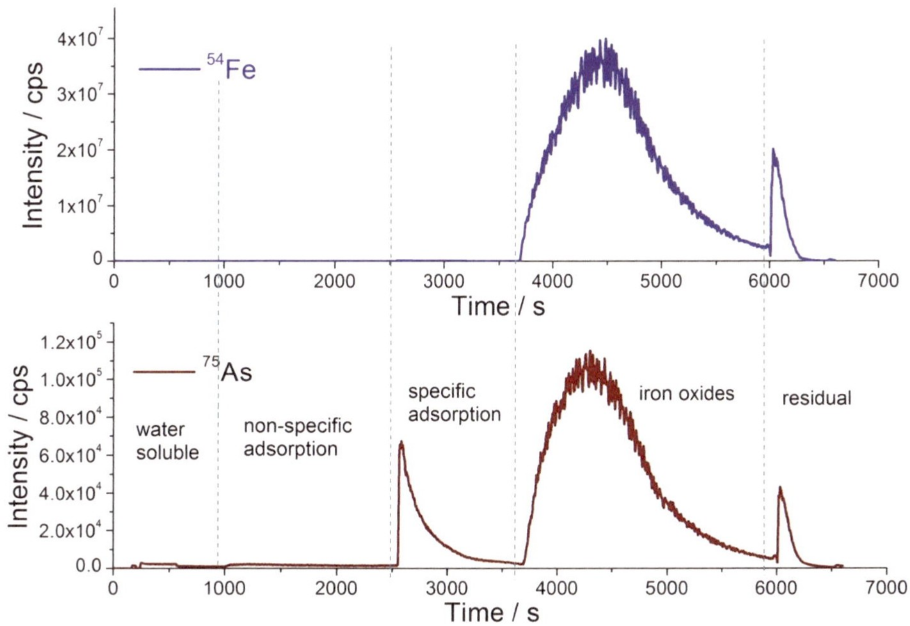

Figure 9.
Sequential analysis of $^{75}As(III)$ and ^{54}Fe in ferrihydrite.

In the second work, an extraction electrospray ionization-hydride generation-microwave plasma torch-mass spectrometry (EESI-HG-MPT-MS) method was proposed by Wu et al. for online speciation analysis of As from raw bio-matrices [20]. In this work, the MPT was coupled to an organic mass spectrometer for sensitive detection of inorganic arsenic species. The overall setup was mainly composed of five parts, which are detailed in **Figure 10**. A mixture of $CH_3OH/H_2O/HAc$ (V:V:V = 50:49:1) was used as an eluent reagent. Raw samples were extracted online by a syringe pump and an applied high voltage of 4 kV. The As-containing effluent was then either directly ionized for organic As detection or for the detection of inorganic As converted into volatile arsenic hydrides under certain pH conditions. It is worth noting that, in this work, a novel sample introduction device that can be used for both gas and solution samples was prepared by inserting a glass pipette into a larger quartz tube (i.d. 0.25 cm). When the analytes reached the tip of the inner glass pipette, a stable electrospray plume was formed, and therefore organic As could be analyzed directly.

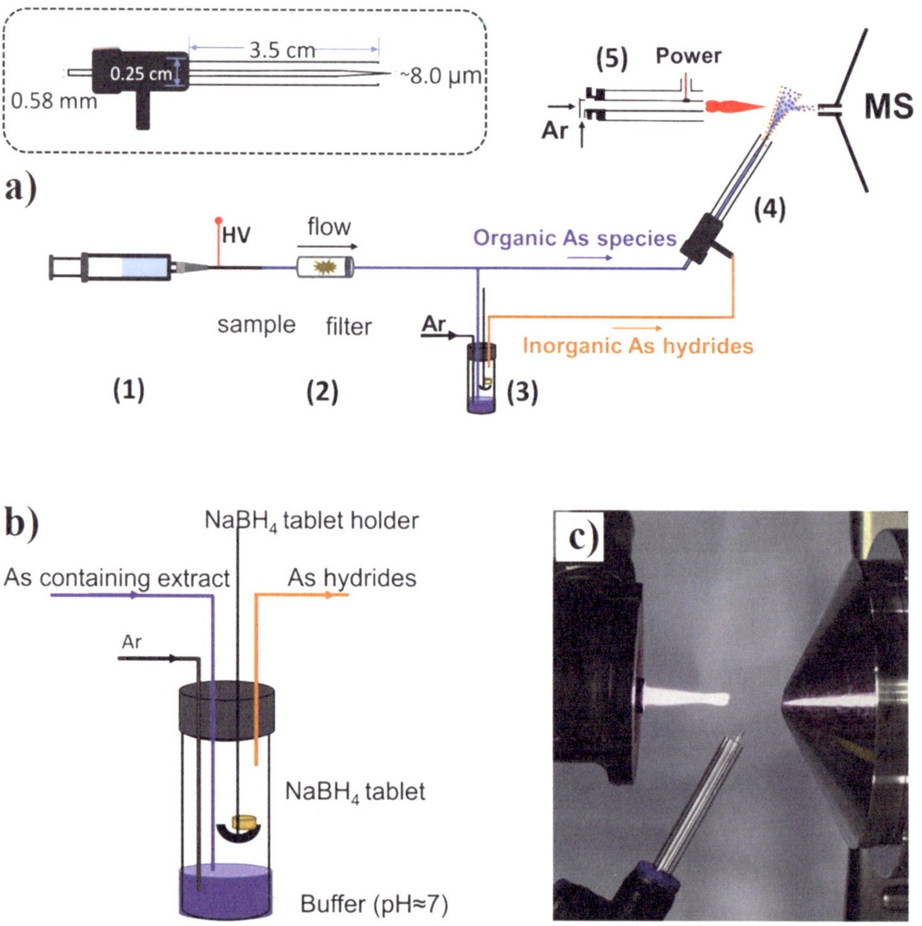

Figure 10.
Sequential analysis of organic and inorganic arsenic species. (a) Schematic diagram of the EESI-HG-MPT-MS, which consists of a (1) syringe pump, (2) sample chamber, (3) generator of hydride As, (4) sample introduction device, (5) MPT and mass spectrometer; (b) schematic diagram of the generator of hydride As; (c) picture of the home-made ion source.

On the other hand, inorganic As was determined via ionization by MPT in the form of m/z 204 ($H_3AsO_4 + NO_3^-$). This method enabled direct obtaining of As speciation in raw biological samples, which avoided multistep extraction and sample preservation, therefore potential redox processes, interconversion, or mass loss could be effectively alleviated. Ambient ionization by MPT allows sensitive detection of inorganic As on a common organic mass spectrometer. The developed method may play a role in different fields such as analytical chemistry, environmental chemistry, and bioanalysis.

3. Conclusions and Future Directions

3.1 Conclusions

In summary, a novel strategy for direct MS analysis of metal speciation in geological samples was proposed by regulating the interaction process between reagents/energy and geological samples, and a sequential ionization device was home-built which can be directly coupled to the commercial mass spectrometer (e.g., ICP-MS, ESI-MS) for the direct MS analysis of geological samples. Rapid sequential characterization of metal speciation in multiple geological samples, such as minerals, fossils, soil, and bio-tissue has been quantitatively achieved based on the sequential analytical strategy and ionization device, without complex sample pretreatment. Compared with traditional methods (e.g., BCR method, Tessier method), the present method offers significant advantages such as high analysis speed, low sample consumption, high recovery, simple operation, low energy and labor consumption, and rich chemical information. These results demonstrate the power of the present method for obtaining information on element abundance, occurrence, and paragenetic association, which is vitally essential for geochemistry.

3.2 Future directions

3.2.1 Sequential analysis of the distribution of metallic constituents in geological samples

Micro-area analysis of geological samples, such as minerals, can provide information on the concentration and distribution of the elements, which is essential for studying the metallogenic mechanism and understanding the structure. In general, laser ablation-inductively coupled plasma-mass spectrometry (LA-ICP-MS) [35], second ion mass spectroscopy (SIMS) [36], proton-induced X-ray emission spectroscopy (PIXRES) [37], etc. were widely used in the micro-area analysis of minerals with point-by-point scanning, which has the advantage of high spatial resolution and disadvantage of time-consuming. Thus, a novel analytical strategy for rapid micro-area analysis is needed. For instance, using a sequential ionization device (**Figure 1**) coupled with MS, the speciation (S) and concentration (C) of the metal in the geological samples can be obtained sequentially with time (T) by MS to establish the relationship Y(S, C) = F(T). Meanwhile, the Microstructure morphology imaging system in the sequential ionization device can record the change of morphology (M) in real time to establish the relationship Y(M) = F(T). By combining the relations the Y(S, C) = F(T) and Y(M) = F(T), a relationship between (S, C) and (M) could be established as Y(S, C) = F(M), which would be the distribution of metal speciation.

3.2.2 Sequential analysis of metallic and organic constituents in geological samples

Besides metallic constituents, organic constituents are also important for geological samples [38]. For example, biomarkers, such as alkanes, aromatic hydrocarbons, terpenoids, steroids, fatty acids, porphyrins, and so on, are helpful in identifying the source of organic matters, inferring the redox conditions of depositional and early diagenetic stages, and in reflecting thermal maturity of sediments. Thus, a method is urgent that could simultaneously analyze the metallic and organic constituents, which would not only provide information on metallic and organic constituents but could also establish the relationship between the metallic and organic constituents. Recently, we have started research to simultaneously analyze the metallic and organic constituents in fossils, soil, PM2.5, tissue, alloy, etc. by sequential ionization MS.

Acknowledgements

This work was financially supported by the National Natural Science Foundation of China (No. 21864001, 21727812, 21705016).

Conflict of interest

The authors declare no conflict of interest.

Author details

Jiaquan Xu[1] and Huanwen Chen[1,2]*

1 Jiangxi Key Laboratory for Mass Spectrometry and Instrumentation, East China University of Technology, Nanchang, P. R. China

2 School of Pharmacy, Jiangxi University of Chinese Medicine, Nanchang, P. R. China

*Address all correspondence to: chw8868@gmail.com

References

[1] Knut B. Relationships between depositional environments, burial history and rock properties. Some principal aspects of diagenetic process in sedimentary basins. Sedimentary Geology. 2014;**301**:1-14. DOI: 10.1016/j.sedgeo.2013.12.002

[2] Fisher RV, Schmincke HU. Pyroclastic rocks and tectonic environment. In: Fisher RV, Schmincke HU, editors. Pyroclastic Rocks. Berlin: Springer; 1984. pp. 383-409. DOI: 10.1007/978-3-642-74864-6_14

[3] Du Y, Ma T, Xiao C, Liu YJ, Chen LZ, Yu HT. Water-rock interaction during the diagenesis of mud and its prospect in hydrogeology. International Biodeterioration & Biodegradation. 2018;**128**:141-147. DOI: 10.1016/j.ibiod.2017.03.020

[4] Clayton RN. Isotopes: From earth to the solar system. Annual Review of Earth and Planetary Sciences. 2007;**35**:1-19. DOI: 10.1146/annurev.earth.35.092006.145059

[5] Belova VV. Development of solvent extraction methods for recovering rare earth metals. Theoretical Foundations of Chemical Engineering. 2017;**51**:599-609. DOI: 10.1134/S004057951605002X

[6] Voegelin A, Weber FA, Kretzschmar R. Distribution and speciation of arsenic around roots in a contaminated riparian floodplain soil: Micro-XRF element mapping and EXAFS spectroscopy. Geochimica et Cosmochimica Acta. 2007;**71**:5804-5820. DOI: 10.1016/j.gca.2007.05.030

[7] Javed MB, Kachanoski G, Siddique T. A modified sequential extraction method for arsenic fractionation in sediments. Analytica Chimica Acta. 2013;**787**:102-110. DOI: 10.1016/j.aca.2013.05.050

[8] Wu D, Sun S, He M, Wu Z, Xiao J, Chen XD, et al. As (V) and Sb (V) Co-adsorption onto ferrihydrite: Synergistic effect of Sb (V) on As (V) under competitive conditions. Environmental Science And Pollution Research. 2018;**25**:14585. DOI: 10.1007/s11356-018-1488-2

[9] Wu D, Pichler T. Preservation of co-occurring As, Sb and Se species in water samples with EDTA and acidification. Geochemistry: Exploration, Environment, Analysis. 2016;**16**:117-125. DOI: 10.1144/geochem2015-369

[10] Popp M, Hann S, Koellensperger G. Environmental application of elemental speciation analysis based on liquid or gas chromatography hyphenated to inductively coupled plasma mass spectrometry-a review. Analytica Chimica Acta. 2010;**668**:114-129. DOI: 10.1016/j.aca.2010.04.036

[11] Chen BB, Peng L, He M, Wang C, Bin H. Identification of cadmium containing metabolites in HepG2 cells after treatment with cadmium-selenium quantum dots. Chinese Chemical Letters. 2023;**34**:107262. DOI: 10.1016/j.cclet.2022.02.067

[12] Sahuquillo A, López-Sánchez JF, Rubio R, Rauret G, Thomas RP, Davidson CM, et al. Use of a certified reference material for extractable trace metals to assess sources of uncertainty in the BCR three-stage sequential extraction procedure. Analytica Chimica Acta. 1999;**382**:317-327. DOI: 10.1016/S0003-2670(98)00754-5

[13] Tessier A, Campbell PGC, Bisson M. Sequential extraction procedure for the speciation of particulate trace metals. Analytical Chemistry. 1979;**51**:844-851. DOI: 10.1021/ac50043a017

[14] Yang JQ, Chen YW, Shi KL, Hu KS, Li R, Gao XQ, et al. Stability of selenium and its speciation analysis in water using automatic system separation and HR-ICP-MS measurement. Chinese Chemical Letters. 2022;**33**:3444-3450. DOI: 10.1016/j.cclet.2022.03.002

[15] Bogusz A, Oleszczuk P. Effect of biochar addition to sewage sludge on cadmium, copper and lead speciation in sewage sludge-amended soil. Chemosphere. 2020;**239**:124719. DOI: 10.1016/j.chemosphere.2019.124719

[16] Wu DB, Yang SP, Li FL, Zhu TG, Chen HW. Online sequential fractionation analysis of arsenic adsorbed onto ferrihydrite by ICP-MS. Analytical Chemistry. 2020;**92**:14309-14313. DOI: 10.1021/acs.analchem.0c03516

[17] Song LL, Li H, Li T, Xu JQ, Chen HW. Sequential speciation analysis of heavy metals in drinking water pipe scales by mass spectrometry. Chemical Research in Chinese Universities. 2022;**38**:774-781. DOI: 10.1007/s40242-022-2131-9

[18] Xu JQ, Zhu TG, Chingin K, Liu YH, Zhang H, Chen HW. Sequential formation of analyte ions originated from bulk alloys for ambient mass spectrometry analysis. Analytical Chemistry. 2018;**90**:13832-13836. DOI: 10.1021/acs.analchem.8b04309

[19] Xu JQ, Zhong DC, Chingin K, Song LL, Chen HW. Chemical profiling of bulk alloys using micro-electrochemical probe mass spectrometry. Analytical Chemistry. 2019;**91**:8304-8309. DOI: 10.1021/acs.analchem.9b01056

[20] Wu DB, Li D, Dong LL, Li GL, Wang L, Tang ZY, et al. Sensitive online speciation analysis of arsenic in biological samples by ambient mass spectrometry. Journal of Analytical Atomic Spectrometry. 2022;**37**:2103-2110. DOI: 10.1039/D2JA00075J

[21] Xu JQ, Xu SR, Xiao YP, Chingin K, Lu HY, Yan RH, et al. Quantitative determination of bulk molecular concentrations of β-agonists in pork tissue samples by direct internal extractive electrospray ionization-mass spectrometry. Analytical Chemistry. 2017;**89**:11252-11258. DOI: 10.1021/acs.analchem.7b00517

[22] Song LL, Chingin K, Wang M, Zhong DC, Chen HW, Xu JQ. Polarity-specific profiling of metabolites in single cell by probe electrophoresis mass spectrometry. Analytical Chemistry. 2022;**94**:4175-4182. DOI: 10.1021/acs.analchem.1c03997

[23] Lu HY, Zhang H, Chingin K, Wei YP, Xu JQ, Ke MF, et al. Sequential detection of lipids, metabolites, and proteins in one tissue for improved cancer differentiation accuracy. Analytical Chemistry. 2019;**91**:10532-10540. DOI: 10.1021/acs.analchem.9b01507

[24] Xu JQ, Li FL, Xia F, Zhu TG, Wu DB, Chingin K, et al. High throughput online sequential extraction of natural rare earth elements and determination by mass spectrometry. Science China: Chemistry. 2021;**64**:642-649. DOI: 10.1007/s11426-020-9928-6

[25] Li FL, Li H, Yang ZQ, Huang TD, Wu DB, Wang SL. Rapid online fractionated analyses of rare earth elements in a dinosaur fossil by mass spectrometry. Journal of Analytical